Essential Plant Nutrients

M. Naeem • Abid A. Ansari • Sarvajeet Singh Gill
Editors

Essential Plant Nutrients

Uptake, Use Efficiency, and Management

Springer

Editors
M. Naeem
Department of Botany
Plant Physiology Section
Aligarh Muslim University
Aligarh, India

Abid A. Ansari
Department of Biology
Faculty of Science
University of Tabuk
Tabuk, Saudi Arabia

Sarvajeet Singh Gill
Centre for Biotechnology
Maharshi Dayanand University
Rohtak, Haryana, India

ISBN 978-3-319-58840-7 ISBN 978-3-319-58841-4 (eBook)
DOI 10.1007/978-3-319-58841-4

Library of Congress Control Number: 2017946075

Printed on acid-free paper

This Springer imprint is published by Springer Nature
The registered company is Springer International Publishing AG
The registered company address is: Gewerbestrasse 11, 6330 Cham, Switzerland

Contents

Contributors

Faraht Abbas Department of Environmental Sciences and Engineering, Government College University, Faisalabad, Pakistan

Neama Abdalla Plant Biotechnology Department, Genetic Engineering Division, National Research Center, Dokki, Giza, Egypt

Muhammad Adrees Department of Environmental Sciences and Engineering, Government College University, Faisalabad, Pakistan

Tariq Aftab Plant Physiology Section, Department of Botany, Aligarh Muslim University, Aligarh, India

Zaheer Akbar Department of Environmental Sciences and Engineering, Government College University, Faisalabad, Pakistan

Md. Khairul Alam School of Veterinary and Life Sciences, Murdoch University, Perth, WA, Australia

Akbar Ali Department of Botany, Dayalbagh Educational Institute, Agra, India

Plant Physiology Section, Department of Botany, Aligarh Muslim University, Aligarh, India

Arlene Asthana Ali Department of Biotechnology, Jamia Millia Islamia, New Delhi, India

Qasim Ali Department of Botany, Government College University, Faisalabad, Pakistan

Shafaqat Ali Department of Environmental Sciences and Engineering, Government College University, Faisalabad, Pakistan

Abdullah H. Al-Saeedi College of Agricultural and Food Sciences, King Faisal University, Al-Hassa, Saudi Arabia

Tarek Alshaal Soil and Water Department, Faculty of Agriculture, Kafrelsheikh University, Egypt

Claudia Álvarez Universidad Nacional de Colombia, Medellin, Colombia

Megahed Amer Soils, Water and Environment Research Institute (SWERI), Agricultural Research Center, Giza, Egypt

Abid A. Ansari Department of Biology, Faculty of Science, University of Tabuk, Tabuk, Saudi Arabia

Felipe Aquea Laboratorio de Bioingeniería, Facultad de Ingeniería y Ciencias, Universidad Adolfo Ibáñez, Santiago, Chile

Muhammad Saleem Arif Department of Environmental Sciences and Engineering, Government College University Faisalabad, Faisalabad, Pakistan

Muhammad Arslan Ashraf Department of Botany, Government College University Faisalabad, Faisalabad, Pakistan

Muhammad Ashraf Department of Soil & Environmental Sciences, University College of Agriculture, University of Sargodha, Sargodha, Pakistan

Lohit K. Baishya ICAR Research Complex for NEH Region, Medziphema, Nagaland, India

K.P. Baiyeri Department of Crop Science, University of Nigeria, Nsukka, Nigeria

Kaushik Batabyal Department of Agricultural Chemistry and Soil Science, Bidhan Chandra Krishi Viswavidyalaya, Nadia, West Bengal, India

Laboratory of Plant Stress Responses, Faculty of Agriculture, Kagawa University, Kagawa, Japan

Prem Kumar Dantu Department of Botany, Dayalbagh Educational Institute, Agra, India

Tariq Ahmad Dar Centre for Biodiversity Studies, School of Biosciences and Biotechnology, Baba Ghulam Shah Badshah University, Rajouri, India

Ayman El-Ghamry Soil and Water Department, Faculty of Agriculture, Mansoura University, Mansoura, Egypt

Hassan El-Ramady Soil and Water Department, Faculty of Agriculture, Kafrelsheikh University, Egypt

S. Kala Soil Chemistry and Fertility, IISWC, Research Centre, Kota, Rajasthan, India

Tamer Elsakhawy Agriculture Microbiology Department, Soil, Water and Environment Research Institute (SWERI), Sakha Agricultural Research Station, Agriculture Research Center (ARC), Giza, Egypt

Masayuki Fujita Laboratory of Plant Stress Responses, Faculty of Agriculture, Kagawa University, Kagawa, Japan

Aliaa Gad Agriculture Microbiology Department, Soil, Water and Environment Research Institute (SWERI), Sakha Agricultural Research Station, Agriculture Research Center (ARC), Giza, Egypt

Sarvajeet Singh Gill Stress Physiology and Molecular Biology Lab, Centre for Biotechnology, MD University, Rohtak, Haryana, India

Arti Goel Amity Institute of Microbial Biotechnology, Amity University Noida, Noida, Uttar Pradesh, India

Viraj Gunarathne Environmental Chemodynamics Project, National Institute of Fundamental Studies, Kandy, Sri Lanka

Muhammad Zulqarnain Haider Department of Botany, Government College University, Faisalabad, Pakistan

Eman Hamad Agriculture Microbiology Department, Soil, Water and Environment Research Institute (SWERI), Sakha Agricultural Research Station, Agriculture Research Center (ARC), Giza, Egypt

Mirza Hasanuzzaman Molecular Biotechnology Group, Center of Molecular Biosciences, Tropical Biosphere Research Center, University of the Ryukyus, Nishihara, Okinawa, Japan

Department of Agronomy, Faculty of Agriculture, Sher-e-Bangla Agricultural University, Dhaka, Bangladesh

Tanvir ul Hassan Dar Centre for Biodiversity Studies, School of Biosciences and Biotechnology, Baba Ghulam Shah Badshah University, Rajouri, India

Israr Ul Hassan Department of Mathematics and Sciences, College of Arts and Applied Sciences, Dhofar University, Salalah, Oman

Zaid ul Hassan Department of Environmental Sciences and Engineering, Government College University, Faisalabad, Pakistan

Md. Shahadat Hossain Laboratory of Plant Stress Responses, Faculty of Agriculture, Kagawa University, Kagawa, Japan

Afzal Hussain Department of Environmental Sciences and Engineering, Government College University, Faisalabad, Pakistan

Cristian Ibañez Laboratorio de Biología Molecular y Bioquímica, Departamento de Biología, Universidad de La Serena, La Serena, Chile

Mohd. Idrees Department of Mathematics and Sciences, College of Arts and Applied Sciences, Dhofar University, Salalah, Oman

Claudio Inostroza-Blancheteau Núcleo de Investigación en Producción Alimentaria (NIPA-UCT), Escuela de Agronomía, Facultad de Recursos Naturales, Universidad Católica de Temuco, Temuco, Chile

Muhammad Iqbal Department of Environmental Sciences and Engineering, Government College University, Faisalabad, Pakistan

M. Irfan Qureshi Department of Biotechnology, Jamia Millia Islamia, New Delhi, India

K.S. Kartikha Soil Chemistry and Fertility CPCRI, Regional Station, Vittal, Karnataka, India

Katalin Sárdi Department of Crop Production and Soil Science, University of Pannonia, Keszthely, Hungary

Rizwana Kausar Soil and Water Testing Laboratory for Research, Sargodha, Punjab, Pakistan

M. Masroor A. Khan Plant Physiology Section, Department of Botany, Aligarh Muslim University, Aligarh, India

Juan D. Leon Universidad Nacional de Colombia, Medellin, Colombia

Jubayer Al Mahmud Laboratory of Plant Stress Responses, Faculty of Agriculture, Kagawa University, Miki-cho, Kita-gun, Kagawa, Japan

Department of Agroforestry and Environmental Science, Faculty of Agriculture, Sher-e-Bangla Agricultural University, Dhaka, Bangladesh

Manish Mathur Central Arid Zone Research Institute, Jodhpur, Rajasthan, India

Sonia Mayakaduwa Environmental Chemodynamics Project, National Institute of Fundamental Studies, Kandy, Sri Lanka

B.P. Meena Agronomy, IISS, Bhopal, Madhya Pradesh, India

Gyanendranath Mitra Faculty of Agriculture, Department of Agricultural Chemistry, Soil Science and Biochemistry, Orissa University of Agriculture and Technology (OUAT), Bhubaneswar, India

Felipe Moraga Laboratorio de Bioingeniería, Facultad de Ingeniería y Ciencias, Universidad Adolfo Ibáñez, Santiago, Chile

Ahmed Mosa Soil and Water Department, Faculty of Agriculture, Mansoura University, Egypt

Muhammad Salman Mubarik Centre of Agricultural Biochemistry and Biotechnology (CABB), University of Agriculture, Faisalabad, Pakistan

M. Naeem Plant Physiology Section, Department of Botany, Aligarh Muslim University, Aligarh, India

Kamrun Nahar Department of Agricultural Botany, Faculty of Agriculture, Sher--e-Bangla Agricultural University, Dhaka, Bangladesh

Laboratory of Plant Stress Responses, Faculty of Agriculture, Kagawa University, Miki-cho, Kita-gun, Kagawa, Japan

Bouzid Nedjimi Laboratory of Exploration and Valorization of Steppe Ecosystem, Faculty of Science of Nature and Life, University of Djelfa, Djelfa, Algeria

Hirosuke Oku Molecular Biotechnology Group, Center of Molecular Biosciences, Tropical Biosphere Research Center, University of the Ryukyus, Okinawa, Japan

Alaa E.-D. Omara Agriculture Microbiology Department, Soil, Water and Environment Research Institute (SWERI), Sakha Agricultural Research Station, Agriculture Research Center (ARC), Giza, Egypt

Nelson W. Osorio Universidad Nacional de Colombia, Medellin, Colombia

Laura Osorno Universidad Nacional de Colombia, Medellin, Colombia

Anisur Rahman Department of Agronomy, Faculty of Agriculture, Sher-e-Bangla Agricultural University, Dhaka, Bangladesh

Laboratory of Plant Stress Responses, Faculty of Agriculture, Kagawa University, Kagawa, Japan

I. Rashmi Soil Chemistry and Fertility, IISWC, Research Centre, Kota, Rajasthan, India

Nasir Rasool Department of Chemistry, Government College University, Faisalabad, Pakistan

Zed Rengel Soil Science and Plant Nutrition, School of Earth and Environment, The University of Western Australia, Crawley, WA, Australia

Marjorie Reyes-Díaz Departamento de Ciencias Químicas y Recursos Naturales, Facultad de Ingeniería y Ciencias, Universidad de La Frontera, Temuco, Chile

Center of Plant, Soil Interaction and Natural Resources Biotechnology, Scientific and Technological Bioresource Nucleus (BIOREN), Universidad de La Frontera, Temuco, Chile

Muhammad Riaz Department of Environmental Sciences and Engineering, Government College University Faisalabad, Faisalabad, Pakistan

Muhammad Rizwan Department of Environmental Sciences and Engineering, Government College University, Faisalabad, Pakistan

Dibyendu Sarkar Department of Agricultural Chemistry and Soil Science, Bidhan Chandra Krishi Viswavidyalaya, Nadia, West Bengal, India

Sher Muhammad Shahzad Department of Soil & Environmental Sciences, University College of Agriculture, University of Sargodha, Sargodha, Pakistan

Tarek Shalaby Horticulture Department, Faculty of Agriculture, Kafrelsheikh University, Kafr El-Shaikh, Egypt

College of Agricultural and Food Sciences, King Faisal University, Al-Hassa, Saudi Arabia

K.C. Shinogi Agriculture Extension, IISS, Bhopal, India

Abhay Shirale Soil Chemistry and Fertility, IISS, Bhopal, Madhya Pradesh, India

Minu Singh Department of Biotechnology, Jamia Millia Islamia, New Delhi, India

Teg Bahadur Singh Department of Botany, Dayalbagh Educational Institute, Agra, India

Moin Uddin Botany Section, Women's College, Aligarh Muslim University, Aligarh, India

F.D. Ugese Department of Crop Production, University of Agriculture, Makurdi, Nigeria

Meththika Vithanage Environmental Chemodynamics Project, National Institute of Fundamental Studies, Kandy, Sri Lanka

Faculty of Applied Sciences, University of Sri Jayewardenepura, Nugegoda, Sri Lanka

Tahira Yasmeen Department of Environmental Sciences and Engineering, Government College University Faisalabad, Faisalabad, Pakistan

About the Editors

M. Naeem is assistant professor in the Department of Botany at Aligarh Muslim University, India. For more than a decade, he has devoted his research to improving the yield and quality of commercially important medicinal and aromatic plants (MAPs). His research focuses on escalating the production of MAPs and their active principles using a novel and safe technique involving radiation-processed polysaccharides as well as the application of potent PGRs. To date, he has successfully run three major research projects as the principal investigator, two of which were sanctioned by the Department of Science & Technology, New Delhi. He also completed another fascinating research project on *Catharanthus roseus* awarded by CSTUP, Lucknow. Dr. Naeem has published more than 80 research papers in reputable national and international journals as well as six books. He has also participated in various national and international conferences and acquired life memberships to various scientific bodies in India and abroad. Based on his research contributions, Dr. Naeem has been awarded a research associateship from the Council of Scientific and Industrial Research, New Delhi; a Young Scientist Award (2011) from the State Government of Uttar Pradesh; a Fast Track Young Scientist Award from the Department of Science & Technology, India; a Young Scientist of the Year Award (2015) from the Scientific and Environmental Research Institute, Kolkata; and a Rashtriya Gaurav Award (2016) from the International Friendship Society, New Delhi.

Abid A. Ansari is assistant professor in the Department of Biology, Faculty of Science, University of Tabuk, Saudi Arabia. Dr. Ansari's research work is concerned with phytoremediation and eutrophication. Dr. Ansari has to his credit a number of research articles of national and international repute, ten edited books, and a number of book chapters on varied aspects of his field of research. He has been awarded Scientist of the Year 2014 and Environmentalist of the Year 2011 by the National Environmental Science Academy, India, and Research Excellence 2016 by the University of Tabuk. He has also participated in various national and international conferences and acquired memberships to various scientific bodies, including the Saudi Biological Society, the International Phytotechnology Society, and the Society for Ecological Restoration.

Sarvajeet Singh Gill is assistant professor at the Centre for Biotechnology, MD University, Rohtak, India. He has made significant contributions towards abiotic stress tolerance in crop plants.

Dr. Sarvajeet Singh Gill's research includes abiotic stress tolerance in crop plants, reactive oxygen species signalling and antioxidant machinery, gene expression, helicases, crop improvement, transgenics, nitrogen and sulphur metabolism, and plant fungal symbiotic interactions. Together with Dr. Narendra Tuteja at the International Centre for Genetic Engineering and Biotechnology (ICGEB), New Delhi, he worked on plant helicases and discovered a novel function of plant MCM6, PDH45 and p68 in salinity stress tolerance that will help improve crop production at suboptimal conditions. A recipient of the Junior Scientist of the Year Award 2008 from the National Environmental Science Academy, Sarvajeet Gill has edited several books and has a number of research papers, review articles, and book chapters to his name.

Part I
Concepts of Plant Nutrients Uptake

Chapter 1
Essential Plant Nutrients and Recent Concepts about their Uptake

Gyanendranath Mitra

Abstract Plants acquire a number of mineral nutrients essential for their metabolism and growth from soil or any other rooting medium. The nutrients have to get through the plasma membrane of root hair cells for use in plant metabolism. According to recent concepts this process is strictly regulated by large groups of genes, which are specific for each nutrient. These genes produce m-RNA transcripts which translate sets of transporter proteins specific for each nutrient. The transporter proteins are lodged inside minute pores located on the plasma membrane. They regulate passage of each nutrient into the cytoplasm. A large number of metabolic enzymes are up- or down-regulated in response to deficiency or sufficiency of plant nutrients. Amino acids, plant growth regulators, intermediate metabolites, and the nutrients themselves are involved in the induction or repression of transporter encoding genes as well as post-translation modification of transporter proteins.

Keywords Nutrient uptake • Plant growth • Nutrient transporters • Abiotic stress

1.1 Essential Plant Nutrients

Plants take up several mineral elements in their ionic forms from the soil or any other growth medium for their metabolism and growth. Some of these elements are called essential since absence or low concentrations of them interfere with plant metabolism and growth and show characteristic deficiency symptoms, which can be corrected by their application. So far 18 elements, C, H, O, N, P, K, Ca, Mg, S, Fe, Zn, Cu, Mn, B, Cl_2, Mo, Co and Ni, have been considered essential for plant nutrition (NRCCA 2010). They are further classified into Macro- and micronutrients.

This article is a brief account on the subject and mostly based on the information given in the book *Regulation of Nutrient Uptake by Plants: A Biochemical and Molecular Approach* by Gyanendranath Mitra, Springer, 2015, with updates from recent research publications.

G. Mitra (✉)
Faculty of Agriculture, Department of Agricultural Chemistry, Soil Science and Biochemistry, Orissa University of Agriculture and Technology (OUAT), Bhubaneswar, India
e-mail: gnmitra@gmail.com

M. Naeem et al. (eds.), *Essential Plant Nutrients*,
DOI 10.1007/978-3-319-58841-4_1

1. **The Macronutrients**, applied in larger quantities to the plants, consist of
   (a) Structural elements: C, H, and O
   (b) Primary nutrients: N, P and K
   (c) Secondary nutrients: S, Ca and Mg.
2. **The Micronutrients**, applied in small quantities to plants, consist of: Zn, Fe, Mn, Cu, B, Mo, Cl^-, Co and Ni.

1.1.1 Beneficial Plant Nutrients

They are not essentially required for all the plants. Some of them are essential for some of the plants but others are beneficial to a few plants, and animals who consume these plants. These include Na, Co, Si, Se and V.

Sodium (Na^+) is essential for *halophytes*, which accumulate salt in vacuoles to maintain turgor and growth. A few of the C4 plants (except corn and sorghum) need Na^+ essentially for specific functions, such as in the concentration of CO_2. Other beneficial effects of Na^+ include greener and glossy leaves due to increased cuticular wax formation and improvement of taste and texture of crops (Brownell and Crossland 1972). Silicon strengthens the stem and provides protection to plants from biotic and abiotic stress. Cobalt is involved in nitrogen fixation by root nodule bacteria and other diazotrophs. Consumption of Selenium rich crop plants such as cabbage, mustard, onion and broccoli provides protection to human beings against cancer and heart disease. The importance of V is due to the discovery in 1980 that it can act as an insulin-mimetic agent.

1.1.2 Non-essential Plant Nutrients

Plants often survive in hostile ionic environment in mineral-rich soils. In their ionic form,

$$Al^{3+}, Au^{+}, Cd^{2+}, Cu^{+}, Cu^{2+}, Co^{2+}, Cr^{3+}, Fe^{2+}, Fe^{3+},$$
$$Hg^{2+}, Mn^{2+}, Ni^{2+}, Pb^{2+}, Sn^{2+}, \mathrm{W}^{6+} \text{ and } Zn^{2+}$$

become toxic to plants at different threshold concentrations. Among them, Zn, Fe, Mn, Cu, B, Mo, Co and Ni are micronutrients and essential for plant growth at low concentrations but become toxic beyond a threshold concentration.

Elements such as Cr, V, W, As, Ag, Hg, Sb, Cd, Pb and U with no known function in plant metabolism have been found to be taken up by plants (Godbold and Hüttermann 1985; Breckle 1991; Nies 1999; Schützendübel and Polle 2002). If the soil or growth medium is rich in one or more of the non-essential elements, these elements are likely to be taken up by plants to a tolerable, or sometimes toxic concentrations. These elements may be of geological origin or accumulated in the soil due to anthropogenic causes. Soils around mining sites or nearer to solid waste dumps from mineral-based industries may contain elements toxic to plants. Use of untreated effluents from industries, use of sewage and sludge from urban centres and use of polluted water as source of irrigation and atmospheric deposits of radioactive isotopes from any overground nuclear activities on soils and plants are other sources of non-essential elements. When food crops are grown on these soils the non-essential elements are likely to be taken up by the plants and may enter into the food chain of man and animals (Greger 2004; Kabta-Pendias and Szdeke 2015). Higher plants have built-in cellular mechanisms for metal detoxification and tolerance to protect them from uptake of these toxic nutrients, such as (1) restriction of metal movement to roots by mycorrhizal association, (2) binding the metals to cell wall and to root exudates, (3) reduce influx across plasma membrane, (4) active efflux into apoplast, (5) scavenging by root border cells, (6) chelation in cytosols by various legands, and (7) transport of accumulated metals to the vacuole. Further an elaborate membrane transport system regulates movement of metal ions across plasma membrane (Mitra 2015).

1.1.3 Radioactive Nuclides

Radioactive nuclides are a part of the terrestrial environment emanating from radioactive substances present in the earth's crust and from cosmic rays. Recently there has been enrichment of specific nuclides in the environment due to (1) manufacture and testing of nuclear weapons, (2) extensive construction of nuclear power plants, (3) commercial fuel reprocessing (4) nuclear waste disposal, (5) Uranium mining and enrichment and (6) nuclear accidents. (Major accidents happened in Chernobyl in USSR, 26th April, 1986, caused by explosion in nuclear power plant due to operational error and in Daiichi Fukushima, Japan, on 11th March, 2011, due to meltdown of nuclear power plant damaged by Tsunami).

The radioactive nuclides released by nuclear weapon tests include

$$^{140}Ba,^{14}\mathrm{C},^{141}Ce,^{144}Ce,^{137}Cs,^{55}Fe,^{3}\mathrm{H},^{131}\mathrm{I},^{54}Mn,$$
$$^{239}Pu,^{241}Pu,^{103}Ru,^{106}Ru,^{125}Sb,^{89}Sr,^{91}\mathrm{Y},^{96}Zr.$$

Some of these and/or their daughter nuclides are released in other operations as described above. The four most harmful radio-nuclides released due to Chernobyl disaster were ^{131}I ($t_{1/2}$ = 8.02 days, causes thyroid cancer), ^{134}Cs ($t_{1/2}$ = 2.07 years, accumulates in heart), ^{137}Cs ($t_{1/2}$ = 30.2 years) and ^{90}Sr ($t_{1/2}$ = 28.8 years, accumulates in bones). The radioactive nuclides monitored from Fukushima Daiichi explosions were ^{131}I and ^{137}Cs. The regulatory levels fixed by Japan were 2 Bq/g for ^{131}I and 0.5 Bq/g for ^{137}Cs. There were soil contaminations with these two nuclides. Soils of a large area of eastern and north-eastern Japan were contaminated with ^{137}Cs. Chernobyl accident data have shown that ^{137}Cs adsorbed on the top soil layer can remain there for long years making the soil unfit for crop production (Yasunari et al. 2011).

Soon after the Chernobyl disaster four square kilometres of pine forest directly downwind of the reactor turned red and died. The radiation level caused by Chernobyl disaster is still very high and 30 km around the factory has been declared as 'Zone of alienation'. It may take 20,000 years to become fit for human habitation. The area however has reverted to become a natural forest and overrun by wildlife due to lack of competition from humans for space and resources. This indicates that plants and animals can survive in a relatively high radiation zone. A study was conducted on progeny of *Arabidopsis* plant collected from Zone of alienation with different levels of contamination. The study indicated a significantly higher resistance of progeny *Arabidopsis* plants to mutagens. There was increased expression of radical scavenging genes *CAT1* and *FSD3* and DNA repair genes *RAD1 and RAD51-like* in these plants (Kavalchuk et al. 2004).

According to World Nuclear Association (2015) the human environment has always been radioactive and accounts for 85% of annual radiation dose, 2.4 mSv/year. The radiation dose received from all nuclear activities accounts for less than 1%.

1.2 Recent Concepts about Nutrient Uptake by Plants

Globally arable soils are deficient in one or more of plant nutrients. The concentration of plant nutrients in soil solution depends on characteristic of the soil, local climatic conditions, nutrient removal due to increased intensity of cropping and management practices such as excessive or less fertiliser use, inadequate irrigation and drainage. The agronomic field operations also change the nutrient profile of the soil.

The requirements of plants for nutrients change with their growth stages, which do not often match with nutrients available in soil solutions. The nutrients available may be in excess or deficient. It has recently been found that plants adopt special mechanisms to acquire nutrients to meet their needs irrespective of their concentration in soil solution.

Plants take up mineral nutrients for their metabolism and growth. As a first step the mineral nutrients need to be transported across plasma membrane of root hair cells into cytoplasm for use in plant metabolism. Recent research indicates that this process is under strict genetic control. There are different groups of genes for each nutrient, which encode transporter proteins whose functions include acquiring specific nutrient from the soil solution and transporting them across plasma membrane of root hair cells for use in plant metabolism. There are different sets of genes, which are induced due to deficiency or sufficiency of a plant nutrient. They produce mRNA transcripts for translation of transporter proteins. Induction or repression of these genes is caused by amino acids, plant growth regulators, intermediate metabolites or nutrients themselves (Orsel et al. 2002; Hammond et al. 2004; Rodriguez-Navarro and Rubio 2006; Miller et al. 2008).

A large number of genes, which are involved in encoding transporter proteins for uptake of different nutrients, have been identified for a number of plants. The amino acid sequence and structure of corresponding transporter proteins and their mechanism of action have been reported.

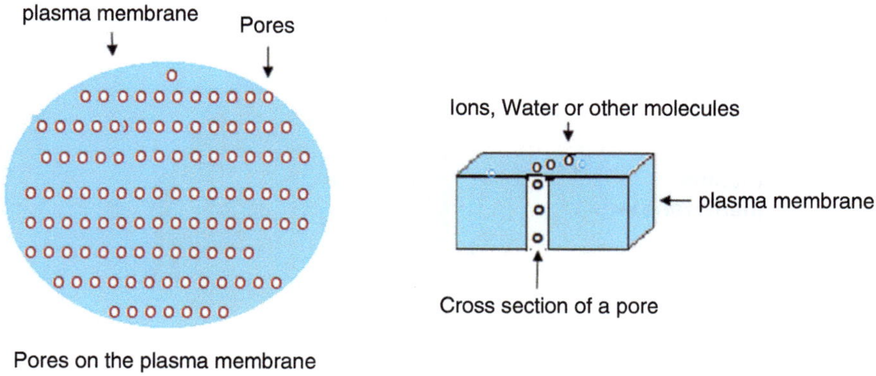

Fig. 1.1 Pores on plasma membrane and cross section of a pore (adapted from Mitra 2015)

The transporter proteins have been classified as (1) Chanel proteins and (2) Transporters.

1.2.1 Ion Channels and Transporters

There are a large number of pores on the plasma membrane of cells, which allow passage of specific nutrient ions or molecules and do not allow any others to pass through them. Such selectivity is ensured by large molecules of intrinsic trans-membrane transporter proteins with fixed topology, which are lodged across the pores of the plasma membrane (Fig. 1.1). They are too large to move across the membrane. The transporter proteins consist of

1. Channel proteins and
2. Transporters (Carriers) proteins.

1.2.1.1 Channel Proteins

Channel proteins are large molecules with multiple trans-membrane α-helices. They alternate between open and closed conformations (gating). There is conformational change of the channel protein due to any one of the extrinsic factors, such as (1) changes in membrane potential (2) binding of a small regulatory molecule or (3) membrane stretch (e.g. via link to the cytoskeleton) (Dubyak 2004; Rainer 2012). These factors determine if the channel is in a gated state (open for transport) or closed state (incapable of ion transport). The extrinsic factors control the accessibility of ions to the pore domain, which acts as a pathway for movement of ions from one side of the membrane to the other side. Since there are no energetic interactions, between channel protein and the transported ion, the rate of transport of ion is fast. There is probably no binding site within the pore to restrict their movement. Even if

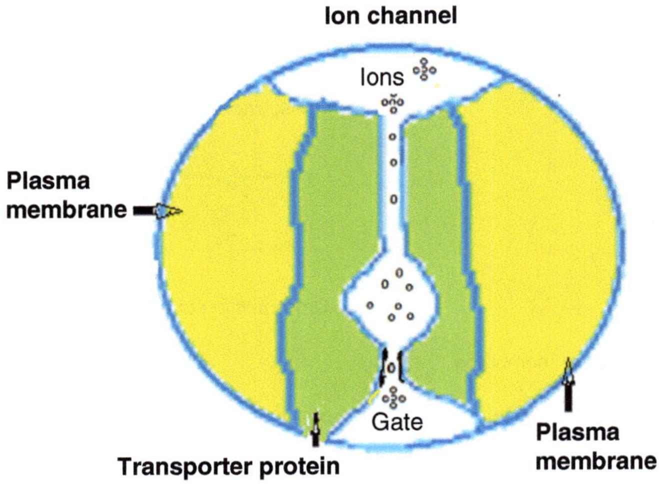

Fig. 1.2 Ion channel, with transporter proteins lining the pore of plasma membrane (adapted from Mitra 2015)

they exist they are shallow and separated by small free energy barriers (Roux et al. 2011).

All channels mediate passive transport of ions down their chemical or electrochemical gradient across the membrane due to difference in concentrations of ions on each side of the membrane as well as any electrical potential across the membrane (Fig. 1.2).

1.2.1.2 Ion Transporters (Carriers)

Transporter proteins are 'vectoral' enzymes (Dubyak 2004). Their functioning involves (1) a selective recognition/binding of the ion to be transported, (2) conformational changes in carrier protein due to binding of the ion and (3) physical movement of the ion across the membrane caused by such conformational changes. Ion transporters can catalyse movement of ions against their electro-chemical gradient (not ion channels) deriving energy from ATP hydrolysis. There are three types of ion transporters:

1. **Uniporters**: They transport one type of ion across the membrane, e.g.: P-type ATPases, $Ca2^{+}$-ATPase.
2. **Symporters (co-transporters)**: They transport more than one type of ion across the plasma membrane, e.g.: NRTs ($2H^{+}/NO_3^{-}$ co-transport), TaHKT1 (K/Na co-transporter).
3. **Anti-porters (Exchangers)**: There is exchange of one ion for the other, which moves in opposite directions, e.g.: CHX (K^{+}/H^{+} anti-porter), CAX (Ca^{2+}/H^{+} anti-porter) (Fig. 1.3).

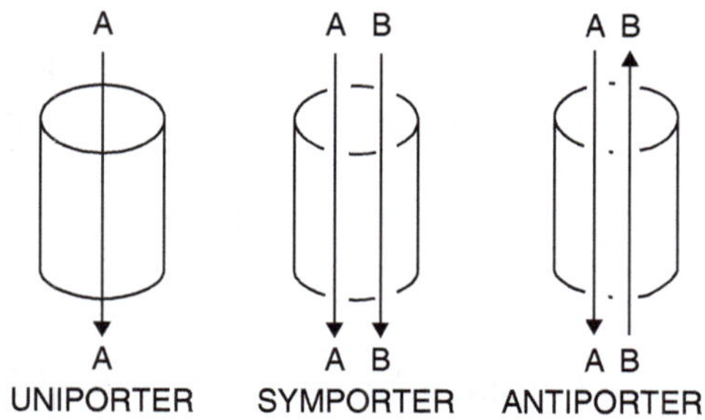

Fig. 1.3 Types of transporters: Uniporters, Symporters and Antiporters (adapted from Mitra 2015)

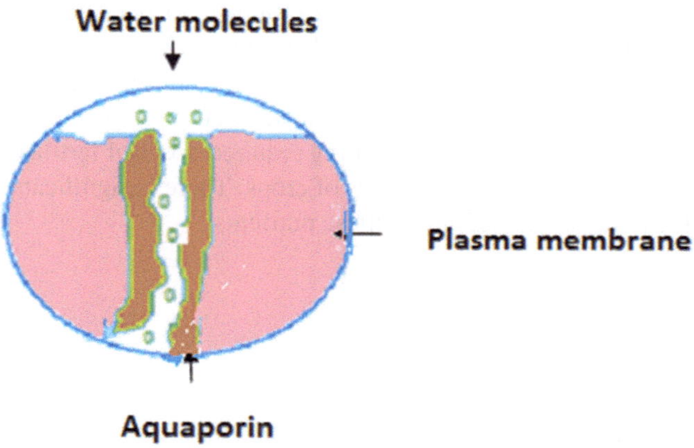

Fig. 1.4 Water channel, with aquaporins lining the walls of the pore across plasma membrane, (adapted from Mitra 2015)

1.3 Mechanism of Uptake of Water

Water is taken up into cells through water channels. Aquaporins, a large protein family found in both eukaryotes and prokaryotes, control movement of water through the narrow channels located on the plasma membrane. Molecules with proper orientation are selectively allowed to pass through the channel. Ions carrying charges such as (H^+) and (H_3O^+) are prevented from passage. Smaller uncharged molecules such as glycerol and urea are allowed passage through the channel. Glycerol molecules, which are larger than water molecules, appear to move in a single file through the amphipathic channel where NPA motifs play a critical role (Chaumont et al. 2000) (Fig. 1.4).

Plant genomes contain a large number of aquaporin (AQP) genes to cope up with adverse water regimes encountered by them during their growth period. *Arabidopsis* has 38 AQP genes of 2–3 kb size (Johanson et al. 2001; Quigley et al. 2002), maize 33 (Chaumont et al. 2000), barley 23 (Katsuhara et al. 2002), rice 34 (Nguyen et al. 2013), wheat 35 (Forrest and Bhave 2008) and soybean 66 (Zhang et al. 2013) AQP genes.

The proteins coded by AQP genes belong to a major intrinsic protein (MIP) family. MIPs are classified into five sub-families such as PIP (plasma membrane intrinsic protein), TIP (tonoplast intrinsic protein), NIP (Nod 26-like intrinsic protein), SIP (small basic intrinsic protein), and XIP. (Recently, XIP genes have been reported to be involved in the transport of a wide range of hydrophobic solutes (Venkatesh et al. 2015)). The members of the families are not confined to the location as indicated by their names. They may be found elsewhere.

1.4 Primary Nutrient (N, P and K)

The primary nutrients, Nitrogen (N), Phosphorus (P) and Potassium (K), are required in relatively large quantities by plants for their metabolism and growth. Arable soils do not contain sufficient reserve of these nutrients to meet the requirements of crops grown on the same patch annually. Crop removal of these nutrients needs to be replenished every year by adding required doses of fertilisers containing these nutrients to harvest optimum yields of crops. There is significant reduction of crop yield due to deficiency of any of these nutrients.

1.4.1 Uptake of Nitrogen (N)

Surface soils generally contain 0.03–0.4% of total N, 95% of which is in the organic form. Plants contain about 1–6% of N of their dry weight. Nitrogen is primarily taken up by plants as NO_3^- or NH_4^+ ions. Unfertilised soils may contain NO_3^- at a concentration of <1 mM but application of fertilisers may raise it >70 mM.

Concentration of N is more or less constant within cytoplasm. Nitrate concentration in cytoplasm is limited to about 2–5 mM and 5–75 mM inside vacuole (Miller and Smith 1996). NH_4^+ is toxic and is not allowed to accumulate within the plants.

Nitrogen is a constituent of amino acids, proteins, purine and pyrimidine rings of nucleic acids, chlorophyll and enzymes. All of these compounds are involved in plant metabolism and growth. Adequate nitrogen supply results in vigorous growth of plants and the leaves turn dark green due to high photosynthetic activity. Deficiency of N causes impaired photosynthesis and degradation of chloroplasts. This appears first as yellowing of older leaves while growing leaves remain green.

1.4.1.1 Mechanism of N Uptake

Plants take up N primarily as NO_3^- and NH_4^+. Acquisition, uptake, transport and redistribution of both of them are under strict genetic control (Siddiqi et al. 1990; King et al. 1993). The primary event of NO_3^- uptake is its transport through plasma membrane of root epidermal and cortical cells. This is carried out by a favourable H^+ (proton) electrochemical gradient maintained by the plasma membrane (PM) H^+-ATPases (Proton pumps) (Miller and Smith 1996; Quaggiotti et al. 2003; Sperandio et al. 2014). PM H^+-ATPase activity maintains membrane potential ($\Delta\Psi$) and proton motive force (Δp) necessary for ion transport. Both for high and low affinity transport system NO_3^- uptake takes place by symport of $2H^+/NO_3^-$ (Crawford and Glass 1998).

1.4.1.2 Nitrate Transport Genes

There are four families of genes involved in transport of NO_3^- across plasma membrane in *Arabidopsis* such as (1) NRT1 (Nitrate Transporter1/Peptide Transporter family, 53 members), (2) NRT2 (7 members), (3) CLC (Chloride channel, 7 members) and (4) SLAC1/SLAH (Slow Anion Channel-Associated 1 homologues, 5 members) (Krapp et al. 2014). The four families have a total of 73 genes out of which 60 are from NRT1/PTR and NRT2 families. Out of 35 genes characterised 24 are nitrate transporters (Krapp et al. 2014).

1.4.1.3 Nitrate Transporters (NRTs)

There are three different nitrate transport systems in plants. When the external NO_3^- concentration is high (1–50 mM), an essentially unregulated and constitutively expressed Low Affinity Transport system (LATS) operates (Crawford and Glass 1998). A High Affinity Transport (HATS) system operates, when external NO_3^- concentration is low (<0.2 mM). Some of them are constitutively expressed (cHATS) and others induced by NO_3^- (iHATS) (Fig. 1.5).

The NRT1 genes encode low affinity transporters (LATS), when the NO_3^- concentration in the soil is high (>1 mM, Orsel et al. 2002). The NRT2 genes encode high affinity nitrate transporters at low NO_3^- concentration (<0.2 mM). Some of the NRT2 genes are inductive (iHATS) and others constitutive (cHATs). AtNRT1;1 (CHL1) is a dual affinity nitrate transporter, switched off and on by phosphorylation/dephosphorylation of threonine T101 in its polypeptide chain (Liu et al. 1999). The CBL (calcineurin B-like) inter-

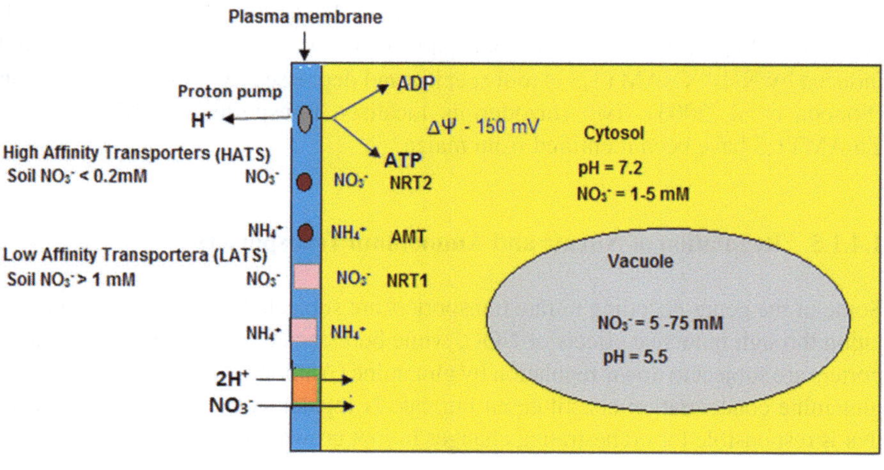

Fig. 1.5 Mechanism of nitrogen uptake by plants (adapted from Mitra 2015)

acting protein kinase, CIPK23 (SnRK3;23), phosphorylates T101 under low nitrate conditions, allowing NRT1;1 to act as a high affinity nitrate transporter (Ho et al. 2009). Dephosphorylated NRT1;1 is a low affinity nitrate transporter.

Nitrate transport in *Arabidopsis* is carried out by two transporters from NRT1 family, AtNRT1;1 and AtNRT1;2, and two from NRT2 family, AtNRT2;1 and AtNRT2;2. When external NO_3^- concentration is low, NRT2;1 proteins localised on the plasma membrane constitute the major component of HATs (72%) activity (Li 2006). It requires a second protein NAR2 for its stability. AtNRT2;1 and AtNAR2;1 form a tetramer with two subunits each, which constitute the active NO_3^- transporter (Yong et al. 2010).

In higher plants *NRT2* genes isolated so far are preferentially expressed in the roots (Tsay et al. 2007). OsNRT1 expressed in epidermal cells of rice roots is a homologue of *Arabidopsis* AtNRT1;1(CHL1) (Lin et al. 2000). Four HATs OsNRT2;1, OsNRT2;2, OsNRT2;3 and OsNRT2;4 and two NAR proteins OsNAR2;1 and OsNAR2;2 have been isolated from rice (Feng et al. 2011; Sperandio et al. 2014). In maize Zm NRT2;1 is involved in influx activity and ZmNRT2;2 in xylem loading process (Trevisan et al. 2008).

OPTs and PTRs: Some members of NRT1 are peptide transporters called OPTs (oligopeptide transporters) involved in transport of tetra- and penta-peptides. Some of the OPTs transport glutathione, glutathione conjugates, phytochelatins and metals (Tsay et al. 2007). PTRs are di- and tri-peptide transporters. Seven families of NRT1/PTRs have been identified. They transport nitrate, di- and tri-peptides, auxins and carboxylates (Fan et al. 2014). During seed germination of barley, HvPTR1 expressed in plasma membrane of scutellar epithelial cells, transport peptides produced by hydrolysis of endosperm storage proteins to the developing embryo.

1.4.1.4 Ammonium Transporters

The AMT1 family of high affinity NH_4^+ transporters contains five members in *Arabidopsis*, of which AtAMT1;1, AtAMT1;2 and AtAMT1;3 have been studied in detail. In rice four NH_4^+ transporter genes have been identified (Suenaga et al. 2003). OsAMT1;1 is expressed in roots and shoots. OsAMT1;2 is root specific and induced by NH_4^+. OsAMT1;3 is root specific and depressed by nitrogen application (Sonoda et al. 2003). Two rhizodermis localised transporters ZmAMT1;1 and ZmAMT1;3 have been identified from maize.

1.4.1.5 Regulation of Nitrate and Ammonium Transporters

Some of the genes encoding nitrate transporters are subjected to transcriptional regulation through inductive effects of NO_3^-, while both encoding NO_3^- and NH_4^+ transporters are subject to down-regulation by glutamine (Anthony et al. 2002). Nitrate and glutamine concentration constitute an intricate N regulatory network at the root tip that is responsible for orchestrating changes in root growth rate and root architecture.

Nitrate stimulates primary root growth, both directly and by antagonising inhibitory effect of glutamine, which stimulates root branching (Walch-Liu and Forde 2008).

Ammonium transporters are oligomeric proteins. They undergo conformational coupling among monomers for ammonium uptake. This provides a mechanism for tight regulation of ammonium transporters. Rapid shut off mechanism is required to prevent toxic accumulation of NH_4^+. Application of higher levels of NH_4^+ blocks NO_3^- uptake by roots.

1.4.1.6 Biotechnological Approach to Increase N-Use Efficiency

The current biotechnological approach to improve NUE includes manipulation of genes involved in (1) N-uptake, (2) N-assimilation and (3) N-translocation to the edible/useful parts of the crops.

No correlations have yet been observed between over-expression of nitrate transporters and NUE. Over-expression of either the NR or the NiR gene in plants increases mRNA levels, and often affects N uptake but does not seem to increase the yield or growth of the plants regardless of the nitrogen source available.

Egami et al. (2012) introduced fungal glutamate dehydrogenase (*gdh A*) gene encoding NADP(H) dependent glutamate dehydrogenase from *A. nidulans* into potato. The GDH potato had higher photosynthetic rate irrespective of N-supply and resulted in higher tuber yield and NUE. A similar claim was made by (Naohiro) Aoki et al. (2009) of the same group for rice and potato (Mitra 2015).

Shrawat et al. (2008) introduced a barley AlaAT (alanine aminotransferase) cDNA driven by a rice tissue-specific promoter (OsAnt1) into rice plants. The transgenic plants had significant increases in the biomass and grain yield as compared to control plants when plants were well supplied with nitrogen. Significant progress is yet to be made in improving NUE through genetic manipulation.

1.4.2 Uptake of Phosphate (Pi)

Total P in surface soils varies from 0.005 to 0.15%, 50% of it in organic form. Pi (phosphate) content of plants is in the range of 0.05–0.5% of their dry weight. Phosphorus is primarily taken up by plants in the forms of phosphate ions. The forms of phosphate ions available to plants are HPO_4^{2-}, $H_2PO_4^-$ and PO_4^{3-} based on the pH of rhizosphere. At pH 7.2 $H_2PO_4^- \approx HPO_4^{2-}$, above pH 7.2 $HPO_4^{2-} > H_2PO_4^-$, but below pH 7.2, $H_2PO_4^-$ dominates and is more than HPO_4^{2-}. The concentrations of phosphate ions, HPO_4^{2-}, $H_2PO_4^-$, in soil solution are very low (0.1–10 μM). Plant uptake of HPO_4^{2-} is much slower than $H_2PO_4^-$ (Mitra 2015). Cellular Pi content is in the range of 2–20 mM (Bieleski 1973; Schachtman et al. 1998).

Phosphorus is a constituent of high energy compounds such as nucleoside triphosphates (ATP, GTP, CTP, UTP), phosphoenolpyruvate, creatinine phosphate, etc., which supply energy to drive endergonic metabolic reactions. Energy storage and energy transfer are the major biochemical functions of the high energy phosphorylated compounds.

A common symptom of Pi deficiency in plants is dark green or purple shoot due to anthocyanin accumulation. Phosphate deficiency causes induction of enzymes involved in synthesis of anthocyanins (Vance et al. 2003; Fang et al. 2009), which protects nucleic acids from UV damage and chloroplast from photo-inhibitory damage (Zeng et al. 2010).

Phosphate deficiency results in starch accumulation in the cells. Low cellular Pi removes allosteric inhibition of the enzyme ADP Glc-pyrophosphorylase involved in starch biosynthesis in cells. Pi deficiency up-regulates some of the glycolytic bypass enzymes such as pyrophosphate (PPi)-dependent phosphofructokinase, PPi-phosphoenol pyruvic kinase, pyruvate phosphate dikinase and the tonoplast H^+-pyrophosphatase (Plaxton and Podesta 2006). Glycolysis appears to be bypassed by avoiding those reactions requiring P. Under severe Pi deficiency a large decline of ATP and ADP (up to 80%) and other nucleoside phosphates occurs. Plants respond by adopting alternate metabolic pathways (Plaxton and Tran 2011). Moderate Pi deficiency causes significant reduction in glutamine synthetase and nitrate reductase enzymes and affects amino acid metabolism and N-assimilation (Calderon-Vazquez et al. 2008).

Mechanisms of Pi Uptake by Plants: Plants respond to Pi deficiency through (1) morphological adaptations, (2) metabolic changes and (3) genetic responses.

1. ***Morphological Adaptation of Plants due to Pi-Deficiency***: Under conditions of Pi-deficiency plants adapt themselves suitably through modification of their roots and shoots so as to acquire more Pi from soil and use them frugally to support plant growth. Rhizosphere is a critical region around roots, where intense interactions among plant roots, soil and microorganisms take place. Deficiency of Pi has profound effects on root growth and its architecture. These are modified suitably to explore a larger volume of soil so as to absorb more P to meet the P-demand of plants. Pi-deficiency causes delayed leaf development, reduction in number of leaves and leaf expansion, decreased photosynthetic capacity, stunted growth (reduced auxiliary shoot emergence and elongation), impaired flower development and an increased root/shoot ratio of the plants (Vance 2010). Pi from lower and older leaves translocates to newer leaves. Enhanced uptake of Pi by roots and translocation to shoots results in excess Pi accumulation in older leaves and may cause chlorosis and necrosis of leaf tips due to Pi-toxicity.
2. ***Metabolic Changes***: Plant roots exude a variety of organic compounds under normal conditions of growth. These include: sugars, organic acids, amino acids, growth hormones, phenolics, proteins etc., which affect rhizosphere chemistry and alter plant-microbe interaction, allelopathy and nutrient acquisition by plants.

Excretion of organic acids in response to Pi deficiency lowers pH in the rhizosphere by 2–3 units than the bulk of the soil. This may increase dissolution of sparingly soluble soil-P (Marschner 1995). While the protons excreted through organic acids lower the pH, the carboxylate anions react with Fe^{3+}, Al^{3+} and Ca^{2+} present in insoluble compounds of Pi-containing minerals. They form chelates with the cations and release Pi for uptake by the plants. This results in an increase of soil solution Pi concentration by about 1000-fold (Plaxton and Tran 2011). Under conditions of Pi-deficiency, plants recycle P from older tissues to new tissues. Plants also remobilize from non-essential uses to essential uses.

Genetic Response to Phosphate Deficiency: Phosphate deficiency results in coordinated induction of hundreds of genes encoding enzymes, which maximise capacity of plants to acquire phosphate more efficiently from external sources and reprioritize internal use of phosphorus (Plaxton and Tran 2011).

Phosphate Transporters: Plants have both low and high affinity transport systems encoded by corresponding genes. The low affinity transport systems are constitutive and operate at higher Pi concentration. High affinity phosphate transporters are located primarily in plasma membrane of root hairs and operate at low Pi concentration. The high affinity transporters are induced when Pi is deficient.

Mechanism of Phosphate Transport: The transporters, which are $H_2PO_4^-/H^+$ symporters, move Pi against the steep concentration gradient of about 10,000-fold or higher (concentration of Pi in soil solution ≈ 0.1–10 μM and cellular concentration inside roots ≈ 2–20 mM) through active transport with energy derived from ATP. The movement from root surface to xylem is symplastic and is at a rate of about 2 mM h^{-1} (Bieleski 1973). Transport of Pi to above ground parts is through xylem flow and to cells in tissues through symplastic transport. Movements of Pi through plasma membrane into cells and into vacuole within cells are carried out by $H_2PO_4^-/H^+$ symporters with energy derived from ATP.

Genes Involved in Pi Acquisition and Transportation: Genes of four transporter families, PHT1, PHT2, PHT3 and PHT4, are found in *Arabidopsis*. Members of PHT1 gene family are expressed in root epidermal cell and the encoded transporters are located on the plasma membrane (Lin et al. 2009). They are high affinity $H_2PO_4^-/H^+$ symporters and function to acquire Pi from the rhizosphere.

Early Genes and Late Genes: Genes that respond to P deficiency can be grouped into 'Early genes' that respond rapidly and often non-specifically to Pi deficiency, or 'Late genes' that impact on the morphology, physiology or metabolism of plants upon prolonged Pi deficiency (Vance et al. 2003, Hammond et al. 2004).

PHO Regulon Genes: There is a Pi-starvation-inducible rescue system in plants with their promoter region, the PHO regulon genes, under a common regulatory system (Goldstein et al. 1988). The Pi-responsive genes, TPSI1 from tomato and Mt4 from *Medicago truncatula*, have *cis*-regulatory elements 'GCACG (G/T)' in their binding sites. The AtPHR1 (phosphate starvation response 1) gene from *Arabidopsis* has a motif, a *cis*-element 'GNATATNC' (P1BS, PHR1 specific binding sequence, *cis*-element 'GNATATNC'), which is shared by several Pi-responsive genes.

Late Genetic Response to Pi Deficiency: PHR1 is involved in coordinated regulation of many 'late' Pi starvation genes, such as RNases, phosphatases, TPSI/Mt4 family (Franco-Zorrilla et al. 2004, Hammond et al. 2004) and OPSI1 (Wasaki et al. 2006), which have PHR1 binding sites. PHR1 binds as a dimer to the promoter of 'late' Pi-starvation genes. Most of the Pi taken up by roots is subsequently transported through xylem to shoots.

Sugar Signalling: Müller et al. (2007) have identified 149 transcripts of Pi induced genes, which are regulated by the interaction between Pi deficiency and sucrose availability. Many of these genes encode proteins involved in carbohydrate metabolism and P re-mobilisation.

Micro RNA (miRNA): MicroRNAs (miRNAs) containing 19–25 nucleotides are found in all animals and plants but not in fungi. They are post-transcriptional regulators encoded by specific genes, several at a time or by some portions of the introns of genes, whose m-RNA they regulate. They either completely destroy the m-RNA if their sequences exactly match (usually in plants) or repress the translation of m-RNA if there is a partial match. In the later case several of them simultaneously bind to the UTR (un-translated) region of m-RNA. In plants they may target the coding region itself (He and Hannon 2004).

miRNA and Phosphate Deficiency: Phosphate deficiency causes up-regulation of miR399, which decreases rapidly on Pi addition (Fujii et al. 2005, Bari et al. 2006). Over-expression of *Arabidopsis* miR399 in tomato results in increased accumulation of Pi. There is also augmented excretion of acid phosphatases and protons by roots, which facilitates Pi acquisition from soil (Gao et al. 2010). Homologues of miR399 have been found in rice, tomato, common bean (*Phaseolus vulgaris)* and *Medicago truncatula* (Kuo and Chiou 2011).

Methods Adapted to Improve Phosphate Use Efficiency: Phosphate (Pi) use efficiency (PUE) of crops is generally low (15–20%) due to various soil and plant related factors. Some of the methods adopted to improve PUE are as follows: (1) Growing suitable plant associations with high and low Pi-uptake capacities, (2) facilitation of Pi availability by one crop to the other through rhizosphere acidification, (3) manipulating expression of genes enabling growth in low-P environments.

1.4.3 Uptake of Potassium (K)

Potassium content of soils is in the range of 0.5–2.5%. Plants contain about 2–10% of K of their dry weight. Cytoplasmic concentration of K^+ is maintained at approximately 100 mM, although vacuole may contain 20–200 mM of K^+ (Gierth and Mäser 2007). Apoplastic concentration of K^+ may vary between 10 and 200 mM and may increase up to 500 mM (White and Karley 2010, Wang et al. 2013).

Potassium activates about 60 enzymes involved in various metabolic processes, such as photosynthesis, protein synthesis, oxidative metabolism etc., and improves quality and stress tolerance of crops. It is also involved in osmo-regulation, turgor driven movements and maintenance of the plasma membrane potential. Within the cytosol, K^+ ion neutralises the soluble and insoluble macromolecular anions and stabilises pH at about 7.2, which is optimal for most enzymatic reactions (Marschner 1995).

Genes involved in K uptake: A large number of proteins encoded by their corresponding genes are involved in K^+ transport in plants. According to Transporter Classification (TC) System there are two major pathways for K^+ acquisitions by plants, Class 1: Channels and other low-affinity systems effective at concentrations near 1 mM and above, Class 2: High-affinity systems operate at K^+ concentrations in the μM range. Channels also contribute to high-affinity K^+ uptake.

Pathways of K^+ Acquisitions by Plants: The pathways fall into several distinct categories as follows: (a) K^+ channels consisting of three families: Shaker-type channels, KCO channels (a total of 15 genes in *Arabidopsis*) and cyclic-nucleotide-gated channels (CNGC, 20 genes in *Arabidopsis*) (Very and Sentenac 2002). (b) Trk/HKT transporters [Na^+/K^+ symporter] (Schachtman 2000), one gene in *Arabidopsis*; (c) KUP/HAK/KT transporters [H^+/K^+ symporter] (Kim et al. 1998); 13 genes in *Arabidopsis.* (d) K^+/H^+ antiporter homologue; six genes in *Arabidopsis*; and (e) glutamate receptors (GLRs; 20 genes in *Arabidopsis*) (Very and Sentenac 2002).

Shaker Channels: Plant voltage-gated channels belonging to the Shaker family participate in K^+ uptake from the soil solution, long-distance K^+ transport in the xylem and phloem, and K^+ fluxes in guard cells during stomatal movements (Chérel 2004). Hyper-polarisation-activated channels are inwardly rectifying (i.e. mediating an inward K^+ current). Depolarisation-activated channels are outwardly rectifying (allowing K^+ efflux from the cell).

SKOR and GORK Channels: The Shaker-type channels are further subdivided into SKOR and GORK channels (both depolarisation-activated), KAT channels and AKT channels. AKT channels contain an ankyrin-binding motif, which is lacking in KAT type channels (Maser et al. 2001). Once K^+ has been taken up, its secretion into the root xylem for delivery to the shoot involves the outwardly rectifying SKOR channel, which could mediate the delivery of up to 50% of the K^+ in the xylem sap. Outwardly rectifying Shaker, GORK, is expressed in guard cells. AKT and KAT channels: *AKT1* gene encodes an inwardly rectifying channel, which plays a role in K^+ uptake from the soil solution. AKT2 has a role in the control of the phloem cell membrane potential, and in the regulation of sucrose loading/unloading into/from the phloem sap. The AKT2 channel also accounts for about 50% of K^+ permeability of mesophyll cells, AKT1 being responsible for the remaining 50%. Inwardly rectifying Shakers, KAT1 and KAT2, are expressed in guard cells.

KCO channels and cyclic-nucleotide-gated channels (CNGC): KCO channels are components of slow vacuolar (SV) double-pore K^+ channels located in the tonoplast.

Trk/HKT Transporters (TC: 2·A·38): HKT (High Affinity K^+ transporters) proteins of plants are part of the Trk superfamily of cation transporters and are topologically related to K^+ channels. Plant HKT amino acid sequences group into two subfamilies, and the genes of subfamily one have longer introns than those of subfamily two (Platten et al. 2006). Subfamily two exclusively contains monocot genes; subfamily one includes monocot genes and all known *HKTs* from dicots.

KUP/HAK/KT Transporters (TC: 2·A·72): All plant genomes contain genes encoding KUP (potassium uptake permeases)/HAK (high affinity potassium transporters)/KT (potassium transporters) (given different acronyms by different research groups) transporters (not found in *Protista* and *Animalia*). All KT/KUP/HAK transporters can be grouped into four distinct clusters. Plants have Cluster I or Cluster II transporters. Cluster III genes are found only in Arabidopsis and rice. Cluster IV is the smallest in number, which comprises only of four rice genes. Cluster I transporters have high affinity for K^+ and play a key role in potassium acquisition, when K^+ availability is low. Cluster II transporters facilitate low-affinity K^+ transport complementing potassium channels. These transporters are localised in the tonoplast and facilitate K^+ efflux from the vacuole. Under conditions of K^+ deprivation, export of K^+ from the vacuole is mediated by a K^+/H^+ symporter with a 1:1 stoichiometry for the maintenance of K^+ homeostasis.

K^+/H^+ Antiporter Homologues: Also known as *CHX* (Cation/H+ eXchanger), a member of the family, AtCHX17 expressed in the cortex and epidermis of the mature root is involved in K^+ acquisition and homeostasis rather than Na^+ transport. **Glutamate receptors (GLR)**: GLRs are non-selective cation channels involved in Ca^{2+} influx and are differentially activated by amino acids especially glutamic acid and glycine. All of the 20 genes of *Arabidopsis* are grouped into three clades and expressed in roots.

1.5 Secondary Nutrients

Calcium (Ca), magnesium (Mg) and sulphur (S) are considered as the secondary nutrients for plants. Soils in many locations contain reasonable amounts of these nutrients. Hence farmers do not apply these nutrients on a regular basis to their crops as NPK.

1.5.1 Uptake of Calcium (Ca^{2+})

Calcium is the fifth most abundant element and constitutes about 3.5% of the earth's crust. Most of the soils are moderate to rich in Ca^{+2} (0.7–1.5%), except strongly acidic tropical soils (0.1–0.3%) and calcareous soils in arid and semi-arid regions (1–30%). There is an equilibrium among solution Ca^{+2}, exchangeable Ca^{+2} and mineral Ca^{+2}.

$$\text{Mineral } Ca^{2+} \rightarrow \text{Solution } Ca^{2+} \rightleftarrows \text{Exchangeable } Ca^{2+}.$$

Healthy plants growing with adequate Ca^{2+} supply have a Ca^{+2} content of 0.1–5% of their shoot dry weight. A steady supply of 1–10 mM Ca^{2+} is required for normal plant growth (Gilroy et al. 1993). Plants have been classified according to their Ca^{2+} requirements: 'Calcifuges', which grow in acid soils with low Ca^{2+} and capacity to tolerate toxic concentrations of Fe, Al and Mn (*Crassulaceae*, *Brassicaceae* and *Fabaceae*); 'Calcicoles', which grow in calcareous soils with high Ca^{2+} and tolerance to deficiency of P and Fe (*Apiales* and *Asterrales*).

Acidic pectin residues (galacturonic acid) of cell wall bind Ca^{2+} and impart rigidity to it. The carboxylic groups of galacturonic acid of cell wall pectins in the middle lamella are responsible for root CEC. Cell wall provides an enormous reservoir of Ca^{2+} (10 μM–10 mM) in contrast to cytosol (100–200 nM). High concentration of Ca^{2+} (0.1–1.0 mM) is required on the outer surface of plasma membrane to maintain its structural and functional integrity (Hepler 2005).

Since Ca^{2+} at higher concentration is cytotoxic, its movement through phloem is slow and its distribution unequal within the plants. The older leaves contain more Ca^{2+} than the younger ones. Most of the Ca^{2+} transported through xylem (300 μm and 16·5 mm) is sequestered and locally deposited. Excessive Ca^{2+} within cytoplasm, which, move to vacuole, are deposited as insoluble salts of oxalic, phosphoric and phytic acids.

Calcium regulates various fundamental processes such as cytoplasmic streaming, thigmotropism, gravitropism, cell division, cell elongation, cell differentiation, cell polarity, photomorphogensis, and plant defence and stress responses (Song et al. 2010). Cytoplasmic concentration of Ca^{2+} needs to be strictly regulated at nano molar (nM) range (100–200 nM). Cytoplasmic streaming is permitted at a low Ca^{2+} concentration of 0.1 μM but an elevated concentration of 1 μM inhibits the process (Hepler 2005).

Various abiotic stresses such as cold, heat, salinity, drought, osmotic and oxidative stresses, and physical stimuli such as touch and swaying of the plants by wind cause transient perturbations of cytosolic Ca^{2+} concentration, which are restored to basal levels within minutes (White 2003, Reddy et al. 2011).

1.5.1.1 Genes Involved in Ca^{2+} Homeostasis

Calcium homeostasis in cytoplasm is achieved through regulation of influx/efflux of Ca^{2+} ion by (1) calcium channels, (2) Ca^{2+}/H^+ antiporters and (3) Ca-ATPases.

Influx of Ca^{2+} ions takes place through voltage gated channels, such as (1) hyper polarisation activated cation channels (HACC), (2) depolarisation activated cation channels (DACC), (3) voltage independent cation channels (VICC), (4) ligand gated channels (cyclic nucleotide gated channels, CNGCs) and (5) glutamate receptors (GLRs).

Efflux of Ca^{2+} from cytosol is carried out by (1) Ca/H^+ antiporters (CAX1), which mediate a high affinity low turnover efflux and found in tonoplasts of *Arabidopsis* (CAX1-CAX11 & MHX), VCAX1 in mungbean, OsCAX1a in rice, GmCAX1 in soybean and (2) P-type Ca-ATPase, which mediate a low affinity high capacity efflux of Ca^{2+}. There are two types of Ca-ATPases: (a) P2A found in ER, Golgi and Endosomes and (b) P2B in ER, plasma membrane, plastid envelop.

1.5.1.2 Ca^{2+} Regulated Gene Expression in Response to some Specific Abiotic Stress

Drought: Exogenous Ca^{2+} has been reported to enhance drought resistance, inhibit synthesis of activating oxides, protect the structure of plasma membrane, maintain normal photosynthesis and regulate the metabolism of plant hormones. Cellular

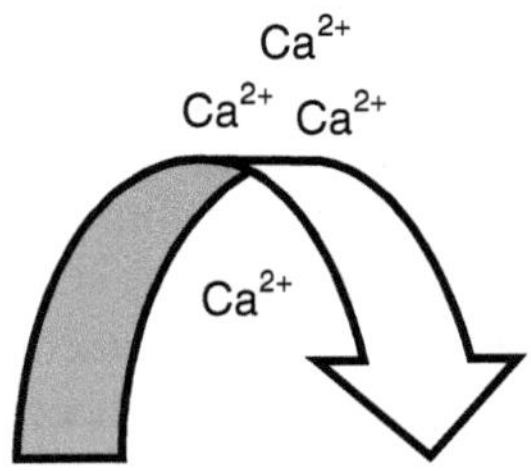

Fig. 1.6 High cytosolic Ca^{2+} on the convex side (adapted from Mitra 2015)

Ca^{2+} as a second messenger transmits drought signal and induces physiological response to water stress.

Cold: Ca^{2+} permeable channel proteins have been reported to be primary temperature sensors in plants and are involved in plant response to cold stress. A large number of genes of CBF regulon are induced during the process of cold acclimation.

Heat: Plants in response to higher temperature synthesise heat shock proteins (HSPs), a number of which has been characterised. Their transcription is tightly regulated by TFs. Elevation of cellular Ca^{2+} due to heat changes expression of several genes including Ca^{2+} sensors.

Mechanical Stimuli, such as touch and wind, induce elevation of cytosolic Ca^{2+} concentration (Braam 2005). Different types of mechanical stimuli induce distinct type of Ca^{2+} response in *Arabidopsis* roots. Touch stimuli induce monophasic elevation of cytosolic Ca^{2+} concentration at the touch site. Bending induces biphasic transient elevation of cytosolic Ca^{2+} concentration on the convex (stretching) side (Fig. 1.6).

1.5.2 Uptake of Magnesium (Mg^{2+})

Mg^{2+} has the smallest ionic radius, highest charge density and largest hydrated radius. There is a 400-fold difference between volumes of hydrated and non-hydrated states (Li et al. 2001, Geberta et al. 2009). Earth's crust contains about 1.93% of Mg. The Mg content of soil may vary from 0.1% coarse-textured humid soils to 4% in fine-textured soils from arid or semi-arid region. Mg^{2+} ions unlike Ca^{2+} are more susceptible to leaching since they are not as strongly adsorbed to clay minerals or organic matter due to their large hydrated radius.

Mg^{2+} concentration in crops varies from 0.1 to 0.4%. The critical limit of Mg^{2+} in dry banana leaves has been reported to be 0.3% and of coconut 0.2% (14th fond) (Mitra 2006). The free Mg^{2+} level in the cytosol is strictly regulated due to its role in photosynthesis and on membrane ionic currents (Shaul 2002). The concentration of Mg^{2+} in the metabolic pool of leaf cells (cytoplasm and chloroplast) is reported to be 2–10 mM (Leigh and Wyn-Jones 1986). Free Mg^{2+} concentration is considerably less. About 90% of Mg^{2+} is complexed with cytoplasmic ATP.

Mg^{2+} acts as cofactor of many enzymes, such as RNA polymerase, ATPases, protein kinases, phosphatases, carboxylases and glutathione synthetase. It is required for aggregation of ribosomes and is the central atom of chlorophyll molecule. During the process of chlorophyll formation, insertion of Mg^{2+} into the porphyrin structure is catalysed by Mg^{2+}-chelatase (Walker and Weinstein 1991; Papenbrock et al. 2000). Mg^{2+} is involved in both light and dark reactions of photosynthesis. Mg^{2+} deficient leaves are therefore highly photo-sensitive (Shaul 2002). Mg^{2+} deprivation elicits rapid Ca^{2+} uptake and activates Ca^{2+}/calcineurin signalling (Wiesenberger et al. 2007). Low magnesium concentrations may become a limiting factor for functional intracellular communication (Geberta et al. 2009).

Mg Transporters: CorA (cobalt resistant phenotype of bacterial mutants) is involved in the transport of Mg^{2+} across biological membranes and has unique structures (Moomaw and Maguire 2008; Geberta et al. 2009). CorA appears to be a constitutive gene since it is not transcriptionally regulated. CorA homologue proteins have been found in all living organism including plants. It has been observed that grasses and cereals treated with Al^{3+} show Mg^{2+} deficiency (Tan et al. 1991) and application of higher levels of Mg^{2+} can alleviate Al^{3+} toxicity (Tan et al. 1991; Matsumoto 2000). It has also been shown that Al^{3+} inhibits Mg^{2+} uptake by roots (Rengel and Robinson 1989). The hydrated radius of Mg^{2+} and Al^{3+} are similar (Bose et al. 2011). At milli-molar concentration Mg^{2+} can effectively compete with Al^{3+} for the same binding sites of the roots.

1.5.3 Uptake of Sulphur (S)

Soil derives S from the S bearing minerals, S-containing fertilisers and from atmosphere, which is enriched by S containing gases primarily SO_2 from industrial emissions. About 30–40% of S in top soil at Rothamsted, UK, is derived from atmospheric deposition (Zhao et al. 2001). Globally soils of many countries are deficient in S and need application of S to meet the crop demand for maintaining their yield and quality parameters (McGrath et al. 1996). S deficiency is generally found in coarse-textured alluvial soils (Entisols and Inceptisols), red and lateritic soils (Alfisols) and Vertisols. Application of S-fertilisers in optimum doses does not have any residual effect since clay minerals do not bind sulphate and it is leached out of soil. Annual application of S as sulphate >50 kg S ha^{-1} for more than 150 years in the Broadbalk Experiments in Rothamsted did not result in any build up of S in the soil (Zhao et al. 2001).

Sulphur (S) is an essential plant nutrient and is considered as the fourth major nutrient after N, P and K (TSI 2008). It is also of importance in human and animal nutrition. The total S content of plant tissues has been reported to be 0.5–1.5% of the dry weight of the plants (Zhao et al. 1993; Marschner 1995; Burandt et al. 2001). Sulphur is a constituent of various organic plant constituents.

Sulphur is a constituent of amino acids cysteine and methionine, which are involved in maintaining protein structure and conformation. It is a constituent of coenzymes and prosthetic groups such as lipoic acid, coenzyme A and thiamine.

Sulphur compounds are involved in response to abiotic and biotic stress, such as glutathione in the detoxification of active oxygen species. Sulphur plays an important ecological role in defence against herbivores and pathogens.

Effects of S on Yield and Quality of Crops: Field experiments at Rothamsted show that yield loss due to S-deficiency in oilseed rape can be up to 70% and in cereals up to 50% (Zhao et al. 2001). Oil seed crops generally have a higher requirement of S as compared to other crops. S-deficiency also affects quality of crops. Under limiting S availability, wheat grains accumulate low sulphate storage proteins such as ω-gliadin and high molecular weight subunits of glutenin at the expense of S-rich proteins. Such changes in protein composition affect dough rheology. The oil content of oil seeds is reported to increase due to S-application as follows: sunflower, 3.8%; linseed, 6%; soybean, 9.2%; mustard, 9.2%; and groundnut, 11.3% (TSI, 2008).

Plant Sulphate Transporters: Plasma membrane sulphate transport is probably a pH-dependent proton coupled co-transport involving $3H^+/SO_4^{2-}$ stoichiometry. The sulphate transporter protein expressed in the plasma membrane of root cells consists of a single polypeptide chain of around 70–74 kD. A large number of sulphate transporter genes have been identified from *Arabidopsis,* rice and other plants.

In *Arabidopsis* the gene family consists of 14 isoforms, subdivided into five groups. Wheat, *Brassica oleracea* and rice have similar gene groups and are close homologues with similar functions. Group-1, high affinity transporters, is located in the plasma membrane. Group-2, low affinity transporters, is located in the plasma membrane. Group-3, of unknown function, may be associated with hetero-dimer association (Kataoka et al. 2004a). Group-4 is involved in efflux of sulphate across tonoplast of vacuole into cytoplasm (Kataoka et al. 2004b). Group-5, a member, Sultr 5;2 is probably an intracellular transporter involved in Mo metabolism in *Arabidopsis* and named as mot1 (Shinmachi et al. 2010).

Regulation of Sulphate Uptake: S uptake is closely coordinated with N and C metabolism. There is an induction of the genes of high affinity sulphate transporters due to addition of sucrose. Nitrogen deficiency strongly reduces sulphur uptake and consequently a significant reduction in accumulation of transcripts of high affinity sulphate transporters AtSultr1;1 and AtSultr1;2. However S-deficiency does not decrease total N content although there is an increase in O-acetyl serine (OAS), which is the precursor for synthesis of cysteine and has a role in the regulation of sulphate uptake and reduction (Hawkesford 2003). Addition of OAS (*O*-acetyl serine) to plants with adequate supply of sulphate leads to an increase in mRNA levels of sulphate transporters, sulphate uptake rates and tissue content of glutathione and cysteine. While sulphate, cysteine and glutathione act as negative regulator of sulphate transporter gene expression, OAS overrides such effects and acts as a positive regulator (Smith et al. 1997). The set of genes regulated under S-deficient conditions differs considerably between leaves and roots (Hirai et al. 2003).

miRNA (microRNA): Expression of miR395 is significantly up-regulated during S-deficiency. Two families of genes involved in sulphate metabolism are targeted by miR395. (1) The APS genes coding ATP sulphurylase isoforms: ATPS1, ATPS3, ATPS4 and (2) The genes of low affinity sulphate transporters, Sultr2;1, which are located in the xylem parenchyma cells of roots and shoots. Sultr2;1 is cleaved by miR395 (Liang et al. 2010).

1.6 The Micronutrients

Zinc (Zn), Iron (Fe), Manganese (Mn), Copper (Cu), Boron (B), Molybdenum (Mo), Chloride (Cl^-), Cobalt (Co) and Nickel (Ni) are considered as micronutrients. They are applied in small quantities to plants but are essential for plant growth. When present in less than optimal quantities they show characteristic deficiency symptoms. However they become toxic to the plants beyond a threshold concentration.

1.6.1 Uptake of Zinc (Zn^{2+})

Lithosphere contains about 80 ppm of Zn. The Zn content of soil is within a range of 10–300 ppm. Zinc deficiency is generally observed in low land soils, since low redox potential causes precipitation of Zn^{2+} as zinc sulphide, zinc carbonate and zinc oxy-hydroxides. In aerobic soils Zn^{2+} is readily available to plants on cation exchange sites. However calcareous soils tend to become Zn deficient even under aerobic conditions. Short-term flooding of these soils may make Zn readily available at an intermediate redox potential (Rose et al. 2013).

Zinc (Zn) is essential for plant growth. Zinc concentration less than 15–20 μg in leaves per gram of dry leaf tissues leads to Zn deficiency. About one third of world's population suffers from mild Zn deficiency. Zinc deficiency causes impaired brain development, dysfunction of the reproductive system, anorexia, immune disorders, hair loss, skin lesions and loss of taste and smell (Fernandez 2007; Song et al. 2010).

Zinc is an essential catalytic component of over 300 enzymes such as alkaline phosphatase, alcohol dehydrogenase, carbonic anhydrase and Cu–Zn superoxide dismutase. Zn plays an important role in transcriptional and post-transcriptional processes, in protein degradation and protein-protein interactions (Broadley et al. 2007, Song et al. 2010). Several transcriptional regulatory proteins have motifs stabilised by Zn, such as Zn finger, Zn cluster and RING finger domains.

Mechanisms of Zn Uptake by Plants: Rice varieties with higher efficiency for Zn uptake secrete citrate and malate at an increased rate in response to Zn deficiency in soil. The amount of citrate and malate excreted is however not enough to mobilise Zn from Zn deficient soil (Rose et al. 2013). Zn deficiency in barley plants is reported to induce synthesis and secretion of mugeniec acids, which are effective in Zn uptake from the soil (Suzuki et al. 2006).

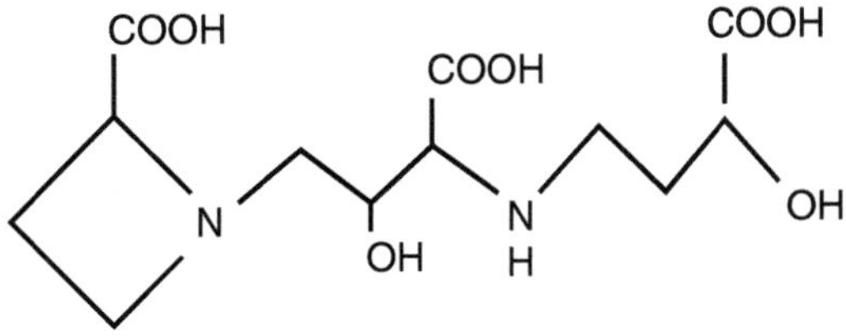

Mugineic acid

Zn^{2+}-Transporters: It is essential to maintain Zn^{2+} homeostasis within plants and their various organs at an acceptable physiological limit since Zn is toxic beyond a threshold concentration. This is carried out by a coordinated expression of Zn^{2+} transporters, which are involved in Zn^{2+} uptake from the soil, translocation of Zn^{2+} to various organs and tissues, in intracellular sequestration and transport to vacuole.

The transporters involved belong to several groups, such as HMAs (heavy metal transporters) of P-type ATPase superfamily (HMA2 and HMA4); PCRs (plant cadmium resistance transporters (PCR2); MTPs (metal transporter proteins) of CDF family and the ZIP (ZRT and IRT like protein family) (ZIP1, ZIP3 and ZIP7).

TFs (transcription factors) have been reported to be involved in molecular control of Zn^{2+} homeostasis in plants under Zn^{2+} deficiency such as bZIPs (basic-region leucine zipper), bZIP19 and bZIP23.

1.6.2 Uptake of Iron

Iron constitutes about 5% of earth's crust. Most of the soils around the world are rich in iron. Iron in soil is present in the form of an amorphous Fe $(OH)_3$ precipitate, which is the immediate source of iron uptake by plants. Availability of Fe to plant roots depends on redox potential and pH of the soil. Plant tissue concentration of 1–5 μM Fe is considered sufficient and a concentration below 1 μM is likely to cause deficiency. A concentration above 10 μM may cause toxicity with reduction of growth parameters (Mitra et al. 2009). However these limits may vary considerably among different plant species and their genotypes.

Fe is a constituent of a number of enzymes and proteins (ferritin, farataxin, Fe–S cluster, etc.). Several flavoproteins contain firmly bound Fe^{3+} ion such as succinic $FADH_2$ dehydrogenases, dihydroorotic acid dehydrogenase, and xanthine and aldehyde oxidases. Iron is a structural component of a number of molecules with porphyrin ring such as cytochromes, hemes, ferrichromes and leghemoglobin. Most of the Fe-containing enzymes are involved in oxidation-reduction reactions in respiration and photosynthesis.

Mechanism of iron uptake by plants: There are two distinct iron-uptake systems based on the response of plants to Fe deficiency, Strategy-I and Strategy-II.

Strategy-I plants: These include all dicots and non-*graminaceous* monocots. Fe deficiency causes a decrease in rhizosphere pH of these plants to facilitate release of Fe^{3+} ion from insoluble sources. The sparingly soluble ferric iron is then reduced at the root surface by membrane-resident NADPH-dependent ferric chelate reductase. Reduced ferrous iron is absorbed into root cells by the high affinity Fe^{2+}-transporter, IRT1, a member of the ZIP metal transporter family. IRT1-like Fe^{2+}-transporters have been isolated from several dicotyledonous species.

Strategy-II plants: These are limited to *graminaceous* monocots. These plants release mugineic acid-family (MAs) phytosiderophores to the rhizosphere, where they solubilise sparingly soluble iron by chelation. The chelated complex is then absorbed into the roots. Rice plants use MAs to acquire Fe from the rhizosphere. Synthesis of MAs and uptake of MA chelated iron are strongly induced under iron deficient conditions.

Iron Transporters: The transporters involved in Fe uptake are (1) IRTs of ZIP family (AtIRT1, AtIRT2; OsIRT1, OsIRT2 etc.) involved in high affinity iron uptake by roots, (2) Nramps (AtNramp1, AtNramp3, AtNramp4, OsNramp5, LeNramp1, AhNramp1 etc. found in various plants) involved in Fe transport, (3) ABC transporter [AtABCB25 (AtATM3) is a mitochondrial ABC transporter involved in biogenesis Fe-S clusters in plants], (4) H^+-ATPase (expressed in the root epidermis release protons to the rhizosphere, which lowers pH and make iron more soluble. Fe deficiency up-regulates the H^+-ATPases, AHA1, AHA2 and AHA7 in the root epidermis) and (5) the YSL transporter (YS1 found in *Poaceae* roots is a proton coupled symporter of Fe(III)-PS complexes).

1.6.3 Uptake of Manganese (Mn^{2+})

Earth's crust contains an average of 1000 $\mu g\ g^{-1}$ of Mn. Total Mn in soils varies between 20 and 3000 $\mu g\ g^{-1}$ with an average of about 600 $\mu g\ g^{-1}$.Mn in the form of its oxides and hydroxides is coated on soil particles along with iron oxides and other constituents. Mn can exist in various oxidation states (0, II, III, IV, VI and VII). In biological systems Mn occurs preferably in the oxidation states of II, III and IV (Guest et al. 2002). The concentration of Mn^{2+} in soil is controlled primarily by MnO_2 and ranges within 1 ppm.

$$MnO_2 + 4\mathrm{H}^+ + 2\mathrm{e}^- = Mn^{2+} + 2\mathrm{H_2O}.$$

About 90% of solution Mn^{2+} exists as organic complexes.

Mn is a component of photosynthetic proteins and enzymes. Its deficiency especially in chloroplast affects water splitting mechanism of Photo-system II (PS-II), which provides electrons for photosynthetic electron transport (Buchanan et al. 2000). Mn is involved in some of the essential metabolic processes, such as photosynthesis, respiration, activation of hormones (IAA through IAA oxidase, Burnell 1988) and synthesis of amino acids (Millaleo et al. 2010).

Mn is a cofactor of about 35 enzymes such as Mn-superoxide dismutase, Mn-catalase, pyruvate carboxylase and phosphoenolpyruvate carboxykinase. Mn is essential for synthesis of chlorophyll, ATP synthesis, bio-synthesis of fatty acids, acyl lipids and proteins, synthes tyrosine and secondary products, flavonoids and lignin. It is involved in RuBP carboxylase reactions and biosynthesis of isoprenoids and assimilation of nitrate. Mn is involved in some of the essential metabolic processes, such as photosynthesis, respiration, activation of hormones, IAA through IAA oxidase and synthesis of amino acids (Millaleo et al. 2010).

Mechanism of Mn^{2+} Uptake: Cellular Mn^{2+} uptake is carried out by a number of transporter proteins.

Mn Transporters: The gene families involved in Mn transport include the following:

(1) Cation/H^+ antiporter: The *Arabidopsis* AtCAX1 is a vacuolar high affinity Ca^{2+}/H^+ antiporter. AtCAX2 has low affinity for Ca^{2+} and possibly transports Mn^{2+} and Cd^{2+} across the tonoplast (Hirschi et al. 1996, 2000). (2) AtNramp1, AtNramp3 and AtNramp4 can transport Fe, Mn and Cd. (3) ZIP5, ZIP6 and ZIP9 transport Mn alone. (4) The proteins of CDF family are involved in efflux of transitional metal cations, Zn^{2+}, Cd^{2+}, Co^{2+}, Ni^{2+} or Mn^{2+} from cytoplasm to outside of the cell or into sub-cellular compartments to maintain metal homeostasis and tolerance to their toxic effects (Hall and Williams 2003; Hanikenne et al. 2005).

1.6.4 Uptake of Copper (Cu^{2+})

Copper content of earth's crust is about 50–70 μg g^{-1}. Total Cu content of soils varies between 1 and 40 μg g^{-1} with an average of 9 μg g^{-1}. Copper (Cu^{2+}) is chemically adsorbed to the surface of clays, and Fe, Al and Mn oxides. It is one of the divalent cations, which is most strongly adsorbed to oxides of Fe and Al, forming Cu-O-Fe and Cu-O-Al surface bonds. It also forms such bonds with clay minerals. In soil solution the dominant ion at pH below 7.0 is Cu^{2+} and above pH 7.0 Cu $(OH)_2$. The solubility of Cu^{2+} decreases with an increase in pH (Lindsay 1979).

$$Cu^{2+} + 2\mathrm{H_2O} = Cu(OH)_2 + 2\mathrm{H}^+.$$

Copper concentration in plant tissues is about 1–5 μg g^{-1} of dry weight and in leaves 5–20 μg g^{-1} of dry weight. Under physiological conditions, copper exists as Cu^+ and Cu^{2+} forms. The Cu^{2+} form is often bound by N in histidine side chain and Cu^+ preferably to S in cysteine or methionine. Both deficiency and toxicity of Cu adversely affect crucial physiological processes in plants. It is essential that Cu concentrations in tissues and cells need to be controlled within a narrow physiological range. This involves uptake of Cu from soil, transport to different parts of the plants and regulation of its concentration in tissues, cells and intra-cellular organelles.

Cu is involved in photosynthetic and respiratory electron transport chain, cell wall lignification, Vit-C metabolism, ethylene perception, carbohydrate metabolism, nitrogen fixation, fatty acid de-saturation and hydroxylation, protection from oxidative stress, pathogen defence and biogenesis of molybdenum cofactor. Cu is an integral part of plastocyanin a protein involved in photosynthetic electron transport and of cytochrome *c* oxidase involved in respiratory electron transport chain. Deficiency of Cu reduces plastocyanin biosynthesis, which affects PSI electron transport. Several enzymes need Cu ion as a cofactor such as polyphenol oxidases, ascorbate oxidase, diamine oxidases and laccase. The detoxification of superoxide radicals are carried out by the enzyme Cu/Zn superoxide dismutase. Cu is involved in oxidative phosphorylation, iron mobilisation, signalling transcription and protein trafficking machinery (Yruela 2005, Pilon et al. 2006, Puig et al. 2007). Redox reactions between Cu^{2+} and Cu^{+} can catalyse production of highly toxic hydroxyl radicals (HO′), which fragments Cu/Zn SOD and causes damage to cell membranes, nucleic acids, proteins and other bio-molecules. Damage to cell membrane causes ion leakage such as K^{+} ion leakage from the excised roots of *Agrostis capillaris*. Oxidative damage caused by Cu toxicity interferes with various cellular processes such as photosynthesis, pigment synthesis and number of other metabolic processes resulting in stunted growth of plants. Cu toxicity causes destruction of thylakoid structure in chloroplast and significant modification of protein and lipid composition of thylakoid membranes.

Genes involved in Cu Uptake: A wide range of gene families and proteins have been identified, which are involved in Cu homeostasis. Some of them are (1) copper transporter proteins (COPT) (transport of Cu from exterior into cytoplasm), (2) P-type ATPases (trans-membrane transport of Cu), (3) the YSL transporters (Cu delivery from vascular tissues to seeds along with iron (Yruela 2009)) and (4) CCH (copper chaperones), involved in symplastic Cu transport through plasmodesmata.

miRNA: When there is excessive Cu, transcription of miR398 is a down-regulated as otherwise it would have cleaved mRNA of CSD1 and CSD2, which are involved in detoxification of Cu-induced oxidative stress.

1.6.5 Uptake of Boron (B)

Boron occurs on earth's crust in most of the igneous rocks at a concentration less than 10 µg g^{-1}. The primary B mineral in soil is Tourmaline, a relatively insoluble borosilicate. Total B concentration in soils is mostly around 7–80 µg g^{-1}. About 95% of soil B is not available to plants. About 0.1 µg g^{-1} of B in soil solution is considered adequate for most of the monocots (Havlin et al. 2007). Availability of B decreases with an increase in pH above pH 6.3–6.5. Liming acid soils may cause temporary B deficiency. High Ca^{2+} concentration in alkaline soils or recently over limed soils negatively affects B availability. Application of higher doses of K-fertilisers may also affect B availability. H_3BO_3 is the preferred form in which roots absorb B. Boron concentration in monocots varies between 6 and 18 µg g^{-1} and in dicots from 20 to 60 µg g^{-1}. Boron deficiency occurs at <20 µg g^{-1} B in mature leaf tissues. B is not readily translocated from older to younger leaves.

Boron is essential for cell wall structure and functions. Primary cell wall of higher plants consists of cellulose, hemicelluloses (xyloglucan and arabinoxylan) and pectic polysaccharides. The major components of pectic matrix are homogalacturonan (HG), rhamnogalacturonan-I (RG-I) and rhamnogalacturonan-II (RG-II). Boron forms cross links in pectic polysaccharides through borate-diol bonding of two rhamnogalacturonan-II (RG-II) molecules in the cell wall. RG-II is present in primary cell wall.

Boron Transporters: Two types of B-transporters have been identified in *Arabidopsis thaliana*, which are involved in B-uptake under conditions of B deficiency, NIP5;1 and BOR1. NIP5;1 gene is up-regulated under B limiting conditions. NIP5;1 protein is localised in the plasma membrane. BOR1 encodes a plasma membrane protein, which is involved in boron efflux from cells (Takano et al. 2002).

1.6.6 Uptake of Molybdenum (Mo)

The average molybdenum content of earth's crust is about 2 μg g^{-1} and of soil 0.2–5 μg g^{-1}. $CaMoO_4$ is the major form in which Mo is found in both acidic and alkaline soils. Various forms of Mo found in soil solution are MoO_4^{-2}, $HMoO_4^-$ and H_2MoO_4 depending upon pH. Availability of Mo increases with increase in soil pH. Mo is strongly adsorbed by Fe/Al oxides and may form insoluble compounds, which become unavailable to plants.

$$2Fe^{3+} + 3MoO4^{-2} = Fe_2\left(MoO_4\right)_3.$$

Plants containing <0.2 μg g^{-1} of dry weight are likely to show Mo deficiency. However this concentration may vary in different plant species.

Mo is taken up by plants as MoO_4^{-2}, which is then used for synthesis of pterin-based Mo cofactor (Moco). Mo in Moco is covalently bound to two S atoms of a unique tricyclic pterin moiety known as molybdo-pterin. All forms of life contain Mo enzymes, which are involved in global cycling of C, S and N. All of these enzymes are activated by Moco except nitrogenase. Moco becomes unstable when it is dissociated from the protein part of the enzymes (Basu and Burgmayer 2011).

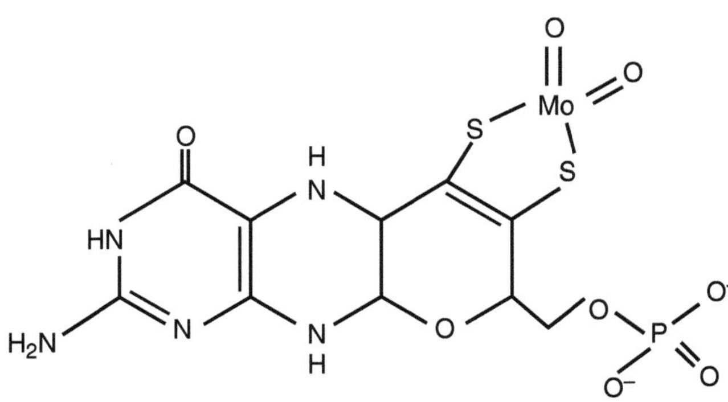

Structure of Moco (adapted from Mitra 2015).

Mo Transporter: A member of group-5 sulphate transporter, Sultr 5;2 is probably an intracellular transporter involved in Mo (molybdenum) metabolism in *Arabidopsis* and is named as MOT1. MOT1 is expressed in all tissues of wheat (Shinmachi et al. 2010). MOT1 is a high affinity Molybdate transporter specific for Mo and allows plant to take up Mo from the scarce resource of Mo in soil.

Mo deficiency affects N and S metabolism in a manner different from N and S deficiency. There is a significant accumulation of phosphate in plants due to Mo deficiency.

1.6.7 Uptake of Nickel (Ni^{2+})

Ni constitutes about 3% of earth's crust. Total concentration of Ni in soil is around 5–500 mg kg^{-1} with an average of 50 mg kg^{-1}. Arable soils contain 3–1000 mg kg^{-1} of Ni (Liu et al. 2012). Nickel becomes available to plants in the form of Ni^{+2} ions. Nickel is readily oxidised in soil and becomes unavailable to plants above pH 6.7. Nickel concentration > 10 μg g^{-1} is generally considered toxic to sensitive plants.

Nickel is a constituent of enzyme urease (Dixon et al. 1975), which is present in a large number of plants. It has been observed in several plants that Ni deficiency results in accumulation of toxic concentration of urea in the leaves due to depression of urease activity. Nickel acts as a cofactor of enzyme urease and is essential for conversion of urea into NH_4^+ for use by plant tissues.

$$NH_2 \cdot CO \cdot NH_2 + 3\mathrm{H_2O} \underset{\text{Urease}}{\overset{\text{Ni}}{=}} 2NH_4^+ + CO_2 + 2OH^-.$$

Studies with ^{63}Ni on wheat plants showed that there was quick movement of ^{63}Ni from labelled part of roots to newly developed roots and transiently to expanding younger leaves. There was also rapid redistribution of ^{63}Ni from older to younger leaves indicating a high mobility through phloem (Page and Feller 2005).

Nickel Transporters: AtIRT1 (Iron Regulated Transporter1), a member of ZIP family, is possibly the major transporter involved in high affinity iron uptake by roots of *Arabidopsis*. It has been suggested that AtIRT1 also transports Ni (Nishida et al. 2011).

1.6.8 Uptake of Chloride (Cl^-)

Broyer et al. (1954) established Chlorine (Cl_2) in the form of Cl^- ion as an essential plant micronutrient in 1954. Chloride concentration in earth's crust is around 0.02–0.05%. It occurs in igneous and metamorphic rocks. Sea water contains about 3.5% of NaCl (Na^+, 1.08%; Cl^-, 1.94%; SO_4^{-2}, 0.27%; Mg^{2+}, 0.13%; Ca^{2+}, 0.04% and K^+,

0.04%) (Munns 2002). Chloride is added to soil along with potassic fertiliser KCl, which constitutes 92% of world potassium consumption in agriculture. Chlorine exists as chloride (Cl^-) ion in soil solution and is in the range of 0.5 ppm in acid soils and more than 6000 ppm in saline and sodic soils.

Concentration of chloride in higher plants is usually 0.2–2.0% but may go up to 10% in saline soils (Fixen 1993).

Chloride (Cl^-) is involved in (1) osmotic adjustments (cell turgor and stomatal movements and regulation of water loss), (2) transport of nutrients in plants (Ca, Mg and K), (3) increases yield of cereal crops, (4) regulation of function of several enzymes, (5) photosynthesis and (6) reduces disease attack (suppresses stem rot, sheath blight in rice, common root rot, stripe rust in wheat, downy mildew in pearl millet, stalk rot in corn etc.). Chloride constitutes an important part of nutrient management for crops like coconut and oil palm.

Minimum concentration of chloride in plant tissue essential for biochemical reactions is about 100 mg kg^{-1} of dry wt. Chloride is a major osmotically active ion in the vacuole and is involved in turgor and osmo-regulation. Chloride is reported to act as an inhibitor of malic enzyme. Chloride acts as an activator of a number of enzymes. Chloride dependent α-amylases, angiotensin converting enzymes (ACE) and photosystem-II are activated by bound chloride.

1.6.8.1 Chloride Transporters and Chloride Channel Gene (CLC)

CCC (Cation Chloride Transporters) family of proteins occur in animals, plants, fungi and bacteria. They are KCl/NaCl symporters. The generalised transport reaction of CCC symporters is:

$$\left\{Na^{+}\ \text{or}\ \mathrm{K}^{+}+Cl^{-}\right\}(\text{out}) \rightleftarrows \left\{Na^{+}\ \text{or}\ \mathrm{K}^{+}+Cl^{-}\right\}(\text{in}).$$

Other transporters involved in chloride channels are (1) Al^{3+}-activated Malate Transporter and (2) NRTs (Nitrate Transporters).

References

Aoki, N., Tomohito, E., Akira, K., Tomomi, A., Masataka, W., Haruto, S., Ryu, O. (2009). Genetic modification of nitrogen use efficiency in potato and rice by introducing fungal glutamate dehydrogenase, American Society of Plant Biologist. Plant Biol, 2009, Honolulu, Hawaii, Abs #P39004.

Anthony, D. M., Glass, D., Britto, T. D., Kaiser, B. N., James, R., Kinghorn, J. R., Kronzucker, H. J., Anshuman, A., Okamoto, M., Rawat, S., Siddiqi, M. Y., Unkles, S. F., & Vidmar, J. J. (2002). The regulation of nitrate and ammonium transport systems in plants. *Journal of Experimental Botany, 53*(370), 855–864.

Bari, R., Pant, B. D., Stitt, M., & Scheible, W.-R. (2006). PHO2, MicroRNA399, and PHR1 define a phosphate-signaling pathway in plants. *Plant Physiology, 141*(3), 988–999.

Basu, P., & Burgmayer, S. J. N. (2011). Pterin chemistry and its relationship to the molybdenum cofactor. *Coordination Chemistry Reviews, 255*(9–10), 1016–1038.
Bieleski, R. L. (1973). Phosphate pools, phosphate transport, and phosphate availability. *Annual Review of Plant Physiology, 24*, 225–252.
Bose, J., Babourina, O., & Rengel, Z. (2011). Role of magnesium in alleviation of aluminium toxicity in plants. *Journal of Experimental Botany, 62*, 2251–2264.
Braam, J. (2005). In touch: Plant responses to mechanical stimuli. *The New Phytologist, 165*, 373–389.
Breckle, C. W. (1991). Growth under heavy metals. In Y. Waisel, A. Eshel, & U. Kafkafi (Eds.), *Plant roots: The hidden half* (pp. 351–373). New York, NY: Marcel Dekker.
Brownell, P. F., & Crossland, C. J. (1972). The requirement of sodium as a micronutrient by species having C_4 dicarboxylic photosynthetic pathway. *Plant Physiology, 49*, 794–797.
Broadley, M. R., White, P. J., Hammond, J. P., Zelko, I., & Lux, A. (2007). Zinc in plants. *The New Phytologist, 173*, 677–702.
Broyer, T. C., Carlton, A. B., Johnson, C. M., & Stout, P. R. (1954). Chlorine-a micronutrient element for higher plants. *Plant Physiology, 29*(6), 526–532.
Buchanan, B., Grusen, W., & Jones, R. (2000). *Biochemistry and molecular biology of plants*. Maryland: American Society of Plant Physiologists, 1367 pp.
Burandt, P., Papenbrock, J., Schmidt, A., Bloem, E., Haneklaus, S., & Schnug, E. (2001). Genotypical differences in total sulfur contents and cysteine desulf-hydrase activities in *Brassica napus* L. *Phyton (Horn, Austria), 41*, 75–86.
Burnell, J. (1988). The biochemistry of manganese in plants. In R. D. Graham, R. J. Hannam, & N. J. Uren (Eds.), *Manganese in soil and plants* (pp. 125–137). Dordrecht, The Netherlands: Kluwer Academic Publishers.
Calderon-Vazquez, C., Ibarra-Laclette, E., Caballero-Perez, J., & Herrera-Estrella, I. (2008). Transcript profiling of *Zea mays* roots reveals gene responses to phosphate deficiency at the plant- and species-specific levels. *Journal of Experimental Botany, 59*, 2479–2497.
Chaumont, F., Barrieu, F., Jung, R., & Chrispeels, M. J. (2000). Plasma membrane intrinsic proteins from maize cluster in two sequence subgroups with differential aquaporin activity. *Plant Physiology, 122*, 1025–1034.
Chérel, I. (2004). Regulation of K^+ channel activities in plants: From physiological to molecular aspects. *Journal of Experimental Botany, 55*, 337–351.
Crawford, N. M., & Glass, A. D. M. (1998). Molecular and physiological aspects of nitrate uptake in plants. *Trends in Plant Science, 3*(10), 367–407.
Dixon, N. E., Gazzola, C., Blakel, R. L., & Zerner, Y. B. (1975). Jack-bean urease (E.C.3.5.1.5.3.). A metallo-enzyme—A simple biological role for nickel. *Journal of the American Chemical Society, 97*, 4131–4133.
Dubyak, G. R. (2004). Ion homeostasis, channels and transporters: An update on cellular mechanisms. *Advances in Physiology Education, 28*(1–4), 143–154.
Egami, T., Wakayama, M., Aoki, N., Sasaki, H., Kisaka, H., Miwa, T., & Olsugi, R. (2012). The effects of introduction of a fungal glutamate dehydrogenase (*gdhA*) gene on photosynthetic rate, biomass, carbon and nitrogen contents in transgenic potato. *Plant Biotechnology, 29*, 57–64.
Fan, X., Xie, D., Chan, J., Lu, H., Xu, Y., Ma, C., & Xu, G. (2014). Over expression of OsPTR6 in rice increased plant growth at different nitrogen supplies but decreased nitrogen use efficiency at high ammonium supply. *Plant Science, 227*, 1–11.
Fang, Z. Y., Shao, C., Meng, Y. J., Wu, P., & Chen, M. (2009). Phosphate signalling in *Arabidopsis* and *Oryza sativa*. *Plant Science, 176*, 170–180.
Feng, H., Yan, M., Li, B., Shen, Q., Miller, A. J., & Xu, G. (2011). Spatial expression and regulation of rice high affinity nitrate transporters by nitrogen and carbon status. *Journal of Experimental Botany, 62*, 2319–2332.
Fernandez, M. (2007). *Application of Biotechnology for functional foods*. PEW: Initiative on Food and Biotechnology.
Fixen, P. E. (1993). Crop response to chloride. *Advances in Agronomy, 50*, 107–150.
Forrest, K. L., & Bhave, M. (2008). The PIP and TIP aquaporins in wheat form a large and diverse family with unique gene structures and functionally important features. *Functional & Integrative Genomics, 8*(2), 115–133.

Franco-Zorrilla, J. M., González, E., Bustos, R., Linhares, F., Leyva, A., & Paz-Ares, J. (2004). The transcriptional control of plant responses to phosphate limitation. *Journal of Experimental Botany, 55*, 285–293.

Fujii, H., Chiou, T. J., Lin, S. I., Aung, K., & Zhu, J. K. (2005). A miRNA involved in phosphate-starvation response in *Arabidopsis*. *Current Biology, 15*, 2038–2043.

Gao, N., Su, Y., Min, J., Shen, W., & Shi, W. (2010). Transgenic tomato over expressing ath-miR399d has enhanced phosphorus accumulation through increased acid phosphatase and proton secretion as well as phosphate transporters. *Plant and Soil, 334*, 123–136.

Geberta, M., Meschenmosera, K., Svidováb, S., Weghuberb, J., Schweyenb, R., Eiflera, K., Lenza, H., Weyanda, K., & Knoopa, V. (2009). A root-expressed magnesium transporter of the MRS2/MGT Gene family in *Arabidopsis thaliana* allows for growth in low-Mg^{2+} environments. *Plant Cell, 21*(12), 4018–4030.

Gierth, M., & Mäser, P. (2007). Potassium transporters in plants–involvement in K^+ acquisition, redistribution and homeostasis. *FEBS Letters, 581*(12), 2348–2356.

Gilroy, S., Bethke, P. C., & Jones, R. L. (1993). Calcium homeostasis in plants. *Journal of Cell Science, 106*, 453–462.

Godbold, D. L., & Hüttermann, A. (1985). Effect of zinc, cadmium and mercury on root elongation of *Picea abies* (karst.) seedlings, and the significance of these metals to forest die-back. *Environmental Pollution, 38*, 375–381.

Goldstein, A. H., Baertlein, D. A., & McDaniel, R. G. (1988). Phosphate starvation inducible metabolism in *Lycopersicon esculentum* I. Excretion of acid phosphatase by tomato plants and suspension-cultured cells. *Plant Physiology, 87*, 711–715.

Greger, M. (2004). *Uptake of nuclides by plants*. Tech Report TR-04-14, Department of Botany, Stockholm University.

Guest, C., Schulze, D., Thompson, I., & Huber, D. (2002). Correlating manganese X-ray absorption near-edge structure spectra with extractable soil manganese. *Soil Science Society of America Journal, 66*, 1172–1181.

Hall, J. L., & Williams, L. E. (2003). Transition metal transporters in plants. *Journal of Experimental Botany, 54*(393), 2601–2613.

Hammond, J. P., Broadle, M. R., & White, P. J. (2004). Genetic responses to phosphorus deficiency. *Annals of Botany, 94*(3), 323–332. doi:10.1093/aob/mch156.

Hanikenne, M., Motte, P., MCS, W., Wang, T., Loppes, R., & Matagne, R. F. (2005). A mitochondrial half-size ABC transporter is involved in cadmium tolerance in *Chlamydomonas reinhardtii*. *Plant, Cell & Environment, 28*, 863–873.

Havlin, J. L., Tisdale, S. L., Beaton, J. D., & Nelson, W. L. (2007). *In soil fertility and fertilizers*. New Delhi: Prentice Hall (India).

Hawkesford, M. J. (2003). Transporter gene families in plants: The sulphate transporter gene family-redundancy or specialization? *Physiologia Plantarum, 117*, 155–165.

He, L., & Hannon, G. J. (2004). Micro RNAs: Small RNAs with a big role in gene regulation. *Nature Reviews. Genetics, 5*, 522–531.

Hepler, P. K. (2005). Calcium: A central regulator of plant growth and development. *Plant Cell, 17*(8), 2142–2155.

Hirai, M. Y., Fujiwara, T., Awazuhara, M., Kimura, T., Noji, M., & Saito, K. (2003). Global expression profiling of sulfur-starved *Arabidopsis* by DNA macroarray reveals the role of O-acetyl-L-serine as a general regulator of gene expression in response to sulfur nutrition. *The Plant Journal, 33*(4), 651–663.

Hirschi, K. D., Zhen, R.-G., Cunningham, K. W., Rea, P. A., & Fink, G. R. (1996). CAX1, an H^+/Ca^{2+} antiporter from *Arabidopsis*. *PNAS, 93*, 8782–8786.

Hirschi, K. D., Korenkov, V. D., Wilganowski, N. L., & Wagner, G. J. (2000). Expression of *Arabidopsis CAX2* in tobacco, altered metal accumulation and increased manganese tolerance. *Plant Physiology, 124*, 125–134.

Ho, C.-H., Lin, S.-H., H-C, H., & Tsay, Y.-F. (2009). CHL1 functions as a nitrate sensor in plants. *Cell, 138*(6), 1184–1194.

Johanson, U., Karlsson, M., Johansson, I., Gustavsson, S., Sjovall, S., Fraysse, L., Weig, A. R., & Kjellbom, P. (2001). The complete set of genes encoding major intrinsic proteins in *Arabidopsis* provides a framework for a new nomenclature for major intrinsic proteins in plants. *Plant Physiology, 126*, 1358–1369.

Kabta-Pendias, A., & Szdeke, B. (2015). *Trace elements in abiotic and biotic environment*. Boca Raton: CRC Press.

Kataoka, T., Hayashi, N., Yamaya, T., & Takahashi, H. (2004a). Root-to-shoot transport of sulfate in *Arabidopsis*: Evidence for the role of SULTR3;5 as a component of low-affinity sulfate transport system in the root vasculature. *Plant Physiology, 136*, 4198–4204.

Kataoka, T., Watanabe-Takahashi, A., Hayashi, N., Ohnishi, M., Mimura, T., Buchner, P., Hawkesford, M. J., Yamaya, T., & Takahashi, H. (2004b). Vacuolar sulfate transporters are essential determinants controlling internal distribution of sulfate in *Arabidopsis*. *Plant Cell, 16*, 2693–2704.

Katsuhara, M., Akiyama, Y., Koshio, K., Shibasaka, M., & Kasamo, K. (2002). Functional analysis of water channel in barley roots. *Plant & Cell Physiology, 43*, 885–893.

Kavalchuk, I., Abramov, V., Pogribny, I., & Kovalchuk, O. (2004). Molecular aspects of plant adaptation to life in the Chernobyl zone. *Plant Physiology, 135*, 357–363.

Kim, E. J., Kwak, J. M., Uozumi, N., & Schroeder, J. I. (1998). *AtKUP1*: An *Arabidopsis* gene encoding high-affinity potassium transport activity. *Plant Cell, 10*, 51–62.

King, B. J., Siddiqi, M. Y., Ruth, T. J., Warner, R. H., & Glass, A. D. M. (1993). Feedback regulation of nitrate influx in barley roots by nitrate, nitrite, and ammonium. *Plant Physiology, 102*, 1279–1286.

Krapp, A., David, L. C., Chardin, C., et al. (2014). Nitrate transport and signalling in *Arabidopsis*. *Journal of Experimental Botany, 65*(3), 789–798.

Kuo, H.-F., & Chiou, T.-J. (2011). The role of micro RNAs in phosphorus deficiency signaling. *Plant Physiology, 156*(3), 1016–1024.

Leigh, R. A., & Wyn-Jones, R. G. (1986). Cellular compartmentation in plant nutrition: The selective cytoplasm and the promiscuous vacuole. In B. Tinker & A. Lauchli (Eds.), *Advances in plant nutrition 2* (pp. 249–279). New York: Praeger Scientific.

Li, F. (2006). *Potassium and water interaction, International workshop on soil potassium and K-fertiliser interaction*, Agricultural College, Guangxi University, Nanning Guangxi-530005, November 14.

Li, L., Tutone, A. F., Drummond, R. S. M., Gardner, R. C., & Luan, S. (2001). A novel family of magnesium transport genes in *Arabidopsis*. *Plant Cell, 13*, 2761–2775.

Liang, G., Yang, F., & Yu, D. (2010). Micro RNA395 mediates regulation of sulphate accumulation and allocation in *Arabidopsis thaliana*. *The Plant Journal, 62*, 1046–1057.

Lin, C. H. M., Koh, S., Stacey, G., SM, Y., Lin, T. Y., & Tsay, Y. F. (2000). Cloning and functional characterization of a constitutively expressed nitrate transporter gene, OsNRT1, from rice. *Plant Physiology, 122*, 379–388.

Lin, W.-Y., Lin, S.-I., & Chiou, T.-J. (2009). Molecular regulators of phosphate homeostasis in plants. *Journal of Experimental Botany, 60*(5), 1427–1438. doi:10.1093/jxb/ern303.

Lindsay, W. L. (1979). *Solubilities of common zinc minerals in soils, chemical equilibria in soils*. New York: Wiley.

Liu, K. H., Huang, C. Y., & Tsay, Y. F. (1999). CHL1 is a dual-affinity nitrate transporter of *Arabidopsis* involved in multiple phases of nitrate uptake. *Plant Cell, 11*, 865–874.

Liu, G. E., Simonne, H. & Li, Y. (2012). *Nickel nutrition in plants, HS1191, one of a series of the horticultural sciences, Florida cooperative extension service*, Institute of Food and Agricultural Sciences, University of Florida, Original publication date June 2011.

Marschner, H. (1995). *Mineral nutrition of higher plants* (2nd ed.). London: Academic.

Maser, P., Thomine, S., Schroeder, J. I., Ward, J. M., Hirschi, K., Sze, H., et al. (2001). Phylogenetic relationships within cation transporter families of *Arabidopsis*. *Plant Physiology, 126*, 1646–1667.

Matsumoto, H. (2000). Cell biology of aluminum toxicity and tolerance in higher plants. *International Review of Cytology, 200*, 1–46.

McGrath, S. P., Zhao, F.J., Withers, P. J. A. (1996). Development of sulfur deficiency in crops and its treatment. In *Proceedings of the Fertiliser Society, No. 379*. Peterborough: The Fertiliser Society.

Millaleo, R., Reyes-Diaz, M., Ivanov, A. G., Mora, M. L., & Alberdi, M. (2010). Manganese as essential and toxic element for plants: Transport, accumulation and resistance mechanisms. *Journal of Soil Science and Plant Nutrition, 10*(4), 470–481.

Miller, A. J., & Smith, S. J. (1996). Nitrate transport and compartmentation in cereal root cells. *Journal of Experimental Botany, 47*, 843–854.

Miller, A. J., Fan, X., Shen, Q., & Smith, S. J. (2008). Amino acids and nitrate as signals for the regulation of nitrogen acquisition. *Journal of Experimental Botany, 59*(1), 111–119.

Mitra, G. N. (2006). *Nutrient management of crops in soils of Orissa*. India: IFFCO.

Mitra, G. N. (2015). *Regulation of nutrient uptake by plants—A biochemical and molecular approach*. India: Springer.

Mitra, G.N., Sahu, S.K., Nayak, R.K. (2009). Characterization of iron toxic soils of Orissa and ameliorating effects of potassium on iron toxicity. *Proceedings of the IPI-OUAT-IPNI International Symposium, Bhubaneswar, Orissa, India*, Vol I: Invited Papers. p. 215.

Moomaw, A. S., & Maguire, M. E. (2008). The unique nature of Mg^{2+} channels. *Physiology (Bethesda), 23*, 275–285.

Müller, R., Morant, M., Jarmer, H., Nilsson, L., & Nielsen, T. H. (2007). Genome-wide analysis of the *Arabidopsis* leaf transcriptome reveals interaction of phosphate and sugar metabolism. *Plant Physiology, 143*, 156–171.

Munns, R. (2002). Comparative physiology of salt and water stress. *Plant, Cell and Environment, 25*, 239–250.

Nguyen, M. X., Moon, S., & Jung, K. H. (2013). Genome-wide expression analysis of rice aquaporin genes and development of a functional gene network mediated by aquaporin expression in roots. *Planta, 238*(4), 669–681.

Nies, D. H. (1999). Microbial heavy-metal resistance. *Applied Microbiology and Biotechnology, 51*, 730–750.

Nishida, S., Tsuzuki, C., Kato, A., Aisu, A., Yoshida, J., & Mizuno, T. (2011). AtIRT1, the primary iron uptake transporter in the root, mediates excess nickel accumulation in *Arabidopsis thaliana* plant. *Cellular Physiology, 52*(8), 1433–1442.

NRCCA. (2010). *North east region certified crop adviser study resources*. USA: Cornell University.

Orsel, M., Filleur, S., Fraisier, V., & Daniel-Vedele, F. (2002). Nitrate transport in plants: Which gene and which control? *Journal of Experimental Botany, 53*(370), 825–833.

Page, V., & Feller, U. (2005). Selective transport of zinc, manganese, nickel, cobalt and cadmium in the root system and transfer to the leaves in young wheat plants. *Annals of Botany, 96*, 425–434.

Papenbrock, J., Mock, H. P., Tanaka, R., Kruse, E., & Grimm, B. (2000). Role of magnesium chelatase activity in the early steps of the tetrapyrrole biosynthetic pathway. *Plant Physiology, 122*, 1161–1169.

Pilon, M., Abdel-Ghany, S. E., Cohu, C. M., Gogolin, K. A., & Ye, H. (2006). Copper cofactor delivery in plant cells. *Current Opinion in Plant Biology, 9*, 1–8.

Platten, J. D., et al. (2006). Nomenclature for HKT transporters, key determinants of plant salinity tolerance. *Trends in Plant Science, 11*, 372–374.

Plaxton, W. C., & Podesta, F. E. (2006). The functional organisation and control of plant respiration. *Critical Reviews in Plant Sciences, 25*, 159–198.

Plaxton, W. C., & Tran, H. T. (2011). Metabolic adaptations of phosphate-starved plants. *Plant Physiology, 156*(3), 1006–1015.

Puig, S., Andres-Colas, N., Garcia-Molina, A., & Penarrubia, L. (2007). Copper and iron homeostasis in *Arabidopsis*: Response to metal deficiencies, interactions and biotechnological applications. *Plant, Cell & Environment, 30*, 271–290.

Quaggiotti, S., Ruperti, B., Borsa, P., Destro, T., & Malagoli, M. (2003). Expression of a putative high-affinity NO_3^- transporter and of an H^+-ATPase in relation to whole plant nitrate transport physiology in two maize genotypes differently responsive to low nitrogen availability. *Journal of Experimental Botany, 54*(384), 1023–1031.

Quigley, F., Rosenberg, J. M., Shachar-Hill, Y., & Bohnert, H. J. (2002). From genome to function: The *Arabidopsis* aquaporins. *Genome Biology, 3*(1), res0001.1–res0001.17.

Rainer, H. (2012). Ion channels in plants. *Physiological Reviews, 92*(4), 1777–1811.

Reddy, A. S. N., Ali, G. S., Celesnik, H., & Day, I. S. (2011). Coping with stresses: Roles of calcium and calcium/Calmodulin-regulated Gene expression. *Plant Cell Journal, 23*(6), 2010–2032.

Rengel, Z., & Robinson, D. L. (1989). Competitive aluminum ion inhibition of net magnesium ion uptake by intact *Lolium multiflorum* roots. *Plant Physiology, 91*, 1407–1413.

Rodriguez-Navarro, A., & Rubio, F. (2006). High-affinity potassium and sodium transport systems in plants. *Journal of Experimental Botany, 57*, 1149–1160.

Rose, T. J., Impa, S. M., Rose, M. T., Pariasca-Tanaka, J., Mori, A., Heur, S., Johnson-Beebout, S. E., & Wissuwa, M. (2013). Enhancing phosphorus and zinc acquisition efficiency in rice: A critical review of root traits and their potential utility in rice breeding. *Annals of Botany, 112*(2), 331–345. doi:10.1093/aob/mcs217.

Roux, B., Berneche, S., Egwolf, B., et al. (2011). Ion selectivity in channels and transporters. *The Journal of General Physiology, 137*(5), 415–426.

Schachtman, D. P. (2000). Molecular insights into the structure and function of plant K^+ transport mechanisms. *Biochimica et Biophysica Acta, 1465*, 127–139.

Schachtman, D. P., Reid, R. J., & Ayling, S. M. (1998). Phosphorus uptake by plants: From soil to cell. *Plant Physiology, 116*, 447–453.

Schützendübel, A., & Polle, A. (2002). Plant responses to abiotic stresses: Heavy metal-induced oxidative stress and protection by mycorrhization. *Journal of Experimental Botany, 53*(372), 1351–1365.

Shaul, O. (2002). Magnesium transport and function in plants: The tip of the iceberg. *Biology of Metals, 15*, 309–323.

Shinmachi, F., Buchner, P., Stroud, J. L., Parmar, S., Zhao, F.-J., McGrath, S. P., & Hawkesford, M. J. (2010). Influence of sulfur deficiency on the expression of specific sulfate transporters and the distribution of sulfur, selenium, and molybdenum in wheat. *Plant Physiology, 153*(1), 327–336.

Shrawat, A. K., Carroll, R. T., DePauw, M., Taylor, G. J., & Good, A. G. (2008). Genetic engineering of improved nitrogen use efficiency in rice by the tissue-specific expression of alanine aminotransferase. *Plant Biotechnology Journal, 6*(7), 722–732.

Siddiqi, M. Y., Glass, A. D. M., Ruth, T. J., & Rufty Jr., T. W. (1990). Studies of the uptake of nitrate in barley, I. Kinetics of $^{13}NO_3^-$ influx. *Plant Physiology, 93*, 1426–1432.

Smith, F. W., Hawkesford, M. J., Ealing, P. M., Clarkson, D. T., van den Berg, P. J., Belcher, A. R., & Warrilow, A. G. S. (1997). Regulation of expression of a cDNA from barley roots encoding a high affinity sulphate transporter. *The Plant Journal, 12*, 875–884.

Song, W. -Y., Choic, K. S., Kimb, D. Y., Geislera, M., Park, J., Vincenzettia, V., Schellenberg, M., Kim, S. H., Limb, Y. P., Nohe, E. W., Leeb, Y., & Martinoia, E. (2010). *Arabidopsis* PCR2 is a zinc exporter involved in both zinc extrusion and long-distance zinc transport. *The Plant Cell, 22*(7), 2237–2252.

Sonoda, Y., Ikeda, A., Saiki, S., von Wiren, N., Yamaya, T., & Yamaguchi, J. (2003). Distinct expression and function of three ammonium transporter genes (OsAMT1;1-1;3) in rice. *Plant & Cell Physiology, 44*(7), 726–734.

Sperandio, M. V. L., Santos, L. A., de Araujo, O. J. L., Braga, R. P., Coelho, C. P., Nogueira, E. M., Fernandes, M. S., & de Souza, S. R. (2014). Response of nitrate transporters and PM H^+-ATPase expression to nitrogen flush on two upland rice varieties contrasting in nitrate uptake kinetics. *Australian Journal of Crop Science, 8*(4), 568–576.

Suenaga, A., Moriya, K., Sonoda, Y., Ikeda, A., von Wiren, N., Hayakawa, T., et al. (2003). Constitutive expression of a novel-type ammonium transporter OsAMT2 in rice plants. *Plant & Cell Physiology, 44*, 206–211.

Suzuki, M., Takahashi, M., Tsukamoto, T., Watanabe, S., Matsuhashi, S., Yazaki, J., Kishimoto, N., Kikuchi, S., Nakanishi, H., Mori, S., & Nishizawa, N. K. (2006). Biosynthesis and secretion of mugineic acid family phytosiderophores in zinc-deficient barley. *The Plant Journal, 48*, 85–97.

Takano, J., Noguchi, K., Yasumori, M., Kobayashi, M., Gajdos, Z., Miwa, K., Hayashi, H., Yoneyama, T., & Fujiwara, T. (2002). *Arabidopsis* boron transporter for xylem loading. *Nature, 420*(6913), 337–340.

Tan, K., Keltjens, W. G., & Findenegg, G. R. (1991). Role of magnesium in combination with liming in alleviating acid-soil stress with the aluminum-sensitive sorghum genotype CV323. *Plant and Soil, 136*, 65–72.
The Sulphur Institute. (2008). *Sulphur in Indian agriculture*. Washington, DC: The Sulphur Institute.
Trevisan, S., Borsa, P., Botton, A., Varotto, S., Malagoli, M., Ruperti, B., & Quaggiotti, S. (2008). Expression of two maize putative nitrate transporters in response to nitrate and sugar availability. *Plant Biology (Stuttgart, Germany), 10*, 462–475.
Tsay, Y. F., Chiu, C. C., Tsai, C. B., Ho, C. H., & Hsu, P. K. (2007). Nitrate transporters and peptide transporters. *FEBS Letters, 581*, 2290–2300.
Vance, C. P. (2010). Sugars, and MicroRNAs: Quaternaries in phosphate acquisition and use. *Plant Physiology, 154*(2), 582–588.
Vance, C. P., Uhde-Stone, C., & Allan, D. L. (2003). Phosphorus acquisition and use: Critical adaptations by plants for securing a non-renewable resource. *The New Phytologist, 157*, 423–447.
Venkatesh, J., TW, Y., Gaston, D., & Park, S. W. (2015). Molecular evolution and functional diversity of X-intrinsic protein gene in plants. *Molecular Genetics and Genomics, 290*(2), 443–460.
Very, A.-A., & Sentenac, H. (2002). Cation channels in the *Arabidopsis* plasma membrane. *Trends in Plant Science, 7*, 168–175.
Walch-Liu, P., & Forde, B. G. (2008). Nitrate signaling mediated by the NRT1.1 nitrate transporter antagonizes L-glutamate-induced changes in root architecture. *The Plant Journal, 54*(5), 820–828.
Walker, C. J., & Weinstein, J. D. (1991). Further characterisation of magnesium chelatase in isolated developing cucumber chloroplasts–substrate specificity, regulation, intactness, and ATP requirements. *Plant Physiology, 95*, 1189–1196.
Wang, M., Zheng, Q., Shen, Q., & Guo, S. (2013). The critical role of potassium in plant stress response. *International Journal of Molecular Sciences, 14*, 7370–7390.
Wasaki, J., Shinano, T., Onishi, K., et al. (2006). Transcriptomic analysis indicates putative metabolic changes caused by manipulation of phosphorus availability in rice leaves. *Journal of Experimental Botany, 57*(9), 2049–2059. doi:10.1093/jxb/erj158.
White, P. J. (2003). Calcium in plants. *Annals of Botany, 92*(4), 487–511.
White, P., & Karley, A. (2010). Potassium. In R. Hell & R. R. Mendel (Eds.), *Cell biology of metals and nutrients* (pp. 199–224). Berlin/Heidelberg: Springer.
Wiesenberger, G., Steinleitner, K., Malli, R., Graier, W. F., Vormann, J., Schweyen, R. J., & Stadler, J. A. (2007). Mg^{2+} deprivation elicits rapid Ca^{2+} uptake and activates Ca^{2+}/calcineurin signaling in *Saccharomyces cerevisiae*. *Eukaryotic Cell, 6*, 592–599.
World Nuclear Association. (2015). *Nuclear radiation and health effects*, Retrieved from world-nuclear.org, London.
Yasunari, T. J., Stohl, A., Hayans, R. S., et al. (2011). Cesium-137 deposition and contamination of Japanese soils due to Fukushima nuclear accident. *PNAS, 108*(49), 19530–19534.
Yong, Z., Kotur, Z., & Glass, A. D. M. (2010). Characterization of an intact two-component high-affinity nitrate transporter from *Arabidopsis* roots. *The Plant Journal, 63*, 739–748.
Yruela, I. (2005). Copper in plants. *Brazilian Journal of Plant Physiology, 17*, 145–146.
Yruela, I. (2009). Copper in plants: Acquisition, transport and interactions. *Functional Plant Biology, 36*(5), 409–430.
Zeng, X. Q., Chow, W. S., LJ, S., Peng, X. S., & Peng, X. X. (2010). Protective effect of supplemental anthocyanins on *Arabidopsis* leaves under high light. *Physiologia Plantarum, 138*, 215–225.
Zhang, D. Y., Ali, Z., Wang, C. B., Xu, L., Yi, J. X., et al. (2013). Genome-wide sequence characterization and expression analysis of major intrinsic proteins in soybean (*Glycine max* L.) *PloS One, 8*(2), e56312.
Zhao, F. J., Bilsborrow, P. E., Evans, E. J., & Syers, J. K. (1993). Sulphur turnover in the developing pods of single and double low varieties of oilseed rape (*Brassica napus* L.) *Journal of the Science of Food and Agriculture, 62*, 111–119.
Zhao, F.J., McGrath, S.P., Hawkesford, M.J. (2001). Sulphur nutrition and the sulphur cycle Institute of Arable Crops, Research report, Rothamsted Experimental Station 2000–2001.

Chapter 2
Nutrient Uptake, Removal, and Cycling in *Eucalyptus* Species

Akbar Ali, M. Naeem, Tariq Ahmad Dar, Mohd. Idrees, M. Masroor A. Khan, Moin Uddin, Prem Kumar Dantu, and Teg Bahadur Singh

Abstract *Eucalyptus* (family Myrtaceae), represented by over 900 species distributed throughout the world, are well known for their essential oils. Nutrient supply is treated as a dynamic variable which is related to relative growth rate. Plants take up the nutrients from the soil and use them for various metabolic processes. The genetic constitution of plants plays a superior role in determining the way assimilates are partitioned between roots, stems, and leaves. Different growth stages of the trees are reflected in processes of change that contribute to the control of nutrient demand, storage, and distribution. The amount of litterfall and nutrient return depends on the degree of biomass production and nutrient uptake in the stand. Cycling of nutrient is an important aspect as significant amounts of nutrients are returned to the soil through different mode and become available for cycling.

Keywords *Eucalyptus* • Biomass • Mineral nutrients • Accumulation • Cycling

---

A. Ali (✉)
Department of Botany, Dayalbagh Educational Institute, Agra 282005, India

Plant Physiology Section, Department of Botany, Aligarh Muslim University, Aligarh 202002, India
e-mail: akbar.agra.alig@gmail.com

M. Naeem • M.M.A. Khan
Plant Physiology Section, Department of Botany, Aligarh Muslim University, Aligarh 202002, India

T.A. Dar
Centre for Biodiversity Studies, School of Biosciences and Biotechnology, Baba Ghulam Shah Badshah University, Rajouri 185234, India

M. Idrees
Department of Mathematics and Sciences, College of Arts and Applied Sciences, Dhofar University, P. O. Box 2509, Salalah 211, Oman

M. Uddin
Women's College, Botany Section, Aligarh Muslim University, Aligarh 202002, India

P.K. Dantu • T.B. Singh
Department of Botany, Dayalbagh Educational Institute, Agra 282005, India

M. Naeem et al. (eds.), *Essential Plant Nutrients*,
DOI 10.1007/978-3-319-58841-4_2

2.1 Introduction

The genus *Eucalyptus* comprises of more than 900 species and various hybrids and varieties. These are well known for their essential oils. The *Eucalyptus* genus because of its high growth rate is presented as great soil nutrients demanding (Câmara et al. 2000). The nutrients absorption by eucalyptus trees is essential to their development and reproduction. Most metabolic reactions are not possible without mineral elements' presence. Different growth stages of the plant are reflected in processes of change that contribute to the control of nutrient demand, storage, and distribution in trees. Deficiency of important nutrients for tree metabolism determine according level its weakening (and therefore susceptibility increase to plagues and diseases), poor growth, not reaching sexual maturity or even death (Foelkel 2005). Among many roles of plant nutrients are constituents of plants' tissues, catalysts in various reactions, osmotic regulators, constituents of buffer systems, and regulators of membrane permeability (Pallardy 2008). Tree nutrition is important throughout development; the nutritional demands are larger from planting to canopy closure because this early stage of growth is predominantly the chlorophyll-producing tissues formation (Gómez 2006).

The circulation of nutrients in forests and plantations has received increasing attention over the past two to three decades with the major emphasis on the biological cycle between plant and soil of nutrient uptake, nutrient reabsorption before leaf senescence, nutrient use efficiency, etc. (Ovington 1965; Attiwill 1981; Bargali et al. 1992; Bisht 1993). Nutrient use efficiency can be estimated as the amount of aboveground net primary production per kilogram of nutrient taken up (Binkley et al. 1992).

The gross annual supply of a given nutrient from the soil is equal to the net annual demand of the stand. It is assumed that the total amount of the nutrient returned annually in the biogeochemical cycle from forest stand to soil is again available to the stand. The balance between the input of a nutrient in rainfall and the output in stream water (the geochemical cycle) is also considered as an available amount of nutrient. All these processes are ecologically important because plant nutrient uptake and nutrient recovery from senescing leaves are major components of ecosystem nutrient cycles.

In the beginning the gross productivity proportion of the plants is highest in the crown (leaves and branches), with high nutrient concentrations. Redistribution of nutrients linked to leaf senescence is small during this period and great amounts of nutrients are absorbed from the soil. This stage is characterized by increased nutrient accumulation rates, which peak during the crown closing phase (Attiwill 1981; Grove et al. 1996). The second stage includes crown closing when leaf biomass is stable or slightly decreasing and the heartwood generates most of the primary gross production. The heartwood nutrient content is low and this leads to a decrease in the nutrient accumulation rate of the tree. This is the phase of maximum soil exploitation by the fine roots and cycling processes are very intense. During the third growth

Table 2.1 Percentage of nutrients at different age in *E. grandis* plantation (Source: Bellote et al. 1980)

Age (years)/% of the total extracted nutrients							
Nutrients	1	2	3	4	5	6	7
N	21.6	34.6	47.7	60.8	73.9	86.9	100
P	21.8	28.1	46.6	69.7	90	99.9	100
K	13.4	27.8	42.2	56.7	85.6	85.6	100
Ca	8.2	15.9	32.7	49.5	83.2	83.2	100
Mg	1.1	18.7	36.2	53.8	89	89	100
S	9.9	24.9	39.9	54.9	85	85	100

stage of the tree, the greatest part of gross primary productivity is associated to the maintenance of the produced biomass (Miller 1984; Grove et al. 1996).

Annual nutrient uptake = annual nutrient increment + annual nutrient return.
Annual nutrient return = $\sum$ (aboveground litterfall biomass of tree each month in the stand × nutrient concentration in litterfall each month.
Cycling coefficient (%) = (annual nutrient return/annual nutrient uptake) × 100%.

2.2 Nutrient Uptake

Plants take up the nutrients from the soil and use them for various metabolic processes. Some of the plant parts such as leaves, branches, seeds, fruits, and roots are returned to the soil during the plant's growth. The nutritional factor can be expressed as a "flux density" or the amount of nutrient available per unit of time and unit of area.

In the beginning the gross productivity proportion of the plants is highest in the crown (leaves and branches), with high nutrient concentrations. The processes regulating nutrient transfers that require basic knowledge are as follows:

1. The geochemical cycle (in the soil)
2. The biogeochemical cycle (between tree and soil)
3. The biochemical cycle (inside the trees)

The eucalyptus can achieve a high biomass production on a low nutrient uptake. The nutrient use efficiency of eucalyptus increased with the decrease of the availability of the nutrient in the soil (FAO 2011). The majority of the nitrogen, phosphorus, and potassium were found in the eucalyptus trees. Laclau et al. (2000) showed that more than half of the nitrogen and 35% of the potassium in a stand of *Eucalyptus alba* hybrids had accumulated in the tree biomass by the time the stand reached 2 years. Bellote et al. (1980) reported the *Eucalyptus grandis* on the evolution of the nitrogen, phosphorus, potassium, calcium, magnesium, and sulfur content as function of age (1–7 years). The data nutrient content for each year shows as a percentage of totals extracted at the end of 7 years (Table 2.1).

All nutrients showed a maximum accumulation at 7 years old (increasing trend with increasing age) with the exception of phosphorus, which at 6 years had already accumulated 99.9% of the total extracted. By the fourth year more than 50% of the total of all nutrients have been removed except calcium, with an accumulation at that age of 49.5%. Schumacher and Caldeira (2001) reported the nutrient content in different compartments of forest biomass in 4-year-old *E. globulus*. The tree stem represents 69% of total biomass, which contained only 29% of the nitrogen, 28% of the phosphorus, 40% of the potassium, 12.5% of the calcium, and 34% of the magnesium. However leaves with 15% of the total biomass (larger canopy proportion than in other studies because are young trees in this instance) contained 60% of the nitrogen, 45% of the phosphorus, 33% of the potassium, 23% of the calcium, and 29.5% of the magnesium. On the other hand, bark was the element that accumulated 55.3% of the total calcium, representing only about 9% of total biomass.

George (1984) reported that 120 t ha^{-1} of total biomass in *Eucalyptus* plantation constituted 69% aboveground biomass and accumulated 42, 50, and 35% N, P, and K of total uptake amounting to 168, 47, and 104 kg ha^{-1}, respectively. Nitrogen status of plants affects growth mainly via light interception by strongly influencing the partitioning of assimilates between roots, stem, and leaves (Agren and Ingestad 1987; Ingestad and Agren 1988). The concentration of potassium but generally not of nitrogen and calcium is significantly lower in heartwood than in sapwood in both eucalyptus and conifers (Turner 1981; Wright and Will 1958).

The decreasing concentrations of Ca and Mg in the bark with increasing tree age may indicate a reduced availability of these nutrients for plants due to exhaustion of their exchangeable forms in the soil, as well as retranslocation of nutrients from this tissue as trees grew older. The reduction of K, Ca, and Mg concentrations in the stem wood, however, may have been caused by stronger supply restrictions by the soil, just as much as by higher internal cycling rates stimulated by higher heartwood production rates with increasing age (Reis and Barros 1990). Plants are opportunists and, when possible, take up more nutrients than actually required for growth of new tissues. A positive correlation between the tissue content of minerals and plant growth is generally observed also under field conditions (Aronsson and Elowson 1980; Pereira et al. 1989; Snowdon and Benson 1992; Rytter and Ericsson 1993; Nilsson and Wiklund 1994, 1995).

2.3 Nutrient Removal

The percentage return of nutrients varies with species, site conditions, plant age, etc. (Sugur 1989). Nutrient withdrawal from senescent leaves to stems and roots is certainly an important mechanism by which plants make more efficient use of a limited nutrient resource. Between one-third and two-thirds of the leaf nitrogen and phosphorus contents can in this way be transported back to woody tissues and stored for use in the coming season (Bernier 1984; Ericsson 1984; Pereira et al. 1989; van den Driessche 1984). Stem harvest with de-barking in the intensively fertilized treatment would result in removal of 127 kg ha^{-1} nitrogen, 13 kg ha^{-1} phosphorus, and 65 kg ha^{-1} potassium reported by Rudhe (2014). Similar values have been estimated by Laclau et al. (2000) where stem harvest with de-barking of *E. alba* hybrids for pulpwood resulted in exports of 82 kg ha^{-1} nitrogen, 23 kg ha^{-1} phosphorus, and 31 kg ha^{-1} potassium. Without knowing all sources of nutrient input to the ecosystem, it is difficult to know if the removal will result in a depletion of nutrients in the long term. Similarly reported by many researchers (Qui et al. 2011; Carneiro et al. 2008; Brady and Weil 2007; Gonçalves et al. 2004), keeping as much biomass as possible at the site could be important to sustain the productivity. As the result indicates, much of the available plant nutrients are received from the litter and the biomass, and if too large amounts of nutrients are removed from the ecosystem, depletion of nutrients in the soil could be expected.

2.4 Nutrient Cycling

Nutrient dynamics in soil-plant system are studied under a common term called "nutrient cycling." Cycling of nutrient is an important aspect as significant amounts of nutrients are returned to the soil through litterfall and become available for cycling.

Cycling processes in forest ecosystems have been defined (Switzer and Nelson 1972) in terms of:

1. The biogeochemical cycle between plant and soil
2. The biochemical cycle of internal transfer within the forest stand
3. The geochemical cycle of import export to the forest ecosystems

In the process of internal biological cycle, there are two important pathways: The first is tree to soil—along this nutrients are transported to forest floor via leaf/litterfall, twig, fruits, and flower drop, etc. The other pathway is movement of nutrient elements from soil to tree through the activity of root systems.

Nutrient cycling in forest plantations can be defined as the exchange of nutrient elements between the living and nonliving components of the forest ecosystem. The uptake of nutrients by plant roots, their incorporation into living tissue, and the release of nutrients from organic matter through decomposition causes nutrients to cycle within terrestrial ecosystems. Therefore, the nutrient cycling process involves nutrient uptake and storage in vegetation in perennial tissues, litter production, litter decomposition, nutrient transformation by soil fauna and flora, nutrient input from the atmosphere, and the weathering of primary minerals and nutrient export from the site by harvest and other removals, leaching, erosion, and gaseous transfers.

Nutrient retranslocation from senescing or dead tissue (leaves, heartwood, or bark) to growing tissues is one of the ways to increase utilization efficiency of limited nutrients by trees. Nutrient retranslocation from leaves means the largest contribution to the biogeochemical cycle of nitrogen, phosphorus, and potassium (mobile nutrients), followed in importance by retranslocation due to duraminization process and in lesser extent by redistribution from the outer bark (Grove et al. 1996). Nutrient retranslocation proportion due to leaf senescence depends on the initial concentration and mobility of nutrient in the phloem. Eucalyptus trees are particularly efficient in phosphorus retranslocation in wood during heartwood formation, finding for various species concentrations from five to thirty times higher in sapwood than in heartwood. Potassium generally follows the same behavior as phosphorus, while nitrogen has smaller concentration differences between sapwood and heartwood. Calcium retranslocation in duraminization process shows significant differences depending on species and site. Therefore in cases where the internal calcium redistribution is small, the tree requires continuous absorption from the soil of this nutrient (Turner and Lambert cited by Grove et al. 1996).

The depletion of soil nutrients is one of the most commonly cited criticisms associated with the eucalyptus (Jagger and Pender 2003). In contrast to other trees commonly used, such as *Leucaena and Acacia*, *Eucalyptus* species do not fix nitrogen from the atmosphere as the leguminous species do. According to some authorities, monoculture plantation forestry may affect soil chemical properties in two ways: nutrient depletion from the soil into the tree components and change in the chemical status of the soil surface as the litter layer is dominated by leaf fall derived from one species. Forest management practices can aggravate or reduce the magnitude of these effects. For example, the nutrient cycle at a site can be improved by mixing *Eucalyptus* species with nitrogen-fixing species such as *Acacia*.

Forrester et al. (2005) have compared monocultures of *E. globulus* (E) and *Acacia mearnsii* (A) and mixtures of these species planted in a species replacement series: 100% E, 75% E + 25% A, 50% E + 50% A, 25% E + 75% A, and 100% A. The result indicated that highest and the lowest aboveground biomass was obtained from 50:50 mixtures and monoculture, respectively.

The concentrations in the physiologically active leaves and litterfall leaves differed after 4.5 years, indicating the magnitude of biochemical nutrient cycling that occurs during the process of leaf senescence. Values of −43.8%, −63.6%, −53.3%, +24.0%, −22.2% were found for N, P, K, Ca, and Mg, respectively, showing the release sequence: P > K > N > Mg and the absence of Ca mobility. The same sequence was observed for leaves and leaf litter (consisting mainly of leaves) after 2.5 years, (−62.9%, −56.7%, +10.6%, and −18.0% for P, K, Ca, and Mg, respectively) (Leite et al. 1998). For *E. diversicolor* and *E. marginata*, the nutrient percentage retranslocated from senescent to younger leaves was −56.0%, −71.1%, −63.0%, +30.0%, and −4.0% for N, P, K, Ca, and Mg, respectively (Grove et al. 1996). Attiwill (1981) reported that the remobilized N and P quantities of senescent eucalyptus leaves are greater than in other annual or woody species.

There are other minor routes like stem flow, crown/canopy washed, and through rainfall in which the nutrients are dissolved in rains as atmospheric input into forest soil. Nutrient remobilization (on a mass base) from the bark and stem wood has the greatest contribution to the biochemical cycle in eucalyptus trees (Grove et al. 1996).

The nonmobile nutrients in the phloem, similar to Ca in the trunk, are retransferred outward during heartwood formation or from the outer bark to tissues in growth process. This process is not observed in leaves, where Ca is retained in the aging tissues (Grove et al. 1996). Calcium remobilization in trunk and branches seems to be the greatest source of Ca in many species with marked differences of Ca concentrations between heart- and sapwood or between inner and outer bark, particularly when Ca absorption by the roots is limited. A large fraction of the nutrients required for growth of new tissues can be met by internal cycling of nutrients. The ability to use the same nutrients over and over again, before they are finally discarded in litter or immobilized in heartwood, is a characteristic which is particularly well developed in trees. Although trees are very efficient at cycling nutrients from old to young tissues, sooner or later part of the nutrients contained in the living biomass will be lost in the litterfall.

The contribution of biochemical cycling to gross annual demand is the amount of a given nutrient which is moved from the sapwood before its transition to physiologically inactive heartwood, plus the amount of that nutrient which is moved from that portion of the stand which is to fall as litter. The external geochemical cycle encompasses input of nutrient elements by various means such as precipitation, weathering of rocks, biological N_2 fixation, fertilization, etc., and output of nutrients through harvest, forest fire, volatilization, leaching, drainage, erosion, etc.

The nutrients that most frequently limit forest growth are nitrogen (N) and phosphorus (P) (Fisher and Binkley 2000). Nutrient availability can alter growth rate through changes in dry mass partitioning, in specific leaf area, or in the assimilation rate per unit leaf area (Kirschbaum et al. 1992; Madeira et al. 2002; Sands et al. 1992; Xu et al. 2002).

References

Agren, G. I., & Ingestad, T. (1987). Root:shoot ratio as a balance between nitrogen productivity and photosynthesis. *Plant, Cell & Environment, 10*, 479–486.

Aronsson, A., & Elowson, S. (1980). Effects of irrigation and fertilization on mineral nutrients in scots pine needles. *Ecological Bulletins, 32*, 219–228.

Attiwill, P. M. (1981). Energy, nutrient flow, and biomass. In *Australian forest nutrition workshop productivity in perpetuity* (pp. 131–144). Melbourne: CSIRO Publishing.

Bargali, S. S., Singh, R. P., & Singh, S. P. (1992). Structure and function of an age series of eucalypt plantations in central Himalaya. II. Nutrient dynamics. *Annals of Botany, 69*, 413–421.

Bellote, A. F. J., Sarruge, J. R., Haag, H. P., & de Oliveira, G. D. (1980). Extração e exportação de nutrientes pelo *Eucalyptus grandis* Hill ex Maiden em função da idade: 1-macronutrientes. *IPEF, 20*, 01–23.

Bernier, B. (1984). *Nutrient cycling in populus: A literature review with implications in intensively-managed plantations. IEA/ENFOR, Report 6* (p. 46). Ottawa: Canadian forest service.

Binkley, D., Dunkin, K. A., De Bell, D., & Ryan, M. G. (1992). Production and nutrient cycling in mixed plantations of *Eucalyptus* and *Albizia* in Hawaii. *Forest Science, 38*, 393–408.

Bisht, K. (1993). Growth of *Quercus leucotrichophora A. Camus* and *Pinus roxburghii* Sarg. Seedlings in relation to nutrient and water. *Proceedings of the Indian National Science Academy, 59*, 71–78.

Brady, N. C., & Weil, R. R. (2007). *The nature and properties of soils* (14th ed.). Upper Saddle River, NJ: Prentice Hall.

Câmara, C. D., de Paula, L. W., & Vieira, S. A. (2000). Corte raso de uma plantação de *Eucalyptus saligna* de 50 anos: impactos sobre a ciclagem de nutrientes em uma microbacia experimental. *IPEF, Scientia Forestalis, 57*, 99–109.

Carneiro, M., Fabião, A., Martins, M. C., Fabião, A., Abrantes da Silva, M., Hilário, L., & Lousã m, M. M. (2008). Effects of harrowing and fertilization on understory vegetation and timber production of a *Eucalyptus globules* Labill. Plantation in Central Portugal. *Forest Ecology and Management, 255*, 591–597.

van den Driessche, R. (1984). Nutrient storage, retranslocation and relationships of stress to nutrition. In G. D. Bowen & N. EKS (Eds.), *Nutrition of plantation forest* (pp. 181–210). London: Academic Press.

Ericsson, T. (1984). *Nutrient cycling in willow. IEA/ENFOR, Report 5* (p. 32). Ottawa: Canadian forestry service.

FAO. (2011). Eucalyptus in east africa, socio-economic and environmental issues. In G. Dessie & T. Erkossa (Eds.), *Planted forests and trees working paper 46/E, forest management team, forest management division*. Rome: FAO.

Fisher, R. F., & Binkley, D. (2000). *Ecology and management of forest soils* (3rd ed.). New York, NY: Wiley.

Foelkel, C. (2005). *Minerais e nutrientes das árvores dos eucaliptos: Aspectos ambientais, fisiológicos, silviculturais e industriais acerca dos elementos inorgânicos presentes nas árvores. Eucalyptus Online Book & Newsletter*. Retrieved from http://www.eucalyptus.com.br/capitulos/capitulo_minerais.pdf.

Forrester, D., Bauhus, J., & Cowie, A. L. (2005). Nutrient cycling in a mixed-species plantation of *Eucalyptus globulus* and *Acacia mearnsii*. *Canadian Journal of Forest Research, 35*, 2942–2950.

George, M. (1984). Nutrients removal from eucalyptus plantations by harvesting wood. *Indian Forester, 110*, 61–67.

Gómez, C. (2006). *Detección de limitantes nutritivos en Eucalyptus a través de ensayos en macetas*. Retrieved from Monografias.com. http://www.monografias.com/trabajos38/limitantes-nutritivos/limitantesnutritivos2.shtml.

Gonçalves, J. L. M., Stape, J. L., Laclau, J. P., Smethurst, P., & Gava, J. L. (2004). Silvicultural effects on the productivity and wood quality of eucalypt plantations. *Forest Ecology and Management, 193*, 45–61.

Grove, T. S., Thomson, B. D., & Malajczuk, N. (1996). Nutritional physiology of eucalypts: Uptake, distribution and utilization. In P. M. Attiwill & M. A. Adams (Eds.), *Nutrition of eucalypts* (pp. 77–108). Australia: CSIRO.

Ingestad, T., & Agren, G. (1988). Nutrient uptake and allocation at steady-state nutrition. *Physiologia Plantarum, 72*, 450–459.

Jagger, P., & Pender, J. (2003). The role of trees for sustainable management of less-favored lands: The case of *Eucalyptus* in Ethiopia. *Forest Policy and Economics, 5*, 83–95.

Kirschbaum, M. U. F., Bellingham, D. W., & Cromer, R. N. (1992). Growth analysis of the effect of phosphorous nutrition on seedlings of *Eucalyptus grandis*. *Australian Journal of Plant Physiology, 19*, 55–66.

Laclau, J. P., Bouillet, J. P., & Ranger, J. (2000). Dynamics of biomass and nutrient accumulation in a clonal plantation of eucalyptus in Congo. *Forest Ecology and Management, 128*, 181–196.

Leite, F. P., Barros, N. F., Novais, R. F., & Fabres, A. S. (1998). Acúmulo e distribuição de nutrientes em *Eucalyptus grandis* sob diferentes densidades populacionais. *Revista Brasileira de Ciência do Solo, 22*, 419–426.

Madeira, M., Fabiao, A., & Steen, E. (2002). Mass and chemical composition of roots in a *Eucalyptus globulus* plantation after one year under different water and plant nutrient regimes (in preparation).

Miller, H. G. (1984). Dynamics of nutrient cycling in plantation ecosystems. In J. R. Aldous (Ed.), *Wood for energy: The implications for harvesting, utilization and marketing* (pp. 137–146). Edinburgh: Institute of chartered foresters.

Nilsson, L. O., & Wiklund, K. (1994). Nitrogen uptake in a Norway spruce stand following ammonium sulphate application, fertigation, irrigation, drought and nitrogen-free-fertilization. *Plant and Soil, 164*, 221–229.

Nilsson, L. O., & Wiklund, K. (1995). Nutrient balance and P, K, mg, S and B accumulation in a Norway spruce stand following ammonium sulphate application, fertigation, irrigation, drought and N-free fertilization. *Plant and Soil, 168–169*, 437–446.

Ovington, J. D. (1965). Organic production, turnover and mineral cycling in woodlands. *Biological Reviews, 40*, 295–336.

Pallardy, S. G. (2008). *Physiology of Woody plants* (3rd ed.p. 454). New York: Elsevier.

Pereira, J. S., Linder, S., Araujo, M. C., Pereira, H., Ericsson, T., Borralho, N., & Leal, L. C. (1989). Optimization of biomass production in *Eucalyptus globulus* plantations. In J. S. Pereira & J. J. Landsberg (Eds.), *Biomass productivity by fast growing trees* (pp. 101–121). Dordrecht, The Netherlands: Kluwer Academic Publishers.

Qui, S., Bell, R. W., Hobbs, R. J., & Mc Comb, A. J. (2011). Estimating nutrient budgets for prescribed thinning in a regrowth eucalyptus forest in south-west Australia. *Forestry, 85*, 01–11.

Reis, M. G. F., & Barros, N. F. (1990). Ciclagem de nutrientes emplantios de eucalipto. In N. F. Barros & R. F. Novais (Eds.), *Relação solo-eucalipto* (pp. 265–302). Viçosa, MG: Folha de Viçosa.

Rudhe, J. (2014). Where has the fertilizer gone? Closing the nutrient budget for a eucalyptus fertilization experiment in Southern China. Master Thesis. Southern Swedish forest Research Centre, Swedish University of Agricultural Sciences.

Rytter, L., & Ericsson, T. (1993). Leaf nutrient analysis in *Salix viminalis* (L.) energy forest stands growing on agricultural land. *Journal of Plant Nutrition and Soil Science, 156*, 349–356.

Sands, P. J., Cromer, R. N., & Kirschbaum, M. U. F. (1992). A model of nutrient response in *Eucalyptus grandis* seedlings. *Australian Journal of Plant Physiology, 19*, 459–470.

Schumacher, M. V., & Caldeira, M. V. W. (2001). Estimativa da biomassa e do conteúdo de nutrientes de um povoamento de *Eucalyptus globulus* (Labillardière) sub-espécie *maidenii. Ciência Florestal, 11*, 45–53.

Snowdon, P., & Benson, M. L. (1992). Effects of combinations of irrigation and fertilization on the growth and above-ground biomass production of *Pinus radiata. Forest Ecology and Management, 52*, 87–116.

Sugur, G. V. (1989). Litter production and nutrient cycling of different species under plantation conditions. *My Forest, 25*, 43–49.

Switzer, G. L., & Nelson, L. E. (1972). Nutrient accumulation and cycling in loblolly pine (*Pinus taeda* L.) plantation ecosystems: The first twenty years. *Soil Science Society of America Proceedings, 36*, 143–147.

Turner, J. (1981). Nutrient supply in relation to immobilization in biomass and nutrient removal in harvesting. In *Proceedings Australian Forest nutrition workshop. Productivity in perpetuity* (pp. 263–275). Canberra, Australia: Australian Forestry Council.

Wright, T. W., & Will, G. M. (1958). The nutrient content of scots and Corsican pines growing on sand dunes. *Forestry, 31*, 13–25.

Xu, D., Dell, B., Malajczuk, N., & Gong, M. (2002). Effects of P fertilisation on productivity and nutrient accumulation in a *Eucalyptus grandis E. urophylla* plantation in southern China. *Forest Ecology and Management, 161*, 89–100.

Chapter 3
Biochar's Influence as a Soil Amendment for Essential Plant Nutrient Uptake

Viraj Gunarathne, Sonia Mayakaduwa, and Meththika Vithanage

Abstract Biochar has recently become an interesting option for soil management in terms of nutrients depleted lands, which is now emerging as an increasing global concern. Since biochar is derived from biomass, they are high in carbon and may contain a range of plant macro- and micronutrients. In addition, the physical microstructure of biochar may crucially influence the role of biochar on plant nutrient uptake determining access to mineralized elements by soil solution, microorganisms, and plant roots. The beneficial use of biochar as a soil amendment in terms of increased crop yield and improved soil quality has been reported. This book chapter extensively discusses the influential nutrients in biochars and their effects on plant nutrient uptake. Further, alteration of the mechanism of nutrient uptake via biochar modification and the effect on nutrient transformation in soil have been reviewed. Biochar impacts on nutrient uptake by different plants under different environmental and soil conditions are not fully understood yet. This chapter will provide insights for future research directions in order to establish an effective biochar-plant nutrient interaction.

Keywords Yield improvement • Soil nutrients • Biochar • Soil reclamation • Fertilizer

---

Viraj Gunarathne and Sonia Mayakaduwa are co-first authors

V. Gunarathne • S. Mayakaduwa
Environmental Chemodynamics Project, National Institute of Fundamental Studies, Kandy, Sri Lanka

M. Vithanage (✉)
Environmental Chemodynamics Project, National Institute of Fundamental Studies, Kandy, Sri Lanka

Faculty of Applied Sciences, University of Sri Jayewardenepura, Nugegoda, Sri Lanka
e-mail: meththikavithanage@gmail.com; meththika@sjp.ac.lk

M. Naeem et al. (eds.), *Essential Plant Nutrients*,
DOI 10.1007/978-3-319-58841-4_3

3.1 Introduction

The threats of nutrient depleted soils that are associated with poor agricultural practices, deforestation, overgrazing, and industrialization are in growing global concern. Increasing soil-nutrient depletion leading to plant nutrient deficiencies has been reported elsewhere (Sanchez 2002). Shortage of essential plant nutrients and subsequent yield reduction may create severe impacts on food security and economic development in the world. The traditional way to enhance soil nutrients is the application of mineral fertilizers (Sanchez 2002). However, due to surface runoff and leaching, nutrient concentration drops in soil, which motivates the farmers for repetitive application of chemical fertilizers. At the same time, the soil may have been enriched in chemical fertilizers. However, plant uptake is minimal due to low bioavailability. These can cause detrimental effects on the soil biology and the environment (Bah et al. 2014).

Although several options have been proposed to overcome the issue of excessive application of nutrients, low availability, and nutrient loss, beneficial solutions are minimal (Eghball and Power 1999; Withers et al. 2001). Biochar has become the most recent interest for soil nutrient management inclusive contaminated soils other than many of its application for environmental remediation and carbon sequestration (Atkinson et al. 2010; Sohi et al. 2010). Biochar technology has received attention in soil research due to its extraordinary potential for improving soil structure and plant nutrient availability (Glaser et al. 2002). Several researchers have documented effective plant responses under biochar amendment (Steiner et al. 2007; Major et al. 2010; Uzoma et al. 2011). Biochar can positively influence plant nutrient uptake directly as a result of its nutrient content and release characteristics, as well as indirectly via enhanced sorption of nutrients (Lehmann et al. 2003a); increase in soil pH (Rondon et al. 2007); improved soil cation exchange capacity (Liang et al. 2006); increased soil physical properties (Chan et al. 2008a), including an increase in water retention (Laird et al. 2010); and alteration of soil microbial populations and functions (Pietikäinen et al. 2000). Furthermore, biochar amendments may facilitate efficient use of fertilizers by retention and thereby reducing nutrient leaching from soil (Lehmann et al. 2003a). Nevertheless, modern agriculture rarely uses biochar for its agronomic value regarding crop response, and at the same time, soil health benefits are yet to be quantified.

In this chapter, we focus on explaining the possible impacts of biochar on plant nutrient uptake, existing information on biochar nutrient properties, the effect on nutrient transformation, and biochar characteristics that determine plant nutrient availability. Moreover, future research potentials on modifications to biochar that may improve plant nutrient uptake, long-term biochar stability, and subsequent plant nutrient responses are highlighted.

3.2 Biochar

Biochar refers to the carbon-rich solid coproduct of pyrolysis, which is the thermal degradation of biomass under oxygen-limited conditions. The origin of biochar is connected to the slash and char techniques used by ancient farmers in Amazon River basin area. They created biochar that was referred to as Terra Preta, by incomplete combustion of plant debris in pits, and had found that Terra Preta could retain soil fertility for centuries (Marris 2006). During last few decades, biochar received increasing attention due to its potential for the significant reduction of atmospheric greenhouse gas levels (Lehmann 2007), remediation of contaminated soil/water (Beesley et al. 2010), and improving soil productivity (Steiner et al. 2008). Despite its organic origin, biochar is recalcitrant to decomposition due to its highly aromatic carbon structure. Thus, biochar has been used for agriculture, in order to increase nutrient availability (Asai et al. 2009), cation exchange capacity (Chan et al. 2008a), and soil water-holding capacity (Masulili et al. 2010) in soil. In particular, biochar has a high surface area and porosity compared to other chemical or biological amendments enabling adsorb/retain nutrients/water or to provide habitats for beneficial soil microorganisms (Glaser et al. 2002; Lehmann et al. 2006).

- Production and physicochemical properties

Different biomass varying from agricultural residues to municipal solid waste can be used as feedstocks for biochar production (Cao et al. 2014). Recently, it has been identified that pyrolytic conversion of organic materials to biochar is an alternative waste management option (Jayawardhana et al. 2016a, b). During pyrolysis, limited oxygen supply may prevent the complete combustion and thereby inhibit the carbon volatilization and ash production to a great extent. Heat released from the pyrolysis facilitates the volatilization of hydrogen and oxygen along with some of the carbon within the biomass. The resulted carbonaceous material may consist of poly-aromatic hydrocarbons with oxygenated functional groups (Warnock et al. 2007). The physical microstructure of biochar may crucially affect the role of biochar on plant nutrient uptake determining access to mineralized elements by soil solution, microorganisms, and plant roots. Biochar pore size distribution is highly variable encompassing nano- (<0.9 nm), micro- (<2 nm) to macropores (>50 nm) (Downie et al. 2009) determining surface area. Due to the porous nature, biochar provides habitats for beneficial soil organisms including mycorrhizae and bacteria. Moreover, the porosity and surface area of biochar may create critical effects on its nutrient retention capacity by surface binding of both cations and anions to its surfaces. The polycyclic aromatic structure of biochar hinders biological decomposition and chemical oxidation, which explains its persistence over centuries in the environment (Glaser et al. 2000).

However, the parameters as feedstock type, pyrolysis temperature, residence time, and heating rate are crucial in determining physicochemical characteristics of biochar produced. Yet, some researchers had suggested that pyrolysis temperature and feedstock might pose the greatest effect on biochar quality (Kloss et al. 2012). It has been reported that an increase in pyrolysis temperature increases biochar's pH, BET surface area, and carbon content (Demirbas 2004). Moreover, the effects of feedstock type on biochar surface area, pores, elemental composition, and functional groups have been highlighted (Sohi et al. 2010; Sun et al. 2014).

- Nutrient availability and concentrations

Since biochars are derived from biomass, they are high in carbon and may contain a range of plant macronutrients (nitrogen (N), phosphorous (P), calcium (Ca), magnesium (Mg), potassium (K), and sulfur (S)) and micronutrients (copper (Cu), iron (Fe), manganese (Mn), and zinc (Zn)) (Chan and Xu 2009; Hossain et al. 2011). Researchers have shown that the nutrient content of the biochar is attributed to the feedstock type (Chan et al. 2008a). Particularly, total P and N content were found to be higher in biochar derived from a feedstock of animal origin (e.g., sewage sludge, broiler liter) than those from plants (e.g., wood/green waste) (Chan et al. 2008a). However, nutrient elements in feedstock tend to mineralize, co-stabilize with carbon, or volatilize to form condensable products during pyrolysis. For instance, potassium and K are largely conserved after converting into inorganic forms, whereas nitrogen is volatilized in proportion to carbon or associated with carbon in the resident fraction (Chan and Xu 2009). Both P and K vaporize at pyrolysis temperatures above 760 °C, whereas magnesium and calcium are lost above 1107 and 1240 °C, respectively. Thus, recent studies have suggested that the biochar produced at low temperatures is suitable for agricultural uses, whereas high-temperature-derived biochar can be effectively used to contaminant adsorption in soils (Agrafioti et al. 2013). Hence, nutrient element concentrations in feedstock materials are not a reliable measure of biochar nutrient value (Angst and Sohi 2013).

Furthermore, the total elemental composition of nutrients does not necessarily reflect the availability of these nutrients to plants. Although phosphorus is mainly found in the ash fraction, pH-dependent reactions and presence of chelating substances often control its solubilization (DeLuca et al. 2015). Availability of magnesium is similar to that of phosphorous. However, according to some research findings, magnesium can be partially volatilized into gaseous or condensable liquid fractions that are not available for plants (Angst and Sohi 2013). In contrast, K availability in biochar is typically high, and researchers confirmed the increase in K uptake by plants after biochar applications (Lehmann et al. 2003a; Chan et al. 2008a). Furthermore, studies have indicated a low nitrogen availability as most of the nitrogen in biochar is present as heterocyclic nitrogen (Knicker et al. 1996). Nevertheless, heterocyclic nitrogen has found to be less resistant than generally assumed as parts of them seemed to be available for plants (De la Rosa et al. 2011). Table 3.1 summarizes nutrient element composition of biochars derived from different feedstocks.

Moreover, nutrient availability in biochar also depends on the environmental conditions and soil type where biochar is used as an amendment. Eventually, even fixed nutrients may break down with time releasing small quantities to the soil solution (Lehmann et al. 2006). As long as the soil is dry, some nutrients remain trapped within mineral layers, and once the soil gets wet, they are released to the soil solution (Esposito 2013).

Table 3.1 Nutrient element composition of some biochars

Feedstock Type	Production conditions	Nutrients (g Kg^{-1})						References
		N	P	K	Mg	Ca	Fe	
Broiler liter	700 °C steam activated	6	48	30	–	–	–	Lima and Marshall (2005)
Coconut shell	500 °C	9.4	73	–	–	–	–	Tsai et al. (2006)
Rice straw	400 °C	9.8	1.3	41	0.010	0.010	0.341	Naeem et al. (2014)
	500 °C	8.5	1.4	48	0.013	0.011	0.521	
	500 °C	13.2	37	–	–	–	–	Tsai et al. (2006)
What straw	400 °C	9.4	3.0	32	0.006	0.008	0.259	Naeem et al. (2014)
	500 °C	8.5	3.4	36	0.007	0.009	0.422	
Pine woodchip	–	0.7	770	2.10	–	–	–	Brantley et al. (2015)
Green waste	450 °C	1.8	–	8.19	1.340	1.600	–	Chan et al. (2008a)
Saw dust	500 °C	3.2	0.30	3.40	1.500	37	1.400	Lentz and Ippolito (2012)
Miscanthus straw	700 °C	–	0.82	18.33	1.520	–	–	Prendergast-Miller et al. (2014)
Sewage sludge	450 °C	<0.03	1.31	2.47	–	–	–	Liu et al. (2014)
Grape pomace	300 °C	–	3.63	164.75	4.860	1.130	1.750	Manolikaki and Diamadopoulos (2016)
Rice husk	300 °C	–	1.80	14.51	0.650	0.040	0.080	Manolikaki and Diamadopoulos (2016)
Dairy manure	300 °C	–	5.39	14.95	8.757	20.185	–	Rajkovich et al. (2012)
	500 °C	–	3.94	14.94	8.498	18.505	–	Rajkovich et al. (2012)
Food waste	300 °C	–	5.87	13.02	3.337	28.177	–	Rajkovich et al. (2012)
	500 °C	–	7.52	21.34	4.461	53.779	–	Rajkovich et al. (2012)
Paper waste	300 °C	–	0.83	2.79	2.428	258	–	Rajkovich et al. (2012)
	500 °C	–	0.82	3.34	2.739	289	–	Rajkovich et al. (2012)
Poultry	300 °C	–	26.41	40.01	8.914	157.531	–	Rajkovich et al. (2012)
	500 °C	–	30.56	28.11	10.436	204.205	–	Rajkovich et al. (2012)

- Role as a soil amendment

Terra Preta had been used to boost soil fertility and improve soil quality as a soil amendment at least 2000 years ago. To date, these soils have found to be highly fertile compared to other soils in the region containing as much as four times more organic matter in the top layer of the soil (Filiberto and Gaunt 2013). Biochar's greater resistance to microbial decay than other soil organic matter is resulted by its particular chemical structure (Smernik et al. 2002), whereas the high nutrient retention is derived from specific chemical and physical properties such as high charge density (Liang et al. 2006; Lehmann et al. 2003b). Thus, biochar has been highlighted as more stable than any other amendment to soil, and they have a nutrient availability beyond a fertilizer effect. Consequently, researchers have indicated that biochar is not comparable with other types of compost or manure that is used for the improvement of soil properties as it is much more efficient than any other organic soil amendment in improving soil quality (Lehmann and Joseph 2015). The beneficial use of biochar as a soil amendment in terms of increased crop yield and improved soil quality has been reported (Major et al. 2010; Haefele et al. 2011). Further, a review of previous research indicated a wide range of biochar application rates (0.5–135 ton/ha of biochar) as well as a huge range of plant responses (−29–324%) (Glaser et al. 2002). According to several researchers, such impacts are attributed to direct nutrient addition or nutrient retention by biochar and their effects on soil pH (Rondon et al. 2007; Yamato et al. 2006). It has been observed that along with crop yield, soil organic carbon, soil pH, and total nitrogen also increased after biochar amendment (Zhang et al. 2012). In addition, an increase in the water-holding capacity of soil was reported after biochar amendments, whereas nutrient leaching was found to be decreased (Glaser et al. 2002; Sohi et al. 2009). Biochar's critical role in increasing retention of nutrient and thereby reducing their leaching has been widely investigated (Eghball and Power 1999; Lehmann et al. 2003a; Steiner et al. 2008). The retention of cations is enhanced by the high surface charge density of biochar, whereas high surface area, internal porosity, and the presence of both polar and nonpolar surface sites facilitate the retention of organic and associated nutrients (Atkinson et al. 2010).

Moreover, biochar is being used for soil remediation as the extent of oxygen containing carboxyl, hydroxyl, and phenolic surface functional groups in biochar effectively binds soil contaminants reducing their mobility in soils (Sohi et al. 2010). Particularly, organic contaminants have been proven to adsorb into the carbonaceous fraction of biochar through electrostatic attractions and polar or hydrophobic interactions (Lehmann et al. 2006; Mayakaduwa et al. 2016a, b). Biochar produced from woodchips and cotton straw could decrease the dissipation of organic pesticides including chlorpyrifos, carbofuran, and fipronil from soil reducing their bioavailability (Sun et al. 2014; Chan and Xu 2009). Besides, PAHs and steroid hormones in soil had been effectively remediated using biochar so that the risk of the pollutants leaching to groundwater and entering food chains could reduce (Chan et al. 2008a; Hossain et al. 2011). Several researchers have confirmed that even a small amount of biochar addition may appreciably reduce the accumulation of organic contaminants in soil (Lentz and Ippolito 2012; Prendergast-Miller et al. 2014). On the other hand, biochar can stabilize heavy metals in contaminated soils that are not biodegradable and persistent for a long time. The retention of heavy metals as Pb, Cd, and Ni by alkaline soil amended with broiler litter biochar has been reported. Bandara et al. (2017b) reported immobilization of Ni, Mn, and Cr in woody biochar-amended serpentine soil. Furthermore, the potential of tea waste biochar on immobilizing Cr in tannery waste contaminated soils has been found to be in high potential (Vithanage et al. 2016). It has been suggested that the mineral composition of biochar as phosphates and carbonates may be crucial in the stabilization of heavy metals due to the reason that they may precipitate with heavy metals reducing their bioavailability (Bandara et al. 2017b). Moreover, cation release and metal complexation may also involve in heavy metal removal in contaminated soil (Liu et al. 2014; Manolikaki and Diamadopoulos 2016).

Additionally, biochar amendments in soil have been highlighted as a possible mean of reducing atmospheric greenhouse gas levels (Bandara et al. 2017a). Annual net emissions of CO_2, N_2O, and CH_4 had reduced by 12% after application of woody biochar (Bailey et al. 2011). The enhanced nitrogen retention by biochar may prevent or limit the production of N_2O (Du et al. 2014). Further, biochar amendments may slow carbon and nitrogen release while sequestering carbon in soil (Jin et al. 2016). Such effects including chemical, physical (abiotic), or microbiological (biotic) interactions between biochar and soil have not been clearly explained (Yanai et al. 2007). Nevertheless, a recent study has successfully used biochar with co-inoculation of fungi and bacteria to enhance enzyme activities that may be crucial in organic matter decomposition and nutrient cycling (Bandara et al. 2017a). Also, Bailey et al. (2011) indicated the potential stimulation of soil enzymes by switchgrass biochar although their effects on enzymes can be variable.

3.3 Biochar Effects on Plant Nutrient Uptake

Although evaluation of plant responses to biochar may be complicated, researchers have indicated the biochar-induced changes in nutrient concentrations in soil and plant tissues (Lehmann et al. 2011). A very high rate of nutrient uptake by maize grown in infertile acidic soils under field conditions has been reported after a single biochar application (Major et al. 2010). The authors have suggested that biochar may increase crop growth through enhanced pH in studied acidic soil or high base cation retention in the root zone. It has been reported that large surface area, high negative surface charge, and greater charge density of biochar determine its higher capacity of cation adsorption per unit carbon than other kinds of soil organic matters (Liang et al. 2006). In addition, microbiological or soil physical mechanisms driven by biochar may contribute to enhance crop growth. Similarly, increased concentrations of several nutrients in plant tissues following biochar applications have been highlighted (Masto et al. 2013). Again, the influences depend on the production variables of biochar and complex physiochemical properties which may involve in the biochar-soil-plant interacting system (Masto et al. 2013). Besides biochar

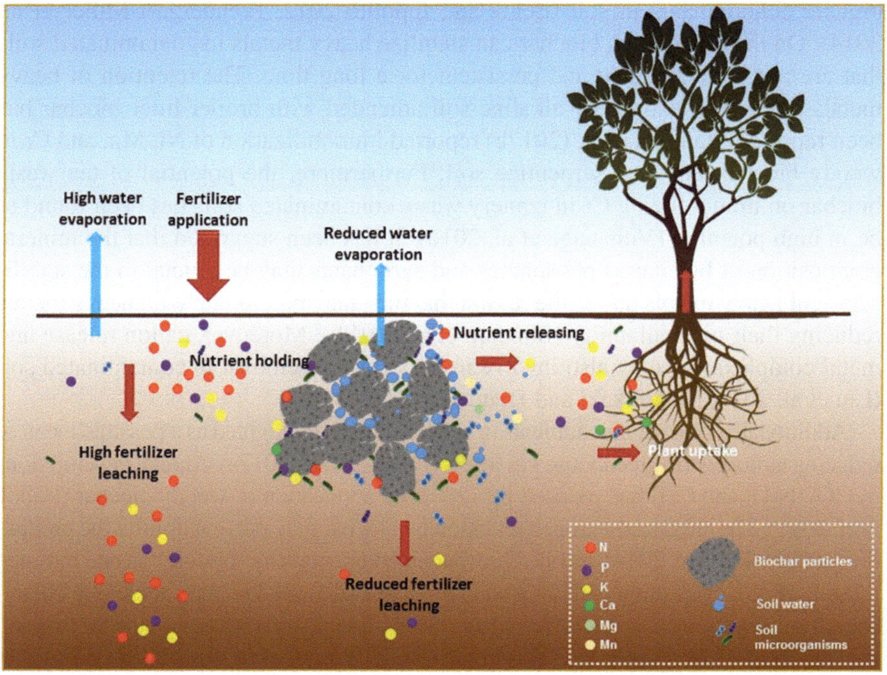

Fig. 3.1 Possible effects of biochar on plant nutrient uptake

characteristics, their interactions with climate, fertilization status, and soil type may affect on uncertainty in how biochar influence on nutrient uptake mechanisms (Biederman and Harpole 2013). Moreover, Biederman and Harpole (2013) highlighted that biochar's dark color alters thermal dynamics in plant-soil interactions that may lead to an influence on nutrient uptake. Biochar amendments can reduce the bulk density in soils leading to increase root penetration that allows the uptake of nutrients from soil solution (Glaser et al. 2002; Lehmann and Joseph 2015). Besides, biochar-induced soil water permeability and water-holding capacity raise the amount of available water for plants (Glaser et al. 2002; Asai et al. 2009). Thus, elevated plant-available fraction of soil water via biochar application may pose direct impacts for plant nutrient uptake. Biochar CEC is another crucial parameter that might pose modifications on nutrient uptake. Although such types of interactions have not been studied yet, it has been suggested that biochar can slow cation loss by inducing a shift in soil water nutrient transport from bypass to matrix flow (Biederman and Harpole 2013). However, such possible physiochemical reactions of biochar on plant nutrient uptake should be properly identified through further research. Figure 3.1 depicts the possible biochar effects on plant nutrient uptake.

3.4 Macronutrients

Nitrogen (N), phosphorous (P), calcium (Ca), magnesium (Mg), potassium (K), and sulfur (S) are considered as plant macronutrients because plants require them in relatively high amounts (>0.1% of dry mass). Those macronutrients are essential for plants to complete their life cycles. Generally, plant roots absorb the ionic forms of those mineral nutrients, which are dissolved in the soil solution. Plant-available form of calcium, magnesium, and potassium is Ca^{2+}, Mg^{2+}, and K^{+}, respectively, whereas nitrogen as NO_3^- or NH_4^+, phosphorous as PO_4^{3-}, and sulfur as SO_4^{2-} (Maathuis 2009).

3.4.1 Nitrogen

In crop cultivation, N is considered as the key annual input regarding fertilizer for crop nutrition, inorganic or organic. Although biochar may not supply N directly, as N is eliminated during pyrolysis or integrated into stable aromatic structures, they interact with soil mineral N that is accessible for plants and depends on the dynamic balance between microbial usage, plant uptake, and mineralization. It has been reported that the total N recovery in crops is higher in charcoal amended plots (18.1%) in comparison to compost treatments (16.5%). Steiner et al. (2008) reported the increased N retention by charcoal amendments more than compost. Application of poultry litter biochar without N fertilizer had resulted in yield increase of radish from 42 to 96% in comparison with control, indicating the enhanced N availability and uptake (Chan et al. 2008b). These researchers have proved that biochar additions significantly increased plant N concentrations. Even at low biochar application rate (10 ton/ha), plant N uptake increased from 41 to 45%, compared to control and N uptake increased further with increasing application rate. Similarly, research findings of Uzoma et al. (2011) indicated that the rate of biochar application improved the rate of N uptake in maize. Researchers have suggested that enhanced N uptake at higher biochar addition rates can be attributed to the increased K, since K is considered as the counter cation accompanying the uptake of N as nitrate ions (Chan et al. 2008b). Moreover, aged biochar influences N availability in a different manner than fresh biochar. Particularly, aged biochar-containing oxygenated functional groups (e.g., carboxyl and hydroxyl) have more capacity to absorb NH_4^+ than fresh biochar (Zheng et al. 2013).

Additionally, the leaching of N or lose as gaseous N in agriculture fields after fertilizer use has become a major limitation for improving production by increasing nitrogen utilization efficiency (Yanai et al. 2007; Ding et al. 2010). Smith and Tibbett (2004) had used sewage sludge as a soil amendment and found that loss of N by NH_3 volatilization or NO_3^- leaching may limit benefits of sludge amendments. Such nutrient losses can be mitigated by amending the soil with biochar which increases the N fertilizer use efficiency (Uzoma et al. 2011). Decreased N losses can be explained by either electrostatic adsorption to exchange sites in biochar facilitating enhanced retention of NH_4^+ (Steiner et al. 2008) or absorption of NH_3 to biochar relying on the biochar surface area and their acid functional group content (Clough et al. 2013). Biochar application levels between 10 and 20% by weight have been shown to reduce NH_4^+ losses in soils (Lehmann et al. 2003a). Another researcher observed the similar results in radish plants (Major et al. 2010). Slavich et al. (2013) showed that feedlot manure biochar had enhanced agronomic nitrogen use efficiency by 23% and thus increased total pasture productivity by 11%. It was observed that N uptake of corn plants was increased by 15% after biochar application with recommended fertilizers (Rajkovich et al. 2012). Similarly, an increased uptake of N by several crops grown in soils amended with biochar and N fertilizer was reported (Van Zwieten et al. 2010a). Nevertheless, biochar application might involve for the limitation of soil N availability for plants in N-deficient soils. It had been suggested that biochar's high C/N ratio might pose negative effects on N immobilization (Lehmann et al. 2003a; Asai et al. 2009).

3.4.2 Phosphorous

In soil, more than 80% of the P remains immobile and unavailable for plant uptake as a result of adsorption, precipitation, or conversion to the organic form. Yet, numerous studies commented that the biochar application increases plant-available P in soil (Asai et al. 2009; Yamato et al. 2006). Application of biochar to the root zone of the P-deficient soil increased plant growth by 59% and P uptake by 73% (Shen et al. 2016). Lehmann et al. (2003a, b) also revealed that increasing biochar application rates also increase the P concentration and uptake in plants. In addition, an increase in grain yield has been recorded from after addition of biochar to rice fields with low available P (Asai et al. 2009). Researchers have explained that microbial biomass is crucial for organic P to be bioavailable and biochar-amended soils are rich in microbial biomass carbon (Lehmann et al. 2011; Masto et al. 2013). High microbial biomass carbon starts to get high amounts of ortho-P for its metabolic functions, leading to having high concentrations of bioavailable P in soil (Masto et al. 2013). On the other hand, P uptake by plants may depend on the association between plants and mycorrhizal fungi which secrets extracellular phosphatases and P-solubilizing organic acids making organic P plant available. Several researchers revealed that biochar encourages mycorrhizal colonization of plant roots by facilitating habitats for them and thereby indirectly promote P solubility (Warnock et al. 2007; Gul and Whalen 2016). Another assumption is that nutrients in biochar increase the production of P-solubilizing organic acids. Deb et al. (2016) have stated that this effect is greater in nutrient poor soils than fertile soils. In addition, enhanced P uptake by maize grain with the application of cow manure biochar had been attributed to the increased P availability dynamics as a result of increased soil pH by biochar (Uzoma et al. 2011). Enhanced soil pH may facilitate increase alkaline extracellular phosphatase activities. For instance, corncob biochar had increased alkaline phosphatase activity ~2 to ~3 times (Du et al. 2014), while swine manure biochar had contributed to a 28.5% increase in alkaline phosphomonoesterase activity (Jin et al. 2016).

Moreover, organic amendments including manure, compost, and sludge have long been applied to assure P sustainability and thereby to increase crop productivity. However, the high mobility of P in organic amendments may not only limit the nutritional benefits but also cause serious environmental issues like eutrophication (Dai et al. 2016). According to previous studies, after manure application, 78% of input P had released to the top soil, whereas biochar application had released almost 1% (Dai et al. 2016). In biochar production, most P fractions become stable during pyrolysis. As a result, biochar may provide a long lasting P source to crop fields (Dai et al. 2015). In addition, P fertilizers are commonly used in crop cultivations to increase yield, and P leaching similar to N is significantly experienced. Thus, the addition of fertilizers along with biochar might be an effective measure that should be confirmed by further evaluations.

3.4.3 Potassium

Plants are known to uptake potassium (K) from relatively dilute soil solutions. However, as biochar can increase soil CEC, thereby they can increase the ability of soil to hold K and store them in the soil for plant uptake. In addition, biochar may inherently contain exchangeable K for plant uptake. One year after biochar application, K content in plant biomass had increased by 57%, whereas manure application had increased 43% during the same period (Lentz and Ippolito 2012). A great availability of K in soil, soon after biochar application, has been reported (Cheng et al. 2008). Further, K uptake by maize grain was significant after the application of cow manure biochar (Uzoma et al. 2011). Several researchers suggested that increased K availability in soil could be attributed to enhanced soil pH by biochar (Manolikaki and Diamadopoulos 2016; Smider and Singh 2014). The increase in soil pH may force on less available K^+ that remains strongly attached to the clay particles to be released into the soil solution. An increase of rice and cowpea biomass by the K provided from biochar has been reported (Lehmann et al. 2003a). Biochar produced from plant biomass increased K uptake in common bean (Rondon et al. 2007). In addition, enhanced concentrations of K in legume biomass had been reported after addition of grass-derived biochar. In the same study, it had been found that available K in biochar applied treatment soils was even exceeding concentrations in the treatments that received K fertilizer (Oram et al. 2014). The authors had highlighted that decrease in net nitrification resulted reduced N uptake and thereby enhanced K uptake as there could be a competition in legume plants for N and K. Particularly, fresh biochar is considered to have available K that can be rapidly taken by plants (Karer et al. 2013). However, some researchers have suggested that high availability of K for plants with biochar may not persist beyond the year after application (Steiner et al. 2007).

3.4.4 Calcium

The potential of Ca uptake by plants is also related to root cation exchange capacity. Organic soil amendments and some clay can adsorb Ca^{2+} as they have negatively charged sites on their surfaces. Then the soil has the potential of exchanging Ca^{2+} with plant root. A significant increase in exchangeable Ca level and enhanced Ca uptake after addition of cow manure was reported (Uzoma et al. 2011). Nevertheless, calcium becomes readily available in the soil after biochar application; biochar has a greater negative surface charge, charge density, and higher surface area than other organic amendments (Somebroek 1993). On the other hand, Ca content in biochar may replace monomeric Al species on soil mineral or soil organic matter exchangeable sites enhancing Ca availability for plants (Novak et al. 2009). According to some research findings, excess Ca levels in the soil after harvesting indicates that Ca release from biochar may exceed even plant requirements (Ma and Matsunaka 2013). A field trial done by Chan et al. (2008a, b), during 4 years of the period with 0, 8 and 20 ton/ha of biochar application rates indicated increased available Ca over time. Further, the available Ca concentrations increased over time, from 101 to 320% and up to 30 cm depths suggesting leaching of Ca is minimum with the application of biochar (Major et al. 2010). However, it was reported that significant increases in plant-available Ca could be observed only at biochar applications higher than 50 ton/ha without fertilizer application (Chan et al. 2008a). Furthermore, biochar, rich in cations as K^+, Na^+, and NH_4^+, may directly or indirectly depress Ca uptake particularly at low Ca concentrations.

3.4.5 Magnesium

Magnesium (Mg) amount that can be uptaken by plants in soil depends on soil pH, and it becomes less available under low pH conditions. Since most of the biochar applications enhance soil pH, magnesium can be easily available for plant uptake. Uzoma et al. (2011) observed the significantly high level of exchangeable Mg in biochar-amended soil than control (Uzoma et al. 2011). Consequently, they found that cow manure biochar is responsible for an increment of Mg concentration in maize grain, which was attributed to increasing of exchangeable Mg in soil with higher biochar application rates. In contrast, some researchers reported that the addition of biochar reduced the uptake of Mg and reduced the yield of corn silage (Lentz and Ippolito 2012).

3.4.6 Sulfur

It has been estimated that up to 95% of S present in agricultural soils remains as sulfate esters or sulfonates being unavailable to the plant. Such organic forms need conversion into inorganic forms via desulfurization carried out by microbes. Thus, some research hypothesis indicates that biochar amendment provides refuges for such microbial populations and thereby enhance S mobilization allowing plants to uptake more S (Fox et al. 2014). It has been reported that compost amendment enhanced available S, thereby S uptake (Chowdhury et al. 2000). Nevertheless, no detailed studies have assessed biochar effect on S uptake yet. Although studies have revealed the changes in biochar that might increase S availability for plants, some studies indicated the decreases of available S observed after adding of even small amounts (i.e., 0.36–0.5%) of biochar to the field (Namgay et al. 2010). Increased soil pH after biochar amendments may negatively affect S oxidation, biochar might add S uptake inhibitors to the soil, or they inhibit microbial activities of S oxidation. Furthermore, organic amendments with high C/S ratio (e.g., rice husk) have been found to result in severe S deficiency of plant due to S immobilization in soil (Chowdhury et al. 2000).

3.5 Micronutrients

Micronutrients such as boron (B), chlorine (Cl), copper (Cu), iron (Fe), manganese (Mn), molybdenum (Mo), nickel (Ni), and zinc (Zn) can be vital for the normal healthy plant growth (Alloway 2008). Since biochar is mentioned as an ideal amendment for metal retention, they may pose impacts on such nutrient uptake. The mechanisms involved in metal retention can be attributed to biochar-induced soil cation exchange capacity, acid neutralization in soil, and biochar's high specific surface area (Beesley et al. 2010; Asai et al. 2009). In addition, it is worthwhile to mention that enhanced soil pH due to biochar additions can cause micronutrient deficiencies which occur at high pH (>6). The application of biochar was reported to reduce the availability of Cd and Zn in mine-contaminated soil thereby decreasing uptake rate of Zn by Jack bean (*Canavalia ensiformis*) and *Mucuna aterrima* plants (Puga et al. 2015). Similarly, the exchangeable Zn concentrations decreased marginally (from 13 to 10 mg/kg) with increasing biochar application rates indicating high Zn sorption capacity of biochar (Jayawardhana et al. 2016a). Another study had shown that B and Zn contents in wheat plant tissues decreased after biochar applications which could be explained by high adsorption capacity of biochar as well as enhanced soil pH leading to precipitate Zn and make B less available (Kloss et al. 2012). As well, compost amendments had contributed to decreasing Zn availability by improving soil porosity, particle size distribution, and cracking patterns allowing the formation of stable water aggregates and thereby limiting the dispersion (Park et al. 2011). Moreover, a combination of biochar and manure had synergistic positive effect for

Mn availability in soil (Lentz and Ippolito 2012). Some other researchers found that addition of biochar can stimulate or inhibit the activity of microorganisms which affect the availability of Mn by alterations in microorganism population and activity (Meek et al. 1968; Abou-Shanab et al. 2003). Novak et al. (2009) observed that Mn concentration in biochar-amended soil has increased, whereas those in biochar leachates have decreased. This can be attributed to the great retention of Mn with different organic and inorganic forms during pyrolysis (Novak et al. 2009; Amonette and Joseph 2009). Further, reduction of Cu uptake by corn silage after biochar application was observed (Lentz and Ippolito 2012). Nevertheless, some researchers reported that the concentration of Cu was not affected significantly by the addition of pecan shell biochar (Novak et al. 2009). Reduction of Cu availability had been reported after biosolid amendments indicating the formation of inorganic metal complexation leading to immobilize available Cu (Park et al. 2011). The Ni uptake in spinach had increased at 3% biochar application compared to the control and decreased significantly at 5%. However decreased Ni uptake had resulted in an increase in biomass production from 29 to 36% (Jayawardhana et al. 2016b). Further, uptake of Fe also had decreased with biochar amendments which may be due to the precipitation of Fe thereby reducing its mobility into phloem cells for long distance translocation (Kloss et al. 2012). However, the low uptake efficiency of the micronutrients after biochar additions suggests that they may prevent toxicity accumulations in plants.

3.6 Effect of Biochar on Nutrient Transformation in Soil

Specific biochar characteristics including large surface area, highly porous structure, and cation exchange capacities may affect nutrient transformation processes in soil. However, only a limited research attention has been given to the influence of biochar on such processes (DeLuca et al. 2015). Biochar additions to soil may alter soil microbial populations or provide habitat for them those are actively giving a contribution to transformations of nutrients including N, P, or S (DeLuca et al. 2015). Net nitrification in acidic forest soil had significantly increased after biochar application which may be due to the reason that autotrophic nitrifying bacteria may favor less acidic soil conditions (Warnock et al. 2007). Further, Rondon et al. (2007) showed a considerable increase of nitrogen fixation in common beans after biochar amendments. Comparably, Dai et al. (2016) have reported an increase in N mineralization nearly two times higher than the control after biochar application in a lettuce plantation. They have indicated the positive correlation of N mineralization with biochar H/C ratio and explained that since less recalcitrant biochar with high H/C ratio can enhance mineralization as they easily decompose releasing N into the mineral pool (Dai et al. 2016). On the contrary, reduced N mineralization has been reported by several studies (Masto et al. 2013; Dai et al. 2015). The authors have shown that the high C/N ratio of the biochar may inhibit N mineralization potential which is likely to depend on the biochar feedstock. Nevertheless, biochar derived from wood or N-limited feedstock along with high C/N ratios tends to immobilize nitrogen by converting their organic forms to inorganic (Lentz and Ippolito 2012). In addition, having a similar cycle to N, S in the soil also may be significantly influenced by biochar (DeLuca et al. 2015). Biochar would increase soil pH declining populations of autotrophic microorganisms involved in organic S oxidation as they favor low pH conditions. Thus, oxidation and mineralization of sulfur may reduce after biochar applications (Jayawardhana et al. 2016a).

A number of studies have proven that biochar modifies soil pH particularly increasing soil pH in acidic soils (Lehmann et al. 2003a; Manolikaki and Diamadopoulos 2016). As soil pH strongly determines precipitation reactions of P, biochar applications to soil may convert P into insoluble pools. Further, ionic P interactions with Al^{+3}, Fe^{+3}, and Ca^{+2} can be altered or adsorbed organic molecules onto biochar that may act as chelating agents precipitating P. For example, simple organic/phenolic acid or complex proteins/carbohydrates have the potential of sorbing to the hydrophobic or charged biochar surface, chelating Al^{+3}, Fe^{+3}, and Ca^{+2}, and thereby they can modify the P solubility (DeLuca et al. 2015). Further, as biochar ages, cation exchange capacities are altered by increasing negative charge sites and decreasing positive charge sites. Hence, aged biochar may reduce the availability of Al^{+3} and Fe^{+3} in soil promoting the recycling of labile P fractions (DeLuca et al. 2015). Moreover, biochar is reported to influence on P mineralization and phosphatase enzyme activities (Jin et al. 2016). To illustrate, research findings has confirmed that the biochar may enhance phosphatase activity that hydrolyzes organic P and converts them into different inorganic forms (Oram et al. 2014).

3.7 Environmental Considerations

As discussed earlier, biochar effects on soil and plants not only depend on the quality of biochar but also on the soil characteristics (e.g., soil pH, texture, organic matter). Most of the research efforts that have been taken were concerned on highly weathered infertile tropical soils (Sohi et al. 2010; Glaser et al. 2002; Blackwell et al. 2009). Though such studies had revealed the positive effects of biochar on both soil and plants, these effects might be somewhat attributed to the depletion of Al toxicity in rhizosphere resulted by enhanced soil pH (Kuka et al. 2013). Thus, the same effects cannot be expected from other soils in different climatic regions (Major et al. 2010). Soil researches in temperate regions have demonstrated that biochar effect on plant and soil is very small, short lived, or undesirable (Jones et al. 2012; Kloss et al. 2014). Research evidence convinced that biochar might even reduce plant growth which may be due to the unfavorably high pH of biochar (Van Zwieten et al. 2010b).

Although biochar has been reported as an effective multifunctional soil amendment, it is essential to establish rigorously monitored supply networks and to ensure that feedstocks come from sustainably managed lands and waste materials. If not properly monitored, the production of biochar could lead to deforestation and processing of non-sustainable feed stocks, exacerbating the problems of decreasing biodiversity and increasing carbon emissions. In many cases, feedstock waste materials are not much valuable, or their disposal may demand cost. However, it is crucial to ensure that contaminants present in the pyrolysis feedstock (e.g., sewage sludge/municipal waste) are eliminated or modified to become more or less available in the biochar product. In the case of such types of feedstock, evaluation of phytotoxicity is an important consideration. On the other hand, biochar itself may serve as a source of combustion-related contaminants such as poly-nuclear aromatic hydrocarbons (PAHs) and dioxins which are produced during the production process. Kookana et al. (2010) revealed that biochar amendment to soil could potentially lead to accumulation of contaminants residues in soil. Furthermore, the maximum amount of biochar that can be applied for a sustainable crop production is questionable. In spite the fact that numerous biochar research has shown increased crop yield with increasing biochar additions, some researchers have stated that biomass production and crop yield decrease at high biochar concentrations (Kloss et al. 2012). Overall, biochar characteristics, as well as specific soil productivity constraints, are indispensable factors to be taken into account before biochar application.

3.8 Remarks

Biochar impacts on plant nutrient uptake deserve a greater attention as no studies directly asses such interactions. Although increased crop production after biochar application has been reported, the explanation for these benefits has not been fully described, and neither the quantitative variability in nutrient uptake and influence on soil microbiology nor the durability of the effects has been specified. According to previous work, biochar is significantly variable in composition and availability of nutrients depending upon feedstock material and pyrolysis conditions. Further, negative surface charge and CEC of biochar may increase with biochar aging. Also, the labile organic carbon in biochar and its intrinsic nutrient supply may be depleted throughout the aging process. Thus, the changes in biochar-nutrient properties and effects on nutrient cycling in biochar-mediated soil over biochar aging should be studied in detail.

However, reported effects on the plant nutrient uptake are not directly attributed to the nutrient composition of biochar but may depend on indirect mechanisms between biochar and soil. Thus, further research is needed to identify and quantify indirect nutrient aspects of biochar produced from different feedstock under different pyrolysis conditions and how the nutrient retention capacities may vary with time. At the same time, strong interest is present at this moment to modify biochar using different physical and chemical techniques to be applied in soil (e.g., sulfur char for saline soil). However, it is important to determine how such modifications influence the macro- and micronutrient availability as well. Moreover, studies on biochar impacts on soil microbial populations, and their activities that may determine plant nutrient uptake are limited. Some hypotheses have been put forward explaining that biochar may provide habitats for microbes, supply protection from predators, or provide a substrate for nutrient requirements. However, the underlying reason why biochar stimulates microbial activities should be fully assessed. On the other hand, soil quality parameters are vital to contribute to the biochar performance, and nutrient status in soil and biochar-nutrient interactions in soil may fluctuate with time. In addition, research opportunities exist to evaluate responses of different plants to different biochar. Hence, long-term data in relation to specific soil parameters and specific plants is critical to promote biochar use in plant productivity. Such research gaps are key challenges to address in order to establish an effective approach on biochar-induced plant nutrient uptake.

References

Abou-Shanab, R., et al. (2003). Rhizobacterial effects on nickel extraction from soil and uptake by *Alyssum murale*. *New Phytologist, 158*(1), 219–224.

Agrafioti, E., et al. (2013). Biochar production by sewage sludge pyrolysis. *Journal of Analytical and Applied Pyrolysis, 101*, 72–78.

Alloway, B. J. (2008). *Micronutrient deficiencies in global crop production*. Netherlands: Springer Science & Business Media.

Amonette, J. E., & Joseph, S. (2009). Characteristics of biochar: Microchemical properties. *Biochar for Environmental Management: Science and Technology, 33*.

Angst, T. E., & Sohi, S. P. (2013). Establishing release dynamics for plant nutrients from biochar. *GCB Bioenergy, 5*(2), 221–226.

Asai, H., et al. (2009). Biochar amendment techniques for upland rice production in Northern Laos: 1. Soil physical properties, leaf SPAD and grain yield. *Field Crops Research, 111*(1), 81–84.

Atkinson, C. J., Fitzgerald, J. D., & Hipps, N. A. (2010). Potential mechanisms for achieving agricultural benefits from biochar application to temperate soils: A review. *Plant and Soil, 337*(1–2), 1–18.

Bah, A., et al. (2014). Reducing runoff loss of applied nutrients in oil palm cultivation using controlled-release fertilizers. *Advances in Agriculture, 2014*, 285387.

Bailey, V. L., et al. (2011). Reconciling apparent variability in effects of biochar amendment on soil enzyme activities by assay optimization. *Soil Biology and Biochemistry, 43*(2), 296–301.

Bandara, T., et al. (2017a). Role of woody biochar and fungal-bacterial co-inoculation on enzyme activity and metal immobilization in serpentine soil. *Journal of Soils and Sediments, 17*(3), 665–673.

Bandara, T., et al. (2017b). Efficacy of woody biomass and biochar for alleviating heavy metal bioavailability in serpentine soil. *Environmental Geochemistry and Health, 39*(2), 391–401.

Beesley, L., Moreno-Jiménez, E., & Gomez-Eyles, J. L. (2010). Effects of biochar and greenwaste compost amendments on mobility, bioavailability and toxicity of inorganic and organic contaminants in a multi-element polluted soil. *Environmental Pollution, 158*(6), 2282–2287.

Biederman, L. A., & Harpole, W. S. (2013). Biochar and its effects on plant productivity and nutrient cycling: A meta-analysis. *GCB Bioenergy, 5*(2), 202–214.

Blackwell, P., Riethmuller, G., & Collins, M. (2009). Biochar application to soil. *Biochar for Environmental Management: Science and Technology, 1*, 207–226.

Brantley, K. E., et al. (2015). Pine woodchip biochar impact on soil nutrient concentrations and corn yield in a silt loam in the Mid-Southern US. *Agriculture, 5*(1), 30–47.

Cao, C. T., et al. (2014). Biochar makes green roof substrates lighter and improves water supply to plants. *Ecological Engineering, 71*, 368–374.

Chan, K. Y., & Xu, Z. (2009). Biochar: Nutrient properties and their enhancement. *Biochar for Environmental Management: Science and Technology, 1*, 67–84.

Chan, K., et al. (2008a). Agronomic values of greenwaste biochar as a soil amendment. *Soil Research, 45*(8), 629–634.
Chan, K., et al. (2008b). Using poultry litter biochars as soil amendments. *Soil Research, 46*(5), 437–444.
Cheng, C.-H., Lehmann, J., & Engelhard, M. H. (2008). Natural oxidation of black carbon in soils: Changes in molecular form and surface charge along a climosequence. *Geochimica et Cosmochimica Acta, 72*(6), 1598–1610.
Chowdhury, M. A. H., et al. (2000). Microbial biomass, S mineralization and S uptake by African millet from soil amended with various composts. *Soil Biology and Biochemistry, 32*(6), 845–852.
Clough, T. J., et al. (2013). A review of biochar and soil nitrogen dynamics. *Agronomy, 3*(2), 275–293.
Dai, L., et al. (2015). Immobilization of phosphorus in cow manure during hydrothermal carbonization. *Journal of Environmental Management, 157*, 49–53.
Dai, L., et al. (2016). Biochar: A potential route for recycling of phosphorus in agricultural residues. *GCB Bioenergy, 8*(5), 852–858.
De la Rosa, J., et al. (2011). Molecular composition of sedimentary humic acids from South West Iberian Peninsula: A multi-proxy approach. *Organic Geochemistry, 42*(7), 791–802.
Deb, D., et al. (2016). Variable effects of biochar and P solubilizing microbes on crop productivity in different soil conditions. *Agroecology and Sustainable Food Systems, 40*(2), 145–168.
DeLuca, T. H., et al. (2015). Biochar effects on soil nutrient transformations. *Biochar for Environmental Management: Science and Technology, 2*, 421–454.
Demirbas, A. (2004). Effects of temperature and particle size on bio-char yield from pyrolysis of agricultural residues. *Journal of Analytical and Applied Pyrolysis, 72*(2), 243–248.
Ding, Y., et al. (2010). Evaluation of biochar effects on nitrogen retention and leaching in multilayered soil columns. *Water, Air, & Soil Pollution, 213*(1–4), 47–55.
Downie, A., Crosky, A., & Munroe, P. (2009). Physical properties of biochar. *Biochar for Environmental Management: Science and Technology*, 13–32.
Du, Z., et al. (2014). Consecutive biochar application alters soil enzyme activities in the winter wheat–growing season. *Soil Science, 179*(2), 75–83.
Eghball, B., & Power, J. F. (1999). Composted and noncomposted manure application to conventional and no-tillage systems: Corn yield and nitrogen uptake. *Agronomy Journal, 91*(5), 819–825.
Esposito, N. C. (2013). *Soil nutrient availability properties of biochar*. MSc Thesis, Faculty of California Polytechnic State University, San Luis Obispo.
Filiberto, D. M., & Gaunt, J. L. (2013). Practicality of biochar additions to enhance soil and crop productivity. *Agriculture, 3*(4), 715–725.
Fox, A., et al. (2014). The role of sulfur-and phosphorus-mobilizing bacteria in biochar-induced growth promotion of *Lolium perenne*. *FEMS Microbiology Ecology, 90*(1), 78–91.
Glaser, B., et al. (2000). Black carbon in density fractions of anthropogenic soils of the Brazilian Amazon region. *Organic Geochemistry, 31*(7), 669–678.
Glaser, B., Lehmann, J., & Zech, W. (2002). Ameliorating physical and chemical properties of highly weathered soils in the tropics with charcoal-a review. *Biology and Fertility of Soils, 35*(4), 219–230.
Gul, S., & Whalen, J. K. (2016). Biochemical cycling of nitrogen and phosphorus in biochar-amended soils. *Soil Biology and Biochemistry, 103*, 1–15.
Haefele, S., et al. (2011). Effects and fate of biochar from rice residues in rice-based systems. *Field Crops Research, 121*(3), 430–440.
Hossain, M. K., et al. (2011). Influence of pyrolysis temperature on production and nutrient properties of wastewater sludge biochar. *Journal of Environmental Management, 92*(1), 223–228.
Jayawardhana, Y. et al. (2016a) *Detection of benzene in landfill leachate from Gohagoda dumpsite and its removal using municipal solid waste derived biochar.*
Jayawardhana, Y., et al. (2016b). Chapter 6: Municipal solid waste biochar for prevention of pollution from landfill leachate. In *Environmental materials and waste* (pp. 117–148). Amsterdam: Academic Press.

Jin, Y., et al. (2016). Manure biochar influence upon soil properties, phosphorus distribution and phosphatase activities: A microcosm incubation study. *Chemosphere, 142*, 128–135.
Jones, D., et al. (2012). Biochar-mediated changes in soil quality and plant growth in a three year field trial. *Soil Biology and Biochemistry, 45*, 113–124.
Karer, J., et al. (2013). Biochar application to temperate soils: Effects on nutrient uptake and crop yield under field conditions. *Agricultural and Food Science, 22*(4), 390–403.
Kloss, S., et al. (2012). Characterization of slow pyrolysis biochars: Effects of feedstocks and pyrolysis temperature on biochar properties. *Journal of Environmental Quality, 41*(4), 990–1000.
Kloss, S., et al. (2014). Biochar application to temperate soils: Effects on soil fertility and crop growth under greenhouse conditions. *Journal of Plant Nutrition and Soil Science, 177*(1), 3–15.
Knicker, H., et al. (1996). 13C-and 15N-NMR spectroscopic examination of the transformation of organic nitrogen in plant biomass during thermal treatment. *Soil Biology and Biochemistry, 28*(8), 1053–1060.
Kookana, R. S., Yua, X.-Y., & Yinga, G.-G. (2010). *"Black is the new green": The blue shades of biochar, in 19th World Congress of Soil Science*. Australia: Brisbane.
Kuka, K., et al. (2013). Investigation of different amendments for dump reclamation in Northern Vietnam. *Journal of Geochemical Exploration, 132*, 41–53.
Laird, D. A., et al. (2010). Impact of biochar amendments on the quality of a typical Midwestern agricultural soil. *Geoderma, 158*(3), 443–449.
Lehmann, J. (2007). A handful of carbon. *Nature, 447*(7141), 143–144.
Lehmann, J., & Joseph, S. (2015). *Biochar for environmental management: Science, technology and implementation*. Routledge: Taylor & Francis.
Lehmann, J., et al. (2003a). Nutrient availability and leaching in an archaeological Anthrosol and a Ferralsol of the Central Amazon basin: Fertilizer, manure and charcoal amendments. *Plant and Soil, 249*(2), 343–357.
Lehmann, J., et al. (2003b). Soil fertility and production potential. In *Amazonian dark earths* (pp. 105–124). New York: Springer.
Lehmann, J., Gaunt, J., & Rondon, M. (2006). Bio-char sequestration in terrestrial ecosystems–a review. *Mitigation and Adaptation Strategies for Global Change, 11*(2), 395–419.
Lehmann, J., et al. (2011). Biochar effects on soil biota–a review. *Soil Biology and Biochemistry, 43*(9), 1812–1836.
Lentz, R., & Ippolito, J. (2012). Biochar and manure affect calcareous soil and corn silage nutrient concentrations and uptake. *Journal of Environmental Quality, 41*(4), 1033–1043.
Liang, B., et al. (2006). Black carbon increases cation exchange capacity in soils. *Soil Science Society of America Journal, 70*(5), 1719–1730.
Lima, I. M., & Marshall, W. E. (2005). Granular activated carbons from broiler manure: Physical, chemical and adsorptive properties. *Bioresource Technology, 96*(6), 699–706.
Liu, T., Liu, B., & Zhang, W. (2014). Nutrients and heavy metals in biochar produced by sewage sludge pyrolysis: Its application in soil amendment. *Polish Journal of Environmental Studies, 23*(1), 271–275.
Ma, Y. L., & Matsunaka, T. (2013). Biochar derived from dairy cattle carcasses as an alternative source of phosphorus and amendment for soil acidity. *Soil Science & Plant Nutrition, 59*(4), 628–641.
Maathuis, F. J. (2009). Physiological functions of mineral macronutrients. *Current Opinion in Plant Biology, 12*(3), 250–258.
Major, J., et al. (2010). Maize yield and nutrition during 4 years after biochar application to a Colombian savanna oxisol. *Plant and Soil, 333*(1–2), 117–128.
Manolikaki, I., & Diamadopoulos, E. (2016). Ryegrass yield and nutrient status after biochar application in two Mediterranean soils. *Archives of Agronomy and Soil Science*. doi:10.1080/03650340.2016.1267341.
Marris, E. (2006). Putting the carbon back: Black is the new green. *Nature, 442*(7103), 624–626.
Masto, R. E., et al. (2013). Biochar from water hyacinth (Eichornia crassipes) and its impact on soil biological activity. *Catena, 111*, 64–71.

Masulili, A., Utomo, W. H., & Syechfani, M. (2010). Rice husk biochar for rice based cropping system in acid soil 1. The characteristics of rice husk biochar and its influence on the properties of acid sulfate soils and rice growth in West Kalimantan, Indonesia. *Journal of Agricultural Science, 2*(1), 39–47.
Mayakaduwa, S., et al. (2016a). Insights into aqueous carbofuran removal by modified and non-modified rice husk biochars. *Environmental Science and Pollution Research*, 1–9.
Mayakaduwa, S., et al. (2016b). Equilibrium and kinetic mechanisms of woody biochar on aqueous glyphosate removal. *Chemosphere, 144*, 2516–2521.
Meek, B. D., MacKenzie, A., & Grass, L. (1968). Effects of organic matter, flooding time, and temperature on the dissolution of iron and manganese from soil in situ. *Soil Science Society of America Journal, 32*(5), 634–638.
Naeem, M. A., et al. (2014). Yield and nutrient composition of biochar produced from different feedstocks at varying pyrolytic temperatures. *Pakistan Journal of Agricultural Sciences, 51*(1), 75–82.
Namgay, T., Singh, B., & Singh, B. P. (2010). Influence of biochar application to soil on the availability of As, Cd, Cu, Pb, and Zn to maize (*Zea mays* L.) *Soil Research, 48*(7), 638–647.
Novak, J. M., et al. (2009). Impact of biochar amendment on fertility of a southeastern coastal plain soil. *Soil Science, 174*(2), 105–112.
Oram, N. J., et al. (2014). Soil amendment with biochar increases the competitive ability of legumes via increased potassium availability. *Agriculture, Ecosystems & Environment, 191*, 92–98.
Park, J. H., et al. (2011). Role of organic amendments on enhanced bioremediation of heavy metal (loid) contaminated soils. *Journal of Hazardous Materials, 185*(2), 549–574.
Pietikäinen, J., Kiikkilä, O., & Fritze, H. (2000). Charcoal as a habitat for microbes and its effect on the microbial community of the underlying humus. *Oikos, 89*(2), 231–242.
Prendergast-Miller, M., Duvall, M., & Sohi, S. (2014). Biochar–root interactions are mediated by biochar nutrient content and impacts on soil nutrient availability. *European Journal of Soil Science, 65*(1), 173–185.
Puga, A., et al. (2015). Biochar application to a contaminated soil reduces the availability and plant uptake of zinc, lead and cadmium. *Journal of Environmental Management, 159*, 86–93.
Rajkovich, S., et al. (2012). Corn growth and nitrogen nutrition after additions of biochars with varying properties to a temperate soil. *Biology and Fertility of Soils, 48*(3), 271–284.
Rondon, M. A., et al. (2007). Biological nitrogen fixation by common beans (*Phaseolus vulgaris* L.) increases with bio-char additions. *Biology and Fertility of Soils, 43*(6), 699–708.
Sanchez, P. A. (2002). Soil fertility and hunger in Africa. *Science, 295*(5562), 2019–2020.
Shen, Q., et al. (2016). Can biochar increase the bioavailability of phosphorus? *Journal of Soil Science and Plant Nutrition, 16*(2), 268–286.
Smernik, R. J., et al. (2002). Determination of T 1ρ H relaxation rates in charred and uncharred wood and consequences for NMR quantitation. *Solid State Nuclear Magnetic Resonance, 22*(1), 50–70.
Smider, B., & Singh, B. (2014). Agronomic performance of a high ash biochar in two contrasting soils. *Agriculture, Ecosystems & Environment, 191*, 99–107.
Smith, M., & Tibbett, M. (2004). Nitrogen dynamics under *Lolium perenne* after a single application of three different sewage sludge types from the same treatment stream. *Bioresource Technology, 91*(3), 233–241.
Sohi, S., et al. (2009). Biochar, climate change and soil: A review to guide future research. *CSIRO Land and Water Science Report, 5*(09), 17–31.
Sohi, S., et al. (2010). A review of biochar and its use and function in soil. *Advances in Agronomy, 105*, 47–82.
Somebroek, W. (1993). Amounts, dynamics and sequestering of carbon in tropical and subtropical soils. *Ambio, 22*, 417–426.

Steiner, C., et al. (2007). Long term effects of manure, charcoal and mineral fertilization on crop production and fertility on a highly weathered Central Amazonian upland soil. *Plant and Soil, 291*(1–2), 275–290.

Steiner, C., et al. (2008). Nitrogen retention and plant uptake on a highly weathered central Amazonian Ferralsol amended with compost and charcoal. *Journal of Plant Nutrition and Soil Science, 171*(6), 893–899.

Sun, Y., et al. (2014). Effects of feedstock type, production method, and pyrolysis temperature on biochar and hydrochar properties. *Chemical Engineering Journal, 240*, 574–578.

Slavich, P. G., Sinclair, K., Morris, S. G., Kimber, S. W. L., Downie, A., & Van Zwieten, L. (2013). Contrasting effects of manure and green waste biochars on the properties of an acidic ferralsol and productivity of a subtropical pasture. *Plant and Soil 366*(1–2):213–227

Tsai, W., Lee, M., & Chang, Y. (2006). Fast pyrolysis of rice straw, sugarcane bagasse and coconut shell in an induction-heating reactor. *Journal of Analytical and Applied Pyrolysis, 76*(1), 230–237.

Uzoma, K., et al. (2011). Effect of cow manure biochar on maize productivity under sandy soil condition. *Soil Use and Management, 27*(2), 205–212.

Van Zwieten, L., et al. (2010a). Effects of biochar from slow pyrolysis of papermill waste on agronomic performance and soil fertility. *Plant and Soil, 327*(1–2), 235–246.

Van Zwieten, L., et al. (2010b). A glasshouse study on the interaction of low mineral ash biochar with nitrogen in a sandy soil. *Soil Research, 48*(7), 569–576.

Vithanage, M., et al. (2016). Potential of biochar and synthetic iron oxides for chromium immobilization in tannery waste polluted soil. *Soil and Groundwater Pollution Remediation, 3*(1), 45–58.

Warnock, D. D., et al. (2007). Mycorrhizal responses to biochar in soil–concepts and mechanisms. *Plant and Soil, 300*(1–2), 9–20.

Withers, P. J., Clay, S. D., & Breeze, V. G. (2001). Phosphorus transfer in runoff following application of fertilizer, manure, and sewage sludge. *Journal of Environmental Quality, 30*(1), 180–188.

Yamato, M., et al. (2006). Effects of the application of charred bark of *Acacia mangium* on the yield of maize, cowpea and peanut, and soil chemical properties in South Sumatra, Indonesia. *Soil Science & Plant Nutrition, 52*(4), 489–495.

Yanai, Y., Toyota, K., & Okazaki, M. (2007). Effects of charcoal addition on N2O emissions from soil resulting from rewetting air-dried soil in short-term laboratory experiments. *Soil Science & Plant Nutrition, 53*(2), 181–188.

Zhang, A., et al. (2012). Effects of biochar amendment on soil quality, crop yield and greenhouse gas emission in a Chinese rice paddy: A field study of 2 consecutive rice growing cycles. *Field Crops Research, 127*, 153–160.

Zheng, H., et al. (2013). Impacts of adding biochar on nitrogen retention and bioavailability in agricultural soil. *Geoderma, 206*, 32–39.

Part II
Plant Nutrients Use Efficiency

Chapter 4
Nutrients Requirement of Medicinal Plants of Dhofar Region of Oman

Mohd. Idrees, Israr Ul Hassan, M. Naeem, Tariq Aftab, and M. Masroor A. Khan

Abstract Dhofar region of Oman has unique position in Arabic peninsula due to Kharif (monsoon) season. During the season, the area receives sufficient water required for development of various type of vegetation including medicinal plants. Medicinal plants of xerophytic nature can survive for a long period. Some of them are perennial like Frankincense and Myrrh, both are considered as the indicators of the region. The other medicinal plant of region includes olive, camel grass, *Nerium*, *Aloe vera*, Eshvarak, rubber bush, etc. The whole region is covered by calcareous soil which has approximately 15% $CaCO_3$. The soil is very poor in nutrient, and all major nutrients like nitrogen, phosphorus, and potassium are found in very low amount or almost absent. Fertilizer application is generally needed to satisfy plant requirements of nutrient for obtaining profitable production. Most of the previous studies on soil are related soil amendment dealt with water scarcity. Little research has been conducted on nutrient application in both timing and doses of application in subtropical arid climates, such as that of Oman. However, some scientist tried to explain the requirement of nutrients of regional medicinal plants including the problem related to absorption of the nutrients, leaching, maintaining soil carbon level, nutrient carbon balances, and fertilizer effect. We tried to collect the information of previous work related to the role of nutrient in development of medicinal plants of Dhofar region and presented in this chapter briefly.

Keywords Mineral fertilizers • Nutrient strategies • Soil profile • Medicinal plants

M. Idrees (✉) • I.U. Hassan
Department of Mathematics and Sciences, College of Arts and Applied Sciences, Dhofar University, P. O. Box 2509, Salalah 211, Oman
e-mail: midress@du.edu.om

M. Naeem • T. Aftab • M.M.A. Khan
Plant Physiology Section, Department of Botany, Aligarh Muslim University, Aligarh 202002, India

M. Naeem et al. (eds.), *Essential Plant Nutrients*,
DOI 10.1007/978-3-319-58841-4_4

4.1 Introduction

The Sultanate of Oman is situated on the southeastern shore of the Arabian Peninsula. The tactical topographical position of the Sultanate made it one of main points of the region in the international maritime trade sector. Business relationships with other countries are due to high plant diversity of Dhofar region. The Dhofar (Ẓufār) Governorate lies in Southern Oman, on the eastern border with Yemen. Historically the region was the chief source of frankincense in the world. Dhofar is directly exposed to the southeast monsoon from mid-June to mid-September, which is known as the Khareef, and hence has a subtropical climate. Due to its climatic condition, it has high diversity of green plants including herbs and shrubs. Dhofar's momentarily rainy climate contrasts abruptly with the adjoining unproductive Empty Quarter Desert. The Salalah plain was once a sound cultivated area with a cultured irrigation system. Rainfall on the seashores and on the inner plains ranges from 20 to 100 mm (0.8–3.9 in.) a year and falls during mid- and late winter. Rainfall in the mountains, particularly over Jebel Akhdar, is much higher and may reach 900 mm (35.4 in.). Development of crop growth and productivity in low-fertility nonsaline soils is rather easy to achieve and largely depend on economics.

Supply of mineral nutrition, according to the plant needs, is important one among various field technique to enhance production. Dose, form, and time of nutrition supply plays important role to increase growth and development. In case of medicinal plants, supply of nutrition can affect the accumulation of secondary metabolites or active molecules. Six macroelements (N, P, K, Mg, Ca, S) and seven microelements (Fe, Mn, Zn, B, Cu, Mo, Cl) are indispensable for normal physiological and biochemical parameters of the plants. An adequate mineral nutrition in the initial phase of plant growth and development is the most important. Early reports have established that supply of both organic and inorganic nourishments augmented growth and yield of crop (Ghanta and Mitra 1993; El-Sheikh 2013a, b). It is a known fact that supply of salt in the root zone can disturb the capacity of plants to take the nutrients from soil and might be the reason of cell toxicity which can reduce plant development (Munns 2002; Parida and Das 2005; Massa et al. 2009). Therefore, an improvement of the saline-root zone (Burn et al. 2001; Massa et al. 2009) can counter salinity stress. Grattan and Grieve (1999) favored supply of big quantities of manures in the root zone to regulate the stress on growth and metabolism of the plant. This chapter summarized nutrient status and requirements of medicinal plants of Oman in general and of Dhofar region in particular as well as contributions of various workers on same issue.

4.2 Salalah Plain

Salalah plain is seaside expanse of the southern province of Dhofar. Dhofar is the only area in Oman to that has advantage of rainfall from the southern monsoon Kharif. The usual 110 mm rainfall is recorded annually; however, sometime it can be recorded from 70 to 360 mm. The period of 2 months (July–August) is normally the "wet" period. Some of the spring water is utilized by falaj (irrigation system) to provide irrigation water for parts of the plain. Recharge is by underflow from the mountains and from the springs. Modern irrigation techniques are in operation in large commercial farms mainly for the production of forage crops such as alfalfa and rhodes grass. *Boswellia sacra*, *Cammiphora myrrha*, *Azadirachta indica*, *Catharanthus roseus*, and *Aloe vera* are medicinal plants of the region.

4.3 Dhofar Jebel

The Jebel Mountain forms a separate agroclimatic zone. The Jebel Mountain received highest rainfall in the region which might be up to 600–700 mm that favors survival of various plant species.

4.3.1 Texture and Nutrient Status of Soil of Dhofar Region

More than 90% of land in Oman is inappropriate for cultivation; most is barren land, stony, or sandy and infertile. Only a minor portion (about 1.4 million ha) is used for cultivation of wild grasses for fodder. Agricultural soils are located on plains, coastal areas, and mountain (The Oman Soil Atlas 1992). Oman lacks truly fertile agriculture soils. Its best soils are the alluvial soils washed down from the mountain in the interior and along coast (El-Sheikh 2013a, b). Information collected from soil assessments on the basis of requirement of particular crop or plant species can be used for nutrient recommendations. Soil survey of 1990–1991 recognized 38 types of soils in various region of Oman. The soils of Dhofar region are bristly coarse (sandy or coarse loamy) with weak water-holding capacity. The soil pH is soberly to strong alkaline and the content of organic matter is very less. These qualities made most soils nonsaline to slightly saline with low fertility. The major components of soil exchange complex are calcium and magnesium that provide average cation exchange capacity to topsoil. Moderate quantity of potassium and phosphorus are available in the soil. Various reports indicate that organic matter is available in higher quantity in comparison to other region but still less than that of international standards. Approximately 0.6% organic carbon content was found in topsoil with minor quantities of iron, manganese, and copper. For cultivation of any crop, soil organic matter whether it is fresh or composted has important role in nutrient status of the soil and thus has effect on plant growth and productivity (Duong et al. 2011). Organic matter plays major role in soils with low fertility where it can improve physical chemical and biological features by providing soil organism and nutrients (Anderson and Domsch 1980; Marinari et al. 2000; Sall et al. 2003; Nyberg et al. 2006). Soil microorganisms obtained from microbial biomass are responsible to decompose organic matter to provide nutrient to soil (Benbi and Richter 2002;

Jedidi et al. 2004). Several factors like type and quality of soil, pH, temperature, and moisture have direct or indirect effect on the process of decomposition and mineralization of organic matter (Joergensen 2000; Hartz et al. 2000; Griffin et al. 2005; Flavel and Murphy 2006).

4.4 Flora of the Oman

For nearly two and a half millennia ago, botanists, naturalists, and travelers had explored flora of Oman. The interest of the world explorer/taxonomist started in 420 BC, when frankincense was exported from Dhofar to Europe via Egypt. P.R.M. Aucher-Eloy, a Frenchman, a well-known taxonomist and plant collector had visited Oman in 1938 and collected more than 250 species from various parts of Oman, among these species, some were considered new at that time. The cloud forest of the Dhofar Mountains is classified as a center of plant diversity in the Arabian Peninsula, with approximately 750 species, of which some 60 are endemic (Radcliffe-Smith 1980; Miller 1994). At present more than 12,000 species of flowering plant are present in Oman. Dhofar region works as reserve area for some plant species and has nearabout 50 members of endemic plant species which belong to two endemic genera. Some important medicinal plants species are the following (Ghazanfar and Al-Sabahi 1993):

Boswellia sacra, Commiphora myrrha, Olea europaea, Cymbopogon schoenanthus, Nerium oleander, Aloe vera, Rhazya stricta, Calotropis procera, Cassia italica, Carthamus tinctorius, Capparis spinosa, Carica papaya, Ricinus communis, Ocimum basilicum, Medicago sativa, Thymus vulgare, Acacia ehrenbergiana, Azadirachta indica, Acacia nilotica, Moringa peregrine, Piper nigrum, Myrtus communis, Nigella sativa, Zingiber officinale, etc.

4.5 Nutrient Management

The reports of various researchers on nutrients management for various medicinal plants are described briefly in the following paragraphs.

*4.5.1 Frankincense (***Boswellia sacra***) and Myrrh (***Commiphora myrrha***)*

The major species of important medicinal plants of Dhofar region are *Boswellia sacra* and *Commiphora myrrha*, commonly known as frankincense and myrrh (*Commiphora myrrha*), respectively. Frankincense and myrrh are the members of same family (Burseraceae) and both are shrub-like trees. The resin of both plants has been used medicinally and for spiritual practices for over 5000 years. Myrrh is obtained from gum resin of *Commiphora sp.* and has various uses in medicinal, cosmetic, and aroma industry. Both as a resin or essential oil, frankincense is useful to provide relax for both the body and mind. The species of *Boswellia sacra* is an indicator of the Dhofar region in Oman and is found in plenty on the steep, precariously eroding slopes in the mountains of Dhofar. *B. sacra* can survive in the most life-threatening conditions and frequently nurtures on rocky hills, commonly in calcareous soil. Generally calcareous soils have near about 15% $CaCO_3$ in the soil that might be presented as powdery, nodules, and crusts forms. Calcareous soils are wildly distributed in drier areas of Oman. Calcareous soil, managed with sufficient water and nutrient supply, has high prospective efficiency for local crop. The soil texture and nutrient status of Dhofar soil indicated in Table 4.1.

Data showed in Table 4.1 indicated that calcareous soils have very low organic matter content and other nutrient. Reports on nutrient management of calcareous soils suggested different strategy for each plant. Nitrogen is one of the significant elements that play an important role in plant growth and metabolism. It is present in the structure of amino acids and hence the proteins. Plants cannot absorb nitrogen directly from the atmosphere. Calcareous soil has more than 15% $CaCO_3$ which directly or indirectly affect the chemistry and availability of nitrogen (N), phosphorus (P), magnesium (Mg), potassium (K), manganese (Mn), zinc (Zn), and iron (Fe). It is better to apply nitrogen fertilizer just before planting and after sprouting in split doses up to plant establishment. Application of nitrogen through side-dressing to the growing crop is an efficient way of nitrogen application. Most of calcareous soils has negligible amount of phosphorous. The soluble form of phosphorus like triple superphosphate, monoammonium phosphate, diammonium phosphate, and ammonium polyphosphate can be effective on calcareous soils. The optimum dose of phosphorus for frankincense tree is not available in the literature, but it should be according to time of plant development. It must be near about 500 times during a developing stage to complete the nutritional requirements (Diez et al. 1992). Excess exogenous supply of phosphorus can cause deficiency of iron and zinc. Zinc sulfate can be used to avoid the deficiency of zinc.

Table 4.1 Status and requirement of nutrients of Dhofar soil

S. N.	Parameters	Status/requirement
1	Soil type	Calcareous soils
2	Sand	45–50%
3	Silt	35–40%
4	Clay	10–15%
5	pH	9–10.5
6	$CaCO_3$	15%
7	Organic matter	Very low/exogenous supply require
8	Nitrogen	Very low/exogenous supply require
9	Phosphorus	Very low/exogenous supply require
10	Zinc	Very low/exogenous supply require
11	Iron	Very low/exogenous supply require

4.5.2 Oleaeuropaea

The olive (*Olea europaea*) is a short perennial tree that flourishes mostly in Mediterranean region but also cultivated other parts of the region with suitable climate. Olive trees (*Olea europaea*) easily grow in calcareous soil (rich in calcium carbonate), which is generally found in slopes, mountains, and coastal areas (Jose et al. 2006). Martínez et al. (2006) conducted an experiment to study the pattern of nutrient loss at mountains slopes for *Olea europaea*. Martínez et al. (2006) used three different type of soil as follows:

1. *Non-tillage with barley strips of 4 m width (BS)*
2. *Conventional tillage (CT)*
3. *Non-tillage without plant strips (NT)*

They realized about total NPK losses (sediments and runoff) in BS type were around 0.87, 0.07, and 0.72 kg/ha, in CT type 1.82, 0.11 and 0.97 kg/ha and in NT type 3.15, 0.29 and 2.45 kg/ha, respectively. Further they noticed that quantity of nutrients was higher in runoff surface than that of standard in CT and NT regarding water quality for NO_3, N–NH_4 and soluble P. The effect of slurry fertilization on two olive (*Olea europaea* L.) groves has been studied by Monge et al. (2000) for one season. They used land sprayed slurry at 15 and 30 m^3/ha for non-irrigated olive stand while 30 and 60 m^3/ha used for irrigated olive. They obtained higher concentration of micronutrient in young leaves than old leaves. Monge et al. (2000) concluded that supply of slurry has not any significant effect on accumulation of nutrients and plant growth. Osman (2010) studied effect of mineral, bio-NPK soil application on cultivars of young olive trees, and foliar fertilization on leaf and shoot chemical composition. Results obtained by Osman (2010) indicated that Coronaiki cultivar of olive has higher concentration of minerals in leaves than that of Manzanillo olive. Coronaiki cultivar showed high C/N ratio than that of other cultivars. Further, he noticed that supply of NPK fertilizer enhanced amino acid, carbohydrate, and mineral in treated olive.

4.5.3 Cymbopogon schoenanthus

Cymbopogon schoenanthus L. Spreng is a sweet-smelling herb used in raw salads and in cooking of several recipes. The properties of *Cymbopogon schoenanthus* like antioxidant activity and chemical composition were evaluated by Khadria et al. (2008). Mohamed (2005) used fertilizer like urea and chicken manure on growth and metabolism yield of *Cymbopogon schoenanthus*. He observed that there is no significant effect of treatment on flowering, growth, and yield.

4.5.4 Nerium oleander

Nerium oleander belongs to dogbane family Apocynaceae. Oleanders are very tolerant of a various soil types. Plants can survive in soil with higher concentration of sodium, chloride, and salts. *Nerium oleander* can also tolerate high pH and alkalinity. The tolerant capacity of oleander makes it evergreen plant of Gulf region. However, oleander can grow easily without proper supply of fertilizer, but application of nitrogen fertilizer may be useful for some cultivar at the time of sowing. Generally, very low amount of phosphorus is recommended for oleander to avoid excess of phosphorus in soil. Vijaiananth and Rameshkumar (2012) observed that supply of NPK enhanced growth parameters and yield of oleander. Pedrinho et al. (2007) reported linear effect of nitrogen fertilizer on growth and accumulation of glycosides.

4.5.5 Aloe vera

Aloe vera is a member of family Liliaceae and has medicinal properties (Anselm 2004; Aloe India 2007; International Aloe Science Council 2007). Egbuchua and Enujeke (2015) observed that fertilizers containing nitrogen and phosphorus had significant effect on overall performance of *A. vera*. Growth parameters including leaf numbers, leaf length, and fresh weight significantly increased in plants treated with nitrogen and phosphorus at the rate of 75 and 40 kg/ha. Further, data indicated that no significant effect was found for combined doses of nitrogen and phosphorus. Tawaraya et al. (2007) investigated the effect of colonization of arbuscular mycorrhizal on nutrient (N and P) uptake and development of *A. vera*. Results obtained by Tawaraya et al. (2007) indicated that colonization of arbuscular mycorrhizal enhanced uptake of nitrogen and phosphorus and hence increased growth and metabolism of *A. vera*. Nutrient like phosphorus and nitrogen played major role in somatic growth and production of *A. vera* (Pareek et al. 1999), signifying the importance of P and N sustenance for growth of *A. vera*. It is a well-known fact that *Aloe vera* is capable to bear xerophytic conditions that is why it can flourish on saline soil of sea coast and on tropical and subtropical areas of world (Rahi et al. 2013). Under salt condition, the additional supply of silicate (Si) to nutrient combination reduced the concentrations of sodium (Na) and chloride (Cl) in leaves, stem, and roots of *A. vera* though it enhanced potassium content and ratio of potassium and sodium (K/Na) significantly. The additional supply of silicate enhanced the capability of root and stem to absorb and translocate potassium, respectively, in *A. vera* (Xu et al. 2006). Salighehdar et al. (2013) observed growth parameters increased significantly in *A. vera* received nutrient solution containing highest amount of nitrogen. Hazrati et al. (2012) studied the role of different fertilizers containing nitrogen on various physiological and morphological properties of *A. vera*. Their findings suggested that supply of nitrogen significantly increased leaf parameters such as number area and thickness of leaves in *A. vera*. The gradual increase in nitrogen supply played positive role on the accumulation of aloin concentration.

4.5.6 Catharanthus roseus

Periwinkle (*Catharanthus roseus* L.) is a perennial plant and can easily grow in tropical and subtropical arid climate such that of Oman, belonging to family Apocynaceae. However, there is no reference on *Catharanthus roseus* in available literature related to Oman region, but some workers conducted experiments to find out its requirement of nutrients in subtropical arid condition similar to Oman. Van Iersel et al. (1999) found that nitrogen content in shoot was correlated to the shoot dry mass of *Catharanthus roseus* seedlings. At the same time, P and K had minute or no effect on shoot growth; henceforth, P and K can be decreased in fertilization for the cultivation of *C. roseus*. Pre- and posttransplant growth of *C. roseus* was upgraded equally by supplement of nitrogen in pretransplant fertilizer (Van Iersel et al. 1999). The concentrations (0.25 mM) of P and K are required for the development of best-quality seedlings of *C. roseus*. The best outcome of NPK fertilization and improved vinblastine content were noted when potassium was not supplied with chloride form. As opposite to this, Smith et al. (1987) noted more intra-cell collection of catharanthine and other indole alkaloids of *C. roseus*. Biosynthesis and accumulation of secondary plant metabolites were affected by environmental factors (De Luca and St. Pierre 2000). Combination of foliar and soil drenching application of nutrient delivered not only a rise in yield but also improved absorption of nutrients by leaves, thus decreasing the risk of losses by leaching, fixing, volatilization, and pollution of ground water by nitrates (Narkiewicz-Jodko et al. 1989).

4.6 Conclusion

The beneficial effect of mineral supply to soil according to the crop requirement to increase growth has been well known in cultivation for more than two centuries. The situation of Arab region especially Oman is somewhat different from rest of the world regarding the study of mineral nutrition for plants in general and particular for medicinal plants. A large area of fertile soil is covered with calcareous soils that support the vegetations capable to tolerate $CaCO_3$. However, some soil amendment reports are available, but most of these are related to water-holding capacity and soil tolerance. Only very few plants species, i.e., *Boswellia sacra*, *Olea europaea*, *Aloe vera*, and *Cammiphora myrrha*, are explored in terms of mineral nutrition.

References

Aloe India. (2007). *The plant of immortality*. Retrieved from http://aloeupflorida.com/careofaloevera.html. pp. 1–4.

Anderson, J. P. E., & Domsch, K. H. (1980). Quantities of plant nutrients in the microbial biomass of selected soils. *Soil Science, 130*, 211–216.

Anselm, A. (2004). *Nature power 3rd edition Anselm Adodo publications* (p. 288). Nigeria: OSB Ewu-Esan.

Benbi, D., & Richter, J. (2002). A critical review of some approaches to modelling nitrogen mineralization. *Biology and Fertility of Soils, 35*, 168–183.

Burn, R., Sttimbreno, A., & Couve, C. (2001). Recycling of nutrients solutions for rose (*Rosa hybrida*) in soil-less culture. *Acta Horticulturae, 554*, 183–191.

De Luca, V., & St. Pierre, B. (2000). The cell and developmental biology of alkaloid biosynthesis. *Trends in Plant Science, 5*(4), 168–173.

Diez, J. A., MaC, C., & Vallejo, A. (1992). Controlling phosphorus fixation in calcareous soils by using coated diammonium phosphate. *Fertilizer Research, 31*, 269–274.

Duong, T. T. T., Penfold, C., & Marschner, P. (2011). Amending soils of different texture with six compost types: Impact on soil nutrient availability, plant growth and nutrient uptake. *Plant and Soil, 354*, 197–209.

Egbuchua, C. N., & Enujeke, E. C. (2015). Growth indices of *Aloe vera* as influenced by nitrogen and phosphorus fertilizers in oxisols of rain forest zone Nigeria. *GJBB, 4*, 45–49.

El-Sheikh, M. A. (2013a). Population structure of woody plants in the arid cloud forests of Dhofar, southern Oman. *Acta Botanica Croatica, 72*, 97–111.

El-Sheikh, M. A. (2013b). Weed vegetation ecology of arable land in Salalah, southern Oman. *Saudi Journal of Biological Sciences, 20*, 291–304.

Flavel, T. C., & Murphy, D. V. (2006). Carbon and nitrogen mineralization rates after application of organic amendments to soil. *Journal of Environmental Quality, 35*, 183–193.

Ghanta, P. K., & Mitra, S. K. (1993). Effect of micronutrients on growth, flowering, leaf nutrient content and yield of banana cv. Giant governor. *Crop Research, 6*, 284–287.

Grattan, S. R., & Grieve, C. M. (1999). Salinity-mineral nutrient relations in horticultural crops. *Scientia Horticulturae, 78*, 127–157.

Griffin, T. S., He, Z., & Honeycutt, C. W. (2005). Manure composition affects net transformation of nitrogen from dairy manures. *Plant and Soil, 273*, 29–38.

Hartz, T. K., Mitchell, J. P., & Giannini, C. (2000). Nitrogen and carbon mineralization dynamics of manures and composts. *Horticultural Science, 35*, 209–212.

Hazrati, S., Tahmasebi Sarvestani, Z., & Amin, S. (2012). The effect of differential nitrogen fertilization on morphological and physiological traits of *Aloe vera* plants. *International Research Journal of Applied and Basic Sciences., 3*, 682–687.

International Aloe Science Council. (2007). Retrieved from http://www.iasc.org/aloe.html.

Jedidi, N., Hassen, A., Cleemput, O. V., & M'Hiri, A. (2004). Microbial biomass in a soil amended with different types of organic wastes. *Waste Management & Research, 22*, 93–99.

Joergensen, R. G. (2000). Ergosterol and microbial biomass in the rhizosphere of grassland soils. *Soil Biology and Biochemistry, 32*, 647–652.

José, R., Francia, M., Víctor, H. D. Z., & Armando, M. R. (2006). Environmental impact from mountainous olive orchards under different soil-management systems. *Science of the Total Environment, 358*, 46–60.

Khadria, A., Serralheirob, M. L. M., Nogueirab, J. M. F., Mohamed, N., Samira, S., & Araújob, M. E. M. (2008). Antioxidant and antiacetylcholinesterase activities of essential oils from *Cymbopogon schoenanthus* L. Spreng. Determination of chemical composition by GC–mass spectrometry and 13C NMR. *Food Chemistry, 109*, 630–637.

Marinari, S., Masciandaro, G., Ceccanti, B., & Grego, S. (2000). Influence of organic and mineral fertilizers on soil biological and physical properties. *Bioresource Technology, 72*, 9–17.

Martínez R. A., Durán Z. V. H., Francia J. R. (2006). Soil erosion and runoff response to plant-cover strips on semi-arid slopes (SE Spain), *Land Degrad. Dev. 17*, 1–11.

Massa, D., Mattson, N. S., & Lieth, H. J. (2009). Effects of saline root environment (NaCl) on nitrate and potassium uptake kinetics for rose plants: A Michaelis Menten modelling approach. *Plant and Soil, 318*, 101–115.

Miller, G. M. (1994). Dhofar fog oasis, Oman and Yemen. In S. D. Davis, V. H. Heywood, & A. C. Hamilton (Eds.), *Centres of plant diversity* (Vol. 1, pp. 309–311). Gland: World Wildlife Fund, International Union for Conservation of Nature.

Mohamed, M. H. A. (2005). *Effect of irrigation intervals and fertilizer types on the growth, herbage yield and oil content of camel's hay (Cymbopogon schoenanthus (L.) Spreng.)*. M.S. Dissertation. University of Khartoum.

Monge, E., Val, J., Espada, J. L., Orús, F., & Betran, J. (2000). Effects of organic wastes on olive mineral nutrition and its influence on fruit quality, preliminary results for macronutrients. *Acta Horticulturae*. doi:10.17660/ActaHortic.2000.512.20.

Munns, R. (2002). Comparative physiology of salt and water stress. *Plant, Cell & Environment, 25*, 239–250.

Narkiewicz-Jodko, J., Tomczak, J., & Nowosielski, O. (1989). *Nawozy dolistno-ochronne w ochronie warzyw przed szkodnikami (plant protecting foliar fertilizers in pest control on vegetables)* (pp. 75–82). Skierniewice: Biul Warzywniczy.

Nyberg, K., Schnürer, A., Sundh, I., Jarvis, Å., & Hallin, S. (2006). Ammonia oxidizing communities in agricultural soil incubated with organic waste residues. *Biology and Fertility of Soils, 42*, 315–323.

Osman, S. M. (2010). Effect of mineral, bio-NPK soil application of young olive trees and foliar fertilization on leaf and shoot chemical composition. *Research Journal of Agriculture and Biological Sciences, 6*, 311–318.

Pareek, O. P., Sharma, B. D., Nath, V., Singh, R. S., & Bhargawa, R. (1999). Effect of nitrogen and phosphorus fertilizers and organic manure on growth and yield of Indina aloe (*Aloe barbadensis* mill.) *Annals of Arid Zone, 38*, 85–86.

Parida, A. K., & Das, A. B. (2005). Salt tolerance and salinity effects on plants: A review. *Ecotoxicology and Environmental Safety, 60*, 324–349.

Pedrinho, R. D., Fernandes, L. P. K., Favero, S., Solon, S., Matias, R., & Sayumi, Y. D. (2007). Efeito da adubação nitrogenada na biomassa e no teor de glicosídeos cardioativos da ornamental *Nerium oleander* L. *Ensaios e Ciência, 11*, 59–66.

Radcliffe-Smith, A. (1980). The vegetation of Dhofar in: Scientific results of the Oman Flora and Fauna survey 1977 (Dhofar). *The Journal of Oman Studies, 2*, 59–86.

Rahi, T. S., Singh, K., & Singh, B. (2013). Screening of sodicity tolerance in *Aloe vera*: An industrial crop for utilization of sodic lands. *Industrial Crops and Products, 44*, 528–533.

Salighehdar, F., Sedaqat-Hoor, S., & Olfati, J. (2013). Effects of four nutrient solutions on vegetative traits of *Aloe vera* L. cv. Austin at six harvest periods. *EJGCST, 4*, 15–27.

Sall, S. N., Masse, D., Bernhard, R. F., Guisse, A., & Chotte, J. L. (2003). Microbial activity during the early stage of laboratory decomposition of tropical leaf litters: The effect of interactions between litter quality and exogenous inorganic nitrogen. *Biology and Fertility of Soils, 39*, 103–111.

Ghazanfar, S.A., & Al-Sabahi, A. M. A. (1993). Medicinal plants of northern and central Oman (Arabia). *Economic Botany, 47*, 89–98.

Smith, J. I., Smart, N. J., Kurz, W. G. W., & Mistra, M. (1987). The use of organic and inorganic compounds to increase the accumulation of indole alkaloids in *Catharanthus roseus* (L.) G. Don on cell suspension cultures. *Journal of Experimental Botany, 38*(14), 1501–1506.

The Oman Soil Atlas. (1992). *UNDP/FAO Project-OMA/87/011 Soil Survey and Land Classification*.

Tawaraya, K., Turjaman, M., & Ekamawanti, H. A. (2007). Effect of arbuscular mycorrhiza colonization on nitrogen and phosphorus uptake and growth of Aloe vera L. *Horticultural Science, 42*(7), 1737–1739.

Van Iersel, M. W., Beverly, R. B., Thomas, P. A., Latimer, J. G., & Mills, H. A. (1999). Nitrogen, phosphorus and potassium effects on pre- and post-transplant growth of salvia and vinca seedlings. *Journal of Plant Nutrition, 22*(9), 1403–1413.

Vijaiananth, A., & Rameshkumar, S. (2012). Effect of different levels of pruning along with NPK on growth and yield of nerium (*Nerium oleander* L.) *International Journal of Current Agricultural Science, 2*, 13–16.

Xu, C. X., Liu, Y. L., Zheng, Q. S., & Liu, Z. P. (2006). Silicate improves growth and ion absorption and distribution in *Aloe vera* under salt stress. *Journal of Plant Physiology and Molecular Biology, 32*, 73–78.

Chapter 5
Quantitative Attributes of Nutrient Uptake and Use Efficiency

Manish Mathur and Arti Goel

Abstract Nutrient uptake and their utilization efficiencies are the two central contemplations for identification and assessment of most suitable cultivars and for their ideal cultivation practices. Several endeavors have formularized quantitative attributes for these two important traits both at ecosystem as well as at species level across the globe under control conditions as well as on natural habitats. However, much confusion has been caused by the use of different terminology. Even the application of the most common terms, such as nutrient use efficiency, may differ among authors, leading to fundamental disagreement over interpretation of the same data. Thus, the present chapter is shaped with some important scattered attributes like definition of nutrient uptake and use efficiency, concept and levels of their measurements (ecosystem and species levels), technical terminology, and quantitative aspects for nutrient use and uptake efficiencies. Twenty-seven quantitative approaches with some supportive parameters like construction cost, resorption, and mechanistic models have also been described. The present chapter will give updated and comparative tools to researchers to facilitate their experiment layout, quantitative evaluation of their data sets, and their interpretation.

Keywords Nutrient measurement • Nutrient utilization • Nutrient uptake • Technical attributes

5.1 Defining Nutrient Uptake and Use Efficiency

Efficiency concepts in plant mineral nutrition have been characterized in light of the procedure by which plants acquire, transport, store, and use the nutrient in order to produce dry matter or grain, at low or high level of nutrient supply (Ciarelli et al. 1998). The definition of nutrient efficiency changes enormously, and it's a helpful

M. Mathur (✉)
Central Arid Zone Research Institute, Jodhpur, Rajasthan 342003, India
e-mail: eco5320@gmail.com; ravi_mm2099@yahoo.com

A. Goel
Amity Institute of Microbial Biotechnology, Amity University Noida,
Noida, Uttar Pradesh 201313, India

M. Naeem et al. (eds.), *Essential Plant Nutrients*,
DOI 10.1007/978-3-319-58841-4_5

parameter to measure the capacity of a plant to acquire and utilize nutrients for production of timber, crops, or forages. Bridgham et al. (1995) defined nutrient uptake efficiency as the proportion of available soil nutrients acquired by plants and nutrient use efficiency as total net primary productivity per quantity of nutrients acquired during the same periods. Rathke et al. (2006) had characterized it as the ratio of seed yield to the amount of nutrient in the aboveground parts of plant, while Tittonell et al. (2007) defined nutrient use efficiency as dry matter produced (kg) by each kg of nutrient uptake. Weih et al. (2011) have depicted it as the process of carbon gain and loss in connection with the procedure related to gain and loss of the major growth-limiting nutrients.

With respect to yield parameters, nutrient efficiency has been characterized as the capacity to deliver a high plant yield in a soil or other media that would somehow limit the production of a standard line (Graham 1984; Gourley et al. 1994). Then again, it is by and large characterized as aggregate plant biomass produced per unit nutrient absorbed, which is comparable to the reciprocal of nutrient concentration in whole plant (Baliger et al. 1990). Further, nutrient acquisition efficiency and nutrient internal utilization efficiency are the two noteworthy segments of plant nutrient use efficiency related to capacity of plant to gain nutrient from the dirt and to plant internal ability to produce yield units per unit nutrient in plant (Good et al. 2004).

Overall nutrient use efficiency in plants is governed by the inundation of ions from the soil to the root surface and by the influx of ions to roots followed by their transport to the shoots and remobilization to plant organ (Jothimani et al. 2007). There are two ways that nutrient use efficiency of the individuals composing the system can influence nutrient use efficiency of the whole system. The first is through its influence on competitive interaction among species. Plant with high nutrient use efficiency should be able to tolerate lower nutrient availabilities: thus, they should be effective competitors in diverse communities where nutrients are in short supply (Tilman et al. 1997). A system made up of such individuals should therefore have a higher productivity per unit of nutrient supplied by the soil than one made up of individuals with low nutrient use efficiencies. The second way that plant nutrient use efficiency can influence ecosystem nutrient use efficiency is through its influence on litter nutrient return (Hobbie 1992). Vitousek (1982) developed the theory of nitrogen use efficiency, and according to it, plant communities on the N-poor soils are less productive but more efficient in their use of N than communities on N-rich soils.

5.2 Concept and Levels of their Measurement

The concept of nutrient use efficiency is central to understanding ecosystem functioning because it is the step in which plants can influence the return of nutrients to soil pool and the quality of the litter (Yuan et al. 2006). Ecologist studied nutrient use efficiency at three scales, most commonly those of the leaf and whole plant (Mathur 2013). At the leaf level, nutrient use efficiency is the ratio of photosynthetic rate to concentration of nutrient in the leaf lamina (Field and Mooney 1986); at the plant level, it is the ratio of growth to nutrient uptake (Hirose 1975). Ecosystem nutrient use efficiency depends on two component indexes: (a) plant-level nutrient use efficiency (i.e., the NPP of the individuals that make up the system per unit of nutrient taken up by them) and (b) uptake efficiency (i.e., total uptake by the individuals that make up the system per unit of nutrient supplied by the soil) (Shaver and Melillo 1984). Plant adaptation to resource availability is often expressed as resource use efficiency or ratio of production to uptake of certain limited resources (Garnier and Aronson 1998; Pastor and Bridgham 1999; Binkley et al. 2004).

Theory suggested that nutrient efficiency increases unimodally with declining soil resources. Yuan et al. (2006) have experimentally demonstrated that whole-plant nitrogen use, uptake, and response increased monotonically with decreasing soil nitrogen and water, being higher on infertile (dry) habitat than on fertile (wet) habitats. They further explained that nutrient use efficiency differed among sites and was often affected by trade-off between net productivity (NP) and mean residence time (MRT, the ratio between the average nutrient pool and the annual nutrient uptake or loss), where plants and communities had adapted in a way to maximize either NP or MRT, but not both concurrently. Further, comparisons of species (Vazquez de Aldana and Berendse 1997; Eckstein and Karlsson 2001), community (Nakamura et al. 2002; Silla and Escudera 2004), and ecosystem levels (Hiremath and Ewel 2001) have suggested that nitrogen use efficiency increases as soil N availability declines. Most of the studies related to utilization efficiency of various nutrients so far exclusively addressed the species in lowland mires, subarctic tundra, or tropical rain forest (Yuan et al. 2006), while few such studies have also been conducted in the semiarid region of the world, particularly northern China (Yuan et al. 2005).

5.2.1 Ecosystem-Level Approaches

By definition, the ecosystem is the system developed through interaction between biotic and abiotic components. From nutrient use efficiency point of view, whole-ecosystem experimental manipulation indicates an alteration of an ecosystem in which either the entire habitat itself is experimentally altered or a subset of the habitat is altered, the response to which is expected to be ecosystem wide (Fayle et al. 2015). Studies have examined the efficiency of plant nutrient uptake and use at multiple levels, from leaves to ecosystem (Garnier and Aronson 1998; Hiremath 1999; Aerts and Chapin 2000; Yuan et al. 2003). Vegetation structure in natural ecosystems is determined by both biotic factors (e.g., competition, facilitation, and predation) (Hacker and Gaines 1997) and abiotic ones (Valverde et al. 1996; Hahs et al. 1999) although the effect of these factors depends on the habitats where plants grow (McAuliffe 1994). A complete assessment of ecosystem productivity and nutrient dynamics requires measuring above- and belowground patterns of biomass increment, turnover, and nutrient content. In fact, a proper evaluation of NUE requires data at the whole-plant level, because patterns of aboveground NUE are not necessarily similar to the whole-plant NUE (Aerts and Chapin 2000; Mathur 2015).

Generally environmental components that impose restrictions on plant development (e.g., water stress, high temperatures, and salinity) assume vital roles in the community structure (Interli and Kilham 2001). For instance, in arid environments, water availability and low levels of nutrients in soil are the main environmental factors affecting the development of plant communities and consequently plant abundance and productivity (Flores and Briones 2001). Ecological theory suggests that species in a mixture may exploit resources of a site more completely and efficiently than a single species would be able to do, leading to greater overall productivity.

To accomplish greater productivity in mixed stands, the species constituting the stands need to show contrasts in their requirements (niches) and the way they utilize site assets and (or) positively affect the growth of each other. This concept of niche separation implies that if two species are too similar in their requirements, they would eventually compete intensely to exclude the other. The mixed stands may experience less intense interspecific than intraspecific light competition as a consequence of the differences in shade tolerance among species. Such a stratified canopy would, in theory, maximize the use of light because of increased light interception and light use efficiency (Kelty 1992), leading to greater total productivity than in pure stands (Smith et al. 1997). This type of response has been found in studies by Montagnini et al. (1995), Montagnini (2000), DeBell et al. (1997), Man and Lieffers (1999), Garber and Maguire (2004), and Mathur (2015).

Hiremath and Ewel (2001) have examined the impact of monoculture and polyculture types on ecosystem nutrient use efficiency. They have inferred that three species significantly influenced nutrient use efficiency by whole ecosystem in monoculture, while in polyculture the additional life-form significantly increased nutrient uptake and uptake efficiencies. They demonstrate that ecosystem nutrient use efficiency is an outcome not only of the characteristics of the species or life-forms that comprise the system but also of factors that affect soil nutrient supply. They considered ecosystem nutrient use efficiency as the proportion of net primary productivity to soil nutrient supply. They further summarized how the ecosystem-level nutrient use efficiency increases with the richness of species or life-forms in a community and what are the relative roles of species traits and nutrient supplying capacity of the soil in determining nutrient use efficiency. Nutrient use efficiency at ecosystem level revealed that NUE increase with richness of species or life-forms in a community. However, this connection at species level has not been worked out, especially in semiarid areas where resources are released in pulses.

Tilman et al. (1997) in the context of diversity and ecosystem productivity and Hooper (1998) in the context of diversity and nutrient retention have suggested that ecosystem nutrient use efficiency depends on the identity of the species making up the system and not on a greater diversity of species per se. A complete evaluation of ecosystem productivity and nutrient dynamics requires measuring above- and belowground patterns of biomass increment, nutrient content, and turnover. In fact, a proper evaluation of nutrient use efficiency requires data at the whole-plant level, because patterns of aboveground utilization efficiency are not necessarily similar to whole-plant utilization efficiency (Aerts and Chapin 2000).

Be that as it may, most of the previous studies have focused exclusively on the nutrient utilization efficiency (NUE) of fine-litter production (Vitousek 1984; Silver 1994; Yasumura et al. 2002; Paoli et al. 2005). Destructive methods of NUE evaluation are generally criticized; therefore, nutrient solution techniques have been used as important tools in short-term experiments to select and identify nutrient-efficient plant (Furlani and Furlani 1988; Ciarelli et al. 1998; Speher and Souza 1999). In addition, these nondestructive techniques might be useful for those species which have low germplasm availability in nature (Furlani et al. 2002). Ecosystem-level approaches for nitrogen and phosphorus efficiency using litter experiments were conducted by workers like Mirmanto et al. (1999), Sayer et al. (2006), Kaspari et al. (2008), Sayer and Tanner (2010), Sayer et al. (2012), Ashford et al. (2013), Homeier et al. (2012), and Turner et al. (2013). Kobe et al. (2005) have utilized global data set with regard to resorption efficiency in green and senescence leaves, and they have summarized that functional relationships between nutrient concentrations in senesced and green leaves were indistinguishable for evergreen and deciduous species, but evergreens tended to have lower concentrations in both senesced and green leaves. General relationships (across species and environments) between senesced and green leaf nutrient concentrations and broad, intraspecific variation in nutrient concentrations suggest that variation in resorption efficiency and proficiency could be a result (and not necessarily a cause) of species sorting across fertility gradients.

Resource pulses are thought to be important drivers of community dynamics in desert ecosystem. Plants in arid and semiarid systems are primarily limited by water, but nutrient may be co-limiting. Summer precipitation pulses can affect ecophysiological parameters directly by increasing soil water availability and also by effects on nutrient availability (Synder et al. 2004). Many short-term isotopic and ecophysiological studies have demonstrated that desert plant species, functional groups, and life stages differ in the capacity to use summer resource pulse (Schwinning et al. 2002; Ivans et al. 2003). Mathur (2005) reported fluctuation in net negative and positive associations among annuals as well as among perennials in 12 different successional stages during pulse, inter-pulse, and non-pulse events. One whole-plant NUE study from semiarid grassland in northern China was conducted by Yuan et al. (2006). Studies of Sala et al. (1989) and Paruelo and Sala (1995) from Patagonian Steppe suggested a tight link among three key elements of the functioning of ecosystems: species composition and N and water cycles. Consequently, changes in any of these three elements will have consequences on the other two. The tight link between N and water capturing strategies suggest that changes in the availability of one resource may have rapid and direct effects on the other. Anthropogenic changes in the composition of plant types will have effects on N and water economy. If climate change results in drier conditions in arid and semiarid regions, deep soil water resources will be reduced. Under these conditions, results from this study predict decreased shrub density and a less conservative nitrogen economy for the affected ecosystems. Removal of the woody plants would convert the vegetation to one dominated by grasses with short N residence times in biomass and therefore high N turnover. Therefore, a reduction in the shrub component would have atmospheric consequences because N atoms are most vulnerable to being lost from the ecosystem via gaseous loss or leaching when in inorganic form. Sala et al. (2012) have studied the contrasting nutrient-capture strategies in shrubs and grasses of a Patagonian arid ecosystem. Their studies have revealed that shrub and grass species have similar nitrogen use efficiency but that they achieve it through opposite mechanism. Shrubs have a conservative N economy absorbing annually only small fractions of their N content, whereas grasses have a more open N economy.

Norris et al. (2007) have quantified nutrient use efficiency for assessment of altered ecosystem nitrogen dynamics as a consequence of land cover changes in tall grass prairie. They measure and compared major pools and fluxes of nitrogen in red cedar forests and adjacent native grasslands. They reported higher ecosystem-level nitrogen use efficiency (ANPP, leaf litter N) in forest compared to grasslands and concluded that differences related to biogeochemistry and both nitrogen and carbon storage throughout areas of the eastern central plain where coverage of red cedars is increasing.

Mangroves are a diverse group of plants and are an ecological entity with little phylogenetic association. This may lead to many intrinsic differences among coexisting species in nutrient uptake and nutrient use efficiency, with significant differences observed between species in their response to nutrient availability (McKee 1993; Lovelock and Feller 2003), which may be partially responsible for differential distribution of species (zonation) observed in mangrove landscapes (Feller et al. 2003). However, convergent evolution has led to similar adaptations among mangrove species in traits such as water relations (Ball 1988; Macinnis-Ng et al. 2004) and architecture (Tomlinson 1986). Interspecific differences in nutrient use efficiency have been observed between mangrove species (Lovelock and Feller 2003) and are also modified by plant interactions with environmental variables (Martin et al. 2010). For example, phosphorus use efficiency differed among mangrove species and decreased with increased nutrient availability and salinity (Martin et al. 2010). PNUE decreases with increasing salinity because under highly saline conditions, mangroves achieve higher photosynthetic water use efficiency by increasing N leaf content in order to maximize photosynthetic carbon gain when stomatal conductance is low.

5.2.2 Species-Level Approaches

It is generally hypothesized that a single plant species is unable to take full advantage of any environment, due to variables such as mineral and nutritional needs, root depth, and metabolic by-products. Monoculture practice in the same area for extended periods depletes resources required by that plant species while neglecting available resources other plant species can take advantage of. Perennial polyculture is a relatively new model in farming practices that aims to reduce many of the ecological costs of farming by mimicking natural ecosystems. The idea behind a "reengineered ecosystem" is to decrease soil erosion by wind and water by providing year-round ground cover. Perennial polyculture means creating a diverse group of plants that are intermixed in the field. Most modern farming relies on monocultures, vast fields of a one plant seeded over and over again (Glover and John 2010). There are many skeptics of perennial polyculture as well. Some studies have found that while perennial polyculture may seem a brilliant solution to agricultural dilemmas, over time, a dominant species emerges, essentially converting the field back to a monoculture system (Griffitha et al. 2011). In this case, the cost of regulating the polyculture field outweighs any benefits gained. Gama-Rodrigues et al. (2007) have concluded that growth and yield in mixed-tree stands were higher than in pure stands owing to the combination of species with complementary ecophysiological attributes, consequently improving the efficiency in nutrient use and cycling.

The separate consideration of nutrient uptake efficiency, nutrient response efficiency, and nutrient use efficiency can yield important insights into the competitive abilities of different species along fertility gradients (Pastor and Bridgham 1999). Yuan et al. (2006) have reported monotonic increase in nitrogen use efficiency with decreasing soil nitrogen and water availability. Further, they reported that interspecific variation in NRE of two *Stipa* species does not confirmed the model developed by Pastor and Bridgham (1999). In their experiment, nutrient response efficiency increased significantly with declining soil N and water availability and, thus, suggested that unimodal response efficiency (Pastor and Bridgham 1999) may not be applicable within species in temperate grasslands of northern China. Yuan et al. (2008) have quantified nitrogen use efficiency in *Stipa krylovii* by partitioning it into productivity (A) and the mean residence time (MRT). Their experiment suggested that these three parameters, i.e., A, MRT, and NUE, were significantly related to the soil N supply, while a negative relationship between A and MRT exists for this species, i.e., plant with higher A had lower MRT.

$$A = NPP / N_{Pool}$$

$$MRT = N_{Pool} / N_{Loss}$$

$$NUE = A \times MRT = NPP / N_{Loss}$$

NPP is the net aboveground biomass; N_{Pool} is the N content in biomass and N_{Loss} quantified by multiplying total liter production by its N concentration. Plant species differ in the efficiency with which they acquire and utilize phosphorus. Phosphate acquisition efficiency (PAE) relates to the different extents to which plants are able to mobilize phosphorus from poorly soluble sources or to take up the soluble phosphorus available in the soil solution. Higher plants show several morphological and physiological adaptations that enable them to acquire phosphorus from sparingly soluble phosphorus soil fraction. These adaptations include root system enlargement, arbuscular mycorrhiza establishment, increased organic acid exudation, rhizosphere acidification, increase production of phosphatases, and enhanced phosphate uptake rate (Narang et al. 2000). Mathur (2015) had studied the impact of stand types (monoculture and polyculture) on phosphorus acquisition efficiency and their subsequent internal utilization efficiency were determined at species level, in a seratonious endangered plant, *Blepharis sindica*. He hypothesized that within semiarid plant communities, stand types influence nutrient dynamics of plants and have two directions and consequently two effects: firstly, it affects the nutrient uptake efficiency and secondly its utilization efficiency through their various bottom-up and top-down factors. Species-level phosphorus acquisition and internal utilization efficiency and their relation with bottom-up and top-down factors for *Corchorus depressus* were carried out by Mathur (2014).

Henderson and Jose (2012) have quantified the effect of water and nutrient availability on nutrient content ($kg{\cdot}ha^{-1}$), resorption efficiency (%), resorption proficiency (g nutrient/kg leaf litter (senesced tissue)), and leaf- and canopy-level nutrient use efficiency of nitrogen (N), phosphorus (P), and potassium (K) for *Populus deltoides*, *Quercus pagoda*, and *Platanus occidentalis*. They concluded that aboveground nutrient content, nutrient resorption efficiency and proficiency, and leaf- and canopy-level nutrient use efficiency were not influenced by increased nitrogen availability. Additionally, they also concluded that maximum biomass production was not necessarily tied to maximum nutrient input. Production as well as nutrient requirements are species specific and may include a compensatory mechanism providing sufficient resources available from the site, to deter nutrient imbalance.

5.3 Factors Associated with Them

Effective uptake can be achieved through four different mechanisms:

(a) Temporal partitioning such that one species takes up nutrients at a time when others do not
(b) Spatial partitioning such that one species takes up nutrients from portions of the habitat that are inaccessible to other species
(c) Uptake of nutrients in different proportions
(d) Uptake of different forms of the same nutrient (e.g., plant root absorb P as either $H_2PO_4^-$)

Nitrogen use efficiency, in particular, has been the subject of a large number of studies (Vitousek 1982; Shaver and Melillo 1984; Aerts and de Caluwe 1994; Bridgham et al. 1995; Yuan et al. 2006). The fundamental tenet of NUE theory is that plant communities on N-poor soils are less productive but more efficient in their use of N than communities on N-rich soils. Comparison of species (Vazquez de Aldana and Berendse 1997; Eckstein and Karlsson 2001), community (Nakamura et al. 2002; Silla and Escudera 2004), and ecosystem level (Hiremath and Ewel 2001) has shown that NUE increases as soil N availability declines. The trade-off between resource acquisition and conservation has been formalized in the leaf economics spectrum (LES), which shows that relationships exist among several key traits across a broad range of species and different climates (Funk 2013). Plant species with low leaf mass per unit area (LMA), high rates of carbon assimilation, high leaf nitrogen (N) content, and short leaf life span occupy one end of the spectrum (fast return on investment), while plant species with high LMA, low rates of carbon assimilation, low leaf N content, and long leaf life spans occupy the other (slow return on investment). With respect to invasion, several researchers have suggested that invasive species are positioned closer to the fast-return end of the LES (Leishman et al. 2010; Penuelas et al. 2010; Ordonez and Olff 2012). This "fast-return" strategy seems at odds with an ability to tolerate low-resource conditions, as species adapted to low-resource systems often display slow growth, resource use efficiency, high LMA, high tissue construction cost, and long-lived tissues (Craine 2009).

In particular, high nitrogen use efficiency considered advantageous under conditions of low soil fertility, as it entails high biomass production per unit of N uptake (Aerts and Chapin 2000). In addition, NUE is a useful index that integrates plant physiological and morphological responses along nutrient available gradients. Woody perennials exhibit a characteristic time course of phosphorus acquisition and internal P redistribution during their life cycle (Fageria et al. 2011; White and Veneklaas 2012). Phosphorus efficiency can be divided into P acquisition efficiency (PAE) and P utilization efficiency (PUE). PAE refers to the ability of plants to take up P from soils, and PUE is the ability to produce biomass or yield using the acquired P (Wang et al. 2010). Enhancing P efficiency in plants can be achieved through improving P acquisition and/or utilization.

However, the contribution from PAE or PUE to plant P efficiency varies with species and environmental conditions (Wang et al. 2010). The main mechanisms related to increased P acquisition efficiency are the root morphology and architecture, which is based on the increase in root volume and area; the genotype ability to associate with soil rhizosphere microorganisms, which would facilitate soil P acquisition; and the genotype capacity to secrete organic compounds in the rhizospheres (like phosphatases and organic acids), which would release P from organic or inorganic soil sources. A higher P internal utilization efficiency has been attributed to a higher grain yield per unit of P in the grain (quotient of utilization) and to higher ability to transfer nutrient from shoot to grain, called P harvest index (Baligar and Fageria 1997).

Several workers found varying K efficiency among different plant species and genotypes. For example, El Dessougi et al. (2010, 2011) showed that wheat and sugar beet had a higher agronomic K efficiency than barley. Sadana and Claassen (1999) showed, in a pot experiment, that sugar beet is more K efficient than wheat and maize. Hence, varying K efficiency may be due to variations in internal requirements of the plant or use efficiency (Trehan and Claassen 1998; Zhang et al. 1999; Fageria et al. 2001). The internal requirement is the K concentration in plants needed to produce a certain proportion of the maximum yield, for example, 90% as used by Foehse et al. (1988). Other reasons for efficiency could be the K uptake ability of the plants, i.e., acquiring K from the soil or solution and accumulating it in the shoots. This depends on the density of roots and on the efficiency of the single roots to take up K or the influx (El Dessougi et al. 2002; Bhadoria et al. 2004). Low shoot growth rate and/or high root length/shoot weight ratio and a high uptake rate per unit root are the major mechanisms employed for K efficiency by crops like *Triticum aestivum*, *Hordeum vulgare*, *Lolium perennial*, *Brassica napus*, *Beta vulgaris*, *Helianthus annuus*, and *Zea mays* (Dessougi et al. 2010).

Proper N application timing and rates are critical for meeting crop needs and indicate considerable opportunities for improving N use efficiency (NUE) (Dhugga and Waines 1989; Blankenau et al. 2002). Reports have shown that split N application in the later stages was effective in attaining higher N uptake efficiency (Kumari et al. 2000; Ashraf and Azam 1998). NUE, grain yield produced per unit of N supply, is a complex trait comprising N uptake efficiency (NupE) and N utilization efficiency (NUTE) (Moll et al. 1982; Ortiz-Monasterio et al. 1997). NupE reflects the ability of the plants in obtaining N, while NUTE reflects the efficiency with which the crop utilizes N in the plant for the synthesis of grain yield.

Tran and Tremblay (2000) reported that wheat NupE was lower in the early applications at planting and tillering than application in the later crop growth stage. Asplund et al. (2016) studied the functional traits associated with nitrogen use efficiency in bread wheat (*Triticum aestivum ssp. aestivum*); their partial least squares regression have identified six traits, namely, leaf chlorophyll of the top leaf at stem elongation, grain ear^{-1}, ear pot^{-1}, straw biomass pot^{-1}, days between emergence and anthesis, and days between emergence and completed senescence, and these are crucial for overall NUE.

Zeugin et al. (2010) from central Panama forest had linked the tree diversity with phosphorus and nitrogen acquisition with specific objectives such as the following: does species richness and species composition affect the size of N and P pool in the trees? And if so, is this caused by selection or complementarity? To what extent do the environmental variables contribute to the explanation of the N and P pool patterns in the plantation? And are N and P use efficiencies affected by the tree species and/or the species richness? They reported a positive but not consistent net effect of biodiversity on the N and P pools, mainly explained by complementarity effects. They also observed significantly lower PUE in the three species mixtures compared to monoculture or the high diverse plots. They have interpreted their results with view of competition and facilitation between species and for other resources. Formerly, Richards et al. (2010) have provided hypotheses and measurements related to the proportion of resource supply captured section and resource use efficiency section.

5.4 Technical Attributes of Nutrient Use Efficiencies

5.4.1 Partial Factor Productivity (PFP)

This is a basic productivity expression, figured in units of crop yield per unit of nutrient applied. It is easily calculated for any farm that keeps records of inputs and yields. It can likewise be figured at the regional and national level, provided reliable statistics on input use and crop yields are available. However, partial factor productivity values vary among crops in different cropping systems, because crops differ in their nutrient and water needs. A comparison between crops and rotations is particularly difficult if it is based on fresh matter yields, since these differ greatly depending on crop moisture contents (e.g., potato vs. cereals). Thusly, geographic areas with various cropping systems are difficult to compare with this indicator.

5.4.2 *Agronomic Efficiency (AE)*

This is figured in units of yield increment per unit of nutrient applied. It more closely reflects the direct production impact of an applied fertilizer and relates directly to economic return. The computation of AE requires knowledge of yield without nutrient input so is only known when research plots with zero nutrient input have been implemented on the farm. If it is calculated using data from annual trials rather than long-term trials, NUE of the applied fertilizer is often underestimated because of residual effects of the application on future crops. Estimating long-term contribution of fertilizer to crop yield requires long-term trials.

5.4.3 *Partial Nutrient Balance (PNB)*

It is the simplest form of nutrient recovery efficiency, usually expressed as nutrient output per unit of nutrient input (a ratio of "removal to use"). Less frequently, it is reported as "output minus input." PNB can be measured or estimated by crop producers as well as at the regional or national level. Often the assumption is made that a PNB close to one suggests that soil fertility will be sustained at a steady state. However, since the balance calculation is a partial balance and nutrient removal by processes, such as erosion and leaching, is usually not included, using a PNB of one as an indicator of soil fertility sustainability can be misleading, particularly in regions with very low indigenous soil fertility and low inputs and production, such as sub-Saharan Africa.

Likewise, all nutrient inputs are rarely included in the balance calculations, thus, the modifier, partial, in the term. Biological N fixation, recoverable manure nutrients, biosolids, irrigation water, and the atmosphere can all be nutrient sources in addition to fertilizer. Values well below one, where nutrient inputs far exceed nutrient removal, might suggest avoidable nutrient losses and thus the need for improved NUE; attainable values, however, are cropping system and soil specific. A PNB greater than one means more nutrients are removed with the harvested crop than applied by fertilizer and/or manure, a situation equivalent to "soil mining" of nutrients. This situation may be desired if available nutrient contents in the soil are known to be higher than recommended. However, in cases where soil nutrient concentration is at or below recommended levels, a PNB > 1 must be regarded as unsustainable (Brentrup and Palliere 2010). Over the transient and on individual homesteads, PNB can show substantial fluctuations due to cash flow and market conditions, especially for P and K. Longer-term assessment of PNB over several years is therefore more useful.

5.4.4 *Apparent Recovery Efficiency (RE)*

Apparent recovery efficiency is one of the more complex forms of NUE expressions and is most commonly defined as the difference in nutrient uptake in aboveground parts of the plant between the fertilized and unfertilized crop relative to the quantity of nutrient applied. It is often the preferred NUE expression by scientists studying the nutrient response of the crop. Like AE, it can only be measured when a plot without nutrient has been implemented on the site but in addition requires measurement of nutrient concentrations in the crop. And, like AE, when calculated from annual response data, it will often underestimate long-term NUE.

5.4.5 Internal Utilization Efficiency (IE)

It is defined as the yield in relation to total nutrient uptake. It varies with genotype, environment, and management. A very high IE suggests deficiency of that nutrient. Low IE suggests poor internal nutrient conversion due to other stresses (deficiencies of other nutrients, drought stress, heat stress, mineral toxicities, pests, etc.).

5.4.6 Physiological Efficiency (PE)

This is defined as the yield increase in relation to the increase in crop uptake of the nutrient in aboveground parts of the plant. Like AE and RE, it needs a plot without application of the nutrient of interest to be implemented on the site. It also requires measurement of nutrient concentrations in the crop and is mainly measured and used in research.

5.4.7 Resorption

Nutrient resorption amid senescence, through its role in reducing nutrient loss, is an important component of plant nutrient conservation strategy (Aerts and Chapin 2000). There are two essential parameters that can be utilized to evaluate this procedure: nutrient resorption proficiency (the level to which a nutrient is reduced during senescence, RP) and nutrient resorption efficiency (the proportional nutrients withdrawn from senescing tissues prior to abscission, RE). RP is generally negatively related to nutrient availability (Killingbeck 1996; Wright and Westoby 2003; Rejmankova 2005; Richardson et al. 2005), while RE shows inconsistent patterns in response to nutrient availability in soils (Aerts 1996; Huang et al. 2012). The varied relationships between RE and nutrient availability reported in previous studies (e.g., Singh 2004; Yuan et al. 2005) may be attributed to the different methods for estimating nutrient resorption. The most generally utilized strategies incorporate leaf mass-based (e.g., Oleksyn et al. 2003; Quested et al. 2003; Rejmankova 2005; Cardenas and Campo 2007) and leaf area-based (e.g., Renteria et al. 2005; Kazakou et al. 2007) estimations (leaf level).

However, substantial mass loss and leaf shrinkage have been observed, resulting in underestimates in RE by using these leaf-level methods. For example, mass losses ranging from 1.5 to 57% among 4 graminoid species (Vernescu et al. 2005), 24–55% among 18 herbaceous species in southern France (Kazakou et al. 2007), and 8–38% among 10 graminoid species at Dutch and Swiss wetland sites (Güsewell 2005), as well as considerable leaf shrinkage (up to 20%) during senescence (van Heerwaarden et al. 2003), have been observed. Thus, van Heerwaarden et al. (2003) suggested that leaf cohorts can be a better approach. This approach has been widely used in some studies (e.g., Güsewell 2005; Aerts et al. 2007) but still yielding inconsistent results. For perennial herbaceous species, their entire aboveground portions die off completely, and nutrients in these tissues will be transported to roots or rhizomes for storage during senescence (Tagliavini et al. 1998). Thus, leaf level-based (such as leaf mass, leaf area, etc.) estimates might not be appropriate for nutrient resorption estimation at a whole-plant level.

Resorption proficiency (the amount of nutrient remaining in senesced leaves) and resorption efficiency (ration of nutrients resorbed relative to green leaf concentrations) are the two related terms. Foliar nutrient resorption, a mechanism by which plants can recycle nutrients internally, also varies with latitude. Tropical trees have lower N resorption and greater P resorption than temperate and boreal trees (Yuan and Chen 2009). However, there is marked variability in foliar nutrient resorption within plant communities (Mayor et al. 2014), particularly for P (Reed et al. 2012). This reflects the wide range in P availability and taxonomic diversity in tropical forests (Alvarez-Clare et al. 2013).

Previous studies mostly focused on the nutrient resorption from leaves, and the contributions of non-leaf tissues resorption have largely been overlooked (Freschet et al. 2010). Yuan et al. (2005) suggested that taking all of the aboveground portions as a nutrient pool to calculate nutrient resorption might be appropriate because nutrients resorbed from non-leaf tissues can be included when estimating nutrient resorption. However, this hypothesis has not yet been tested in field experiments. It is assumed that many N-limited ecosystems are shifting toward P limitation because of increased atmospheric N deposition and climate change. Such shift can lead to lower P concentrations both in green and in senescing leaves, due to growth dilution of the P in leaves.

The process of nutrient resorption decreases the possibility of mineral loss in litter fallen on the forest floor, and afterward, the withdrawn nutrients are redeposited in developing tissues such as leaves or productive structures (like seeds) or stocked for further usage. Resorption is an integrated part of the well-predicted process of leaf senescence and seems to occur in many species (Wright and Westoby 2003). Mineral resorption parameters are essential for determining the correctness of ecological system and biogeochemical patterns (Vergutz et al. 2012; Liu et al. 2014; Bilgin et al. 2016). The usability of nitrogen (N) and phosphorous (P) is of special significance for the functioning of most terrestrial plant associations. Since resorption contributes to decreasing a plant's dependence on available mineral supply, it could potentially be as substantial to a plant's mineral economy as absorption by roots and is one of the most significant strategies used by plants to preserve minerals (Chapin and Kedrowski 1983; Aerts 1996; Killingbeck 1996; Aerts and Chapin 2000; McGroddy et al. 2004; Ratnam et al. 2008).

It is estimated that about 50% of leaf N and P is recycled through resorption across a broad range of perennial life models (Aerts 1996; Aerts and Chapin 2000; Ratnam et al. 2008). Resorption proficiency is a parameter used for measuring nutrient resorption. Resorption proficiency appears to be more responsive than resorption efficiency to mineral availability. Some studies revealed that N fertilization resulted in higher N contents in the litter of many species (Van Heerwaarden et al. 2003). Resorption efficiency can be calculated as the percentage of nitrogen (NRE) and phosphorus (P-RE), and it was recovered from senescing leaves (Orgeas et al. 2002; Rejmankova 2005; Bilgin et al. 2016). Nitrogen resorption proficiency (N-RP) and phosphorus resorption proficiency (P-RP) were the concentration of nutrient in senescent leaves (Killingbeck 1996).

5.4.8 Construction Cost

Construction cost is a quantifiable measure of energy demand for biomass production (Song et al. 2007). It has been defined as the amount of glucose required to provide carbon skeletons, reductant, and energy for the synthesis of organic compounds. In general, low construction cost is associated with high relative growth rates; even small differences in construction cost can lead to substantial differences in growth rate. Leaf construction cost per unit of mass (CC_{mass} equivalent to gram glucose per gram dry mass) is calculated through

$$CC_{mass}\left[\frac{(0.06968\Delta Hc-0.065)\times(1-ash)+7.5\left(k\frac{N}{14.0067}\right)}{E_G}\right]$$

In this equation, k is the oxidation state of N substrate (+5 for nitrate or −3 for ammonium), and E_G is the growth efficiency. E_G can be estimated to be 0.087 across species. This parameter for *Mikania micrantha*, *Lantana camara*, *Urena lobata*, and I in relation to NUE was quantified by Song et al. (2007), while Petritan et al. (2010) quantified this for *Fagus sylvatica*, *Acer pseudoplatanus*, and *Fraxinus excelsior.* Geng et al. (2013) have quantified this attribute for *Alternanthera philoxeroides* by using parameters like mineral concentration (Min) and acid alkalinity:

$$CC_{mass}=\left[-1.041+5.077\times\frac{c}{1000-Min}\right]\times\frac{1000-Min}{1000}+\left(5.325x\frac{N_{org}}{1000}\right)$$

$$\textit{and } Min = Ash - AAx30 + Nitrate$$

5.4.9 Mechanistic Models

A mechanistic model assumes that a complex system can be understood by examining the workings of its individual parts and the manner in which they are coupled. Potassium uptake and supply level of the soil can be described by a mechanistic model (Claassen and Steingrobe 1999). The model calculates the diffusive and convective transport of nutrients toward the root under consideration of sorption and desorption processes. The uptake rate into the root is calculated by Michaelis-Menten kinetics. Applying a nutrient uptake model to wheat plants, it was shown that where calculated transport and uptake agreed well with measured data, the high K uptake efficiency of wheat was mainly due to its large root system. Dessougi et al. (2010) found that sugar beet had fewer roots, but it could realize much higher influx than calculated by the model. Samal et al. (2010) have quantified potassium uptake efficiency and dynamics in the rhizosphere of maize, wheat, and sugar beet by using mechanistic model. Plant parameters for model calculation are as follows:

Mean root radius: Mean root radius (r_0) calculated from fresh root weight (FRW) in gram and root length (RL) in cm, assuming a specific gravity for roots of 1 gcm^{-3}:

$$r_0 = \sqrt{\frac{FRW}{\pi \times RL}}$$

Average half distance between neighboring roots (r_1) can be quantified through

$$r_1 = \sqrt{\frac{Soil\,volum\left(cm^3\right)}{\pi \times root\,lenght\left(cm\right)}}$$

Water influx: Assuming that young plant has exponential root growth, water influx (ν_0) is quantified through

$$\nu_0 = \frac{T_2 - T_1}{RA_2 - RA_1} \times \frac{\ln\left(\frac{RA_2}{RA_1}\right)}{t_2 - t_1}$$

where $T_2 - T_1$ is the amount of water (cm^3) transpired between t_1 and t_2, and RA is the root surface area (cm^2). Total evapotranspiration loss of water is determined from the water loss from the pots with plants minus the water loss from the unplanted pots.

Relative shoot and root growth rate:

$$RGR = \frac{\ln \times \left(\frac{SDW_2}{SDW_1}\right)}{t_2 - t_1}$$

$$k = \frac{\ln \times \left(\frac{RL_2}{RL_1}\right)}{t_2 - t_1}$$

In the above equations, SWD is the shoot dry weight in gram and is the mean of three replications, t is the time of harvest in seconds, and RL is the root length in cm. Further, relationships between net K influx (I_n) and K concentration at the root surface (CL_0) determine the K uptake kinetics, and this relation can be described by modified Michaelis-Menten function:

$$I_{\mathrm{n}} = \frac{I_{\max}\left(C_{L0} - C_{L\min}\right)}{K_{\mathrm{m}} + C_{L0} - C_{L\min}}$$

Maximum net influx (I_{max}) was obtained from the influx measured from the treatment with the highest K level for each crop. This gives the maximum net K influx of plants well supplied with K for the growing conditions. Plants of lower K supply may have a higher I_{max}, but to which extend this may be important and will be studied through a sensitivity analysis on I_{max}. As the influx was calculated per cm of root, it was recalculated per cm^2 of total root surface area including the surface area of root hairs per cm root. Root surface area (RSA) and root hair surface area (RHSSA) are quantified through

$$RSA = 2\pi r_0 RL$$

where r_0 is the root radius and RL is root length,

$$RHSA = 2\pi rh_0 RHL$$

where rh_0 is the root hair radius (0.0005 cm as reported by author), and RHL is the total root hair length per cm of root. Samal et al. (2010) took RHL value from Hofbauer (1990 see Samal et al. 2010). Minimum solution concentration (C_{Lmin}) is the concentration at which net influx equals zero, and they took this from Meyer (1993). The Michaelis constant (K_m) is the difference between concentration at which influx is half of I_{max} and C_{Lmin}. The values were taken from Meyer (1993 see Samal et al. 2010).

Net K influx: Potassium influx is the net amount of K taken up per unit root length (or root surface area) per unit time. K influx (I_n) quantified through

$$I_{\mathrm{n}} = \frac{U_2 - U_2}{RL_2 - RL_1} x \frac{\ln\times\left(\frac{RL_2}{RL_1}\right)}{t_2 - t_1}$$

where U is K content in μmol plant^{-1}, RL is root length per plant in cm, t is time of harvest in seconds, and the subscripts 1 and 2 refer to the first and second harvest, respectively.

5.5 Quantitative Aspects for Nutrient Use Efficiency

Much confusion has been caused by the use of different terminologies. Even the application of the most common terms, such as nutrient use efficiency, may differ among authors, leading to fundamental disagreement over interpretation of the same data

(Knops et al. 1997). Therefore, it is important to choose an appropriate index. Furthermore, because of a general lack of comprehensive data, Vitousek (1982, 1984), Bridgham et al. (1995), Hiremath and Ewel (2001), Fioretto et al. (2003), Perez et al. (2003), and Pavon et al. (2005) used leaf litter and its nutrient content as alternative measures of net primary production and nutrient uptake, respectively. The evaluation of NUE is useful to differentiate the plant species, genotypes, and cultivars for their ability to absorb and utilize nutrient for maximum yields. The NUE is based on (1) uptake efficiency (acquisition from soil, influx rate into roots, influx kinetics, radial transport in roots are based on root parameters per weight or length, and uptake is also related to the amount of the particular nutrient applied or present in soil), (2) incorporation efficiency (transport to shoot and leaves are based on shoot parameters), and (3) utilization efficiency (based on remobilization, whole plant, i.e., root and shoot parameters).

5.5.1 Nutrient Use Efficiency (NUE) (Vitousek 1982)

$$NUE = \frac{NPP}{N_{ac}}$$

where NPP = net primary productivity; N_{ac} = nutrient available in soil. NUE (Berendse and Aerts 1987):

$$NUE = \frac{p}{N_{ac}}$$

where P = productivity (g cm^{-2} year^{-1}); R_{ac} = rate of resource uptake or acquisition by plant (g N m^{-2} year^{-1}.

5.5.2 Nitrogen Use Efficiency Traits

Moll et al. (1982) and Ortiz-Monasterio et al. (1997) reported formulas for N use efficiency and its component traits:

$$N\,uptake\,efficiency\left(\%\right) = \frac{Ntf\left(kg\,ha^{-1}\right)}{Ns\left(kg\,ha^{-1}\right)} \times 100$$

where Ntf = total aboveground N content at maturity of fertilized treatment; Ntc = total aboveground N content at maturity of control treatment; Ns = N supplied.

$$N\,biomass\,production\,efficiency\left(kg\,kg^{-1}N\right) = \frac{TDWf\left(kg\,ha^{-1}\right) - TDWc\left(kg\,ha^{-1}\right)}{Ntf\left(kg\,ha^{-1}\right) - Ntc\left(kg\,ha^{-1}\right)}$$

where TDWf = total dry weight of fertilized treatment; TDWc = total dry weight of control treatment.

$$N\ utilization\ efficiency\left(kg\, kg^{-1} N\right)$$
$$= Harvest\ index \times N biomass production\ efficiency$$
$$N\ use\ efficiency_{grain\ yield}\left(kg\ kg^{-1} N\right)$$
$$= N\ uptake\ efficiency \times N\ utilization\ efficiency$$
$$= \frac{GDWf\left(kg\ ha^{-1}\right) - GDWc\left(kg\, ha^{-1}\right)}{Ns\ \left(kg\, ha^{-1}\right)}$$

where GDWf = grain dry weight of fertilized treatment; GDWc = grain dry weight of control treatment.

5.5.3 Nutrient Efficiency Ratio (NER) (Gerloff and Gabelman 1983)

$$NER = \frac{\left(Units\ of\ Yield, kg\right)}{\left(Units\ of\ elements\ in\ tissue, kg\right)} kg\, kg^{-1}$$

NER is generally utilized to differentiate genotypes into efficient and inefficient nutrient use.

5.5.4 Physiological Efficiency (PE)

$$PE = \frac{\left(Yield\ F\ kg - Yield\ C,\ kg\right)}{\left(Nutrient\ uptake\ F,\ kg - Nutrient\ uptake\ C,\ kg\right)} = kg\ kg^{-1}$$

where *F* is the plant receiving fertilizer and *C* is the plant receiving no fertilizer. Factors like genotypes, environmental and management practices govern the ability of a species to transform nutrients acquired from fertilizer into economic yield (grain). Low PE suggests suboptimal growth (nutrient deficiencies, drought stress, heat stress, mineral toxicities, and pests).

5.5.5 Agronomic Efficiency (AE kg Yield Increase per kg Nutrient Applied)

It is the product of nutrient recovery from mineral or organic fertilizer (RE) and the efficiency with which the plant uses each additional unit of nutrient (PE). AE depends on management practices that affect RE and PE:

$$AE = \frac{(Yield\,F\,kg - Yield\,C,kg)}{(Quantity\ of\ nutrient\ applied,\ kg)} = kg\,kg^{-1}$$

$$AE = \frac{Nutrient\,uptake\,F, kg - Nutrient\,uptake\,C, kg(Yield\,F\,kg - Yield\ C,kg)}{(Quantity\,of\ nutrient\ applied, kg)} = \times 100 = \%$$

5.5.6 Agro-Physiological Efficiency (APE)

APE is defined as the economic yield (grain) obtained per unit of nutrient absorbed:

$$APE = \frac{(Yield\,F, kg - Yield\,C, kg)}{(Nutrient\,uptake\,F, kg\,Straw + grain) - (Nutrient\,uptake\,C, kg\,Straw + grain)} = kg\,kg^{-1}$$

5.5.7 Apparent Nutrient Recovery Efficiency (ANR)

$$ANR = \frac{(R_{ac} - R_{\min}) \times NPP}{N_{ac}}$$

where NPP = net primary productivity; N_{ac} = nutrient available in soil; R_{ac} = rate of resource uptake or acquisition by plant (g N m^{-2} $year^{-1}$). ANR reflects the plant's ability to acquire applied nutrient from soil.

5.5.8 Apparent Nutrient Use Efficiency (Murthy et al. 2010)

$$Apparent\ nutrient\ use\ efficiency(\%) = \frac{Uptake\,in\,treated\ plot - uptake\,in\,control\ plot}{Fertilizer\,dose} \times 100$$

5.5.9 Apparent Nitrogen Use Efficiency

Sistani et al. (2010) have quantified this parameter by applying the following formula for *Cynodon dactylon*:

$$ANUE = \left[\frac{(total\,N\,uptake\ of\ \mathrm{a}\,litter - total\,N\,uptake\,of\ the\,control)}{total\,N\,applied\,as\,litter}\right]$$

5.5.10 Apparent Nutrient Recovery (Paul et al. 2014)

ANR reflects the plant's ability to acquire applied nutrient from soil. RE depends on the congruence between plant demand and nutrient release from fertilizer. RE is affected by the application method (amount, timing, placement, *N* form) and factors that determine the size of the crop nutrient sink (genotype, climate, plant density, abiotic/biotic stresses):

$$Apparent\ Recovery = \frac{Nutrient\ uptkae\ in\ fertilizer\ treatment\left(Un\right) - Nutrient\ uptake\ in\ control\ \left(Uc\right)}{Ammount\ of\ nutrient\ applied\left(Fa\right)} \times 100$$

Apparent Zn recovery efficiency (ARE-Zn) can be quantified as (Fageria and Baligar 2005)

$$Apparent\ Zn\ recovery\ efficiency = \left[\left(\frac{Zu\ uptake\ in\ grain + straw\ in\ mg\ pot\ at\ higher\ zn\ level - zn\ uptake\ in\ grain + straw\ in\ mg\ pot\ at\ lower\ zn\ lever}{Quantity\ of\ Zn\ applied\ in\ mg\ pot^{-1}}\right)\right] \times 1$$

Nutrient harvest index can be quantified through

$$Nutrient\ harvest\ index = \frac{\left(Nutrient\ uptake\ in\ grain\right)}{\left(Total\ uptake\ by\ crop\right) \times 100}$$

5.5.11 Apparent Nutrient Balance (Liu et al. 2003 and Salam et al. 2014)

Apparent nutrient balance = (nutrient uptake by crop) - (nutrient additions).

Sorgona et al. (2006) have quantified nitrogen use efficiency by using three parameters, viz., (a) nitrogen efficiency ratio as total plant dry weight divided by total nitrogen accumulation (g TDW mg^{-1} N), (b) nitrogen utilization efficiency quantified as total plant dry weight divided by nitrogen concentration (g^2 TDW mg^{-1} N), and (c) nitrogen uptake efficiency calculated as total nitrogen accumulation divided by root dry weight (mg N g^{-1} RDW). They also described NER trend in response to nitrate availability:

$$NER\left(g\,TDW\,mg^{-1}\,N\right) = A_{NER} + B_{NER}\,e^{C_{NER}\left(NO_3^-\right)}$$

where A_{NER} (g TDW mg^{-1} N) indicates the NER value at the equilibrium and B_{NER} (g TDW mg^{-1} N) indicates the difference between the maximum NER value and A_{NER}, while C_{NEWR} (mM^{-1}) is the decay constant that indicates the nitrate concentration in which the half maximum NER is obtained. Tateno and Takeda (2010) have quantified nitrogen uptake and nitrogen use efficiency at leaf level, leaf litter, aboveground, belowground, and whole stand level. They quantified *N* concentration at the stand level by using basal area of species and it can be described as

$$N\,concentration\,at\,the\,stand\,level = \sum \frac{\left(N_i \times BA_i\right)}{\sum BA_i}$$

N_i is the *N* concentration of species I, and BAi is the total basal area of species i in each plot. And *N* resorption efficiencies of leaves are calculated as (green leaf *N* content-leaf litter *N* content)/green leaf *N* content × 100. Fageria et al. (2010) have quantified nitrogen harvest index by using the following equation:

$$Nitrogen\,harvest\,index = \frac{Nitrogen\,uptake\,in\,grain}{Nitrogen\,uptake\,in\,grain\,and\,shoot}$$

Similarly, zinc harvest index can be calculate through

$$Zn\,HI\left(mg\,pot^{-1}\right) = \frac{Zn\,uptake\,in\,grain}{Zn\,uptake\,in\,grain + straw}$$

5.5.12 Nutrient Uptake Efficiency (NupE) (Shaver and Melillo 1984; Bridgham et al. 1995 and Pastor and Bridgham 1999)

$$NupE = \frac{N_{ac}}{N_{av}}$$

where N_{ac} = is the amount of nutrient uptake by plant; N_{av} = is the available nutrient in soil.

5.5.13 Net Primary Productivity (NPP) (Bridgham et al. 1995)

$$NPP = \frac{\left(R_{ac} - R_{\min}\right) \times NPP_{\max}}{R_{ac} - R_{\min + \propto}}$$

where R_{min} is the estimated minimum resource requirement for positive growth, NPP_{max} is the production asymptote, and ∝ is the half-saturation constant with respect to resource.

5.5.14 Nutrient Use Efficiency (NUE)

$$NUE = \frac{\left(R_{ac} - R_{\min}\right) \times NPP_{\max}}{\left(R_{ac} - R_{\min + \propto}\right) \times N_{ac}}$$

This equation was parameterized using data for NPP and resource and nonlinear regression to estimate $R_{\min}$, $NPP_{\max}$, and ∝ using appropriate software. Utilization efficiency (UTE kg kg^{-1} N) and nitrogen use efficiency (NUE kg kg^{-1} N) were assessed for upland rice varieties in northwestern Amhara region of the Ethiopia (Lake et al. 2015):

$$UPE = \left[\frac{BNYF - BNY0}{Nf}\right] \times 100$$

$$UTE = \left[\frac{GYf - GY0}{BNYf - BNY0}\right]$$

$$NUE = UPE \times UTE$$

$$GNY = \left(\frac{GNC \times GY}{100}\right)$$

$$SNY = \frac{SNC \times \left(BY - GY\right)}{100}$$

$$BNY = \left(GNY + SNY\right)$$

$$NHI\left(\%\right)\left[\frac{GNY}{BNY} \times 100\right]$$

where BNYf = biomass *N* yield in *N*-fertilized plot (kg ha^{-1}); BNY0 = biomass *N* yield in non-*N*-fertilized plot (kg ha^{-1}); Nf = *N* fertilizer applied (kg ha^{-1}); GYf = grain yield in *N*-fertilized plot (kg ha^{-1}); GY0 = grain yield in non-*N*-fertilized plot (kg ha^{-1}); GNC = grain *N* concentration (%); GY = grain yield (kg ha^{-1}); SNC = straw N concentration (%); and BY = biomass yield (kg ha^{-1}).

5.5.15 Nutrient Response Efficiency (Bridgham et al. 1995)

NRE is the product of nutrient use efficiency (P/R_{ac}) and nutrient uptake efficiency, which is a proportion of potentially available nutrient in the environment (R_{ac}, g nutrient m^{-2} $year^{-1}$) that plant actually taken up (R_{ac}/R_{av}):

$$NRE = \frac{P}{R_{av}} = \frac{P}{R_{ac}} \times \frac{R_{ac}}{R_{av}} = \frac{A}{L_{n}} \times \frac{R_{ac}}{R_{av}}$$

Thus, nutrient response efficiency depends on productivity (A), the residence time of the plant nutrient pool ($1/L_n$), the nutrient uptake rate (R_{ac}), and the abilities of different plants to compete for available nutrient (R_{ac}/R_{av}).

5.5.16 *Productivity at a Given Level of Resource Availability [P(R$_{av}$)]*

Production at a given resource availability can be approximated by a Michaelis-Menten equation for resource uptake, if one assumes a direct coupling between uptake of a limiting resource and production:

$$P_{R_{av}} = R_{av} \times \left(\frac{P}{R_{av}} \right) + 0$$

5.5.17 *Leaf-Level Potential Photosynthetic Nutrient Use Efficiency (Kuijk and Anten 2009)*

(PPNUE = light-saturated photosynthetic rate\leaf nutrient content).

At leaf level, NUE is often quantified as the ratio between photosynthetic capacity and leaf nitrogen (nutrient) content. PPNUE = P_0/N_0, where P_0 is the maximum photosynthetic rate (μmol C m^{-2} s^{-1}) of young fully exposed leaves and N_0 is the nitrogen content per unit area of young fully exposed leaf (mmol N m^{-2}). According to Henderson and Jose (2012), leaf-level and canopy-level nutrient use efficiency is calculated by using the following formula:

$$\textit{Leaf level nutrient use efficiency}\left(\mathrm{g\,g^{-1}}\right) = \left(\frac{1}{\textit{leaf level nutrient content}\ \mathrm{g\,g^{-1}}} \right) \times (1 - \textit{resorption})$$

$$\textit{Resportion}(\%) = \left(\frac{Foliar_{(live)} - Foliar_{(litter)}}{Foliar_{(live)}} \right) \times 100$$

Canopy-level nutrient use efficiency (CNUE) can be calculated by using aboveground biomass produced in year divided by the peak production (peak foliar production was determined from monthly leaf litter collection) and nutrient content of green leaves for each species in each treatment (Henderson and Jose 2012), and this can be expressed as CNUE (Mg kg^{-1}) = Mg/($kg_{foliage}$ × kg kg^{-1} (nutrient)).

5.5.18 Whole-Canopy NUE (mol C Mmol N^{-1})

$$Canopy-NUE=\left(P_{daily}\times LLS\right)/\left(\left(N_{total}\right)\times\left(1-r\right)\right)$$

where the numerator is the whole-canopy daily carbon gain (P_{daily}, mol C g leaf-mass^{-1} day^{-1}) multiplied by the leaf life span (LLS) in days. The denominator is the amount of nitrogen invested in a gram lead over its life span and the lost from the plant, i.e., the product of whole-canopy foliar nitrogen content (N_{total}, mmol g^{-1}) and the fraction of nutrients not resorbed before leaf abscission [(1 − r), in which r indicates resorption]. The above equation can be divided into two parts, i.e., whole-canopy photosynthetic nitrogen productivity (PNP, mol C mmol N^{-1} day^{-1}) and mean residence time (MRT) of nitrogen in a plant (the inverse of nitrogen turnover when plant growth is at steady state): PNP = $P_{daily} \backslash N_{total}$

$$MRT=LLS/\left(1-r\right)$$

According to Hiremath and Ewel (2001), MRT is not useful for intraspecific variation. However, this formula is based on the calculation for cumulative PNUE in Hiremath and Ewel (2000) with the main exception that carbon gain and nitrogen content are determined for whole canopy, thus including leaves of different age.

5.5.19 Nutrient Influx Rate (in) (Baligar 1986)

$I_n = [(U_2 - U_1)\ (T_2 - T_1) \times [I_nL_2 - I_n L_1)/L_2 - L_1)]$.

where I_n = nutrient influx rate (pmoles cm^{-1} s^{-1}); U = given ion quantity in a whole plant (mmoles\plants); and L_2 and L_1 = relative root length.

5.5.20 Phosphorus Use Efficiency Index (EI) (Siddiqi and Glass 1981)

$$EI=\left[\left(total\ DM\right)2\right]/\,total\,P\,content$$

5.5.21 Partial Nutrient Balance (PNB)

$$PNB=UH/\mathrm{F}$$

where UH = quantity of nutrient removed in harvest crop portion to the quantity of nutrient applied. The primary goal of this measure of efficiency is to determine how close a system is to one. A PNB value close to 1 indicates that mass balance exists—

nutrient application to a unit of land approximately equals nutrient removal. Such a balance is necessary for the fertility level of a system to be sustained.

5.5.22 *Agronomic Efficiency of a Profitable Maintenance Application (Jat et al. 2013)*

$$F_{maint}\, r\, Y_{\mathrm{F}}$$

where F_{maint} is the maintenance rate, r is the rate of nutrient removal per harvested crop unit, and Y_{F} is the fertilized crop yield. When efficiencies are examined, it is often difficult to know how to interpret them. How much efficiency can reasonably be expected?

5.5.23 *Nutrient Re-translocation Efficiency (Finzi et al. 2001 and Biswas and Khan 2011)*

$$NRE(\%) = \left\{ \frac{Nutrient\ in\ green\ leaves - Nutrient\ in\ leaf\ litter}{Nutrient\ in\ green\ leaves} \right\} \times 100$$

The relative importance of P acquisition efficiency and P internal utilization efficiency in the P use efficiency (PUE) was investigated according to Moll et al. (1982). This methodology was developed to investigate the relative importance of two variables (PAE and PUTIL), obtained experimentally, over a third variable (phosphorus use efficiency) which is obtained by the multiplication of PAE and PUTIL. This information could be valuable to utilize weight for each of two variables (PAE and PUTIL) that to be used in selection program related to P use efficiency. The same is also exercised for two variables related to P internal utilization efficiency and PHI.

5.5.24 *Phosphorus Acquisition Efficiency (PAE)*

$$PAE = \frac{Phosphorus\ in\ plant}{Phosphorus\ in\ soil}$$

5.5.25 *Phosphorus Internal Utilization Efficiency (PUTIL)*

$$PUTIL = \frac{Seed\ dry\ matter\ produced\,(seed\ biomass)}{Phosphorus\ in\ plant}\,k$$

5.5.26 Quotient of Phosphorus Utilization (QUTIL)

$$QUTIL = \frac{Seed\ biomass}{Phosphorus\ in\ seed}$$

5.5.27 Phosphorus Harvest Index

$$PHI = \frac{Phosphorus\ in\ seed}{Phosphorus\ in\ plant}$$

Mathur's (2014) study on species-level phosphorus acquisition and internal utilization efficiency of desert woody perennial *Corchorus depressus* revealed that for PUE at low P environmental conditions, PAE was 10–37 times more important than PUTIL. Similarly for PUTIL, QUTIL was 17–56 times more important than PHI. Kaiser-Meyer-Olkin test (0.597) and Bartlett's test of sphericity indicated appropriate use of factor analysis (PCA) and significant correlation between variables in the present study, respectively. Ordination analysis showed lack of correlation between PAE and PUTIL, indicating that selection of one of these should not affect the other. Percent cover of plant, seed output, root length, and root collar diameter exhibit various relationships with different P variables. Regression analysis between P variables and community factors revealed that at species level, dominance of *C. depressus* reflects its higher P acquisition efficiency; however, as the community diversity (richness, Shannon and Wiener index and evenness) increases, the P internal utilization and PUE were inhibited. It can be concluded that P internal utilization and PUE of *C. depressus* are largely influenced by temporal factors (increase and decrease of community diversity during pulse and non-pulse events, respectively). As a result, this species achieved effective nutrient use through temporal partitioning, through which it fulfilled P requirements during low resource availability. Among the soil variables, soil nitrogen supported PUTIL, PUE, and QUTIL, while soil moisture and soil pH favored PHI and soil N, and PHI showed negative relationship.

Overall crop nitrogen use efficiency as suggested by Weih et al. (2011), and according to them, can be broken down into three major components: NUE = $U_N \times E_{N,g} \times C_{N,g}$. The mean *N* uptake efficiency (U_N) is the mean *N* content during major growth period per unit *N* content in seed grain (g N $(gN)^{-1}$). In the calculation of the U_N, the seed *N* (Ns) can be replaced by fertilizer or soil *N* to obtain a measure of the *N* uptake and, ultimately, NUE on basis of fertilizer and or soil *N*. The grain-specific *N* efficiency ($E_{N,g}$) reflects the grain biomass produced per unit of mean plant *N* content during major growth period (g biomass $(g\ N)^{-1}$), and the grain *N* concentration ($C_{N,g}$) reflects N conservation or storage (g N $(g\ biomass)^{-1}$). The grain *N* concentration of other relevant components, to obtain a NUE measure expressing the yield of the component of the interest per unit of seed N (or soil N, if seed N is replaced by soil N):

$$U_N = \frac{N_i'}{N_S} + \frac{\frac{N_i'}{N_S}\left(e^{R_{N-t}} - 1\right)}{2}, where\ R_N \frac{\ln N_f' - \ln N_i'}{t}$$

This divides the U_N into N uptake efficiency prior to major growth period and the relative N uptake rate during major growth period (R_N); $E_{N,g} = E_{N\,x\text{-}r}B\ g$. This allows the analysis of relationships between E_N, g, whole-plant N efficiency, and biomass allocation to grains (i.e., harvest index):

$$C_{N,g} = C_N \frac{r_{N,g}}{r_{B,g}}$$

The measures refer to ground area basis and year, e.g., N content of seed grains in g^{-2} $year^{-1}$, if not specified differently. The components of the above equation can be quantified by using:

Plant biomass (g) at final harvest stage (B)
Biomass (g) of produced grains at final stage (B_g)
Plant N concentration (N/B; g g^{-1}) at final harvest stage (C_N)
Grain N concentration (N_g/B_g; g g^{-1}) at final harvest (C_N·g)
Plant N (B/N' g g^{-1}) efficiency (E_N)
Grain-specific N (B_g/N' g g^{-1}) efficiency (E_N,g)
Plant N content (g) at final harvest (N)
N content (g) of seed grain (N_s)
N content (g) of produced grain at final stage (N_g)
Mean plant N content during major growth period
N_i' which is the initial plant N content at the beginning of major growth period (g)
N_f' which is the final plant N content at the end of the major growth period (g)
Mean relative N accumulating rate during major growth period (R)
Grain biomass allocation at final harvest, i.e., equal to the harvest index (rB g)
Grain N allocation at final harvest (r_N g)
Duration of major growth period (days or weeks; t)
Mean N uptake efficiency during major growth period per N content in seed grain (U_N)

These parameters are quantified as follows: divides the U_N into the N uptake efficiency prior to major growth period and the relative N uptake rate during major growth period (RN); allows the analysis of relationships between $E_N \times g$ whole-plant N efficiency and biomass allocation to grain (i.e., harvest index); and facilitates insight into N conservation aspects. The measures refer to ground area basis and year, e.g., N content of seed grains in gm^{-2} $year^{-1}$; if not specified differently according to Xu et al. (2014), some agronomic indices commonly used to describe NUE include agronomic efficiency (AE, kg crop yield increase per kg nutrient applied), apparent recovery efficiency (RE, kg nutrient taken up per kg nutrient applied), and partial factor productivity (PFP, kg crop yield per kg nutrient applied) (Dobermann 2007; Liu et al. 2011; Xu et al. 2014):

$$AE_X = \frac{(Y - Y_0)}{Fx}$$

$$RE_X = \frac{(U - U_0)}{Fx}$$

$$PFP_X = \frac{(Y)}{Fx}$$

where X is the nutrient (N, P, or K); F is the amount of fertilizer applied (kg ha^{-1}); Y (kg ha^{-1}) is the yield with optimum management (OPT) or farming practices (FP); Y_0 is the yield (kg ha^{-1}) in the control treatment with no N, P, or K; U is the total plant nutrient uptake in aboveground biomass at maturity in an OPT or FP plot (kg ha^{-1}); and U_0 is total plant nutrient uptake in aboveground biomass at maturity in a plot with no N, P, or K(kg ha^{-1}).

5.6 Conclusion

Nutrient use efficiency is often misunderstood or misinterpreted if it is discussed as an isolated issue and not in the context of the efficiency of the total production system. It is important to remember that efficiency and economic viability of total food production system are objectives within which the various components need to be optimized to achieve overall goals. The present chapter enlisted 27 indices and 12 formulas additionally. Further review of these quantitative approaches raveled that mathematical expression of impact of competition, facilitation, surrounding of alien species, root morphology, and root architecture need to be incorporated in NUE studies. Among the nutrients, N, P, and K have largely being explored, and mathematical expression of their nutrient efficiency traits was sufficiently addressed; however, studies related to zinc, iron, and other important micronutrient need to be conducted.

References

Aerts, R. (1996). Nutrient resorption from senescing leaves of perennials: Are there general patterns? *Journal of Ecology, 84*, 597–608.

Aerts, R., & Chapin, F. S. (2000). The mineral nutrition of wild plants revisited a re-evaluation of processes patterns. *Advances in Ecological Research, 30*, 1–67.

Aerts, R., & de Caluwe, H. (1994). Nitrogen use efficiency of *Carex* species in relation to nitrogen supply. *Ecology, 75*(8), 2362–2372.

Aerts, R., Cornelissen, J. H. C., Van, L., & Callagha, T. V. (2007). Climate change has only a minor impact on nutrient resorption parameters in high-altitude petland. *Oecologia, 151*(1), 132–139.
Alvarez-Clare, S., Mack, M. C., & Brooks, M. (2013). A direct test of nitrogen and phosphorus limitation to net primary productivity in a lowland tropical wet forest. *Ecology, 94*, 1540–1551.
Ashford, O. S., Foster, W. A., Turner, B. L., Sayer, E. J., Sutcliffe, L., & Tanner, E. V. J. (2013). Litter manipulation the soil arthropod community in a low tropical rainforest. *Soil Biology and Biochemistry, 62*, 5–12.
Ashraf, M., & Azam, F. (1998). Fate interaction with soil N of fertilizer ^{15}N applied to wheat at different growth stages. *Cereal Research Communications, 26*, 397–404.
Asplund, L., Bergkvist, G., & Weih, M. (2016). Functional traits associated with nitrogen use efficiency in wheat. *Acta Agriculture Scinavica. Section B, Soil and Plant Science, 66*(2), 153–169.
Baligar, V. C. (1986). Interrelationships between growth and nutrient uptake in alfafa and corn. *Journal of Plant Nutrition, 9*, 1391–1404.
Baligar, V. C., & Fageria, N. K. (1997). Nutrient use efficiency in acid soils nutrient management plant use efficiency. In A. C. Monitz, A. M. C. Furlani, N. K. Fageria, C. A. Rosolem, & H. Cantarelles (Eds.), *Plant-soil interaction at low PH sustainable agricultural forestry production* (pp. 75–93). Brazil: Brazilian Soil Science Society Campinas.
Baliger, V. C., Duncan, R. R., & Fageria, N. K. (1990). Soil-plant interaction on nutrient use efficiency in plants. In V. C. Baliger & R. R. Duncan (Eds.), *Crops as enhancers of nutrient use* (pp. 351–373). San Diego: Academic Press.
Ball, M. C. (1988). Ecophysiology of mangroves. *Trees Structure and Function, 2*, 129–142.
Berendse, F., & Aerts, R. (1987). Nitrogen use efficiency: A biologically meaningful definition? *Functional Ecology, 1*, 293–296.
Bhadoria, P. S., EL Dessougi, H., Liebersbach, H., & Claasen, N. (2004). Phosphorus uptake kinetics, size of root system and growth of maize and groundnut in solution culture. *Plant and Soil, 262*, 327–336.
Bilgin, A., Zeren, Y., & Guzel, S. (2016). Foliar N P resorption nutrient N, P, C S contents of Vaccinium arctostaphylos L. *Vaccinium myrtillus* L. from east black sea region of Turkey. *Turkish Journal of Botany, 40*, 137–146.
Binkley, D., Stape, J. L., & Ryan, M. G. (2004). Thinking about efficiency of resource use in forests. *Forest Ecology and Management, 193*, 5–16.
Biswas, S., & Khan, D. K. (2011). Major nutrient dynamics of two plant species at matha protected forest of Purulia, West Bengal, India. *International Journal of Environmental Sciences, 2*(1), 210–219.
Blankenau, K., Olfs, H. W., & Kuhlmann, H. (2002). Strategies to improve the use efficiency of mineral fertilizer nitrogen applied to winter wheat. *Journal of Agronomy and Crop Science, 188*, 146–154.
Brentrup, F., & Palliere, C. (2010). *Nitrogen use efficiency as an agro-environmental indicator. OECD workshop "Agri-environmental indicators: Lessons learned and future directions" 23–26 march*. Switzerland: Leysin.
Bridgham, S. D., Pastor, J., McClaugherty, C., & Richardson, C. J. (1995). Nutrient use efficiency: a litterfall index, a model, a test along a nutrient-availability gradient in North Carolina pearls. *The American Naturalist, 145*, 1–21.
Cardenas, I., & Campo, J. (2007). Foliar nitrogen phosphorus resorption decomposition in the nitrogen-fixing tree *Lysiloma microphyllum* in primary secondary seasonally tropical dry forests in Mexico. *Journal of Tropical Ecology, 23*, 107–113.
Chapin, F. S., & Kedrowski, R. A. (1983). Seasonal-changes in nitrogen phosphorus fractions autumn retranslocation in evergreen deciduous taiga trees. *Ecology, 64*, 376–391.
Ciarelli, D. M., Furlani, A. M. C., Dechen, A. R., & Lima, M. (1998). Genetic variation among maize genotypes for P-uptake P-use efficiency in nutrient solution. *Journal of Plant Nutrition, 21*, 2219–2229.
Claassen, N., & Steingrobe, B. (1999). Mechanistic simulation models for a better understanding of nutrient uptake from soil. In Z. Rengel (Ed.), *Mineral nutrition of crops—Fundamental mechanisms and implications* (pp. 327–367). New York: Haworth.

Craine, J. M. (2009). *Resource strategies of wild plants*. Princeton: Princeton University Press.
DeBell, D. S., Cole, T. G., & Whitesell, C. D. (1997). Growth, development yield in pure mixed sts of eucalyptus Albizia. *Forest Science, 43*, 286–298.
Dessougi, H. I., Classen, N., & Steingrobe, B. (2010). Potassium efficiency of department crops grown on sy soil under controlled conditions. *University of Khartoum Journal of Agricultural Sciences, 18*(3), 310–334.
Dhugga, K. S., & Waines, J. G. (1989). Analysis of nitrogen accumulation use in bread durum wheat. *Crop Science, 29*, 1232–1239.
Dobermann, A. (2007). Nutrient use efficiency–measurement management. In A. Krauss et al. (Eds.), *Fertilizer best management practice: General principles, strategy for their adoption voluntary initiatives vs regulations. IFA International workshop on fertilizer best management practices, Brussels, Belgium, 7–9 march 2007* (pp. 1–28). Paris, France: International Fertilizer Industry Association.
Eckstein, R. L., & Karlsson, P. S. (2001). Variation in nitrogen-use efficiency among within sub-arctic graminoids herbs. *New Phytology, 150*, 641–651.
EL Dessougi, H., Claassen, N., & Steingrobe, B. (2002). Potassium efficiency mechanism of wheat, barley and sugar beet grown on a K fixing soil under controlled conditions. *Journal of Plant Nutrition and Soil Science, 165*, 732–737.
EL Dessougi, H. I., Claasseen, N., & Steingrobe, B. (2010). Potassium efficiency of different crops grown on a sandy soil under controlled conditions. *University of Khartoum Journal of Agricultural Sciences, 18*(3), 310–334.
EL Dessougi, H. I., Claasseen, N., & Steingrobe, B. (2011). Potassium efficiency of wheat and sugar beet evaluated under field conditions. *University of Khartoum Journal of Agricultural Sciences, 19*(1), 26–48.
Fageria, N. K., & Baligar, V. C. (2005). Growth components zinc recovery efficiency of upl rice genotypes. *Pesquisa Agropecuária Brasileira, 40*(12), 1211–1215.
Fageria, N. K., Barbosa, M. P., & Filho, D. C. J. G. C. (2001). Potassium use efficiency in common bean genotypes. *Journal of Plant Nutrition, 24*, 1937–1945.
Fageria, N. K., Morais, O. P., & Santos, A. B. (2010). Nitrogen use efficiency in upl rice genotypes. *Journal of Plant Nutrition, 33*, 1696–1711.
Fageria, N. K., Baligar, V. C., & Jones, C. A. (2011). *Growth mineral nutrition of field crops*. Boca Raton: CRC Press.
Fayle, T. M., Turner, E. C., Basset, Y., Ewers, R. M., Reynolds, G., & Novotny, V. (2015). Whole ecosystem experimental manipulations of tropical forest. *Trends in Ecology & Evolution, 30*(6), 334–346.
Feller, I. C., Whigham, D. F., McKee, K. L., & Lovelock, C. E. (2003). Nitrogen limitation of growth nutrient dynamics in a disturbed mangrove forest, Indian River lagoon, Florida. *Oecologia, 134*, 405–414.
Field, C. B., & Mooney, H. A. (1986). The photosynthesis-nitrogen relationship in wild plants. In T. J. Givnish (Ed.), *On the economy of plant from function* (pp. 25–55). New York: Cambridge University press.
Finzi, A. C., Allen, A. S., DeLucia, E. H., Ellsworth, D. S., & Schlesinger, W. H. (2001). Forest litter production, chemistry decomposition of following two years of free-air CO_2 enrichment. *Ecology, 82*(2), 470–484.
Fioretto, A., Papa, S., & Fuggi, A. (2003). Litterfall and litter decomposition in a low Mediterranean shrubland. *Biology and Fertility of Soils, 39*, 37–44.
Flores, J., & Briones, O. (2001). Plant life-form germination in a Mexican inter-tropical desert: Effects of soil water potential temperature. *Journal of Arid Environments, 47*, 485–497.
Freschet, G. T., Cornelissen, J. H. C., van Logtestijn, R. S. P., & Aerts, R. (2010). Substantial nutrient resorption from leaves, stems roots in a subarctic flora: What is the link with other resource economics traits? *New Phytologist, 186*, 879–889.
Foehse, D., Claassen, N., & Jungk, A. (1988). Phosphorus efficiency of plants. *Plant and Soil, 110*, 101–109.

Funk, L. J. (2013). The physiology of invasive plants in low-resource environments. *Conservation Physiology, 1*, 1–17.

Furlani, A. M. C., & Furlani, P. R. (1988). *Composition pH of nutrient solutions for physiological studies plant selection under adverse nutritional conditions, Boletim Tecnico* (Vol. 121, p. 34). Campinas: Campinas Instituto Agronomico.

Furlani, A. M. C., Furlani, P. R., Tanaka, R. T., Mascarenhas, H. A. A., & Delgado, M. D. P. (2002). Variability of soybean germplasm in relation to phosphorus uptake and use efficiency. *Scientia Agricola, 59*, 529–536.

Gama-Rodrigues, A. C., Barros, N. F., & Comedford, N. B. (2007). Biomass nutrient cycling in pure mixed sts of native tree species in southeastern Bahia, Brazil. *Revista Brasileira de Ciência do Solo, 31*, 287–298.

Garber, S. M., & Maguire, D. A. (2004). Stand productivity development in two mixed-species spacing trials in the Central Oregon cascades. *Forest Science, 501*, 92–105.

Garnier, E., & Aronson, J. (1998). Nitrogen use efficiency from leaf to stand level: Clarifying the concept. In H. Lambers, H. Poorter, & v. V. MMU (Eds.), *Inherent variation in plant growth. Physiological mechanisms and ecological consequences* (pp. 515–538). Leiden, The Netherlands: Backhuys Publishers.

Geng, X., Jian, S., Li, B., & Pan, X. (2013). Do higher resource capture ability utilization efficiency facilitate the successful invasion of exotic plant? A case study of *Alternanthera philoxeroides. American Journal of Plant Sciences, 4*, 1839–1845.

Gerloff, G. C., & Gabelman, W. H. (1983). Genetic basis of inorganic plant nutrition. In A. Lauchli & R. L. Bieleski (Eds.), *Inorganic plant nutrition, encyclopedia of plant physiology* (Vol. 5, pp. 453–480). New York: Springer.

Glover, J. D., & John, P. R. (2010). Perennial grains food security for the future. *Issues in Science and Technology, 26*, 41–48.

Good, A. G., Sharawat, A. K., & Muench, D. G. (2004). Can less yield more? Is reducing nutrient input into the environment compatible with maintaining crop production? *Trends in Plant Science, 9*, 597–605.

Gourley, C. J. P., Allan, D. L., & Russelle, M. P. (1994). Plant nutrient efficiency a comparison of definition suggested improvement. *Plant and Soil, 158*, 29–37.

Graham, R. D. (1984). Breeding characteristic in cereals. In P. B. Tinker & A. Lauchli (Eds.), *Advances in plant nutrient* (pp. 57–90). New York: Praeger.

Griffitha, A. P., Francis, M., Epplin, S. D., & Fuhlendorf Robert, G. A. (2011). Comparison of perennial polycultures monocultures for producing biomass for biorefinery feedstock. *Agronomy Journal, 103*, 617–627.

Güsewell, S. (2005). Nutrient resorption of wetl graminoids is related to the type of nutrient limitation. *Functional Ecology, 19*, 344e35.

Hacker, S. D., & Gaines, S. D. (1997). Some implications of direct positive interactions for community species diversity. *Ecology, 78*, 1990–2003.

Hahs, A., Enright, N. J., & Thomas, I. (1999). Plant communities, species richness their environmental correlates in the sheaths of Little Desert National Park, Victoria. *Australian Journal of Ecology, 24*, 249–257.

Henderson, D. E., & Jose, S. (2012). Nutrient use efficiency of three fast growing hardwood species across a resource gradient. *Open Journal of Forestry, 2*(4), 187–199.

Hiremath, A. J. (1999). *Nutrient use efficiency in simplified tropical ecosystem*. Ph.D. Dissertation, University of Florida, Gainesville, FL. 184 P.

Hiremath, A. J., & Ewel, J. J. (2000). Control of ecosystem nutrient use efficiency bottom up or top down? In A. R. Berkowitz & C. Brewer (Eds.), *Proceedings of the symposium nutrient cycling*. Cambridge: Harvard University.

Hiremath, A. J., & Ewel, J. J. (2001). Ecosystem nutrient use efficiency, productivity, nutrient accrual in model tropical communities. *Ecosystems, 4*, 669–682.

Hirose, T. (1975). Relations between turnover rate, resource utility, structure of some plant populations: A study in the matter budgets. *Journal of Faculty of Science, University of Tokyo, Section III: Botany, 11*, 355–407.
Hobbie, S. E. (1992). Effect of plant species on nutrient cycling. *Trends in Ecology & Evolution, 7*, 336–339.
Homeier, J., Hertel, D., Camenzind, T., Cumbicus, N. L., Maraun, M., Martinson, G. O., Poma, L. N., et al. (2012). Tropical Andean forests are highly susceptible to nutrient inputs—Rapid effects of experimental N P addition to an Ecuadorian montane forest. *PloS One, 7*, e47128.
Hooper, D. U. (1998). The role of complementarily competition in ecosystem responses to variation in plant diversity. *Ecology, 79*, 704–719.
Huang, J. Y., HL, Y., Wang, B., Li, L. H., Xiao, G. J., & Yuan, Z. Y. (2012). Nutrient resorption based on different estimations of five perennial herbaceous species from the grassland in inner Mongolia China. *Journal of Arid Environments, 78*, 1–8.
Interli, S. J., & Kilham, S. S. (2001). Limiting resources the regulation of diversity in phytoplankton communities. *Ecology, 82*, 1270–1282.
Ivans, C. Y., Leffler, A. J. U., Spaulding, J. M., Stark, R. J., & Caldwell, R. M. M. (2003). Root responses nitrogen acquisition by *Artemisia tridentata and Agropyron desertorum* following small summer rainfall events. *Oecologia, 134*, 317–324.
Jat, M. L., Satyanarayana, T., Majumdar, K., Parihar, C. M., Jat, S. L., Tetarwal, J. P., Jat, R. K., & Saharawat, Y. S. (2013). Fertilizer best management practices for maize systems. *Indian Journal of Fertilizer, 9*(4), 80–94.
Jothimani, S., Pian, B. J., & Thiyagarajan, M. (2007). Nutrient use efficiency in crop plants. *Agrobios Newsletter, 8*, 28–31.
Kaspari, M., Garacia, M. N., Harmus, K. E., Santana, M., Wright, S. J., & Yavitt, J. B. (2008). Multiple nutrients limit litterfall decomposition in a tropical forest. *Ecology Letters, 11*, 35–43.
Kazakou, E., Garnier, E., Navas, M. L., Roumet, C., Collin, C., & Laurent, G. (2007). Components of nutrient residence time the leaf economics spectrum in species from Mediterranean old-fields differing in successional status. *Functional Ecology, 21*, 235–245.
Kelty, M. J. (1992). Comparative productivity of monocultures mixed species system. In M. J. Kelty, B. C. Larson, & C. D. Oliver (Eds.), *The ecology silviculture of mixed species forests* (pp. 125–141). Dordrecht, The Netherlands: Kluwer Academic Publishers.
Killingbeck, K. T. (1996). Nutrients in senesced leaves: Keys to the search for potential resorption resorption proficiency. *Ecology, 77*, 1716–1727.
Knops, J. M. H., Koening, W. D., & Nash, T. N. (1997). On the relationship between nutrient use efficiency fertility in forest ecosystem. *Oecologia, 110*, 550–556.
Kobe, R. K., Lepczyk, C. A., & Iyer, M. (2005). Resorption efficiency decreases with increasing green leaf nutrients in a global data set. *Ecology, 86*(10), 2780–2792.
Kuijk, M., & Anten, N. P. R. (2009). Whole canopy nitrogen use efficiency of pioneer species in early secondary forest succession in Vietnam. *Ecological Research, 24*(4), 811–820.
Kumari, K., Sharma, M., Sharma, R. K., & Balloli, S. S. (2000). Effect of late application of nitrogen on yield protein content of wheat. *Annals of Agricultural Research, 21*(1), 288–291.
Lake, T., Belete, K., & Bayu, W. (2015). Genotypic variation trait association in nitrogen use efficiency of upl rice varieties in north-western amhara region, Ethiopia. *International Journal of Research and Review, 2*, 124–133.
Leishman, M. R., Thomson, V. P., & Cooke, J. (2010). Native exotic invasive plants have fundamentally similar carbon capture strategies. *Journal of Ecology, 98*, 28–42.
Liu, X. J., Wanga, J. C., Lua, S. H., Zhanga, F. S., & Zenga, X. Z. (2003). Effects of non-flooded mulching cultivation on crop yield, nutrient uptake nutrient balance in rice–wheat cropping systems. *Field Crops Research, 83*, 297–311.
Liu, X., He, P., Jin, J., Zhou, W., Sulewski, G., & Phillips, S. (2011). Yield gaps, indigenous nutrient supply, nutrient use efficiency of wheat in China. *Agronomy Journal, 103*, 1452.
Liu, C., Liu, Y., Guo, K., Wang, S., & Yang, Y. (2014). Concentrations resorption patterns of 13 nutrients in different plant functional types in the karst region of south-western China. *Annals of Botany, 113*, 873–885.

Lovelock, C. E., & Feller, I. C. (2003). Photosynthetic performance resource utilization of two mangrove species coexisting in a hypersaline scrub forest. *Oecologia, 134*, 455–462.

Macinnis-Ng, C., McClenahan, K., & Eamus, D. (2004). Convergence in hydraulic architecture, water relations primary productivity amongst habitats across seasons in Sydney. *Functional Plant Biology, 31*, 429–439.

Man, R., & Lieffers, V. J. (1999). Are mixtures of aspen white spruce more productive than single species stands? *The Forestry Chronicle, 753*, 505–512.

Martin, K. C., Bruhn, D., Lovelock, C. E., Feller, I. C., Evans, J. R., & Ball, M. C. (2010). Nitrogen fertilization enhances water-use efficiency in a saline environment. *Plant, Cell & Environment*. doi:10.1111/j.1365-3040.2009.02072.x.

Mathur, M. (2005). *Ecology prospecting of some medicinal plants of aphrodisiac potential*. Doctoral Thesis, Jai Narayan Vyas University, Jodhpur, India.

Mathur, M. (2013). Characterization of whole plant nutrient utilization efficiency under heterogenous environmental conditions. *Asian Journal of Bio Science, 3*, 14–21.

Mathur, M. (2014). Species level phosphorus acquisition internal utilization efficiency their relation with bottom up top down factors. *Applied Ecology and Environmental Research, 12*(1), 123–142.

Mathur, M. (2015). Assessment of stand types for phosphorus acquisition and internal utilization efficiency in a desert plant *Blepharis sindica* (T. Andres). *Arid Ecosystems, 5*(2), 102–111.

Mayor, J. R., Wright, J. S., & Turner, B. L. (2014). Species-specific response of foliar nutrients to long term nitrogen and phosphorus additions in a lowland tropical forest. *Journal of Ecology, 102*, 36–44.

McAuliffe, J. R. (1994). Landscape evolution, soil formation, ecological patterns processes in Sonoran Desert Bajadas. *Ecological Monographs, 64*, 111–148.

McGroddy, M. E., Daufresne, T., & Hedin, L. O. (2004). Scaling of C:N:P stoichiometry in forests worldwide: Implications of terrestrial Redfield-type ratios. *Ecology, 85*, 2390–2401.

McKee, K. L. (1993). Soil physicochemical patterns mangrove species distribution—Reciprocal effects? *Journal of Ecology, 81*, 477–487.

Meyer, D. (1993). *Effizienz von Kulturpflanzen bei der Nutzung des nichtaustauschbaren Kaliums von Böden*. Dissertation, Universität Göttingen.

Mirmanto, E., Proctor, J., Green, J., Nagy, L., & Suriantata. (1999). Effects of nitrogen phosphorus fertilization in a low evergreen rainforest. *Philosophical Transactions of the Royal Society of London. Series B: Biological Sciences, 354*, 1825–1829.

Moll, R. H., Kamprath, E. J., & Jackson, W. A. (1982). Analysis interpretation of factors, which contribute to efficiency of nitrogen utilization. *Agronomy Journal, 74*, 562–556.

Montagnini, F. (2000). Accumulation in above-ground biomass soil storage of mineral nutrients in pure mixed plantations in a humid tropical l. *Forest Ecology and Management, 134*, 257–270.

Montagnini, F., Gonzalez, E., Porras, C., & Reingans, R. (1995). Mixed pure forest plantations in the humid neotropics: A comparison of early growth, pest damage establishment costs. *Commonwealth Forestry Review, 744*, 306–314.

Murthy, K. R., Raveendra, H. R., & Reddy, M. T. B. (2010). Effect of chromolaena parthenium as green manure their compost on yield, uptake nutrient use efficiency on typic paleustalf. *European Journal of Biological Sciences, 4*(1), 41–45.

Nakamura, T., Uemura, S., & Yabe, K. (2002). Variation in nitrogen use traits within between five Carex species growing in the low mires of northern Japan. *Functional Ecology, 16*, 67–72.

Narang, R. M., Bruene, A., & Altman, T. (2000). Analysis of phosphate acquisition efficiency in different Arabidopsis accessions. *Plant Physiology, 124*, 1786–1899.

Norris, M. D., Blair, J. M., & Johnson, J. C. (2007). Altered ecosystem nitrogen dynamics as a consequence of l cover changes in tallgrass prairie. *The American Naturalist, 158*(2), 432–445.

Oleksyn, J., Reich, P. B., Zytkowiak, R., Karolewski, P., & Tjoelker, M. G. (2003). Nutrient conservation increases with latitude of origin in European *Pinus sylvestris* populations. *Oecologia, 136*, 220–235.

Ordonez, A., & Olff, H. (2012). Do alien plant species profit more from high resource supply than natives? A trait-based analysis. *Global Ecology and Biogeography, 22*, 648–658.

Orgeas, J., Ourcival, J. M., & Bonin, G. (2002). Seasonal spatial patterns of foliar nutrients in cork oak *Quercus suber* L. growing on siliceous soils in Provence France. *Plant Ecology, 164*, 201–211.

Ortiz-Monasterio, J. I., Sayre, K. D., Rajaram, S., & Mc-Mahon, M. (1997). Genetic progress in wheat yield nitrogen use efficiency under four nitrogen rates. *Crop Science, 37*, 898–904.

Paoli, G. D., Curran, L. M., & Zak, D. R. (2005). Phosphorus efficiency of aboveground productivity along a nutrient gradient in Bornean low rainforest a test of the unimodal nutrient response efficiency hypothesis. *Ecology, 86*, 1548–1561.

Paruelo, J. M., & Sala, O. E. (1995). Water losses in the Patagonian steppe: A modeling approach. *Ecology, 76*, 510–520.

Pastor, J., & Bridgham, S. D. (1999). Nutrient efficiency along nutrient availability gradient. *Oecologia, 18*, 50–58.

Paul, T., Bisht, P. S., Pey, P. C., Barman, A., Bhattacharya, S., Saha, T., & Nath, C. P. (2014). Rice productivity nutrient use efficiency as influenced by nutrient management in long term rice Oryza Sativa-wheat *Triticum aestivum* system. *Annals of Agricultural Research New Series, 35*(2), 171–175.

Pavon, N. P., Briones, O., & Flores-Rivas, J. (2005). Litterfall production and nitrogen content in an intertropical semi-arid Mexican scrub. *Journal of Arid Environments, 60*, 1–13.

Penuelas, J., Sardans, J., Llusia, J., Owen, S. M., Carnicer, J., Giambelluca, T. W., Rezende, E. L., Waite, M., & Niinemets, U. (2010). Faster return on 'leaf economics' different biogeochemical niche in invasive compared with native plant species. *Global Change Biology, 16*, 2171–2185.

Perez, C. A., Carmona, M. R., & Armesto, J. J. (2003). Non-symbiotic nitrogen fixation, net nitrogen mineralization, and denitrification in evergreen forest of Chilloe Island, Chille: A comparison with other temperate forests. *Gayana, 60*, 25–33.

Petritan, A. M., Lupke, B. V., & Petritan, I. C. (2010). A comparative analysis of foliar chemical composition leaf construction cost of beech *Fagus sylvatica*, sycamore maple *Acer pseudoplatanus* ash *Fraxinus excelsior* saplings along a light gradient. *Annals of Forest Science, 67*, 610–618.

Quested, H. M., Cornelissen, J. H. C., Press, M. C., Callaghan, T. V., Aerts, R., Trosien, F., Riemann, P., Gwynn-Jones, D., Kondratchuk, A., & Jonasson, S. E. (2003). Decomposition of sub-arctic plants with differing nitrogen economies: A functional role for hemiparasites. *Ecology, 84*, 3209–3221.

Rathke, G. W., Behrens, T., & Diepenbrock, W. (2006). Integrated nitrogen management strategies to improve seed yield, oil content nitrogen efficiency of winter oilseed rape *Brassica napus* L.: A review. *Agriculture, Ecosystems and Environment, 117*, 80–108.

Ratnam, J., Sankaran, M., Hanan, N. P., Grant, R. C., & Zambatis, N. (2008). Nutrient resorption patterns of plant functional groups in a tropical savanna: Variation functional significance. *Oecologia, 157*, 141–151.

Reed, S. C., Townsend, A. R., Davidson, E. A., & Cleveland, C. C. (2012). Stocichiometric patterns in foliar nutrient resorption across multiple scales. *New Phytologist, 196*, 173–180.

Rejmankova, E. (2005). Nutrient resorption in wet macrophytes: Comparison across several regions of different nutrient status. *The New Phytologist, 167*, 471–482.

Renteria, L., Jaramillo, V. J., Martinez-Yrizar, A., & Perez-Jimenez, A. (2005). Nitrogen phosphorus resorption in trees of a Mexican tropical dry forest. *Trees, 19*, 431–441.

Richards, A. E., Forrester, D. I., Bauhus, J., & Scherer-Lorenzen, M. (2010). The influence of mixed tree plantations on the nutrition of individual species: A review. *Tree Physiology, 30*, 1192–1208.

Richardson, S. J., Peltzer, D. A., Allen, R. B., & Mcglone, M. S. (2005). Resorption proficiency along a chronosequence: Responses among communities within species. *Ecology, 86*, 20–25.

Sadana, U. S., & Claassen, N. (1999). Potassium efficiency and dynamics in the rhizosphere of wheat, maize and sugar beet evaluated by a mechanistic model. *Journal of Plant Nutrition, 22*(6), 939–950.

Sala, O. E., Golluscio, R. A., Lauenroth, W. K., & Soriano, A. (1989). Resource partitioning between shrubs grasses in the Patagonian steppe. *Oecologia, 81*, 501–505.

Sala, O. E., Golluscio, R. A., Lauenorth, W. K., & Roset, P. A. (2012). Contrasting nutrient-capture strategies in shrubs grasses of a Patagonian arid ecosystem. *Journal of Arid Environments, 82*, 130–135.

Salam, M. A., Solaiman, A. R. M., Karim, A. J. M. S., & Saleque, M. A. (2014). System productivity, nutrient use efficiency apparent nutrient balance in rice-based cropping system. *Archives of Agronomy and Soil Science, 60*(6), 747–764.

Samal, D., Koval, J. L., Steingrobe, B., Sadana, U. S., Bhadoria, P. S., & Claasseen, N. (2010). Potassium uptake efficiency and dynamics in the rhizosphere of maize (*Zea mays* L.), wheat (*Triticum aestivum* L.), and sugar beet (*Beta vulgaris* L.) evaluated with a mechanistic model. *Plant and Soil, 332*, 105–121.

Sayer, E. J., & Tanner, E. V. J. (2010). Experimental investigation of the importance of litterfall in low semi-evergreen tropical forest nutrient cycling. *Journal of Ecology, 98*, 1052–1062.

Sayer, E. J., Tanner, E. V. J., & Lacey, A. L. (2006). Effects of litter manipulation on early-stage decomposition meso-arthropod abundance in a tropical moist forest. *Forest Ecology and Management, 229*, 285–293.

Sayer, E. J., Wright, J. S., Tanner, V. J., Yavitt, J. B., Harms, K. E., & Powers, J. S. (2012). Variable responses of low tropical forest nutrient status to fertilization litter manipulation. *Ecosystems, 15*, 387–400.

Schwinning, S., Davis, K., Richardson, L., & Ehleringer, J. R. (2002). Deuterium enriched irrigation indicates different from rain use in shrub/grass species of the Colorado plateau. *Oecologia, 130*, 345–355.

Shaver, G. R., & Melillo, J. M. (1984). Nutrient budget of marsh plants: Efficiency concepts relation to availability. *Ecology, 65*, 1491–1510.

Siddiqi, M. Y., & Glass, A. D. M. (1981). Utilization index: A modified approach to the estimation and comparison of nutrient utilization efficiency in plants. *Journal of Plant Nutrition, 4*, 289–302.

Silla, F., & Escudera, A. (2004). Nitrogen use efficiency trade-offs between N productivity mean residence time at organ, plant population level. *Functional Ecology, 18*, 511–521.

Silver, W. L. (1994). Is nutrient availability related to plant nutrient use in humid tropical forest? *Oecologia, 98*, 336–343.

Singh, A. (2004). Effect of fertilization on N P resorption efficiency of selected leguminous non-leguminous tropical trees planted on coal mine spoil. *Journal of the Indian Institute of Science, 84*, 173–182.

Sistani, K. R., Adeli, A., & Tweolde, H. (2010). Apparent use efficiency of nitrogen phosphorus from litter applied to bermudagrass. *Communications in Soil Science and Plant Analysis, 41*(15), 1873–1884.

Smith, D. M., Larson, B. C., Kelty, M. J., & Ashton, P. M. S. (1997). *The practice of silviculture: Applied forest ecology*. New York: Wiley.

Song, L., Guan-Yan, N. I., Chen, B., & Peng, S. (2007). Energetic cost of leaf construction in the invasive weed *Mikania micrantha* H. B. K. its co-occurring species: Implication for invasiveness. *Botanical Studies, 48*, 331–338.

Sorgona, A., Abenavoli, M. R., Gringeri, P. G., & Cacco, G. A. (2006). Comparison of nitrogen use efficiency definitions in citrus rootstocks. *Scientia Horticulturae, 109*, 389–393.

Speher, C. R., & Souza, L. A. C. (1999). Selecting soyabean *Glycine max* L. Merril tolerant to low-calcium stress in short term hydroponics experiment. *Euphytica, 106*, 35–38.

Synder, K. A., Donovan, L. A., James, J. J., Tiller, R. L., & Richards, J. H. (2004). Extensive summer water pulses do not necessarily lead to canopy growth of great basin northern Mojave Desert shrubs. *Oecologia, 141*, 325–334.

Tagliavini, M., Millard, P., & Quartieri, M. (1998). Storage of foliar-absorbed nitrogen remobilization for spring growth in young nectarine *Prunus persica var. nectarina* trees. *Tree Physiology, 18*, 203–207.

Tateno, R., & Takeda, H. (2010). Nitrogen uptake nitrogen use efficiency above below ground along a topographic gradient of soil nitrogen availability. *Oecologia, 163*, 793–804.

Tilman, D., Lehman, C. L., & Thomas, K. T. (1997). Plant diversity ecosystem productivity theoretical considerations. *Proceedings of the National Academy of Sciences of the United States of America, 94*, 1856–1861.

Tittonell, P., Zingore, S., Van Wijk, M. T., Corbeels, M., & Giller, K. E. (2007). Nutrient use efficiencies crop responses to N, P manure applications in Zimbabwean soils: Exploring management strategies across soil fertility gradients. *Field Crops Research, 100*, 348–368.

Tomlinson, P. B. (1986). *The botany of mangroves*. Cambridge: Cambridge University Press; 419 p.

Tran, T. S., & Tremblay, G. (2000). Recovery of N-15 labeled fertilizer by spring bread wheat at different N rates and application times. *Canadian Journal of Soil Science, 80*, 533–539.

Trehan, S. P., & Claassen, N. (1998). External K requirement of young plants of potato, sugar beet and wheat in flowing solution culture resulting from different internal requirements and uptake efficiency. *Potato Research, 41*, 229–237.

Turner, B. L., Yavitt, J. B., Harm, K. E., Garcia, M. N., Romero, T. E., & Wright, I. J. (2013). Seasonal changes treatment effects on soil inorganic nutrients following a decade of fertilizer addition in a lowl tropical forest. *Soil Science Society of America Journal, 77*, 1357. –1369.

Valverde, P. L., Zavala-Hurtado, J. A., Montaña, C., & Ezcurra, E. (1996). Numerical analyses of vegetation based on environmental relationships in the southern Chihuahuan Desert. *The Southwestern Naturalist, 41*, 424–433.

Van Heerwaarden, L. M., Toet, S., & Aerts, R. (2003). Nitrogen phosphorus resorption efficiency proficiency in six subarctic bog species after 4 years of nitrogen fertilization. *Journal of Ecology, 91*, 1060–1070.

Vazquez de Aldana, B. R., & Berendse, F. (1997). Nitrogen–use efficiency in six perennial greases from contrasting habitats. *Functional Ecology, 11*, 619–626.

Vergutz, L., Manzoni, S., Porporato, A., Novais, R. F., & Jackson, R. B. (2012). Global resorption efficiencies concentrations of carbon nutrients in leaves of terrestrial plants. *Ecological Monographs, 82*, 205–220.

Vernescu, C., Coulas, J., & Ryser, P. (2005). Leaf mass loss in wetland graminoids during senescence. *Oikos, 109*, 187–195.

Vitousek, P. M. (1982). Nutrient cycling nitrogen use efficiency. *The American Naturalist, 119*, 553–572.

Vitousek, P. M. (1984). Litter fall, nutrient cycling nutrient limitation in tropical forests. *Ecology, 65*, 285–298.

Wang, X., Shen, J., & Liao, H. (2010). Acquisition or utilization, which is more critical for enhancing phosphorus efficiency in modern crops? *Plant Science, 179*, 302–306.

Weih, M., Asplund, L., & Bergkvist, G. (2011). Assessment of nutrient use in annual perennial crops: A functional concept for analyzing nitrogen use efficiency. *Plant and Soil, 339*, 513–520.

White, P. J., & Veneklaas, E. J. (2012). Nature nurture: The importance of seed phosphorus content. *Plant and Soil, 357*, 1–8.

Wright, I. J., & Westoby, M. (2003). Nutrient concentration, resorption lifespan: Leaf traits of Australian sclerophyll species. *Functional Ecology, 17*, 10–19.

Xu, X. P., He, P., Qiu, S. J., Pampolino, M. F., Zhao, S. C., Johnston, A. M., & Zhou, W. (2014). Estimating a new approach of fertilizer recommendation across small-holder farms in China. *Field Crops Research, 163*, 10–17.

Yasumura, Y., Hikosaka, K., Matsui, K., & Hirose, T. (2002). Leaf-level nitrogen-use efficiency of canopy understory species in a beech forest. *Functional Ecology, 16*, 826–834.

Yuan, Z., & Chen, H. Y. H. (2009). Global trends in senesced leaf nitrogen and phosphorus. *Global Ecology and Biogeography, 18*, 532–542.

Yuan, Z. Y., Li, L. H., & Huang, J. H. (2003). On plant nutrient use efficiency. I. Some aspects in reassessing plant nutrient use efficiency. LiCS. *Advances in Plant Sciences*, 187–200.

Yuan, Z. Y., Li, L. H., Han, X. G., Wan, S. Q., & Zhang, W. H. (2005). Variation in nitrogen economy of two Stipa species in the semiarid region of northern China. *Journal of Arid Environments, 61*(1), 13–25.

Yuan, Z. Y., Li, L. H., Han, X. G., Chen, S. P., Wang, Z. W., Chen, Q. S., & Bai, W. M. (2006). Nitrogen response efficiency increased monotonically with decreasing soil resource availability a case study from a semiarid grassland in northern China. *Oecologia, 148*, 564–572.

Yuan, Z. Y., Chen, H. Y., & Ling, L. H. (2008). Nitrogen use efficiency: Does a trade-off exist between the N productivity the mean residence time within species? *Australian Journal of Botany, 56*, 272–277.

Zeugin, F., Potvin, C., Jansa, J., & Scherer-Lorenzen, M. (2010). Is the tree diversity an important driver for phosphorus nitrogen acquisition of a young tropical plantation? *Forest Ecology and Management, 260*, 1424–1433.

Zhang, G., Jingxing, C., & Tirrore, E. A. (1999). Genotypic variations for potassium uptake and utilization efficiency in wheat. *Nutrient Cycling in Agro Ecosystems, 54*, 41–48.

Chapter 6
Nutrient Use Efficiency

Dibyendu Sarkar and Lohit K. Baishya

Abstract Sustainable crop production delivering high yield to meet ever increasing demand for food requires optimum use of fertilizers (nutrients) by crops. Since fertilizers are costly both financially and environmentally, increasing efficiency of applied fertilizers is a prerequisite. Nutrient use efficiency (NUE) may be defined as yield per unit fertilizer input or in terms of recovery of applied fertilizer. The most appropriate expression of NUE is, however, determined by the question being asked and often by the spatial or temporal scale of interest for which reliable data are available. Analysis of global temporal trends in NUE for N, P and K showed that partial factor productivity (units of crop yield per unit of applied nutrient) and partial nutrient balance (units of nutrient uptake per unit of applied nutrient) for Africa, North America, Europe and the EU-15 are trending upwards, while in Latin America, India and China, they are trending downwards. Typical NUE values for crops are best set locally within the appropriate cropping system, soil, climate and management contexts. Development of new cultivars with higher NUE, coupled with the best management practices, will contribute to sustainable agricultural systems. Improvement in NUE thus requires simultaneous consideration of multiple aspects like crop improvement as well as crop management involving a multidisciplinary approach.

Keywords Partial factor productivity • Partial nutrient balance • Agronomic efficiency • Apparent nutrient recovery • Fertilizer • N use efficiency • P use efficiency • K use efficiency

---

D. Sarkar (✉)
Department of Agricultural Chemistry and Soil Science, Bidhan Chandra Krishi Viswavidyalaya, Mohanpur, Nadia, West Bengal 741252, India
e-mail: dsarkar04@rediffmail.com

L.K. Baishya
ICAR Research Complex for NEH Region, Medziphema, Nagaland 797 106, India
e-mail: lkbaishya@rediffmail.com

M. Naeem et al. (eds.), *Essential Plant Nutrients*,
DOI 10.1007/978-3-319-58841-4_6

6.1 Introduction

World population is expected to grow from 6.0 billion in 1999 to 9.0 billion by 2050. It is estimated that the world will need 60% more cereal production between 2000 and 2050 (FAO 2009) to meet the demand for food required to feed future generations. Sustainable crop production delivering high yield to meet ever-increasing demand is a major challenge for the agricultural production system. Fertilizers have been used by the farmers to harvest the desired crop yield. However, imbalanced and/or excessive use of fertilizers in crop production remains a great concern. Nutrient mining is a major cause for poor crop yield in developing countries, particularly in Africa. Therefore, increasing agricultural production as well as nutrient use efficiency (NUE) continues to be a major challenge for world agriculture. Nutrient use efficiency is largely impacted by fertilizer management as well as soil, plant and environmental factors. This indicates potential for nutrient losses to the surrounding environment from agroecosystems and is important for evaluating the health of crop production systems. Application of fertilizers to crops is costly—in terms of financial cost of production and transport and also in terms of environmental pollution (e.g. eutrophication, C foot print). These necessitates for increased NUE for economic viability and environmental sustainability of agroecosystem.

6.2 Concept of Nutrient Use Efficiency

Nutrient use efficiency may be defined in terms of production outputs relative to nutrient inputs or in terms of recovery of applied nutrients. Estimation of NUE is based on (a) nutrient uptake efficiency (acquisition from soil, influx rate into roots, transport in roots) and (b) utilization efficiency (transport and remobilization in shoots and leaves based on plant parameters). Evaluation of NUE is useful to differentiate soils for their capacity to supply nutrient to crops and to differentiate crop species and cultivars for their ability to absorb and utilize nutrient for producing maximum yields. Definitions commonly used in agronomic research to evaluate the efficiency of applied nutrient in the form of fertilizer (Cassman et al. 2002; Dobermann 2007) are described here.

6.2.1 Partial Factor Productivity (PFP)

Partial factor productivity is defined as units of crop yield per unit of applied nutrient. It can be easily estimated by the farmers who keep records of inputs and outputs of their farm. If the information on inputs (fertilizer used) and outputs (crop yield) are available, it can also be calculated at regional and national levels. The values of PFP vary among crops across the cropping systems, because crops differ in their nutritional requirements based on soil fertility, fertilizer application and water demands.

6.2.2 Partial Nutrient Balance (PNB)

Partial nutrient balance is a simple form of estimation of NUE, usually expressed as nutrient output per unit of nutrient input (a ratio of nutrient removal to nutrient use). The value of a PNB close to one assumes that soil fertility will be sustained at a steady state. Values well below one, where nutrient inputs far exceed nutrient removal, might suggest avoidable nutrient losses and thus the need for improved NUE. A PNB greater than one means more nutrients are removed with the harvested crop than applied by fertilizer and/or manure, a situation equivalent to 'soil mining' of nutrients. This situation may be desired if available nutrient contents in the soil are known to be higher than recommended.

6.2.3 Agronomic Efficiency (AE)

Agronomic efficiency is defined as units of yield increase per unit of applied nutrient. It shows the direct impact of applied fertilizer on crop yield. Estimation of AE requires information of yield without nutrient input so is only known when research plots without nutrient input have been included in the study. Calculation of AE using data from short-term (e.g. annual, biennial) trials rather than long-term trials often underestimates NUE because residual effects of fertilizer application are overlooked in short-term trials.

6.2.4 Apparent Nutrient Recovery Efficiency (ANR)

Apparent nutrient recovery efficiency is a complex form of NUE index and is defined as the difference in nutrient uptake in above-ground parts of the plant grown in the fertilized and unfertilized plots relative to the quantity of nutrient applied. It is often preferred by researchers studying crop response to applied nutrient. Like AE, it can only be calculated when a plot without nutrient has been maintained on the experimental site but in addition requires determination of nutrient concentrations in the crop.

6.2.5 Internal Utilization Efficiency (IUE)

Internal utilization efficiency is defined as the yield in relation to total nutrient uptake. It varies with crop/species, environment and management. A very high IUE suggests deficiency of that nutrient, whereas low IUE suggests poor internal nutrient conversion due to stresses (e.g. deficiencies and toxicities of other nutrients, biotic and abiotic stresses).

6.2.6 Physiological Efficiency (PE)

Physiological efficiency is defined as the increased yield in relation to the increase in crop uptake of nutrient. Similar to AE and ANR, it requires a plot without application of the nutrient. It also requires determination of nutrient concentrations in the crop and is mainly used by researchers.

Because each of the indices discussed above has different interpretations, fertilizer research should include measurements of several indices to understand the factors governing nutrient uptake and fertilizer efficiency to evaluate different management strategies. Interpretation must also consider other crop management factors. For example, AE and ANR are not appropriate indices of NUE when comparing cropping practices such as crop establishment methods or different water management regimes. In this case the crop yield in control plots (Y_0) differs significantly because of these management practices. In these cases, PFP is more appropriate index for making comparisons. Likewise, comparisons of ANR and PE among genotypes should use quality varieties, which are adapted to particular growth conditions. Caution is required when using AE, ANR or PE for assessing trends in NUE in long-term experiments, because depletion of indigenous soil nutrient resources in permanent nutrient omission plots (0-N, 0-P or 0-K plots) will lead to overestimation of the true NUE in fertilized plots (Dobermann 2007).

Other indices are sometimes used (Gourley et al. 1993; Huggins and Pan 1993), but they have no additional advantages for understanding efficient fertilizer management. More detailed studies on the fate of nutrients in agroecosystems often involve isotopes, which are particularly useful for understanding loss, immobilization, fixation and release mechanisms. Determination of ANR in research plots is usually done by the calculation described in Table 6.1. An alternative method for estimation of ANR for N involves using the ^{15}N isotope as a tracer in the fertilizer to determine the proportion of fertilizer applied that was taken up by the crop. The two methods are usually related; however, ANR determined by the ^{15}N method will usually be lower than the difference estimates (Table 6.1) due to cycling of the ^{15}N through microbial processes in soil (Cassman et al. 2002). Tracers are more useful when recovery is measured in the soil as well as in the plant, particularly in the long term. Ladha et al. (2005) reported a range of 5.7–7.1% ANR for N from several

Table 6.1 Summary of commonly used concepts of nutrient use efficiency

Definition	Formula
Partial factor productivity (PFP) is units of crop yield per unit of applied nutrient	$PFP = Y_F/F$ or $PFP = (Y_0/F) + AE$
Partial nutrient balance (PNB) is units of nutrient uptake per unit of applied nutrient	$PNB = U_F/F$
Agronomic efficiency (AE) of applied nutrient is units of crop yield increase per unit of applied nutrient	$AE = (Y_F - Y_0)/F$ or $AE = ANR \times PE$
Apparent nutrient recovery efficiency (ANR) is increase in units of nutrient uptake per unit of applied nutrient	$ANR = (U_F - U_0)/F$
Internal utilization efficiency (IE) of a nutrient is units of crop yield per unit of nutrient uptake	$IE = Y_F/U_F$
Physiological efficiency (PE) of applied nutrient is increase in units of crop yield per unit increase in nutrient uptake from fertilizer	$PE = (Y_F - Y_0)/(U_F - U_0)$

F, amount of nutrient applied (kg ha^{-1}); Y_F and Y_0, crop yield with and without applied nutrient (kg h^{-1}), respectively; U_F and U_0, total nutrient uptake in above ground biomass at maturity (kg ha^{-1}) in fertilized and unfertilized control plot, respectively

studies where ^{15}N was used to estimate N recovery by five subsequent crops, excluding the first growing season.

6.3 Status of Nutrient Use Efficiency

Nutrient use efficiency in terms of PFP and PNB for N differs considerably in regions (Table 6.2). The highest values of the two estimates occur with the lowest N rates for the regions of Africa and Eastern Europe/Central Asia. These regions also characterized by the lowest average cereal yields and PNB values for N are much greater than one, which indicated mining of N from native soil N pools, and such cereal production systems may not be sustainable. Even with the extremely high average PNB in sub-Saharan Africa, the region experiences great intercountry variability with generally higher values in the east and lower values in the central and western part of the continent (Smaling et al. 1997). Farms having good access to resources will have PNB values often less than one (nutrient input exceeds removal), while those with fewer resources will be greater than one. East Asia shows the lowest PNB for N (0.46) at the highest average N input rate. This suggests the potential for improving NUE while maintaining productivity.

Research plots typically generate comprehensive information on nutrient uptake and removal at crop harvest for fertilized and unfertilized plots, enabling estimation of all the NUE indices (Table 6.1). Research reports often include measurements of more than one NUE expression, because each of those indices addresses different questions and has different interpretations. Estimation of NUE in research plots is

Table 6.2 Regional partial factor productivity and partial nutrient balance for N applied to cereals along with average N rates and crop yield during 1999–2002/2003 (adapted from Dobermann and Cassman 2005; Fixen et al. 2015)

Region	N rate kg ha^{-1}	Cereal yield Mg ha^{-1}	Grain[a] N kg ha^{-1}	PFP for N	PNB for N
Sub-Saharan Africa	9	1.1	17	122	1.80
Eastern Europe, Central Asia	25	2.1	32	84	1.30
Oceania	48	1.9	29	40	0.59
Latin America	55	2.9	44	53	0.79
South Asia	58	2.4	36	41	0.62
Southeast Asia	65	3.2	48	49	0.74
West Asia, North Africa	68	2.3	35	34	0.51
Northeast Asia (Japan, S. Korea)	89	6.1	92	69	1.03
North America	112	5.1	77	46	0.68
Western Europe	113	5.5	83	49	7.30
East Asia (China, Vietnam, Korea DPR)	155	4.8	72	31	0.46
World	70	3.1	47	44	0.66

[a]Assuming 1.5% N in cereal grain; PFP for N, units grain yield per unit of applied N; PNB for N, units grain N per unit of applied N

generally greater than those in farmers' field (Dobermann 2007). Differences in agronomic management contribute to such differences between research plots and farmers' fields.

Whether trials are in farmers' fields or on research stations, high-yield cereal systems tend to have higher AE than those with lower yield levels. This should not be surprising since the higher nutrient requirements of crops at high yield levels are likely to exceed the nutrient supplying ability of unfertilized control plots to a greater extent than at lower yield levels. This increases the difference between the yield of the fertilized and unfertilized crop. Additionally, a crop with a faster nutrient accumulation rate may reduce the potential for nutrient losses from the production field. While analysing data from diverse summaries of NUE of cereal systems elsewhere, Fixen et al. (2015) reported that 36% variability in AE for N could be explained simply by average grain yield.

6.3.1 *Nitrogen Use Efficiency*

Ladha et al. (2005) conducted an extensive review of 93 published studies and reported an average ARN for N of 55%, which compares well to other global estimates of 50% by Smil (1999) and 57% by Sheldrick et al. (2002) and to

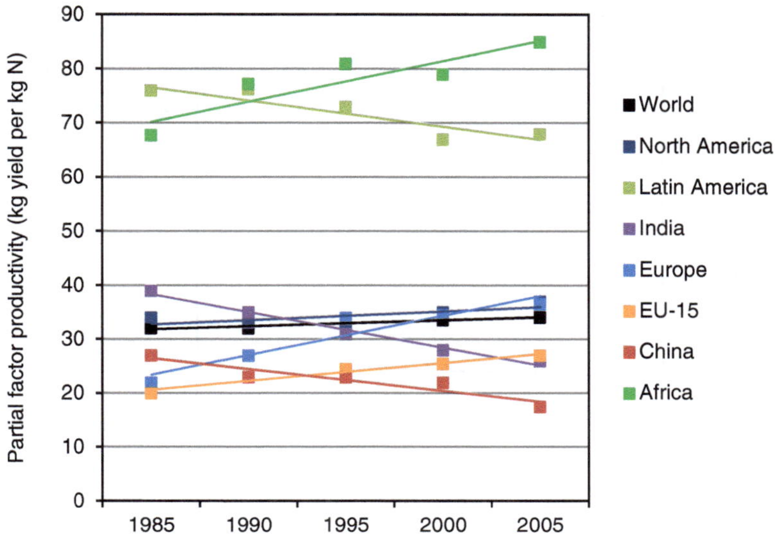

Fig. 6.1 Regional partial factor productivityPartial factor productivity (PFP):for N for N over a 25-year (1983–2007) period (adapted from Fixen et al. 2015)

estimates reported from the USA and Canada of 56% (Howarth et al. 2002) and 52% (Janzen et al. 2003), respectively. Fixen et al. (2015) followed a systematic approach in estimating regional trends in NUE around the world over a 25-year period (1983–2007). With the available information (FAO 2012; IFA 2012), they were constrained to estimate PFP and PNB only. World PFP and PNB for N levels have shown a very slight increase over the 25-year period. Regional temporal trends in PFP are in most cases similar to PNB for N, but trends among global regions clearly differ (Figs. 6.1 and 6.2). Africa and Latin America in 1985 had by far the highest PFP and PNB for N values but with trends in opposite directions. The values of PFP for N show that both these regions have extremely high productivity per unit of applied N fertilizer. However, the excessive PNB values for Africa show that it is becoming more dependent on non-fertilizer sources to balance crop removal of N, an unsustainable situation. In contrast, Latin America has maintained very high productivity per unit of N but has also moved towards a more sustainable nutrient balance. In general, PNB and PFP for Africa, North America, Europe and the EU-15 are trending upwards, while Latin America, India and China are trending downwards. It is interesting to note that PNB for Europe during the last decade appears to have levelled off at around 70% and that PNB for Latin America, India and China has been declining at about the same rate for the 25-year period.

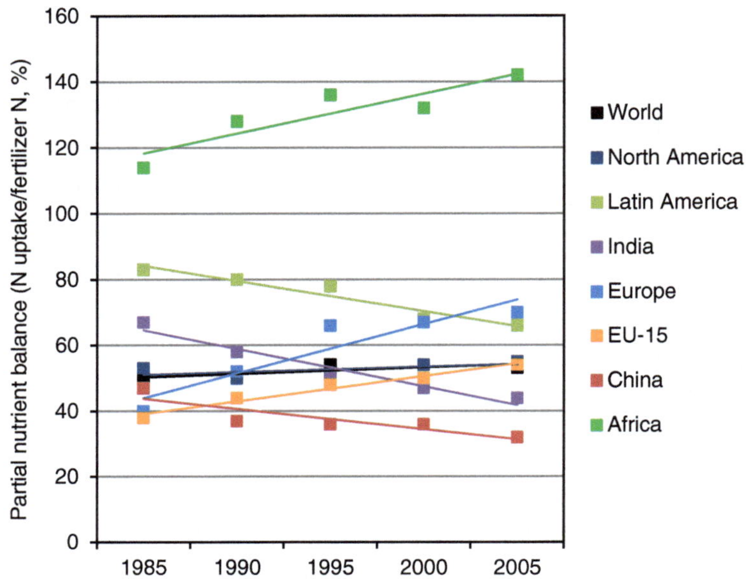

Fig. 6.2 Regional partial nutrient balance for N over a 25-year (1983–2007) period (adapted from Fixen et al. 2015)

6.3.2 Phosphorus Use Efficiency

While most of the benefits and recovery of N fertilization occur during the year of application, much of the benefits of P and K application on many soils were realized in subsequent years (Syers et al. 2008). Appropriate evaluation of the current status and long-term trends of P use efficiency needs to consider these residual effects. Short-term AE, ANR and PFP for P are usually best interpreted within the context of current soil fertility status and associated PNB, which indicates future soil fertility status, if the current PNB remains unchanged. Global PFP and PNB for P have increased over a 25-year period (1983–2007) with PFP in the last 5-year period (2003–2007) approaching 195 kg production per kg P and PNB approaching 70% (Figs. 6.3 and 6.4). Regionally, Africa has markedly separated itself from all other regions in terms of both PFP and PNB. During 1983–1987, Africa, India and China had nearly identical PNB of around 90% but moved in opposite directions over the 25-year period with PNB in Africa doubling to 180%, while China and India dropped to approximately 50%. The PNB values for Africa indicate extreme mining of soil P, while the values in China and India indicate that soil P levels should be increased. These figures do not take into account changes in the use of local rock phosphate, but there is no evidence that this was significant. There is a paucity of reliable information on the use of rock phosphate as a direct application fertilizer in Africa, but various sources indicate that amounts used have remained very low. Average application rates at the country level are less than 0.5 kg ha^{-1}, even for

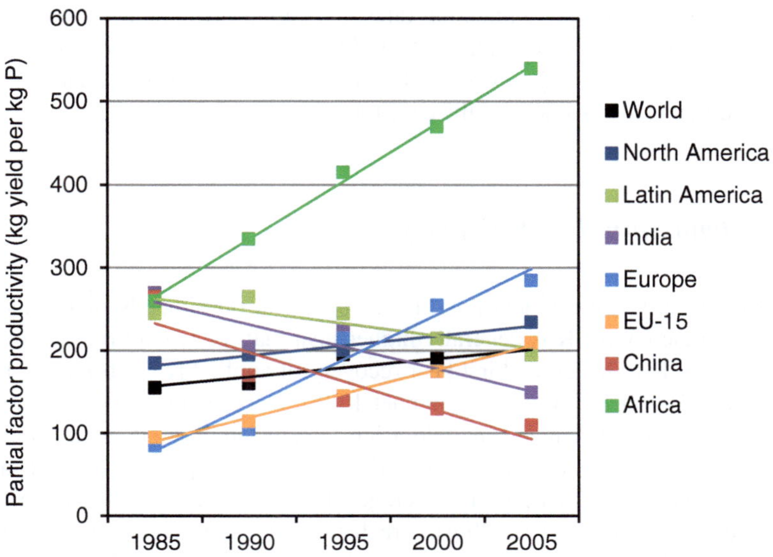

Fig. 6.3 Regional partial factor productivity for P over a 25-year (1983–2007) period (adapted from Fixen et al. 2015)

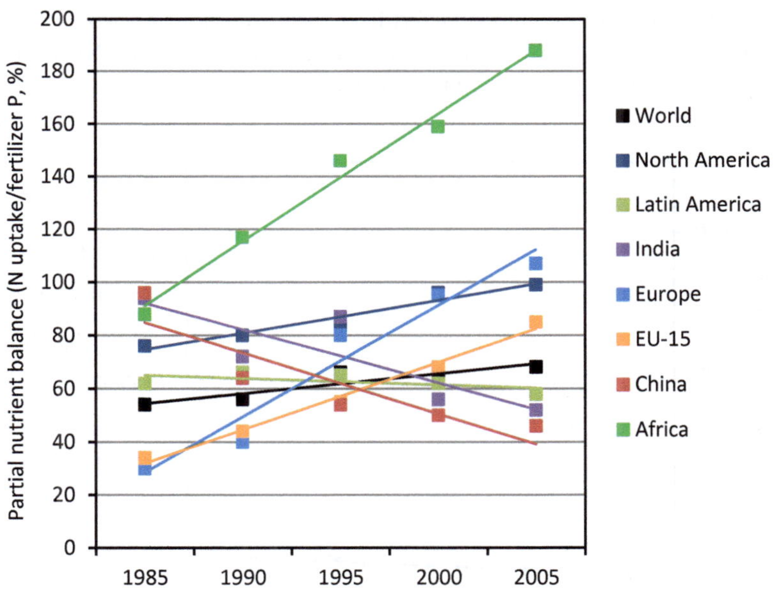

Fig. 6.4 Regional partial nutrient balance for P over a 25-year (1983–2007) period (adapted from Fixen et al. 2015)

countries with the highest application rates, indicating insignificant P contribution from rock phosphates. In general, PNB and PFP for Africa, North America, Europe and the EU-15 are trending upwards while for Latin America, India and China are trending downwards, just as was the case for N.

6.3.3 Potassium Use Efficiency

Information on K use efficiency is limited than that on either N or P. First year recovery efficiency for K is generally higher than that for P, with the exception of some strongly K-fixing soils. First year recovery of applied K has been reported in the range of 20–60% (Baligar and Bennet 1986). Dobermann (2007) summarized average recovery efficiencies in field trials conducted in Asia in the range of 38–51%. Jiyun (2012) summarized field trials on cereal crops in China and showed ANR for K in the 25–32% range and average AE values of 8–12. Dobermann (2007) suggested that AE for K of 10–20 were realistic targets for cereals on soils that are low in available K content. As with N and P, world PFP and PNB for K have increased over the 25-year (1983–2007) period, with PFP in the last 5-year period (2003–2007) approaching 145 kg of production per kg K and PNB approaching 140% (Figs. 6.5 and 6.6). Globally, non-forage crops were removing 40% more K than was being applied as commercial fertilizer during this 5-year period. Regionally, across the 25-year period, China underwent the greatest change in PNB from about 500–100%. For Africa, both PFP and PNB increased markedly across the 25 years

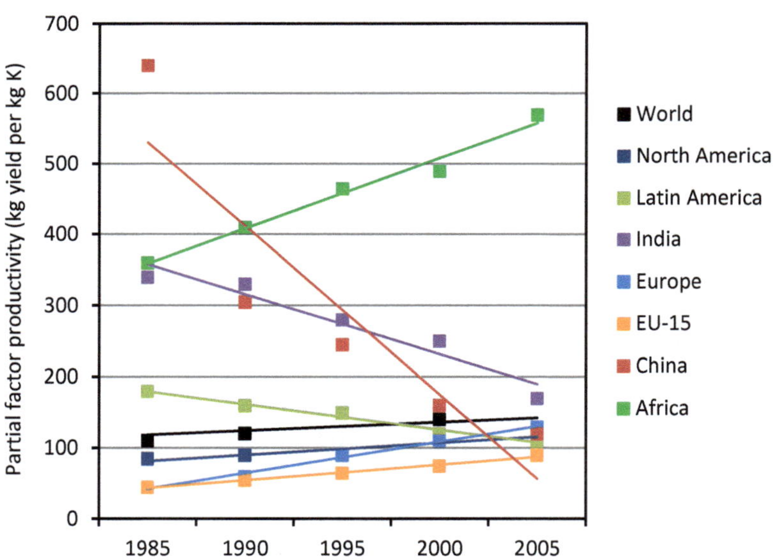

Fig. 6.5 Regional partial factor productivity for K over a 25-year (1983–2007) period (adapted from Fixen et al. 2015)

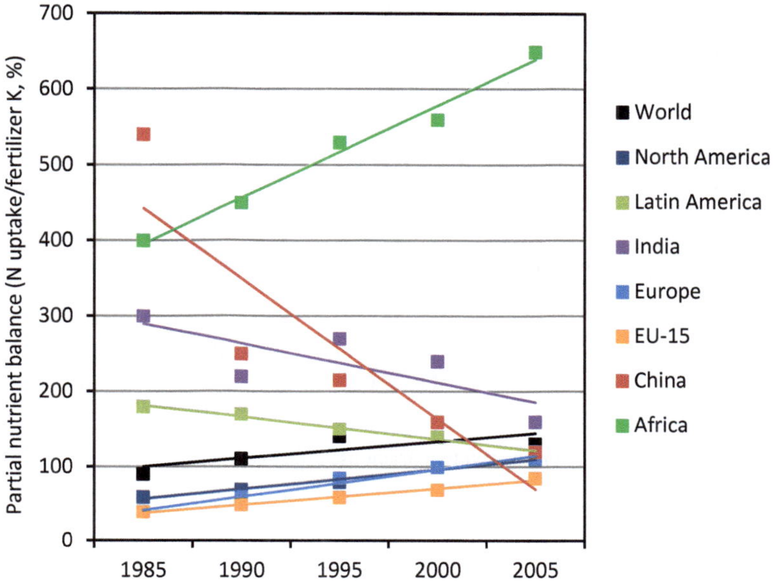

Fig. 6.6 Regional partial nutrient balance for K over a 25-year (1983–2007) period (adapted from Fixen et al. 2015)

with a PNB of 600% during 2003–2007 indicating that crops removed more than six times the amount of K that was applied as fertilizer. In general, PNB and PFP for Africa, North America, Europe and the EU-15 are trending upwards in K, while Latin America, India and China are trending downwards, just as was the case for N and P. The absence of forage crop production and K removal in these NUE estimates impacts some regions much more than others and should be kept in mind in comparing the absolute values of the expressions. Differences in temporal trends (slopes of the lines) are likely to be more reliable.

6.4 Factors Influence Nutrient Use Efficiency

Studying the basic mechanisms which influence the efficiency of applied nutrient through fertilizers is essential. Nutrient use efficiency in any cropping system is a function of capacity of soil to supply adequate levels of nutrients in plant available form and ability of plant to acquire, transport in roots and shoot and to remobilize to other parts of the plant. Therefore, NUE is governed by a complex set of factors including (a) soil properties, (b) crop characteristics, (c) climate and (d) management issues (Fig. 6.7).

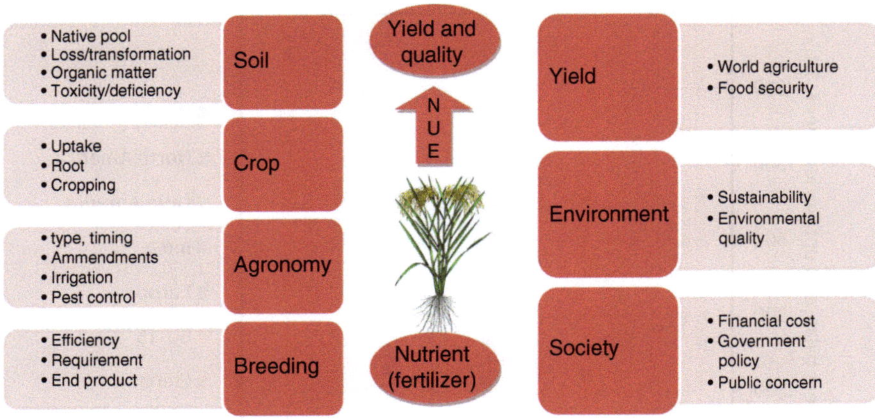

Fig. 6.7 Factors and issues for improving nutrient use efficiency

6.4.1 Soil Factors

Soil factors that largely influence NUE through nutrient transformation (adsorption/desorption, mineralization/immobilization) and their losses (leaching and volatilization) are texture, organic matter (OM) content, cation exchange capacity, pH, moisture, temperature, aeration, compaction, flora and fauna. Nutrient use efficiency and crop production in any agroecosystem are thus controlled by nutrient supplying capacity of soil which is governed by chemical, physical and biological qualities of the soil.

6.4.1.1 Nutrient Availability in Soil

Crop response to applied nutrient is dependent on the capacity of soil to supply that nutrient to meet the crop requirement. Chemical tests have been used over decades to estimate available nutrient status in soils to predict the probability of obtaining an economic response to applied nutrient. Soil testing for available nutrient helps in adjusting the quantity of fertilizer to be applied to crop and in improving NUE. Some nutrient pools not estimated by conventional soil tests may supply considerable amounts of nutrients during crop growth. In sandy soils, mica may supply adequate amount of K even though the soil tests indicate low available K. Crop responses to K on these soils are not expected in short period but may be recorded in long term.

Concentrations of heavy metals (e.g. Cd, Cr, Ni, Pb, Cu, Zn, As, Co and Mn) in some agricultural soils have been increasing due to the use of soil amendments, pesticides and other anthropogenic activities like indiscriminate industrialization and urbanization (Alloway 1995; Ha et al. 2014). These heavy metals, if present at excess levels, pose phytotoxicity and can reduce plant growth, nutrient uptake and eventually NUE (Marschner 1995; Baligar et al. 2001). The availability of these heavy metals to plants is influenced by soil pH, temperature, redox potentials, ligand-exchange mechanisms and composition of soil solution.

6.4.1.2 Nutrient Loss and Transformation

The amount of available nutrients estimated by soil tests may not be entirely available to crops because of their loss through leaching, volatilization, denitrification and transformation to unavailable forms. Leaching loss is important for NO_3–N, because it is not held by exchange sites in soil. Leaching loss of nutrients, in general, occurs in light-textured soil coupled with heavy rain. Leaching of Ca, K, Mg, B and SO_4^{2-} is common in acid soil. Volatilization of ammonia in high soil pH is considerable when urea is applied at the surface. Denitrification loss of N mainly occurs under submerged rice cultivation, particularly at higher temperature, and in the presence of easily decomposable OM. Transformation of available nutrients into unavailable forms is largely influenced by soil characteristics.

6.4.1.3 Soil Organic Matter

Soil OM increases NUE by improving physical, chemical and biological properties of soil, protecting surface soil from erosion and providing reservoir of plant nutrients. In tropics, the maintenance of soil OM is very difficult because of its rapid decomposition due to high temperature. Land preparation for conventional cultivation with ploughing disturbs the soil affecting the distribution and stability of soil aggregates. This facilitates oxidation of OM in soils and results in a decrease in its contents. In cultivated soils prevalent cropping system and associated agronomic practices influence the level at which organic C stabilize in a particular agroecosystem.

6.4.1.4 Chemical Constraints in Soil

Salinity, acidity and low OM content in soil are the major chemical constraints which result in deficiency/toxicity of essential nutrients. Acidic soils are characterized by the presence of phytotoxic levels of Al, Mn, Fe and H and deficient levels of N, P, K, Ca, Mg, Mo, B and Zn to support good plant growth (Patiram 2007). These factors are largely responsible for poor crop growth and lower NUE. Salinity reduces availability of N, P, K or essential ions due to high concentrations of Ca, reduces uptake of K and Ca because of high Na concentrations (in sodic soils), and reduces uptake of NO_3–N due to excess levels of SO_4^{2-} and Cl^- in the soil (Ogle et al. 2004). Deficiencies/toxicities of essential minerals can affect root length, thickness, surface areas, density, root hairs, root growth (in terms of biomass) and root to shoot ratios and in turn affect nutrient uptake and NUE.

6.4.1.5 Physical Conditions of Soil

Despite adequate nutrient supply, unfavourable physical conditions such as high bulk density, presence of hard pan in upper layers, poor structure and texture, surface sealing and crusting, high or low water holding capacity, water logging and extreme drying or poor aeration can lead to poor crop growth and also reduce NUE. These factors can affect mineralization and immobilization, adsorption and precipitation mechanisms, leaching, runoff and gaseous losses via denitrification and ammonification (Bhattacharyya et al. 2015). A greater influence of sodic soils on crop growth may be from the degradation of soil physical properties rather than from soil chemical properties. Physical deterioration of sodic soils can be either crusting or hardpan formation, which contributes to reduced permeability of roots, consequently decreasing water and nutrient uptake.

The basic requirements for crop growth in terms of physical soil conditions of soil are adequate soil moisture and aeration, optimum soil temperature and freedom from mechanical stress. Soil aeration is a vital component for active root growth and its function. Changes in the oxidative and reductive state of soil have an influence on the availability of plant nutrients. Similarly, soil temperature plays a vital role in the growth and function of plant roots and availability of plant nutrients. Soil aeration and temperature largely influence efficiency of applied nutrients by influencing plant, soil parameters and microbial activities. Soil moisture influences root growth, plant nutrient absorption and NUE. Nutrient absorption is influenced directly by soil moisture and indirectly by the effect of water on plant metabolic activities, soil aeration and concentration of nutrients in soil solution. If soil moisture becomes a limiting factor during the critical stage of crop growth, fertilizer application may adversely affect crop yield and nutrient recovery. The efficiency of applied nutrient depends on the rate of applied nutrient and the amount of water supply. These necessitate defining optimum nutrient requirement for crops in relation to available rainfall and irrigation water.

6.4.1.6 Biological Conditions

Rhizobia, diazotrophic bacteria and mycorrhizae in the rhizosphere are beneficial in improving root growth by fixing atmospheric N_2, suppressing pathogens, producing phytohormones to facilitate uptake of less mobile nutrients such as P and micronutrients and in mobilization/solubilization of unavailable nutrients. Ladha et al. (1996) reported that free-living and/or associated phototrophs and heterotrophs in irrigated rice can fix from 50 to 100 kg N ha^{-1}, contributing to the increased supply and efficiency of N. Arbuscular mycorrhizal fungi form a beneficial symbiosis with roots, thereby increases root surface area, which assists roots to explore larger soil volumes to bring more ions closer to the roots and contributing to higher nutrient inflow. Primary benefits of arbuscular mycorrhizae are enhanced acquisition of mineral nutrients; plant tolerance to soil acidity, salinity and alkalinity; and increased ability of plants to withstand or exclude elements toxic to plant growth. Other benefits for the host plants are the improvement in water uptake and the resistance to drought. Phosphorus uptake in plant is most affected by arbuscular mycorrhizal interaction, but it can also directly increase the uptake of Zn, Cu and N (Abbott and Robson 1984).

Weeds compete with crop plants for water, nutrients and sunlight, thereby reduce crop yield and consequently NUE. Allelopathic interactions of weed and crop plants are quite common. Infections of diseases and insects also reduce crop yield and NUE. Soil-borne pathogens such as actinomycetes, bacteria, fungi, nematodes and viruses present in the rhizosphere lead to pathogenic stress and bring changes in the morphology and physiology of roots and shoots that reduces plants' ability to absorb nutrients effectively. Diseases and insects that infect plant leaves reduce photosynthetic activity resulting in lower utilization of absorbed nutrients (Fageria 1992). Plant diseases are greatly influenced by environmental factors, including deficiencies and/or toxicities of essential elements. The severity of obligate and facultative parasites on plants is influenced by the level of N and P available to plants. Lack of Ca, Mg, Zn, B, Mn, Mo, Ni, Cu, Fe and Si is known to induce various diseases in plants (Graham and Webb 1991).

6.4.2 Crop Factors

6.4.2.1 Nutrient Uptake

Genetic variability of crop has been reported to cause the differences in NUE and nutrient uptake. Such differences in growth and NUE in plants have been related to differences in absorption, translocation, shoot demand, dry matter production per unit of nutrient absorbed and environmental interactions. Overall NUE in plants is governed by the flux of ions from the soil to the root surface and by the influx of ions into the roots followed by their transport to the shoots and remobilization to plant organs. Plants and microorganisms exude inorganic and organic substances that may alter soil pH as well as directly influence nutrient availability through transformation. Root exudates can also exert a direct influence on rhizosphere microorganisms. Plants and microorganisms compete for uptake of available nutrients in the rhizosphere. Most plant species can increase their capacity to access nutrients by altering root morphology (increasing surface area by growing long, thin roots with numerous, and long, root hairs) and by changing the capacity and/or affinity of plasma membrane-embedded transporters capable of carrying nutrients into the cytosol.

6.4.2.2 Root Characteristics

Roots are the principal organs for nutrient absorption from soil. The root morphological factors such as length, thickness, surface area and volume have profound effects on the plant's ability to acquire and absorb nutrients in soil. These parameters influence the ability of the roots to penetrate high-density soil layers and to tolerate temperature and moisture extremes and toxicities and deficiencies of elements. Additionally, the ability to modify the rhizosphere pH and the nutrient uptake kinetics is also affected by root morphology. Sauerbeck and Helal (1990) summarized root activities that affect nutrient availability in the rhizosphere as follows: (a) modification of rhizosphere pH; (b) exudation of organic acids, chelators, reductants and oxidants; (c) extracellular enzymes to turn over organically bound nutrients; and (d) providing substrate for microbial biomass.

6.4.2.3 Crop Rotation

The nature of cropping sequence greatly influences fertilizer requirement and its efficiency. Crops have differential feeding capacities on applied as well as native nutrients. The crops like maize and potato require high levels of fertilizer, and they may not fully utilize the applied fertilizer. Because of this some amount of nutrient is left in the soil as residual which can be utilized by the succeeding crop in the sequence. The magnitude of residual effect is dependent on the rate and kind of fertilizer, the cropping, the management regime followed and the soil conditions. Crops have a tendency for luxury consumption of N and K and may not leave any residual effect of excess fertilizer added. The reduced doses of fertilizer application crop may leave the soil in an exhausted condition, and the fertilizer requirement of the succeeding crop may increase. Legumes in the cropping sequence result N-rich root residues in the soil for the succeeding crop and thus reduce its N requirement.

6.4.3 *Climate Factors*

Temperature, solar radiation and precipitation during crop growth influence nutrient availability in soil and plants' ability to utilize the nutrients and produce yield. Soil temperature influences the rate of nutrient release from organic and inorganic reserves and the uptake by roots and subsequent translocation and utilization in plants. Solar radiation has a direct influence on photosynthesis, which in turn influences demand of nutrients for plant. The quality of radiation and crop shading reduces crop growth, N_2 fixation and ion uptake (Fageria 1992). Total rainfall is not as important for crop production and higher NUE as is the distribution of rainfall during the growing season and how fertilizers interact with the water balance at the root zone. To a larger extent, climatic variables cannot be changed, but cultivar selection and crop management must be tailored to prevailing climatic conditions. In a breeding program, it is vital to include physiological traits that improve the plants' ability to tolerate multiple climatic stress factors.

6.4.4 Management Factors

Changes in the soil nutrient reserve and alteration in root systems under different crop management systems might have direct bearing on the nutrient availability, uptake by crops and NUE. The recovery efficiencies of fertilizers are also influenced by amendments such as lime and OM due to their effects in nutrient dynamics. Best management practices such as source, rate, method of application and split application of nutrients should be optimized based on soil, plant and climatic factors to reduce nutrient losses due to leaching, denitrification, ammonia volatilization, runoff and fixation. Minimum tillage, no tillage, conservation tillage and traditional tillage can bring profound changes in soil quality, soil OM and nutrient availability in soils. Rooting pattern, water holding capacity, water infiltration, aeration, soil compaction and soil temperature are also influenced by the type of tillage practices. Crop rotation, use of cover crops and green manure crops are known to improve soil fertility.

6.5 Approaches for Increasing Nutrient Use Efficiency

Increasing NUE requires a diverse set of approaches from agronomy to molecular plant breeding. This includes germplasm screening, transgenic and genetic approaches and a consideration of abiotic and biotic stresses. Improvement in NUE thus requires the simultaneous consideration of multiple aspects of crop performance, involving a multidisciplinary approach.

6.5.1 Agronomic Approach

Optimization of agronomic managements for a crop under any agroecosystems is prerequisite for improving NUE.

6.5.1.1 Soil Amelioration

Addition of OM to soil maintains good soil physical, chemical and biological conditions which largely governs NUE. The SOM helps to maintain good aggregation and increase water holding capacity and exchangeable K, Ca and Mg. It also reduces P fixation and leaching of nutrients and decreases toxicities of Al and Mn. Best management practices such as addition of crop residues, green manure, compost, animal manure, use of cover crops, reduced tillage and avoiding burning of crop residues can significantly improve the level of soil OM and contribute to the sustainability of the cropping systems and higher NUE. Results from long-term fertilizer experiments have shown that the integration of organic manures with chemical fertilizers can maintain high productivity and organic C status in soil (Mandal et al. 2007). There is a significant increase in soil organic C due to incorporation of rice or wheat straw into the soil instead of removing or burning it.

Liming is an effective way to correct soil chemical constraints. It improves the availability of Ca, Mg, Mo, P, soil structure and cation exchange capacity. The fixation of atmospheric N_2 by free-living and symbiotic organisms like rhizobium is increased. Potential toxicity of Al and Mn is reduced. Lime has very low mobility in soil, and when surface applied it does not reduce the acidity of sub-surface soil horizons. Contrary to lime, gypsum ($CaSO_4$) has a greater downward movement, and when applied to the surface, it can still impact and reduce the acidity of the subsoil (Ritchey et al. 1980). Downward movement of Ca in soil has resulted in increased rooting depth and in higher uptake rates of N, Ca, Cu, P and Mn by corn (*Zea mays* L) grown in Cerrado acid Oxisol of Brazil (Sousa et al. 1992). Reduction of subsoil acidity usually leads to deeper rooting and higher water and mineral uptake by plants.

6.5.1.2 Improving Physical Conditions of Soil

Tillage, mulching, irrigation, addition of OM and amendments like gypsum in sodic soil are major management techniques that create suitable physical condition for crop growth. Changes in the soil nutrient reserve and alteration in root systems under different tillage systems might have direct bearing on the nutrient availability and its use efficiency for crops. Tillage practices such as conventional, conservation and no-tillage can make changes in soil OM, nutrient concentrations, bulk density, water holding capacity, soil temperature and soil aeration. Greater availability of P, Ca, K and N has been reported under no-tillage than those under conventional tillage (Ismail et al. 1994; Mahboubi et al. 1993). Minimum tillage enhances root growth for barley (*Hordeum vulgare* L) and oat (*Avena sativa* L) cropping systems (Ehlers et al. 1983; Ellis et al. 1977). Minimum tillage has also been reported to increase root weight, length and density, increasing the nutrient and water use efficiencies (Adkinson 1990). Such improved root parameters contributed to higher yields and uptake efficiencies of N, P, Ca, S, Cu, Fe and Zn. Crop rotation and use of cover crops and green manure crops are known to improve soil fertility and physical properties and to minimize pest and weed problems (Delgado et al. 1999).

6.5.1.3 Fertilizer Management

While applying fertilizers selection of the *right source of nutrients* for application at the *right rate*, at the *right time* and in the *right place* is important for achieving adequate nutrient use efficiency. Fertilizer source, rate, timing and placement are interdependent and are interlinked with other agronomic management practices involved in a crop production. Interactions among nutrient source, rate, time and place are also considered in managing fertilizers for controlling soil fertility.

6.5.1.3.1 Right Source of Fertilizers

Any nutrient applied through fertilizers must be in plant available forms or in a form that changes into a plant available form in the soil. The source of fertilizer should be selected based on (a) soil physical and chemical properties (e.g. application NO_3–N to flooded rice soils or broad casting urea on soil with high pH should be avoided), (b) synergisms among nutrient elements and sources should be understood (e.g. P × Zn interaction, liming vs. P, Zn and B availability, fertilizer complementing manure), (c) blending compatibility of fertilizer materials must be recognized because certain combinations of fertilizer sources absorb, (d) granule size should be similar to avoid product segregation, (e) benefits and sensitivities to nutrient element in fertilizers also need to be appropriately assessed, (f) accompanying ion/element in the fertilizer along with target nutrient element (e.g. chloride ions accompanying K in muriate of potash (KCl) are beneficial to maize but can be detrimental to tobacco and some fruits; sources of P fertilizer may contain plant available Ca and S) and (g) effects of nonnutritive elements (e.g. natural deposits of rock phosphate are enriched in several metals like Cd).

Since nitrates are easily leached and lost by denitrification, retardation of nitrification of ammonium-containing or ammonium-producing fertilizers by using nitrification inhibitors in submerged rice field helps in increasing N use efficiency. Chemical nitrification inhibitors that are widely used are N-serve [2-chloro-6-(trichloromethyl) pyridine], AM [2-amino-4-chloro-6-methylpyrimidine], DCD (dicyandiamide), 4-amino-1,2,4-6-triazole-HCl (ATC), 3,4-dimethylpyrazole phosphate (DMPP), carbon disulphide (CS2) and ammonium thiosulphate [$(NH_4)_2S_2O_3$]. Neem (*Azadirachta indica*) cake-coated urea was shown to have nitrification inhibiting properties. Application of PPD (phenyl phosphorodiamidate) and NBPT [*N*-(*n*-butyl) thiophosphoric triamide] with urea has been suggested to reduce the rate of urea hydrolysis and improve its efficiency (Prasad and Power 1995). Nitrification inhibitor DCD has the potential to reduce N_2O emissions and increase NUE of irrigated systems of barley (Delgado and Mosier 1996). Slow- and controlled-release fertilizers have added advantages in increasing nutrient recovery by plants, lowering N_2O and NH_3 emissions and NO_3-N leaching from cropping systems. Slow-release N fertilizers such as isobutylidenediurea (IBDU), crotonaldehydediurea (CDU) and polyolefin-coated urea (POCU) are currently available in the market. Further research and development is needed to continue developing new products that can increase the recovery of fertilizers while maintaining and or increasing yields and protecting the environment.

6.5.1.3.2 Right Rate of Fertilizers

Scientific methods are involved to evaluate soil nutrient supply to crops to work out fertilizer application rates. Soil and plant analysis and crop response studies are useful tools in this regard. Quantity and plant availability of nutrients in all available indigenous nutrient sources such as manure, composts, biosolids, crop residues and irrigation water need to be determined. In estimating fertilizer application rates, consideration of NUE is important because some loss of nutrients from soil-plant system is unavoidable. If the removal of nutrients from a cropping system exceeds applied amount for years together, loss of soil fertility may occur. This aspect also needs to be considered while working out fertilizer application rates. For nutrients unlikely to be retained in the soil, the most economic rate of application is where the last unit of nutrient applied is equal in value to the increase in crop yield it generates (law of diminishing returns). For nutrients retained in the soil, their residual effect to future crops should be considered. Assessment of probabilities of predicting economically optimum rates is crucial.

6.5.1.3.3 Right Timing of Fertilizers

Fertilizers should be applied to match the crop nutrient demand, which depends on the date of planting, crop growth characteristics and sensitivity to nutrient deficiencies at particular growth stages. For example, to increase B use efficiency, the time of supplying B to the crops should synchronise with the stages of their growth when the requirements for B are most critically important (Sarkar et al. 2007). Such need for B varies for different crops and their stages of growth. In the case of grain crops like cereals and oilseeds, the supply of B is more important during reproductive development rather than during the vegetative growth stage. Whereas in the case of root crops like potato, tapioca (*Manihot esculenta* C.) and sugar beet (*Beta vulgaris* L.) where translocation of photosynthates from source to sink is needed for a longer period, a steady and prolonged supply of B throughout their growth period may be necessary. Application of nutrient late in the cropping season may be associated with its accumulation in the economic produce of the crop (e.g. grain) rather than in the leaves or stalk. Therefore, nutrient application late in the cropping season may increase both quality and quantity of grain yield. For example, top dressed of N at tillering stage of wheat has been found to increase both yield and protein content of wheat, especially at low levels of soil N.

Fertilizer timing is also governed by dynamics of nutrient supply in soil. Mineralization of soil OM makes some nutrient elements available to crops, but if mineralization does not occur during the critical crop growth stages requiring that nutrient, deficiencies may limit productivity. Similarly timing of weather factors influencing nutrient loss will govern fertilizer timing (e.g. leaching losses tend to be more frequent during monsoon season). Logistic of field operations will also influence the timing of fertilizer application (e.g. multiple applications of nutrients may or may not combine with those of crop protection products, and nutrient applications should not delay time-sensitive operations such as planting).

6.5.1.3.4 Placement of Fertilizers

Placement of fertilizer needs to ensure that nutrients are adequately intercepted by roots. An example is the band placement of P fertilizer for wheat, ensuring sufficient nutrition of the young seedling, increasing yield substantially, even though amounts applied and taken up are small. Fertilizer placement should take into account that soils vary in nutrient supplying capacity and nutrient loss potential. Logistic of soil preparation also needs to be recognized, and fertilizer placement should fit the needs of tillage system. In conservation tillage systems, subsurface fertilizer applications need to ensure that soil coverage by crop residues is maintained. The efficiency of the applied fertilizers can be improved considerably if the rooting habits of various plants during early growth stages are known. This is particularly true for immobile nutrients. If a plant produces taproot system early, fertilizer can best be placed directly below the seed. On the other hand, if lateral roots are formed early, side dressing of fertilizer would be efficient. Mycorrhizal fungi often associated with plant roots increase the ability of plants to absorb nutrients particularly under low soil fertility. In tropical regions the major soil problems in rain-fed systems that affect crop production are low soil fertility, salinity, alkalinity, acidity, Fe toxicity and P and Zn deficiencies. Nutrient use efficiency in these soils is improved by fertilizer placement and timing to critical crop growth stages (Baligar et al. 2001).

6.5.1.4 Integrated Nutrient Management

Integrated plant nutrient management (IPNM) is another approach to enhance NUE. The basic concept of IPNM is the maintenance of soil fertility and health for sustained crop production on long-term basis and use of chemical fertilizers along with different organic sources available at the farm to meet nutrient requirement. Consequently, the major focus is on the management of soil OM and plant nutrients through integrated use of mineral fertilizers, green manures, sewage sludge and food industry wastes. In the case of legumes as a green manure or a dual-purpose crop (grain + green manure) such as mung bean (*Vigna radiata*) and cowpea (*Vigna unguiculata*) residue left after one picking of pods applied to rice can contribute 40–120 kg N ha^{-1} in rice-wheat system and thus reduce fertilizer N application to crops. Legume residues decompose fairly fast under subtropical and tropical conditions especially under submerged conditions and make N readily available to the rice crop. Similarly, application of different organic manures can supply 40–120 kg N ha^{-1} in different cereal-based cropping systems. Apart from soil, climate and cropping, N supply from organic sources depends on their C:N ratio and rate of application.

6.5.1.5 Site-Specific Nutrient Management

Site-specific nutrient management is a dynamic field specific to nutrient management in particular cropping season to optimize the supply and demand of nutrient based on its cycling through soil-plant system. Dobermann et al. (2002) reported that field-specific management of macronutrients increased nutrient uptake, yield and NUE in irrigated rice farms in Asia. Site-specific technology in the future might help to develop sound management systems that lead to reduced fertilizer inputs, thereby improving efficiency of costly fertilizer input and the degradation of the environment.

6.5.2 Crop Improvement Approach

6.5.2.1 Breeding

Most of the breeding programmes focussed on yield improvements and protection of that yield with resistance to biotic and abiotic constraints. Nutrient use efficiency itself has seldom been a target. However, increased yield without increase in applied nutrient will result in increased NUE; therefore, the target of increasing yield has resulted in increased NUE. When nutrient supply from soil is suboptimal (e.g. acid- and salt-affected soils), the efficiency with which mineral nutrients are used by plants is important in overall NUE. Breeding programmes should consider plant characteristics such as the ability to produce near maximum yields at low nutrient levels and extensive root systems efficient in exploring large soil volumes to produce cultivars with high NUE that can contribute to sustainability and environmental protection. Conventional plant breeding has enhanced N use efficiency in rice cultivars (Fischer 1998), but new methods are needed that can advance how specific traits are identified and pass from one cultivar to the other or from one species to another.

Among macronutrients, breeding for adaptation to low-nutrient environments has advanced mainly for P and to some extent N and K. In the case of micronutrients, some breeding efforts have been reported for Fe, Mn, Zn, Cu and B (Rengel 2005). Breeding for such nutrient-efficient genotypes adapted to low-input agricultural ecosystems should be prioritized. Growing nutrient-efficient crop genotypes on soils of low nutrient availability represents an environmental friendly approach that would reduce land degradation by minimizing application of fertilizers to crop production (Rengel 2001). Such novel approach would reduce the danger of exhaustion of soil nutrient resources through 'land mining' at least for P and micronutrients because the total soil supply of these elements is sufficient for hundreds of years of sustainable cropping by new, efficient genotypes that can acquire the soil nutrient pools not available to inefficient genotypes.

6.5.2.2 Genetic Control

Targeted manipulation of key genes controlling NUE would be an ideal approach, and considerable effort has been expended in identifying and testing possible key steps improving NUE. For NO_3 assimilation attention has focussed on primary assimilation, for example, nitrate reductase and glutamine synthetase (McAllister et al. 2012). Another key enzyme showing promise is alanine amino transferase. This enzyme catalyses the transfer of an amino group from glutamate to oxaloacetate and has been shown to increase biomass accumulation at suboptimal N inputs when overexpressed in a tissue-specific manner in *Brassica napus*, thus improving NUE (Good et al. 2007).

Because of limiting physiological processes, application of ever-increasing fertilizer to crop does not necessarily result in a linear increase in yield or uptake of nutrients. Targeting these limiting processes is key areas for crop improvement for yield and NUE (Gaju et al. 2011). The relative importance of these processes varies during crop development. To increase NUE through genetic approach, the primary targets are root, canopy and grain (Hawkesford 2012).

6.5.2.2.1 Root

Target traits are early root proliferation, shallow root proliferation to capture applied fertilizer and deep roots and their proliferation to access nutrient reserves at lower depth. Cluster roots (Neumann and Martinoia 2002) are designed to facilitate local soil acidification and to enable access to soil nutrient pools not normally plant available, particularly phosphate. Efficient root systems also require efficient mechanisms for uptake of nutrients from soil. Uptake from the soil depends upon cell membrane ion transporters, production of exudates or extracellular compounds to facilitate nutrient uptake. Transporters are involved within the plant to optimize translocation or storage of nutrients in roots, shoots and storage organs. Direct breeding for root characteristics is extremely difficult and has seldom been attempted. Laboratory screens are usually limited to seedlings, often in the absence of soil, and field screens involve cumbersome excavations or invasive and laborious coring techniques.

A study on P use efficiency in *Brassica* indicated high efficiency in current commercial varieties, most likely principally due to breeding for yield; however, P use efficiency and root trait QTLs were found to be associated in a mapping population (Hammond et al. 2009).

6.5.2.2.2 Canopy

Nutrient uptake during vegetative development is required to make and maintain a photosynthetically active canopy, a prerequisite for yield optimization. Early uptake of nutrients, especially N, helps canopy development and eventual closure. Availability of N will determine the canopy size. Maximal light capture produces photosynthate destined for storage organs such as the grain. As crop matures, the canopy often senesces, decreasing photosynthetic capacity but facilitating remobilization of minerals, including N, to grain. Breeding targets may be early canopy closure, a closed canopy with a lower need for N, a canopy with improved photosynthetic activity (Parry et al. 2011) or extended canopy longevity. Delaying canopy senescence during storage organ generation (e.g. grain filling) may be beneficial for higher yield but will impact negatively on the remobilization processes, and sink tissues may suffer from nutrient deficiency.

6.5.2.2.3 Grain

Although adequate nutrition is required for optimum grain yield, both to produce photosynthesising canopy and then to support sink organ growth, a large sink will also have a higher mineral nutrient demand. This demand is driven by the requirements for sink synthesis but is also required for nutritional quality (mineral nutrient content). However, decreased mineral nutrient content in modern wheat grain compared to older varieties is the fact due to increased grain yield but with reduced redistribution of mineral to the grain (Fan et al. 2008). These necessitate urgent need for improved mineral nutrient partitioning in grain, particularly minerals important for human and animal diets (e.g. Fe, Zn and Se). Future breeding programmes thus need to consider acquisition of nutrients by crop and their redistribution to grain in addition to crop yield.

6.5.2.3 Selection of Genotypes

Plant genotypes differ in the mechanisms for acquisition of nutrients from environments with low nutrient availability. Genotypes with higher NUE may have an increased capacity (a) to exploit the soil (large root surface area), (b) to convert non-available nutrient forms into available forms and/or (c) to take up nutrients across the plasma membrane. Therefore, the key approach for breeding crops for high NUE is to identify and exploit variation in existing germplasm. Recent screening experiments indicated considerable variation in key NUE traits, for example, in wheat (Gaju et al. 2011) and barley (Bingham et al. 2012), and for P use efficiency (Hammond et al. 2009). Use of mapping populations and QTL analysis enables the dissection of sub-traits contributing to the overall complex traits such as NUE. Yield, nutrient uptake and NUE all have multiple identifiable QTL, and some of these overlap with QTL contributing to other traits such as anthesis date and plant height

(Bogard et al. 2011). Where overlaps exist, this indicates a link between the traits and possibly the same gene or allele contributing to that complex quantitative trait. For example, in a study on wheat, 233 QTLs were identified for traits related to NUE and yield (Laperche et al. 2007). De-convoluting these complex multigene traits is a need of the hour, which will facilitate breeding for higher NUE.

6.6 Conclusions

Increasing NUE in plants is important for enhancing yield and quality of crops, reducing fertilizer input cost and environmental sustainability. The NUE in plants should be clearly defined and carefully selected to reflect the end use. Plant species and cultivars within species differ in absorption and utilization of nutrients, and such differences are attributed to morphological, physiological and biochemical processes in plants and their interaction with climate, soil, fertilizer and management practices. An improved NUE in plants can be achieved by careful manipulation of plant, soil, fertilizer, environmental and management practices. There is a great need for a multidisciplinary effort of plant breeders and physiologists, agronomists and soil scientists to formulate an effective way to increase NUE in plants.

References

Abbott, L. K., & Robson, A. D. (1984). The effect of mycorrhizae on plant growth. In C. L. Powell & D. J. Bagyaraj (Eds.), *A mycorrhizae* (pp. 113–130). Boca Raton, FL: CRC Press.

Adkinson, D. (1990). Influence of root system morphology and development on the need for fertilizer and efficiency of use. In V. C. Baligar & R. R. Duncan (Eds.), *Crops as enhancers of nutrient use* (pp. 411–451). San Diego, CA: Academic Press.

Alloway, B. J. (Ed.). (1995). *Heavy metals in soils* (2nd ed.). New York: Wiley.

Baligar, V., & Bennet, O. (1986). Outlook on fertilizer use efficiency in the tropics. *Fertilizer Research, 10*, 83–96.

Baligar, V. C., Fageria, N. K., & He, Z. L. (2001). Nutrient use efficiency in plants. *Communications in Soil Science and Plant Analysis, 32*, 921–950.

Bhattacharyya, R., Ghosh, B. N., Mishra, P. K., Mandal, B., Rao, C. S., Sarkar, D., et al. (2015). Soil degradation in India: Challenges and potential. *Sustainability, 7*, 3528–3570.

Bingham, I. J., Karley, A. J., White, P. J., Thomas, W. T. B., & Russell, J. R. (2012). Analysis of improvements in nitrogen use efficiency associated with 75 years of spring barley breeding. *European Journal of Agronomy, 42*, 49–58.

Bogard, M., Jourdan, M., Allard, V., Martre, P., Perretant, M. R., Ravel, C., et al. (2011). Anthesis date mainly explained correlations between post-anthesis leaf senescence, grain yield, and grain protein concentration in a winter wheat population segregating for flowering time QTLs. *Journal of Experimental Botany, 62*, 3621–3636.

Cassman, K. G., Dobermann, A., & Walters, D. T. (2002). Agroecosystems, nitrogen-use efficiency, and nitrogen management. *Ambio, 31*, 132–140.

Delgado, J. A., & Mosier, A. R. (1996). Mitigation alternatives to decrease nitrous oxides emissions and urea-nitrogen loss and their effect on methane flux. *Journal of Environmental Quality, 25*, 1105–1111.

Delgado, J. A., Sparks, R. T., Follett, R. F., Sharkoff, J. L., & Riggenbach, R. R. (1999). Use of winter cover crops to conserve soil and water quality in the San Luis Valley of south Central Colorado. In R. Lal (Ed.), *Soil quality and soil erosion* (pp. 125–142). Boca Raton, FL: CRC Press.

Dobermann, A. (2007). Nutrient use efficiency—Measurement and management. In *IFA international workshop on fertilizer best management practices* (pp. 1–28). Brussels, Belgium: IFA.

Dobermann, A., & Cassman, K. G. (2005). Cereal area and nitrogen use efficiency are drivers of future nitrogen fertilizer consumption. *Sci China., 48*, 745–758.

Dobermann, A., Witt, C., Dawe, D., Abdulrachman, S., Gines, H. C., Nagarajan, R., et al. (2002). Site-specific nutrient management for intensive rice cropping systems in Asia. *Field Crops Research, 74*, 37–66.

Ehlers, W., Kope, U., Hess, F., & Bohm, W. (1983). Penetration resistance and root growth of oats in tilled and untilled loess soil. *Soil and Tillage Research, 3*, 261–275.

Ellis, P. B., Elliott, J. G., Barness, B. T., & Howse, K. R. (1977). Comparison of direct drilling, reduced cultivation and ploughing on the growth of cereals. 2 spring barley on a sandy loam soil: Soil physical conditions and root growth. *The Journal of Agricultural Science, 89*, 631–642.

Fageria, N. K. (1992). *Maximizing crop yields*. New York: Marcel Dekker.

Fan, M.-S., Zhao, F.-J., Fairweather-Tait, S. J., Poulton, P. R., Dunham, S. J., McGrath, S. P., et al. (2008). Evidence of decreasing mineral density in wheat grain over the last 160 years. *Journal of Trace Elements in Medicine and Biology, 22*, 315–324.

FAO. (2012). FAOSTAT. FAO Statistics Division. http://faostat3.fao.org.

FAO (Food and Agriculture Organization of the United Nations). (2009). FAOSTAT. FAO Statistics Division. http://faostat3.fao.org.

Fischer, K. S. (1998). Toward increasing nutrient use efficiency in rice cropping systems: The next generation of technology. *Field Crops Research, 56*, 1–6.

Fixen, P., Brentrup, F., Bruulsema, T. W., Garcia, F., Norton, R., & Zingore, S. (2015). Nutrient/fertilizer use efficiency: Measurement, current situation and trends. In P. Drechsel, P. Heffer, H. Magen, R. Mikkelsen, & D. Wichelns (Eds.), *Managing water and fertilizer for sustainable agricultural intensification* (1st ed., pp. 8–38). Paris, France: IFA, IWMI, IPNI and IPI.

Gaju, O., Allard, V., Martre, P., Snape, J. W., Heumez, E., LeGouis, J., et al. (2011). Identification of traits to improve the nitrogen-use efficiency of wheat genotypes. *Field Crops Research, 123*, 139–152.

Good, A. G., Johnson, S. J., De Pauw, M., Carroll, R. T., Savidov, N., Vidmar, J., et al. (2007). Engineering nitrogen use efficiency with alanine aminotransferase. *Canadian Journal of Botany, 85*, 252–262.

Gourley, C. J. P., Allan, D. L., & Russele, M. P. (1993). Defining phosphorus efficiency in plants. In N. J. Barrow (Ed.), *Plant nutrition—From genetic engineering to field practice* (pp. 363–366). Dordrecht, Netherlands: Kluwer Academic Publishers.

Graham, R. D., & Webb, M. J. (1991). Micronutrients and disease resistance and tolerance in plants. In J. J. Mortvadt, F. R. Cox, L. M. Shuman, & R. M. Welch (Eds.), *Micronutrients in agriculture, Soil Science Society American book series* (Vol. 4, 2nd ed., pp. 329–370). Madison, WI: Soil Science Society America. Inc..

Ha, H., Olson, J. R., Bian, L., & Rogerson, P. A. (2014). Analysis of heavy metal sources in soil using Kriging interpolation on principal components. *Environmental Science & Technology, 48*, 4999–5007.

Hammond, J. P., Broadley, M. R., White, P. J., King, G. J., Bowen, H. C., Hayden, R., et al. (2009). Shoot yield drives phosphorus use efficiency in *Brassica oleracea* and correlates with root architecture traits. *Journal of Experimental Botany, 60*, 1953–1968.

Hawkesford, M. J. (2012). Improving nutrient use efficiency in crops. In *eLS*. Chichester: John Wiley & Sons, Ltd. doi:10.1002/9780470015902.a0023734.

Howarth, R. W., Boyer, E. W., Pabich, W. J., & Galloway, J. N. (2002). Nitrogen use in the United States from 1961-2000 and potential future trends. *Ambio, 31*, 88–96.

Huggins, D. R., & Pan, W. L. (1993). Nitrogen efficiency component analysis - an evaluation of cropping system differences in productivity. *Agronomy Journal, 85*, 898–905.

IFA (International Fertilizer Industry Association). (2012). IFA statistics. http://www.fertilizer.org/Statistics.

Ismail, I., Blevins, R. L., & Frye, W. W. (1994). Long-term no tillage effects on soil properties and continuous corn yields. *Soil Science Society of America Journal, 58*, 193–198.

Janzen, H. H., Beauchemin, K. A., Bruinsma, Y., Cambell, C. A., Desjardins, R. L., Ellert, B. H., et al. (2003). Th, E.G. the fate of nitrogen in agroecosystems: An illustration using Canadian estimates. *Nutrient Cycling in Agroecosystems, 67*, 85–102.

Jiyun, J. (2012). Changes in the efficiency of fertilizer use in China. *Journal of the Science of Food and Agriculture, 92*, 1006–1009.

Ladha, J. K., Kundu, D. K., Angelo-van Coppenolle, M. G., Peoples, M. B., Carangal, V. R., & Dart, P. J. (1996). Legume productivity and soil nitrogen dynamics in lowland rice-based cropping systems. *Soil Science Society of America Journal, 60*, 183–192.

Ladha, J. K., Pathak, H., Krupnick, T. J., Six, J., & van Kessel, C. (2005). Efficiency of fertilizer nitrogen in cereal production: Retrospects and prospects. *Advances in Agronomy, 87*, 85–156.

Laperche, A., Brancourt-Hulmel, M., Heumez, E., Gardet, O., Hanocq, E., Devienne-Barret, F., et al. (2007). Using genotype x nitrogen interaction variables to evaluate the QTL involved in wheat tolerance to nitrogen constraints. *Theoretical and Applied Genetics, 115*, 399–415.

Mahboubi, A. A., Lal, R., & Faussey, N. R. (1993). Twenty-eight years of tillage effects on two soils in Ohio. *Soil Science Society of America Journal, 57*, 506–512.

Mandal, B., Majumder, B., Bandyopadhyay, P. K., Hazra, G. C., Gangopadhyay, A., Samantaray, R. N., et al. (2007). The potential of cropping systems and soil amendments for carbon sequestration in soils under long-term experiments in subtropical India. *Global Change Biology, 13*, 357–369.

Marschner, H. (1995). *Mineral nutrition of higher plants*. San Diego, CA: Academic Press.

McAllister, C. H., Beatty, P. H., & Good, A. G. (2012). Engineering nitrogen use efficient crop plants: The current status. *Plant Biotechnology Journal, 10*, 1011–1025.

Neumann, G., & Martinoia, E. (2002). Cluster roots—An underground adaptation for survival in extreme environments. *Trends in Plant Science, 7*, 162–167.

Ogle, D., Majerus, M., & St. John, L. (2004). Plants for saline to sodic soil conditions. http://www.plant-materials.nrcs.usda.gov/pubs/idpmstn5465.pdf.

Parry, M. A. J. P., Reynolds, M., Salvucci, M. E., Raines, C., Andralojc, P. J., Zhu, X., et al. (2011). Raising yield potential of wheat. II. Increasing photosynthetic capacity and efficiency. *Journal of Experimental Botany, 62*, 453–468.

Patiram. (2007). Management and future research strategies for enhancing productivity of crops on the acid soils. *Journal of the Indian Society of Soil Science, 55*, 411–420.

Prasad, R., & Power, J. F. (1995). Nitrification inhibitors for agriculture, health and environment. *Advances in Agronomy, 54*, 233–281.

Rengel, Z. (2001). Genotypic differences in micronutrient use efficiency in crops. *Communications in Soil Science and Plant Analysis, 32*, 1163–1186.

Rengel, Z. (2005). Breeding crops for adaptation to environments with low nutrient availability. In M. Ashraf & P. J. C. Harris (Eds.), *Abiotic stresses: Plant resistance through breeding and molecular approaches* (pp. 239–276). New York: The Haworth Press.

Ritchey, K. D., Sousa, D. M. G., Laboto, E., & Correa, O. (1980). Calcium leaching to increase rooting depth in a Brazilian savannah Oxisols. *Agronomy Journal, 72*, 40–44.

Sarkar, D., Mandal, B., & Kundu, M. C. (2007). Increasing use efficiency of boron fertilizers by rescheduling the time and methods of application for crops in India. *Plant and Soil, 301*, 77–85.

Sauerbeck, D. R., & Helal, H. M. (1990). Factors affecting the nutrient efficiency in plants. In N. E. Balsam, M. Dambroth, & B. C. Loughman (Eds.), *Genetic aspects of plant mineral nutrition* (pp. 11–17). Dordrecht, Netherlands: Kluwer Academic Publisher.

Sheldrick, W. F., Syers, J. K., & Lindgard, J. A. (2002). Conceptual model for conducting nutrient audits at the national, regional, and global scales. *Nutrient Cycling in Agroecosystems, 62*, 61–72.

Smaling, E. M., Nandwa, S. M., & Janssen, B. H. (1997). Soil fertility in Africa is at stake. In P. Sanchez & R. Buresh (Eds.), *Replenishing soil fertility in Africa: Special publication* (Vol. 51, pp. 47–62). Madison, WI: Soil Science Society of America.

Smil, V. (1999). Nitrogen in crop production: An account of global flows. *Global Biogeochemical Cycles, 13*, 647–662.

Sousa, D. M. G., Lobato, E., Ritchey, K. D., & Rein, T. A. (1992). Response of annual crops and leucaena to gypsum in the Cerrado. In *2nd seminar on the use of gypsum in agriculture* (Vol. 2, pp. 277–306). Brazil: IBRAFOS.

Syers, J. K., Johnson, A. E., & Curtin, D. (2008). *Efficiency of soil and fertilizer phosphorus use: Reconciling changing concepts of soil phosphorus behaviour with agronomic information, FAO-fertilizer and plant nutrition bulletins* (Vol. 18). Rome: Food and Agriculture Organization of the United Nations.

Chapter 7
Understanding the Dynamics of Phosphorus Starvation and Plant Growth

Tariq Ahmad Dar, Moin Uddin, Akbar Ali, M. Masroor A. Khan, and Tanvir ul Hassan Dar

Abstract Phosphorus is one of the essential macronutrients required in relatively large quantities by the plants for normal growth and development and to complete their life cycle. It is an important constituent of biomolecules like nucleic acids, phospholipids, enzymes and adenosine triphosphate. Phosphate signaling allows higher plants to respond and adapt to the phosphate-deficient conditions efficiently. Phosphorus deficiency in the soil produces responses and adaptive changes in the plants like changes in root morphology and architecture, improved uptake and utilization of P, metabolic changes, exudation of organic acids, and numerous enzymes for the solubilization of the inorganic and organic reserves of P in the rhizosphere (phosphate mobilization). Therefore, the understanding of the proper mechanisms of the adaptation of plants to low P availability will help in the selection and breeding to improve productivity under P-limited environments. The present review gives an overview of the plant responses to P-limited environments and the developments made so far in this area of study.

Keywords Plant development • Root architecture • Phosphate mobilization and transporters • Phosphate signalling

---

T.A. Dar (✉) • T. ul Hassan Dar
Centre for Biodiversity Studies, School of Biosciences and Biotechnology, Baba Ghulam Shah Badshah University, Rajouri 185234, India
e-mail: dartariq21@gmail.com

M. Uddin
Botany Section, Women's College, Aligarh Muslim University, Aligarh 202002, India

A. Ali
Department of Botany, Dayalbagh Educational Institute, Agra 282005, India

Plant Physiology Section, Department of Botany, Aligarh Muslim University, Aligarh 202002, India

M.M.A. Khan
Plant Physiology Section, Department of Botany, Aligarh Muslim University, Aligarh 202002, India

M. Naeem et al. (eds.), *Essential Plant Nutrients*,
DOI 10.1007/978-3-319-58841-4_7

7.1 Introduction

Phosphorus is one of the essential macronutrients required for normal plant growth and development constituting about 0.2% of plant dry matter (Harrison et al. 2002). It plays several essential roles in plants and is the key constituent of important biomolecules like nucleic acids, phospholipids, enzymes, and adenosine triphosphate (ATP) (Marschner 2012). Plants obtain phosphorus from the soil in the form of inorganic phosphate ($H_2PO_4^-$) (Vance et al. 2003). Although the amount of phosphorus present in the soil may be large, it is fixed into other forms which are not taken by the plants ultimately resulting in phosphorus deficiency. Hence, P is one of the most unavailable and unapproachable macronutrients in the soil (Vance et al. 2003) and often limits plant growth. Therefore different types of phosphate fertilizers are added to the soil to overcome P deficiency.

7.2 Response of Plants to Phosphorus Deficiency

To overcome the phosphorus deficiency, plants adapt different mechanisms to improve the acquisition of phosphorus from the soil (Fig. 7.1).

7.2.1 Changes in Root System

The root design plays an important role in inorganic phosphorus (Pi) acquisition. The root systems with more surface area are able to explore a given volume of soil more effectively and hence much efficient Pi acquisition (Lynch 1995). Under Pi deficiency one of the most common response of plants is the increase in root growth instead of shoot growth, which results in increasing root-to-shoot dry weight ratio (Hermans et al. 2006), and this occurs as a result of change in the distribution pattern of photosynthates between roots and shoots which is brought about by plant hormones (Nacry

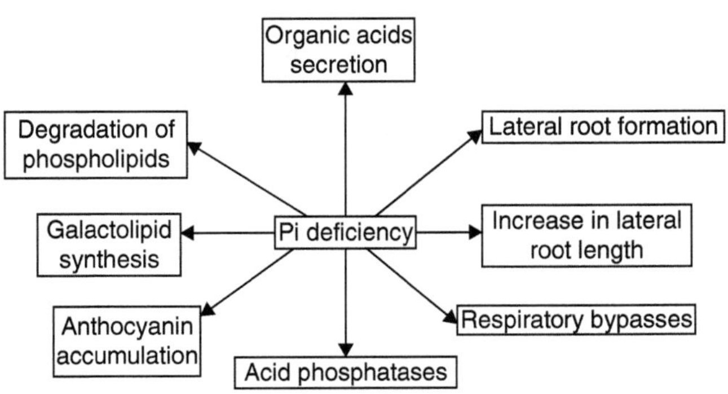

Fig. 7.1 Morphological and physiological responses of plants to phosphorus deficiency

et al. 2005), sugar signaling (Vance 2010), and nitric oxide in white lupin (Wang et al. 2010). It has been found in *Arabidopsis* that under Pi starvation conditions, there is inhibition of primary root growth (Lopez-Bucio et al. 2003), while the growth of lateral roots is stimulated (Lopez-Bucio et al. 2003; Nacry et al. 2005). The plants growing under P-deficient conditions have much branched root system with enhancement in the number and length of root hairs (Bates and Lynch 1996; Ma et al. 2001) which significantly enhances surface area for the efficient Pi acquisition (Bates and Lynch 2001a, b). Such plants are also known to have increased lateral root development and are much more efficient in acquiring Pi from the soil to maintain their growth (Zhu et al. 2010). The modifications in the root system in response to deficiency of P are known to be controlled by the plant growth regulators especially auxins, ethylene, and cytokinins (Casson and Lindsey 2003). Ethylene is known to be a positive regulator of root hair development (Michael 2001) and controls root design by showing molecular cross-talk with auxin. The production of ethylene is known to be stimulated in response to the P deficiency (Borch et al. 1999) which may be the reason for the enhanced root hair formation in P-deficient plants (Michael 2001). The inhibition of primary root is mediated by the genes, low phosphate root 1 (LPR1) and low phosphate root 2 (LPR2), which are known to encode two multicopper oxidase enzymes, and the mutants of these two genes lpi1 and lpi2 do not have a typical primary root inhibition induced by Pi starvation, and both genes are suggested to be required in the regulation of root development under Pi starvation (Guimil and Dunand 2006).

7.2.2 Increasing Phosphate Mobilization and Its Utilization

Only inorganic form of phosphorus (Pi) is available to plant cells because of the presence of Pi transporters located on the plasma membrane, while as due to lack of transporters, organic forms like phosphomonoesters and nucleic acids cannot be translocated. Therefore mobilization of these organic forms is needed to release Pi which is brought about by hydrolases like acid phosphatases and ribonucleases. During Pi-deficient conditions, there is increased secretion of phosphatases which ultimately results in mobilization of phosphorus and its subsequent uptake by the plant. Plants grown in Pi-limited environment have upregulation of the genes encoding purple acid phosphatases (PAPs) and ribonucleases (RNases). The secreted acid phosphatases from plants have high activity and transcription under Pi-deficient conditions (Li et al. 2002). The genes encoding these acid phosphatases are known to be upregulated under P deficiency like the expression of *AtsAPase* gene in *Arabidopsis*, which is enhanced in P deficiency (Haran et al. 2000), and the expression of purple acid phosphatase genes *AtPAP11* and *AtPAP12* increased in P-deficient *Arabidopsis* suspension cells (Li et al. 2002). There are about 29 purple acid phosphatase (PAP) genes identified in *Arabidopsis* genome, and about 11 of these are upregulated (Misson et al. 2005) during Pi-deficient conditions which confirm the vital role of acid phosphatases during phosphorus deficiency. Ribonucleases are known to cause mobilization and release of the organic P in soil for the ready availability of the plants (Bariola et al. 1994; Duff et al. 1994). In *Arabidopsis* about three genes have been identified which produce S-like RNAases which bring about Pi mobilization, namely, *AtRNS1, AtRNS2, and AtRNS3* (Taylor and Green 1991; Bariola et al. 1994). According to Marschner (2012), roots cause acidification of the rhizosphere which results in dissolution of sparingly available soil P. The factors which determine the pH of rhizosphere include cation/anion uptake ratios and nitrogen assimilation. The availability of ammonium to plant roots decreases pH as the supply of nitrates causes increase in pH of rhizosphere. In response to P deficiency, the plants of lupin are known to stimulate the release of protons and citrate exudation by roots accompanied with the inhibition of nitrate uptake (Shen et al. 2005). The effects of the release of organic acids on Pi acquisition have been well documented (Vance et al. 2003; Hinsinger et al. 2005).

Roots are also known to exudate a variety of molecules in response to Pi deficiency including P-mobilizing compounds such as organic acids in *Arabidopsis* (Narang et al. 2000) and rice (Begum et al. 2005), protons, and phosphatases. Pi-deficient rape typically releases malic acid near its root tips or at sites in contact with insoluble rock phosphate. The organic acids are produced during TCA cycle, and it has been found that the enzymes involved in TCA cycle like citrate synthase, phosphoenolpyruvate carboxylase, and malate dehydrogenase are upregulated under Pi-deficient conditions which ultimately results in the production and release of more organic acids into the rhizosphere (Wu et al. 2003). Organic acids like malate and citrate are involved in the release of Pi from Al–, Fe–, and Ca–P complexes (Ryan et al. 2001). Plant growth-promoting rhizobacteria including phosphorus solubilizing bacteria (PSB) lead to enhancement of the P acquisition by increasing solubilization of Pi to plants (Richardson et al. 2009). They are known to impart this effect through acidification of soil surrounding the roots and release of enzymes like phosphatases and phytases (Zhang et al. 2010), or they may produce carboxylates like oxalate, citrate, and gluconate (Jones and Oburger 2011).

7.2.3 Regulation of Inorganic Phosphate (Pi) Transporters

A range of genes are involved in mediating adaptations to Pi deficiency through the regulation of Pi acquisition, remobilization of internal phosphorus, Pi transport processes, and changes in the metabolism and signaling of Pi (Fang et al. 2009). Phosphate transporter (PT) is a major group of genes which are induced in different plant species under P deficiency (Miller et al. 2001). During Pi deficiency the ability of the roots to uptake Pi is enhanced (Mimura et al. 1998), and the mechanism involved is the induction of Pi transporters. It has been revealed by experiments that the Pi deficiency-induced genes are regulated by the concentration of Pi inside the plant and not the Pi status of the soil surrounding the roots (Shane et al. 2003). The regulatory cascades involving transcriptional factors are known to coordinate the changes in gene expression. Several transcription factors have been identified by the transcriptional profiling methods that are increased under Pi-deficient conditions (Wu et al. 2003). The expression of the genes encoding Pi transporters of the Pht1 subfamily is increased along with the increase in the size of the plant root system, thereby increasing the surface area (Smith et al. 2003). With the proper understanding about the regulatory signaling cascades, genes involved in plant responses to P deficiency, and the identification of transcription factors, it would be possible to develop crops with improved phosphorus utilization efficiency (PUE).

7.2.4 Changes in Metabolism

In response to a number of environmental and developmental signals including Pi deficiency, accumulation of anthocyanins in different plant tissues is reported (Steyn et al. 2002), and the genes involved in the biosynthetic and regulatory pathway of anthocyanins are upregulated during Pi deficiency (Steyn et al. 2002; Misson et al. 2005; Morcuende et al. 2007). It is also believed that Pi starvation-induced anthocyanin accumulation requires *PHR1* (Rubio et al. 2001). The cellular respiratory pathways are also adjusted in response to Pi deficiency (Fang et al. 2009). Prolonged deficiency of inorganic phosphate reduces the cytoplasmic Pi, ATP, and ADP which ultimately results in the disturbance of the normal glycolytic and oxidative phosphorylation pathways. To overcome such situation, plants modify these respiratory pathways by bypassing the ADP-/ATP-/Pi-dependent reactions (Theodorou and Plaxton 1993; Plaxton 1996). The enzyme activity data in Pi-starved *Brassica nigra* suspension cells has revealed that the glycolytic pathways are modified with induction of the activities of the enzymes like phosphoenolpyruvate (PEP) phosphatases, UDP glucose pyrophosphorylase, PEPC, and non-phosphorylating NADP-dependent glyceraldehyde 3-phosphate dehydrogenase, while the metabolic byproduct PPi serves as an alternative donor of energy to withstand the shortage of ATP concentration. Also in response to low Pi, normal oxidative phosphorylation pathway is modified. Lipid composition of plant membranes undergoes drastic changes during Pi starvation with decrease in the content of phospholipids and an increase of non-phosphorous lipids in several species, including photosynthetic bacteria (Benning et al. 1993) and *Arabidopsis* (Essigmann et al. 1998). Sulfolipids are known to increase in thylakoid membranes (Essigmann et al. 1998), while the galactolipids increase in extraplastidic membranes (Hartel et al. 2000) as well as in thylakoid membranes in *Arabidopsis* in response to Pi starvation. The degradation of phospholipids in response to Pi starvation is consistent with the upregulation of phospholipase C and phospholipase D genes accompanied with the downregulation of the genes like phosphoethanolamine N-methyltransferases involved in biosynthesis of phospholipids. Auxin signaling is believed to be essential for the induction of MGD2/3 which is involved in the synthesis of galactolipid digalactosyl diacylglycerol (DGDG) during Pi starvation.

7.3 Phosphate Signaling

The mechanism of sensing the level of inorganic phosphate by plants is not clearly understood yet; however, the knowledge in this field is growing rapidly. The concept that plants sense the Pi concentration comes from the Pi sensing and signaling systems in unicellular organisms like *Escherichia coli* (Torriani 1990; Wanner 1993) and *Saccharomyces cerevisiae* (Persson et al. 2003; Mouillon and Persson 2006) in which PhoR and PHO81 act as sensors of Pi deficiency and regulate downstream signaling networks, respectively. There is also the possibility of the existence of the similar Pi sensing systems in higher plants which has been supported by using phosphite ($H_2PO_3^-$), an analog of Pi. Application of phosphate in *Arabidopsis* represses the typical Pi deficiency responses like anthocyanin accumulation, promotion of lateral roots and root hairs, alteration in the composition of lipids of membranes, and upregulation of Pi deficiency-inducible genes (Ticconi et al. 2001; Varadarajan et al. 2002) as well as root proteoid formation in white lupin (Gilbert et al. 2000) The Pi deficiency is believed to trigger the transcriptional, biochemical, and physiological changes that ultimately result in better absorption of P from the soil and improves the P use efficiency (Amtmann et al. 2005; Hammond and White 2008). One of the transcription factors, miRNAs, is believed to be responsible for the homeostasis and signaling of Pi through the identification and characterization of PSR miRNAs under P-deficient conditions (Kuo and Chiou 2011).

References

Amtmann, A., Hammond, J. P., Armengaud, P., & White, P. J. (2005). Nutrient sensing and signalling in plants: Potassium and phosphorus. *Advances in Botanical Research, 43*, 209–257.

Bariola, P. A., Howard, C. J., Taylor, C. B., Verburg, M. T., Jaglan, V. D., & Green, P. J. (1994). The Arabidopsis ribonuclease gene RNS1 is tightly controlled in response to phosphate limitation. *The Plant Journal, 6*(5), 673–685.

Bates, T. R., & Lynch, J. P. (1996). Stimulation of root hair elongation in *Arabidopsis thaliana* by low phosphorus availability. *Plant, Cell & Environment, 19*, 529–538.

Bates, T. R., & Lynch, J. P. (2001a). Root hairs confer a competitive advantage under low phosphorus availability. *Plant and Soil, 236*, 243–250.

Bates, T. R., & Lynch, J. P. (2001b). The efficiency of *Arabidopsis thaliana* (Brassicaceae) root hairs in phosphorus acquisition. *American Journal of Botany, 87*, 964–970.

Begum, H. H., et al. (2005). The function of a maize-derived phosphoenolpyruvate carboxylase (PEPC) in phosphorus-deficient transgenic rice. *Soil Science & Plant Nutrition, 51*, 497–506.

Benning, C., Beatty, J. T., Prince, R. C., & Somerville, C. R. (1993). The sulfolipid sulfoquinovosyldiacylglycerol is not required for photosynthetic electron transport in *Rhodobacter sphaeroides* but enhances growth under phosphate limitation. *Proceedings of the National Academy of Sciences of the United States of America, 90*, 1561–1565.

Borch, K., Bouma, T. J., Lynch, J. P., & Brown, K. M. (1999). Ethylene: A regulator of root architectural responses to soil phosphorus availability. *Plant, Cell & Environment, 22*, 425–431.

Casson, S. A., & Lindsey, K. (2003). Genes and signalling in root development. *The New Phytologist, 158*, 11–38.

Essigmann, B., Guler, S., Narang, R. A., Linke, D., & Benning, C. (1998). Phosphate availability affects the thylakoid lipid composition and the expression of SQD1, a gene required for sulfolipid biosynthesis in *Arabidopsis thaliana. Proceedings of the National Academy of Sciences of the United States of America, 95*, 1950–1955.

Fang, Z., Shao, C., Meng, Y., Wu, P., & Chen, M. (2009). Phosphate signaling in *Arabidopsis* and *Oryza sativa. Plant Science, 176*, 170–180.

Gilbert, G. A., et al. (2000). Proteoid root development of phosphorus deficient lupin is mimicked by auxin and phosphonate. *Annals of Botany, 85*, 921–928.

Guimil, S., & Dunand, C. (2006). Patterning of Arabidopsis epidermal cells: Epigenetic factors regulate the complex epidermal cell fate pathway. *Trends in Plant Science, 11*, 601–609.

Hammond, J. P., & White, P. J. (2008). Sucrose transport in the phloem: Integrating root responses to phosphorus starvation. *Journal of Experimental Botany, 59*, 93–109.

Haran, S., Logendra, S., Seskar, M., Bratanova, M., & Raskin, I. (2000). Characterization of *arabidopsis* acid phosphatase promoter and regulation of acid phosphatase expression. *Plant Physiology, 124*, 615–626.

Harrison, M. J., Dewbre, G. R., & Liu, J. (2002). A phosphate transporter from *Medicago truncatula* involved in the acquisition of phosphate released by arbuscular mycorrhizal fungi. *Plant Cell, 14*, 2413–2429.

Hartel, H., Dormann, P., & Benning, C. (2000). DGD1-independent biosynthesis of extraplastidic galactolipids after phosphate deprivation in Arabidopsis. *Proceedings of the National Academy of Sciences of the United States of America, 97*, 10649–10654.

Hermans, C., Hammond, J. P., White, P. J., & Verbruggen, N. (2006). How do plants respond to nutrient shortage by biomass allocation? *Trends in Plant Science, 11*, 610.

Hinsinger, P., Gobran, G. R., Gregory, P. J., & Wenzel, W. W. (2005). Rhizosphere geometry and heterogeneity arising from root-mediated physical and chemical processes. *The New Phytologist, 168*, 293–303.

Jones, D. L., & Oburger, E. (2011). Solubilization of phosphorus by soil microorganism. In E. K. Buenemann, A. Oberson, & E. Frossard (Eds.), *Phosphorus in Action* (pp. 169–198). New York: Springer.

Kuo, H. F., & Chiou, T. J. (2011). The role of microRNAs in phosphorus deficiency signaling. *Plant Physiology, 156*, 1016–1024.

Li, D., et al. (2002). Purple acid phosphatases of *Arabidopsis thaliana*. Comparative analysis and differential regulation by phosphate deprivation. *The Journal of Biological Chemistry, 277*, 27772–27781.

Lopez-Bucio, J., Cruz-Ramırez, A., & Herrera-Estrella, L. (2003). The role of nutrient availability in regulating root architecture. *Current Opinion in Plant Biology, 6*, 280–287.

Lynch, J. P. (1995). Root architecture and plant productivity. *Plant Physiology, 109*, 7–13.

Ma, Z., Bielenberg, D. G., Brown, K. M., & Lynch, J. P. (2001). Regulation of root hair density by phosphorus availability in *Arabidopsis thaliana. Plant, Cell & Environment, 24*(4), 459–467.

Marschner, H. (2012). *Mineral nutrition of higher plants* (3rd ed.). Great Britain: Elsevier Science Ltd.

Michael, G. (2001). The control of root hair formation: Suggested mechanisms. *Journal of Plant Nutrition and Soil Science, 164*, 111–119.

Miller, S. S., Liu, J., Allan, D. L., Menzhuber, C. J., Fedorova, M., & Vance, C. P. (2001). Molecular control of acid phosphatise secretion into the rhizosphere of proteoid roots from phosphorus-stressed white lupin. *Plant Physiology, 127*, 594–606.

Mimura, T., Reid, R., & Smith, F. (1998). Control of phosphate transport across the plasma membrane of *Chara corallina. Journal of Experimental Botany, 49*, 13–19.

Misson, J., Raghothama, K. G., Jain, A., Jouhet, J., Block, M. A., Bligny, R., & Doumas, P. (2005). A genome-wide transcriptional analysis using *Arabidopsis thaliana* Affymetrix gene chips determined plant responses to phosphate deprivation. *Proceedings of the National Academy of Sciences of the United States of America, 102*, 11934–11939.

Morcuende, R., Bari, R., Gibon, Y., Zheng, W., Pant, B. D., Blasing, O., & Scheible, W. R. (2007). Genome-wide reprogramming of metabolism and regulatory networks of *Arabidopsis* in response to phosphorus. *Plant, Cell & Environment, 30*, 85–112.

Mouillon, J. M., & Persson, B. L. (2006). New aspects on phosphate sensing and signalling in *Saccharomyces cerevisiae. FEMS Yeast Research, 6*(2), 171–176.

Nacry, P., Canivenc, G., Muller, B., Azmi, A., Van Onckelen, H., Rossignol, M., & Doumas, P. (2005). A role for auxin redistribution in the responses of the root system architecture to phosphate starvation in *Arabidopsis. Plant Physiology, 138*, 2061–2074.

Narang, R. A., Bruene, A., & Altmann, T. (2000). Analysis of phosphate acquisition efficiency in different Arabidopsis accessions. *Plant Physiology, 124*, 1786–1799.

Persson, B. L., Lagerstedt, J. O., Pratt, J. R., Pattison-Granberg, J., Lundh, K., Shokrollahzadeh, S., & Lundh, F. (2003). Regulation of phosphate acquisition in *Saccharomyces cerevisiae*. *Current Genetics, 43*, 225–244.

Plaxton, W. C. (1996). The organization and regulation of plant glycolysis. *Annual Review of Plant Physiology and Plant Molecular Biology, 47*, 185–214.

Richardson, A. E., Barea, J. M., McNeill, A. M., & Prigent-Combaret, C. (2009). Acquisition of phosphorus and nitrogen in the rhizosphere and plant growth promotion by microorganism. *Plant and Soil, 321*, 305–339.

Rubio, V., Linhares, F., Solano, R., Martín, A. C., Iglesias, J., Leyva, A., & Paz-Ares, J. (2001). A conserved MYB transcription factor involved in phosphate starvation signaling both in vascular plants and in unicellular algae. *Genes & Development, 15*(16), 2122–2133.

Ryan, P. R., Delhaize, E., & Jones, D. L. (2001). Function and mechanism of organic anion exudation from plant roots. *Annual Review of Plant Biology, 52*(1), 527–560.

Shane, M., De Vos, M., De Roock, S., & Lambers, H. (2003). Shoot P status regulates cluster-root growth and citrate exudation in *Lupinus albus* grown with a divided root system. *Plant, Cell & Environment, 26*, 265–273.

Shen, J., Li, H., Neumann, G., & Zhang, F. (2005). Nutrient uptake, cluster root formation and exudation of protons and citrate in *Lupinus albus* as affected by localized supply of phosphorus in a split-root system. *Plant Science, 168*, 837–845.

Smith, F. W., Mudge, S. R., Rae, A. L., & Glassop, D. (2003). Phosphate transport in plants. *Plant and Soil, 248*, 71–83.

Steyn, W. J., Wand, S. J. E., Holcroft, D. M., & Jacobs, G. (2002). Anthocyanins in vegetative tissues: A proposed unified function in photoprotection. *New Phytologist, 155*, 349–361.

Taylor, C. B., & Green, P. J. (1991). Genes with homology to fungal and S-gene RNases are expressed in *Arabidopsis thaliana*. *Plant Physiology, 96*, 980–984.

Ticconi, C. A., Delatorre, C. A., & Abel, S. (2001). Attenuation of phosphate starvation responses by phosphite in Arabidopsis. *Plant Physiology, 127*(3), 963–972.

Theodorou, W. C., & Plaxton, M. E. (1993). Metabolic adaptations of plant respiration to nutritional phosphate deprivation. *Plant Physiology, 101*, 339–334.

Torriani. (1990). From cell membrane to nucleotides: The phosphate regulation *in Escherichia coli*. *BioEssays, 12*, 371–376.

Vance, C. P. (2010). Quantitative trait loci, epigenetics, sugars, and microRNAs: Quaternaries in phosphate acquisition and use. *Plant Physiology, 154*, 582–588.

Vance, C. P., Uhde-Stone, C., & Allan, D. L. (2003). Phosphorus acquisition and use: Critical adaptations by plants for securing a non-renewable resource. *The New Phytologist, 157*, 423–447.

Varadarajan, D. K., et al. (2002). Phosphite, an analog of phosphate, suppresses the coordinated expression of genes under phosphate starvation. *Plant Physiology, 129*, 1232–1240.

Wang, B. L., Tang, X. Y., Cheng, L. Y., Zhang, A. Z., Zhang, W. H., Zhang, F. S., Liu, J. Q., Cao, Y., Allan, D. L., Vance, C. P., et al. (2010). Nitric oxide is involved in phosphorus deficiency-induced cluster-root development and citrate exudation in white lupin. *The New Phytologist, 187*, 1112–1123.

Wanner, B. L. (1993). Gene regulation by phosphate in enteric bacteria. *Journal of Cellular Biochemistry, 51*, 47–54.

Wu, P., Ma, L., Hou, X., Wang, M., Wu, Y., Liu, F., & Deng, X. W. (2003). Phosphate starvation triggers distinct alterations of genome expression in *Arabidopsis* roots and leaves. *Plant Physiology, 132*, 1260–1271.

Zhang, F., Shen, J., Zhang, J., Zuo, Y., Li, L., & Chen, X. (2010). Rhizosphere processes and management for improving nutrient use efficiency and crop productivity: Implications for China. *Advances in Agronomy, 107*, 1–32.

Zhu, J., Zhang, C., & Lynch, J. P. (2010). The utility of phenotypic plasticity of root hair length for phosphorus acquisition. *Functional Plant Biology, 37*, 313–322.

Chapter 8
Response Pattern of Selected Tropical Perennials to Organic and Inorganic Fertilizers Based on Empirical Data

K.P. Baiyeri and F.D. Ugese

Abstract Tropical perennials have immense nutritional, social and economic significance. However, the environment of their growth often limits their productivity and ultimate utilization. The ability of tropical soils to support plant growth is severely restricted by such properties as poor structure and drainage, low inherent fertility and nutrient imbalances. Socio-economic pressures often lead to more intensive soil cultivation and widespread adoption of inappropriate practices that further degrades the soil. Crop response under such circumstances is generally suboptimal. These soils merit peculiarly appropriate nutrient management options. These include development and utilization of fertilizer technologies uniquely suited for the tropics rather than direct transfer of those intended for the temperate environment and application of organic manures wholly or in combination with inorganic fertilizers. Available evidence indicates the favourable impact of organic manure application on soil properties leading to improved plant response in terms of productivity, produce quality, including postharvest attributes. This is suggestive of the strategic role of organic sources of nutrients in improving the ability of tropical soils to support plant growth and productivity indefinitely while also ensuring human health and environmental quality. In this chapter, empirical data on response pattern of some selected tropical perennials to organic and inorganic fertilizers are presented and discussed.

Keywords Perennials plants • Fertilizers • Tropical environment • Growth responses

K.P. Baiyeri (✉)
Department of Crop Science, University of Nigeria, Nsukka, Nigeria
e-mail: paul.baiyeri@unn.edu.ng

F.D. Ugese
Department of Crop Production, University of Agriculture, Makurdi, Nigeria

M. Naeem et al. (eds.), *Essential Plant Nutrients*,
DOI 10.1007/978-3-319-58841-4_8

8.1 The Tropical Environment and Crop Productivity

According to Amberger (2006), the tropics geographically refer to the areas lying between latitude 23.5°N and 23.5°S. In other words, they are areas lying between the tropic of Cancer and that of Capricorn. Average temperature in the tropics exceeds 18 °C throughout the year. Tropical soils have many weaknesses as far as agricultural productivity is concerned. They are dominated by Oxisols and Ultisols, which cover approximately 51% of the total tropical landscape. These soils are characterized by high weathering and drainage, low cation exchange capacity (CEC) and lack the capacity to adsorb large quantities of K from applied fertilizers on their exchange sites. They also have low pH, base status and available P (Baligar and Bennett 1986; Kaddar et al. 1984). The usefulness of these soils in crop production is limited by high acidity levels and low fertility exacerbated by intense leaching of essential nutrients and lack of primary minerals. The Ultisols and Oxisols generally have problems bordering on Al toxicity, poor nutrient content, nutrient imbalances and multiple nutrient deficiencies (Sanchez et al. 1997). In addition, tropical soils are loose in structure, highly drained and low in essential elements needed for normal plant functions (Baligar and Bennett 1986). The use of acid-tolerant crops on such soils, though commendable, is not sustainable since soil reserves of essential minerals are so low that it can be quickly depleted. Thus, a more integrated approach is better such as use of suitable genotypes, appropriate soil amendments and sound management practices.

With such traits as described above, application of fertilizers to tropical soils to improve crop productivity is imperative. However, mineral fertilizer use is presenting daunting challenges of different dimensions. The fertilizer technology in use in the tropics has been directly transferred from the temperate region for which it was initially and specifically developed functions (Baligar and Bennett 1986). Its suitability is therefore questionable. Even when these fertilizers are produced, the rising cost of importation and transportation is an obstacle to profitable utilization in many parts of the tropics.

8.2 Fertilizer Use and Crop Response in the Tropics

Global increase in food production in the last 50 years or so has been attributed to agricultural intensification with the use of mineral fertilizers as an important component (FAO 2006). This has helped to feed the population that has increased by close to 90% within the period while per capital area dropped from 0.42 to 0.25 ha. For obvious reasons, a large proportion (70–80%) of future increases in food production in developing countries are still expected from land intensification rather than increase in area under cultivation. Of the about fivefold increase in the use of fertilizers since 1960 (FAO 2006), contribution by fertilizer N in per capita food production has been estimated at 40% (Smil 2002), representing the most significant contribution in this regard.

It is evident that the dramatic increase in mineral fertilizer use which stemmed from high population pressure and quest for better nutrition was based on the perceived failure of existing agricultural practices to meet such demands. Traditional practices were dependent mainly on organic matter for the supply of needed nutrients. Although organic sources still constitute an important component of soil nutrient globally, their contribution is less than that of mineral fertilizers. Currently, animal manures are known to contribute only 11% of total N required for global food production (Smil 1999).

Fertilizer use efficiency, which refers to the extent to which applied nutrients are taken up by the crop in question (Singh and Ryan 2015), is generally low in the tropics. In contrast to temperate crops, fertilizer nutrient recovery by tropical crops is lower (Baligar and Bennett 1986). The efficiency of added N, P and K is about 50%, 10% (or less) and 20–40%, respectively. Losses of N are due largely to ammonia volatilization, leaching and denitrification; P losses are mostly due to fixation, while K is lost mainly by leaching. A twin problem to the low fertilizer use efficiency in tropical areas is the high cost of production of mineral fertilizer (FAO 1977). In developing countries, 64% of the energy used in agriculture is channelled towards fertilizer production of which almost all of it is monopolized by N production. Even in the developed countries, 35% of the energy used in agriculture is still expended on fertilizer manufacture with N production gulping the lion share. With the rising cost of fossil fuels, the strain on developing economies is considerable. This is even more worrisome when viewed in the context of prevailing poverty levels particularly among the farming communities in most of these regions. Unfortunately it has been difficult to find a less costly process of N fixation than the Haber-Bosch process (Nelson 1974), a situation that has endured to the present time.

In the tropics, subsistence farming is still widespread, with farmers scarcely affording mineral fertilizers. In some of these regions, recourse is still made to land fallowing for soil fertility restoration. However, with increasing population pressure, fallow periods have become too short to make any reasonable impact on soil fertility build up (Amberger 2006). Continuous cropping on soils without replenishing nutrients lost through crop harvest results in nutrient mining, a condition that leads to declining yields and degraded soils (Amberger 2006; Singh and Ryan 2015).

8.3 The Consequences of Mineral Fertilizer Use

As noted earlier, drastic increases in global food production and better nutrition can be ascribed largely to the use of fertilizers. However, their use particularly that of N has been linked with some issues bordering on soil health and productivity as well as overall environmental quality.

8.3.1 Soil Acidification

One of the undesirable effects of nitrogenous fertilizers is the soil acidifying effect. Acidification imposes serious limitation on the supply of phosphorus to plants and the effectiveness of phosphate fertilizers (Kotschi 2013). The major synthetic nitrogen fertilizers, namely, urea, ammonium sulphate, ammonium nitrate, monoammonium phosphate and diammonium phosphate, have high acidity index (Hart 1998). Acidity index defines the quantity of lime that must be added to soil to neutralize 1 kg of N fertilizer (Kotschi 2013). Of the fertilizers mentioned above, ammonium sulphate has the highest acidity index of 1.10, followed by urea (0.71), while diammonium phosphate has the least (0.37). Closely associated with soil acidification is Al toxicity prevalent under highly acidic soils and deficiency of a number of nutrients such as P, Ca and Mg (Akinrinde 2006). Al toxicity alone (Akinrinde 2006) or in combination with P deficiency (Amberger 2006) has been indicated as the most important factor limiting crop performance in acid soils.

Soil acidification is a global issue. However, acidifying effect of synthetic N on soils in the tropics is viewed more seriously. This is because severe leaching and weathering which characterize the humid tropics are known to worsen soil acidification problems. Secondly, the practice of liming which tends to address soil acidity problems is not widespread in the tropics (Kotschi 2013). This could be due to lack of awareness or the high cost of liming materials.

8.3.2 Environmental Quality Deterioration

Rising global temperatures have been linked to greenhouse gas emissions especially CO_2 (Oyebade et al. 2010). Agriculture has been reported to account for 12% (Smith and Reuther 2007) or 15% (Leakey 2009) of all greenhouse gas emissions. Close to half (47%) of this proportion is due to clearing and burning of forest land for farming purposes, while one-third (17%) is due to nitrous oxide released into the atmosphere from N fertilization (Smith and Reuther 2007).

Apart from the industrial production of N fertilizers which generates significant quantities of CO_2, the industrial process itself is energy demanding, requiring 1.2% of the world's energy demand. The manufacture of synthetic N gulps up to 90% of the energy used in the fertilizer industry (Kongshaugh 1998). Besides, the production of nitric acid, an important component in the synthesis of N fertilizer, releases considerable amounts of nitrous oxide into the atmosphere (Kotschi 2013). Low soil pH as enhanced by synthetic N application also promotes the production of nitrous oxide through the processes of nitrification and denitrification (Singh and Ryan 2015).

It has been reported that globally, more synthetic N is used than actually needed. Nitrogen takes up three-quarters of mineral fertilizer use globally (Kotschi 2013). Excessive application of synthetic N fertilizers encourages leaching loses of nitrate and cations to groundwater and surface water bodies impacting negatively on their quality. Subsoil oxidation of pyrite by leached nitrates could release sulphate and trace elements such as nickel, copper, lead, etc. (Singh and Ryan 2015).

8.3.3 Controversial Role in Organic Matter Depletion

The role of applied synthetic N fertilizers in organic matter decomposition is at best controversial. Understandably, application of N has been noted to enhance plant growth leading to greater biomass that subsequently increases organic matter content of the soil (Singh and Ryan 2015). On the other hand, decomposition of organic matter has been linked with the presence of synthetic N in the soil, the rate of decomposition being proportional to the amount of N applied to the soil. Khan et al. (2009) and Mulvaney et al. (2009) demonstrated from long-term experiments across the globe the organic matter-reducing effect of continuous N application. This position has however being hotly disputed by Ladha et al. (2011) and Geiseller and Scow (2014) who have faulted the premise (time-response ratio) on which such conclusions were drawn. By adopting the time by fertilizer response approach, they demonstrated increases in organic matter content of soils receiving doses of synthetic N over time. Closely linked to this is the role of mineral N fertilizers on soil biological life. Predominant evidence seems to suggest temporary reduction in biological life after application of synthetic N. The situation however normalizes after few weeks. On the whole, biological life is better enhanced in fertilized than non-fertilized plots (Geiseller and Scow 2014). From all indications, the last is yet to be heard about the real impact of synthetic N on the decomposition of soil organic matter. At present, it might be necessary applying some caution in the use of synthetic N as we await more definitive conclusions from scientific investigations.

8.4 Considerations for Soil Nutrient Supply in the Tropics

Fertilizer application in the tropics must take cognizance of pertinent issues if it must ensure sustainable agricultural productivity as well as human and environmental well-being. Some of such issues deemed relevant are briefly stated below:

8.4.1 The Uniqueness of Tropical Soils

It has been noted earlier that a significant proportion of tropical soils have high acidity, low native fertility status, loose structure and high weathering capacity and are easily drained. Fertilization will have to take into account the need to improve soil structure, reduce acidity and increase nutrient content for enhanced productivity (Kotschi 2013). Inorganic fertilizers make more meaning when applied in combination with organic manure under degraded soil conditions.

8.4.2 Climatic Variables

Under tropical conditions, the vagaries of climate take their toll on nutrient efficiency. Prevailing high temperatures accelerate organic matter decomposition (Webster and Wilson 1980) and its logical depletion as well as soil weathering. High amounts of rainfall equally significantly contribute to leaching losses of soluble nutrients.

8.4.3 Low Fertilizer Efficiency

As indicated earlier, there is low rate of applied nutrient recovery, particularly N. On a general note, fertilizer nutrient recovery by crops is lower in the tropics compared to the temperate environment (Baligar and Bennett 1986). This is because tropical soils are different from temperate soils and need fertilizers specifically prepared to suit their conditions. Presently, fertilizers developed have been for the temperate regions. It is obvious that to increase nutrient recovery rate in tropical soils, suitable fertilizer materials and application methods must be adopted. In other words, fertilizer materials and application methods customized for tropical areas will drastically reduce nutrient wastage and enhance plant uptake and overall performance. In acid soils N loses are dominated by denitrification and leaching, while in alkaline soils, ammonium volatilization predominates (Baligar and Bennett 1986).

8.4.4 Economic Constraints

FAO (2006) has identified obstacles to fertilizer use as including, among others, the discouraging crop produce price relative to fertilizer price and the limited purchasing power of small-scale resource-poor farmers. At present, there is no viable alternative to the Haber-Bosch process of synthetic N production, which is very costly in terms of materials and energy (Nelson 1974; Kotschi 2013). This adds to the cost of fertilizers when they eventually arrive. Obviously, fertilizer materials that are cheap and readily available would make more sense at least from the economic standpoint.

8.5 Description of Selected Tropical Perennials

8.5.1 Musa *Species*

Banana and plantain belong to the family *Musaceae*. While bananas refer to all members of the genus *Musa*, plantains are considered a subset of bananas. The latter are not necessarily cooked before consumption particularly when ripe, but the former are almost exclusively cooked before being consumed. Originating primarily in Malaysia, Indonesia, the Philippines and Papua New Guinea, *Musa* species are now cultivated throughout the tropics and some parts of the subtropics. They are the leading fruit in terms of global production and trade (Daniells et al. 2011). *Musa* species are considered a rich energy source with carbohydrate making up 22 and 32% of fruit weight of banana and plantain, respectively (Honfo et al. 2011). In sub-Saharan Africa, they make appreciable contribution to food security as they provide more than 25 and 10% of the daily intake of carbohydrates and calories, respectively, to more than 70 million people (IITA 2000).

Daniells et al. (2011) describe *Musa* plant as a giant perennial monocotyledonous herb reaching heights of 2–9 m. In the mature state, it consists of an underground rhizome having roots and suckers, a pseudostem and a bunch with several fruits. The rhizome is considered to be the plant's true stem (Hartmann et al. 1981). The pseudostem consists of overlapping leaf sheaths and could support a canopy of 10–15 expanded leaves. Flowering occurs after vegetative growth phase, eventually resulting to individual fruits and bunches.

8.5.2 Citrus Species

Citrus species (*Citrus sinensis*) probably originated in North Eastern India, Burma and the surrounding areas from where the species crossed into China, the North East and to the Mediterranean region. From the Mediterranean, it was carried to other parts of the world. The production of citrus particularly the sweet orange cultivars is more heavily concentrated in subtropical and warm temperate areas. Tropical areas are also producing sizeable quantities of other citrus species (Davies et al. 1994).

The global production of citrus, which is about 102 million metric tonnes per year, is much greater than that of other tropical and subtropical fruits, including banana, mango, apple, pear and peach. The subtropical regions produce much of the best quality fruits with Brazil and the United States supplying up to 45% of the total world orange production. The United States on the other hand accounts for up to 40% of grapefruit production in the world (FAO 2004).

Citrus species are small- to medium-sized evergreen trees, reaching a height of 5–15 m. The leaves are unifoliate, margins entire with petioles that are most often winged. The leaves, oval and glossy, are alternately arranged on the shoot. The flowers, which are strongly scented, occur singly (solitary) or in small clusters.

The fruit of citrus is a specialized berry (hesperidium) with segmented inner fleshy parts with an outer rind. The fruit exocarp and mesocarp have a leathery texture. The fruit is more or less spherical or somewhat elongated. Each of the easily separated segments or carpels or sections has many juice cells and several seeds covered by a leathery skin (exocarp) containing many oil glands. Some, like the tangelo, have no seeds, while some have only few seeds. Rind colour may be yellow to orange when ripe (Baiyeri and Ugese 2011).

8.5.3 Passiflora edulis

The passion fruit is a woody perennial climbing plant with tendrils. The leaves are green, alternate, finely toothed and glossy on the upper side, duller on the underside, oval in shape and 7.5–20 cm long. Fully developed leaves have three deep lobes. The leaves, young stems and tendrils have red or purple taints. Flowers are fragrant, solitary, 5–10 cm in diameter, positioned at the nodes. They have three bracts, green in colour; five sepals, greenish white in colour; and five white petals. Flowers generally are predominantly purple. Fruits are oblong, 4–7.5 cm in diameter, green brown, dark purple or light yellow in colour, with a smooth but tough rind. The juicy flesh is yellow or orange and could hold up to 250 seeds. The seeds are small, hard brown or black in colour.

The plant has a shallow root system (Baiyeri and Ugese 2011). In most parts of the world, its major utilization is as fruit juice (Ani and Baiyeri 2008). Apart from human and animal nutrition, passion fruit is credited with a number of industrial and medicinal uses (Morton 1987).

8.5.4 Carica papaya

It is a soft-wooded evergreen tree, 5–10 m tall, usually unbranched. The leaves occur at the top position of the trunk and are spirally arranged. Characteristically, the tree has compound or palmately lobed leaves with each leaf having seven lobes. Usually, leaves are large, 50–70 cm in diameter and borne on exceptionally long hollow petioles. Scars of former leaves and fruits are normally noticeable on lower sections of the trunk (Baiyeri and Ugese 2011).

The plant is dioecious and has small unisexual flowers occurring in leaf axils. Fruits could be spherical or elongated, measuring 15–45 cm long and 10–30 cm wide, and could weigh up to 9 kg. Many shiny black or greyish-black seeds are contained in the large central cavity of the fruit. When ripe the skin of the fruit becomes amber or orange in colour. Production statistics from 2002 to 2010 have placed Nigeria fourth after India, Brazil and Indonesia, contributing 6.79% of global production (Evans and Ballen 2012). The ripe fresh fruits are consumed directly or in fruit salads. They are used in making soft drinks, jams, ice cream flavouring or canned in syrup, while unripe fruit is used as a vegetable. Papain, processed from plant latex, is useful as a meat tenderizer, for manufacture of chewing gum and as a digestive aid. Papaya also has a number of other medicinal values (Olarewaju 2004).

8.5.5 Vitellaria paradoxa

The shea butter tree (*Vitellaria paradoxa*, C. F. Gaertn Synonym: *Butyrospermum parkii*) belongs to the family Sapotaceae and is indigenous to the West African savannah within which it is widely distributed (Opeke 1987; Keay 1989). The species is naturally distributed from Upper Gambia to Uganda, but dispersal to other parts of the world has been made possible by man (FAO 1989). In Nigeria, *V. paradoxa* is found scattered through the Guinea and Sudan savannah zones although occurrence has been reported in isolated parts of Western Nigeria (Keay 1989).

Vitellaria paradoxa has great nutritional and socio-economic significance. The fleshy mesocarp, which constitutes 50–80% of the fruit, is sweet and edible when ripe, while the white kernel of the nut is rich in oil known as shea butter. In Europe, the oil is utilized in the cosmetic industry (Vickery and Vickery 1969) and in chocolate production and in pastry (Boffa et al. 1996). Locally, shea butter is used for cooking, particularly in Northern Nigeria (Keay 1989), and has been rated the second most important oil crop in Africa after oil palm, assuming greater importance in those areas climatically unsuitable for oil palm production (ICRAF 2000). It is also locally used as illuminant and in soap, candle and pomade preparations (Awoleye 1995).

The tree is generally small to medium in size, 7–25 m tall with a short bole. The bark is thick, rough deeply fissured, greyish in colour and fire resistant. The simple, leathery, glossy leaves are oblong or obovate-oblong, glabrous when mature with a wavy margin. Leaves are arranged spirally in dense clusters of 20–30 at the tips of branches. The flowers are yellow green or creamy white and fragrant and occur in clusters at the end of twigs. Fruit is a berry, round or oval shaped, normally 10–60 g, but occasionally larger especially if multiple seeded. Fruit colour is green, turning yellowish green when mature although colour change may not be drastic enough to be easily noticed. The nut, embedded in the mesocarp is brown, smooth, and shiny, with a testa and a hilum. *Vitellaria paradoxa* has a long taproot, while most lateral roots are concentrated at the top 10 cm depth but could extend outwards to a distance of 20 m. Secondary lateral roots can grow vertically downwards to the same depth as the taproot.

8.6 Fertilizer Response Patterns of Selected Perennials

8.6.1 Musa *Species*

Banana and plantains are heavy feeders and easily deplete nutrient resources after just few years of cultivation (Wilson 1987; Baiyeri and Tenkouano 2007). Application of large amounts of external nutrient inputs is therefore required for sustainable production (Hossner and Juo 1999). In West Africa in particular, banana production is dominated by organic manure-based cultivation with plots receiving organic matter and nutrients from household refuse (Swennen 1990). It is obvious that to maintain acceptable levels of production, proper nutrient management practices are critical.

8.6.1.1 Mining Abilities of *Musa* Genotypes

An insight into the nutrient mining abilities of *Musa* genotypes is critical to promoting informed fertilizer recommendations and nutrient management for enhanced productivity.

Existing fertilizer recommendations in *Musa* production as, for instance, that by van Ee (1999) appear to be largely blanket with little or no consideration for genotypic variations. On the contrary, *Musa* genotypes have been shown to exhibit marked variation in nutrient mining abilities. A study by Ortese et al. (2011) to determine nutrient absorption of six *Musa* genotypes by measuring residual nutrient concentration at specific growth stages was carried out in Nigeria. The cultivars, namely, Agbagba, PITA 22, Nsukka Local, FHIA 17, Fougamou and BITA 7, exhibited marked variation in the extent and pattern of nutrient absorption (Table 8.1). Of the six macronutrients (N, P, K, Ca, Mg, S) investigated, only N did not show significant change in mean residual content with advance in age of the crop even though it tended to increase with age. Residual values of P and K were highest at the fourth month of growth of previous crop, decreasing thereafter. In contrast, residual concentrations of Ca, Mg and S increased progressively from the fourth to the sixth month of growth of banana and plantain genotypes.

Table 8.1 Effect of six *Musa* genotypes and duration of their growth on mean values of residual nutrient concentration (mg/100 g) in growth medium previously cropped with the six *Musa* genotypes at Nsukka, Nigeria

Genotype	N	P	K	Ca	Mg	S
Agbagba	1.2	53.4	159.7	2731.4	1461.6	10.3
PITA 22	1.1	19.6	180.7	1206.7	269.8	9.6
Nsukka local	0.9	35.2	184.7	1129.5	524.5	6.4
FHIA 17	1.2	23.3	91.5	747.5	595.7	5.2
Fougamou	1.4	33.2	135.3	1327.1	2477.1	5.4
BITA 7	1.3	22.8	84.7	969.8	414.0	6.2
Control	1.9	18.3	8.15	1900.0	346.7	11.2
LSD (0.05)	0.1	1.1	5.1	65.7	34.2	0.5
Growth duration (months)						
4	1.2	42.7	307.8	631.0	291.4	2.7
5	1.3	27.6	32.7	1737.0	1086.3	9.1
6	1.3	17.9	21.5	1922.9	1232.0	11.5
LSD (0.05)	NS	0.7	3.4	43.0	22.4	0.3

NS no significant difference
Adapted from Ortese et al. (2011)

Higher residual amounts of nutrients implied lower root absorption and vice versa. As such genotypes are expected to absorb more of P and K after the fourth month of growth, while more active absorption of Ca, Mg and S could be expected to occur at this stage (fourth month). By implication, timing of application of particularly single fertilizers could be tailored towards coinciding with the period of more active root absorption. Fortunately, P and K had similar pattern of uptake by the crop, while uptake of N did not vary significantly from the fourth to the sixth month. Thus the compound fertilizer NPK could still be applied following this principle. This aspect of nutrient application timing is most critical where inorganic fertilizers are involved. Organic matter by its nature undergoes continuous mineralization releasing nutrients for crop use. The issue of application timing is therefore less critical. The latter is also less susceptible to leaching losses of nutrients. Thus correct timing of fertilization is imperative in curtailing nutrient loses, enhancing plant uptake and promoting crop productivity.

In the study reported above (Ortese et al. 2011), it was even more interesting that the absorption of certain nutrients by the banana and plantain genotypes was not uniform (Fig. 8.1a, b). In media that were previously grown to Fougamou and

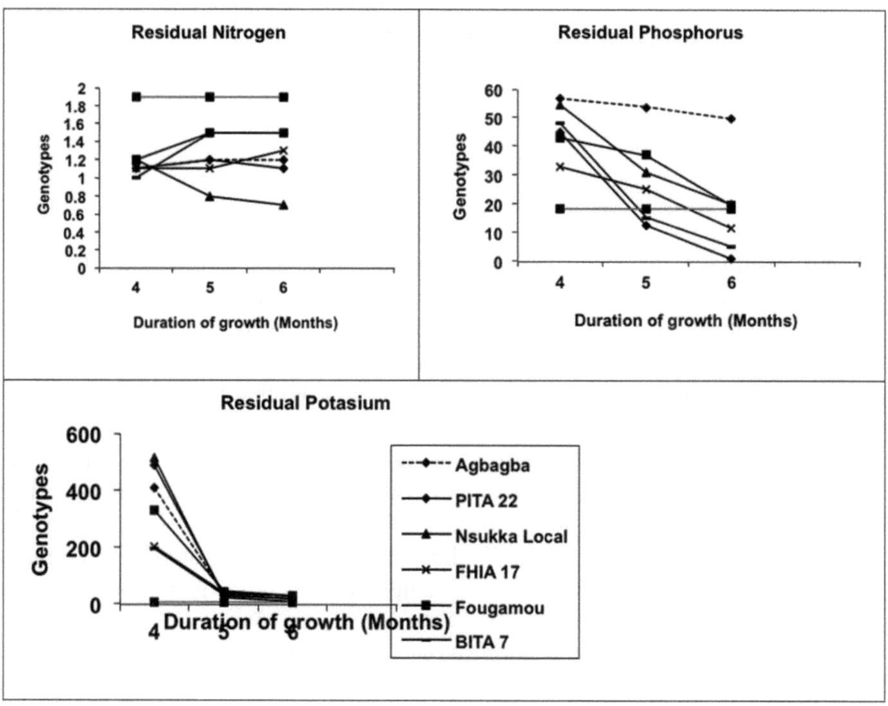

Fig. 8.1 (**a**) Residual concentration (mg/100 g) of N, P and K of growth media previously cropped with six *Musa* genotypes. Adapted from Ortese et al. (2011). (**b**) Residual concentration (mg/100 g) of Ca, Mg and S of growth media previously cropped with six *Musa* genotypes. Adapted from Ortese et al. (2011)

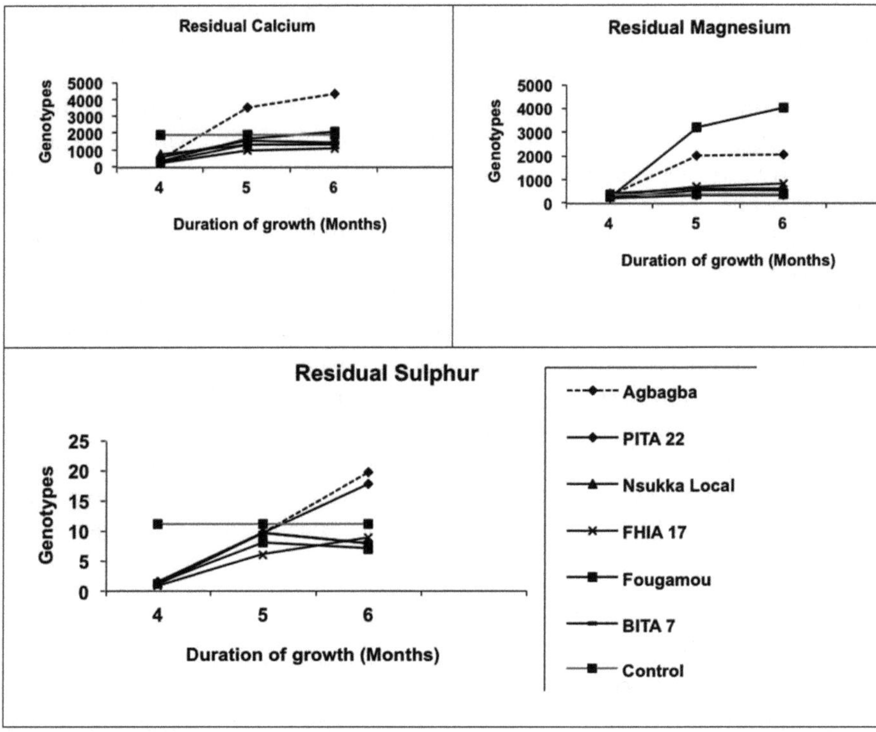

Fig. 8.1 Continued

BITA 7, residual N increased up to the fifth month and then levelled off. However, in media earlier cropped with FHIA 17, slight increases in N content were observed after the fifth month. In the rest of the cultivars, N concentration had an inverse relationship with age of previous crop. All genotypes exhibited fairly similar response in terms of P and K. In respect of Ca, even though a general increase in quantity was observed with age of previous crop, Agbagba had the highest values, while FHIA 17 had the least. Thus Agbagba could be viewed as having the least desire for Ca in sharp contrast to FHIA 17. Fougamou and Agbagba also showed greater aversion in absorbing Mg relative to other genotypes. Agbagba, PITA 22 and to some extent FHIA 17 showed reticent absorption of S throughout the period of evaluation, while the behaviour of the other genotypes was not consistent.

The overall picture has great implications for decision-making in *Musa* production. Fertility status of prospective sites for *Musa* cultivation, determined through soil test, could be used to identify or select the best cultivar for planting on such sites or the amount of particular nutrients to be added. Generally, cultivars with shy feeding ability with prospects of high yield will be greatly valued as they could reduce or eliminate the cost of additional fertilizer input. This would be a great relief to resource-poor farmers. It is therefore needful that the nutrient absorption abilities of different genotypes be investigated, particularly of the more popular varieties.

8.6.1.2 Fertilizers and Placement Method

Although heavy deployment of external nutrient inputs is a necessary requirement for sustainable *Musa* production (Hossner and Juo 1999), proper placement is equally important to enhance uptake, reduce losses and enhance crop performance. In a screen house trial, Baiyeri and Tenkouano (2008) evaluated three manure placement methods on root growth and nutrient uptake of a plantain hybrid—PITA 14. A control treatment with no manure application was included for comparison. In the first treatment (T1), full dose of manure was applied as top dressing. In the second treatment (T2), the dose was applied as bottom dressing that is 20 cm below surface of the pot and covered with soil substrate, while the third placement (T3) was a split combination whereby 50% of manure was top-dressed and the remaining 50% bottom dressed. The control treatment (T4) had no manure application. The substrate used for the screen house evaluation consisted of 60 kg of topsoil with the equivalent of 20 t/ha poultry manure (about 7.5 kg) filled into cylindrical pots of dimensions 80 × 60 cm. Effect on root and shoot growth, biomass production and distribution and nutrient uptake were assessed at 3 and 5 months after transplanting (MAT).

Results indicated significant effect of manure placement on root and shoot growth although specific leaf area (SLA) was not affected, but its values for the control were inferior to those of manure treatments. Surface application triggered impressive growth 3 MAT, but this could not be sustained up to the fifth month. At this later period, T3, which had split application of manure at the top and bottom, showed superior growth and was followed by T2, T1 and the control (T4) in that order. Biomass production followed similar pattern with T1 taking the lead 3MAT but being overtaken by T3 later (5 MAT). Plants that received no manure application had longer and more numerous roots, which could not confer any advantage on shoot growth as it showed the least shoot growth. Split placement of manure facilitated greater uptake of N, P and K with higher accumulation of the three elements in the leaves and highest amounts of N in the roots. Plants had the highest amounts of P and K in the roots when manure was top-dressed. Expectedly, control plants had the least amounts of NPK in their leaves and roots. Plants with high amounts of NPK in their roots also showed high biomass production. This is exciting when considered in relation to yield. Although yield was not assessed in the study under consideration, positive significant relationship between growth and yield has already been established in *Musa* species (Stover and Simmonds 1987; Baiyeri and Mbah 1994; Baiyeri et al. 2000).

Field evaluation conducted later using the same treatments employed in the screen house with inclusion of inorganic fertilizer (urea) provided further insights into its influence on growth, crop phenology, dry matter partitioning and yield (Baiyeri et al. 2011a, b). While effect of manure placement was marginal in the plant crop, the ratoon crop showed significant growth response to manure placement. The effect of black sigatoka disease was severely attenuated in plants that received doses of manure. Similarly, plants that received organic manure exhibited earlier flowering and fruit maturity compared to those that received either inorganic fertilizer or no fertilizer at all (Table 8.2). Split placement of manure led to 6 and 25 days earlier flowering and 9 and 24 days earlier maturity compared with top dressing the manure and bottom dressing singly, respectively. The number of day's differential in flowering and maturity occasioned by placement methods is too large to be ignored and could make all the difference even in economic terms when early harvest may attract premium prices.

In the plant crop, bunch yield and yield components were statistically similar but found to be superior to those of control plants that received no fertilizer application. In the ratoon crop however, split placement of manure at the top and bottom of the planting holes triggered production of bunches and fruits with heaviest weights. Fruits were also longer under this treatment. Inorganic fertilizer or lack of manure produced fruits that were very light and with small circumference. Such results serve to confirm the popular notion that organic fertilizers are better than the inorganic types in tropical soils as the former, apart from releasing nutrients more slowly and over a longer period, also improve soil physical characteristics. Nitrogen loses due to volatilization are expected to be more severe when the material is top-dressed. Volatilization could account for up to 50% losses in nitrogen (Mattila and Joki-Tokola 2003). With split placement of manure, volatilization losses are minimized. Besides, bottom placement of manure could attract lower nitrification and mineralization due to lower temperatures and oxygen levels at such depths (Baiyeri et al. 2011a). This method, apart from being more productive, is also more sustainable.

Dry matter partitioning pattern normally responds to manure placement. In the case under consideration (Baiyeri et al. 2011b), split placement of manure facilitated more dry matter allocation to above-ground portion of the plant. However, genotypic differences in the particular manner of above-ground allocation of assimilate became evident. Thus while BITA 3 partitioned a greater percentage of its assimilate to the pseudostem and leaves, PITA 14 allocated a greater proportion of its photosynthate to the edible portion. Little wonder that this manure placement method led to higher economic yields subsequently.

8.6.1.3 The Role of Mycorrhiza

Beneficial effect of ectomycorrhiza on growth of *Musa* species has been strongly suspected. In a study, Ortese et al. (2011) observed that when maize was grown on media previously cropped to two *Musa* genotypes (PITA 22 and BITA 7) on which ectomycorrhiza growth had been earlier noticed, the maize plants showed luxuriant growth by exhibiting remarkable greenness. Mycorrhizal association with plant roots could either be intracellular as in the case of arbuscular mycorrhizal fungi (AMF) or extracellular as in ectomycorrhizal fungi. In either case, it increases active absorptive surface area of roots and stimulates greater nutrient and water uptake

Table 8.2 Effect of manure placement method and genotype on the phenology, bunch yield and yield components of plantain

	Days to flowering		Days to fruit maturation		Bunch yield (t/ha)		Number of fruits/bunch		Fruit weight (g)	
Treatment	Plant crop	Ratoon crop	Plant crop	Ratoon crop	Plant crop	Ratoon crop	Plant crop	Ratoon crop	Plant crop	Ratoon crop
Manure placement										
T1	245	481	362	588	23.89	33.49	112.80	126.50	152.30	179.70
T2	264	499	377	610	20.50	32.98	96.30	131.20	140.40	173.60
T3	239	469	353	577	23.30	37.29	111.20	125.60	146.50	208.10
T4	323	489	411	593	22.14	27.73	104.90	108.50	129.60	173.20
T5	307	494	403	593	16.27	26.84	89.70	129.1	118.90	155.40
$LSD_{(0.05)}$	36.64	NS	22.14	NS	4.45	5.23	NS	14.83	21.71	22.86
Genotype										
BITA 3	276	480	366	564	19.76	34.49	94.30	136.7	137.30	212.40
PITA 14	226	479	342	600	24.71	26.14	114.80	100.60	153.60	175.70
PITA 23	324	500	436	613	19.20	29.94	99.80	135.20	121.80	145.80
$LSD_{(0.05)}$	28.34	NS	17.15	17.44	3.45	4.05	17.00	11.48	16.82	17.71

NS no significant difference
Adapted from Baiyeri et al. (2011a, b)

even in moisture stress areas. Among other benefits, it also increases host plant disease resistance (Naher et al. 2013; Zachée et al. 2008). Beneficial effects of mycorrhiza have been reported in fruit crops such as banana (Grant et al. 2005) and citrus (trifoliate orange) (Wu and Zou 2012). With the potential of mycorrhiza to improve growth and upgrade yield of *Musa* genotypes, it merits greater studies towards deliberate colonization of plant roots by these organisms. The impoverished nature of most tropical soils further justifies this consideration.

8.6.2 Nutrition of **Passiflora edulis**

The yellow passion fruit is a shallow-rooted perennial, which climbs by means of tendrils and is reputed as a heavy feeder (Knight and Sauls 1994). Because of its shallow roots, it does not explore nutrients and moisture at deeper levels of the soil, making adequate nutrition and moisture supply a prime concern. Cultivation of yellow passion fruit is still at its early stage in Nigeria although its prospects are bright. As such a comprehensive package for its nutrition is yet to be developed. However initial experience with poultry manure is exciting and promising. Ani and Baiyeri (2011) found progressive increase in growth aspects of the plant as poultry manure rate increased from 0 to 15 t/ha. Poultry manure application also shortens the time to flowering. This is attributable to the high content of N and K in the manure (Thomson and Troeh 1978) and their release over a longer period of time. Nakasone and Paull (1999) and Araujo et al. (2006) have indicated that N and K are the most important nutrients in passion fruit.

Apart from growth and flowering phenology, addition of poultry manure has marked effect on fruiting, fruit yield and juice content. Generally, increase in poultry manure rates show at least corresponding marginal increases in yield attributes such as number of fruits, pulp weight, weight and volume of juice and juice yield index (Table 8.3). However, 10 t/ha poultry manure seems to give highest values of weight and volume of fruit juice (Ani and Baiyeri 2011). As pointed out earlier, passion fruit is a heavy feeder, and its shallow root system restricts the ability of its roots to exploit growth resources at lower soil depths. As such poor nutrition could have a telling effect on its performance. In the case cited above, control plants did not only delay flowering but also failed to bear fruits in the first season of harvest. On the contrary, where poultry manure was applied, there was better fruit and juice yield in the first and second seasons. Generally adequate plant nutrition impacts favourably on flowering, fruit set and final yield (Araujo et al. 2006; Mohammed 2002).

One of the most prized products from passion fruit is the fruit juice. The quality of the juice is often determined from the field (Wills et al. 1998) since after harvest no amount of postharvest treatment can improve poor quality of produce (Seymour et al. 1993). Addition of poultry manure, apart from its effect on growth, reproduction and yield of passion fruit also exert significant influence on juice quality. Ani and Baiyeri (2008) found increase in juice pH and protein content with increasing levels of poultry manure (0–15 t/ha). Fat content and total soluble solids did not

Table 8.3 Effect of poultry manure on chemical composition and antinutrient content of fruit juice of yellow passion fruit in the first (December 2005) and second (August 2006) harvest periods in Nigeria

Manure rate (t/ha)	Juice pH	Titratable acidity	Sugar	Vit. A	Vit. C	Total phenol	Crude protein	Fat	TSS	Moisture	Calcium oxalate	Hydrogen cyanide	Phytate	Tannin
		(%)		(mg/100 mL)			(%)				(mg/100 g)			
First year of harvest														
5	3.11	3.75	6.91	5.30	15.60	0.777	2.19	1.00	78.55	21.45	1.407	0.0029	0.673	0.240
10	3.09	3.95	7.96	5.33	16.40	0.781	2.77	0.50	77.73	22.24	2.110	0.0031	0.723	0.310
15	2.93	4.35	8.93	5.70	17.40	0.747	3.36	1.00	79.04	20.88	1.485	0.0035	0.830	0.700
$LSD_{(0.05)}$	0.08	0.10	1.27	0.18	0.54	0.410	0.027	NS	NS	0.62	0.221	0.0003	0.001	0.099
Second year of harvest														
0	2.72	1.90	5.47	8.53	14.41	0.600	1.65	1.74	21.17	85.57	1.250	0.0009	0.625	0.293
5	2.74	2.07	5.91	8.50	24.53	0.300	2.31	3.75	20.30	86.33	1.250	0.0007	0.343	0.000
10	2.75	2.33	6.44	9.67	25.63	0.400	2.54	5.75	20.43	86.10	1.250	0.0004	0.286	0.000
15	2.77	2.83	7.78	12.67	30.95	0.467	3.50	6.63	20.63	86.30	1.250	0.0000	0.200	0.000
$LSD_{(0.05)}$	0.01	0.13	0.33	1.85	0.70	0.02	0.054	0.13	0.21	0.57	NS	0.0001	0.035	0.217

TSS total soluble solids, *NS* no significant difference
Adapted from Ani and Baiyeri (2008)

vary significantly in the first season of harvest but did show such variation in the second season (Table 8.3). Titratable acidity and vitamin A and C contents also increased with level of poultry manure. Antinutritional factors did not respond in any consistent manner to addition of the manure. In the first season, concentration of hydrogen cyanide and phytate increased with increase in poultry manure rate, but in the second season of harvest, there was an inverse relationship. Tannin content had linear relationship with poultry manure rate in the first season, but in the second, tannin presence was not detected in the juice of fruits that received poultry manure application. What is particularly important is that none of the antinutritional factors were found in concentrations considered harmful to human health in both seasons.

Similarly, addition of poultry manure did not increase juice pH and titratable acidity beyond acceptable levels specified by Morton (1987) and Nakasone and Paull (1999) as acceptable. Soil moisture content could play a decisive role in the seasonal variation in juice quality traits. Thus generally manure application improved juice content and quality indices with 15 t/ha rate found most suitable for juice quality (Ani and Baiyeri 2008). It is interesting that this rate produced the highest fruit yield that was statistically at par with that of 10 t/ha. For economic reasons, Ani and Baiyeri (2011) recommended 10 t/ha.

Effect of fertilizer is far reaching and has great impact on seed quality through its influence as an important component of the environment of mother plant (Cardwell 1954). Seed quality could be assessed by considering such traits as seed weight or germination/seedling parameters. Fertilization of mother plants of yellow passion fruit has been found to produce seeds with superior attributes compared to the non-fertilized plants (Baiyeri et al. 2011c). In the present case, a combination of organic and inorganic fertilizers was found to exert the best influence on seed and resulting seedling quality. Thus the treatment 20 t/ha poultry manure +50 kg N/ha + 150 kg K/ha has been found to be better than either 30 t/ha or 20 t/ha poultry manure or inorganic fertilizer alone (200 kg N/ha + 600 kg K/ha) or the various other combinations of poultry manure and inorganic fertilizer. Mother plants that received a combination of 20 t/ha poultry manure, 50 kg N/ha and 150 kg K/ha produced seeds that recorded best germination in terms of amount and speed of germination, emergence and early seedling growth attributes.

8.6.3 Nutrition of Citrus Species

Citrus trees are grown on a wide range of soil types. Consequently, availability and inherent levels of nutrients may vary widely. In most areas, supplemental nutrients are needed to obtain commercially acceptable growth and yields. Citrus species do benefit from application of external nutrient inputs. But generally specific rate of nutrients to be applied to citrus trees is dependent on such factors as temperature and rainfall regimes, growing region, fertility status of soil, age of tree and bearing status (Davies and Albrigo 1994). Generally a total of 12 nutrient elements comprising six macronutrients, N, P, K, Ca, Mg and S, and six micronutrients, Mn, Cu, Zn, B, Fe and Mo, are needed by citrus apart from C, H and O. This is particularly needed when citrus is grown on impoverished soil. In soils adjudged suitable for citrus production, only N and K could be supplemented because of their leachable nature (Davies and Albrigo 1994). In addition N and K represent the largest component of nutrients removed from the soil by the fruit on an annual basis (Alva et al. 2006). Fruit quality of citrus is greatly influenced by nutrition and fertilization practices with N and K having greater influence in situations where other elements are also available and do not pose limitations on tree performance (Embleton et al. 1978).

Growth region also affects the amount of supplemental nutrition required. Regions with high rainfall and temperatures are more likely to lose nutrients from the soil due to leaching or volatilization. Crop load is also closely tied to fertilization programmes for mature trees since many nutrients are removed by harvesting the fruit. For example, a tonne of 'Valencia' oranges contains 1.31 kg N, 0.91 kg P and 1.8 kg K (Smith and Reuther 2007). Therefore, an orchard producing 50 tonnes ha^{-1} would require 66, 9.5 and 90 kg ha^{-1} of N, P and K, respectively, to replenish nutrients lost during harvest. This estimate does not account for losses due to leaching or volatilization or utilization of nutrients for vegetative growth (Gmitter and Hu 1990). Although a useful advice is to use NPK mixed fertilizer 12-8-18, or something close to the formula (Samson 1980), another rule is to fertilize in such a way that nutrient losses per unit of harvested product are compensated for. It must however be noted that since nutrient uptake efficiency ranges from 20 to 50% in citrus groves, nutrient application must exceed the minimum requirement of the tree. However, split application of 2–3 times per year has been advocated for, to avoid placing a large amount of fertilizer in the soil and thereby encouraging leaching loses (Alva et al. 2006).

Even though citrus produced in Nigeria does not enter international trade, the country occupies the ninth position among the top 10 citrus producer countries (UNCTAD 2010). There is evidence that farmers in Benue State, recognized as Nigeria's most important citrus zone (Avav and Uza 2002), have some level of familiarity with citrus fertilization. About 98% of growers apply fertilizer in their orchards due to its positive influence on tree growth, fruit yield and quality. Of the proportion that applies fertilizer, the greatest number (68.3%) use NPK compound fertilizers, while only 5% use organic manure (Ortese et al. 2012). A limitation in fertilizer application types employed by farmers is the highly restricted number of nutrients involved. Among fertilizer types used, only NPK and of course poultry manure can supply up to three nutrient elements. At present, the use of poultry manure in citrus cultivation is not widespread among Benue farmers. In the face of rising cost of inorganic fertilizer and even non-availability, farmers need to be encouraged to utilize different sources of organic manure to increase yields.

An important scientific study on citrus (sweet orange) fertilization in Nigeria (Ortese 2014) has noted the pattern of response to fertilization and the influence of previous orchard management level on such response. Effect of fertilization becomes more pronounced after the third year of fertilizer application. Compared to other treatments, namely, control (no fertilizer application), 600 kg/ha NPK 15:15:15, 300 kg/ha NPK 15:15:15 + 5 t/ha poultry manure and 600 kg/ha NPK 15:15:15 + 20 t/

ha rice hull mulch, 10 t/ha poultry manure had higher fruit weight, peel weight and residue weight, but this was significantly so only with respect to the control. The 10 t/ha poultry manure generally favoured best fruit metric traits and fruit yield and reduced fruit weight loss although this was marginal.

Nutritional quality of juice was significantly ($p < 0.05$) influenced by farm management during each of the 3 years of evaluation. After 1 year of fertilizer treatment, well-managed orchard had the highest ash, potassium, moisture, sodium, sugar, volume of fruits juice and vitamin C contents in the fruits. By the second year, poorly managed orchard gave fruits that had the highest acidity, potassium, sugar and total soluble solids, whereas fairly managed orchard produced fruits that had the highest sodium, pH and vitamin C. Results of fertilization on fruit quality traits were not consistent from year to year. However, it was evident that 10 t/ha poultry manure, 300 kg/ha NPK 15:15:15 + 5 t/ha poultry manure and 600 kg NPK/ha + 20 t/ha rice hull mulch enhanced fruit juice quality traits. It is obvious that while fertilization has great potential in increasing yields and improving quality of fruit in Nigeria, more extensive evaluations will have to be done for more definitive conclusions.

8.6.4 Nutrition of **Vitellaria paradoxa**

Vitellaria paradoxa is a highly valued tree of the African savannah. Some of the reasons discouraging its cultivation are the slow growth rate and long gestation period (Umali and Nikiema 2002). The extremely slow growth rate has been considered as an added impetus to a comprehensive fertilization package for the species (Ugese et al. 2012). Presently fertilizer recommendation appears very scanty. For instance, Umali and Nikiema (2002) have recommended application of 2.5 kg ammonium sulphate, 1.5 kg calcium phosphate and 1.5 kg potassium chloride for ten trees.

For effective development of fertilization package to address nutrient requirements of plants, the first stage is a proper understanding of nutrient deficiency symptoms (Chweya 1982). This is a necessary step since for every plant species; expression of deficiency of particular nutrients is specific (Wallace 1961). Ugese et al. (2012) investigated the effect of macronutrient deficiencies on growth, dry matter and foliar attributes of shea seedlings. Deficiency of all the nutrients produced plants with smaller leaf areas (Table 8.4). Fewer number of leaves resulted from seedlings starved of N, P and K. Apart from S, dry matter content of normal seedlings was generally higher than that of nutrient-deficient seedlings. The absence of N and P in sand culture led to stunting of seedlings, while deficiency of all nutrients except Ca produced some noticeable foliar colour change. Imbalances in foliar nutrient concentration were observed with deficiency of each nutrient. Normal growth and dry matter production of seedlings with adequate provision of the macronutrients points to the critical role that fertilization could play in the general performance of this species.

Table 8.4 Effect of macronutrient deficiencies on growth of *Vitellaria* seedlings

Nutrient solution	Number of leaves	Leaf area (cm^2)	Plant height (cm)	Total shoot length (cm)	Stem girth (cm)	Total root length (cm)
Complete	5.7	29.2	6.3	13.0	0.32	71.3
– Nitrogen	4.3	13.5	5.2	11.8	0.25	54.8
– Phosphorus	3.7	12.9	4.0	10.5	0.29	73.7
– Potassium	4.5	20.0	5.4	11.4	0.31	48.3
– Calcium	4.7	24.2	5.9	12.2	0.32	59.3
– Magnesium	4.8	15.0	3.4	10.1	0.26	48.5
– Sulphur	5.5	19.5	5.2	11.1	0.29	31.5
$LSD_{(0.05)}$	1.3	3.5	1.0	2.2	0.05	7.2

Adapted from Ugese et al. (2012)

Seedling growth in *Vitellaria paradoxa* has been shown to be strongly influenced by potting media. For seedling plants, emergence speed and growth can be enhanced by suitable growth medium (Baiyeri 2003). Growth medium is considered the most important determinant of seedling quality (Baiyeri and Mbah 2006). In the shea tree (*Vitellaria paradoxa*), Ugese et al. (2011) found out that among growth media used in raising shea seedlings, medium 1:2:3 (rice hull/poultry manure/river sand) gave better seedling performance than the rest of the media. Even medium (2:3:1 rice hull/poultry manure/river sand) with lower bulk density, higher water holding capacity and total porosity and which had better nutritional content did not perform as much. Some of the above-mentioned physical properties reportedly favoured seedling performance of African breadfruit (Baiyeri and Mbah 2006) and nutmeg (Abirami et al. (2010). However, some of the physical properties of the medium in question particularly its high water holding capacity may have presented less than optimum conditions for shea seedling growth (Ugese et al. 2011). *Vitellaria paradoxa* is typically a savanna species and does not tolerate waterlogged conditions. This underscores a basic principle in formulating growth media for plant species. Thus not only nutrient content is taken into consideration but also the physical conditions of the medium. This is informative for proper management of this important species even in the field.

8.6.5 *Nutrition of* Carica papaya

Carica papaya is very responsive to fertilizers, and yields could be significantly influenced by fertilizer application (Rice et al. 1986). However, fertilizer recommendation in papaya does not seem to be so definite. Samson (1980) reports that in Queensland, 1250 kg NPK 12–34-12/ha is broadcast and ploughed in, while 40 kg N/ha is given every 2 months thereafter. Organic manure application in planting holes and annually is also recommended. In Australia, 0.9–1.4 kg of NPK 8-12-6 or equivalent per tree per annum at three split applications is recommended. In addition, the species hardly responds to higher levels of fertilization by giving vegetative growth at the expense of fruit yield as is the case with many other tree crops (Rice et al. 1986). As such it is capable of utilizing heavy doses of fertilizer to give reasonable output. General recommendation by van Ee (1999) is based on plant uptake which fertilizer application is supposed to compensate for. Fruit harvest of 20 t/ha removes from the soil 35.4 kg N, 9.2 kg P_2O_5, 102.2 kg K_2O, 9.8 kg CaO and 6 kg MgO per year. It is easily observable that nutrient removal is dominated by N and K. In Nigeria, fertilization of papaya is not well defined probably because the plant is hardly grown on plantation level. However one or few stands are maintained around the household with ample supplies of domestic refuse. With the uptake of large quantities of N and K by the crop, organic matter, particularly poultry manure, is likely to benefit the crop immensely since it is dominated by N and K (Thomson and Troeh 1978).

Results of early trials on papaya fertilization in Nigeria are quite promising. Olajide (2016) evaluating 12 accessions of papaya at three levels of poultry manure (0, 5, 10 t/ha) observed positive effect on tree growth attributes with increasing performance as poultry manure rate increased (Table 8.5). Flowering and fruiting phenology were also affected by application of poultry manure. Days to flower budding, anthesis, first fruit set and first fruit maturity were reduced with poultry manure application. Proportion of plants that bore fruits increased with increase in rate of manure. There was no significant difference between fruit number and yield of plants that received 5 and 10 t/ha of poultry manure although the higher rate tended to give numerically higher values. It was however noteworthy that plants that received zero application of poultry manure failed to bear fruit altogether within the time under consideration. This underscores the need for proper nutrition of plants for enhanced productivity. In *Passiflora edulis*, poor nutritional status undermined the ability of plants to bear fruits on time (Ani and Baiyeri 2011). Unfortunately poultry manure rates evaluated had little influence on post-harvest behaviour of papaya fruits.

8.7 Conclusions and Future Outlook

There is convincing evidence of differential nutrient mining abilities of *Musa* cultivars with agreeable implications for fertilization timing, nutrient management and crop yield. It is evident that placement of manure at the top and bottom of planting hole has more favourable effect on growth, phenology and yield of *Musa* species. The clear superiority of organic manure as against inorganic fertilizers in influencing the performance of banana and plantains is obvious. As such given the fragile nature of tropical soils, greater attention needs to be given to this nutrient source in the supply of growth resources in *Musa* cultivation. Combinations of organic and inorganic fertilizers for profitable cultivation of this species will however need to be considered in cases where organic manure is scarce. What is apparent is that for high productivity levels to be sustained with soil and consumer health in focus, organic manure must be involved in production of banana and plantains. The impact of various fertilizer combinations on storability and quality attributes of fruits would need to be assessed. Similarly early signs of ectomycorrhizal benefits in *Musa*

Table 8.5 Effect of poultry manure rates on growth (15 months after transplanting), flowering, fruiting phenology and yield of papaya grown at Nsukka, Nigeria

Poultry manure (t/ha)	Plant height (cm)	Growth vigour	No of leaves	Stem circumf (cm)	Days to flower budding	Days to anthesis	Days to first fruit set	Days to first fruit maturity	No of fruits 11 months after fruit set	Yield (t/ha)
0	56.5	3.2	5.7	5.0	79.0 (7.2)[a]	0.0 (0.7)	0.0 (0.7)	0.0 (0.7)	0.0 (0.7)	0.0
5	167.6	3.4	15.6	15.0	71.6 (8.5)	44.2 (6.7)	362.3 (19.1)	511.3 (22.6)	5.7 (2.3)	1.3
10	178.1	3.7	16.4	16.7	70.0 (8.4)	42.7 (6.6)	330.1 (18.1)	479.6 (21.8)	6.6 (2.6)	1.3
$LSD_{(0.05)}$	7.84	0.27	0.84	0.68	1.24	0.71	0.52	0.36	NS	NS

Adapted from Olajide (2016)

[a]Values in brackets were square root transformed and used for ANOVA

production need to be more closely investigated owing to the serious promise it holds in beefing up crop yields at low cost.

Passiflora edulis, a new entrant with promising future in Nigeria, requires reasonable levels of fertilizer for optimum performance. The species has shown favourable response to fertilization with poultry manure. Generally, manure application hastens growth and reproduction and increases yield of the crop. Application rates of 10 and 15 t/ha of poultry manure have been found most suitable for yield of passion fruit. However, best fruit juice quality indices are obtained with application of 15 t/ha. Thus in situations in which fruit juice quality is not too critical, production can be executed at manure rates of 10 t/ha which appears more economical than the 15 t/ha. A combination of 20 t/ha poultry manure, 50 kg N/ha and 150 kg K/ha has been adjudged the best for quality seed production. It might be necessary to assess a wider array of combinations of nutrient or nutrient source combinations for optimum seed quality of *Passiflora edulis.*

Like banana/plantains and *Passiflora edulis*, citrus species respond to fertilization. Response, particularly to organic fertilizers, may not be as dramatic as in the other species. Usually, marked impact becomes more evident after the third year of application. However, it impacts favourably on fruit yield, juice content and quality. Poultry manure at the rate of 10 t/ha gives marginally higher fruit weight, peel weight and residue weight over varying combinations of poultry manure and NPK fertilizer. This rate (10 t/ha poultry manure) also favour best fruit metric traits and fruit yield and minimal fruit weight loss. Generally, fruit juice quality indices are improved when plants are supplied with combinations of poultry manure and NPK fertilizer in varying proportions. It is clear that in citrus, the issue of fertilization is a bit more complex as such factors as management history, age of plantation and season could play critical roles in determining response to applied external nutrient inputs. Accordingly, this has to be considered in future efforts at enhancing citrus nutrition. Additionally, effect of application of poultry manure or other organic amendments could last beyond 3 years. This aspect needs to be conclusively investigated to better guide nutrient management in this important species.

Vitellaria paradoxa is an invaluable tree species but is not yet fully domesticated. Although it benefits from fertilization and its slow growth rate is an added impetus to nutrient addition, such practice is still at infancy. Nutrient deficiency symptoms as found in the species are a clue to the important role proper nutrition that could play in the proper functioning and ultimate productivity of this species. Preliminary results suggest the need for proper balance between nutrient content and physical properties of growth medium for optimum tree performance. Considerations as indicated here are expected to guide further work in this regard to arrive at informed fertilization package for *Vitellaria*. In addition, fertilization of adult trees for fruit yield and quality needs to be investigated.

Carica papaya, though a pleasant fruit crop, is not yet widely cultivated at plantation level particularly in Africa. This scenario may change as societies come to better appreciate and exploit the industrial potentials of this species as is the case with the industrial extraction of papain in Tanzania (Rice et al. 1986). It however has credentials as a suitable fertilization candidate. This includes its capacity to take up large quantities of manure without readily tilting towards vegetative growth at the expense of fruit yield. Applications of 0–10 t/ha of poultry manure have already shown favourable impact on vegetative growth, reproductive development and fruit yield. Since the species has capacity to absorb large quantities of fertilizer without adverse effect on fruit yield, it might be needful to experiment with higher doses of this manure. The suitability of other manures in the fruit yield of papaya also needs to be explored. Unfortunately, poultry manure did not exert beneficial effect on post-harvest behaviour of papaya fruits such as delay in fruit ripening or reduced incidence of rot. The highly climacteric nature of this fruit is one of the key reasons limiting its distant marketing and hence plantation establishment. Solutions to this challenge may however have to be sought using other approaches.

The overall picture tilts towards greater emphasis on utilization of organic sources of nutrients in the cultivation of tropical perennials. These organic materials are more suitable in improving the physical and chemical properties of the largely fragile and weak soils of the tropics and thus ensuring soil health and productivity. These fertilizers, compared to inorganic types, have superior influence on crop growth, yield and produce quality. They are also cheaper, relatively more accessible and environmentally friendlier. They therefore deserve the focus of attention in the cultivation of fruits in the tropics.

References

Abirami K., Rema J., Mathew P. A., Srinivasan V., & Hamza S. (2010). Effect of different propagation media on seed germination, seedling growth and vigour of nutmeg (Myristica fragrans Houtt.). *Journal of Medicinal Plants Research 4*, 2054–2058.

Akinrinde, E. A. (2006). Strategies for improving crops' use-efficiencies of fertilizer nutrients in sustainable agricultural systems. *Pakistan Journal of Nutrition, 5*, 185–193.

Alva, A. K., Mattos, D. J., Paramasivam, M. N., & Sajwan, K. S. (2006). Potassium management for optimizing citrus production and quality. *International Journal of Fruit Science, 6*, 1–43.

Amberger, A. (2006). *Soil fertility and plant nutrition in the tropics and subtropics*. Paris: IFA and IPI.

Ani, J. U., & Baiyeri, K. P. (2008). Impact of poultry manure and harvest season on juice quality of yellow passion fruit (Passiflora edulis var. flavicarpa Deg.) in the sub-humid zone of Nigeria. *Fruits 63*(4), 239–247.

Ani, I. U., & Baiyeri, K. P. (2011). The effects of poultry manure rates and cropping year on phenology and yield response of yellow passion fruit (*Passiflora edulis* var flavicarpa Deg) in sub-humid zone of Nigeria. *Proceedings of the 29th Annual Conference of the Horticultural Society of Nigeria, 24th–29th July 2011.*

Araujo, R., Bruckner, C. H., Martinez, H. E. P., Saloao, L. C. C., Alvarez, V. H., Souza, A., Pereira, W. E., & Hizumi, S. (2006). Quality of yellow passion fruit (*Passiflora edulis* Sims F Var Flavicarpa Deg) as affected by potassium nutrition. *Fruits, 61*, 109–115.

Avav, T., & Uza, D. V. (2002). Agriculture. In A. L. Pigeonniere (Ed.), *Africa atlases: Nigeria.* Paris: Les Editions J A.

Awoleye, F. (1995). Effects of seed sources on the growth of seedlings of *Vitellaria paradoxum* in the southern Guinea savanna of Nigeria. *Nigerian Journal of Botany, 8*, 65–69.

Baiyeri, K. P. and Mbah, B. N. (1994). Growth and yield correlation in false horn plantain (Musa AAB cv. Agbagba) in a sub-humid zone of Nigeria. *MusAfrica, 5*, 3–4.

Baiyeri, K. P. (2003). Evaluation of nursery media for seedling emergence and early seedling growth of two tropical tree species. *Moor Journal of Agricultural Research, 4*, 60–65.

Baiyeri, K. P., & Mbah, B. N. (2006). Effects of soilless and soil based nursery media on seedling emergence, growth and response to water stress of African breadfruit (*Treculia Africana* Decne). *African Journal of Biotechnology, 5*, 1405–1410.

Baiyeri, K. P., & Tenkouano, A. (2007). Manure placement influenced growth and dry matter yield of a plantain hybrid. *African Crop Science Conference Proceedings, 8*, 385–390.

Baiyeri, K. P., & Tenkouano, A. (2008). Fruit characteristics and ripening pattern of ten *Musa* genotypes in a sub-humid environment in Nigeria. *Fruits, 63*, 3–9.

Baiyeri, K. P., & Ugese, F. D. (2011). Tropical fruits and vegetables: Physical properties. In J. Glínski, J. Horabik, & J. Lipiec (Eds.), *Encyclopedia of Agrophysics*. Poland: Springer, AA Dordrecht.

Baiyeri, K. P., Mbah, B. N., & Tenkouano, A. (2000). Yield components of triploid and tetraploid Musa genotypes in Nigeria. *HortScience, 35*, 1338–1343.

Baiyeri, K. P., Ndukwe, O. O., & Tenkouano, A. (2011a). Manure placement method influenced growth, phenology and bunch yield of three Musa genotypes in a humid zone of southern Nigeria. *Proceedings of the 29th Annual Conference of the Horticultural Society of Nigeria, 24th–29th July 2011.*

Baiyeri, K. P., Ndukwe, O. O., & Tenkouano, A. (2011b). The effect of manure placement methods on dry matter partitioning pattern in three Musa genotypes grown in a humid agro-ecology of southern Nigeria. *Proceedings of the 29th Annual Conference of the Horticultural Society of Nigeria, 24th–29th July 2011.*

Baiyeri, K. P., Ugese, F. D., & Uchendu, T. O. (2011c). The effects of previous fertilizer treatments on passion fruit seed quality, and seedling emergence and growth qualities in soilless media. *Journal of Agricultural Technology, 7*, 1397–1407.

Baligar, V. C., & Bennett, O. L. (1986). Outlook on fertilizer use efficiency in the tropics. *Fertilizer Research, 10*, 83–96.

Boffa, J. M., Yameogo, G., Nikiema, P., & Taonda, J. B. (1996). What future for the shea tree? *Agroforestry Today, 8*, 5–9.

Cardwell, V. B. (1954). Seed germination and crop production. In *Physiological basis of crop growth and development*. Madison: American Society of Agronomy/Crop Science Society of America.

Chweya, J. A. (1982). Nutrient deficiency symptoms in kale (*Brassica oleracea* Var Acephala). *East African Agricultural and Forestry Journal, 48*, 15–18.

Daniells, J., Englberger, I., & Lorens, A. (2011). Farm and forestry production and marketing profile for banana and plantain (Musa spp.) In C. R. Elevitch (Ed.), *Specialty crops for Pacific Island Agroforestry*. Holuolua, Hwai'I: Permanent Agriculture Resources (PAR). Retrieved May 10, 2016 from http://agroforestry.net.scps.

Davies, F. S., Maurer, M. Z., & Albrigo, I. G. (1994). Reclaimed wastewater for irrigation of citrus in Florida. *HortTechnology, 3*, 163–167.

Davis, F. S., & Albrigo, L. G. (1994). Citrus. Wallington: CAB International.

van Ee, S. (1999). *Fruit growing in the tropics* (2nd ed.). Agrodok-series no. 5. Agromisa foundation.

Embleton, T. W., Jones, W. W., Cree, C. B., & Garber, M. J. (1978). Effects of N on Valencia Production. *California Citrog, 45*, 22–24.

Evans, E. A., & Ballen, F. H. (2012). *An overview of global papaya production, trade and consumption*. UF/IFAS Extension: University of Florida.

FAO. (1977). *The state of food and agriculture*. Rome: FAO.

FAO. (1989). *Utilization of tropical foods*. Tropical oilseed paper no 47.

FAO. (2004). *Food and Agriculture organization Faostat-statistical database*. Retrieved May 21, 2004 from http://apps.www.fao.org

FAO. (2006). *Fertilizer use by crop*. FAO fertilizer and plant nutrition bulletin no 17.

Geiseller, D., & Scow, K. M. (2014). Long-term effects of mineral fertilizers on soil microorganisms – A review. *Soil Biology and Biochemistry, 75*, 54–63.

Gmitter, C. J., & Hu, U. M. (1990). The influence of cultivar and high nitrogen and potassium fertilization on fruit quality traits of young orange trees. *Proceedings of the Florida State Horticultural Society, 106*, 8–12.

Grant, C., Bittman, S., Montreal, M., Plenchette, C., & Morrel, C. (2005). Soil and fertilizer phosphorus: Effects on plant P supply and mycorrhizal development. *Canadian Journal of Plant Science, 85*, 3–14.

Hart, J. (1998). *Fertilizer and lime materials*. Oregon State Extension Services.

Hartmann, H. T., Flocker, W. J., & Kofranek, A. M. (1981). *Plant science: Growth, development and utilization of cultivated plants*. New Jersey: Prentice-Hall.

Honfo, F. G., Tenkouano, A., & Coulibaly, O. (2011). Banana and plantain-based foods consumption by children and mothers in Cameroon and southern Nigeria: A comparative study. *African Journal of Food Science, 5*, 287–291.

Hossner, L. R., & Juo, A. S. R. (1999). *Soil nutrient management for sustained food crop production for upland farming systems in the tropics*. Taipei: Food and fertilizer technology center – an international center for farmers in Asia and Pacific Region.

ICRAF. (2000). *International Centre for Research in agroforestry*. Agroforestree Database.

IITA. (2000). *Annual report, project 2*. Ibadan, Nigeria: International Institute of Tropical Agriculture.

Kaddar, T., Russel, D. A., & Cooke, G. W. (1984). *The vital role of potassium fertilizers in tropical agriculture*. Alabama: International Fertilizer Development Center.

Keay, R. W. J. (1989). *Trees of Nigeria*. Oxford: Clarendon Press.

Khan, S. A., Mulvaney, R. L., Ellsworth, T. R., & Boast, C. W. (2009). The myth of nitrogen fertilization for soil carbon sequestration. *Journal of Environmental Quality, 36*, 1821–1832.

Knight, R. J., & Sauls, J. W. (1994). *The passion fruit*. Gainesville, Fla: University of Florida, Florida Cooperative Extension Service Fact Sheet.

Kongshaugh, G. (1998). *Energy consumption and greenhouse gas emissions in fertilizer production*. Marraketsch, Maroko: IFA Technical Conference.

Kotschi, J. (2013). *A soiled reputation: Adverse impacts of mineral fertilizers in tropical agriculture*. WWF Germany: Heinrich Böll Stiftung.

Ladha, J. K., Reddy, C. K., Padre, A. T., & van Kessel, C. (2011). Role of nitrogen fertilization in sustaining organic matter in cultivated soils. *Journal of Environmental Quality, 40*, 1756–1766.

Leakey, R. R. B. (2009). Agroforestry: A delivery mechanism for multi-functional agriculture. In L. R. Kellimore (Ed.), *Handbook on agroforestry: Management practices and environmental impact*. New York: Nova Science Publishers.

Mattila, P. K., & Joki-Tokola, E. (2003). Effect of treatment and application technique of cattle slurry on its utilization by ley. *Nutrient Cycling in Agroecosystems, 65*, 221–230.

Mohammed, J. M. (2002). Imbalance in nutrient supply as a threat to sustainable crop production. In A. Kraus (Ed.), *International fertilizer correspondent no 8*. International Potash Institute.

Morton, J. F. (1987). Passionfruit. In J. F. Morton (Ed.), *Fruits of warm climates*. Miami, Florida.

Mulvaney, R. L., Khan, S. A., & Ellsworth, T. R. (2009). Synthetic nitrogen fertilizers deplete soil nitrogen: A global dilemma for sustainable cereal production. *Journal of Environmental Quality, 38*, 2295–2314.

Naher, U. A., Othman, R., & Panhwar, Q. A. (2013). Beneficial effects of mycorrhizal association for crop production in the tropics – A review. *International Journal of Agriculture and Biology, 15*, 1021–1028.

Nakasone, N. Y., & Paull, R. E. (1999). *Tropical fruits*. New York: CABI Publishing Company.

Nelson, L. B. (1974). *Fertilizers for all-out food production*. ASA special publication no 23.

Olajide, K. (2016). *Agronomic evaluation of 12 pawpaw accessions grown under three manure rates in derived savanna agro*-ecology. M.Sc. Thesis, Department of Crop Science, University of Nigeria, Nsukka.

Olarewaju, J. D. (2004). Promotion of medicinal horticultural plants. *Proceedings of 22nd Annual Conference of Horticultural Society of Nigeria, 4th–9th July 2004*.

Opeke, L. K. (1987). *Tropical tree crops*. Ibadan: Spectrum Books Ltd.

Ortese, E. (2014). *Studies on orchard management and aspects of agronomic intervention to sweet orange (Citrus sinensis) production in Benue State, Nigeria*. Ph.D. Thesis, Department of Crop Science, University of Nigeria, Nsukka.

Ortese, E., Baiyeri, K. P., & Ugese, F. D. (2011). Residual macronutrient concentration and follower maize (*Zea mays*) crop performance in soilless growth medium previously cropped with six *Musa* genotypes. *Journal of Crop Science and Biotechnology, 14*, 321–327.

Ortese, E., Baiyeri, K. P., & Ugese, F. D. (2012). Demographic features of citrus producers and agronomic management of the crop in Benue state, Nigeria. *PAT, 8*, 180–190.

Oyebade, B. A., Aiyeloja, A. A., & Ekeke, B. A. (2010). Sustainable agroforestry potentials and climate change mitigation. *Advances in Environmental Biology, 4*, 58–63.

Rice, R. P., Rice, L. W., & Tindall, H. D. (1986). *Fruit and vegetable production in Africa*. London: Macmilan.

Samson, J. A. (1980). *Tropical fruits*. London: Longman Group Limited.

Sanchez, P., Izac, A. M., Buresh, R., Shepherd, K., Soule, M., Mokwunye, U., Palm, C., Woomer, P., & Nderitu, C. (1997). Soil fertility replenishment in Africa as an investment in natural resource capital. In R. J. Buresh, P. A. Sanchez, & F. Calhoun (Eds.), *Replenishing soil fertility in Africa*. Madison: Soil Science Society of America.

Seymour, G. B., Taylor, E., & Tucker, G. A. (1993). *Biochemistry of fruits ripening*. London: Chapman and Hall.

Singh, B., & Ryan, J. (2015). *Managing fertilizers to enhance soil health*. Paris: IFA. Retrieved from www.fertilizer.org/Library

Smil, V. (1999). Nitrogen in crop production: An account of global flows. *Global Biogeochemical Cycles, 13*, 647–662.

Smil, V. (2002). Nitrogen and food production: Proteins for human diets. *Ambio, 31*, 126–131.

Smith, P. F., & Reuther, W. M. (2007). Observations on boron deficiency in citrus. *Indvs, 31*, 5–7.

Stover, R. H., & Simmonds, N. W. (1987). Bananas. In *Tropical agricultural series* (3rd ed.). New York: John Wiley and Sons.

Swennen, R. (1990). Plantain cultivation under West African conditions: a reference manual. International Institute for Tropical Agriculture, Ibadan, Nigeria, Amarin Printing Group Co. Ltd. Thailand. 24p.

Thomson, L. M., & Troeh, F. R. (1978). *Soils and soil fertility* (4th ed.). McGraw Hill: Publication in Agriculture.

Ugese, F. D., Baiyeri, K. P., & Mbah, B. N. (2011). Nursery media influences growth of seedlings of the shea nut tree (*Vitellaria paradoxa* C. F. Gaertn.) *Afr J Plant Sci Biotech, 5*, 56–59.

Ugese, F. D., Baiyeri, K. P., & Mbah, B. N. (2012). Expressions of macronutrient deficiency in seedlings of the shea butter tree (*Vitellaria paradoxa* C.F. Gaertn.) *Journal of Agricultural Technology, 8*, 1051–1058.

Umali, B. E., & Nikiema, A. (2002). *Vitellaria paradoxa* C. F. Gaertn. Record from Protabase. In L. P. A. Oyen & R. H. M. J. Lemmens (Eds.), *PROTA*. The Netherlands: Wageningen.

UNCTAD. (2010). *United Nations conference on trade and development*. Retrieved from http://www.Unctad. Org/infocomm.

Vickery, M. L., & Vickery, B. (1969). *Plant products of tropical Africa*. London: Macmillan.

Wallace, T. (1961). *The diagnosis of mineral deficiencies in plants* (2nd ed.). NJ: Chemical Publishing Co Inc.

Webster, C. C., & Wilson, P. N. (1980). *Agriculture in the tropics* (2nd ed.640 pp). London, UK: Longman Group.

Wills, R. B. H., McGlasson, W. B., Graham, D., & Daryl, J. (1998). *An introduction to physiology and handling of fruits, vegetables and ornamentals*. Wallingford: CABI.

Wilson, G. F. (1987). Status of bananas and plantains in West Africa. In G. J. Persley & E. A. De Langhe (Eds.), *Banana and plantain breeding strategies, ACIAR Proceedings* (Vol. 21, pp. 28–32).

Wu, Q. S., & Zou, Y. N. (2012). Evaluating effectiveness of four inoculation methods with A*rbuscular mycorrhizal* fungi on trifoliate orange seedlings. *International Journal of Agriculture and Biology, 14*, 266–270.

Zachée, A., Bekolo, N., Bime, N. D., Yalen, M., & Godswill, N. (2008). Effect of mycorrhizal inoculums and urea fertilizer on diseases development and yield of groundnut crops (*Arachis hypogaea* L.) *African Journal of Biotechnology, 7*, 2823–2827.

Chapter 9
Unravelling the Impact of Essential Mineral Nutrients on Active Constituents of Selected Medicinal and Aromatic Plants

Minu Singh, Arlene Asthana Ali, and M. Irfan Qureshi

Abstract Humans for centuries have used plants to relieve discomfort and treat various health ailments. Medicinal herbs are used throughout developed and developing countries as home remedies, over-the-counter drug products and raw materials for the pharmaceutical industry, and represent a considerable proportion of the universal drug market. The medicinal value of plant depends on the nature of plant constituents, known as active principal or active constituent, present in it. Active constituents are those chemical substances, which are exclusively responsible for remedial activity of plant and serve as lead compounds in drug discovery and design. Traditional systems of medicine, modern medicines, folk medicines, food supplements, nutraceuticals, pharmaceutical intermediates and synthetic drugs are invariably dependent on the proportionate presence of active constituents found in plants. Nutrition plays an important role in the growth and development of all crop plants. The contribution of macro- and micronutrients in building indispensable organic compounds and in almost all plant life processes shows the noteworthy and diversified role of these minerals in the modification of plant metabolism. Severity or insufficiency of these minerals causes varied effects in plant metabolism. The role and contributions of various mineral elements can be revealed through their regulatory role played in metabolism of medicinal and aromatic plants. Secondary plant metabolism is a function of concentrations of minerals in the soil. Important mineral elements present in soil are transferred to plant areas where their need arises, thereafter governing various physiological activities. Therefore, biosynthesis and accumulation of these bioactive molecules in a plant system are broadly dependent on the availability and accessibility of mineral elements in the soil. This review chapter is an attempt to understand how essential mineral nutrients affect active constituents of selected medicinal and aromatic plants viz. fennel (*Foeniculum vulgare*), mentha (*Mentha arvensis, Mentha piperita, Mentha citrata*), lemongrass (*Cymbopogon flexosus*), Artemisia (*Artemisia annua*), turmeric (*Curcuma longa*), ginger (*Zingiber officinale*), periwinkle (*Catharanthus roseus*), Aloe vera (*Aloe barbadensis*).

Keywords Mineral nutrients • MAPs • Nutraceuticals • Plant metabolism

---

M. Singh (✉) • A.A. Ali • M. Irfan Qureshi
Department of Biotechnology, Jamia Millia Islamia, New Delhi, India
e-mail: minusingh5@gmail.com

M. Naeem et al. (eds.), *Essential Plant Nutrients*,
DOI 10.1007/978-3-319-58841-4_9

9.1 Introduction

Medicinal and aromatic plants (MAPs) have been used since approximately 5000 BC for their preservative and medicinal properties, in addition to enhancing the aroma and flavor of foods (Silviya et al. 2016). They are important factors in sustainable development, environmental protection and public health. A major segment of the world flora includes MAPs which are the source of raw materials used in the pharmaceutical, fragrance, cosmetic, flavour and perfumery industries (Bogers et al. 2006; Miguel et al. 2004). Aromatic and medicinal plants play a significant role in the life of people and are present in innumerable forms. Their use continues undiminished today, and according to the World Health Organization (WHO), nearly 80% of the planet's population, especially in developing countries, still depends on plant produced medicines for their healthcare (Collin 2006; Craker and Gardner 2006; Craker et al. 2003; Gurib-Fakim 2006). Interest in MAPs is increasing all over the world because of their lesser side effects as compared to the synthetic drugs besides cost-effectiveness and easy availability (Newman and Cragg 2007; Smillie and Khan 2010). Promoters of natural products have the opinion that herbal medicines are safe enough, and even if the desirable therapeutic response is not achieved, their use is not dangerous to health because of their natural origin. Now, the medicinal plants as a whole occupy a stable position in modern medicine, since the pharmaceutical industry is showing special interest in using or synthesizing natural substances extracted from the plants. The quality of MAPs can be primarily estimated by their active constituent(s) that produce a desirable therapeutic response.

9.2 Active Constituents and Their Significance for Medicinal Purposes

Active constituents are chemicals produced by plants in a vast diversity of more than 200,000 structures (Hartman 2007). The active constituents are often referred to as "secondary metabolites" of which there are several classes including alkaloids, flavonoids, coumarins, glycosides, gums, polysaccharides, phenols, tannins, terpenes and terpenoids (Gottlieb 1990; Grindberg et al. 2007; Nunnery et al. 2010; Okwu 2004). Moreover, aromatic plants contain odorous, volatile, hydrophobic and highly concentrated compounds called essential oils (or volatile or ethereal oils) which are complex mixtures of secondary metabolites consisting of low-boiling-point phenylpropenes and terpenes. These are obtained from various parts of the plant such as flowers, buds, seeds, leaves, twigs, bark, wood, fruits and roots. Researchers have revealed that terpenes are synthesized via the mevalonic pathway from precursor acetyl-CoA, while phenolic compounds are aromatic substances formed via the shikimic acid pathway or the mevalonic pathway. Moreover, nitrogen-containing active constituents (alkaloids) are synthesized primarily from aliphatic amino acids derived from the tricarboxylic acid pathway or aromatic acids derived from the shikimic acid pathway, via acetyl-CoA, mevalonic acid. Contrary to primary metabolites, secondary metabolites are not essential for growth processes but facilitates the plant in adapting to the environment, e.g., by serving as feeding deterrents against herbivores, protective agents against pathogens or abiotic factors, pollinator attractants, antioxidants, or chemical signals. Reports also suggest their specific role in various parasitic, pathogenic or symbiotic interactions (Croteau et al. 2000; Wink 2003). These active constituents owing to their bioactivity play an important role in the modern pharmacy and medicine. They range from general antioxidant agents which provide some protection against cancer or cardiovascular disease incidence. They appear to neutralize free radicals, inhibit enzymes that activate carcinogens, and activate enzymes that detoxify carcinogens (Mathai 2000; Meagher and Thomson 1999). They also demonstrate efficacy against specific human ailments such as depression, anxiety, or microbial infection, etc. (Percival 2000; WHO 1999). Active constituents play a significant role in modulation of hormone metabolism and detoxification of enzymes and in stimulation of the immune system (Rao 2003). They are also helpful in treatment of diabetes and high blood pressure (Modak et al. 2007; Tabassum and Ahmad 2011). Despite the use of MAPs and demand for plant-derived compounds, their availability is a major blockage in supplying the pharmaceutical needs. The biosynthesis of secondary metabolites, though controlled genetically, is strongly affected by soil and climatic and agricultural factors (Briskin 2000). Therefore, it is logical that their production can be stimulated or altered using physical or chemical environmental triggers, including the nutritional changes.

9.3 Regulatory Role of Some Essential Mineral Nutrients on Plant Metabolism and Active Constituents

Minerals have diversified functions in medicinal plant metabolism. The role and contributions of various mineral elements can be revealed through their regulatory role played in metabolism of MAPs. Severity or scarcity of these causes multifarious effects in plant metabolism. Plants obtain a number of inorganic mineral elements from their environment to ensure successful growth and development of both vegetative and reproductive tissues. These minerals serve numerous functions: as structural components in macromolecules, as cofactors in enzymatic reactions, as osmotic solutes needed to maintain proper water potential, or as ionized species to provide charge balance in cellular compartments. Minerals can be divided into two classes, based on the relative amounts needed for plant growth. The macronutrients include nitrogen (N), potassium (K), calcium (Ca), magnesium (Mg), phosphorus (P) and sulphur (S); these are generally found in plants at concentrations greater than 0.1% of dry tissue weight. The recognized micronutrients include iron (Fe), zinc (Zn), manganese (Mn), copper (Cu), boron (B), chlorine (Cl), molybdenum (Mo) and nickel (Ni); these generally are found at concentrations less than 0.01% of dry tissue weight. These 14 minerals, along with the elements carbon (C), hydrogen (H) and oxygen (O), are broadly accepted as essential for the growth of all plants.

Among various nutrients, N, P, Ca, Mg, P and S are considered to be of prime importance. These nutrients play several important roles in metabolic and regulatory process in plants (Marschner 2002). Fertilizer treatment is known to be one of the main factors increasing the yield and quality of plants. It affects the accumulation and mineralization of organic matter added to the soil and determines plant production potential. Fertilization also determines the concentration of bioactive plant chemicals (Said-Al Ahl et al. 2009) in MAPs. A number of studies reported the effects of agricultural practices on the secondary metabolites in medicinal and aromatic plants (Ashraf et al. 2006; Ozguven et al. 2002; Özgüven et al. 2008; Sekeroglu and Ozguven 2006). Fertilization also influences the volatile oil content and constitution of medicinal and aromatic plants. The production of essential oils not only depends upon the metabolic state and preset developmental differentiation programme of the synthesizing tissue but also is highly integrated with the physiology of the whole plant. Thus, the plant nutrition plays a vital role in enhancing morphological and phytochemical characteristic of plants.

The regulatory roles of some important mineral nutrients are briefly described below.

9.3.1 Nitrogen

Nitrogen is the fourth most abundant element in plants after carbon, hydrogen and oxygen. It is absorbed by plants as nitrate ions (NO_{3-}), ammonium ions (NH_4^+) and urea. As a constituent of all amino acids and proteins (and thus all enzymes), nitrogen serves a central role in cellular metabolism. Additionally, as a component of nucleotides and nucleic acids deoxyribonucleic acid (DNA) and ribonucleic acid (RNA), nitrogen is critical for the transcription, translation and replication of genetic information. Nitrogen also plays an important role in the production of the chlorophyll and is therefore necessary for photosynthesis. Physiological maturity and yield of many crops have been governed by the N supply to the crops.

Several studies conducted on different MAPs show that N fertilization significantly contributes to an enhancement in active constituents and composition of essential oil, through their effect on biomass yield, plant height, leaf area and photosynthetic rate (Arabaci and Bayram 2004; Sreevalli et al. 2004; Ashraf et al. 2006; Sekeroglu and Ozguven 2006; Daneshkhah et al. 2007; Mahfouz and Sharaf-Eldin 2007; Daneshian et al. 2009; Zheljazkov et al. 2010; Hendawy and Khalid 2011). Kiferle et al. (2013) conducted an experiment in hydroponic culture in order to study the response of N nutrition on biomass and rosmarinic acid (RA). The treatment consisted of different NO_3^- concentrations (0.5, 5.0 and 10.0 mol m^{-3}) or NO_{3-}/NH_4^+ molar ratios (1:0, 1:1 and 0:1; total N concentration was 10.0 mol m^{-3}). The concentration of other nutrients was as follows: 1.0 mol m^{-3} P-H_2PO_4, 10.0 mol m^{-3} K^+; 3.0 mol m^{-3} Ca^{2+}; and 1.5 mol m^{-3} Mg^{2+} plus trace elements. Their results revealed that application of 5.0 mol m^{-3} NO_{3-} concentration proved optimum for plant growth and RA production. However, the addition of NH_4^+ to the nutrient solution was found detrimental to both growth and RA production. In another study carried out by Nguyen and Niemeyer (2008) on three cultivars of basil, viz. dark opal, Genovese and sweet Thai, it was found that the lowest level of N (0.1 mM) fertilization resulted in significantly higher rosmarinic and caffeic acid concentrations as compared to higher N level (5.0 mM). Said-Al Ahl et al. (2009) in their study showed that the application of the highest N rate (1.2 g N·10 kg^{-1} of soil) proved the best with respect to herbage yield and essential oil production in *Origanum vulgare* L. Azizi and Dias (2004) conducted a field study to determine the effect of N and P fertilizer on the flavonoids (chlorogenic and isochlorogenic acid, apigenin, biapigenin, rutin, quercetin, isoquercetin and amentoflavone) content of *Hypericum perforatum.* They reported that 125 kg N ha^{-1} and 50 kg P_2O_5 ha^{-1} proved optimum for production of high dry herb yield and flavonoids content in comparison with control treatment. Jabbari et al. (2011) performed a field trial to study the response of *Thymus vulgaris* L. to N and Fe application. Treatments comprised of soil and foliar application of N and P at the rate of 50 kg ha^{-1} and 5 g m^2, respectively. The results obtained indicated that foliar application of N enhanced thymol percentage (2.56%) and thymol (19.56 Kg ha^{-1}), carvacrol (0.53 Kg ha^{-1}) and p-cymene (1.18 Kg ha^{-1}) yield. Similar results were also reported by Barnauskiene et al. (2003) in thyme. They found that thymol, carvacrol and p-cymene were significantly affected by different N fertilizer levels; however the related percentages of these and other compounds in thyme herb subjected to different fertilization doses were not significantly affected. Arabaci et al. (2007) reported that nitrogen fertilizer increased linalool content in the essential oil of *Lavandula hybrida*; nevertheless same condition decreased linalyl acetate content. While Ashraf et al. (2006) reported that concentration of p-cymene in the essential oil of black cumin seed increased on application of nitrogen fertilizer, no change in the level of α-pinene or β-pinene was observed at varying level of nitrogen. Moreover, Puttanna et al. (2010) in their experimental study also noticed a significant increase in rosemary oil yield with N and K application. Nurzyska-Wierdak (2012) reported that foliar feeding of nitrogen in the form of urea contributed to an enhanced concentration of linalool and *epi-α*-cadinol as well as a decreased content of 1,8-cineole, geraniol and eugenol in the oil of *Ocimum basilicum* L. This correlation could be explained by the complicated pathways of biosynthesis of terpene and aromatic compounds present in volatile oils and by the effect of quickly absorbed nitrogen on the biosynthesis of particular oil components.

9.3.2 Phosphorus

Phosphorus is often the limiting element in the soil. It is present in fixed forms as organic and inorganic molecules in the soil and is absorbed by plants as monovalent ($H_2PO_4^-$) or divalent (HPO_4^{-2}) anions. It is a structural component of numerous macromolecules, including nucleic acids, phospholipids, certain amino acids and several coenzymes. It has a significant role in energy transfer via the pyrophosphate bond in ATP, and the attachment of phosphate groups to many different sugars provides metabolic energy in photosynthesis and respiration. P aids in root development, flower initiation and seed and fruit development. Seeds have the highest concentration of P in a mature plant, and P is required in large quantities in young cells, such as shoots and root tips, where metabolism is high and cell division is rapid. P has been shown to reduce disease incidence in some plants and has been found to improve the quality of certain crops.

Ram et al. (2003) reported the increase in essential oil yield of *Mentha arvensis* with the foliar application of phosphate. Similarly Harendra and Yadav (2007) in another study recorded significant increase in the oil content of mustard with increase in P level from 0 to 39.3 kg P_2O_5 ha^{-1}. Thirty percent of oil content was recorded at 39.3 kg which was significant over all other levels including control. Ramezani et al. (2009) recorded a significant increase in essential oil content of basil with P application. Ram et al. (2003) reported the increase in essential oil yield of *Mentha arvensis* with the foliar application of phosphate. Boroomand et al. (2011b) reported significantly higher oil content of basil (2.40%) at 100 Kg P_2O_5 ha^{-1} which was superior over all other P levels. Prasad et al. (2012) noticed a significant amelioration with P application in oil yield of rose-scented geranium and also in the content of citronellol and 10-epi-γ-eudesmol. Lu et al. (2013) examined the response of different levels of P fertilization on root growth and production of bioactive compounds in *Salvia miltiorrhiza*. They found that application of 0.4 g P per plant significantly enhanced the root growth and accumulation of bioactive compounds, viz. danshensu, (0.055%), salvianolic acid B, (4.50%), cryptotanshinone, (0.056%), tanshinone II A (0.127%) and total tanshinone (0.226%) as compared to the other P treatments.

9.3.3 Calcium

Calcium (Ca), one of the essential nutrients for plants, plays a major role in the initiation of many signal transduction processes in higher plant cells, including bud formation, polar growth, gas exchange, regulation, secretion, movements and light and hormone-regulated growth and development. Calcium is involved in cell elongation and cell division, influences the pH of cells and also acts as a regulatory ion in the source-sink translocation of carbohydrates through its effects on cells and cell walls (Hirschi 2004). Calcium is absorbed by plants as a divalent cation (Ca^{2+}) and as a constituent of cell walls in the form of calcium pectate. Numerous investigations confirm the beneficial effect of Ca fertilization on growth and essential oil yield in treated plants (Dordas et al. 2007; Lee and Yang 2005; Mumivand et al. 2011; Suh and Park 2000; Supanjani Tawaha et al. 2005). In *Chrysanthemum coronarium* Supanjani Tawaha et al. (2005) reported that total dried weight, plant height, essential oil content and yield increased significantly by increasing lime application up to 2 t ha^{-1}. Studies of Heidari et al. (2014) also showed that foliar $Ca(NO_3)_2$ application rates significantly augmented essential oil content and yield of *Artemisia dracunculus*. Prasad et al. (2008) in another study found that the herbage yield and oil content and yield of geranium (*Pelargonium* sp.) showed a significant enhancement with the foliar application of calcium chloride. Results of the hydroponic experiment conducted by Alaghemand et al. (2013) in their study revealed that content of hyoscyamine and scopolamine in leaf was enhanced with the Ca. application. Furthermore, Lee and Yang (2005) in their studies also found that $CaCO_3$ significantly increased total plant yield, stem diameter, flower P content and essential oil content and yield of *Chrysanthemum boreale* up to 1.5 t ha^{-1}.

9.3.4 Potassium

Potassium is abundantly present in soluble form in the cytoplasm and in the vacuolar cell sap. It is absorbed by plants as the monovalent cation (K^+). K ion serves to activate number of enzymes, notably those involved in photosynthesis and respiration. It is the most abundant cation in the cytoplasm, and because it is not metabolized, its accompanying anions contribute significantly to the osmotic potential of cells. Thus, potassium functions in plant water relations processes and affects cell extension and growth through the regulation of turgor, leaf gas exchange through the control of stomatal opening/closing and long-distance nutrient flow through pressure-driven phloem translocation. Potassium promotes the translocation of photosynthates (sugars) for plant growth. It has been shown to improve disease resistance in plants, improve the size of grains and seeds and improve the quality of fruits and vegetables. There are several supporting studies which proves that K fertilization have beneficial effect on MAPs, regarding active constituents, essential oil yield and composition (Davies et al. 2009; Ezz El-Din et al. 2010; Jeliazkova et al. 1999; Said-Al Ahl et al. 2009). Lu et al. (2013) reported that K treatment at the rate of 1.25 g K per plant improved growth and accumulation of bioactive compound (danshensu, 0.52% and salvianolic acid B, 4.13%) in *S. Miltiorrhiza*, as compared to the other K treatments. In basil, the essential oil content in herbage was found significantly higher in plants fertilized with the higher K (0.8 g Kdm^{-3}) rather than the lower rate (0.4 g Kdm^{-3}). Also, higher amount of potassium contributed to an enhanced percentage of 1.8-cineole, linalool, eugenol and γ-cadinene in basil oil (NurzyEska-Wierdak and Borowski 2011). In case of *Rosa damascena* Mill., the application of the medium rate of potassium (30 kg ha^{-1}) contributed to maximum essential oil concentration as reported by Daneshkhah et al. (2007). Conversely, a negative effect was observed in the accumulation of cryptotanshinone, tanshinone II A and total tanshinone. Also, findings of some researchers showed that application of K fertilization did not influence the essential oil contents of patchouli (Singh and Ganesha Rao 2009), palmarosa (Singh 2008), rosemary (Singh et al. 2007) and lemongrass (Singh et al. 2005).

9.3.5 Magnesium

Magnesium (Mg) is an indispensable plant nutrient. It has a wide range of key roles in many plant functions. One of the magnesium's well-known roles is in the photosynthesis process, as it is a building block of the chlorophyll, which makes leaves appear green. Magnesium also is a necessary activator for many critical enzymes, including ribulose bisphosphate carboxylase (RuBisCO) and phosphoenolpyruvate carboxylase (PEPC), both essential enzymes in carbon fixation. Thus low amounts of Mg lead to a decrease in photosynthetic and enzymatic activity within the plants. Additionally Mg plays a key role in protein synthesis, phloem loading, phloem loading, partitioning and utilization of photoassimilates and generation of reactive oxygen species. Consequently, many critical physiological and biochemical processes in plants are adversely affected by Mg deficiency, leading to impairments in growth and yield. Magnesium is also crucial in stabilizing ribosome structures; hence, a lack of magnesium causes depolymerization of ribosomes leading to pre-mature aging of the plant.

Szoke et al. (2004) in their study showed that addition of $MgSO_4$ to the solid and liquid MS media, respectively, favourably affects the biomass and essential oil production of chamomile organized cultures and hairy root cultures. The total essential oil content of hairy roots increased with higher Mg concentration, but in contrast, the relative percentile distribution of the main essential oil components (e.g. tr-β-farnesene, α-farnesene) declined while that of other terpenes improved. Zamborine et al. (2016) examined the effects of different fertilizers on biomass, drug yield and essential oil content of lemon balm (*Melissa officinalis* L.) under field conditions. They reported that the treatment with 150 kg ha^{-1} N + 22 kg ha^{-1} P supplemented by 36 kg ha^{-1} Mg resulted in 25–27% higher folium mass compared to both untreated and N + P treated plots. Similarly, supplementary dosages of 36 kg ha^{-1} Mg + 149 kg ha^{-1} K assured by 24% more folium yield compared to untreated and 10% more yield compared to N + P fertilized control. Furthermore, the beneficial effect of Mg fertilization on secondary metabolites of *Phyllanthus niruri* was also reported by Hanudin et al. (2012).

9.3.6 Sulphur

Sulphur is an essential nutrient for plant growth. Although it is considered a secondary nutrient, it is now becoming recognized as the 'fourth macronutrient', along with N, P and K. Sulphur is essential for chlorophyll formation. It is a major constituent of one of the enzymes required for the formation of the chlorophyll molecule. Sulphur is a constituent of the amino acids three amino acids (cysteine, cystine and methionine), which are the building blocks of protein. About 90% of plant S is present in these amino acids. Some of the most important functions of S include its role in specialized peptides, such as glutathione and thioredoxins, in redox reactions, and the role of disulphide bond (S-S) formation in the stabilization of protein structure. Sulphur is necessary for the efficient fixation of nitrogen by the leguminous plants. Aziz et al. (2010a) reported that S fertilization enhanced the essential oil yield and changed the level of secondary metabolites in dragonhead plants. Similar results were observed by Zheljazkov et al. (2008) where S application caused a rise in production of essential oil and eucalyptol content in basil oil.

9.4 Effect of Essential Mineral Nutrients on Active Constituents of Selected MAPs

9.4.1 Fennel

Fennel (*Foeniculum vulgare* Mill) is a medicinal plant which acts as a stimulant, diuretic, carminative, sedative, galactagogic, emmenagogic, expectorant and antispasmodic (Kandil et al. (2002). Fennel fruits are effective in the treatment of diseases like cholera, bile disturbances, nervous disorder, constipation, dysentery and diarrhoea (Leung and Foster 1996). The fruits are also used as a standard medicinal tea for dyspepsia in Germany (EMEA 2008). The essential oil (EO) of fennel is used in perfumery industry, in confectionery and in food flavouring as well. The EO, majorly contains anethole, i.e. 1-methoxy-4-(1-propenyl) benzene, fenchone, i.e. (+) trimethyl (2,2,1) bicycloheptan-2-one and limonene, i.e. 4-isopropenyl-1-methylcyclohexene, terpenic compounds and minor compounds like á-pinene and â-pinene (Mahfouz and Sharaf-Eldin 2007).

Studies of Kandil et al. (2002) demonstrated that ammonium nitrate-treated plants produced the highest limonene content. Anethole content effectively increased using organic nitrogen sources (compost, compost/*Azotobacter* and chicken manure) in comparison to applying ammonium nitrate. However, fenchone and estragole content were found insignificant with different nitrogen fertilization. Generally fertilization with phosphorus sources such as superphosphate and rock-P/S was more effective than rock-P alone in increasing the fenchone and anethole content. Fertilization with feldspar as a potassium source increased anethole content by 43% compared with the anethole content in plants grown with potassium sulphate. Feldspar effectively increased the fenchone content of fennel EO. The speculated reason is feldspar can release K at slower rates than soluble K fertilizers do (Bakken et al. 1997) apart from containing Fe, Mn, Cu, Mn and Zn. Ohallorans et al. (1993) found that fertilization with chicken manure increased both the available N and exchangeable K in the soil. But fennel fruits oil components (limonene, fenchone, estragole and anethole) were not significantly affected by the different treatments of N, P and K (Buntain and Chung 1994; Desmarest 1978). It is to be noted that organic fertilization gave similar fennel fruit oil quality as observed after chemical fertilization. The recommended fertilization dose of *F. vulgare* Mill plants is 40 kg ha^{-1} P and feldspar at the rate of 50 kg ha^{-1} K to get higher yields of fruits and better essential oil quality (Kandil et al. 2002). The highest anethol in fennel essential oil occurred with the half dose of N, P and K (357 kg ammonium sulphate +238 kg calcium super phosphate +60 kg potassium sulphate ha^{-1}) and inoculation with *Bacillus megaterium* (Mahfouz and Sharaf-Eldin 2007).

9.4.2 *Mentha*

Peppermint, an important aromatic plant, is grown for production of essential oil, an important aromatic agent. The use of its essential oil in a number of consumer products such as herbal teas, chewing gum, toothpaste and mouthwashes and in pharmaceuticals, confectionary and aromatherapy (Mint Industry Research Council 2015). The essential oil content and composition of peppermint are important economically for the peppermint industry and for consumers. The volatile oil from peppermint comprises primarily monoterpenes, with less than 2% sesquiterpenes (Croteau et al. 1972a). High menthol (>45%), low menthofuran (<1–3%), low menthone (15–25%) and high menthyl acetate (4–9%) are generally considered as indicators of supreme quality of peppermint oils. Carvone (76.65%) comprises the major component of essential oil of *Mentha spicata*, the other constituents being limonene (9.57%) and 1,8-cineole (1.93%). The trace constituents are Z-β-ocymene, *cis*-muurola-4 (14), 5-diene and bicyclogermacrene (<0.01%) (Chauhan et al. 2009).

It has been known that fertilization affects peppermint essential oil composition and content (Scavroni et al. 2005; Zheljazkov et al. 2009). The optimum N fertilization rate has been decided to be 250 lb./a since there was no significant difference in oil yield between 250 lb./a and higher rates (Mitchell and Farris 1996). The effect of N, P and K on the essential oil of three Bulgarian cultivars was studied (Jeliazkova et al. 1999). With the increased fertilizer rates, an increased essential oil yield (16 to 119% compared to the control) was evident, the highest yield obtained using treatments of N 533.6, P_2O_5 182 and K_2O 240 kg ha^{-1}. The GC study of the oil from different variants revealed that with increasing fertilizer application, the menthol content of cv. *Zefir* and cv. *Tundja* increased, supporting other reports (Marotti et al. 1994). The content of other active constituents did not have a significant change with the tested fertilizer rates. In an attempt to evaluate the effect of N on peppermint productivity, oil content and its composition, Zheljazkov et al. (2009) noticed N at 80 kg ha^{-1} did not have an effect on the oil content, constituents or yield of the oil constituents compared with 0 kg ha^{-1}. The lack of N response to essential oil constituents (−)-menthol, (−)-menthone, (+)-menthofuran and eucalyptol in their study may be ascribed to the residual N in the soil, relatively low N application rate (80 kg ha^{-1}) and the variability within the experiment, which might have shrouded the effect of N.

9.4.3 *Cymbopogon*

The tropical aromatic herb, *Cymbopogon*, commonly known as lemongrass, belongs to the family Poaceae (true grasses). Species well known for their good quality essential oils (EOs) are Lemongrass (*Cymbopogon flexuosus*), wild lemongrass (*C. citratus*), palmarosa (*C. martinii) and* citronella (*C. winterianus*). EOs of most *Cymbopogon* spp. comprises of compounds like citral, geraniol, citronellol, citronellal, linalool, elemol, 1,8-cineole, limonene, b-caryophyllene, methyl heptenone, geranyl acetate and geranyl formate (Ganjewala and Gupta 2013; Ganjewala et al. 2008). Citral is responsible for the characteristic lemon-like fragrance of the plant (Husain 1994). Lemongrass extract, oil, citral and citral derivatives have potent bioactivities like antimicrobial (Kim et al. 1995), allelopathic (Chaimovitsh et al. 2012), antihelmintic (Kumaran et al. 2003), anti-inflammatory (Katsukawa et al. 2010), anticancer (Sharma et al. 2009), antioxidant (Khadria et al. 2008) and insect and mosquito repellent (Kalita et al. 2013).

It was observed that phosphorus nutrition had a positive effect on the formation of EOs in *Cymbopogon nardus and C. winterianus* (Katiyar et al. 2011; Ranaweera and Thilakaratne 1992a). P and Mg have a synergistic effect on oil production when applied together. This positive effect of P and Mg nutrition on the essential oil production could be explained by the involvement of these ions in the biosynthesis of EOs. P nutrition alone or P with Mg also increased the 'total geraniol' in oil indicated by the increase of its constituents: borneol, linalool and methyl isoeugenol (Ranaweera and Thilakaratne 1992b). The increase of total geraniol was always accompanied by the decrease of total hydrocarbons, which indicates the better utilization of monoterpenes in biosynthesis of constituents of total geraniol in oil as a result of phosphorus nutrition. Under soilless conditions, the optimal ratio of N:P:K in the nutrient solution for *C. citratus* Stapf. EO output was found to be 50:35:15 atom % (Mairapetyan et al. 1997). Field experiments were conducted at Mississippi, to evaluate the effects of N (0, 40, 80 and 160 kg N ha^{-1}) (as NH_4NO_3) and S (0, 30, 60 and 90 kg S ha^{-1} as sulphur bentonite, 90% S) on lemongrass biomass productivity, content and yield of EO and its composition (Zheljazkov et al. 2011). N × S interaction affected % neral and yields of *t*-caryophyllene, neral and geranial; N × harvest interaction affected percent and yield of essential oils, % *t*-caryophyllene and neral and yields of *t*-caryophyllene, neral and geranial; and S × harvest interaction affected dry weight yield, % neral, and the yields of neral and geranial. The N × S × harvest interaction effect on % geranial was also significant ($p < 0.05$). The lack of clear trend indicates the presence of factors other than N, S and harvest that affect the concentration of geranial in lemongrass essential oil. The average essential oil content at Verona varied from 0.34 to 0.55% depending on the N × harvest combination, with the highest content obtained from the third harvest (regrowth) at either 40 or 80 kg N ha^{-1}. The interaction of N application rate and harvest also affected geranial yield, with the second harvest, with an N application rate of 160 kg ha^{-1} giving the largest yield. Within each S application rate, due to the greater essential oil content in the third harvest (regrowth), the total amount of essential oil yield from Harvest 1 and Harvest 3 (regrowth) would be greater than the yield from the single harvest system, Harvest 2. Application of Zn combined with Fe (100 ppm each) produced the highest content of EO, soluble sugars and phenolic compounds. There was a proportionate increase of the principal compounds of geranial and neral with increasing levels of ZN or Fe-EDTA as compared with control (Aziz et al. 2010b). Pinzón-Torres et al. (2014) performed a study to analyze the quality and yield of the EO obtained of the *C. martini* crop under three types of fertilizers: Nutrimon® 14-14-14, organic fertilizer and mineral NPK granular fertilizer {(ammoniacal nitrogen 10%, nitric nitrogen (4%), phosphorus (P_2O_5) (14%) and water soluble potassium (K_2O) (14%)}. Relatively high percentages of geraniol were obtained (second harvest organic and 3rd harvest NPK6 + organic, 84%; 2nd harvest NPK2, 83%, and first harvest NPK6, 82%, respectively). The high percentages of geranyl acetate were obtained from the first harvests with organic (33%), green manure + organic (31%), green manure + NPK2 (32%), green manure + NPK4 (30%) and green manure (32%) that supports the advantages of using organic or green manure over mineral fertilizer for *C. martini* crops. Field experiments were conducted by Jemal and Abebe (2016) to investigate the effect of biochar rate application on yield and yield components of *C. citratus* L. The value of C content, available P and total N is high as expected from charred materials. The value of exchangeable cations (K, Ca and Mg) from biochar is medium to high. The highest EO yield was obtained by the application of 15 ton/hectare biochar rate from sugar factory followed by the application of 15 ton/hectare biochar rate from coffee husk. Biochar prevents nutrients from leaching, thereby improving the efficiency of nutrients applied alongside biochar (Major et al. 2010).

9.4.4 *Artemisia*

Artemisia annua L., a traditional Chinese herb of the family Compositae, is a popular medicinal plant. The plant is a storehouse of an array of bioactive compounds including flavonoids, coumarins, steroids, phenolics, purines, lipids, aliphatic compounds, monoterpenoids, triterpenoids and sesquiterpenoids. The principal active compound artemisinin or qinghaosu (0.1–0.6% dry weight) is a sesquiterpene lactone (Davies et al. 2011). Artemisinin is stored in low concentrations (0.1–0.6% dry weight), along with other sesquiterpenoids such as artemisinin, dihydroartemisinic acid, artemisinic acid and arteannuin B in the glandular leaf and inflorescence trichomes (Brisibe et al. 2009; Covello 2008). Artemisinin shows bioactivity against chloroquine-resistant and chloroquine-sensitive strains of *Plasmodium falciparum* and is used to treat celebral malaria (Liu et al. 2003). Semi-synthetically prepared derivatives of artemisinin such as dihydroartemisinin, artesunate, artemether, arteether and artemisinin show unique pharmacological activities against many parasites (Brisibe et al. 2009; Dhingra et al. 2000).

Singh (2000) reported an increase of 26.2 and 40.1% in herb, oil and artemisinin yields compared with control by application of 50 and 100 kg N ha^{-1}, respectively. Artemisinin concentration of the dried leaves notably increased from 6.3 to 27.5 mg 100 g^{-1} with different nitrogen treatments (0, 40, 80 and 120 kg ha^{-1}) (Özgüven et al. 2008). For shoot cultures and leaves, the optimum total nitrogen concentration for artemisinin production using *A. annua* L. was found to be 45 mM (Liu et al. 2003) and 106 mg L^{-1}, respectively, beyond which the artemisinin decreased (Davies et al. 2009). In studies with *A. annua* hairy roots, a high nitrate to ammonia ratio and low nitrogen in the medium favoured artemisinin accumulation (Liu et al. 1997; Wang and Tan 2002). N combined (80 mg kg^{-1} soil) with foliar GA_3 amplified the content and yield of artemisinin in the treated plants over the control (Aftab et al. 2011a). Above 200 mg/L phosphate, there was a significant decrease in the artemisinin content (Liu et al. 2003). Davies et al. (2011) concluded that P application rate of 30 mg P L^{-1} is optimum and avoids economical and environmental wastage. Artemisinin content was also maximized in the plants fertilized with P and inoculated with AM fungi (Kapoor et al. 2007). Ferreira (2007) reported that K deficiency significantly increases the concentration of artemisinin. The artemisinin concentration was significantly higher (75.5%) in K-deficient plants as compared to plants under the complete nutrient treatment. Thus, a mild potassium deficiency is fruitful for *A. annua* farmers to attain gains in artemisinin/ha while saving on K fertilization. But K did not change the leaf artemisinin concentration or leaf artemisinin yield (Davies et al. 2009).

9.4.5 Turmeric

Popularly known as turmeric, the rhizomatous plant *Curcuma longa* L. from the Zingiberaceae family is grown in many warm regions of the world. India fares well in the world turmeric market with 80% of the production and trade share of more than 60% (Turmeric World 2014). A large number of phytochemical compounds are involved in the varied metabolic pathways of this medicinal herb which gives turmeric powder its value in the marketplace (Xie et al. 2009). The terpenoid-rich rhizomes contain carbohydrates (69.4%), fat (5.1%), protein (6.3%) and minerals (3.5%), volatile oil (5.0–6.0%) and oleoresin (7.9–10.4%). Curcuminoids, the characteristic orange–yellow compounds found in turmeric include curcumin, or diferuloylmethane (2.5–6.0%) consisting of 94% curcumin (curcumin I), 6% demethoxycurcumin (curcumin II) and 0.3% bisdemethoxycurcumin (curcumin III) (Chempakam and Parthasarathy 2008). Turmeric finds its use in medicine, condiment, dye and cosmetic. It has tremendous pharmaceutical properties such as anti-inflammatory, antioxidant, anticarcinogenic, antimalarial, antitumour, antibacterial, anti-HIV and anti-Alzheimer (Green and Mitchell 2014; Prasad et al. 2014). The extract is used to treat some diseases like gastritis, ulcers, osteoarthritis, atherosclerosis, cardiac disease and diabetes (Sarker and Nahar 2007).

Nutrients especially N, P and K should be applied during the active growth phase (Fourth and fifth month) of the plant because uptake of nutrients is effective in this period (Sivaraman 2007). Singh et al. (2011) performed two pot experiments to study the response of turmeric to N and P application. Treatments comprised of basal application of different doses of N, viz.0, 30, 60, 90 and 120 kg N ha^{-1} and P, viz. 0, 15, 30, 45 and 60 kg P per ha, respectively. The results obtained showed that application of N and P on curcumin content was found effective over the control. Treatment N_{90} proved optimum and caused an increase of 33.5 and 27.2% at 120 and 180 days after planting (DAP), respectively, over the control, whereas treatment P_{45} significantly increased curcumin content by 21.2 and 19.8% at 120 and 180 DAP, respectively, over the control (P_0). Sadanandan and Hamza (1996c) reported that NPK at 60, 50 and 120 kg ha^{-1} with micronutrients were optimum for varieties

Suvarna, Suguna and Alleppey, whereas NPK at 50, 40 and 100 kg ha^{-1} with micronutrients was optimum for Sudarshana, for rhizome yield and curcumin production. Soil N is the precursor for phenyl propanoid pathway involving curcumin synthesis in turmeric since N forms the structural unit of many proteins (Sandeep et al. 2015). On application of N (maximum 30 kg N ha^{-1}) and P the volatile oil content of *Curcuma amada* Roxb. increased significantly, but it decreased at higher levels of K application. Maximum nonvolatile ether extract (NVEE) was acquired at 45 kg N ha^{-1}. For P and K, a gradual decline in NVEE was observed with increasing levels of the nutrient (Mridula and Jayachandran 2001). In vitro studies of turmeric with different concentrations of mineral nutrients (PO_4^{3-} , Ca^{2+}, Mg^{2+} and KNO_3) revealed that in the high fertilizer treatments, the highest concentrations of curcumin and demethoxycurcumin were obtained with 1.25 mM PO_4^{3-} , 3 mM Ca^{2+}, 1.5 mM Mg^{2+} and 60 mM KNO_3. Reducing KNO_3 concentration to 20 mM at 1.5 mM $Mg^{2+}$and 9 mM Ca^{2+} maximized bisdemethoxycurcumin concentration to 25.0 ± 2.6 μg/g with 9 buds/vessel. In the low input fertilizer treatments, the best induction media for high concentrations of all three curcuminoids (curcumin, demethoxycurcumin and bisdemethoxycurcumin) was with 6.25 mM PO_4^{3-} , 3 mM Ca^{2+}, 1.5 mM Mg^{2+} and 60 mM KNO_3 (El-Hawaz et al. 2016). The quality of turmeric measured in terms of its active constituents did not show any significant variations among these three nutrient management systems, viz. organic nutrient management (ONM), conventional nutrient management (CNM) and integrated nutrient management (INM) (Srinivasan et al. 2016).

9.4.6 Ginger

Ginger, an important spice used all over the world, has strong medicinal and aromatic properties. It belongs to Zingiberaceae family which is characterized by tuberous or non-tuberous rhizomes (Chen et al. 2008). Saponins are characteristic of ginger rhizomes, whereas tannins are less (Adanlawo and Dairo 2007). The major active constituents are the gingerols of which [6] gingerol [5-hydroxy-1-(4-hydroxy-3-methoxy phenyl) decan-3-one] is the most abundant. The volatile oil of ginger rhizome consists of mainly mono and sesquiterpenes: camphene, beta-phellandrene, curcumene, cineole, geranyl acetate, terpineol, terpenes, borneol, geraniol, limonene, linalool, alpha-zingiberene (30–70%), beta-sesquiphellandrene (15–20%), beta-bisabolene (10–15%) and alpha-farnesene. Oleoresin consists zingiberol, the main component responsible for aroma and pungent substances, namely, gingerol, shogaol, zingerone and paradol as well as zingiberene, gingediol, diarylheptanoids, vitamins and phytosterols (Polasa and Nirmala 2003). Ginger has many medicinal properties against a variety of ailments including arthritis, rheumatism, sore throats, muscular aches, constipation, nausea, diarrhoea, hypertension, dementia, infectious diseases and respiratory disorders. Antioxidation, antimicrobial, anticancer, and anti-larval effects are also known (Kalaivani et al. 2012; Zhao et al. 2011). Shogaol tackles neuroinflammatory responses in cortical astrocytes (Shim et al. 2011).

Volatile oil content of ginger increased when nitrogen was applied but up to a certain level. Excess of nitrogen decreased its content (Xizhen et al. 1998). Application of 120 kg N per hectare increased the essential oil content (Singh et al. 2014). The concentration of essential oil was the highest with the treatment N600 and N450 at two different soil fertilities, respectively (Lu-jiu et al. 2010). The ameliorative role of basal application of N on essential oil content could be attributed to the central roles played by N in plants like enhancing the amount of biomass yields per unit area, leaf area development and photosynthetic rate (Sangwan et al. 2001; Sharma and Kumar 2012). Studies on sustained nutritional condition for ginger showed that 50 kg K_2O ha^{-1} besides 20 tonnes of farmyard manure as basal and 25 tonnes green leaf mulch together with NPK at 75, 50 and 50 was optimum for oleoresin production (Sadanandan et al. 2002). The combined treatments M3P3 (bokashi 60%, charcoal husk 20%, coco peat 20% with organic fertilizer concentrations 4.5%) gave the best response to Zingeron level (1.88%) (Soeparjono 2016).

9.4.7 Periwinkle

Madagascar-native periwinkle (*Catharanthus roseus* (L.) G. Don, synonym *Vinca rosea* L., belonging to Apocynaceae family has vast medicinal properties owing to its ability to synthesize a variety of terpenoid indole alkaloids (TIAs), unevenly distributed in all parts of the plant. Vinblastine and vincristine, present in the leaves, are anticarcinogenic indoles, whereas root-derived ajmalicine and serpentine are used as antihypertensives and sedatives. Vindoline and catharanthine found in leaves and in leaves and roots, respectively, have diuretic properties (Hassan et al. 2009; Łata 2007).

Catharanthus roseus needs a high nitrogen supply along with other nutrients to obtain a high content of biologically active substances. The total alkaloid concentration and its yield increased up to a higher level than the herbal yield, i.e. up to 300 N dm^{-3} (Łata and Sadowska 1996a). Janishevskiĭ and Dzhaparidze (1990) witnessed nearly the same results. A rise of herb yield due to fertilization (Rajeswara and Singh 1990) rather than an increase of the alkaloid content is responsible for the increased alkaloid production (Sreevalli et al. 2004). It is to be noted that the total alkaloid content depends on the source of nitrogen. $N–NO_3$ resulted in a 50% increase of the alkaloid concentration, whereas a mixture of $N–NO_3$ and $N–NH_4$ diminished the concentration by 45% (Miranda-Ham et al. 1996). Application of nitrogen as amides and nitrates yielded better results. Foliar application of $N–NH_2$ + P and $N–NH_2$ + Zn increased the alkaloid content by 59% and 75%, respectively (Łata and Sadowska 1996b). The formation of indole alkaloids ajmalicine and serpentine was curbed by nitrogen containing mineral salts with composition identical to MS-medium [KNO_3 (18.9×10^{-3} M) together with $(NH_4)NO_3$(20×10^{-3} M)] or phosphate [KH_2PO_4 (1.2×10^{-3} M)] . For illuminated cultures, highest accumulation of alkaloids, anthocyanins and total phenolics was observed in the absence of phosphate and mineral nitrogen. $CaCl_2$ and $MgSO_4$ salts and other micronutrients and vitamins of the M S-medium did not affect alkaloid accumulation (Knobloch et al. 1982; Knobloch and Berlin 1980). These studies elucidate the role of nitrogen and phosphate containing mineral salts for the formation of secondary compounds by *Catharanthus* cells. The general augment of secondary compounds under medium conditions which limit growth (low auxin, phosphate and nitrogen) advocates the principle that growth and secondary metabolism are regulated antagonistically by competing for the common precursor amino acids for protein synthesis and secondary metabolite production (Vining 1990). Contradictory results were obtained by Mendonça Freitas et al. (2015) in their study where deficiency of N, P, Mg and S reduced ajmalicine concentration. Potassium deficiency increased ajmalicine concentration in *C. roseus* roots. Potassium, when not applied in the chloride form, increased the vinblastine content of *C. roseus*. Contradictorily, higher intracell accumulation of catharanthine and other indole alkaloids was observed when KCl was added to the suspension cultures (Łata 2007). On doubling the concentration of Mg and increasing the concentration of Co by five times of the original nutrient medium, alkaloid biosynthesis in callus tissue increased by 33% and 35%, respectively (Boldyreva and Velichko 2003). One of the many ways by which *C. roseus* plants overcome drought stress was by accumulation of indole alkaloids in shoots and roots triggered by the application of $CaCl_2$ (Jaleel et al. 2007a; Jaleel et al. 2007b). Sprays of B, B and Zn or Zn and Fe led to an 80% increase and a 55% decrease of the total alkaloid content in the roots and leaves, respectively. Zn is a requisite for the synthesis of tryptophan, the precursor of indole alkaloids (Lovkova et al. 2005). Nitrogen plays an important role in the biosynthesis and accumulation of alkaloids in plants. Abdolzadeh et al. (2006) studied to evaluate of the effects of varied sources of nitrogen (2.75, 5.5, 11, 22 and 32 mM) on growth and the total alkaloids content of *C. roseus* including that of vincristine and vinblastine. According to their study, the highest content of amino acids, proteins, total nitrogen, total alkaloids, vincristine and vinblastine was noticed in plant supplied with nitrate plus ammonium. In another study of Hassan et al. (2009), effect of N and K on growth, yield and alkaloid content in *C. roseus* was studied under field conditions. Significant improvement in all the parameters was recorded to the highest extent in plants fertilized with 150 Kg/fed of N and 25 Kg/fed of K.

9.4.8 Aloe vera

Aloe vera (*Aloe barbadensis* Miller) belonging to the family Liliaceae is an industrially important and traditional plant known to produce a gel with medicinal value (Hamman 2008). Short growth period and its high economic value make it quite valuable (Gantait et al. 2014). Water constitutes a considerable part (99–99.5%) of the plant. The rest (0.5–1.0%) of the solid material contain more than 75 different compounds of potent bioactivity including water- and fat-soluble vitamins, minerals, enzymes, simple/complex polysaccharides, phenolic compounds and organic acids (Abdulrahman 2016). Glucomannans and acemannan are the notable carbohydrate polymers present in *Aloe vera* (Eshun and He 2004). Phenolic compounds found are chromone, anthraquinone (including the hydroxyanthracene derivatives, aloin A and B, barbaloin, isobarbaloin and aloe emodin) or anthrone derivatives. Other recognized active components include bradykininase, magnesium lactate and salicylic acid (Atherton 1998; Klein and Penneys 1998). The Aloe latex (aloe), aloe gel (*Aloe vera*) and aloe whole leaf (aloe extract) each with different chemical composition have different therapeutic use. Aloe latex acts as a laxative; Aloe gel heals wounds, burns and frostbites, with addition to possessing anti-inflammatory, antifungal, hypoglycaemic and gastroprotective properties; and aloe extract can be used to combat cancer and AIDS (Mukherjee et al. 2014).

Ji-Dong et al. (2006) and Hazrati (2012) reported that the application of N helped increase the growth and concentration of the phenolic compound aloin of *Aloe vera* plants. In yet another study, the aloin and barbaloin increased in leaf latex with the nitrogen application (Saradhi et al. 2007). Out of the many tested phosphorus levels, a considerably higher gel content of *Aloe vera* was recorded at 150 Kg P_2O_5 ha^{-1} (Boroomand et al. 2011). Tawaraya et al. (2007) from their study found that inoculation of soil with mycorrhizal fungi enhanced the ability of Aloe to acquire P and N from soil thus decreasing application of P and N fertilizer. Cardarelli et al. (2013) performed an experiment represented by pre plant fertilization, reduced-4 g/L and standard-8 g/L [fertilizer containing 60.0 g kg^{-1} N, 65.5 g kg^{-1} P, 24.9 g kg^{-1} K, 12.0 g kg^{-1} Mg, 7.1 g kg^{-1} Ca and trace elements], and post plant fertilization, standard fertilization regime (nutrient solution 120 mg L^{-1} N, 100 mg L^{-1} K, 72 mg L^{-1} Ca, 30 mg L^{-1} Mg and trace elements) and reduced fertilization regime (50% of standard nutrient solution). Doubling the fertilizer dose from 50 to 100% increased plant growth but reduced the leaf bioactive compounds, in *A. barbadensis*, by decreasing the aloin and the b-polysaccharide concentration. This suggests that the secondary metabolism of plant was endorsed by low nutrient availability as observed by Rouphael et al. (2012b). Equivalent nutrient level of N80P40K80 organic sources of fertilizer treatments worked better than inorganic source (Abdulrahman 2016) suggesting the prospect of growing organic *Aloe vera* in a commercial cultivation as a better marketable product.

9.5 Conclusion

Owing to the bioactivity of active constituents, medicinal and aromatic plants (MAPs) have been utilized as therapeutics for thousands of years. Most of these compounds are secondary metabolites produced by the plants only in low amounts from natural sources. Although many studies are focused on the effect of minerals on the overall growth, development and yield of the MAPs, comparatively few studies have been done to evaluate the direct role of minerals on active constituents of these plants. Secondary metabolites which are a source of important pharmaceuticals are used by plants to adapt to the environment. The involvement of macro- and micronutrients in building basic organic compounds and in almost all plant life processes and the results of the presented agronomic studies show the important role of these minerals in the amendments of active constituents. The response to macro- and micronutrients application varied depending on the crop and its variety. The rate of application of macronutrients like nitrogen, phosphorus, potassium, sulphur, calcium, magnesium and microelements and the synergistic or antagonistic interactions between these minerals determine the essential oil yield and its chemical composition. Usually N, P and K nutrients gave better oil yields when applied in some definite combinations rather than individual applications of these major nutrients. Some plants required a specific element in much higher concentration than others, and others were able to tolerate a much higher concentration of an essential element that would prove detrimental to a different species. Hence, fertilization by mineral nutrients influences the biogenesis of secondary plant substances, thereby determining the quantity and composition of active constituents. It could be concluded that an optimal and balanced mineral fertilization, customized for individual MAPs, according to their nutritional requirements and growing conditions, is an important cultivation factor determining the therapeutic and medicinal values of the MAPs. Contradictory results obtained by different authors in general may be ascribed to different experimental conditions, and harvesting dates, to residual presence of the tested nutrients already in the soil or to unawareness of the ontogenesis of the substances. The soil buffer ability for the nutrient might hint us about plant uptake and specific fertilizer requirements. Inoculation of biofertilizers like mycorrhizal fungi or bacteria into the soil could enhance the availability and uptake of fertilizer from the soil, thus decreasing the amount of fertilizer application. Identification of medicinal and aromatic plant varieties which are efficient absorbers of soil fertilizers/mineral nutrients (absorb less mineral nutrients per unit yield or active compounds produced), without compromise of the oil quality, is recommendable to farmers.

References

Abdolzadeh, A., Hosseinian, F., Aghdasi, M., & Sadgipoor, H. (2006). Effects of nitrogen sources and levels on growth and alkaloid content of periwinkle. *Asian Journal of Plant Sciences, 5*, 271–276.

Abdulrahman, E. M. E. (2016). *Impact of nutrients and bio-stimulants on growth and yield of Aloe vera plants*. Doctoral dissertation, Sudan University of Science and Technology.

Adanlawo, I. G., & Dairo, F. A. S. (2007). Nutrient and anti-nutrient constituents of ginger (*Zingiber officinale*, roscoe) and the influence of its ethanolic extract on some serum enzymes in albino rats. *International Journal of Biological Chemistry, 1*, 38–46.

Aftab, T., Khan, M. M. A., Idrees, M., Naeem, M., & Moinuddin, M. (2011a). Optimizing nitrogen levels combined with gibberellic acid for enhanced yield, photosynthetic attributes, enzyme activities and artemisinin content of *Artemisia annua*. *Frontiers of Agriculture in China, 5*, 51–59.

Alaghemand, A., Ghorbanpour, M., Asli, D. E., & Moghaddasian, B. (2013). Calcium fertilization effects on hyoscyamine and scopolamine accumulation in henbane (*Hyoscyamus niger* L.) under hydroponic culture. *European Journal of Experimental Biology, 3*, 228–232.

Arabaci, O., & Bayram, E. (2004). The effect of nitrogen fertilization and different plant densities on some agronomic and technologic characteristic of *Ocimum basilicum* L. (basil). *Journal of Agronomy, 3*, 255–262.

Arabaci, O., Bayram, E., Baydar, H., Savran, A. F., Karagodan, T., & Ozay, N. (2007). Chemical composition, yield and contents of essential oil of *Lavandula hybrida* Reverchon grown under different nitrogen fertilizer, plant density and location. *Asian Journal of Chemistry, 19*, 2184–2192.

Ashraf, M., Qasim, A., & Zafar, I. (2006). Effect of nitrogen application rate on the content and composition of oil, essential oil and minerals in black cumin (*Nigella sativa* L.) seeds. *Journal of the Science of Food and Agriculture, 86*, 871–876.

Atherton, P. (1998). *Aloe vera*: magic or medicine? *Nursing Standard, 12*, 49–52.

Aziz, E. E., El-Danasoury, M. M., & Craker, L. E. (2010a). Impact of sulphur and ammonium sulphate on dragonhead plants grown in newly reclaimed soil. *Journal of Herbs Spices & Medicinal Plants, 16*, 126–135.

Aziz, E. E., Ezz El-Din, A. A., & Omer, E. A. (2010b). Influence of zinc and iron on plant growth and chemical constituents of *Cymbopogon citratus* L. grown in newly reclaimed land. *International Journal of Academic Research, 2*, 278–283.

Azizi, M., & Dias, A. (2004). Nitrogen and phosphorus fertilizers affect flavonoids contents of St. John's wort (*Hypericum perforatum* L.). *Proceedings of 4th International Iran & Russia Conference, Shahrkurd, Iran 458–462*. Retrieved from http://iirc.narod.ru/4conference/Fullpaper/20003.pdf

Bakken, A. K., Gauteb, H., & Myhr, K. (1997). The potential of crushed rocks and mine tailings as slow releasing K fertilizers assessed by intensive cropping with Italian reygrass in different soil types. *Nutrient Cycling in Agroecosystems, 47*, 41–48.

Barnauskiene, R., Rimantas Venskutonis, P., Visykelis, P., & Dambruskiene, E. (2003). Influence of nitrogen fertilizers on the yield and composition of thyme (*Thymus vulgaris*). *Journal of Agricultural and Food Chemistry, 51*, 7751–7758.

Bogers, R. J., Craker, L. E., & Lange, D. (Eds.). (2006). *Medicinal and aromatic plants: Agricultural, commercial, ecological, legal, pharmacological and social aspects (Wageningen UR Frontis series)*. Dordrecht: Springer.

Boldyreva, Y. A., & Velichko, N. A. (2003). Effect of the compounds of mineral nutrition on the *Catharanthus roseus* Callus tissue growth and synthesis of alkaloids. *Biotekhnologiya, 19*, 53–62.

Boroomand, N., Marezi, A., & Sadat, H. G. M. (2011b). Effect of organic and phosphorus on mineral and yield of *Ocimum basilicum. 7th Iranian Congress of Horticultural Science, Isfahan, Iran.*

Boroomand, N., Nakhaei, M., & Sadat, H. G. M. (2011a). Effect of potassium and phosphorous on growth and yield of *Aloe vera* L. *7th Iranian Congress of Horticultural Science, Isfahan, Iran. 375–381.*

Brisibe, E. A., Umoren, U. E., Brisibe, F., Magalhäes, P. M., Ferreira, J. F., Luthria, D., Wu, X., & Prior, R. L. (2009). Nutritional characterisation and antioxidant capacity of different tissues of *Artemisia annua* L. *Food Chemistry, 115*, 1240–1246.

Briskin, D. (2000). Medicinal plants and phytomedicines. Linking plant biochemistry and physiology to human health. *Plant Physiology, 124*, 507–514.

Buntain, M., & Chung, B. (1994). Effects of irrigation and nitrogen on the yield components of fennel (*Foeniculum vulgare* Mill.) *Australian Journal of Experimental Agriculture, 34*, 845–849.

Cardarelli, M., Rouphael, Y., Rea, E., Lucini, L., Pellizzoni, M., & Colla, G. (2013). Effects of fertilization, arbuscular mycorrhiza, and salinity on growth, yield, and bioactive compounds of two aloe species. *HortScience, 48*, 568–575.

Chaimovitsh, D., Stelmakh, R. O., Altshuler, O., Belausov, E., Abu-Abied, M., Rubin, B., Sadot, E., & Dudai, N. (2012). The relative effect of citral on mitotic microtubules in wheat roots and BY2 cells. *Plant Biology, 14*, 354–364.

Chauhan, R. S., Kaul, M. K., Shahi, A. K., Kumar, A., Ram, G., & Tawa, A. (2009). Chemical composition of essential oils in *Mentha spicata* L. Accession [IIIM (J) 26] from north-west Himalayan region, India. *Industrial Crops and Products, 29*, 654–656.

Chempakam, B., & Parthasarathy, V. A. (2008). Turmeric. In V. A. Parthasarathy, B. Chempakam, & T. J. Zachariah (Eds.), *Chemistry of spices* (pp. 97–123). Oxfordshire: CABI.

Chen, I. N., Chang, C. C., Ng, C. C., Wang, C. Y., Shyu, Y. T., & Chang, T. L. (2008). Antioxidant and antimicrobial activity of Zingiberaceous plants in Taiwan. *Plant Foods for Human Nutrition, 63*, 15–20.

Collin, H. (2006). Herbs, spices and cardiovascular disease. In K. V. Peter (Ed.), *Handbook of herbs and spices* (Vol. 3, pp. 126–137). Cambridge: Woodhead Publishing Limited.

Covello, P. S. (2008). Making artemisinin. *Phytochemistry, 69*, 2881–2885.

Craker, L. E., Gardner, Z., & Etter, S. C. (2003). Herbs in American fields: A horticultural perspective of herb and medicinal plant production in the United States, 1903 to 2003. *HortScience, 38*, 977–983.

Craker, L. E., & Gardner, Z. E. (2006). Medicinal plants and tomorrow's pharmacy: An American perspective. In R. J. Bogers, L. E. Craker, & D. Lange (Eds.), *Medicinal and aromatic plants: Agricultural, commercial, ecological, legal, pharmacological and social aspects* (Vol. 17, pp. 29–41). Dordrecht: Springer.

Croteau, R., Burbott, A. J., & Loomis, W. D. (1972a). Biosynthesis of mono-and sesqui-terpenes in peppermint from glucose – ^{14}C and $^{14}CO_2$. *Phytochemistry, 11*, 2459–2467.

Croteau, R., Kutchan, T. M., & Lewis, N. G. (2000). Natural products (secondary metabolites). In B. Buchanan, W. Gruissem, & R. Jones (Eds.), *Biochemistry and molecular biology of plants* (pp. 1250–1318). Rockville: American Society of Plant Physiologists.

Daneshian, A., Gurbuz, B., Cosge, B., & Ipek, A. (2009). Chemical components of essential oils from basil (*Ocimum basilicum* L.) grown at different nitrogen levels. *International Journal of Natural & Engineering Sciences, 3*, 8–12.

Daneshkhah, M., Mohsen, K., & Nikbakht, A. (2007). Effects of different levels of nitrogen and potassium fertilizers on flower yield and essential oil content of *Rosa damascena* Mill. From Barzok of Kashan. *Iranian Journal of Horticultural Science and Technology, 8*, 83–90.

Davies, M., Atkinson, C., Burns, C., Woolley, G., Hipps, N. A., Arroo, R. R. J., Dungey, N., Robinson, T., Brown, P., Flockart, I., Hill, C., Smith, L., & Bentley, S. (2009). Enhancement of artemisinin concentration and yield in response to optimization of nitrogen and potassium supply to *Artemisia annua*. *Annals of Botany, 104*, 315–323.

Davies, M. J., Atkinson, C. J., Burns, C., Arroo, R., & Woolley, J. (2011). Increases in leaf artemisinin concentration in *Artemisia annua* in response to the application of phosphorus and boron. *Industrial Crops and Products, 34*, 1465–1473.

Desmarest, P. (1978). New aspects of fennel cultivation in France. *Acta Horticulturae, 73*, 289–295.

Dhingra, V., Pakki, S. R., & Narasu, M. L. (2000). Antimicrobial activity of artemisinin and its precursors. *Current Science, 78*(6), 709–713.

Dordas, C., Apostolides, G., & Goundra, O. (2007). Boron application affects seed yield and seed quality of sugar beets. *The Journal of Agricultural Science, 145*, 377–384.

El-Hawaz, R., Tharayil, N., Bridges, W., & Adelberg, J. (2016). Mineral nutrition of *Curcuma longa* L. in bioreactors affects subsequent development of curcuminoids following transfer to the greenhouse. *Industrial Crops and Products, 83*, 186–193.

Eshun, K., & He, Q. (2004). *Aloe vera*: A valuable ingredient for the food, pharmaceutical and cosmetic industries-a review. *Critical Reviews in Food Science and Nutrition, 44*, 91–96.

European Medicines Agency (EMEA). (2008). Committee on herbal medicinal products (HMPC). Assessment report on *Foeniculum vulgare* miller. Emea/Hmpc/137426/2006.

Ezz El-Din, A. A., Hendawy, S. F., Aziz, E. E., & Omer, E. A. (2010). Enhancing growth yield and essential oil of caraway plants by nitrogen and potassium fertilizers. *International Journal of Academic Research, 2*, 192–197.

Ferreira, J. F. (2007). Nutrient deficiency in the production of artemisinin, dihydroartemisinic acid, and artemisinic acid in *Artemisia annua* L. *Journal of Agricultural and Food Chemistry, 55*(5), 1686–1694.

Ganjewala, D., & Gupta, A. K. (2013). 8 Lemongrass (*Cymbopogon flexuosus* Steud.) Wats essential oil: Overview and biological activities. *RPMP, 37*, 235–271.

Ganjewala, D., Kumari, A., & Khan, K. H. (2008). Ontogenic and developmental changes in essential oil content and compositions in *Cymbopogon flexuosus* cultivars. In B. N. Prasad & L. Mathew (Eds.), *Recent advance in biotechnology* (pp. 82–92). New Delhi: Excel India Publishers.

Gantait, S., Sinniah, U. R., & Das, P. K. (2014). *Aloe vera*: A review update on advancement of *in vitro* culture. *Journal Acta Agriculturae Scandinavica, Section B—Soil & Plant Science, 64*, 1–12.

Gottlieb, O. R. (1990). Phytochemicals: Differentiation and function. *Phytochemistry, 29*, 1715–1724.

Green, C. E., & Mitchell, S. A. (2014). The effects of blanching, harvest time and location (with a minor look at postharvest blighting) on oleoresin yields, percent curcuminoids and levels of antioxidant activity of turmeric (*Curcuma longa*) rhizomes grown in Jamaica. *Modern Chemistry & Applications, 2*, 2–9.

Grindberg, R. V., Shuman, C. F., Sorrels, C. M., Wingerd, J., & Gerwick, W. H. (2007). Neurotoxic alkaloids from cyanobacteria. In E. Fattorusso (Ed.), *Modern Alkaloids POT-S* (pp. 139–170). Weinheim: Wiley-VCH.

Gurib-Fakim, A. (2006). Medicinal plants: Traditions of yesterday and drugs of tomorrow. *Molecular Aspects of Medicine, 27*, 1–93.

Hamman, J. H. (2008). Composition and applications of *Aloe vera* leaf gel. *Molecules, 13*, 1599–1616.

Hanudin, E., Wismarini, H., Hertiani, T., & Sunarminto, B. H. (2012). Effect of shading, nitrogen and magnesium fertilizer on phyllanthin and total flavonoid yield of *Phyllanthus niruri* in Indonesia soil. *Journal of Medicinal Plants Research, 6*, 4586–4592.

Harendra, K., & Yadav, D. S. (2007). Effect of phosphorus and sulphur levels on growth, yield and quality of Indian mustard (*Brassica juncea*) cultivars. *Indian Journal of Agronomy, 52*, 1541–1157.

Hartman, H. (2007). *Consumer culture and the future of organic usage*. The Hartman Group, Inc. Retrieved from www.hartmangroup.com/products/HB/2006_11_01.html.

Hassan, R. A., Habib, A. A., & El-Din, A. E. (2009). Effect on nitrogen and potassium fertilization on growth, yield and alkaloidal content of periwinkle (*Catharanthus roseus* G. Don). *Medicinal and Aromatic Plant Science and Biotechnology, 3*, 24–26.

Hazrati, S. (2012). Effects of various levels of N on productivity of *Aloe barbadensis* L. and its inhibitory effect on *Trichophyton rubrum*. *Advances in Horticultural Science, 24*(4), 187–190.

Heidari, S., Soltani, F., Azizi, M., & Hadian, J. (2014). Foliar application of Ca and K improves growth, yield, essential oil yield and nutrient uptake of tarragon (*Artemisia dracunculus* L.) grown in Iran. *International Journal of Biosciences, 4*, 323–338.

Hendawy, S. F., & Khalid, K. A. (2011). Effect of chemical and organic fertilizers on yield and essential oil of chamomile flower heads. *Medicinal and Aromatic Plant Science and Biotechnology, 5*, 43–48.

Hirschi, K. D. (2004). The calcium conundrum: Both versatile nutrient and specific signal. *Plant Physiology, 136*, 2438–2442.

Husain, A. (1994). Palmarosa. In *Essential oil plants and their cultivation* (pp. 58–66). Lucknow: Central Institute of Medicinal and Aromatic Plants (CIMAP).

Jabbari, R., Dehaghi, M. A., Sanavi, A. M. M., & Agahi, K. (2011). Nitrogen and iron fertilization methods affecting essential oil and chemical composition of thyme (*Thymus vulgaris* L.) med plant. *Advances in Environmental Biology, 5*, 433–438.

Janishevskiĭ, F. V., & Dzhaparidze, N. M. (1990). The effect of potassium fertilizer form on *Catharanthus roseus*. *Subtropicheskie Kul'tury, 2*, 131–135.

Jaleel, C. A., Manivannan, P., Kishorekumar, A., Sankar, B., Gopi, R., Somasundaram, R., & Panneerselvam, R. (2007a). Alterations in osmoregulation, antioxidant enzymes and indole alkaloid levels in *Catharanthus roseus* exposed to water deficit. *Colloids and Surfaces. B, Biointerfaces, 59*, 150–157.

Jaleel, C. A., Manivannan, P., Sankar, B., Kishorekumar, A., Gopi, R., Somasundaram, R., & Panneerselvam, R. (2007b). Water deficit stress mitigation by calcium chloride in *Catharanthus roseus:* Effects on oxidative stress, proline metabolism and indole alkaloid accumulation. *Colloids and Surfaces. B, Biointerfaces, 60*, 110–116.

Jeliazkova, E. A., Zheljazkov, V. D., Craker, L. E., Yankov, B., & Georgieva, T. (1999). NPK fertilizer and yields of peppermint, *Mentha* x *piperita. Acta Horticulturae, 502*, 231–236.

Jemal, K., & Abebe, A. (2016). Determination of bio-char rate for improved production of lemon grass (*Cymbopogon citratus* L.) *International journal of Advanced Biological and Biomedical Research, 4*, 149–157.

Ji-Dong, W., Zaho-pu, L., Qing-song, Z., Ling, L., & Feng-zhi, P. (2006). Effects of different N levels on seedling growth, nitrate and its secondary metabolites in *Aloe vera* seeding. *Plant Nutrition and Fertilizer Science, 12*, 864–868.

Kalaivani, K., Senthil-Nathan, S., & Murugesan, A. G. (2012). Biological activity of selected Lamiaceae and Zingiberaceae plant essential oils against the dengue vector *Aedes aegypti* L. (Diptera: Culicidae). *Parasitology Research, 110*(3), 1261–1268.

Kalita, B., Bora, S., & Sharma, A. K. (2013). Plant essential oils as mosquito repellent-a review. *International Journal of Research and Development in Pharmacy & Life Sciences, 3*(1), 741–747.

Kandil, M., Ahmed, S., Sator, C., & Schnug, E. (2002). Effect of organic and inorganic fertilisation on fruit and essential oil yield of fennel (*Foeniculum vulgare* Mill.) grown in Egypt. In *Proceeding Fachtagung für Heil- und Gewuerzpflanzen Ahrweiler (im Druck).*

Kapoor, R., Chaudhary, V., & Bhatnagar, A. K. (2007). Effects of arbuscular mycorrhiza and phosphorus application on artemisinin concentration in *Artemisia annua* L. *Mycorrhiza, 17*, 581–587.

Katiyar, R., Gupta, S., & Yadav, K. R. (2011). *Cymbopogon winterianus*: An important species for essential java citronella oil and medicinal values. In *National Conference on Forest Biodiversity: Earth's living treasure FRI Kanpur.*

Katsukawa, M., Nakata, R., Takizawa, Y., Hori, K., Takahashi, S., & Inoue, H. (2010). Citral, a component of lemongrass oil, activates PPARá and ã and suppresses COX–2 expression. *Biochimica et Biophysica Acta, 1801*, 1214–1220.

Khadria, A., Serralheirob, M. L. M., Nogueirab, J. M. F., Neffatic, M., Smitia, S., & Araujob, M. E. M. (2008). Antioxidant and antiacetylcholinesterase activities of essential oils from *Cymbopogon schoenanthus* L. Spreng. Determination of chemical composition by GC–mass spectrometry and 13C NMR. *Food Chemistry, 109*, 630–637.

Kiferle, C., Maggini, R., & Pardossi, A. (2013). Influence of nitrogen nutrition on growth and accumulation of rosmarinic acid in sweet basil (*Ocimum basilicum* L.) grown in hydroponic culture. *Australian Journal of Crop Science, 7*, 321–327.

Kim, J. M., Marshall, M. R., Cornell, J. A., Preston, J. F., & Wei, C. I. (1995). Antibacterial activity of carvacrol, citral, and geraniol against *Salmonella typhimurium* in culture medium and on fish cubes. *Journal of Food Science, 60*, 1364–1368.

Klein, D. K., & Penneys, N. S. (1998). Aloe Vera. *Journal of the American Academy of Dermatology, 18*, 714–720.

Knobloch, K. H., Bast, G., & Berlin, J. (1982). Medium-and light-induced formation of serpentine and anthocyanins in cell suspension cultures of *Catharanthus roseus. Phytochemistry, 21*, 591–594.

Knobloch, K. H., & Berlin, J. (1980). Influence of medium composition on the formation of secondary compounds in cell suspension cultures of *Catharanthus roseus* (L.) G. Don. *Zeitschrift fuer Naturforschung, 35c*, 551–556.

Kumaran, A. M., D'Souza, P., Agarwal, A., Bokkolla, R. M., & Balasubramaniam, M. (2003). Geraniol, the putative anthelmintic principle of *Cymbopogon martini. Phytotherapy Research, 17*, 957–960.

Łata, B. (2007). Cultivation, mineral nutrition and seed production of *Catharanthus roseus* (L.) G. Don in the temperate climate zone. *Phytochemistry Reviews, 6*, 403–411.

Łata, B., & Sadowska, A. (1996a). Effect of nitrogen level in the substrate on yield and alkaloid concentration in *Catharanthus roseus* L. (G.) don. *Folia Horticulturae, 8*, 59–69.

Łata, B., & Sadowska, A. (1996b). Effect of N, P, K, and Zn foliar fertilization on the yield of *Catharanthus roseus* L. (G.) don. *Folia Horticulturae, 8*, 51–58.

Lee, K. D., & Yang, M. S. (2005). Changes in mineral and terpene concentration following calcium fertilization of *Chrysanthemum boreale* M. *Research Journal of Agriculture and Biological Sciences, 1*, 222–226.

Leung, A. Y., & Foster, S. (1996). *Encyclopaedia of common natural ingredients used in food, drugs and cosmetics* (2nd ed.). New York: John Wiley and Sons, Inc.

Liu, C. Z., Guo, C., Wang, Y., & Ouyang, F. (2003). Factors influencing artemisinin production from shoot cultures of *Artemisia annua* L. *World Journal of Microbiology and Biotechnology, 19*, 535–538.

Liu, C. Z., Wang, Y. C., Ouyang, F., Ye, H. C., & Li, G. F. (1997). Production of artemisinin by hairy root cultures of *Artemisia annua* L. *Biotechnology Letters, 19*, 927–930.

Lovkova, M. Y., Buzuk, G. N., Sokolova, S. M., & Buzuk, L. N. (2005). Role of elements and physiologically active compounds in the regulation of synthesis and accumulation of indole alkaloids in *Catharanthus roseus* L. *Applied Biochemistry and Microbiology, 41*, 299–305.

Lu, L., He, C., Jin, Y., Zhang, X., & Wei, J. (2013). Effects of the applications of phosphorus and potassium fertilizers at different growth stages on the root growth and bioactive compounds of *Salvia miltiorrhiza* Bunge. *Australian Journal of Crop Science, 7*, 1533–1543.

Lu-jiu, L. I., Rong-le, L. I., Fang, C. H., Ji-yun, J. I., Jia-jia, W. A., Dian-li, Y. A., & Dong-ping, L. I. (2010). Effect of different N application rates on yield, qualities and N uptake of ginger. *Plant Nutrition and Fertilizer Science, 2*, 019.

Mahfouz, S. A., & Sharaf-Eldin, M. A. (2007). Effect of mineral vs. biofertilizer on growth, yield, and essential oil content of fennel (*Foeniculum vulgare* Mill). *International Agrophysics, 21*, 361.

Mairapetyan, S. K., Tadevosyan, A. H., Alexanyan, S. S., & Stepanyan, B. T. (1997). Optimization of the N: P: K ratio in the nutrient medium of some soilless aromatic and medicinal plants. In II WOCMAP Congress medicinal and aromatic plants, part 3: Agricultural production, post harvest techniques. *Biotechnology, 502*, 29–32.

Major, J., Rondon, M., Molina, D., Riha, S. J., & Lehmann, J. (2010). Maize yield and nutrition during 4 years after biochar application to a Colombian savanna oxisol. *Plant and Soil, 333*, 117–128.

Marotti, M., Piccaglia, R., Giovanelli, E., Deans, S. G., & Eaglesham, E. (1994). Effects of planting time and mineral fertilization on peppermint (*Mentha* x *piperita* L.) essential oil composition and its biological activity. *Flavour and Fragrance Journal, 9*, 125–129.

Marschner, H. (2002). *Mineral nutrition of higher plants* (2nd ed.). New York: Academic Press Inc..

Mathai, K. (2000). Nutrition in the adult years. In L. K. Mahan & S. Escott-Stump (Eds.), *Krause's food nutrition and diet therapy* (Vol. 271, 10th ed., pp. 274–275).

Meagher, E., & Thomson, C. (1999). Vitamin and mineral therapy. In G. Morrison & L. Hark (Eds.), *Medical nutrition and disease* (2nd ed., pp. 33–58). Malden: Blackwell Science.

Mendonça Freitas, M. S., Gama, M. C., Monnerat, P. H., De Carvalho, A. J. C., Lima, T. C., & Vieira, I. J. C. (2015). Induced nutrient deficiencies in *Catharantus roseus* impact ajmalicine bioproduction. *Journal of Plant Nutrition, 39*, 835–841.

Miguel, G., Simoes, M., Figueiredo, A. C., Barroso, J. G., Pedro, L. G., & Carvalho, L. (2004). Composition and antioxidant activities of the essential oils of *Thymus caespititius*, *Thymus camphorates* and *Thymus mastichina*. *Food Chemistry, 86*, 183–188.

Mint Industry Research Council. (2015). Retrieved Oct 15, 2016 from, http://usmintindustry.org/

Miranda-Ham, M. L., Gomez, I., & Loyola-Vargas, V. M. (1996). Effect of inorganic nitrogen source on ammonium assimilation enzymes of *Catharanthus roseus* plants. *Python Buenos Aires, 58*(1/2), 125–133.

Mitchell, A. R., & Farris, N. A. (1996). Peppermint response to nitrogen fertilizer in an arid climate. *Journal of Plant Nutrition, 19*, 955–967.

Modak, M., Dixit, P., Londhe, J., Ghaskadbi, S., & Devasagayam, T. P. A. (2007). Indian herbs and herbal drugs used for the treatment of diabetes. *Journal of Clinical Biochemistry and Nutrition, 40*, 163–173.

Mridula, K. R., & Jayachandran, B. K. (2001). Quality of mango-ginger (*Curcuma amada* roxb.) as influenced by mineral nutrition. *Journal of Tropical Agriculture, 39*, 182–183.

Mukherjee, P. K., Nema, N. K., Maity, N., Mukherjee, K., & Harwansh, R. K. (2014). Phytochemical and therapeutic profile of *Aloe vera*. *Journal of Natural Remedies, 14*, 1–26.

Mumivand, H., Babalar, M., Hadian, J., & Fakhr Tabatabaei, M. (2011). Plant growth and essential oil content and composition of *Satureja hortensis* L. cv. Saturn in response to calcium carbonate and nitrogen application rates. *Journal of Medicinal Plants Research, 5*, 1859–1866.

Newman, D. J., & Cragg, G. M. (2007). Natural products as sources of new drugs over the last 25 years. *Journal of Natural Products, 70*, 461–477.

Nguyen, P. M., & Niemeyer, E. D. (2008). Effects of nitrogen fertilization on the phenolic composition and antioxidant properties of basil (*Ocimum basilicum* L.) *Journal of Agricultural and Food Chemistry, 56*, 8685–8691.

Nunnery, J. K., Mevers, E., & Gerwick, W. H. (2010). Biologically active secondary metabolites from marine cyanobacteria. *Current Opinion in Biotechnology, 21*, 787–793.

NurzyEska-Wierdak, R., & Borowski, B. (2011). Changes in the content and chemical composition of sweet basil essential oil under the influence of fertilization of plants with nitrogen and potassium. *Annales Universitatis Mariae Curie-Skłodowska Sectio DDD, Pharmacia, 24*(3), 133–145.

Nurzyska-Wierdak, R. (2012). Sweet basil essential oil composition: Relationship between cultivar, foliar feeding with nitrogen and oil content. *Journal of Essential Oil Research, 24*, 217–227.

Ohallorans, J. M., Munoz, M. A., & Colberg, O. (1993). Effect of chicken manure on chemical properties of a Mollisil and tomato production. *Journal of Agriculture of the University of Puerto Rico, 77*, 181–191.

Okwu, D. E. (2004). Phytochemicals and vitamin content of indigenous spices of south eastern Nigeria. *Journal of Sustainable Agriculture and Environment, 6*, 30–34.

Ozguven, M., Kirpik, M., & Sekeroglu, N. (2002). Determination of the optimal sowing time and nitrogen fertilization for lavender (*La-vandula angustifolia* Mill.) in the Cukurova conditions, In *Proceedings of the Workshop on Agricultural and Quality Aspects of Me-dicinal and Aromatic Plants, Adana, Turkey* (Vol. 2000, pp. 217–223).

Özgüven, M., Sener, B., Orhan, I., Sekerglu, N., Kirpik, M., Kartal, M., Pesin, I., & Kaya, Z. (2008). Effects of varying nitrogen doses on yield, yield components and artemisinin content of *Artemisia annua* L. *Industrial Crops and Products, 27*, 60–64.

Percival, S. (2000). Use of *Echinacea* in medicine. *Biochemical Pharmacology, 60*, 155–158.

Pinzón-Torres, J. A., Porras, C., Jovany, N., Durán García, D. C., Martínez Morales, J. R., & Stashenko, E. (2014). Green biomass production and quality of essential oils of palmarosa (*Cymbopogon martini* Roxb.) with application of synthesis fertilizers and organic fertilizers. *Acta Agronómica, 63*, 335–342.

Polasa, K., & Nirmala, K. (2003). Ginger: Its role in xenobiotic metabolism. *ICMR Bulletin, 33*, 57–62.

Prasad, A., Chattopadhyay, A., Yadav, A., & Kumari, R. (2008). Variation in the chemical composition and yield of essential oil of rose-scented geranium (*Pelargonium* sp.) by the foliar application of metallic salts. *Flavour and Fragrance Journal, 23*, 133–136.

Prasad, A., Kumar, S., Pandey, A., & Chand, S. (2012). Microbial and chemical sources of phosphorus supply modulate the yield and chemical composition of essential oil of rose-scented geranium (*Pelargonium* species) in sodic soils. *Biology and Fertility of Soils, 48*, 117–122.

Prasad, S., Tyagi, A. K., & Aggarwal, B. B. (2014). Recent developments in delivery, bioavailability, absorption and metabolism of curcumin: The golden pigment from golden spice. *Cancer Research and Treatment, 46*, 2–18.

Puttanna, K., Praksa Rao, E. V. S., Singh, R., & Ramesh, S. (2010). Influence of nitrogen and potassium fertilization on yield and quality of rosemary in relation to harvest number. *Communications in Soil Science and Plant Analysis, 41*, 190–198.

Rajeswara, R. B. R., & Singh, M. (1990). Effect of NPK fertilizers and spacing on periwinkle (*Catharanthus roseus*) under irrigated and rained conditions. *Herba Hungarica, 29*, 1–2.

Ram, M., Singh, R., Ram, D., & Sangwan, R. S. (2003). Foliar application of phosphate increased the yield of essential oil in menthol mint (*Mentha arvensis*). *Australian Journal of Experimental Agriculture, 12*, 1263–1268.

Ramezani, S., Rezaei, M. R., & Sotoudehnia, P. (2009). Improved growth, yield and essential oil content of basil grown under different levels of phosphorus sprays in the field. *Journal of Applied Biological Sciences, 3*, 105–110.

Ranaweera, S. S., & Thilakaratne, W. P. (1992a). Mineral nutrition of *Cymbopogon nardus* (L) Rendle: Part I. Effects of magnesium and phosphorus nutrition on growth and the yield of essential oil. *Vidyodaya Journal of Science, 4*, 201–208.

Ranaweera, S. S., & Thilakaratne, W. P. (1992b). Mineral nutrition of *Cymbopogon nardus* (L) Rendle: Part II. Effects of magnesium and phosphorus nutrition on the fractional composition of essential oil. *Vidyodaya Journal of Science, 4*, 209–219.

Rao, B. (2003). Bioactive phytochemicals in Indian foods and their potential in health promotion and disease prevention. *Asia Pacific Journal of Clinical Nutrition, 12*, 9–22.

Rouphael, Y., Cardarelli, M., Lucini, L., Rea, E., & Colla, E. (2012b). Nutrient solution concentration affects growth, mineral composition, phenolic acids and flavonoids in leaves of artichoke and cardoon. *HortScience, 47*, 1424–1429.

Sadanandan, A. K., & Hamza, S. (1996c). Studies on nutritional requirement of bush pepper (*Piper nigrum* L) for yield and quality. In N. M. Mathew & J. Kuruvilla (Eds.), *Development in plantation crops research* (pp. 223–227). New Delhi: Allied Publishers Ltd.

Sadanandan, A. K., Peter, K. V., & Hamza, S. (2002). Role of potassium nutrition in improving yield and quality of spice crops in India. In N. S. Pasricha, & S. K. Bansal (Eds.), *Potassium for sustainable crop production* (pp. 445–454). Gurgaon, Haryana: Potash Research Institute of India; Switzerland: International Potash Institute.

Said-Al Ahl, H., Hasnaa, A. H., Ayad, S., & Hendawy, S. F. (2009). Effect of potassium humate and nitrogen fertilizer on herb and essential oil of oregano under different irrigation intervals. *Journal of Applied Sciences, 23*, 319–323.

Sandeep, I. S., Nayak, S., & Mohanty, S. (2015). Differential effect of soil and environment on metabolic expression of turmeric (*Curcuma longa* cv. Roma). *Indian Journal of Experimental Biology, 53*, 406–411.

Sangwan, N. S., Farooqi, A. H. A., Shabih, F., & Sangwan, R. S. (2001). Regulation of essential oil production in plants. *Plant Growth Regulation, 34*, 3–21.

Saradhi, V. S. P., Khanam, S., Shivananda, B. G., Vasantha, K. T., & Shivananda, T. N. (2007). Effect of NPK fertilizers on chemical constituents of *Aloe vera* leaves. *Journal of Natural Remedies, 7*, 258–262.

Sarker, S., & Nahar, L. (2007). Bioactivity of turmeric. In P. Ravindran, K. Nirmal Babu, & K. Sivaraman (Eds.), *Turmeric the Genus Curcuma* (pp. 257–296). Boca Raton: CRC Press.

Scavroni, J., Boaro, C. S. F., Marques, M. O. M., & Ferreira, L. C. (2005). Yield and composition of the essential oil of *Mentha piperita* L. (Lamiaceae) grown with biosolid. *Brazilian Journal of Plant Physiology, 17*, 345–352.

Sekeroglu, N., & Ozguven, M. (2006). Effects of different nitrogen doses and plant densities on yield and quality of *Oenothera biennis* L. grown in irrigated lowland and un-irrigated dryland conditions. *Turkish Journal of Agriculture and Forestry, 30*, 125–135.

Sharma, P. R., Mondhe, D. M., Muthiah, S., Pal, H. C., Shahi, A. K., & Saxena, A. K. (2009). Anticancer activity of an essential oil from *Cymbopogon flexuosus*. *Chemico-Biological Interactions, 179*, 160–168.

Sharma, S., & Kumar, R. (2012). Effect of nitrogen on growth, biomass and oil composition of clary sage (*Salvia sclarea* Linn.) under mid hills of north western Himalayas. *Indian Journal of Natural Products and Resoures, 3*, 79–83.

Shim, S., Kim, S., Choi, D. S., Kwon, Y. B., & Kwon, J. (2011). Anti-inflammatory effects of [6]-shogaol: Potential roles of HDAC inhibition and HSP70 induction. *Food and Chemical Toxicology, 49*, 2734–2740.

Silviya, R. M., Dabhi, B. K., Aparnathi, K. D., & Prajapati, J. B. (2016). Essential oils of herbs and spices: Their antimicrobial activity and application in preservation of food. *International Journal of Current Microbiology and Applied Sciences, 5*, 885–901.

Singh, M. (2000). Effect of nitrogen, phosphorus and potassium nutrition on herb, oil and artemisinin yield in *Artemisia annua* under semi-arid tropical conditions. *Journal of Medicinal and Aromatic Plant Sciences, 22*, 368–369.

Singh, M. (2008). Effect of nitrogen and potassium fertilizer on growth, herbage and oil yield of irrigated Palmarosa (*Cymbopogon martinii* [roxb.] wats. Var. motia burk) in a semi-arid tropical climate. *Journal of Agronomy and Crop Science, 54*, 395–400.

Singh, M., & Ganesha Rao, R. S. (2009). Influence of sources and doses of N and K on herbage, oil yield and nutrient uptake of patchouli [*Pogostemon cablin* (Blanco) Benth.] in semi-arid tropics. *Industrial Crops and Products, 29*, 229–234.

Singh, M., Ganesha Rao, R. S., & Ramesh, S. (2005). Effects of nitrogen, phosphorus and potassium on herbage, oil yield, oil quality and soil fertility of lemongrass in a semi-arid tropical region of India. *The Journal of Horticultural Science and Biotechnology, 80*, 493–497.

Singh, M., Ganesha Rao, R. S., & Ramesh, S. (2007). Effects of N and K on growth, herbage, oil yield and nutrient uptake pattern of rosemary (*Rosmarinus officinalis* L.) under semi-arid tropical conditions. *The Journal of Horticultural Science and Biotechnology, 82*, 414–419.

Singh, M., Khan, M. M. A., & Naeem, M. (2011). *Mineral nutrition of turmeric (Curcuma longa L.) - a commercial spice crop*. Germany: Lambert Academic Publishing House.

Singh, M., Khan, M. M. A., & Naeem, M. (2014). Effect of nitrogen on growth, nutrient assimilation, essential oil content, yield and quality attributes in *Zingiber officinale* Rosc. J Saudi Soc. *Agricultural Sciences, 15*, 171–178.

Sivaraman, K. (2007). Agronomy of turmeric. In P. N. Ravindran, K. Nirmal Babu, & K. Sivaraman (Eds.), *Turmeric-The genus Curcuma* (pp. 129–153). Boca Raton: CRC Press.

Smillie, T. J., & Khan, I. A. (2010). A comprehensive approach to identifying and authenticating botanical products. *Clinical Pharmacology and Therapeutics, 87*, 175–186.

Soeparjono, S. (2016). The effect of media composition and organic fertilizer concentration on the growth and yield of red ginger rhizome (*Zingiber officinale* Rosc.) *Agriculture and Agricultural Science Procedia, 9*, 450–455.

Sreevalli, Y., Kulkarni, R. N., Baskaran, K., & Chandrashekara, R. S. (2004). Increasing the content of leaf and root alkaloids of high-alkaloid-content mutants of periwinkle through nitrogen fertilization. *Industrial Crops and Products, 19*, 191–195.

Srinivasan, V., Thankamani, C. K., Dinesh, R., Kandiannan, K., Zachariah, T. J., Leela, N. K., Hamza, S., Shajina, O., & Ansha, O. (2016). Nutrient management systems in turmeric: Effects on soil quality, rhizome yield and quality. *Industrial Crops and Products, 85*, 241–250.

Suh, E. J., & Park, K. W. (2000). Effect of calcium ion in nutrient solution on the content and composition of essential oil of sweet basil in hydroponics. *Journal of Korean Society of Horticultural Science, 41*, 598–601.

Supanjani Tawaha, A. R. M., Yang, M. S., Han, H. S., & Lee, K. D. (2005). Role of calcium in yield and medicinal quality of *Chrysanthemum coronarium* L. *Journal of Agronomy, 4*, 186–190.

Szoke, E., Maday, E., Kiss, S. A., Sonnewend, L., & Lemberkovics, E. (2004). Effect of magnesium on essential oil formation of genetically transformed and non-transformed chamomile cultures. *Journal of the American College of Nutrition, 23*, 763S–767S.

Tabassum, N., & Ahmad, F. (2011). Role of natural herbs in the treatment of hypertension. *Pharmacognosy Reviews, 5*, 30–40.

Tawaraya, K., Turjaman, M., & Ekamawanti, H. A. (2007). Effect of Arbuscular mycorrhizal colonization on nitrogen and phosphorus uptake and growth of *Aloe vera* L. *Horticultural Science, 42*, 1737–1739.

Turmeric World. (2014). Retrieved October 1, 2016 from http://turmericworld.com/production.php

Vining, L. C. (1990). Functions of secondary metabolites. *Annual Review of Microbiology, 44*, 395–427.

Wang, J. W., & Tan, R. X. (2002). Artemisinin production in *Artemisia annua* hairy root cultures with improved growth by altering the nitrogen source in the medium. *Biotechnology Letters, 24*, 1153–1156.

WHO. (1999). *WHO monographs on selected medicinal plants* (Vol. 1). Geneva: WHO.

Wink, M. (2003). Evolution of secondary metabolites from an ecological and molecular phylogenetic perspective. *Phytochemistry, 64*, 3–19.

Xie, Z., Ma, X., & Gang, D. (2009). Modules of co-regulated metabolites in turmeric (*Curcuma longa*) rhizome suggest the existence of biosynthetic modules in plant specialized metabolism. *Journal of Experimental Botany, 60*(1), 87–97.

Xizhen, A., Zhifeng, C., Jingran, Q., Liping, C., & Dewan, Z. (1998). Effects of different levels of supplying fertilizer on ginger quality. *Journal of Shandong Agricultural University, 29*, 183–188.

Zamborine, E. N., Szabo, K., Rajhart, P., Lelik, L., Bernáth, J., & Popp, T. (2016). Effect of nutrients on drug production and essential oil content of lemon balm (*Melissa officinalis* L.) *Journal of Essential Oil Bearing Plants, 18*(6), 1508–1515.

Zhao, X., Yang, Z. B., Yang, W. R., Wang, Y., Jiang, S. Z., & Zhang, G. G. (2011). Effects of ginger root on laying performance and antioxidant status of laying hens and on dietary oxidation stability. *Poultry Science, 90*, 1720–1727.

Zheljazkov, V. D., Cantrell, C. L., Astatkie, T., & Cannon, J. B. (2011). Lemongrass productivity, oil content, and composition as a function of nitrogen, sulfur, and harvest time. *Agronomy Journal, 103*, 805–812.

Zheljazkov, V. D., Cantrell, C. L., Astatkie, T., & Ebelhar, M. W. (2010). Peppermint productivity and oil composition as a function of nitrogen, growth stage, and harvest time. *Agronomy Journal, 102*, 124–128.

Zheljazkov, V. D., Cantrell, C. L., Ebelhar, M. W., Rowe, D. E., & Coker, C. (2008). Productivity, oil content, and oil composition of sweet basil as a function of nitrogen and sulphur fertilization. *Horticultural Science, 43*, 1415–1422.

Zheljazkov, V. D., Cerven, V., Cantrell, C. L., Ebelhar, W. M., & Horgan, T. (2009). Effect of nitrogen, location, and harvesting stage on peppermint productivity, oil content, and oil composition. *Horticultural Science, 44*, 1267–1270.

Part III
Plant Nutrition and Abiotic Stress

Chapter 10
Actions of Biological Trace Elements in Plant Abiotic Stress Tolerance

Mirza Hasanuzzaman, Kamrun Nahar, Anisur Rahman, Jubayer Al Mahmud, Md. Shahadat Hossain, Md. Khairul Alam, Hirosuke Oku, and Masayuki Fujita

Abstract With the increase of global population, the demand for food crops, oil, fiber and other by-product yielding crops is increasing. In contrast to this increasing demand, abiotic stresses hinder the productivity of plants. Abiotic stresses sometimes reduce more than half of the crop yields. To attain global food security, understanding of plant responses to abiotic stresses is crucial because this is the prerequisite for developing approaches/tools for improving plant stress tolerance. Trace elements are nutrients required in small quantities to facilitate a range of physiological functions. These elements stimulate growth but are not essential. Some are essential only for certain plant species or required under a given condition. Trace elements not only improve plant physiological processes and growth but play roles in improving plant stress tolerance. However, the actual physiological functions of trace elements in conferring abiotic stress tolerance are still under study. This chapter focuses

---

M. Hasanuzzaman (✉)
Molecular Biotechnology Group, Center of Molecular Biosciences, Tropical Biosphere Research Center, University of the Ryukyus, 1 Senbaru, Nishihara, Okinawa 903-0213, Japan

Department of Agronomy, Faculty of Agriculture, Sher-e-Bangla Agricultural University, Dhaka 1207, Bangladesh
e-mail: mhzsauag@yahoo.com

K. Nahar
Department of Agricultural Botany, Faculty of Agriculture, Sher-e-Bangla Agricultural University, Dhaka 1207, Bangladesh

Laboratory of Plant Stress Responses, Faculty of Agriculture, Kagawa University, Miki-cho, Kita-gun, Kagawa 761-0795, Japan

A. Rahman
Department of Agronomy, Faculty of Agriculture, Sher-e-Bangla Agricultural University, Dhaka 1207, Bangladesh

Laboratory of Plant Stress Responses, Faculty of Agriculture, Kagawa University, Miki-cho, Kita-gun, Kagawa 761-0795, Japan

M. Naeem et al. (eds.), *Essential Plant Nutrients*,
DOI 10.1007/978-3-319-58841-4_10

on the roles of trace elements emphasizing especially the recent advances on the actions of biological trace elements in plant abiotic stress tolerance.

Keywords Abiotic stress • Trace elements • Antioxidants • Phytoprotectants • Metabolites

10.1 Introduction

With the increase of global population, the demand of food crops, oil, fiber, and other by-product yielding crops is increasing (Condon et al. 2004; Morison et al. 2008). In contrast to this increasing demand, abiotic stresses hinder the productivity of plants. Abiotic stresses sometimes reduce more than half of the crop yields (Wang et al. 2003). Climate prediction models demonstrate the frequency of occurrence of different abiotic stresses including salinity, drought, flooding, extreme temperature, UV radiation, ozone, and metal toxicity that have increased to a great extent, compared to the previous decades (IPCC 2008; Mittler and Blumwald 2010). Under stressful conditions, plants are subjected to various physiological and metabolic changes throughout their life cycles starting from germination to final harvest. These changes include inhibition of germination, decreased growth, less photosynthesis, lower dry matter assimilation, water imbalance, impaired nutrient uptake, and ultimate reduction of yield and quality (Hasanuzzaman et al. 2012a, b, 2013a, b, c, d, 2014a, b, 2015, 2016). To attain global food security, understanding of plant responses to abiotic stresses is crucial because this is the prerequisite for developing approaches/tools for improving plant stress tolerance (Condon et al. 2004; Morison et al. 2008). In response to abiotic stresses, antioxidants, hormones, osmoprotectants, and many other essential metabolites are altered which have major roles to defend abiotic stresses. Traditional crop improvement methods are time-consuming, costly, sometimes uncertain, or unsuccessful and sometimes cause crop loss due to

J.A. Mahmud
Laboratory of Plant Stress Responses, Faculty of Agriculture, Kagawa University, Miki-cho, Kita-gun, Kagawa 761-0795, Japan

Department of Agroforestry and Environmental Science, Faculty of Agriculture, Sher-e-Bangla Agricultural University, Dhaka 1207, Bangladesh

M.S. Hossain • M. Fujita
Laboratory of Plant Stress Responses, Faculty of Agriculture, Kagawa University, Miki-cho, Kita-gun, Kagawa 761-0795, Japan

M.K. Alam
School of Veterinary and Life Sciences, Murdoch University, Perth, WA, Australia

H. Oku
Molecular Biotechnology Group, Center of Molecular Biosciences, Tropical Biosphere Research Center, University of the Ryukyus, 1 Senbaru, Nishihara, Okinawa 903-0213, Japan

evolution or genetic degradation. The use of exogenous phytoprotectants including micronutrient or trace elements is potent approaches to improve plant abiotic stress tolerance (Hasanuzzaman et al. 2011a, b, 2013a, b, c; Nahar et al. 2015a).

Trace elements are nutrients required in small quantities to facilitate a range of physiological functions. These elements stimulate growth but are not essential. Some are essential only for certain plant species or required under a given condition. The micronutrients or trace elements may include iron (Fe), zinc (Zn), copper (Cu), manganese (Mn), molybdenum (Mo), chromium (Cr), cobalt (Co), nickel (Ni), selenium (Se), silicon (Si), etc. Trace elements not only improve plant physiological processes and growth but play roles in improving plant stress tolerance. For instance, Si has been proved to improve plants' tolerance against drought, salinity, high temperature, chilling, UV radiation, nutrient imbalance, and metal toxicity (Raven 2001; Liang et al. 2007; Ma and Yamaji 2008; Hasanuzzaman and Fujita 2011; Ahmed et al. 2012). Plant species supplemented with Se have shown enhanced resistance to salinity (Hawrylak-Nowak 2009; Hasanuzzaman et al. 2011a, b), drought (Hasanuzzaman et al. 2010; Hasanuzzaman and Fujita 2011), extreme temperature (Djanaguiraman et al. 2010; Hawrylak-Nowak et al. 2010), metal toxicity (Hasanuzzaman et al. 2012b), and UV radiation (Yao et al. 2010). Selenium enhances reactive oxygen species (ROS) and oxidative stress tolerance (Hasanuzzaman et al. 2010; Hasanuzzaman and Fujita 2011). Exogenous Zn improved cadmium (Cd) tolerance and increased biomass accumulation by reducing its uptake and Cd-induced membrane damage (Wu and Zhang 2002). Zinc application reduced sodium uptake, enhanced antioxidants, and reduced salt-induced lipid peroxidation and electrolyte leakage (Aktas et al. 2006; Tavallali et al. 2010). Zinc also increased seed yield and thousand kernels weight against drought stress (Monjezi et al. 2012). Exogenous application of boron (B) conferred drought tolerance (Moeinian et al. 2011; Abdel-Motagally and El-Zohri 2016), salt tolerance (Salim 2014), and aluminum (Al) tolerance (Zhou et al. 2015). Salt (Pandya et al. 2004), Cd (Pal'ove-Balang et al. 2006; Peng et al. 2008; Sebastian and Prasad 2015), and drought (Upadhyaya et al. 2012) stresses were alleviated by exogenous manganese (Mn) application. However, the actual physiological functions of trace elements in conferring abiotic stress tolerance are still under study. Considering their importance, this chapter focuses on the roles of trace elements emphasizing specially the recent advances on the actions of biological trace elements in plant abiotic stress tolerance.

10.2 Abiotic Stress: The Worst Enemy for Plants

Unlike animals, plants are sessile organisms and thus are bound to grow in the same habitat fighting with various environmental factors termed as abiotic stress. In the era of climate change, plants are constantly suffering from various abiotic stresses like salinity, drought, extreme temperature, flooding, metal toxicity, air pollution, ozone, UV radiation, and so on. These episodes are becoming more common due to rapid changes in climate. These stressors, alone or in

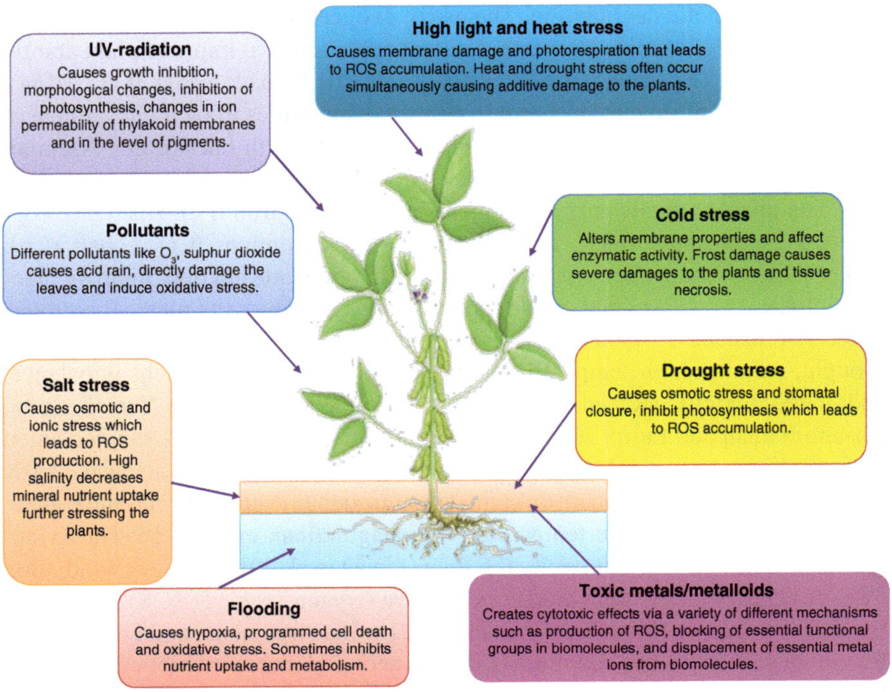

Fig. 10.1 Major abiotic stresses for plants and their general effects (Hasanuzzaman et al. 2016; With Permission from Elsevier)

combination, attack plants, and thus the consequence is often unpredictable and becomes more severe than we expect. These effects are also greatly varied with the type and duration/extent of stresses and also depend on the plant species and genotypes (Pandey 2015). Therefore, it is very difficult to generalize the common effect of abiotic stress in plants. However, the major effects of abiotic stress in plants includes loss of seed germination, growth inhibition, disruption of energy assimilation and photosynthesis, imbalance in metabolism, imbalance in nutrient homeostasis and nutrient uptake, reduction of crop yield, and deterioration in crop quality (Fig. 10.1). One of the most common effects of abiotic stress in plants is oxidative stress which is described separately in a section later in this chapter.

10.3 Plant Responses to Abiotic Stresses

10.3.1 Salinity

Among these abiotic stresses, salinity/salt stress is considered as one of the most destructive stresses which reduces the land area and production of crop (Yamaguchi and Blumwald 2005; Shahbaz and Ashraf 2013). Salinity has terrible effects on the germination, vigor, and yield of the crop (Munns and Tester 2008). Genotype, growth stage, intensity, as well as duration of stress are some of the most important factors which are accountable for the response of the plant to salt stress. When plant exposed to high salinity, it suffers from osmotic stress, ionic specificity, nutritional and hormonal disorders, changing metabolic and physiological processes, and finally oxidative damage (Munns 2002; Zhu 2007). Cellular membrane disorder, photosynthesis reduction, different toxic metabolite synthesis, nutrient unavailability, and ultimate plant death are some common consequences of salt stress in plants (Mahajan and Tuteja 2005). Thus, there are three main ways of affecting plants through salt stress, e.g., by developing osmotic stress through the reduction of soil water potential, creating the ionic imbalance within the cell, lowering the concentration of some ions specially potassium (K^+) and calcium (Ca^{2+}), and ultimately causing sodium (Na^+) and chloride (Cl^-) ions' toxicity. Under salt stress, increased Na^+ interferes with K^+ uptake which is very much important for protein synthesis by acting as binding material, and thus insufficient concentrations of K^+ reduces the protein synthesis (Blaha et al. 2000). Increased Na^+-K^+ ratio and reduced K^+ availability result in the disruption of many enzymatic processes. Two major stresses called osmotic and ionic stresses are associated with salt stress. A usual response of plant to salt stress is growth inhibition which is caused by the changing concentration of soluble salt and osmotic potential of soil solution (Tavakkoli et al. 2011). Plant exposure to salt stress causes inhibition of water uptake and decreases cell expansion and lateral growth due to immediate osmotic stress occurring in the root medium (Munns and Tester 2008). Under salt-induced osmotic stress, development of shoot is severely affected compared to the root. Slower and reduced development of leaf area compared to root might be the reason behind it (Munns and Tester 2008). Impaired chlorophyll (chl) synthesis due to Cl^- ion results in the chlorotic toxicity. Significant reduction of growth and water use efficiency are also resulting from the excess accumulation of Cl^-. Germination is one of the most important phases of plant life cycle determining the final plant density. Germination and establishment of seedling are reported to decrease by salinity and considered as one of the major problems of the salt affected area. Salinity directly acts as a limiting factor for seed germination by delaying the germination time and reducing the final germination percentage and also limiting plant growth (Rahman et al. 2000). Extended salinity inhibited seed germination, while a dormant state was observed when seeds were subjected to lower level of salt stress (Khan and Weber 2008) and it was also revealed that germination process inhibited by salt stress due to lower osmotic potential of the growth medium. Salt stress also negatively affects the metabolism of nucleic acid (Gomes-Filho et al. 2008), proteins (Dantas et al. 2007), and seed food reserve utilization (Othman et al. 2006). An experiment was conducted by Kaveh et al. (2011) with *S. lycopersicum*, and he found significantly reduced germination when subjected to salinity. The gradual retardation and reduced germination of *Hordeum secalinum* were observed with an increase in NaCl concentrations. It was also concluded that 40% and 38% reductions in germination rate occur when treated with 400 and 500 mM NaCl, respectively (Lombardi and Lupi 2006). Bordi (2010) reported that different levels of NaCl treatments significantly reduced the germination percentage of *B. napus*. Some probable reasons of above-mentioned germination inhibition caused by salt stress are ionic imbalance, osmotic stress, and reduced seed water uptake. A study with four rice cultivars also showed remarkable germination reduction when subjected to different salinity levels, and germination was also more significantly reduced in sensitive cultivar compared to tolerant one (Hasanuzzaman et al. 2009). Salt stress primarily affects plant by reducing the rate of the plant growth. Firstly, the water uptake capacity of the plant is reduced, and thus growth rate of plant also reduced quickly due to salinity. There are two phases involved in the salt-induced growth reduction. Osmotic stress is the reason behind the first-phase growth reduction, and second phase of growth reduction is caused by excess accumulation of salt and its toxicity to plants (Munns 2002). The second one ultimately affects the yield by decreasing the supply of photosynthate in plants and causing death of leaves. Delayed flowering and reduced crop yield are the results of this growth reduction caused by salt stress (Munns and Tester 2008). Moreover, alteration of plant water relation also caused by salinity resulting in cell turgor loss. At the same time, water uptake becomes more difficult because of the more negative water potential caused by increased salts present in the root zone (Munns 2002). Stomatal closure is one effective strategy of osmotically stressed plants for managing loss of water (James et al. 2008). Reduction in CO_2 uptake and restricted carboxylation reaction of leaf are executed by closing of stomata. Therefore, the reduction of carbohydrate production takes place during photosynthesis which negatively influences the plant growth and crop yield. The reduction of photosynthesis along with the reduced chl *a*, chl *b*, and carotenoid (Car) pigments were observed under salt stress in rice (Amirjani 2011), mustard (Ahmed et al. 2012), and many other crops (Hasanuzzaman et al. 2012a). Plants when under salt stress become unable to take nutrients and water from the soil due to physiological drought, reduces cell expantion, leaf growth, stomatal closure, and photosynthesis as well. Salt stress causes the premature yellowing, senescence, and ultimate death of the older leaf tissues resulting from the increased concentration of toxic ions which adversely affects photosynthesis (Munns 2002; Munns and Tester 2008). Thus, reduced photosynthesis is a remarkable effect of salt stress (Leisner et al. 2010; Raziuddin et al. 2011). Higher Na^+ and K^+ toxicity in chloroplasts also inhibits the photosynthesis process. It has been revealed that salinity involved a positive growth inhibition which is associated with a marked inhibition of photosynthesis (Fisarakis et al. 2001). Again, salt stress may affect carbon metabolism or photophosphorylation (Sudhir and Murthy 2004). High salt stress remarkably hinders crop yield due to impaired growth and physiological development. In *Vigna radiata*, reduction in pod per plant, seed per pod, and seed weight was observed under high salinity (Nahar and Hasanuzzaman 2009). However, plant tolerance to salt stress is a complicated process, and it depends on the plants age, growth phase, and environment. To endure the stress of salinity, plants have developed well-organized adaptive features at the biochemical level and have antioxidant defense system to fight against the adverse condition. Usually, under salt stress, plants accumulate osmoprotectants of different types like proline (Pro), betaines, glycerols, sucrose, trehalose, etc. (Budak et al. 2013). The mechanisms how plant responds to salt stress are very complex and are still a matter of further research. It is an urgent task for the plant biologists to explore the possible ways of alleviating the damaging effects of salt stress.

10.3.2 Drought

Drought stress remained the greatest constraints for plant productivity throughout the world due to the increasing shortage of water. Drought stress limits plant growth significantly though the reduction of plant growth depends upon the degree of osmotic stress, stages of crop growth, and cultivar (Budak et al. 2013). Reduction in root and shoot growth is greater in susceptible soybean genotype because tolerant genotype can hold much water than susceptible one (Hossain et al. 2014). A study on growth of six bedding plants under drought stress revealed that plant dry weight, leaf area, and plant height differ significantly depending on species. Here, a significant reduction was observed in dry weight of *Cineraria maritima* L, *Petunia x hybrida*, and *Plumbago auriculata* under water-deficit condition, whereas leaf area reduced in *Plumbago* and plant height reduced in *Plumbago* and *Vinca* compared to control (Niu et al. 2006). Sapeta et al. (2013) reported the growth (stem diameter, stem length, leaf number, etc.) of two *Jatropha curcas* was reduced due to drought stress. Similarly, upon rehydration both genotypes showed tolerance to drought. In another study, two varieties of barley exposed to drought (50% field capacity, FC) showed significant reduction in leaf area and leaf appearance rate (Thameur et al. 2012). Reduction in water uptake and energy supply, impaired enzyme activities, loss of turgor, and reduction in cell division and expansion are considered as reasons for inhibited growth under water shortage condition (Taiz and Zeiger 2006; Hasanuzzaman et al. 2014c). Photosynthesis, the most sensitive physiological process to dehydration, is considered as physiological basis of crop yield (Luo et al. 2016). Several factors influence photosynthesis such as leaf area, stomatal opening and closure, water status in the leaf tissues, rate of CO_2 assimilation, electron transport and CO_2 assimilation reactions, and ribulose bisphosphate (RuBP) generation. Alteration in any of these factors upon dehydration condition affects photosynthesis (Ahmad et al. 2014; Hasanuzzaman et al. 2014b). The reduction in photosynthesis under deficit soil moisture condition in potato is due to stomatal closure, reduction in CO_2 assimilation, and excess production of ROS (Li et al. 2015; Obidiegwu et al. 2015). Comparative study between two maize cultivars revealed that photosynthetic activity was higher in drought-tolerant maize cultivar Giza 2 than susceptible Trihybrid 321 cultivar under water shortage condition (Moussa and Abdel-Aziz 2008). Reduction in photosynthetic pigments, changes in gas exchange characteristics, damage to photosystems, and reduced activity of enzymes related to photosynthesis are the common responses under water shortage condition and thus affect photosynthesis (Ashraf and Harris 2013). Marcińska et al. (2013) reported the reduction in chl content and photosynthesis under drought stress in wheat. Accumulation of compatible solutes is a common response of plant under water shortage condition. Sugar, sugar alcohol, glycine betaine, amino acids, and Pro are common compatible solutes in plant involved in osmotic adjustment, ROS scavenging, and protection in protein and cell structure (Budak et al. 2013). In recent years, Pro, glycine betaine, and total free amino acids were found to be increased in plant under dehydration condition to maintain leaf turgor and to improve water uptake in drying soil (Ajithkumar and Panneerselvam 2013). Water shortage significantly affects water relation in plants. Water relation can be understood by some attributes such as relative water content (RWC), leaf water potential, and transpiration rate (Nahar et al. 2017).

10.3.3 Waterlogging

Among the abiotic stresses, excess water or flooding that causes submergence or waterlogging conditions is also harmful and even lethal for plant (Nishiuchi et al. 2012). Submergence is the condition when a plant is completely covered with water or remain under water and subjected to anaerobic from aerobic condition (Jackson and Ram 2003; Nishiuchi et al. 2012). In anaerobic submerged condition, plants also subject to other secondary stresses such as low light, impaired gas exchange, and unavailability of plant nutrients (Ram et al. 1999). On the other hand, water logging is the condition of soil when excess water limits gas diffusion in plant root zone (Setter and Waters 2003). Excess water or waterlogging condition caused by heavy rainfall, poor drainage, and irrigation practices subjected plants in complex changes in several environmental parameters and plant suffered from deficiency of oxygen, light, and nutrient. Deficiency of oxygen occurred by waterlogged condition due to limited gas diffusion and rapid consumption of oxygen by soil microorganism leads to hypoxia and progressively to anoxia within few hours. Due to hypoxia or anoxia, one of the immediate responses of water logging is closing of stomata to avoid water loss that inhibits respiration (García-Sánchez et al. 2007). Inhibited respiration consequently downregulates photosynthetic machinery which decreases chl content (Damanik et al. 2010). Water logging and submergence also decrease leaf water potential, stomatal conductance, and gas exchange which also decrease photosynthetic pigment (Arbona et al. 2008). In addition, under submerged condition, due to oxygen shortage, plant accumulates volatile ethylene which has both toxicity and signaling effect (Steffens et al. 2011). Although the mechanism is still unclear, hypoxic condition also results in oxidative stress through overproduction of ROS (Kumutha et al. 2009; Sairam et al. 2011).

Excess water damages most crops with exception of rice and some other aquatic plants (Bailey-Serres and Colmer 2014). Compared with other plants, rice has some adaptive potentiality for submergence tolerance like formation of aerenchyma and leaf gas film which contribute internal aeration during submergence. Besides these, tolerant rice cultivars survive under complete submergence using special strategies of growth control, namely, quiescence and escape strategy. Though rice can be grown well under flooded and waterlogging conditions, it is not well adapted to sudden and total submergence when sustained for several days (Nishiuchi et al. 2012; Jackson and Ram 2003). The visible injury symptoms in rice seedlings appeared with faster elongation of pseudostem and yellowing of older leaves (Jackson and Ram 2003). Banerjee et al. (2015) showed some biochemical changes in rice seedlings under submergence condition (7 days) like decrease in lipid peroxidation and increase of protein oxidation, phenol content, and flavonoid content. Rice seedlings are also damaged by oxidative stress in post submergence condition due to massive oxygen burst and higher light intensity relative to submerged conditions. At recovery stage,

Ella et al. (2003) showed higher lipid peroxidation and lower chl content in rice seedlings after 7 days of submergence. Short-term (24 h) water logging condition in wheat induced transient N deficiency which reduced growth and yield (Robertson et al. 2009). After 72 h of waterlogging at three-leaf and jointing stage, growth and yield attributes of summer maize decreased in field condition (Ren et al. 2014).

Kumutha et al. (2009) observed higher lipid peroxidation and growth inhibition in pigeon pea seedlings due to activation of DPI-sensitive NADPH oxidase under 4 days of waterlogging condition. It also produced smaller and chlorotic leaves which senesce earlier. Kumar et al. (2013) studied the physiological and morphological response of tolerant and sensitive cultivars varies in *V. radiata* after 3, 6, and 9 days of waterlogging. Growth inhibition and yield reduction were higher in sensitive cultivar (Pusa Baisakhi & MH–1 K–24) compared to tolerant cultivar (T 44 & MH–96–1) because of poor maintenance of leaf respiration rate and lower formation of adventitious root. Waterlogging and post-waterlogging condition induced oxidative stress and decrease photosynthetic rate in citrus seedlings (Hossain et al. 2009).

10.3.4 High Temperature

Increasing global temperatures in last few decades is responsible for severe crop loss in different countries throughout the world (Long and Ort 2010). Global temperature is expected to increase up to 2.5–5.4 °C by year 2100 (Ciscar 2012). Intensity, duration, and the rate of increase of temperature are important factors to create damage effects on plants (Wahid et al. 2007). Reduced germination and biomass, increased tillering, chlorosis, necrosis, early senescence of floral bud, premature death, and senescence of fruit are some common damage effects of high temperature (HT, Wahid et al. 2007). Heat stress during seed development results in structural disintegration and physiological disorders which further reduces germination and vigor, emergence, and seedling establishment (Akman 2009). High temperature hindered seed germination in rice by reducing a group of proteins involved in methionine metabolism, amino acid biosynthesis, energy metabolism, reserve degradation, and protein folding (Liu et al. 2014). Germination of maize is inhibited beyond 37 °C due to mutilation of protein synthesis whereas coleoptile growth is completely stopped at 45 °C (Riley 1981; Akman 2009). Increase of soil temperature beyond 45 °C causes germination failure and epicotyl emergence in sorghum (Peacock et al. 1993). In tomato, seedling emergence stops beyond 30 °C (Camejo et al. 2005). In rice, between 15 and 37 °C germination was 90%. But no germination occurred at 8 and 45 °C (Hartmann et al. 1997). Seed germination of *Phacelia tanacetifolia* inhibited by 2.7%, 5.8%, 84.0%, 89.0%, and 91.5% at temperatures of 20, 25, 30, 35, and 40 °C, respectively, compared to 15 °C (optimum for germination) (Tiryaki and Keles 2012). Lack of plasma membrane reorganization, cytoplasmic ion channel maintenance, and inhibition of the reactivation of the metabolic processes for mobilization of ions from a bound to a free form are responsible for germination reduction under HT in *Phacelia* seeds (Pirovano et al. 1997). High temperature adversely affects soil-plant- water relation, nutrient uptake, photosynthesis, transpiration, and respiration, disrupts source/sink balance, and hinders translocation of assimilate. As a result, growth and development are negatively affected (Prasad et al. 2011). Plants suffer from HT damages are often characterized by decreased water content and accumulation of osmoprotectant molecules. High temperature reduced water content in trifoliate orange seedlings (Fu et al. 2014). An increase of Pro level was noticed in transgenic plant grown under HT condition (Cvikrová et al. 2013). High temperature (31/18 °C and 34/22 °C, day/night) was found to decrease total dry weight and harvest index in wheat (Prasad et al. 2011). Enzymes are thermolabile. Catalytic properties of most enzymes are lost because of which denature of enzyme under HT. Biosynthesis of proteins and enzymes are hampered under HT. Severe HT results in complete denaturation of protein, enzymes, and membrane function and causes cell death (Allakhverdiev et al. 2008). Structural disorganization of thylakoids and swelling of stromal lamellae are HT-induced damages (Zhang et al. 2005). Chlorophyll content decreased by 19% in wheat subjected to HT (31/18 °C and 34/22 °C, day/night) (Prasad et al. 2011). Increase of temperature from 22 to 32 °C decreases photosynthesis rate and transpiration (Zhang et al. 2010). Solubility of oxygen increases more compared to carbon dioxide under HT. Oxygen becomes more concentrated for which oxygenation of RuBP occurs at HT that is termed as photorespiration. Under HT stress, gross photosynthesis is inhibited but respiration and photorespiration increase for which net photosynthesis decreases (Allakhverdiev et al. 2008; Mittler et al. 2012). As HT stress continues with decreasing net photosynthesis, plant begins to use up carbohydrate reserves that cause growth reduction. Inhibition of photosynthesis during HT stress is related to reduction of activity of RuBisCo (Allakhverdiev et al. 2008). The functioning of photosystem II (PS II) is inhibited by HT (Mittler et al. 2012) due to decrease of electron transport, removal of external proteins, and release of calcium and magnesium ions from binding site (Wahid et al. 2007; Zlatev and Lidon 2012). The singlet oxygen is overproduced under HT that damage D1 and D2 proteins (Yoshioka et al. 2006). HT restricts conversion of photosynthetic assimilates into starch that ultimately hampers grain development of cereal or seed development of legume (Stone and Nicolas 1994). Reproductive development has been long recognized as extremely sensitive to heat stress and most important reason for decreasing plant productivity (Thakur et al. 2010). The male gametophyte is more sensitive to HT in contrast to pistil or the female gametophyte (Hedhly 2011). In tomato flowers due to reduction of anthers and pollen viability, the successful fertilization is seriously hampered which further hinders development of the embryo (Barnabás et al. 2008). High temperature reduced soluble sugars in the anther and pollen grains which decreased pollen viability (Ismail and Hall 1999). In barley and *Arabidopsis*, anther development is hampered under HT (30–35 °C) due to arrest of cell proliferation, anomalous vacuoles, and mitochondrial and chloroplast development (Sakata et al. 2010). Decrease of water use efficiency during reproductive development of wheat results in decrease of photosynthesis and leaf area. As a result grain weight and sugar content of kernels decreased (Shah and Paulsen 2003) which also modulated nutritional flour quality (Hedhly et al. 2009). High-temperature stress adversely affects xylem and phloem loading which reduces water transportation and nutrient uptake (Taiz and Zeiger 2006). High-temperature stress (38 °C, 4 days) significantly increased the transpiration rate of tomato plant (Cheng et al. 2009). Increases of temperature may result in yield reduction between 2.5 and 10% in different agronomic species throughout twenty-first century (Hatfield et al. 2011). A decrease of 6.7–10% yield in rice (Peng et al. 2004; Lyman et al. 2013) and 10% reduction of yield in wheat (You et al. 2009) has been predicted with increase of 1 °C temperature in growing periods. Maize pollen viability decreases in temperatures above 35 °C (Dupuis and Dumas 1990). High temperature above 30 °C decreased seed production in soybean (Lindsey and Thomson 2012) and common bean (Porch 2006) grown in tropical humid regions. Seed set on male-sterile, female-fertile soybean plants decreased when day time temperature increased by 30–35 °C (Wiebbecke et al. 2012). In *Citrus sinensis*, temperatures higher than 30 °C increased fruit drop (Cole and McCloud 1985). In wheat, HT (31/18 °C and 34/22 °C, day/night) reduced grain numbers by 56% and individual grain weight by 25% which decreased grain yield significantly (compared to control temperature, 24/14 °C) (Prasad et al. 2011). Failure of fertilization, reduced panicle size, and damaged floral primordia were correlated to 54–64% loss in seed set in sorghum (Jain et al. 2007).

10.3.5 Chilling

In general, chilling stress indicates the low-temperature injury without forming ice crystals, and freezing stress indicates injury due to ice formation within plant tissues. Tolerance to chilling or freezing stresses is dependable to plant species. Chilling temperature reduces crop productivity by affecting several aspects of plant growth and developmental processes (Sanghera et al. 2011). Early growth stages of plant are often adversely affected by chilling stress. Optimum temperature range for germination of rice seed is 20–35 °C. According to Yoshida (1981), the temperature of 10 °C is minimum critical value, and below this temperature, rice seeds did not germinate. In *B. napus*, for 50% germination, it took 3 days at 8 °C, whereas it took 13 days at 2 °C (Angadi et al. 2000). The maximum seed germination and vigor index was reported in *T. aestivum* plants at 20–30 °C, compared to the temperature below 20 °C (Buriro et al. 2011). *Gossypium hirsutum* exposed to low temperature (LT) below 20 °C showed delayed and reduced germination (Krzyzanowski and Delouche 2011). In *T. aestivum*, the germination was drastically reduced at temperature below 8–10 °C (Zabihi-e-mahmoodabad et al. 2011). Chilling temperature causes seedling growth inhibition and abnormal phenotype of seedlings including stem discoloration, leaf yellowing or whitening, white patches/spots, withering, and reduced tillering at their vegetative stage (Nahar et al. 2012). Seedling growth and dry weight of mung bean seedlings decreased under chilling temperature (6 °C). These seedlings also showed abnormal phenotypic appearance (Nahar et al. 2015b). Nahar et al. (2009) observed cold (11 °C) injury symptoms in rice. Stunted growth, leaf chlorosis, irregular number of tiller, and malformed and discolored grain symptoms were commonly observed in chill affected rice plant. Lower temperature reduced growth duration, decreased biomass accumulation, and harvested index, seed number and seed weight of soybean (Calviñoa et al. 2003). Male gametophyte developmental process, meiosis, and tapetal hypertrophy are distorted under chilling stress. Pollen grain becomes distorted and short, anther protein degrades, and pollen tubes are deformed under chill temperature which causes pollen sterility. Fertilization and further development of fruit and seed are hampered due to chill-induced inhibition of male and female gametophyte development which is mainly due to reduced size of reproductive organs and anomalous structures of embryo sac (Jiang et al. 2002). Delayed flowering, bud abscission, sterile flowers, aborted embryo, and unfilled grain are evident in chill affected rice and other grain crop (Jiang et al. 2002; Thakur et al. 2010). Low temperatures (−2° to −5.5 °C) reduced the number of flower bud in *Simmondsia chinensis* (Nelson and Palzkill 1993). Premature flowering development in Chinese cabbage is increased by chilling temperature (Kalisz and Cebula 2001). Distorted panicle initiation and emergence, delayed heading, and sterile and malformed spikelet were noticed in rice genotypes under chilling stress of 11 °C (Nahar et al. 2009). Delayed flowering, delayed blooming, and higher rate of pollen shed and pod abortion were noticed in different genotypes of canola under chilling stress (Miller et al. 2001).

Chilling temperature damages cell membrane by protein denaturation, disrupting protein lipid structure and precipitation, and disrupting membrane permeability (Wang and Li 2006). Chilling stress thus converts the membrane into solid gel phase. Ion leakage, imbalanced anion/cation exchange, disruption of anions and cataions homeostasis, disruption of osmosis, and diffusion processes become the fate of chill affected plant (Farooq et al. 2009). Chill temperature inhibits photosynthesis due to disruption of thylakoid membrane and chloroplast structure. Electron transport, carbon cycle metabolism and stomatal conductance are hampered by chilling temperature. The PS II is considered as the primary target of chill-induced damage. Chilling temperature highly decreases the quantum efficiency of PS II. Chilling temperature restricts RuBisCO regeneration and confines the photophosphorylation (Allen and Ort 2001). Total chl content reduced by 50% in rice due to exposure to LT (15/10 °C, 14 days) (Aghaee et al. 2011). Total chl (a + b) of mung bean plants reduced by 32% and 38%, after 2 and 3 days of chilling temperature (6 °C) exposure, respectively, compared to the control seedlings (Nahar et al. 2015b). Wheat plants exposed to chill temperature of 3 °C (for 48 and 72 h) showed decreased chl, CO_2 assimilation, and transpiration rates (Yordanova and Popova 2007). Dehydration stress may also result from freezing of cell constituents, solutes, and water (Yadav 2010). In temperatures below 0 °C, ice formation is started in intracellular spaces that are also the vital reason for hindering the solute transportation (Thomashow 1999; Yadav 2010). Damaging the root system chill temperature reduces water and nutrient uptake and causes dehydration and osmotic stress (Chinnusamy et al. 2007). Damaged root system and stomatal closure were described as cause of chill-induced dehydration stress (Yadav 2010). Cucumber root showed chill-induced damage in cortical cells. Chill temperature also increased density of cytoplasm and damaged the endoplasmic reticulum (Lee et al. 2002). Decreased root hydraulic conductance, leaf water, and turgor potentials are common effects of chill temperature in sensitive plants (Aroca et al. 2003). During chill-induced dehydration, stress plant metabolic functions are altered which include accumulation of enzymes and isozymes (Hurry et al. 1994). Low-temperature-induced yield reduction is a common occurrence to plants due to its adverse effects on vegetative and reproductive phases (Nahar et al. 2009; Kalbarczyk 2009; Riaz-ud-din et al. 2010). In rice, chilling temperature significantly reduced yield which was due to flower abortion, pollen and ovule infertility, distorted fertilization, poor seed filling, and seed setting (Thakur et al. 2010). Other study revealed that chilling temperature reduced yield by 30–40% in temperate regions (Andaya and Mackill 2003). Rice varieties of BRRI dhan46 and BRRI dhan31 showed 16 and 37% yield reduction, respectively, due to late sowing mediated chilling stress (Nahar et al. 2009). Low temperature induced anomalous seed structure in *B. napus* and *B. rapa*. Low temperature reduced diameter and created white patches, white reticulation, red-brown pigmentation, folded seed, and shriveled seed (Angadi et al. 2000). Seed weight in *G. max* plants decreased by 5% under chill temperature of 13/23 °C (day/night) as reported by Rasolohery et al. (2008). Chilling stress of 11 °C caused yield reduction in maximum genotypes of rice, and only 23 genotypes were screened out among the 244 genotypes which showed higher yield and had better phenotypic appearance in terms of vegetative and reproductive development (Nahar et al. 2009).

10.3.6 Toxic Metals

World population is increasing day by day together with fast industrialization. As a result, big amount of toxic metals is turned out and enter into the environment (Sarma 2011). Increasing environmental pollution caused by toxic metals is becoming a significant problem in the modern world (Sun et al. 2005). Nowadays, losses in agricultural yield are very common because of soil contamination by toxic metals, and it is a serious threat for wildlife and humans health (Sharma and Dubey 2007; Sharma and Dietz 2008). Due to high reactivity of toxic metal, they can easily influence growth, development, and energy synthesis processes of organisms. So, metal toxicity is one of the key abiotic stresses leading to dangerous health effects in plants and animals (Maksymiec 2007). Excessive toxic metal uptake may occur by plants, which possibly will part take in many physiological and biochemical reactions of plants that will be able to break the normal growth of the plant by upsetting absorption, translocation, or their synthetic processes (Hasanuzzaman and Fujita 2012a,b). Plant suffered a lot by toxic metal/metalloids from germination to final seed yield. Seed germination is the primary stage of plant life, and so it is very sensitive to many environmental stresses. Germination of seeds of *Salicornia brachiata* was affected badly by different doses of Cd, Ni, and As (Sharma et al. 2011). In addition, some scientists reported that growth of seedling is very susceptible to heavy metals (Hg, Pb, Cu, Zn, etc.) in comparison to germination of seed, while Cd is the exception, because it disturbed both processes at same concentrations (Li et al. 2005). Different toxic metal/metalloids cause oxidative stress in plants. The generation of ROS is the primary response of plants under metal stress. A variety of metals can directly generate ROS through

Haber-Weiss reactions. Moreover, ROS overproduction and oxidative stress occurrence in plants may occur due to indirect consequence of heavy metal toxicity (Wojtaszek 1997; Mithofer et al. 2004). Lipid peroxidation of cells is the most detrimental effects under heavy metal stress, which can cause biomembrane deterioration directly. Hasanuzzaman et al. (2012b) observed that Cd stress (1 mM $CdCl_2$, 48 h) increased malondialdehyde (MDA) content by 134% and H_2O_2 content by 60% in rapeseed seedlings. Toxic metals/metalloids also altered the activities of antioxidant enzymes of plants as reported by several researchers. Nahar et al. (2016) reported that Cd stress in mung bean plants reduced growth, damaged chl, decreased leaf RWC and altered proline (Pro), and enhanced oxidative damage [lipid peroxidation, H_2O_2 content, $O_2^{\bullet -}$ generation rate] by disrupting activities of nonenzymatic (AsA and GSH) and enzymatic [catalase (CAT), superoxide dismutase (SOD), ascorbate peroxidase (APX), glutathione reductase (GR), mono dehydro-ascorbate (MDHAR), dehydroascorbate reductase (DHAR), glutathione peroxidase (GPX), and glutathione S-transferase (GST)] antioxidants. Similar kinds of physiological disorders were observed in rice plant under Cd stress (Rahman et al. 2015) and wheat plant under arsenic (As stress) (Hasanuzzaman and Fujita 2013).

10.3.7 Ozone

Ozone (O_3) is a heterogenous air pollutant framed in the troposphere layer. In presence of sunlight, O_3 is developed due to oxidation of oxides of nitrogen, volatile organic hydrocarbons, and carbon monoxide those generate primarily from gas motors and burning/blazing of other fossil fuels (Kesselmeier and Staudt 1999). Expanding levels of surrounding ozone may force destructive impacts on human, animals, and plants. The harming influence of O_3 to plant species might be acute or chronic. Acute damage is created by a high concentration of O_3 (>150 ppb) inside a brief timeframe (Gillespie et al. 2011) and brought about foliar injuries and impels an array of biochemical and physiological reactions in plants (Tsukahara et al. 2015). Chronic damage to lower O_3 concentrations is portrayed by decreasing growth and development without noticeable harm (Gillespie et al. 2011). Crops demonstrate differential intra and interspecific phenotypic responses to O_3 (Biswas et al. 2008; Brosche et al. 2010). The obvious indications of O_3 damage have been evaluated in foliage part. Chlorotic or necrotic injuries on leaves, reduced biomass production, and yield reduction are common consequences of O_3 stress (Wilkinson et al. 2012). The reproduction process of plants is greatly hampered due to exposure of tropospheric O_3. Grain characteristics of rice, wheat, and maize crops are negatively affected under O_3 stress (Biswas and Jiang 2011). It is estimated that elevated O_3 can reduce grain yield of different rice cultivars by 4.2–5.2% (Sawada et al. 2016) and yield of soybean by 23–27% (Wang and Mauzerall 2004). The fruits of different types of plants are affected differentially under O_3 stress. Pod development is adversely affected by O_3 which results in reduction of number and size of pod. Tuber number and yield have been demonstrated to decrease as affected by O_3 in tuber crops (Wilkinson et al. 2012). In potato, tuber number and total tuber yield decreased by 38% and 45%, respectively, due to exposure of O_3 (Foster et al. 1983). Pollen germination and tube development are adversely affected by O_3; abscissions of bud or flower and individual ovules or seeds are commonly occurred phenomenon under O_3 stress (Black et al. 2000).

10.3.8 Ultraviolet Radiation

Solar radiation supplies the essential energy for growth and development of plant via photosynthesis but high light and particularly its integral ultraviolet (UV) fraction cause stress potentially leading to severe injury to cellular components of plants even the DNA and proteins (Müller-Xing et al. 2014). The ultraviolet radiation from sunlight is divided into three broad classes: UV-A (320–400 nm), UV-B (290–320 nm), and UV-C (280–100 nm) light. UV-A and UV-B can easily reach to the surface of earth, but UV-C is filtered out in the atmosphere (Biedermann et al. 2011). Stratospheric ozone layer does not affected by the UV-A, because it is not attenuated by ozone layer. The wavelength of UV-C is shorter than 280 nm, and it cannot arrive at ground level, and this is not expected to change. The UV-B radiation gets top most importance by the researcher and geologist, because ozone absorbed it. Among UV radiation types, even though UV-B is only a negligible constituent of the total solar radiation (less than 0.5%), due to its high energy, its potential for causing biological damage is remarkably high, and even little raise could lead to considerable biological damage (Zlatev et al. 2010). Due to large-scale production of anthropogenic pollutants, intensities of UV-B radiation in the sunlight are slowly increasing in the world as a result of depletion is going on in the stratospheric ozone layer. Previous research on UV-B radiation confirmed that it has significant photobiological consequences on growth, development, and other biochemical activities of plants (McNamara and Hill 2000). The intensities of UV-B vary with the angle of sun and thickness of stratospheric ozone layer. So, all organisms of the earth are exposed to UV-B radiation in different intensities. The rate of enhance UV-B intensity depends on latitude, and higher amount of UV-B is recorded in antarctic and arctic regions (Zlatev et al. 2010).

Frequent discharge of chlorofluorocarbons and other pollutants due to anthropogenic activities is increasing the amount of ultraviolet radiation in the earth surface which is the main reason of stratospheric ozone layer depletion (Dai et al. 1997; Sharma et al. 1998). The stratospheric ozone depletion and increased solar UV radiation cause injury to plants and animals (Costa et al. 2002; Ravindran et al. 2010). Sunlight is obligatory for photosynthesis of terrestrial plants, so they are mostly vulnerable to UV radiation (Greenberg et al. 1996; Sharma et al. 1998). The UV-A radiation shows less harmful effect to organism among all the UV radiations. In addition, also UV-C radiation does not show much harmful effects on the biota because it is observed by the oxygen and ozone layer (Barta et al. 2004). The impact of UV radiation on most cultivated plants is negative and unfavorable to plant growth and development (Vyšniauskienė and Rančelienė 2014). UV radiation creates oxidative stress (Costa et al. 2002), which come up from the detrimental effects of ROS, and they react with lipids, proteins, and nucleic acid (Dai et al. 1997). Plants have many enzymatic and nonenzymatic detoxification systems under normal healthy growth condition to efficiently scavenge or detoxify the ROS or their secondary reaction products (Bartling et al. 1993). Nasibi and M-kalantari (2005) carried out an experiment regarding the action of the antioxidant defense system in *Brassica napus*, under different ultraviolet radiation (UV-A, UV-B, and UV-C). This experiment demonstrated that UV-B and UV-C radiation induced oxidative stress. Because, they increased the lipid peroxidation and total ascorbate content of cells. The treatments of UV-B and UV-C significantly increased flavonoids and anthocyanins levels compared with control and UV-A-treated plants. These compounds perform as internal filters and also have an antioxidant action against ROS such as H_2O_2. Protein content increased under UV-B and UV-C, which probably related to the synthesis of different defense proteins such as HSP. On the other hand, treatment of UV-B radiation for 8 days on maize plants (*Zea mays* L. cv. SC. 704, 21 days old) increased lipid peroxidation (MDA content) and decreased RWC. UV-B stress also increased carotenoid content and activity of some enzymes (APX and GPX) Javadmanesh et al. (2012). In addition, plants under adverse conditions decreased their protein content in root tissue but not in shoot tissue. Vyšniauskienė and Rančelienė (2014) reported that compared to the control, the plant height of early potato cultivars reduced after the UV-B exposure. However, fresh and dry weight of leaves, chlorophyll *a* and *b*, and carotenoid contents in potato cultivars remained unaffected after UV-B treatment. Antioxidative enzymes also increased due to exposure of UV-B in potato cultivars which prove the active response of plant to UV-B-induced stress. They also suggested that response and tolerance depend on the plant genotype. Similar kind of reduction in photosynthetic parameters was observed in cotton plant under UV-B stress (Reddy et al. 2003). UV radiation reduces the total mass and yield of different plants (Gao et al. 2004; Liu et al. 2013). A field experiment was carried out with three soybean cultivars [Hai339 (H339), Heinong 35(HN35), and Kennong18 (KN18)] by exposing them to higher UV-B radiation. The UV-B radiation reduced height of plant, dry weight of individual stem, and yield per plant. Pod number per plant was the most responsible component for yield change under UV-B radiation in the 2-year study. UV-B radiation reduced pod number per plant of three soybean cultivars by 43.5% for KN18, 30.4% for HN35, and 29.6% for H339. They also stated that seed number per pod was less affected than the pod number per plant under UV-B light treatment. The UV-B radiation had no noteworthy effect on effective filling period, but seed size was negatively impacted by this radiation, and it reduced 12.3% for all the cultivars of soybean. UV-B radiation decreased the cotyledon cell number, as a result seed size was decreased. No significant effect on cell volume of cotyledon, cell weight, or cell growth rate of soybean varieties was observed under enhanced UV-B radiation (Liu et al. 2013). Similarly Gao et al. (2004) reported that higher UV-B radiation caused a considerable declination in the growth, yield, and quality of maize. Besides the yield and biomass reduction, UV-B radiation altered the nutrient status of plants (Yue et al. 1998).

10.4 Abiotic Stress-Induced Oxidative Stress

Reactive oxygen species are unavoidable by-products derived from the reduction of molecular oxygen (O_2) and include some free radicals: superoxide ($O_2^{\bullet-}$), hydroxyl radical ($OH^{\bullet}$), alkoxyl ($RO^{\bullet}$), and peroxyl ($ROO^{\bullet}$) and some non-radical products, H_2O_2, singlet oxygen ($^1O_2$), etc. (Gill and Tuteja 2010; Sandalio et al. 2013). Chloroplast, mitochondria, and peroxisomes are the main source of ROS production, chloroplast being the main source (Apel and Hirt 2004). Physical and chemical activation is involved in ROS generation. Physical activation involves transfer of energy from photo activated pigment such as excited chl (chl triplet state, ^{3}chl) that involved in the transfer of its excitation energy onto O_2 to make $^1O_2$. Chemical activation involves univalent reduction of dioxygen or triplet oxygen (molecular oxygen) to water; four electrons and four protons are engaged which leave three major ROS, viz. $O_2^{\bullet-}$, H_2O_2 and •OH (Gill and Tuteja 2010; Sandalio et al. 2013). Under normal or nonstress environment, electron flow from excited PS centers is transferred to $NADP^+$ which is reduced to NADPH. This NADPH enters Calvin cycle where it reduces final electron acceptor (CO_2). Calvin cycle recycles NADP (Vašková et al. 2012). But under stress conditions, NADP recycling is impaired, electron transport chain (ETC) can be over loaded, and electron leakage from ferredoxin is converted to O_2, reducing it to $O_2^{\bullet-}$ through Mehler reaction (Elstner 1991). The $O_2^{\bullet-}$ can be converted into more reactive H_2O_2 by SOD activity in outer stromal membrane surface (Gill and Tuteja 2010). In presence of Fe^{2+}, H_2O_2 is transformed into extreme reactive OH• through Fenton reaction (at Fe-S centers) (Gill and Tuteja 2010; Sharma et al. 2012). Peroxisome metabolic processes are chiefly involved in the production of H_2O_2. The $O_2^{\bullet-}$ can also be produced in peroxisome (Noctor et al. 2002). Mitochondrial electron transport chain (ETC) is also an important source of ROS production in plant cells. Mitochondrial ETC and ATP syntheses are coupled strongly, and stress-induced imbalance between these two may lead to over-reduction of electron carriers as a result of which ROS overproduce (Noctor et al. 2007; Blokhina and Fagerstedt 2010). NADPH-dependent electron transport process is involved with Cyt P_{450} and $O_2^{\bullet-}$ generation in the endoplasmic reticulum (Mittler 2002). In cell wall, plasma membrane, and apoplastic area, ROS including $OH^{\bullet}$ and H_2O_2 can be overproduced through ranges of biochemical reactions (Heyno et al. 2011).

Under normal growing (nonstress) condition, ROS is kept in a balance state by well-equipped antioxidant defense system. In contrast, under environmental stress condition, excess ROS generation leads to oxidative stress. The ROS can cause membrane lipid peroxidation, protein oxidation, and DNA damage. Cell structure and biochemical and physiological process are disrupted by ROS-induced oxidative stress and, at extreme state, gradually lead to programmed cell death (Gill and Tuteja 2010; Hasanuzzaman et al. 2012a). ROS-induced oxidative stress is a common phenomenon under different abiotic stresses. The ROS production under salt stress might be increased by three-fold. The level of H_2O_2 can be increased by 30-fold under salt stress (Singh and Flowers 2010). Salinity-induced oxidative stress was documented in different plant species. Salt stress significantly increased H_2O_2 and MDA levels in mung bean seedlings (Nahar et al. 2014). Salt affected wheat seedlings showed high level of H_2O_2 by 60%, compared to control. Lipid peroxidation in those wheat seedlings increased by 27 and 73% at 150 mM and 300 mM NaCl, respectively (Hasanuzzaman et al. 2011a, b). Drought is one of major stresses contributing the major ROS and causing severe oxidative stress within the plants. More than 70% of total H_2O_2 were demonstrated to produce due to photorespiration under drought stress condition (Noctor et al. 2002). Cellular membrane damage was related to significant rise in $O_2^{\bullet-}$ and H_2O_2 in leaves of drought affected *Malus* spp. (Wang et al. 2012). Similar results were demonstrated in mung bean seedlings under PEG-induced drought stress (Nahar et al. 2017). High-temperature raised ROS production in wheat seedlings. The H_2O_2 contents in wheat seedlings were 0.5, 0.58, 0.78, and 1.1 μmol g^{-1} FW in response to 22, 30, 35, and 40 °C (2 h) temperature stresses, respectively (Kumar et al. 2012). Mung bean seedlings exposed drought and high-temperature stress separately or in combined increased H_2O_2 and $O_2^{\bullet-}$ production which was evidenced from higher cellular H_2O_2 and $O_2^{\bullet-}$ contents and increased number of spots of these ROS in leaves, compared to control. The H_2O_2 level, $O_2^{\bullet-}$ generation rate, lipoxygenase (LOX) activity, and MDA contents under combined HT and drought stress were 122, 146, 108, and 120%, respectively, in contrast to control (Nahar et al. 2016a). Low temperature creates imbalance between light absorption and light use due to inhibition of C_3 cycle which enhances ROS production. ROS generation also increased due to increased photosynthetic electron flux to O_2 as well as the over-reduction of respiratory ETC (Hu et al. 2008). Waterlogging or flooding induces hypoxic or anoxic environment. Under waterlogging stress, ROS can be produced when plants go through a hypoxia/anoxia from normoxic conditions and also under the condition when plants return to an aerobic environment (Irfan et al. 2010). Enhanced accumulation of H_2O_2 or lipid peroxidation under anoxic conditions was reported previously (Sairam et al. 2011). Plants under heavy metal exposure cause a significant increase of ROS production and subsequent oxidative stress. Redox-active metals such as Fe, Cu, Cr, V, and Co initiate redox reactions in the cell and directly involved in the formation of $OH^{\bullet}$ via Haber-Weiss and Fenton reactions (Sharma and Dietz 2008). Other metals also generate ROS and cause oxidative stress in different ways (Hasanuzzaman et al. 2012a; Nahar et al. 2016). Heavy metal-dependent activation of LOX also leads to lipid peroxidation (Nahar et al. 2016). Mung bean seedlings exposed Cd stress showed a higher level of ROS generation including H_2O_2 and $O_2^{\bullet-}$, a higher LOX activity, and increased lipid peroxidation (Nahar et al. 2016). Wheat seedlings exposed to As; 0.25 and 0.5 mM showed 41 and 95% increase of H_2O_2 and 58 and 180% increase of lipid peroxidation level, respectively, compared to control (Hasanuzzaman and Fujita 2013). Ozone (O_3) being strong oxidant interacts with apoplastic constituents in generating ROS (Yan et al. 2010). Different studies designated that elevated O_3 highly increased the levels of $O_2^{\bullet-}$, H_2O_2 and lipid peroxidation in plants (Yan et al. 2010; Feng et al. 2011). UV-B radiation also generates ROS and results in damage to proteins, lipids, nucleic acids, and associated enzymes (Du et al. 2011; Singh et al. 2011).

10.5 Highlights of Potent Micronutrients and Trace Elements Associated with Plants' Functions

Micronutrients and trace elements are required in small amount but equally important for plant functions as like macronutrients. Their major functions in plants are summarized in Table 10.1. Plants show some deficiency symptoms due to lack of these nutrients as shown in Fig. 10.2.

Table 10.1 Major functions of trace elements/micronutrients in plants

Elements	Functions
Copper	Essential for carbohydrate and nitrogen metabolism also plays roles in lignin synthesis for cell wall strengthening. It plays an important role in chl synthesis. It also activates some enzymes by playing role as coenzyme
Iron	It plays role in nitrogen reduction and fixation and lignin formation. Important for chlorophyll synthesis and energy transfer
Zinc	Important for energy production, formation of chlorophyll, and plant hormone like indole acetic acid (IAA). Helps in protein synthesis, better seed quality, and uniform crop maturity
Boron	Helps in carbohydrate transport, cell wall formation, pollen formation
Manganese	Important for enzyme activity and other metabolic processes. Important for nitrogen metabolism, pollen tube growth, and root hair formation
Molybdenum	Essential for pollen formation, fruit and grain formation. It can convert nitrate into usable form
Cobalt	Promotes to auxin formation and nitrogen fixation. It is also essential for the development of bud, plant stem, and coleoptiles. It also increases drought resistance to seed
Nickel	Component of some plant enzymes, most notably urease, which metabolizes urea nitrogen into useable ammonia within the plant. It plays role in root nodule growth. Essential part of glyoxalase system
Silicon	Strengthens cell walls, stimulates plant growth. Performs as essential function in healing plants in response to environmental stress
Selenium	It stimulates nitrogen assimilation and regulates the water status of plants. Acts as antioxidant and stress protectants

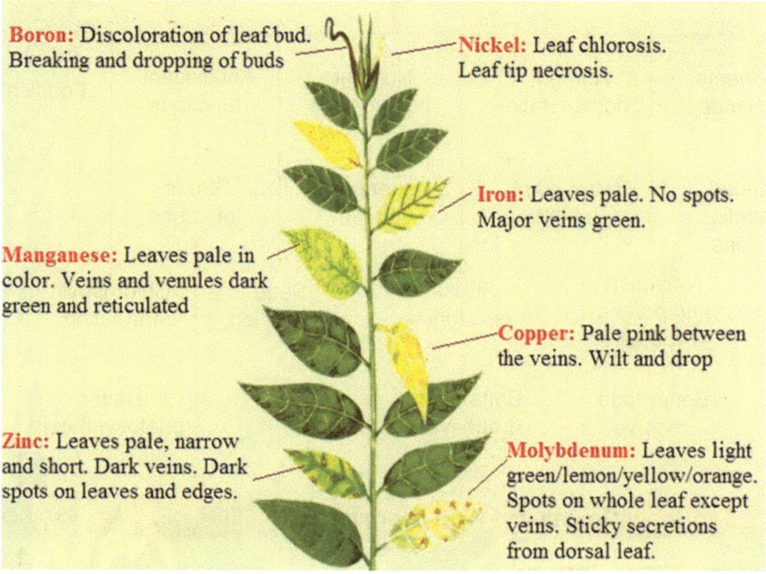

Fig. 10.2 Visualization of deficiency symptoms of major trace elements in plants

10.6 Functional Role of Trace Elements in Conferring Abiotic Stress Tolerance

Although trace elements are required in small amount, their essentialities are the same as macronutrients. Although several trace elements have not yet been identified as essential elements, they have some vital functions in plant growth and development (Fig. 10.3). In addition, these elements have specific metabolic functions and diverse roles in stress tolerance (Table 10.2). However, these elements showed toxicity in plants when exposed to high concentration.

10.6.1 Copper

Due to its presence in multiple oxidation states in vivo, Cu is involved in many physiologic conditions. The role of Cu in plants stress tolerance is well considered due to its functions as structural element of photosynthetic electron transport, cell signaling, and metabolism as well as participation in enzymatic activities (Yruela 2005). It is important that both deficiency and abundance of Cu negatively affect plant processes. For instance, photosynthetic ETC is hampered under both Cu

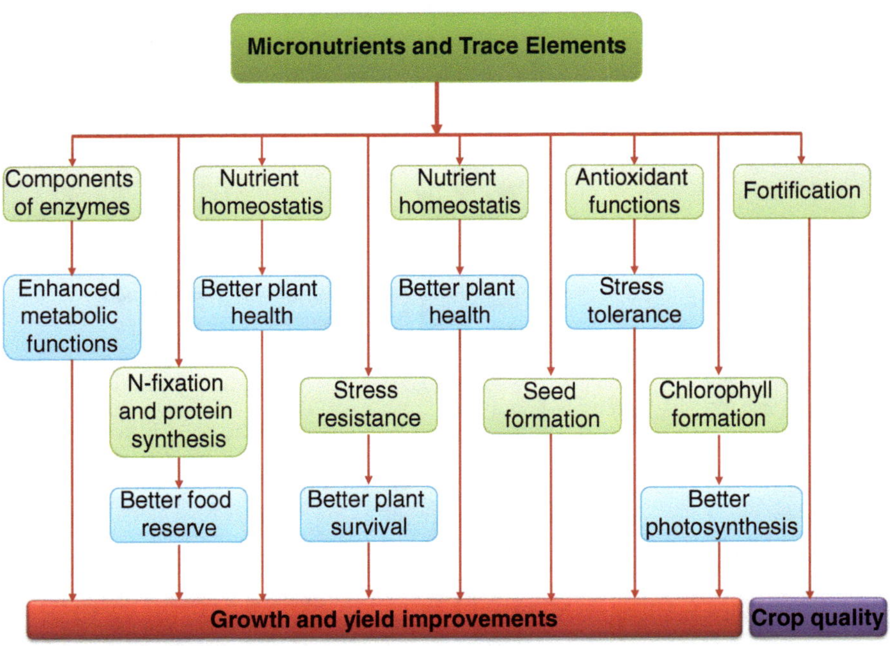

Fig. 10.3 Different aspects of the beneficial roles of trace elements in boosting plant growth, yield, and quality

Table 10.2 Beneficial effects of trace elements supplementation in plants grown under adverse environmental conditions

Plant species	Trace elements	Stress	Beneficial effects	References
Hordeum vulgare	300 μM $ZnSO_4$	0.1–10 μM $CdCl_2$ (10–22 DAS)	• Reduced Cd uptake and MDA • Increased biomass	Wu and Zhang (2002)
Capsicum annuum	2 and 10 mg $ZnSO_4$ kg^{-1} soil	Salt, 0.5 and 1.5%, NaCl, 20 days	• Reduced Na uptake	Aktaş et al. (2006)
Pistacia vera	5–20 mg $ZnSO_4$ kg^{-1} soil	Salt, NaCl 0.8–3.2 g kg^{-1} soil, 100 days	• Reduced lipid peroxidation and LOX activity • Reduced electrolyte leakage • Increased phenolic contents	Tavallali et al. (2010)
Triticum aestivum	Zn and Fe	Drought, stop watering at pollination and seed-filling stage	• Increased seed yield and kernel weight	Monjezi et al. (2012)
T. aestivum	50 mg B L^{-1}	Drought, 50, 75, and 100% irrigation	• Increased chl and Car • Decreased H_2O_2 and Pro content	Abdel-Motagally and El-Zohri (2016)
T. aestivum	B (0, 0.5, 1%)	Drought, irrigation withdrawal	• Improved LAI, CGR, and NAR	Moeinian et al. (2011)
Zea mays	Foliar spray of 50 and 100 ppm boric acid	Salt, 4% NaCl	• Increased plant height, leaf number, shoot dry weight, and grain weight	Salim (2014)
Citrus grandis	2.5 and 20 μM H_3BO_3	Al stress, 1.2 mM $AlCl_3$, 18 week	• Higher dry weight, root citrate secretion, root malate secretion	Zhou et al. (2015)
H. vulgare	2.0 mmol m^{-3} Mn	Salt stress, 125 mM NaCl and 9.6 mM $CaCl_2$, 24 days	• Increased NAR, RGR, photosynthesis, Mn uptake in salt-stressed seedlings	Cramer and Nowak (1992)
H. vulgare	3 mM Mn, 1 WAS	Salinity, 8 dS m^{-1}; two weeks after sowing (WAS)	• Increased dry weight, NAR, RGR, leaf area ratio	Pandya et al. (2004)
Oryza sativa	5 and 25 μM $MnCl_2$	Cd, 5 and 25 μM $CdCl_2$, 7 days	• Increased chl, Car, AsA, and sugar content • Decreased Cd uptake, MDA content, CAT, and POD activity	Sebastian and Prasad (2015)

(continued)

Table 10.2 (continued)

Plant species	Trace elements	Stress	Beneficial effects	References
Z. mays	100 μM $MnSO_4$	Cd, 10 μM $CdCl_2$, 12 h	• Increased root length and decrease Cd uptake	Pal'ove-Balang et al. (2006)
Phytolacca americana	1–5 mM $MnCl_2$	Cd, 10 and 50 μM $CdCl_2$	• Increased root, stem, and leaf dry weight and decreased Cd content	Peng et al. (2008)
Camellia sinensis	50 and 100 μM $MnCl_2$	Drought, Irrigation withdrawal	• Increased dry mass and RWC, total phenolic and AsA content	Upadhyaya et al. (2012)
C. annuum	1.8 mM K_2SiO_3	Salinity, 50 mM NaCl	• Improved growth	Manivannan et al. (2016)
			• Increased photosynthesis rate, stomatal conductance	
			• Decreased electrolyte leakage	
			• Decrease ROS production and lipid peroxidation	
			• Increased activities of antioxidant enzymes	
Anethum graveolens	1.5 mM $Na_2Si_3O_7$	Salinity, 10 dS m^{-1}	• Improved growth	Shekari et al. (2015)
			• Increased K^+ content and reduced Na^+ content	
			• Reduced lipid peroxidation	
			• Increased activities of antioxidant enzymes	
Z. mays	1.5 mM $Na_2O_3Si.5H_2O$	Alkalinity, 25–75 mM Na_2CO_3	• Improved growth	Latef and Tran (2016)
			• Enhancement in leaf RWC and levels of photosynthetic pigments	
			• Increased soluble sugars, soluble proteins, total free amino acids and K	
			• Enhanced activities of SOD, CAT, and POD	

(continued)

Table 10.2 (continued)

Plant species	Trace elements	Stress	Beneficial effects	References
O. sativa	350 kg ha^{-1} of Si fertilizer	Drought, −0.050 or −0.025 MPa	• Increased Pro content • Increased peroxidase activity	Mauad et al. (2016)
Saccharum officinarum	2 mM Ca_2SiO_4	Salinity, 100 mM NaCl	• Reduced tissues Na^+ concentration • Improved K^+ uptake, K^+/Na^+ ratios and Ca^{2+} content • Increased shoot and root dry matter	Ashraf et al. (2010)
Vitis vinifera	4 mM Si $Na_2Si_3O_7$	Salinity, 20 mM NaCl	• Reduced stomatal resistance • Reduced MDA and H_2O_2 contents • Increased APX activity	Soylemezoglu et al. (2009)
Brassica napus	2 and 4 mM K_2SiO_3	Salinity, 300 mM NaCl	• Increased leaf area, leaf fresh weight, seed yield, and photosynthesis • Increased APX and NR activities • Increased chl content	Bybordi (2012)
Z. mays	0.4, 0.8, 1.2, 1.6, 2.0, 2.4, 2.8, and 3.2 mM $Si(OH)_4$	Salinity, 150 mM NaCl	• Improved growth • Increased net CO_2 assimilation rate, stomatal conductance (g_s), transpiration, and leaf sub-stomatal CO_2 concentration	Parveen and Ashraf (2010)
Spartina densiflora	500 µM Na_2SiO_3	Salinity, 171 and 680 mM NaCl	• Improved growth associated with higher net photosynthetic rate, water-use efficiency, and g_s • Reduced tissue Na^+content	Mateos-Naranjo et al. (2013)

(continued)

Table 10.2 (continued)

Plant species	Trace elements	Stress	Beneficial effects	References
Solanum lycopersicum	2.5 mM K_2SiO_3	Drought, 10% PEG	• Improved water balance • Increased growth and photosynthesis • Decreased ROS production • Enhanced activities of SOD and CAT • Increased level of AsA and GSH	Shi et al. (2016)
T. aestivum	Si 50 mg kg^{-1} and 150 mg kg^{-1}	Drought, 50%, 75% and 100% FC	• Increased plant biomass, plant height, and spike weight • Increased tissue Si concentration and uptake	Ahmad et al. (2007)
Z. mays	0.8 mM Na_2SiO_3	Drought, 50% water deficit	• Increased dry mass, tissue nutrient content, water use efficiency • Reduced leaf wilting	Janislampi (2012)
Glycine max	1.70 mM Na_2SiO_3	Drought , −0.5 MPa	• Increased root and shoot dry matter and the ratio of root/shoot • Increased leaf water potential, growth • Increased net photosynthetic rate, *gs*, chl content • Decreased free Pro content, lipid peroxidation, and electrolytes leakage	Shen et al. (2010)
S. bicolor	200 mg L^{-1} Si	Drought, withholding irrigation	• Increased net photosynthetic rate • Decreased shoot to root ratio by increasing root growth	Ahmed et al. (2011)

(continued)

Table 10.2 (continued)

Plant species	Trace elements	Stress	Beneficial effects	References
Oryza sativa	1.5 mM Na_2SiO_3	HT, 39 °C	• Increased number of pollen grain those were with higher diameter • Increased anther dehiscence percentage, pollination, and fertilization	Li et al. (2005)
Z. mays	10 mM K_2SiO_3	Chilling, 2/3 ± 1 °C	• Enhanced photosynthesis • Reduced ROS production • Enhanced AsA and GSH pool	Habibi (2016)
Cucumis sativus	0.1 and 1 mM K_2SiO_3	Chilling (15/8 °C)	• Reduced leaves withering • Increased activities of SOD, GPX, APX, MDHAR, GR, GSH • Increased AsA content • Decreased levels of H_2O_2, $O_2^{\bullet-}$ and MDA	Liu et al. (2009)
T. aestivum	0.1 and 1.0 mM K_2SiO_3	Freezing, −5 °C	• Increased leaf water content • Improved activities of antioxidant enzymes AsA, GSH, SOD, and CAT • Reduced H_2O_2, free Pro, and MDA content	Liang et al. (2008)
C. limon	50, 150 and 250 mg L^{-1} K_2SiO_3	Freezing, 0.5 °C for 28 days	• Increased phenolics and flavonoids concentration • Improved fruit quality • Reduced chilling injury	Mditshwa et al. (2013)

(continued)

Table 10.2 (continued)

Plant species	Trace elements	Stress	Beneficial effects	References
O. sativa	0.6 mM $Si(OH)_4$	Heavy metal (HM), 10 μM $CdCl_2$, 4 days	• Improved plant growth • Improved photosynthesis • Maintenance of nutrient homeostasis • Stimulation of the AsA-GSH pathway	Farooq et al. (2016)
S. nigrum	1.0 mM Na_2SiO_3	HM, 100 μM $CdCl_2$	• Decreased Cd accumulation • Reduced electrolyte leakage • Increased activities of SOD, POD, CAT, and APX	Liu et al. (2013)
O. sativa	1.25, and 2.5 mM Na_2SiO_3	HM, 100 μM $K_2Cr_2O_7$	• Increased seedling height, dry biomass, and soluble protein content • Reduced Cr uptake and translocation • Improved antioxidant defense	Zeng et al. (2011)
B. chinensis	1.5 mM K_2SiO_3	HM, 0.5 and 5 mg L^{-1} $CdCl_2$	• Increased shoot and root biomass • Decreased Cd uptake and root-to-shoot transport • Increased SOD, CAT, APX, reduced MDA, and H_2O_2 concentrations	Song et al. (2009)
C. sativus	1 mM Na_2SiO_3	HM, 100 μM $CdCl_2$	• Reversed chlorosis, protected the chloroplast from disorganization • Increased pigment contents, intercellular CO_2 concentration, *gs* and net photosynthetic rate • Improved water use efficiency	Feng et al. (2010)

(continued)

Table 10.2 (continued)

Plant species	Trace elements	Stress	Beneficial effects	References
Z. mays	1 mM Si as $Si(OH)_4$	HM, 200 or 500 µM $MnSO_4$	• Ameliorated chloroplast damage and photoinhibition • Improved detoxification and compartmentation of Mn	Doncheva et al. (2009)
Picea abies	0.2, 0.5 and 1.0 mM Si	HM, 0.2, 0.5 and 1.0 mM Al	• Ameliorated adverse effects on cell wall thickening, degree of vacuolation, and the degeneration of mitochondria, golgi bodies, endoplasmic reticulum, and nucleus • Reduced cell death	Prabagar et al. (2011)
V. vinifera	4 mM $Na_2Si_3O_7$	HM, 20 mg kg^{-1} H_3BO_3	• Reduced tissue B concentration • Increased activities of CAT and APX • Reduced Pro, H_2O_2, and MDA content	Soylemezoglu et al. (2009)
O. sativa	Si fertilizer ($CaSiO_3$) @ 40 g m^{-2}	UV-B radiation, 250–350 nm	• Improved cell walls of sclerenchyma, vascular bundle sheath, and metaxylem vessel cells, cellulose, non-cellulosic polysaccharides, lignin, and phenolic acids	Goto et al. (2003)
G. max	1.70 mM Si	UV-B radiation, 290–320 nm	• Increased root and shoot dry weight and their ratio • Increased net photosynthetic rate and *gs* • Decreased H_2O_2 content	Shen et al. (2010)

(continued)

Table 10.2 (continued)

Plant species	Trace elements	Stress	Beneficial effects	References
Vigna unguiculata	5–10 μM Na_2SeO_4	Salinity, 100 mM NaCl	• Improved photosynthetic pigments • Increased Pro and soluble sugar content and PAL • Improved growth and yield	Manaf (2016)
L. esculentum	5–10 μM Na_2SeO_3	Salinity, 25–50 mM NaCl	• Improved the integrity of cell membrane • Increased leaf RWC • Enhanced antioxidant defense system	Mozafariyan et al. (2016)
L. esculentum	50 μM Na_2SeO_3	Salinity, 100 mM NaCl	• Improved photosynthetic efficiency • Decreased ROS production • Enhanced antioxidant defense	Diao et al. (2014)
B. napus	25 μM Na_2SeO_4, 48 h	Salinity, 100 and 200 mM NaCl, 48 h	• ncreased AsA and GSH contents, GSH/GSSG ratio, and activities of APX, MDHAR, DHAR, GR, GST, GPX, and CAT • Reduced levels of H_2O_2 and MDA	Hasanuzzaman et al. (2011b)
C. sativus	5, 10, or 20 μM Na_2SeO_4, 11 days	Salinity, 50 mM NaCl, 11 days	• Decreased content of Cl^- in the shoots tissues • Increased Pro accumulation • Enhanced antioxidant defense • Decreased lipid peroxidation	Hawrylak-Nowak (2009)

(continued)

Table 10.2 (continued)

Plant species	Trace elements	Stress	Beneficial effects	References
Rumex patientia x *R. tianshanicus*	1–5 μM Na_2SeO_4, 43 days	Salinity, 100 mM NaCl, 43 days	• Stimulated growth • Increased activities of SOD and POD • Increased accumulation of water-soluble sugar	Kong et al. (2005)
S. vulgare	20 mg L^{-1} Se	Drought, 35–70% FC	• Improved growth • Increased photosynthesis • Increased yield	Aissa et al. (2016)
T. aestivum	15 μg L^{-1} Na_2SeO_4	Drought, 20% FC	• Increased dry matter accumulation • Protection of leaf photochemical event • Increased photosynthesis	Hajiboland et al. (2016)
T. aestivum	7.06 μM Se	Water deficit	• Increased stress tolerance index • Increased biomass	Nawaz et al. (2014)
Trifolium repens	5 μM Na_2SeO_4, 0–5 days	Drought, PEG 6000 (−1.0 MPa), 0–5 days	• Decreased levels of H_2O_2, TBARS, DHA, and GSSG • Increased the levels of AsA and GSH and AsA/DHA and GSH/GSSG ratios • Improved the activities of MDHAR, DHAR, and GR	Wang et al. (2011)
B. napus	15 and 30 g L^{-1} as Na_2SeO_3	Drought, limited irrigation at early stem elongation	• Increased plant height • Increased pod and seed development • Increased yield	Zahedi et al. (2009)
T. aestivum	Se (Na_2SeO_3) 0.5 mg kg^{-1}, 20 days	Drought, 30% FC	• Increased root activity • Increased Pro content • Increased activities of POD and CAT	Xiaoqin et al. (2009a)

(continued)

Table 10.2 (continued)

Plant species	Trace elements	Stress	Beneficial effects	References
T. aestivum	1.0, 2.0, and 3.0 mg kg^{-1} Na_2SeO_3, 20 days	Water stress of 30% FC, 20 days	• Increased root activity • Increased chl, carotenoids and Pro content • Increased activities of POD and CAT activities • Decreased MDA content	Xiaoqin et al. (2009b)
Fagopyrum esculentum	1 g m^{-3} Na_2SeO_4, 7–8 weeks	Reduction of water by 50%, 8 weeks	• Improved *gs*, potential photochemical efficiency of PS II, respiratory potential • Increased yield	Tadina et al. (2007)
Z. mays	20 g $ha^{-1}Na_2SeO_4$	Drought, withholding water	• Enhanced activities of SOD • Reduced MDA content • Improved grain yield	Sajedi et al. (2011)
T. aestivum	0, 25, 50, 75, and 100 μM of Na_2SeO_4, 30 or 60 min	Drought, withholding water for 1 week	• Increased root length and total biomass • Increased stress tolerance index • Increased total sugar content and total free amino acids	Nawaz et al. (2013)
B. napus	25 μM Se (Na_2SeO_4), 48 h	Drought, 10 and 20% PEG-6000, 48 h	• Increased activities of APX, DHAR, MDHAR, GR, GST, GPX, and CAT • Decreased GSSG content, H_2O_2 • Decreased lipid peroxidation	Hasanuzzaman and Fujita (2011)
Z. mays	5–15 μM $Na_2SeO_3 \cdot 5H_2O$	Drought, 25% PEG	• Unregulated AsA-GSH cycle • Decreased ROS production	Yildiztugay et al. (2016)

(continued)

Table 10.2 (continued)

Plant species	Trace elements	Stress	Beneficial effects	References
Z. mays	5–15 μM $Na_2SeO_3 \cdot 5H_2O$	HT (24–44 °C)	• Upregulated AsA-GSH cycle • Decreased ROS production	Yildiztugay et al. (2016)
T. aestivum	2–4 mg L^{-1} Se	HT, 38 ± 2 °C	• Improved growth and photosynthesis • Increased yield attributes and yield • Increased nonenzymatic antioxidants • Enhanced activities of antioxidant enzymes	Iqbal et al. (2015)
S. bicolor	Na_2SeO_4 (75 mg L^{-1}), 7–28 days	HT, 40/30 °C, 7–28 days	• Increased antioxidant enzyme • Decreased ROS levels and membrane damage • Increased antioxidant defense • Increased grain yield	Djanaguiraman et al. (2010)
C. sativus	2.5, 5, 10, or 20 μM Na_2SeO_4	LT, 10 °C/5 °C for 24 h, day	• Improved shoot fresh weight • Increase of Pro content • Reduced MDA level	Hawrylak-Nowak et al. (2010)
T. aestivum	0.5, 1.0, 2.0, 3.0 mg kg^{-1} Na_2SeO_3, 72 h	LT, 4 °C, 72 h	• Increased biomass • Increased chl, anthocyanins, flavonoids, and phenolic compounds • Increased activities of POD and CAT	Chu et al. (2010)

(continued)

Table 10.2 (continued)

Plant species	Trace elements	Stress	Beneficial effects	References
T. aestivum	Se (5 mg Se L^{-1}), 5, 10, and 15 h	LT, 3 or 5 °C, 14 days	• Enhanced growth • Increased chl, anthocyanin, sugar, and Pro contents • Enhanced antioxidant defense system • Decreased membrane damage	Akladious (2012)
S. bicolor	Se (3 and 6 μM L^{-1} Na_2SeO_4), 6 h	LT, 4–8 °C, 7 days	• Enhanced growth • Increased levels of chl, anthocyanin, sugar, Pro, and AsA • Increased enzymatic activities • Diminished lipid peroxidation	Abbas (2012)
C. frutescens	0, 3 and 7 μM Na_2SeO_3	HM, 0.25–0.5 mM $CdCl_2$	• Improved growth • Improved photosynthetic pigments • Decreased Pro • Enhanced CAT activity	Shekari et al. (2016)
B. napus	50 and 100 μM Se (Na_2SeO_4), 24 h	HM. 0.5 and 1.0 mM $CdCl_2$, 48 h	• Increased the AsA and GSH contents, the GSH/GSSG ratio • Increased activities of APX, MDHAR, DHAR, GR, GPX, and CAT • Reduced the MDA and H_2O_2 levels	Hasanuzzaman et al. (2012b)
B. napus	2 μM Na_2SeO_4, 14 days	HM, 400 and 600 μM, $CdCl_2$, 14 days	• Reduced oxidative stress by modulating SOD, CAT, APX, GPX activities • Prevented Cd-induced alteration of DNA methylation pattern	Filek et al. (2008)

(continued)

Table 10.2 (continued)

Plant species	Trace elements	Stress	Beneficial effects	References
Pteris vittata	5, 10 μM of Na_2SeO_4, 5 and 10 days	HM, 150 or 300 μM of Na_2HAsO_4, 10 days	• Improved antioxidant system including thiol and GSH levels • Reduced As uptake	Srivastava et al. (2009)
B. oleracea	1 mg L^{-1} Na_2SeO_3, 10 and 40 days	HM, 1 mg L^{-1} $CdCl_2$, 40 days	• Increased chl content • Improved α-tocopherol level and reduced oxidative damage	Pedrero et al. (2008)
Lolium perenne	1.0, 1.5, 2.0, 5.0 and 10 μM Na_2SeO_3, 20 days	HM, 0.2 mM $AlCl_3$, 20 days	• Improved POD activity • Reduced $O_2^{\bullet -}$ and lipid peroxidation	Cartes et al. (2010)
T. aestivum	1.0 and 2.0 mg Se kg^{-1}, 8 h	UV-B, 40 W, 305 nm, 8 h	• Increased root weight and root activity • Increased flavonoids and Pro content • Increased activities of POD and SOD • Reduced MDA and $O_2^{\bullet -}$	Yao et al. (2009)
Euglena gracilis	10^{-7}, 10^{-8}, 10^{-9} and 10^{-10} M, $Na_2SeO_3 \cdot 5H_2O$, 40 min	UV-A, 320–400 nm, 40 min	• Improved light-enhanced dark respiration and photosynthesis	Ekelund and Danilov (2001)

deficiency and excess Cu conditions (Yruela 2005). Copper is found to be actively involved in many antioxidant enzymes in plants. Growth, protein content, and antioxidant enzymes' activities showed differential responses under different concentrations of copper (0–800 μM) as reported by Gao et al. (2008). Higher activities of enzymes in *Jatropha curcas* seedlings were observed at the concentration of 200–400 μM. However, it was also dependent of plant organs. Azooz et al. (2012) observed the effect of Cu in wheat and found that 2 mM Cu showed optimum growth and biochemical parameters, while no changes were observed up to 10 mM Cu, and thereafter, the growth and biochemical parameters were significantly reduced. The activities of antioxidant enzymes such as CAT, POD, APX, and SOD were increased in dose-dependent manners. This effect of Cu in plants was associated with the biosynthesis of free amino acids and Pro (Azooz et al. 2012).

10.6.2 Zinc

Since 1932, Zinc has been considered as a vital micronutrient for plants because its participation as a functional component of around 200 enzymes and transcription factor involved in biomolecule synthesis and metabolism of nucleic acid and lipid. Furthermore, Zn plays role in biomass production, chl formation, pollen function, and fertilization (Ebrahimian and Bybordi 2011; Hafeez et al. 2013). Zn (at low concentration 1–2 μM) increases plant growth, chl content, and crude protein (Samreen et al. 2013). Role of Zn in conferring drought stress tolerance maintaining water use efficiency was described by Waraich and his group (Waraich et al. 2011a, b). They proposed that Zn can enhance the auxin level directly or indirectly by increasing tryptophan which is a precursor of auxin. Then auxin increased the root growth under drought condition. In addition, Zn decreases the activity of membrane-bound NADPH oxidase; for this reason, ROS generation decline and SOD, CAT, and POD activities increase to protect cell from oxidative damage under drought stress. Cd-induced damage can be restored by using Zn. Wu and Zhang (2002) reported Cd-induced reduction in root and shoot biomass, micronutrient imbalance (Cu, Fe, and Zn), and higher amount of Cd uptake reduced by the application of 300 μmol L^{-1} Zn. Exogenous Zn prevented the Cd uptake by making block, maintained micronutrient (Cu, Fe, and Zn), reduced MDA, and, at the same time, increased antioxidant enzyme activity to protect barley plant from Cd toxicity. Supplemental Zn can reduce salinity-induced oxidative stress by reducing MDA, H_2O_2, and lipoxygenase activities as well as increasing APX and CAT activity in pistachio seedling (Tavallali et al. 2010). Application of Zn increased fresh and dry weight and Pro content and enhanced the activities of CAT, SOD, and GST under salt stress in sunflower leaves (Ebrahimian and Bybordi 2011). Role of Zn in abiotic stress tolerance is still ambiguous. Further study is required to find out mechanism of Zn-induced abiotic stress tolerance and elucidate nature of interaction among Zn and other nutrient elements.

10.6.3 Nickel

Although the biological functions of Ni in plants have been reported since long time, its essentiality has been established recently due to its role as an activator of the enzyme urease (Fabiano et al. 2015). However, its role as protectants of plant stress is rarely available rather than its toxic effects. Recent studies have shown that Ni may activate *OsGLY11.2* (an isoform of glyoxalase I), which is the first line of enzymes in the degradation of cytotoxic methylglyoxal (MG) (Mustafiz et al. 2014), a cytotoxic compound which is produced in high concentration under stress. This MG is detoxified via glyoxalase system where both Gly I and Gly II enzymes are involved. Importance of Ni for the activity of Gly I not only suggest its role in MG detoxification but also the redox state of GSH which is a strong antioxidant and vital player in metal tolerance. Fabiano et al. (2015) studied the role of Ni in the relationship between the MG cycle and GSH homeostasis and stated that Ni may have a key participation in plant antioxidant metabolism, especially in stressful situations (Fabiano et al. 2015). However, further study is necessary to elucidate the actual functions of Ni in conferring stress tolerance to plants.

10.6.4 Boron

Though the physiological role of B in plant is still unclear, it is evident that both the deficiency and toxicity of B resulted in many anatomical, biochemical, and physiological changes in plants (Herrera-Rodríguež et al. 2010). The deficiency of B in plant affects growth and reduces yield, whereas sufficient supply results in better growth and good yield (Shabaan 2010). As an essential element, B is required for many essential functions of plant such as maintaining cell structures and functions as well as cell division, sugar transport, hormone development, respiration, carbohydrate metabolism, membrane transport, and transportation and metabolism of other essential plant nutrients (Shabaan 2010, Herrera-Rodríguež et al. 2010). Boron can also play role in fruit and seed development by increasing fertility, pollen tube growth, and carbohydrate utilization (Blevins and Lukaszewski 1994). Therefore, deficiency of B causes many physiological and biochemical changes in plant that affect plant growth and yield (Shabaan 2010). It is also revealed that exogenous B can develop stress tolerance by regulating physiological and biochemical attributes under environmental stress condition. Foliar application of B improves growth and yield of wheat by increasing chl and Car content and decreasing H_2O_2 and Pro content under drought stress condition (Abdel-Motagally and El-Zohri 2016). Moeinian et al. (2011) also showed that foliar application of B improve growth, yield, and grain quality of *T. aestivum* under drought stress condition by improving leaf area index (LAI), crop growth rate (CGR), and net assimilation rate (NAR). Foliar application of B alleviates salinity and improves growth and yield attributes in maize seedlings under salt stress by regulating phenols, proline, amino acids, and soluble sugar (Salim 2014). Zhou et al. (2015) reported that gene related to ROS and aldehyde detoxification and metabolism, cell transport, Ca signaling, and hormone and gene regulation play role in B-induced alleviation of aluminum toxicity in *Citrus grandis* seedlings.

10.6.5 Manganese

As an essential trace element, manganese (Mn) is associated with various plant metabolic processes. It plays an important role in plant photosynthesis, respiration, and hormone activation. It also takes part in synthesis of protein, lipid, fatty acids, amino acids, ATP, flavonoids, etc. (Lidon et al. 2004; Millaleo et al. 2010). Usually Mn presents in soil as free Mn^{2+} which is only available form for plant and readily taken up by plant via an active transport system in epidermal root cells. Deficiency of Mn is dangerous for plant, because it weakens structural resistance against pathogen and decreased tolerance ability toward abiotic stresses such as heat and drought stress. In addition, Mn deficiency also affects the water-splitting system of photosystem II (PS II) that provides electron directly for photosynthesis which also makes plant vulnerable toward stress (Gherardi and Rengel 2003; Millaleo et al. 2010). On the other hand, the presence of excess Mn in plant root zone also extremely toxic to plant. Higher accumulation of Mn by plant alters various physiological and metabolic processes which causes chlorosis, necrosis, oxidative stress, and growth and yield reduction (Ducic and Polle 2005; Millaleo et al. 2010; Arya and Roy 2011). Considering the beneficial role of lower amount and toxic effect of excess amount, Mn plays dual role in plant system: essential micronutrient and toxic element, respectively (Ducic and Polle 2005; Millaleo et al. 2010).

However, among the beneficial roles that we discussed, Mn also plays role directly against various abiotic stresses. Being a divalent cation, Mn can compete with other divalent toxic ion and reduced their uptaken and transportation within plant as they are uptaken by active transporter and have some common transporter during uptake and transportation (Hirschi et al. 2000; Pittman 2005). In addition, Mn also acts as cofactor in Mn-CAT and Mn-SOD enzymes that participates in plant antioxidant defense system. Although it is not clear but assumed that Mn acts as $O_2^{\bullet-}$ and H_2O_2 scavenger (Ducic and Polle 2005). Moreover, many studies revealed that supplemental Mn plays positive role in conferring stress tolerance under various abiotic stress conditions (Table 10.2). Supplementation of Mn in Cd-treated plant improves growth and chl and Car contents and decrease lipid peroxidation by reducing Cd uptake and enhancing antioxidant defense system (Pal'ove-Balang et al. 2006; Peng et al. 2008; Sebastian and Prasad 2015). Exogenous application of Mn in barley seedlings confer salt stress tolerance and improve growth by increasing net assimilation and photosynthetic rate under salt stress condition (Cramer and Nowak 1992; Pandya et al. 2004). Upadhyaya et al. (2012) reported that Mn plays positive role on post-drought stress recovery in tea by influencing growth and antioxidative response. To observe the protective effect of Mn under salt stress condition, we have grown 12-day-old *O. sativa* seedlings under 150 mM NaCl for 3 and 6 days under controlled environment. Salt stress resulted in marked increase in Na^+, MDA, H_2O_2, $O_2^{\bullet-}$, and Pro and MG content which in turns exhibited chlorosis and growth inhibition in time-dependent manner (Rahman et al. 2016). On the other hand, Mn supplementation (0.5 mM MnSO4) reduced oxidative stress by lowering the levels of MDA and H_2O_2 and enhancing the activities MDHAR, DHAR, SOD, CAT, Gly I, and Gly II and improving the contents of nonenzymatic antioxidants. Exogenous Mn also maintains low Na^+ and high K^+ content and thus maintained ion homeostasis.

10.6.6 Silicon

Although it is not considered as essential for plant function, in rice, Si showed some essential functions, and it is absorbed from soil in amounts that are even higher than those of the essential macronutrients (Datnoff et al. 2001). There are plenty of evidences on the role of Si in plant stress tolerance (Hasanuzzaman et al. 2014b). Beneficial effects of exogenous Si in enhancing abiotic stress resistance are reported in many plant studies (Table 10.2). The protective effect of Si under abiotic stress are mainly due to the deposition of Si as $SiO_2.nH_2O$ in leaves and stems of plants and partly by the interaction between $Si(OH)_4$ and other elements such as Al. It is also suggested that the function of Si in plants might involve mechanical defense rather than physiological changes (Fig. 10.4; Ma and Yamaji 2008; Hasanuzzaman et al. 2014b).

Hydroponically supplemented Si (1.8 mM K_2SiO_3) significantly increased growth and alleviated salinity stress (50 mM NaCl) in *C. annuum* which was due the enhancement of antioxidant defense, improvement of photosynthesis, and maintenance of the nutrient balance (Manivannan et al. 2016). For example, electrolyte leakage potential in salt stress treatment by 38% upon exposure illustrated the NaCl-induced cell membrane damage, which, in turn, was reduced by 33% upon Si supplementation. Similarly, the addition of Si mitigated the oxidative damage by decreasing the MDA content by 29% and H_2O_2 content by 26% which efficiently minimized the oxidative burst. While investigating the beneficial role of Si in salt stressed (10 dS/m) *Anethum graveolens*, Shekari et al. (2015) observed that exogenous Si (1.5 mM) could be able to improve the salt-induced inhibition of growth and

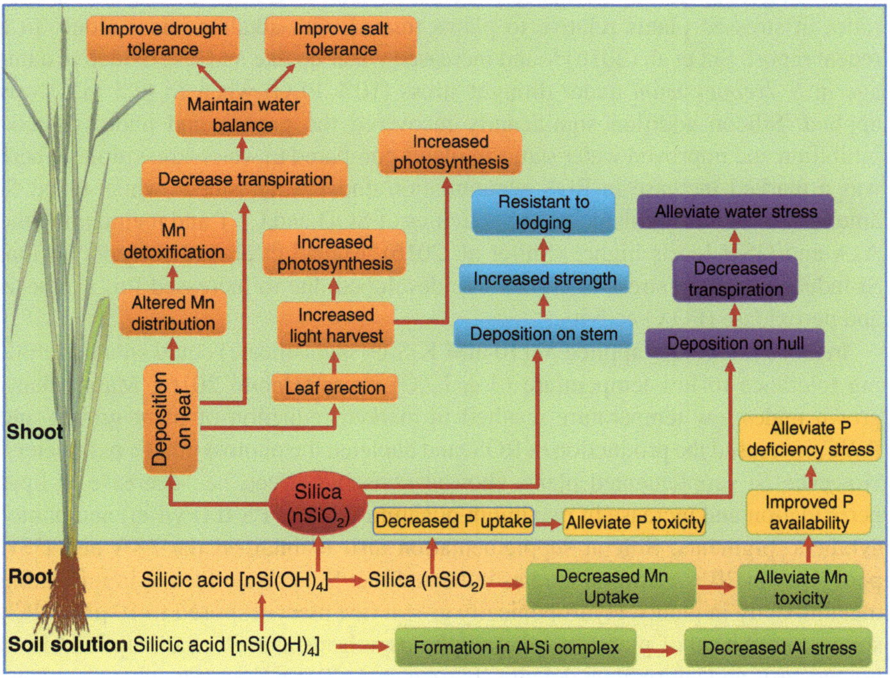

Fig. 10.4 Silicon-induced abiotic stress tolerance in plants (Source: Hasanuzzaman et al. 2014b, with permission from Elsevier)

Na/K ratio. Silicon also reduced lipid peroxidation which was associated with enhanced activities of antioxidant enzymes. Mateos-Naranjo et al. (2013) observed that Si could ameliorate nutrient imbalances in salt affected *Spartina densiflora*. The K^+/Na^+ ratio of leaves of *Spartina densiflora* was greater in Si-treated plants, and these plants also had higher levels of essential nutrients (Si, Al, Cu, Fe, K, and P) in their tissues (Mateos-Naranjo et al. 2013). In *B. napus*, Si (2 and 4 mM) was found to be effective in enhancing antioxidative enzyme activities and photosynthesis (Bybordi 2012). Protective role of Si in drought stress tolerance was reported by many authors. These effects were mainly due to the maintenance of water balance, cellular integrity, and antioxidant defense. However, little information is available on its role in water uptake and in less Si-accumulating plants. Silicon also provided better protection under alkali stress (25–75 mM Na_2CO_3). After 25 days of stress growth parameters, leaf RWC, and the contents of photosynthetic pigments, soluble sugars, total phenols and potassium ion (K^+), as well as potassium/sodium ion (K^+/Na^+) ratio were decreased, while the contents of soluble proteins, total free amino acids, proline, Na^+ and MDA, as well as the activities of SOD, CAT, and POD were increased. On the other hand, seed priming with 1.5 mM $Na_2O_3Si.5H_2O$ improved growth of stressed plants, which was accompanied by the enhancement in leaf RWC and levels of photosynthetic pigments, soluble sugars, soluble proteins, total free amino acids and K^+, as well as activities of SOD, CAT, and POD enzymes. Furthermore, Si supplement resulted in a decrease in the contents of Pro, MDA, and Na^+, which together with enhanced K^+ level led to a favorable adjustment of K^+/Na^+ ratio, in stressed plants relative to plants treated with alkaline stress alone. In a recent report, Shi et al. (2016) found increased water uptake and less oxidative damage in *S. lycopersicum* under drought stress (10% PEG) when Si (2.5 mM) was applied. Silicon addition significantly recovered the growth and photosynthetic inhibition and improved water status in plants compared to water stress alone. There was a marked increase in ROS accumulation under water stress, while added Si ameliorated these by enhancing the activities of SOD and CAT and maintaining the AsA and GSH levels higher (Shi et al. 2016). Mauad et al. (2016) reported that Si-induced drought stress tolerance was developed due to increased Pro synthesis and peroxidase (POX) activity.

In *Z. mays*, foliarly applied Si (10 mM K_2SiO_3 metasilicate) could enhance chilling tolerance to low-temperature (3 ± 1 °C) stress (Habibi 2016). Maize plants grown under low temperature resulted in marked reduction of plant growth and RWC, increased the production of ROS, and depleted the photosynthetic parameters. However, Si supplemented plants showed revisable effect, i.e., decrease in lipid peroxidation and increase in maximum quantum yield of PS II (Fv/Fm) and photosynthetic pigments. Silicon supplementation also maintained the AsA and GSH pool (Habibi 2016). Several studies indicated that Si is also effective in mitigating metal toxicity in plants. Hydroponically grown rice plants exposed to 10 μM $CdCl_2$ showed inhibition of plant growth and photosynthesis as well as disruption of ion homeostasis and antioxidant defense (Farooq et al. 2016). However, when the plants were supplemented with 0.6 mM $Si(OH)_4$, it reversed the effects of Cd by improving plant growth, photosynthesis, nutrient homeostasis, and enhancement of antioxidant defense system, especially AsA-GSH cycle, and thus readjusting cell redox homeostasis (Farooq et al. 2016). In *S. nigrum*, Si-induced Cd stress tolerance was associated with the activation of antioxidant defense and reduction of Cd uptake as reported by Liu et al. (2013). They observed that Cd markedly enhanced the production of ROS and showed higher electrolyte leakage and accumulated higher amount of Cd in shoot and root. In contrary, Si reduced H_2O_2 accumulation and prevented cell death and the electrolyte leakage and H_2O_2 concentration in functional leaves. Si supplementation also enhanced the activities of SOD, CAT POD, and APX (Liu et al. 2013).

10.6.7 Selenium

After the intense research of past few decades, the beneficial roles of Se in plants have been observed, but the question is still unresolved whether Se is an essential micronutrient for plants (Terry et al. 2000). However, it is still recognized as an essential micronutrient (Hasanuzzaman et al. 2014b). In many plant species, Se exerts a positive effect on plant growth, physiology, and productivity under abiotic stress (Hasanuzzaman et al. 2010, 2011b, 2012b, 2014a, b; Table 10.2). One of the roles of Se in exerting beneficial effects on the stress tolerance is the enhancement of the antioxidant capacity (Hasanuzzaman et al. 2014a, b).

Several research results have shown that Se at low concentration provided protection to different plant species against salt stress. Selenium-induced growth promotion and improvement of photosynthesis and antioxidant defense system in salt-treated tomato plants were reported by Diao et al. (2014). Under salt stress (100 mM NaCl), plant growth and photosynthetic attributes were inhibited while Se supplementation (50 μM Na_2SeO_3) reversed the effects. Se treatment also limited the ROS generation which was mainly due to the enhanced activities of SOD, GR, DHAR, MDHAR, and GPX and the redox pool of AsA and GSH (Diao et al. 2014). In pot culture Se (5–10 μM) supplemented *V. unguiculata* showed enhanced tolerance to salt stress (50 mM NaCl) through the highest values of photosynthetic pigments, proline, phenylalanine ammonia lyase (PAL), and total soluble sugar as well as growth and yield attributes (Manaf 2016). In a recent study, we found that exogenous Se (5-10 μM) could alleviate salt (25–50 mM NaCl)-induced damages by improving the integrity of cell membranes and by increasing leaf RWC, and photosynthetic pigments under stress conditions and subsequently increasing plant biomass and yield (Mozafariyan et al. 2016). Under high salt concentration (50 mM NaCl) 10 μM Se performed well than other dose.

Kong et al. (2005) reported that low concentrations (1–5 μM) of Se stimulated growth and enhanced antioxidant enzyme (SOD and POD) activities in leaves of sorrel (*R. patientia* × *R. tianshanicus*) seedlings under salt stress. In contrast, at higher concentrations (10–30 μM), Se showed fewer beneficial effects. In *C. sativus* leaves, Se treatments (5–10 μM) increased the growth, synthesis of photosynthetic pigments, and Pro levels under salt stress (Hawrylak-Nowak 2009). In our recent study, we observed beneficial effects of exogenous Se (25 μM Na_2SeO_4) in salt (100 and 200 mM NaCl)-stressed *B. napus* seedlings (Hasanuzzaman et al. 2011a, b). Selenium treatment increased the components of AsA-GSH cycle and other antioxidant enzymes and maintained the AsA and GSH pool which reduced levels of H_2O_2 and MDA when compared to plants exposed to salt stress alone (Hasanuzzaman et al. 2011a, b). So far the confirmation about which form of Se is more effective is yet to be elucidated. In a recent study Hawrylak-Nowak (2015) reported that Se application, especially in the form of selenite, could upregulate the antioxidant defense in salt-treated lettuce. It was also found that the root growth and increase in photosynthetic pigment were increased upon Se supplementation.

Several plant studies that focused on the protective effects of Se under drought stress indicated that the effects of Se are due to its ability to regulate the water status of plants under water-deficit condition. In wheat, Se supplementation improved dry matter and grain yield under both well watered and a drought condition which was mainly due to enhancement of photosynthesis rate, protection of leaf photochemical events, accumulation of organic osmolytes, and improvement of water use efficiency. Selenium was also able to increase root length and diameter which made the plant able to uptake more water and maintaining higher water content in leaves (Hajiboland et al. 2016). Nawaz et al. (2014) reported that the growth and biomass of wheat seedlings increased under drought due to Se fertigation. Se supplemented seedlings showed higher stress tolerance index and biomass with 7.06 μM Se. Kuznetsov et al. (2003) reported that the addition of 0.1 or 0.25 mM Se caused a 2–6% increase in leaf water content, thereby increasing the drought resistance. The Se-induced improvement in leaf tissue water status was accompanied by a sharp (two to fourfold) inhibition of stress-induced accumulation of Pro and a significant inhibition of POX activity (Kuznetsov et al. 2003). Wang et al. (2011) examined the effect of Se (5 μM Na_2SeO_4) on the AsA-GSH cycle in *Trifolium repens* seedlings subjected to drought. They observed that Se application decreased the lipid peroxidation and H_2O_2 by maintaining the higher AsA and GSH pool higher. Selenium supplementation significantly increased the activities of MDHAR, DHAR, and GR. Among the enzymes, GR showed the highest increase in activity compared to DHAR and MDHAR. In our laboratory, we studied the beneficial role of Se pretreatment (25 μM Na_2SeO_4, 48 h) in *B. napus* seedlings under drought stress (10 and 20% PEG-6000) (Hasanuzzaman and Fujita 2011). Drought-stressed seedlings showed increases in GSH and GSSG content; however, the AsA content increased only under moderate stress (Table 8). The MDHAR and GR activities increased only under moderate stress (10% PEG). The activities of DHAR, GST, and GPX significantly increased at all levels of drought, while CAT activity decreased. Drought stress resulted in a marked increase in the levels of H_2O_2 and MDA. In contrast, Se-pretreated seedlings exposed to drought stress showed a rise in AsA and GSH content and upregulated activities of CAT, APX, DHAR, MDHAR, GR, GST, and GPX when compared with the drought-stressed seedlings without Se. In turn, the Se-treated seedlings showed a considerable decrease in the levels of H_2O_2 and MDA and considerable alleviation of oxidative stress (Hasanuzzaman and Fujita 2011). Very recently, Nawaz et al. (2013) found beneficial role of Se priming in conferring drought stress tolerance. In their experiment, seeds of *T. aestivum* were soaked in distilled water or Na_2SeO_4 solutions (25, 50, 75, and 100 μM) for 30 or 60 min, followed by re-drying and subsequent sowing. Priming with Se significantly increased root length, stress tolerance index, and total biomass of germinated seedlings. Yildiztugay et al. (2016) observed that exogenous Se could upregulate the AsA-GSH cycle in heat-exposed *Z. mays* plants and minimized the overproduction of ROS which in turns protected the plants from oxidative stress induced by drought (25% PEG).

Recent studies also indicated the protective role of Se under HT stress. Yildiztugay et al. (2016) observed that exogenous Se could upregulate the AsA-GSH cycle in heat-exposed *Z. mays* plants and minimized the overproduction of ROS which in turns protected the plants from oxidative stress induced by HT (24–44 °C). Exogenously applied Se also found to be effective under HT stress in spring wheat under field condition. Heat stress (38 ± 2 °C) significantly depleted the antioxidative potential, affected growth, photosynthetic pigments, and grain yield of wheat which were satisfactorily reversed by Se supplementation which helped the wheat plants to increase fertility and hence avoid reduction of grain yield under HT stress (Iqbal et al. 2015). Djanaguiraman et al. (2010) reported that beneficial effect of foliarly applied Se (75 mg L^{-1}) could enhance photosynthesis, membrane integrity, and antioxidant defense which improved the yield components and grain yield of *S. bicolor* plants grown under HT stress (40/30 °C). Se application increased photosynthetic rate and stomatal conductance in HT-stressed plants by 13.2 and 12.4%, respectively, compared to non-supplemented plants under stress, while the $O_2^{\bullet-}$, H_2O_2 and MDA content decreased by 11.5, 35.4 and 28.4%, respectively. Recently, we found that Se could minimize HT-induced damages to *B. napus* seedlings by enhancing antioxidant defense and glyoxalase systems (Hasanuzzaman et al. 2014b). Heat (38 °C, 24 and 48 h)-exposed seedlings exhibited marked decrease in the chl content and increased RWC, MDA, H_2O_2, Pro, and MG contents in time-dependent manners. Selenium supplemented HT-treated seedlings recovered these damages which were evident with decrease level of MDA, H_2O_2, and MG which was correlated with enhanced ecaivities of CAT, GPX, MDHAR, DHAR, GR, Gly I, and Gly II as well as higher redox balance of AsA and GSH (Hasanuzzaman et al. 2014b). Protective role of Se under LT stress has been reported in few plant studies, but the effect was mostly dose dependent. According to Djanaguiraman et al. (2005), Se was able to increase the tolerance of *G. max* plants to LT stress by promoting antioxidant capacity, and it improved growth and developmental processes of that plant under LT. Recently, Abbas (2012) found that SeO_4^{2-} at low concentrations (3 and 6 mg L^{-1}) enhanced growth, levels of chl, anthocyanin, sugar, Pro, and AsA, and enzymatic activities in *S. bicolor* seedlings subjected to LT stress. However, high levels of SeO_4^{2-} (12 mg L^{-1}) resulted in toxic effects. Low levels of SeO_4^{2-} (3 and 6 mg L^{-1}) also diminished lipid peroxidation by enhancing the activities of APX and GPX.

Selenium has been documented to reduce metal toxicity in several research studies. The modes of action were varied and are still unclear; however, some suggested reasons included improvement of the antioxidant defense system, reduction of metal uptake, formation of nontoxic Se-metal complexes, and phytochelatin activity (Vorobets and Mykiyevich 2000; Sun et al. 2010). Moreover, Se is effective at sustaining physiological activities, growth, and developmental processes even in HM toxic environments (Pedrero et al. 2008; Cartes et al. 2010). A study on peeper showed that Se supplementation diminished Cd toxicity on photosynthesis pigment (Shekari et al. 2016). The application of Selenium at 7 μM significantly increased leaf area in the plants grower at 0.25 mM Cd. The application Se at 3 μM with 0.25 mM Cd and Se 3 and 7 μM with 0.5 mM Cd increased the activity of CAT. Selenium 7 μM decreased Pro content of pepper leaves exposed to Cd 0.5 mM (30%). Selenium significantly enhanced antioxidant activity of leaves which was diminished by Cd toxicity. In general, Se has beneficial effect on plant growth and antioxidant enzymes of pepper under Cd stress and nonstress conditions (Shekari et al. 2016). In *B. oleracea*, Cd phytotoxicity resulted in elevated MDA level and decreased photosynthetic pigment and tocopherol concentrations, but Se treatment effectively alleviated these adverse effects (Pedrero et al. 2008). In *B. napus*, Se (2 μM) conferred tolerance to Cd (400 and 600 μM) stress by reducing lipid unsaturation and peroxidation, modulating the activity of antioxidative enzymes (SOD, CAT, APX, GPX), and preventing Cd-induced changes in the DNA methylation pattern (Filek et al. 2008). Sun et al. (2010) showed that enhanced Cd tolerance by Se might be due to removal of Cd from metabolically active cellular sites, induction of Se to scavenge the Cd-induced ROS generated, and the regulation phytochelatin synthesis associated enzymes induced by of Se. In our recent study, we observed that rapeseed seedlings grown under Cd stress (0.5 and 1.0 mM $CdCl_2$) showed substantial increases in MDA and H_2O_2 levels (Hasanuzzaman et al. 2012b). The AsA content of the seedlings decreased significantly upon exposure to Cd stress. The amount of GSH increased only at 0.5 mM $CdCl_2$, while GSSG increased at any level Cd with concomitant decreases in the GSH/GSSG ratio. The activities of antioxidant enzymes also reduced under Cd stress. Importantly, Se-pretreated seedlings exposed to Cd showed increases in the AsA and GSH contents, GSH/GSSG ratio, and the activities of APX, MDHAR, DHAR, GR, GPX, and CAT. However, in most of the cases, pretreatment with 50 μM Se showed better results compared to 100 μM Se. These results indicated that the exogenous application of Se at low concentration increased the tolerance of the plants to Cd-induced oxidative damage by enhancing their antioxidant defenses (Hasanuzzaman et al. 2012b).

Selenium improves plant growth and survival under UV radiation, as reported in several studies. Se (0.01 and 0.05 mg kg^{-1} soil) improved the antioxidative capacity, protected chloroplast enzymes, and increased shoot yield in *Lactuca sativa* under combined UV-B and UV-C stress (Pennanen et al. 2002). Significant increases in the activities of POD and SOD, together with reduced MDA and $O_2^{\bullet-}$levels, were documented in *T. aestivum* under UV-B radiation. Selenium also increased root activity, flavonoid, and Pro contents in this plant (Yao et al. 2010). Recently, we completed a pot experiment with three rice varieties, viz., BRRI dhan45, BRRI dhan47, and Nipponbare grown under different concentration of salt water (50–150 mM). Salt stresses reduced the plant height and tillers $hill^{-1}$, leaf relative water con-

Fig. 10.5 Effect of Se supplementation in mitigating adverse effects of salt stress in different rice cultivars. Plants were subjected to 50 mM NaCl throughout its life cycles with or without 0.5 mM Na_2SeO_3

tent, and chl content in dose-dependent manner. Salt stress also reduced the effective tillers hill^{-1}, number of filled grains panicle^{-1}, 1000-grain weight, grain yield, and straw yield. However, when the plants were supplemented with 0.5 mM Se (sodium selenite, Na_2SeO_3), these observed parameters were significantly increased compared to salt treatment alone. Importantly, the beneficial effect of Se on salt stress tolerance was prominent up to 100 mM NaCl, while it could not be more beneficial above this level of salt concentration (Naim 2015). The response of rice plant to Se and salt stress greatly varied in different cultivars (Fig. 10.5).

10.7 Molecular Approaches to Manipulate the Genes Associated with Trace Elements Actions in Plants

Plant cannot synthesize inorganic molecule including macronutrient, micronutrient, and heavy metals. But plant can uptake inorganic molecule from growing medium using root. To get entry into root or to be translocated, metal ions are needed to cross both cellular and organellar membranes. These membranes contain different transporter proteins which show specificity to particular inorganic elements (Guerinot 2000). For example, IRT1 is a transporter protein which shows specificity to Fe, but Mn and Zn can also be transported using this transporter. Similarly ZIP1, ZIP2, and ZIP3 act as Zn transporter in plant and *AtECA1* for Ca^{2+} and Mn^{2+}, *AtNramp3* for Mn^{2+}, Fe^{2+} and Cd^{2+} in Fe-deficiency conditions, *AtOPT3* for possible transport of Cu^{2+} and Fe^{2+} and Mn^{2+}, and so on (Millaleo et al. 2010). Many genes already identified those encode these transporter proteins. Expression of transporter gene depends on the presence or absence of a particular ion. Regulation of genes at transcriptional and posttranscriptional level controls metal uptake as excessive accumulation metal ions cause toxicity to plants. For instance, sensing the intracellular Zn level, Zn-responsive transcriptional activator protein ZAP1 induces gene expression to uptake Zn. On the other hand, Zn uptake reduced by endocytosis of ZRT1 transporter protein (Guerinot 2000).

Abiotic stress limits ion uptake in plants. The roles of trace element in different types of abiotic stresses are well discussed in previous sections. But with the blessing of modern science, it is possible to manipulate the genes related to trace element uptake. Increasing trace element content in plant by overexpression of gene may enhance tolerance against drought, salinity, and heavy metal toxicity (Table 10.3). Very limited literatures are found in this research area. Li et al. (2011) found that overexpressed Na^+/H^+ antiporter (*AtNHX5*) gene induced extreme dehydration tolerance in paper mulberry. The *AtNHX5*-overexpressing plants survived under drought and salinity, whereas the wild-type (WT) plants could not survive. Under salt stress, *AtNHX5*-overexpressed plant contained higher amounts of Na^+ and K^+ in leaves compared to WT. Higher leaf water content and leaf chl contents, Pro and soluble sugars, and less membrane damage were observed in transgenic plants than the WT plants under both drought and saline conditions. They suggested the possibility of *AtNHX5* gene in improving the tolerance against abiotic stresses in paper mulberry plants. In another study, Sasaki et al. (2016) reported that overexpression of *OsHMA3* showed Cd toxicity tolerance. *OsHMA3*-overexpressed line alleviated the Cd-inhibited growth. Higher amount of Cd found in the roots of *OsHMA3*-overexpressed line than shoots of wild type and vector control line. This result suggests the role of *OsHMA3* in enhancing vacuolar sequestration of Cd in the roots. Furthermore, the *OsHMA3*-overexpressed line constitutively upregulates five transporter genes belonging to the ZIP family. They finally proposed that overexpression of *OsHMA3* is an efficient way to reduce Cd accumulation in the grain and to enhance Cd tolerance in rice.

Table 10.3 Overexpressed transporters that enhance abiotic stress tolerance

Transporter gene	Effects	References
NcZNT1	Overexpression of *NcZNT1* conferred tolerance against high Cd and Zn in *A. thaliana*	Lin et al. (2016)
ZTP29	*Arabidopsis* zinc transporter *ZTP29* was involved in the response to salt stress, may be due to Zn-induced upregulation of the UPR pathway	Wang et al. (2010)
NIP5;1	Enhanced expression of *NIP5;1* in *A. thaliana* promoted root growth under B deficiency condition	Kato et al. (2009)
AtNHX1	Overexpression of this gene enhanced salinity stress tolerance in wheat by reducing Na^+	Xue et al. (2004)
OsNHX1	Overexpressed *OsNHX1* increased K^+ under salinity thus contributed to salt stress tolerance and promoted growth of maize	Chen et al. (2007)
NtCBP4	This gene was responsible for Ni toxicity by reducing Ni^{2+} uptake in transgenic tobacco plant	Arazi et al. (1999)
CAX2	Overexpression of this gene increased Cd and Mn toxicity though transgenic tobacco plant accumulated more Cd and Mn	Hirschi et al. (2000)
CsMTP8	High or low amount of Mn-induced upregulation or downregulation improved Mn homeostasis in cucumber	Migocka et al. (2014)

10.8 Conclusions and Outlooks

The trace elements are not critical for all plants but may improve plant growth and yield through their different beneficial influences on plants' physiological and morphological mechanisms. Essential trace elements or micronutrients are elements necessary for maintaining the life processes in plants. The continous hammering on soil for producing more foods without allowing time for natural replenishment of the nutrients uptaken by plants from the same piece of land has led to evolution of newer elements essential for plants. Trace elements present at very low concentrations in agroecosystems are essential to plant growth and development. Both deficiency and toxicity of trace elements occur in agroecosystems. However, the range of essential to lethal for these elements is rather thin. Application of trace elements in fertilizers is effective in correcting micronutrient deficiencies for crop production, whereas, though costlier, remediation of soils contaminated with metals through phytoremediation appears promising as a cost-effective approach. Stressed plant suffering from due by far but not limited to climatic change/aberrations, seasonal rotations, or geographical positions around the globe are galore (Hasanuzzaman et al. 2014b). Plants also develop physiological and morphological mechanisms for adaptations with the stressful conditions. On reflection of the research works have been conducted so far, it can be said that participation of trace elements in mechanisms of stress tolerance in plants is very vital (Hasanuzzaman et al. 2011b). To be more specific, under adverse environmental conditions, supplementing trace elements have mostly beneficial effects in plant survival, vegetative growth, and reproductive growth and productivity (yield), unless they are in toxic concentration in soil-plant systems. As for example, 50 mg B L^{-1} foliar spray at booting and anthesis stage of *Triticum aestivum* reduces drought stress effects on plants through increasing chl and Car and decreasing H_2O_2 and Pro content (Abdel-Motagally and El-Zohri 2016). Abiotic stress-induced oxidative stress is another phenomena that plants experienced under different stress conditions such as waterlogging, drought, salinity, heat, ozone, UV-B radiation, redox-active metals, etc. but trace elements can mitigate through different ways, for example, by reducing the production of ROS and other radicals. Besides these, the actions of trace elements on a wide range of crops (*Euglena gracilis* to *Oryza sativa*) have been tested in terms of tolerance or avoidance of abiotic stresses, and the results were positive to a great extent.

Trace elements are indispensable part for either physiological, biochemical or molecular approach of stress tolerant mechanism development in plants. To exemplify, *Arabidopsis* zinc transporter *ZTP29* was involved in the response to salt stress through Zn-induced upregulation of the unfolded protein response (UPR) pathway, and overexpression of *AtNHX1* gene reduced Na^+ content in plant and enhanced salinity stress tolerance in wheat. The reviewed research results on role of trace elements on abiotic stress tolerance in plants predict that trace elements are only trace for their presence, but they are huge for their actions and overlooking them as trace elements will results in failure of developing tolerance in plants to abiotic stresses.

However, thorough actions of trace elements on abiotic stress avoidance, tolerance, or resistance mechanisms by plants are still poorly understood for drawing out a full panorama. Foolproof researches focusing on the equilibrium of trace elements in an agroecosystem and elaboration of soil biochemical and molecular approaches that can be used to diagnose trace elements responsible for stress control and to use for future use are urgently needed. Besides these, most of the experiments conducted around the world on actions of beneficial trace elements on plant tolerance of abiotic stresses are confined mostly to laboratory conditions and short-term studies only. Field performances of the biochemical, molecular approaches along with agronomic management practices are also needed to strengthen the science of plants tolerance to abiotic stresses.

Acknowledgments At the very first the authors wish to thank Taufika Islam Anee, Tasnim Farha Bhuiyan, and Mazhar Ul Alam, Laboratory of Plant Stress Responses and Faculty of Agriculture, Kagawa University, Japan, for their critical readings and suggestions in improving the manuscript. We are also highly thankful to Mr. Md. Rumainul Islam, Environmental Engineering Program, Chongqing University, China, for providing us several supporting articles. The first author acknowledges Japan Society for the Promotion of Science (JSPS) for funding in his research. As page limitation precluded us from citing a large number of studies, we apologize to those whose original publications are therefore not directly referenced in this chapter.

References

Allen, D. J., & Ort, D. R. (2001). Impacts of chilling temperatures on photosynthesis in warm-climate plants. *Trends in plant science, 6*(1), 36–42.

Abbas, S. M. (2012). Effects of low temperature and selenium application on growth and the physiological changes in sorghum seedlings. *Journal of Stress Physiology & Biochemistry, 8*, 268–286.

Abdel-Motagally, F. M. F., & El-Zohri, M. (2016). Improvement of wheat yield grown under drought stress by boron foliar application at different growth stages. *Journal of the Saudi Society of Agricultural Sciences*. doi:10.1016/j.jssas.2016.03.005.

Aghaee, A., Moradi, F., Zare-maivan, H., Zarinkamar, F., Irandoost, H. P., & Sharifi, P. (2011). Physiological responses of two rice (*Oryza sativa* L.) genotypes to chilling stress at seedling stage. *African Journal of Biotechnology, 10*, 7617–7621.

Ahmad, F., Rahmatullah, Aziz, T., Maqsood, M. A., Tahir, M. A., & Kanwal, S. (2007). Effect of silicon application on wheat (*Triticum aestivum* L.) growth under water deficiency stress. *Emirates Journal of Food and Agriculture, 19*, 17.

Ahmad, P., Jamsheed, S., & Hameed, A. (2014). *Drought stress induced oxidative damage and antioxidants in plants*. New York: Elsevier.

Ahmed, M., Asif, M., & Goyal, A. (2012). Silicon the non-essential beneficial plant nutrient to enhanced drought tolerance in wheat. In A. Goyal (Ed.), *Crop plant* (pp. 31–48). Rijeka: InTech.

Ahmed, M., Hassen, F., Qadeer, U., & Aslam, M. A. (2011). Silicon application and drought tolerance mechanism of sorghum. *African Journal of Agricultural Research, 6*, 594–607.

Aissa, N., Malagoli, M., & Radhouane, L. (2016). An approach to alleviate the impact of drought stress with selenium amendment. *Iranian Journal of Science and Technology, Transactions A: Science*. (in press).

Ajithkumar, I. P., & Panneerselvam, R. (2013). Osmolyte accumulation, photosynthetic pigment and growth of *Setaria italica* (L) P. Beauv. under drought stress. *Asian Pacific Journal of Reproduction, 2*, 220–224.

Akladious, S. A. (2012). Influence of different soaking times with selenium on growth, metabolic activities of wheat seedlings under low temperature stress. *African Journal of Biotechnology, 11*, 14792–14804.

Akman, Z. (2009). Comparison of high temperature tolerance in maize, rice and sorghum seeds by plant growth regulators. *Journal of Animal and Veterinary Advances, 8*, 358–361.

Aktaş, H., Abak, K., Öztürk, L., & Çakmak, I. (2006). The effect of zinc on growth and shoot concentrations of sodium and potassium in pepper plants under salinity stress. *Turkish Journal of Agriculture and Forestry, 30*, 407–412.

Allakhverdiev, S. I., Kreslavski, V. D., Klimov, V. V., Los, D. A., Carpentier, R., & Mohanty, P. (2008). Heat stress: An overview of molecular responses in photosynthesis. *Photosynthesis Research, 98*, 541–550.

Amirjani, M. R. (2011). Effect of salinity stress on growth, sugar content, pigments and enzyme activity of rice. *International Journal of Botany, 7*, 73–81.

Andaya, V. C., & Mackill, D. J. (2003). QTLs conferring cold tolerance at the booting stage of rice using recombinant inbred lines from a japonica × indica cross. *Theoretical and Applied Genetics, 106*(6), 1084–1090.

Angadi, S. V., Cutforth, H. W., & McConkey, B. G. (2000) Seeding management to reduce temperature stress in *Brassica* species. *Saskatchewan Soils and Crops Proceedings*.

Apel, K., & Hirt, H. (2004). Reactive oxygen species: Metabolism, oxidative stress and signal transduction. *Annual Review of Plant Biology, 55*, 373–399.

Arazi, T., Sunkar, R., Kaplan, B., & Fromm, H. (1999). A tobacco plasma membrane calmodulin-binding transporter confers Ni^{2+} tolerance and Pb^{2+} hypersensitivity in transgenic plants. *The Plant Journal, 20*, 171–182.

Arbona, V., Hossain, Z., López-Climent, M. F., Pérez-Clemente, R. M., & Gómez-Cadenas, A. (2008). Antioxidant enzymatic activity is linked to waterlogging stress tolerance in citrus. *Physiologia Plantarum, 132*, 452–466.

Aroca, R., Vernieri, P., Irigoyen, J. J., Sánchez-díaz, M., Tognoni, F., & Pardossi, A. (2003). Involvement of abscisic acid in leaf and root of maize (*Zea mays* L.) in avoiding chilling induced water stress. *Plant Science, 165*, 671–679.

Arya, S. K., & Roy, B. K. (2011). Manganese induced changes in growth, chlorophyll content and antioxidants activity in seedlings of broad bean (*Vicia faba* L.) *Journal of Environmental Biology, 32*, 707–711.

Ashraf, M., Afzal, R. M., Ahmed, R., Mujeeb, F., Sarwar, A., & Ali, L. (2010). Alleviation of detrimental effects of NaCl by silicon nutrition in salt-sensitive and salt-tolerant genotypes of sugarcane (*Saccharum officinarum* L.) *Plant and Soil, 326*, 381–391.

Ashraf, M., & Harris, P. J. C. (2013). Photosynthesis under stressful environments: An overview. *Photosynthetica, 51*, 163–190.

Azooz, M. M., Abou-Elhamd, M. F., & Al-Fredan, M. A. (2012). Biphasic effect of copper on growth, proline, lipid peroxidation and antioxidant enzyme activities of wheat (*Triticum aestivum* cv. Hasaawi) at early growing stage. *Australian Journal of Crop Science, 6*, 688–694.

Bailey-Serres, J., & Colmer, T. D. (2014). Plant tolerance of flooding stress–recent advances. *Plant, Cell & Environment, 37*, 2211–2215.

Banerjee, S., Dey, N., & Adak, M. K. (2015). Assessment of some biomarkers under submergence stress in some rice cultivars varying in responses. *American Journal of Plant Sciences, 6*, 84–94.

Barnabás, B., Jager, K., & Feher, A. (2008). The effect of drought and heat stress on reproductive processes in cereals. *Plant, Cell & Environment, 31*, 11–38.

Barta, C., Ka'lai, T., Hideg, K., Vass, I., & Hideg, E. (2004). Differences in the ROS-generating efficacy of various ultraviolet wavelengths in detached spinach leaves. *Functional Plant Biology, 31*, 23–28.

Bartling, D., Radzio, R., Steiner, U., & Weiler, E. W. (1993). A glutathione-S transferase with glutathione-peroxidase-activity from *Arabidopsis thaliana*-molecular cloning and functional characterization. *European Journal of Biochemistry, 216*, 579–586.

Biedermann, S., Mooney, S., & Hellmann, H. (2011). Recognition and repair pathways of damaged DNA in higher plants. In C. Chen (Ed.), *Selected topics in DNA repair* (pp. 201–236). Rijeka: InTech.

Biswas, D. K., & Jiang, G. M. (2011). Differential drought induced modulation of ozone tolerance in winter wheat species. *Journal of Experimental Botany, 62*, 4153–4162.

Biswas, D. K., Xu, H., Li, Y. G., Liu, M. Z., & Chen, Y. H. (2008). Assessing the genetic relatedness of higher ozone sensitivity of modern wheat to its wild and cultivated progenitors/relatives. *Journal of Experimental Botany, 59*, 951–963.

Black, V. J., Black, C. R., Roberts, J. A., & Stewart, C. A. (2000). Impact of ozone on the reproductive development of plants. *The New Phytologist, 147*, 421–447.

Blaha, G., Stelzl, U., Spahn, C. M. T., Agrawal, R. K., Frank, J., & Nierhaus, K. H. (2000). Preparation of functional ribosomal complexes and effect of buffer conditions on tRNA positions observed by cryoelectron microscopy. *Methods in Enzymology, 317*, 292–306.

Blevins, D. G., & Lukaszewski, K. M. (1994). Proposed physiologic functions of boron in plants pertinent to animal and human metabolism. *Environmental Health Perspectives, 7*, 31–33.

Blokhina, O., & Fagerstedt, K. V. (2010). Reactive oxygen species and nitric oxide in plant mitochondria: Origin and redundant regulatory systems. *Physiologia Plantarum, 138*, 447–462.

Bordi, A. (2010). The influence of salt stress on seed germination, growth and yield of canola cultivars. *Notulae Botanicae Horti Agrobotanici Cluj-Napoca, 38*, 128–133.

Brosche, M., Merilo, E., Mayer, F., Pechter, P., & Puzorjova, I. (2010). Natural variation in ozone sensitivity among *Arabidopsis thaliana* accessions and its relation to stomatal conductance. *Plant, Cell & Environment, 33*, 914–925.

Budak, H., Kantar, M., & Yucebilgili Kurtoglu, K. (2013). Drought tolerance in modern and wild wheat. *Scientific World Journal*. doi:10.1155/2013/548246.

Buriro, M., Oad, F. C., Keerio, M. I., Tunio, S., Gandahi, A. W., Hassan, S. W. U., & Oad, S. M. (2011). Wheat seed germination under the influence of temperature regimes. *Sarhad Journal of Agriculture, 27*, 539–543.

Bybordi, A. (2012). Effect of ascorbic acid and silicium on photosynthesis, antioxidant enzyme activity, and fatty acid contents in canola exposure to salt stress. *Journal of Integrative Agriculture, 11*, 1610–1620.

Calviñoa, P. A., Sadrasc, V. O., & Andradeb, F. H. (2003). Development, growth and yield of late-sown soybean in the southern pampas. *European Journal of Agronomy, 19*, 265–275.

Camejo, D., Rodríguez, P., Morales, M. A., Dell'amico, J. M., Torrecillas, A., & Alarcon, J. J. (2005). High temperature effects on photosynthetic activity of ´ two tomato cultivars with different heat susceptibility. *Journal of Plant Physiology, 162*, 281–289.

Cartes, P., Jara, A. A., Pinilla, L., Rosas, A., & Mora, M. L. (2010). Selenium improves the antioxidant ability against aluminium-induced oxidative stress in rye grass roots. *The Annals of Applied Biology, 156*, 297–307.

Chen, M., Chen, Q. J., Niu, X. G., Zhang, R., Lin, H. Q., CY, X., Wang, X. C., Wang, G. Y., & Chen, J. (2007). Expression of OsNHX1 gene in maize confers salt tolerance and promotes plant growth in the field. *Plant, Soil and Environment, 53*(11), 490–498.

Cheng, L., Zou, Y., Ding, S., Zhang, J., Yu, X., Cao, J., & Lu, G. (2009). Polyamine accumulation in transgenic tomato enhances the tolerance to high temperature stress. *Journal of Integrative Plant Biology, 51*, 489–499.

Chinnusamy, V., Zhu, J., & Zhu, J. K. (2007). Cold stress regulation of gene expression in plants. *Trends in Plant Science, 12*, 444–451.

Chu, I., Yao, X., & Zhang, Z. (2010). Responses of wheat seedlings to exogenous selenium supply under cold stress. *Biological Trace Element Research, 136*, 355–363.

Ciscar, J. C. (2012). The impacts of climate change in Europe (the PESETA research project). *Climatic Change, 112*, 1–6.

Cole, P., & McCloud, P. (1985). Salinity and climatic effects on the yields of citrus. *Australian Journal of Experimental Agriculture, 25*, 711–717.

Condon, A. G., Richards, R. A., Rebetzke, G. J., & Farquhar, G. D. (2004). Breeding for high water-use efficiency. *Journal of Experimental Botany, 55*, 2447–2460.

Costa, H., Gallego, S. M., & Tomaro, M. L. (2002). Effects of UV-B radiation on antioxidant defense system in sunflower cotyledons. *Plant Science, 162*(6), 939–945.

Cramer, G. R., & Nowak, R. S. (1992). Supplemental manganese improves the relative growth, net assimilation and photosynthetic rates of salt-stressed barley. *Physiologia Plantarum, 84*, 600–605.

Cvikrová, M., Gemperlová, L., Martincová, O., & Vanková, R. (2013). Effect of drought and combined drought and heat stress on polyamine metabolism in proline-over-producing tobacco plants. *Plant Physiology and Biochemistry, 73*, 7–15.

Dai, Q., Yan, B., Huang, S., Liu, X., & Peng, S. (1997). Response of oxidative stress defense system in rice (*Oryza Sativa*) leaves with supplemental UV-B radiation. *Physiologia Plantarum, 101*, 301–308.

Damanik, R. I., Maziah, M., Ismail, M. R., Ahmad, S., & Zain, A. M. (2010). Responses of the antioxidative enzymes in Malaysian rice (*Oryza sativa* L.) cultivars under submergence condition. *Acta Physiologiae Plantarum, 32*, 739–747.

Dantas, B. F., De Sa Ribeiro, L., & Aragao, C. A. (2007). Germination, initial growth and cotyledon protein content of bean cultivars under salinity stress. *Revista Brasileira de Sementes, 29*, 106–110.

Datnoff, L. E., Synder, G. H., & Korndörfer, G. H. (2001). *Silicon in agriculture, studies in plant sciences* (Vol. 40). Dordrecht, Netherlands: Elsevier.

Diao, M., Ma, L., Wang, J., Cui, J., Fu, A., & Liu, H.-Y. (2014). Selenium promotes the growth and photosynthesis of tomato seedlings under salt stress by enhancing chloroplast antioxidant defense system. *Journal of Plant Growth Regulation, 33*, 671–682.

Djanaguiraman, M., Devi, D. D., Shanker, A. K., Sheeba, A., & Bangarusamy, U. (2005). Selenium-an antioxidative protectant in soybean during senescence. *Plant and Soil, 272*, 77–86.

Djanaguiraman, M., Prasad, P. V. V., & Seppanen, M. (2010). Selenium protects sorghum leaves from oxidative damage under high temperature stress by enhancing antioxidant defense system. *Plant Physiology and Biochemistry, 48*, 999–1007.

Doncheva, S., Poschenrieder, C., Stoyanova, Z., Georgieva, K., Velichkova, M., & Barceló, J. (2009). Silicon amelioration of manganese toxicity in Mn-sensitive and Mn-tolerant maize varieties. *Environmental and Experimental Botany, 65*, 189–197.

Du, H., Liang, Y., Pei, K., & Ma, K. (2011). UV radiation-responsive proteins in rice leaves: A proteomic analysis. *Plant & Cell Physiology, 52*, 306–316.

Ducic, T., & Polle, A. (2005). Transport and detoxification of manganese and copper in plants. *Brazilian Journal of Plant Physiology, 17*, 103–112.

Dupuis, L., & Dumas, C. (1990). Influence of temperature stress on in vitro fertilization and heat shock protein synthesis in maize (*Zea mays* L.) reproductive systems. *Plant Physiology, 94*, 665–670.

Ebrahimian, E., & Bybordi, A. (2011). Exogenous silicium and zinc increase antioxidant enzyme activity and alleviate salt stress in leaves of sunflower. *Journal of Food, Agriculture and Environment, 9*, 422–427.

Ekelund, N. G. A., & Danilov, R. A. (2001). The influence of selenium on photosynthesis and "light-enhanced dark respiration" (LEDR) in the flagellate *Euglena gracilis* after exposure to ultraviolet radiation. *Aquatic Sciences, 63*, 57465.

Ella, E. S., Kawano, N., & Ito, O. (2003). Importance of active oxygen-scavenging system in the recovery of rice seedlings after submergence. *Plant Science, 165*, 85–93.

Elstner, E. F. (1991). Mechanisms of oxygen activation in different compartments of plant cells. In E. J. Pell & K. L. Steffen (Eds.), *Active oxygen/oxidative stress and plant metabolism* (pp. 13–25). Rockville, MD: American Society of Plant Physiology.

Fabiano, C. C., Tezotto, T., Favarin, J. L., Polacco, J. C., & Mazzafera, P. (2015). Essentiality of nickel in plants: A role in plant stresses. *Frontiers in Plant Science, 6*, 754. doi:10.3389/fpls.2015.00754.

Farooq, M., Aziz, T., Wahid, A., Lee, D. J., & Siddique, K. H. M. (2009). Chilling tolerance in maize: Agronomic and physiological approaches. *Crop & Pasture Science, 60*, 501–516.

Farooq, M. A., Detterbeck, A., Clemens, S., & Dietz, K.-J. (2016). Silicon-induced reversibility of cadmium toxicity in rice. *Journal of Experimental Botany, 67*(11), 3573–3585. doi:10.1093/jxb/erw175.

Feng, J., Shi, Q., Wang, X., Wei, M., Yang, F., & Xu, H. (2010). Silicon supplementation ameliorated the inhibition of photosynthesis and nitrate metabolism by cadmium (cd) toxicity in *Cucumis sativus* L. *Scientia Horticulturae, 123*, 521–530.

Feng, Z., Pang, J., Kobayashi, K., Zhu, Z. N., & Ort, D. R. (2011). Differential responses in two varieties of winter wheat to elevated ozone concentration under fully open-air field conditions. *Global Change Biology, 17*, 580–591.

Filek, M., Keskinen, R., Hartikainen, H., Szarejko, I., Janiak, A., Miszalski, Z., & Golda, A. (2008). The protective role of selenium in rape seedlings subjected to cadmium stress. *Journal of Plant Physiology, 165*, 833–844.

Fisarakis, I., Chartzoulakis, K., & Stavrakas, D. (2001). Response of sultana vines (*V. vinifera* L.) on six rootstocks to NaCl salinity exposure and recovery. *Agricultural Water Management, 51*, 13–27.

Foster, K. W., Timm, H., Labanauskas, C. K., & Oshima, R. J. (1983). Effects of ozone and sulfur dioxide on tuber yield and quality of potatoes. *Journal of Environmental Quality, 12*, 75–80.

Fu, X. Z., Xing, F., Wang, N. Q., Peng, L. Z., Chun, C. P., Cao, L., Ling, L. L., & Jiang, C. L. (2014). Exogenous spermine pretreatment confers tolerance to combined high-temperature and drought stress in vitro in trifoliate orange seedlings via modulation of antioxidative capacity and expression of stress-related genes. *Biotechnology and Biotechnological Equipment, 28*, 192–198.

Gao, S. R., Yan, M., Cao, W., Yang, S., Wang, F., & Chen, F. (2008). Effects of copper on growth, antioxidant enzymes and phenylalanine ammonia-lyase activities in *Jatropha curcas* L. seedling. *Plant, Soil and Environment, 54*(3), 117–122.

Gao, W., Zheng, Y., Slusser, J. R., Heisler, G. M., Grant, R. H., Xu, J., & He, D. (2004). Effects of supplementary ultraviolet-B irradiance on maize yield and qualities: A field experiment. *Photochemistry and Photobiology, 80*, 127–131.

García-Sánchez, F., Syvertsen, J. P., Gimeno, V., Botia, P., & Pérez- Pérez, J. G. (2007). Responses to flooding and drought stress by two citrus root stock seedlings with different water-use efficiency. *Physiologia Plantarum, 130*, 532–542.

Gherardi, M., & Rengel, Z. (2003). Genotypes of lucerne (*Medicago sativa* L.) show differential tolerance to manganese deficiency and toxicity when grown in bauxite residue sand. *Plant and Soil, 249*, 287–296.

Gill, S. S., & Tuteja, N. (2010). Reactive oxygen species and antioxidant machinery in abiotic stress tolerance in crop plants. *Plant Physiology and Biochemistry, 48*, 909–930.

Gillespie, K. M., Rogers, A., & Ainsworth, E. A. (2011). Growth at elevated ozone or elevated carbon dioxide concentration alters antioxidant capacity and response to acute oxidative stress in soybean (*Glycine max*). *Journal of Experimental Botany, 62*(8), 2667–2678. doi:10.1093/jxb/erq435.

Gomes-Filho, E., Machado Lima, C. R. F., Costa, J. H., da Silva, A. C., Guia Silva, d., Lima, M., de Lacerda, C. F., & Prisco, J. T. (2008). Cowpea ribonuclease: Properties and effect of NaCl-salinity on its activation during seed germination and seedling establishment. *Plant Cell Reports, 27*, 147–157.

Goto, M., Ehara, H., Karita, S., Takabe, K., Ogawa, N., & Yamada, Y. (2003). Protective effect of silicon on phenolic biosynthesis and ultraviolet spectral stress in rice crop. *Plant Science, 164*, 349–356.

Greenberg, B. M., Wilson, M. I., Gerhardt, K. E., & Wilson, K. E. (1996). Morphological and physiological responses of *Brassica napus* to ultraviolet radiation: Photomodification of ribulose 1-5- bis phosphate Carboxilase/oxygenase and potential acclimation processes. *Plant Physiology, 148*, 78–85.

Guerinot, M. L. (2000). The ZIP family of metal transporters. *Biochimica et Biophysica Acta—Biomembranes, 1465*, 190–198.
Habibi, G. (2016). Effect of foliar-applied silicon on photochemistry, antioxidant capacity and growth in maize plants subjected to chilling stress. *Acta Agriculturae Slovenica, 107*, 33–43.
Hafeez, B., Khanif, Y. M., & Saleem, M. (2013). Role of zinc in plant nutrition: A review. *American Journal of Agricultural Economics, 3*, 374–391.
Hajiboland, R., Sadeghzadeh, N., Ebrahimi, N., Sadeghzadeh, B., & Mohammadi, S. A. (2016). Influence of selenium in drought-stressed wheat plants under green house and field conditions. *Acta Agriculturae Slovenica, 105*, 175–191.
Hartmann, H. T., Kester, D. E., Davies, F. T. J., & Geneve, R. L. (1997). *Plant propagation principles and practices* (p. 770). Upper Saddle River, NJ: Prentice Hall.
Hasanuzzaman, M. & Fujita, M. (2011). Exogenous silicon treatment alleviates salinity-induced damage in *Brassica napus L.* seedlings by up-regulating the antioxidant defense and methylglyoxal detoxification system. *Abstract of Plant Biology 2011*, American Society of Plant Biology. Retrieved February 12, 2003, from http://abstracts.aspb.org/pb2011/public/P10/P10001.html/.
Hasanuzzaman, M., & Fujita, M. (2012a). Heavy metals in the environment: Current status, toxic effects on plants and possible phytoremediation. In N. A. Anjum, M. A. Pereira, I. Ahmad, A. C. Duarte, S. Umar, & N. A. Khan (Eds.), *Phytotechnologies: Remediation of environmental contaminants* (pp. 7–73). Boca Raton, FL: CRC Press.
Hasanuzzaman, M., & Fujita, M. (2012b). Selenium and plants' health: The physiological role of selenium. In C. Aomori & M. Hokkaido (Eds.), *Selenium: Sources, functions and health effects* (pp. 101–122). New York: Nova Science Publishers.
Hasanuzzaman, M., & Fujita, M. (2013). Exogenous sodium nitroprusside alleviates arsenic-induced oxidative stress in wheat (*Triticum aestivum* L.) seedlings by enhancing antioxidant defense and glyoxalase system. *Ecotoxicology, 22*, 584–596.
Hasanuzzaman, M., Fujita, M., Islam, M. N., Ahamed, K. U., & Nahar, K. (2009). Performance of four irrigated rice varieties under different levels of salinity stress. *International Journal of Integrative Biology, 6*, 85–90.
Hasanuzzaman, M., Hossain, M. A., & Fujita, M. (2010). Selenium in higher plants: Physiological role, antioxidant metabolism and abiotic stress tolerance. *Journal of Plant Science, 5*, 354–375.
Hasanuzzaman, M., Hossain, M. A., & Fujita, M. (2011a). Nitric oxide modulates antioxidant defense and the methylglyoxal detoxification system and reduces salinity-induced damage of wheat seedlings. *Plant Biotechnology Reports, 5*, 353–365.
Hasanuzzaman, M., Hossain, M. A., & Fujita, M. (2011b). Selenium-induced up-regulation of the antioxidant defense and methylglyoxal detoxification system reduces salinity-induced damage in rapeseed seedlings. *Biological Trace Element Research, 143*, 1704–1721.
Hasanuzzaman, M., Hossain, M. A., da Silva, J. A. T., & Fujita, M. (2012a). Plant responses and tolerance to abiotic oxidative stress: Antioxidant defense is a key factor. In V. Bandi, A. K. Shanker, C. Shanker, & M. Mandapaka (Eds.), *Crop stress and its management: Perspectives and strategies* (pp. 261–316). Berlin: Springer.
Hasanuzzaman, M., Hossain, M. A., & Fujita, M. (2012b). Exogenous selenium pretreatment protects rapeseed seedlings from cadmium-induced oxidative stress by upregulating the antioxidant defense and methylglyoxal detoxification systems. *Biological Trace Element Research, 149*, 248–261.
Hasanuzzaman, M., Nahar, K., & Fujita, M. (2013a). Plant response to salt stress and role of exogenous protectants to mitigate salt-induced damages. In P. Ahmad, M. M. Azooz, & M. N. V. Prasad (Eds.), *Ecophysiology and responses of plants under salt stress* (pp. 25–87). New York: Springer.
Hasanuzzaman, M., Nahar, K., & Fujita, M. (2013b). Extreme temperatures, oxidative stress and antioxidant defense in plants. In K. Vahdati & C. Leslie (Eds.), *Abiotic stress—plant responses and applications in agriculture* (pp. 169–205). Rijeka: InTech. doi:10.5772/54833.
Hasanuzzaman, M., Nahar, K., Fujita, M., Ahmad, P., Chandna, R., Prasad, M. N. V., & Ozturk, M. (2013c). Enhancing plant productivity under salt stress: Relevance of poly-omics. In P. Ahmad,

M. M. Azooz, & M. N. V. Prasad (Eds.), *Salt stress in plants: Signaling, omics and adaptations* (pp. 113–156). New York: Springer.
Hasanuzzaman, M., Gill, S. S., & Fujita, M. (2013d). Physiological role of nitric oxide in plants grown under adverse environmental conditions. In N. Tuteja & S. S. Gill (Eds.), *Plant acclimation to environmental stress* (pp. 269–322). New York: Springer. doi:10.1007/978-1-4614-5001-6_11.
Hasanuzzaman, M., Alam, M. M., Rahman, A., Hasanuzzaman, M., Nahar, K., & Fujita, M. (2014a). Exogenous proline and glycine betaine mediated upregulation of antioxidant defense and glyoxalase systems provides better protection against salt-induced oxidative stress in two rice (*Oryza sativa* L.) varieties. *BioMed Research International, 2014*, 757219. doi:10.1155/2014/757219.
Hasanuzzaman, M., Nahar, K., & Fujita, M. (2014b). Silicon and selenium: Two vital trace elements in conferring abiotic stress tolerance to plants. In A. Parvaiz & S. I. Rasool (Eds.), *Emerging technologies and management of crop stress tolerance, Biological techniques* (Vol. 1, pp. 375–420). New York: Academic Press.
Hasanuzzaman, M., Nahar, K., Gill, S. S., & Fujita, M. (2014c). Drought stress responses in plants, oxidative stress, and antioxidant defense. In N. Tuteja & S. S. Gill (Eds.), *Climate change and plant abiotic stress tolerance* (pp. 209–250). Germany: Wiley-Blackwell. doi:10.1002/9783527675265.ch09.
Hasanuzzaman, M., Nahar, K., & Fujita, M. (2015). Arsenic toxicity in plants and possible remediation. In K. R. Hakeem, M. Sabir, M. Ozturk, & A. Murmet (Eds.), *Soil remediation and plants: Prospects and challenges* (pp. 433–501). Amsterdam: Elsevier.
Hasanuzzaman, M., Nahar, K., Rahman, A., Mahmud, J. A., Hossain, M. S., & Fujita, M. (2016). Soybean production and environmental stresses. In M. Miransari (Ed.), *Environmental stresses in soybean production: Soybean production* (Vol. 2, pp. 61–102). New York: Elsevier.
Hatfield, J. L., Boote, K. J., Kimball, B. A., Ziska, L. H., Izaurralde, R. C., Ort, D., Thomson, A. M., & Wolfe, D. W. (2011). Climate impacts on agriculture: Implications for crop production. *Agronomy Journal, 103*, 351–370.
Hawrylak-Nowak, B. (2009). Beneficial effects of exogenous selenium in cucumber seedlings subjected to salt stress. *Biological Trace Element Research, 132*, 259–269.
Hawrylak-Nowak, B. (2015). Selenite is more efficient than selenate in alleviation of salt stress in lettuce plants. *Acta Biologica Cracoviensia Series Botanica, 57*, 49–54.
Hawrylak-Nowak, B., Matraszek, R., & Szyma'nska, M. (2010). Selenium modifies the effect of short-term chilling stress on cucumber plants. *Biological Trace Element Research, 138*, 307–315.
Hedhly, A. (2011). Sensitivity of flowering plant gametophytes to temperature fluctuations. *Environmental and Experimental Botany, 74*, 9–16.
Hedhly, A., Hormaza, J. I., & Herrero, M. (2009). Global warming and sexual plant reproduction. *Trends in Plant Science, 14*, 30–36.
Herrera-Rodríguež, M. B., González-Fontes, A., Rexach, J., Camacho-Cristóbal, J. J., Maldonado, J. M., & Navarro-Gochicoa, M. T. (2010). Role of boron in vascular plants and mechanisms to boron stresses. *Plant Stress, 4*(2), 115–122.
Heyno, E., Mary, V., Schopfer, P., & Krieger-Liszkay, A. (2011). Oxygen activation at the plasma membrane: Relation between superoxide and hydroxyl radical production by isolated membranes. *Planta, 234*, 35–45.
Hirschi, K. D., Korenkov, V. D., Wilganowski, N. L., & Wagner, G. J. (2000). Expression of *Arabidopsis CAX2* in tobacco altered metal accumulation and increased manganese tolerance. *Plant Physiology, 124*, 125–133.
Hossain, M. M., Liu, X., & Qi, X. (2014). Differences between soybean genotypes in physiological response to sequential soil drying and rewetting. *The Crop Journal, 2*, 1–15. doi:10.1016/j.cj.2014.08.001.
Hossain, Z., López-Climent, M. F., Arbona, V., Pérez-Clemente, R. M., & Gómez-Cadenas, A. (2009). Modulation of the antioxidant system in citrus under waterlogging and subsequent drainage. *Journal of Plant Physiology, 166*, 1391–1404.

Hu, W. H., Song, X. S., Shi, K., Xia, X. J., Zhou, Y. H., & JQ, Y. (2008). Changes in electron transport, superoxide dismutase and ascorbate peroxidase isoenzymes in chloroplasts and mitochondria of cucumber leaves as influenced by chilling. *Photosynthetica, 46*, 581–588.

Hurry, V. M., Malmberg, G., Gardestorm, P., & Oquist, G. (1994). Effects of a short term shift to low temperature and of long term cold hardening on photosynthesis and ribulose-1, 5- bisphosphate carboxylase/oxygenase and sucrose phosphate activity in leaves of winter rye (*Secale cereale* L.) *Plant Physiology, 106*, 983–990.

IPCC. (2008). Climate change and water. In B. C. Bates, Z. W. Kundzewicz, J. Palutikof, & S. Wu (Eds.), *Technical paper of the intergovernmental panel for climate change* (p. 210). Geneva: Secretariat.

Iqbal, M., Hussain, I., Liaqat, H., Ashraf, M. A., Rasheed, R., & Rehman, A. U. (2015). Exogenously applied selenium reduces oxidative stress and induces heat tolerance in spring wheat. *Plant Physiology and Biochemistry, 94*, 95–103.

Irfan, M., Hayat, S., Hayat, O., Afroz, S., & Ahmad, A. (2010). Physiological and biochemical changes in plants under waterlogging. *Protoplasma, 241*, 3–17.

Ismail, A. M., & Hall, A. E. (1999). Reproductive-stage, heat tolerance, leaf membrane thermostability and plant morphology in cowpea. *Crop Science, 39*, 1762–1768.

Jackson, M. B., & Ram, P. C. (2003). Physiological and molecular basis of susceptibility and tolerance of rice. *Annals of Botany, 91(2)*, 227–241.

Jain, M., Prasad, P. V. V., Boote, K. J., Hartwell, A. L., & Chourey, P. S. (2007). Effects of season-long high temperature growth conditions on sugar-to-starch metabolism in developing microspores of grain sorghum (*Sorghum bicolor* L). *Planta, 227*, 67–79.

James, R. A., von Caemmerer, S., Condon, A. G., Zwart, A. B., & Munns, R. (2008). Genetic variation in tolerance to the osmotic stress component of salinity stress in durum wheat. *Functional Plant Biology, 35*, 111–123.

Janislampi, K. W. (2012). Effect of silicon on plant growth and drought stress tolerance. M.S. thesis, Department of Plants, Soils, and Climate, Utah State University. Retrieved February 11, 2013, from http://digitalcommons.usu.edu/etd/1360/.

Javadmanesh, S., Rahmani, F., & Pourakbar, L. (2012). UV-B radiation, soil salinity, drought stress and their concurrent effects on some physiological parameters in maize plant American-Eurasian. *The Journal of Toxicological Sciences, 4*(4), 154–164.

Jiang, Q. W., Kiyoharu, O., & Ryozo, I. (2002). Two novel mitogen-activated protein signaling components, *OsMEK1* and *OsMAP1*, are involved in a moderate low-temperature signaling pathway in Rice. *Plant Physiology, 129*, 1880–1891.

Kalbarczyk, R. (2009). Potential reduction in cucumber yield (*Cucumis sativus* L.) in poland caused by unfavourable thermal conditions of soil. *Acta Scientiarum Polonorum Hortorum Cultus, 8*, 45–58.

Kalisz, A., & Cebula, S. (2001). Direct plant covering and soil mulching in the spring production of some Chinese cabbage cultivars. Effect of temperature on premature bolting. *Folia Horticulturae, 13*, 13–22.

Kato, Y., Miwa, K., Takano, J., Wada, M., & Fujiwara, T. (2009). Highly boron deficiency tolerant plants generated by enhanced expression of *NIP5;1*, a boric acid channel. *Plant & Cell Physiology, 50*, 58–66.

Kaveh, H., Nemati, H., Farsi, M., & Jartoodeh, S. V. (2011). How salinity affect germination and emergence of tomato lines. *Journal of Biological and Environmental Sciences, 5*, 159–163.

Kesselmeier, J., & Staudt, M. (1999). Biogenic volatile organic compounds (VOC): An overview on emission, physiology and ecology. *Journal of Atmospheric Chemistry, 33*, 23–88.

Khan, M. A., & Weber, D. J. (2008). *Ecophysiology of high salinity tolerant plants (tasks for vegetation science)* (1st ed.). Amsterdam: Springer.

Kong, L., Wang, M., & Bi, D. (2005). Selenium modulates the activities of antioxidant enzymes, osmotic homeostasis and promotes the growth of sorrel seedlings under salt stress. *Plant Growth Regulation, 45*, 155–163.

Krzyzanowski, F. C., & Delouche, J. C. (2011). Germination of cotton seed in relation to temperature. *Revista Brasileira de Farmacognosia, 33*, 543–548.

Kumar, P., Pal, M., Joshi, R., & Sairam, R. K. (2013). Yield, growth and physiological responses of mung bean (*Vigna radiata* L) genotypes to waterlogging at vegetative stage. *Physiology and Molecular Biology of Plants, 19*(2), 209–220.

Kumar, R. R., Goswami, S., Sharma, S. K., Singh, K., Gadpayle, K. A., Kuma, N., Rai, G. K., Singh, M., & Rai, R. D. (2012). Protection against heat stress in wheat involves change in cell membrane stability, antioxidant enzymes, osmolyte, H_2O_2 and transcript of heat shock protein. *International Journal of Plant Physiology and Biochemistry, 4*, 83–91.

Kumutha, D., Ezhilmathi, K., Sairam, R. K., Srivastava, G. C., Deshmukh, P. S., & Meena, R. C. (2009). Waterlogging induced oxidative stress and antioxidant activity in pigeon pea genotypes. *Biologia Plantarum, 53*, 75–84.

Kuznetsov, V. V., Kholodova, V. P., Kuznetsov, V. V., & Yagodin, B. A. (2003). Selenium regulates the water status of plants exposed to drought. *Doklady Biological Sciences, 390*, 266–268.

Latef, A. A. A., & Tran, L.-S. (2016). Impacts of priming with silicon on the growth and tolerance of maize plants to alkaline stress. *Frontiers in Plant Science*. doi:10.3389/fpls.2016.00243.

Lee, H., Guo, Y., Ohta, M., Xiong, L., Stevenson, B., & Zhu, J. K. (2002). LOS_2, a genetic locus required for cold responsive transcription encodes a bifunctional enolase. *The EMBO Journal, 21*, 2692–2702.

Leisner, C. P., Cousins, A. B., Offermann, S., Okita, T. W., & Edwards, G. E. (2010). The effects of salinity on photosynthesis and growth of the single-cell C4 species *Bienertia sinuspersici* (Chenopodiaceae). *Photosynthesis Research, 106*, 201–214.

Li, J., Cang, Z., & Jiao, F. (2015). Influence of drought stress on photosynthetic characteristics and protective enzymes of potato at seedling stage. *Journal of the Saudi Society of Agricultural Sciences*. doi:10.1016/j.jssas.2015.03.001.

Li, M., Li, Y., Li, H., & Wu, G. (2011). Overexpression of AtNHX5 improves tolerance to both salt and drought stress in (*Broussonetia papyrifera* L.) vent. *Tree Physiology, 31*, 349–357. doi:10.1093/treephys/tpr003.

Li, W., Khan, M. A., Yamaguchi, S., & Kamiya, Y. (2005). Effects of heavy metals on seed germination and early seedling growth of *Arabidopsis thaliana*. *Plant Growth Regulation, 46*, 45–50.

Liang, Y., Sun, W., Zhu, Y. G., & Christie, P. (2007). Mechanisms of silicon mediated alleviation of abiotic stresses in higher plants: A review. *Environmental Pollution, 147*, 422–428.

Liang, Y., Zhu, J., Li, Z., Chu, G., Ding, Y., & Zhang, J. (2008). Role of silicon in enhancing resistance to freezing stress in two contrasting winter wheat cultivars. *Environmental and Experimental Botany, 64*, 286–294.

Lidon, F. C., Barreiro, M., & Ramalho, J. (2004). Manganese accumulation in rice: Implications for photosynthetic functioning. *Journal of Plant Physiology, 161*, 1235–1244.

Lin, Y. F., Hassan, Z., Talukdar, S., Schat, H., & Aarts, M. G. M. (2016). Expression of the ZNT1 zinc transporter from the metal Hyperaccumulator Noccaea Caerulescens confers enhanced zinc and cadmium tolerance and accumulation to *Arabidopsis thaliana*. *PloS One, 11*(3), e0149750. doi:10.1371/journal.pone.0149750.

Lindsey, L., & Thomson, P. (2012). High temperature effects on corn and soybean. *C.O.R.N Newsletter, 2012*, 23–26.

Liu, B., Liu, X., Li, Y., & Herbert, S. J. (2013). Effects of enhanced UV-B radiation on seed growth characteristics and yield components in soybean. *Field Crops Research, 154*, 158–163.

Liu, J. J., Lin, S. H., PL, X., Wang, X. J., & Bai, J. G. (2009). Effects of exogenous silicon on the activities of antioxidant enzymes and lipid peroxidation in chilling-stressed cucumber leaves. *Agricultural Sciences in China, 8*, 1075–1086.

Liu, S. J., Heng-Heng, X., Wang, W.-Q., Ni, L., Wang, W.-P., Møller, I. M., & Song, S.-Q. (2014). A proteomic analysis of rice seed germination as affected by high temperature and ABA treatment. *Physiologia Plantarum, 154*, 142–161.

Lombardi, T., & Lupi, B. (2006). Effect of salinity on the germination and growth of Hordeum secalinum Schreber (Poaceae) in relation to the seeds after-ripening time. *Atti Soc tosc Sci nat Mem Serie B, 113*, 37–42.

Long, S. P., & Ort, D. R. (2010). More than taking the heat: Crops and global change. *Current Opinion in Plant Biology, 13*, 240–247.

Luo, H. H., Zhang, Y. L., & Zhang, W. F. (2016). Effects of water stress and rewatering on photosynthesis, root activity, and yield of cotton with drip irrigation under mulch. *Photosynthetica, 54*, 65–73. doi:10.1007/s11099-015-0165-7.

Lyman, N. B., Jagadish, K. S. V., Nalley, L. L., Dixon, B. L., & Siebenmorgen, T. (2013). Neglecting rice milling yield and quality underestimates economic losses from high-temperature stress. *PloS One, 8*, e72157.

Ma, J. F., & Yamaji, N. (2008). Functions and transport of silicon in plants. *Cellular and Molecular Life Sciences, 65*, 3049–3057.

Mahajan, S., & Tuteja, N. (2005). Cold, salinity and drought stresses: An overview. *Archives of Biochemistry and Biophysics, 444*, 139–158.

Maksymiec, W. (2007). Signaling responses in plants to heavy metal stress. *Acta Physiologiae Plantarum, 29*, 177–187.

Manaf, H. H. (2016). Beneficial effects of exogenous selenium, glycine betaine and seaweed extract on salt stressed cowpea plant. *Annals of Agricultural Science, 61*, 41–48.

Manivannan, A., Soundararajan, P., Muneer, S., Ho Ko, C., & Jeong, B. R. (2016). Silicon mitigates salinity stress by regulating the physiology, antioxidant enzyme activities, and protein expression in *Capsicum annuum* 'Bugwang'. *BioMed Research International.* doi:10.1155/2016/3076357.

Marcińska, I., Czyczyło-Mysza, I., & Skrzypek, E. (2013). Impact of osmotic stress on physiological and biochemical characteristics in drought-susceptible and drought-resistant wheat genotypes. *Acta Physiologiae Plantarum, 35*, 451–461. doi:10.1007/s11738-012-1088-6.

Mateos-Naranjo, E., Andrades-Moreno, L., & Davy, A. J. (2013). Silicon alleviates deleterious effects of high salinity on the halophytic grass *Spartina densiflora. Plant Physiology and Biochemistry, 63*, 115–121.

Mauad, M., Costa Crusciol, C. A., Nascente, A. S., Filho, H. G., & Lima, G. P. P. (2016). Effects of silicon and drought stress on biochemical characteristics of leaves of upland rice cultivars. *Revista Ciência Agronômica, 47*, 532–539.

McNamara, A. E., & Hill, W. R. (2000). UV-B irradiance gradient affects photosynthesis and pigments but not food quality of perihyton. *Freshwater Biology, 43*, 649–662.

Mditshwa, A., Bower, J. P., Bertling, I., Mathaba, N., & Tesfay, S. Z. (2013). The potential of postharvest silicon dips to regulate phenolics in citrus peel as a method to mitigate chilling injury in lemons. *African Journal of Biotechnology, 12*, 1482–1489.

Migocka, M., Papierniak, A., Maciaszczyk-Dziubińska, E., Poździk, P., Posyniak, E., Garbiec, A., & Filleur, S. (2014). Cucumber metal transport protein MTP8 confers increased tolerance to manganese when expressed in yeast and *Arabidopsis thaliana. Journal of Experimental Botany, 65*(18), 5367–5384. doi:10.1093/jxb/eru295.

Millaleo, R., Reyes-Diaz, M., Ivanov, A. G., Mora, M. L., & Alberdi, M. (2010). Manganese as essential and toxic element for plants: Transport, accumulation and resistance mechanisms. *Journal of Soil Science and Plant Nutrition, 10*, 476–449.

Miller, P., Lanier, W., Brandt, S. (2001). Using growing degree days to predict plant stages. Montana State University Extension Service. SKU MT200103AG.

Mithofer, A., Schulze, B., & Boland, W. (2004). Biotic and heavy metal stress response in plants: Evidence for common signals. *FEBS Letters, 566*, 1–5.

Mittler, R. (2002). Oxidative stress, antioxidants and stress tolerance. *Trends in Plant Science, 7*, 405–410.

Mittler, R., & Blumwald, E. (2010). Genetic engineering for modern agriculture: Challenges and perspectives. *Annual Review of Plant Biology, 61*, 443–462.

Mittler, R., Finka, A., & Goloubinoff, P. (2012). How do plants feel the heat? *Trends in Biochemical Sciences, 37*, 118–125.

Moeinian, M. R., Zargari, K., & Hasanpour, J. (2011). Effect of boron foliar spraying application on quality characteristics and growth parameters of wheat grain under drought stress. *American-Eurasian Journal of Agricultural & Environmental Sciences, 10*(4), 593–599.

Monjezi, F., Vazan, F., & Hassnzadehdelouei, M. (2012). Effects of iron and zinc spray on wheat in drought stress. *Cercetări Agronomice în Moldova, XLVI*(1), 153.

Morison, J. I. L., Baker, N. R., Mullineaux, P. M., & Davies, W. J. (2008). Improving water use in crop production. *Philosophical Transactions of the Royal Society B: Biological Sciences, 363*, 639–658.

Mozafariyan, M., Kamelmanesh, M. M., & Hawrylak-Nowak, B. (2016). Ameliorative effect of selenium on tomato plants grown under salinity stress. *Archives of Agronomy and Soil Science*. doi:10.1080/03650340.2016.1149816.

Moussa, H. R., & Abdel-Aziz, S. M. (2008). Comparative response of drought tolerant and drought sensitive maize genotypes to water stress. *Australian Journal of Crop Science, 1*(1), 31–36.

Müller-Xing, R., Xing, Q., & Goodrich, J. (2014). Footprints of the sun: Memory of UV and light stress in plants. *Frontiers in Plant Science, 5*, 474.

Munns, R. (2002). Comparative physiology of salt and water stress. *Plant, Cell & Environment, 25*, 239–250.

Munns, R., & Tester, M. (2008). Mechanisms of salinity tolerance. *Annual Review of Plant Biology, 59*, 651–681.

Mustafiz, A., Ghosh, A., Tripathi, A. K., Kaur, C., Ganguly, A. K., Bhavesh, N. S., Tripathi, J. K., Pareek, A., Sopory, S. K., & Singla-Pareek, S. L. (2014). A unique Ni^{2+}-dependent and methylglyoxal-inducible rice glyoxalase I possesses a single active site and functions in abiotic stress response. *The Plant Journal, 78*, 951–963.

Nahar, K., Biswas, J. K., & Shamsuzzaman, A. M. M. (2012). *Cold stress tolerance in rice plant: Screening of genotypes based on morphophysiological traits*. Saarbrücken, Germany: Lambert Academic Publishing.

Nahar, K., Biswas, J. K., Shamsuzzaman, A. M. M., Hasanuzzaman, M., & Barman, H. N. (2009). Screening of indica rice (*Oryza sativa* L.) genotypes against low temperature stress. *Botany Research International, 2*, 295–303.

Nahar, K., & Hasanuzzaman, M. (2009). Germination, growth, nodulation and yield performance of three mungbean varieties under different levels of salinity stress. *Green Farming, 2*, 825–829.

Nahar, K., Hasanuzzaman, M., Alam, M. M., & Fujita, M. (2014). Regulatory roles of exogenous glutathione in conferring salt tolerance in mung bean (*Vigna radiata* L.): Implication of antioxidant defense and methylglyoxal detoxification system. *Biologia Plantarum, 59*, 745–756.

Nahar, K., Hasanuzzaman, M., Ahamed, K. U., Öztürk, M., & Fujita, M. (2015a). Plant responses and tolerance to high temperature stress: Role of exogenous phytoprotectants. In K. U. R. Hakeem (Ed.), *Crop production and global environmental issues* (pp. 385–436). Cham, Switzerland: Springer.

Nahar, K., Hasanuzzaman, M., Alam, M. M., & Fujita, M. (2015b). Exogenous spermidine alleviates low temperature injury in mung bean (*Vigna radiata* L.) seedlings by modulating ascorbate-glutathione and glyoxalase pathway. *International Journal of Molecular Sciences, 16*, 30117–30132.

Nahar, K., Hasanuzzaman, M., Alam, M. M., Rahman, A., Suzuki, T., & Fujita, M. (2016). Polyamine and nitric oxide crosstalk: antagonistic effects on cadmium toxicity in mung bean plants through upregulating the metal detoxification, antioxidant defense and methylglyoxal detoxification systems. *Ecotoxicology and environmental safety, 126*, 245–255.

Nahar, K., Hasanuzzaman, M., Alam, M. M., Rahman, A., Mahmud, J. A., Suzuki, T., & Fujita, M. (2017). Insights into spermine-induced combined high temperature and drought tolerance in mung bean: osmoregulation and roles of antioxidant and glyoxalase system. *Protoplasma, 254(1)*, 445–460.

Naim, A. (2015). *Mitigation of salt stress in rice by exogenous application of selenium*. M.S. thesis, Department of Agronomy, Sher-e-Bangla Agricultural University, Dhaka, Bangladesh.

Nasibi, F., & M-Kalantari, K. H. (2005). The effects of uv-a, uv-b and uv-c on protein and ascorbate content, lipid peroxidation and biosynthesis of screening compounds in *Brassica napus*. *Iranian Journal of Science and Technology, 29*, 40–48.

Nawaz, F., Ashraf, M. Y., Ahmad, R., & Waraich, E. A. (2013). Selenium (Se) seed priming induced growth and biochemical changes in wheat under water deficit conditions. *Biological Trace Element Research, 151*, 284–293.

Nawaz, F., Ashraf, M. Y., Ahmad, R., Waraich, E. A., & Shabbir, R. N. (2014). Selenium (se) regulates seedling growth in wheat under drought stress. *Advances in Chemistry*. doi:10.1155/2014/143567.

Nelson, J. M., Palzkill, D. A., & Bartels, P. G. (1993). Irrigation cut-off date affects growth, frost damage, and yield of jojoba. *Journal of the American Society for Horticultural Science, 118*(6), 731–735.

Nishiuchi, S., Yamauchi, T., Takahashi, H., Kotula, L., & Nakazono, M. (2012). Mechanisms for coping with submergence and waterlogging in rice. *Rice, 5*, 2. doi:10.1186/1939-8433-5-2.

Niu, G., Rodriguez, D. S., Circle, M., et al. (2006). Impact of drought and temperature on growth and leaf gas exchange of six bedding plant species under greenhouse conditions. *Hortscience, 41*, 1408–1411.

Noctor, G., De Paepe, R., & Foyer, C. H. (2007). Mitochondrial redox biology and homeostasis in plants. *Trends in Plant Science, 12*, 125–134.

Noctor, G., Veljovic-Jovanovic, S. O. N. J. A., Driscoll, S., Novitskaya, L., & Foyer, C. H. (2002). Drought and oxidative load in the leaves of C3 plants: a predominant role for photorespiration? *Annals of Botany, 89*(7), 841–850.

Obidiegwu, J. E., Bryan, G. J., Jones, H. G., & Prashar, A. (2015). Coping with drought: Stress and adaptive responses in potato and perspectives for improvement. *Frontiers in Plant Science, 6*, 542. doi:10.3389/fpls.2015.00542.

Othman, Y., Al-Karaki, G., Al-Tawaha, A. R., & Al-Horani, A. (2006). Variation in germination and ion uptake in barley genotypes under salinity conditions. *World Journal of Agricultural Sciences, 2*, 11–15.

Pal'ove-Balang, P., Kisova, A., Pavlovkin, J., & Mistrik, I. (2006). Effect of manganese on cadmium toxicity in maize seedlings. *Plant, Soil and Environment, 52*, 143–149.

Pandey, G. K. (2015). *Elucidation of abiotic stress signaling in plants: Functional genomics perspectives* (Vol. 2). New York: Springer.

Pandya, D. H., Mer, R. K., Prajith, P. K., & Pandey, A. N. (2004). Effect of salt stress and manganese supply on growth of barley seedlings. *Journal of Plant Nutrition, 27*, 1361–1379.

Parveen, N., & Ashraf, M. (2010). Role of silicon in mitigating the adverse effects of salt stress on growth and photosynthetic attributes of two maize (*Zea mays* L.) cultivars grown hydroponically. *Pakistan Journal of Botany, 42*, 1675–1684.

Peacock, J. M., Miller, W. B., Matsuda, K., & Robinson, D. L. (1993). Role of heat girdling in early seedling death in sorghum. *Crop Science, 30*, 138–143.

Pedrero, Z., Madrid, Y., Hartikainen, H., & Cámara, C. (2008). Protective effect of selenium in broccoli (*Brassica oleracea*) plants subjected to cadmium exposure. *Journal of Agricultural and Food Chemistry, 56*, 266–271.

Peng, K., Chunling, L., Wuxin, Y., Chunlan, L., Xiangdong, L., & Shen, Z. (2008). Manganese uptake and interactions with cadmium in the hyperaccumulator-*Phytolacca americana* L. *Journal of Hazardous Materials, 154*, 674–681.

Peng, S., Huang, J., Sheehy, J. E., Laza, R. C., Visperas, R. M., Zhong, X., Centeno, G. S., Khush, G. S., & Cassman, K. G. (2004). Rice yields decline with higher night temperature from global warming. *Proceedings of the National Academy of Sciences of the United States of America, 101*, 9971–9975.

Pennanen, A., Xue, T., & Hartikainen, H. (2002). Protective role of selenium in plant subjected to severe UV irradiation stress. *Journal of Applied Botany, 76*, 66–76.

Pirovano, L., Morgutti, S., Espen, L., & Cocucci, S. M. (1997). Differences in the reactivation process in thermosensitive seeds of *Phacelia tanacetifolia* with germination inhibited by high temperature (30°C). *Physiologia Plantarum, 99*, 211–220.

Pittman, J. (2005). Managing the manganese: Molecular mechanisms of manganese transport and homeostasis. *The New Phytologist, 167*, 733–742.

Porch, T. G. (2006). Application of stress indices for heat tolerance screening of common bean. *Journal of Agronomy and Crop Science, 192*, 390–394.

Prabagar, S., Hodson, M. J., & Evans, D. E. (2011). Silicon amelioration of aluminium toxicity and cell death in suspension cultures of Norway spruce (*Picea abies* L.) *Environmental and Experimental Botany, 70*, 266–276.

Prasad, P. V. V., Pisipati, S. R., Momčilović, I., & Ristic, Z. (2011). Independent and combined effects of high temperature and drought stress during grain filling on plant yield and chloroplast EF-Tu expression in spring wheat. *Journal of Agronomy and Crop Science, 197*, 430–441.

Rahman, M. S., Matsumuro, T., Miyake, H., & Takeoka, Y. (2000). Salinity-induced Ultrastructural alterations in leaf cells of rice alternations in leaf cells of rice (*Oryza sativa* L.). *Plant Production Science, 3*, 422–429.

Rahman, A., Mostofa, M. G., Nahar, K., Hasanuzzaman, M., & Fujita, M. (2015). Exogenous calcium alleviates cadmium-induced oxidative stress in rice seedlings by regulating the antioxidant defense and glyoxalase systems. *Brazilian Journal of Botany*. doi:10.1007/s40415-015-0240-0.

Rahman, A., Hossain, M. S., Mahmud, J. A., Nahar, K., Hasanuzzaman, M., & Fujita, M. (2016). Manganese-induced salt stress tolerance in rice seedlings: Regulation of ion homeostasis, antioxidant defense and glyoxalase systems. *Physiology and Molecular Biology of Plants*. doi:10.1007/s12298-016-0371-1.

Ram, P. C., Singh, A. K., Singh, B. B., Singh, V. K., Singh, H. P., Setter, T. L., Singh, V. P., & Singh, R. K. (1999). Environmental characterization of floodwater in eastern India: Relevance to submergence tolerance of lowland rice. *Experimental Agriculture, 35*, 141–152.

Rasolohery, C. A., Berger, M., Lygin, A. V., Lozovaya, V. V., Nelson, R. L., & Dayde, J. (2008). Effect of temperature and water availability during late maturation of the soybean seed on germ and cotyledon isoflavone content and composition. *Journal of Science and Food Agriculture, 88*, 218–228.

Raven, J. A. (2001). Silicon transport at the cell and tissue level. In L. E. Datnoff, G. H. Snyder, & G. H. Korndorfer (Eds.), *Silicon in agriculture* (pp. 41–55). Amsterdam: Elsevier.

Ravindran, K. C., Indrajith, A., Pratheesh, P. V., Sanjiviraja, K., & Balakrishnan, V. (2010). Effect of ultraviolet-B radiation on biochemical and antioxidant defence system in *Indigofera tinctoria* L. seedlings. *International Journal of Engineering Science and Technology, 2*, 226–232.

Raziuddin, Farhatullah, Hassan, G., Akmal, M., Shah, S. S., Mohammad, F., Shafi, M., Bakht, J., & Zhou, W. (2011). Effects of cadmium and salinity on growth and photosynthesis parameters of *Brassica species*. *Pakistan Journal of Botany, 43*, 333–340.

Reddy, K. R., Kakani, V. G., Zhao, D., Mohammed, A. R., & Gao, W. (2003). Cotton responses to ultraviolet-B radiation: Experimentation and algorithm development. *Agricultural and Forest Meteorologyis, 120*, 249–265.

Ren, B., Zhang, J., Li, X., Fan, X., Dong, S., Liu, P., & Zhao, B. (2014). Effects of waterlogging on the yield and growth of summer maize under field conditions. *Canadian Journal of Plant Science, 94*, 23–31.

Riaz-ud-din, Subhani, G. M., Ahmad, N., Hussain, M., & Rehman, A. U. (2010). Effect of temperature on development and grain formation in spring wheat. *Pakistan Journal of Botany, 42*, 899–906.

Riley, G. J. P. (1981). Effects of high temperature on protein synthesis during germination of maize (*Zea mays* L.). *Planta, 151*, 75–80.

Robertson, D., Zhang, H., Palta, J. A., Colmer, T., & Turner, N. C. (2009). Waterlogging affects the growth, development of tillers, and yield of wheat through a severe, but transient, N deficiency. *Crop & Pasture Science, 60*, 578–586.

Sairam, R. K., Dharmar, K., Lekshmy, S., & Chinnusam, V. (2011). Expression of antioxidant defense genes in mung bean (*Vigna radiata* L.) roots under water-logging is associated with hypoxia tolerance. *Acta Physiologiae Plantarum, 33*, 735–744.

Sajedi, N., Madani, H., & Naderi, A. (2011). Effect of microelements and selenium on superoxide dismutase enzyme, malondialdehyde activity and grain yield maize (*Zea mays* L.) under water deficit stress. *Notulae Botanicae Horti Agrobotanici Cluj-Napoca, 39*, 153–159.

Sakata, T., Oshino, T., Miura, S., Tomabechi, M., Tsunaga, Y., Higashitani, N., Miyazawa, Y., Takahashi, H., Watanabe, M., & Higashitani, A. (2010). Auxins reverse plant male sterility caused by high temperatures. *Proceedings of the National Academy of Sciences of the United States of America, 107*, 8569–8574.

Salim, B. B. M. (2014). Effect of boron and silicon on alleviating salt stress in maize. *Middle East Journal of Agricultural Research, 3*(4), 1196–11204.
Samreen, T., Humaira, & Shah, H. U. (2013). Zinc effect on growth rate, chlorophyll, protein and mineral contents of hydroponically grown mungbeans plant (*Vigna radiata*). *Arabian Journal of Chemistry*. doi:10.1016/j.arabjc.2013.07.005.
Sandalio, L. M., Rodríguez-Serrano, M., Romero-Puertas, M. C., & del Ríıo, L. A. (2013). Role of peroxisomes as a source of reactive oxygen species (ROS) signaling molecules. Del Río LA (ed) peroxisomes and their key role in cellular signaling and metabolism. *Sub-Cellular Biochemistry, 69*, 231–255.
Sanghera, G. S., Wani, S. H., Hussain, W., & Singh, N. B. (2011). Engineering cold stress tolerance in crop plants. *Current Genomics, 12*, 30–43.
Sapeta, H., Costa, J. M., Lourenço, T., Marocod, J., Lindee, P. V., & Oliveiraa, M. M. (2013). Drought stress response in *Jatropha curcas*: Growth and physiology. *Environmental and Experimental Botany, 85*, 76–84.
Sarma, H. (2011). Metal hyperaccumulation in plants: A review focusing on phytoremediation technology. *Journal of Environmental Science and Technology, 4*, 118–138.
Sasaki, A., Yamaji, N., & Ma, J. F. (2016). Transporters involved in mineral nutrient uptake in rice. *Journal of Experimental Botany, 67*(12), 3645–3653. doi:10.1093/jxb/erw060.
Sawada, H., Tsukahara, K., Kohno, Y., Suzuki, K., Nagasawa, N., & Tamaok, M. (2016). Elevated ozone deteriorates grain quality of *Japonica* Rice cv. Koshihikari, even if it does not cause yield reduction. *Rice, 9*, 7.
Sebastian, A., & Prasad, M. N. V. (2015). Iron-and manganese-assisted cadmium tolerance in *Oryza sativa* L.: Lowering of rhizotoxicity next to functional photosynthesis. *Planta, 241*, 1519–1528.
Setter, T. L., & Waters, I. (2003). Review of prospects for germplasm improvement for waterlogging tolerance in wheat, barley and oats. *Plant and Soil, 253*, 1–34.
Shabaan, M. M. (2010). Role of Boron in plant nutrition and human health. *American Journal of Plant Physiology, 5*(5), 224–240.
Shah, N., & Paulsen, G. (2003). Interaction of drought and high temperature on photosynthesis and grain-filling of wheat. *Plant and Soil, 257*, 219–226.
Shahbaz, M., & Ashraf, M. (2013). Improving salinity tolerance in cereals. *Critical Reviews in Plant Sciences, 32*, 237–249.
Sharma, A., Gontia-Mishra, I., & Srivastava, A. K. (2011). Toxicity of heavy metals on germination and seedling growth of *Salicornia brachiata*. *Journal of Phytology, 3*, 33–36.
Sharma, P., & Dubey, R. S. (2007). Involvement of oxidative stress and role of antioxidative defense system in growing rice seedlings exposed to toxic levels of aluminium. *Plant Cell Reports, 26*, 2027–2038.
Sharma, P., Jha, A. B., Dubey, R. S., & Pessarakli, M. (2012). Reactive oxygen species, oxidative damage, and antioxidative defense mechanism in plants under stressful conditions. *Journal of Botany, 37*, 1–26.
Sharma, P. K., Anand, P., Sankhalkar, S., & Shety, R. (1998). Photochemical and biochemical changes in wheat seedlings exposed to supplementary ultraviolet-B radiation. *Plant Science, 132*, 21–30.
Sharma, S. S., & Dietz, K. J. (2008). The relationship between metal toxicity and cellular redox imbalance. *Trends in Plant Science, 14*, 43–50.
Shekari, F., Abbasi, A., & Mustafavi, S. H. (2015). Effect of silicon and selenium on enzymatic changes and productivity of dill in saline condition. *Journal of the Saudi Society of Agricultural Sciences*. doi:10.1016/j.jssas.2015.11.006.
Shekari, L., Kamelmanesh, M. M., Mozafariyan, M., & Sadeghi, F. (2016). Role of selenium in mitigation of cadmium toxicity in pepper grown in hydroponic condition. *Journal of Plant Nutrition*. doi:10.1080/01904167.2016.1161773.
Shen, X., Zhou, Y., Duan, L., Li, Z., Eneji, A. E., & Li, J. (2010). Silicon effects on photosynthesis and antioxidant parameters of soybean seedlings under drought and ultraviolet-B radiation. *Journal of Plant Physiology, 167*, 1248–1252.
Shi, Y., Zhang, Y., Han, W., Feng, R., Hu, Y., Guo, J., & Gong, H. (2016). Silicon enhances water stress tolerance by improving root hydraulic conductance in *Solanum lycopersicum* L. *Frontiers in Plant Science, 7*, 196. doi:10.3389/fpls.2016.00196.

Singh, R., Singh, S., Tripathi, R., & Agrawal, S. B. (2011). Supplemental UV-B radiation induced changes in growth, pigments and antioxidant pool of bean (*Dolichos lablab*) under field conditions. *Journal of Environmental Biology, 32*, 139–145.

Singh, R. K., & Flowers, T. J. (2010). Physiology and molecular biology of the effects of salinity on rice. In *Handbook of plant and crop stress* (pp. 899–939). Boca Raton: CRC Press.

Song, A., Li, Z., Zhang, J., Xue, G., Fan, F., & Liang, Y. (2009). Silicon-enhanced resistance to cadmium toxicity in *Brassica chinensis* L. is attributed to Si-suppressed cadmium uptake and transport and Si-enhanced antioxidant defense capacity. *Journal of Hazardous Materials, 172*, 74–83.

Soylemezoglu, G., Demir, K., Inal, A., & Gunes, A. (2009). Effect of silicon on antioxidant and stomatal response of two grapevine (Vitis Vinifera L.) rootstocks grown in boron toxic, saline and boron toxic-saline soil. *Scientia Horticulturae, 123*, 240–246.

Srivastava, M., Ma, L. Q., Rathinasabapathi, B., & Srivastava, P. (2009). Effects of selenium on arsenic uptake in arsenic hyperaccumulator *Pteris vittata* L. *Bioresource Technology, 100*, 1115–1121.

Steffens, B., Geske, T., & Sauter, M. (2011). Aerenchyma formation in the rice stem and its promotion by H_2O_2. *New Phytologist, 190*, 369–378.

Stone, P. J., & Nicolas, M. E. (1994). Wheat cultivars vary widely in their responses of grain yield and quality to short periods of post-anthesis heat stress. *Australian Journal of Plant Physiology, 21*, 887–900.

Sudhir, P., & Murthy, S. D. S. (2004). Effects of salt stress on basic processes of photosynthesis. *Photosynthetica, 42*, 481–486.

Sun, H. W., Ha, J., Liang, S. X., & Kang, W. J. (2010). Protective role of selenium on garlic growth under cadmium stress. *Communications in Soil Science and Plant Analysis, 41*, 1195–1204.

Sun, Q., Wang, X. R., Ding, S. M., & Yuan, X. F. (2005). Effects of exogenous organic chelators on phytochelatins production and its relationship with cadmium toxicity in wheat (*Triticum aestivum* L.) under cadmium stress. *Chemosphere, 60*, 2–31.

Tadina, N., Germ, M., Kreft, I., Breznik, B., & Gaberščik, A. (2007). Effects of water deficit and selenium on common buckwheat (*Fagopyrum esculentum* Moench.) plants. *Photosynthetica, 45*, 472–476.

Taiz, L., & Zeiger, E. (2006). Stress physiology. In L. Taiz & E. Zeiger (Eds.), *Plant physiology* (5th ed., pp. 671–681). Sunderland: Sinauer Associates.

Tavakkoli, E., Fatehi, F., Coventry, S., Rengasamy, P., & McDonald, G. K. (2011). Additive effects of Na^+ and Cl^- ions on barley growth under salinity stress. *Journal of Experimental Botany, 62*, 2189–2203.

Tavallali, V., Rahemi, M., & Eshghi, S. (2010). Zinc alleviates salt stress and increases antioxidant enzyme activity in the leaves of pistachio (*Pistacia vera* L. "Badami") seedlings. *Turkish Journal of Agriculture and Forestry, 34*, 349–359. doi:10.3906/tar-0905-10.

Terry, N., Zayed, A. M., de Souza, M. P., & Tarun, A. S. (2000). Selenium in higher plants. *Annual Review of Plant Physiology and Plant Molecular Biology, 51*, 401–432.

Thakur, P., Kumar, S., Malik, J. A., Berger, J. D., & Nayyar, H. (2010). Cold stress effects on reproductive development in grain crops, an overview. *Environmental and Experimental Botany, 67*, 429–443.

Thameur, A., Lachiheb, B., & Ferchichi, A. (2012). Drought effect on growth, gas exchange and yield, in two strains of local barley Ardhaoui, under water deficit conditions in southern Tunisia. *Journal of Environmental Management, 113*, 495–500. doi:10.1016/j.jenvman.2012.05.026.

Thomashow, M. F. (1999). Plant cold acclimation: Freezing tolerance genes and regulatory mechanisms. *Annual Review of Plant Physiology and Plant Molecular Biology, 50*, 571–579.

Tiryaki, I., & Keles, H. (2012). Reversal of the inhibitory effect of light and high temperature on germination of *Phacelia tanacetifolia* seeds by melatonin. *Journal of Pineal Research, 52*, 332–339.

Tsukahara, K., Sawada, H., Kohno, Y., Matsuura, T., Mori, I. C., & Terao, T. (2015). Ozone-induced rice grain yield loss is triggered via a change in panicle morphology that is controlled by aberrant panicle organization 1 Gene. *PloS One, 10*, e0123308. doi:10.1371/journal.pone.0123308.

Upadhyaya, H., Dutta, B. K., Sahoo, L., & Panda, S. K. (2012). Comparative effect of Ca, K, Mn and B on post-drought stress recovery in tea (*Camellia sinensis* L.) *American Journal of Plant Sciences, 3*, 443–460.

Vašková, J., Vasko, L., & Kron, I. (2012). Oxidative processes and antioxidative metaloenzymes. In M. A. El-Missiry (Ed.), *Antioxidant enzyme*. Rijeka: InTech. doi:10.5772/50995.

Vorobets, N., & Mykiyevich, I. (2000). Single and combined effects of lead and selenium on sunflower seedlings. *Scientific Workshop on Horticulture and Vegetable Growing, 19*, 390.

Vyšniauskienė, R., & Rančelienė, V. (2014). Effect of UV-B radiation on growth and antioxidative enzymes activity in Lithuanian potato (*Solanum tuberosum* L.) cultivars. *Zemdirbyste-Agriculture, 101*(1), 51–56. doi:10.13080/z-a.2014.101.007.

Wahid, A., Gelani, S., Ashraf, M., & Foolad, M. R. (2007). Heat tolerance in plants: An overview. *Environmental and Experimental Botany, 61*, 199–223.

Wang, C. Q., HJ, X., & Liu, T. (2011). Effect of selenium on ascorbate–glutathione metabolism during PEG-induced water deficit in *Trifolium repens* L. *Journal of Plant Growth Regulation, 30*, 436–444.

Wang, L. J., & Li, S. H. (2006). Salicylic acid-induced heat and cold tolerance in relation to Ca^{2+} homeostasis and antioxidant systems in young grape plants. *Plan Sci, 170*, 685–694.

Wang, M., Xu, Q., Yu, J., & Yuan, M. (2010). The putative Arabidopsis zinc transporter ZTP29 is involved in the response to salt stress. *Plant Molecular Biology, 73*(4), 467–479.

Wang, S., Liang, D., Li, C., Hao, Y., Maa, F., & Shu, H. (2012). Influence of drought stress on the cellular ultrastructure and antioxidant system in leaves of drought tolerant and drought-sensitive apple rootstocks. *Plant Physiology and Biochemistry, 51*, 81–89.

Wang, W. X., Vinocur, B., & Altman, A. (2003). Plant responses to drought, salinity and extreme temperatures: Towards genetic engineering for stress tolerance. *Planta, 218*, 1–14.

Wang, X., & Mauzerall, D. L. (2004). Characterizing distributions of surface ozone and its impact on grain production in China, Japan and South Korea: 1990 and 2020. *Atmospheric Environment, 38*, 4383–4402.

Waraich, E. A., Ahmad, R., & Ashraf, M. Y. (2011a). Improving agricultural water use efficiency by nutrient management in crop plants. *Acta Agriculturae Scandinavica Section B—Soil and Plant Science, 61*, 291–304.

Waraich, E. A., Ahmad, R., & Saifullah. (2011b). Role of mineral nutrition in alleviation of drought stress in plants. *Australian Journal of Crop Science, 5*, 764–777.

Wiebbecke, C. F., Graham, M. A., Cianzio, S. R., & Palmer, R. G. (2012). Day temperature influences themale-sterile locus ms9 in soybean. *Crop Science, 52*, 1503–1510.

Wilkinson, S., Mills, G., Illidge, R., & Davies, W. J. (2012). How is ozone pollution reducing our food supply? *Journal of Experimental Botany, 63*, 527–536.

Wojtaszek, P. (1997). Oxidative burst: An early plant response to pathogen infection. *Biochemical Journal, 322*, 681–692.

Wu, F., & Zhang, G. (2002). Alleviation of cadmium-toxicity by application of zinc and ascorbic acid in barley. *Journal of Plant Nutrition, 25*, 2745–2761. doi:10.1081/PLN-120015536.

Xiaoqin, Y., Jianzhou, C., & Guangyin, W. (2009a). Effects of drought stress and selenium supply on growth and physiological characteristics of wheat seedlings. *Acta Physiologiae Plantarum, 31*, 1031–1036.

Xiaoqin, Y., Jianzhou, C., & Guangyin, W. (2009b). Effects of selenium on wheat seedlings under drought stress. *Biological Trace Element Research, 130*, 283–290.

Xue, Z. Y., DY, Z., Xue, G. P., Zhang, H., Zhao, Y. X., & Xia, G. M. (2004). Enhanced salt tolerance of transgenic wheat (*Tritivum aestivum* L.) expressing a vacuolar Na^+/H^+ antiporter gene with improved grain yields in saline soils in the field and a reduced level of leaf Na^+. *Plant Science, 167*, 849–859.

Yadav, S. K. (2010). Cold stress tolerance mechanisms in plants: A review. *Agronomy for Sustainable Development, 30*, 515–527.

Yamaguchi, T., & Blumwald, E. (2005). Developing salt-tolerant crop plants: Challenges and opportunities. *Trends in Plant Science, 10*(12), 615–620.

Yan, K., Chen, W., He, X., Zhang, G., Xu, S., & Wang, L. (2010). Responses of photosynthesis, lipid peroxidation and antioxidant system in leaves of *Quercus mongolica* to elevated O_3. *Environmental and Experimental Botany, 69*, 198–204.

Yao, X., Chu, J., & Ba, C. (2010). Responses of wheat roots to exogenous selenium supply under enhanced ultraviolet-B. *Biological Trace Element Research, 137*, 244–252.

Yao, X., Chu, J., & Wang, G. (2009). Effects of selenium on wheat seedlings under drought stress. *Biological Trace Element Research, 130*, 283–290.

Yildiztugay, E., Ozfidan-Konakci, C., Kucukoduk, M., & Tekis, S. A. (2016). The impact of selenium application on enzymatic and non-enzymatic antioxidant systems in *Zea mays* roots treated with combined osmotic and heat stress. *Archives of Agronomy and Soil Science*. doi:10.1080/03650340.2016.1201810.

Yordanova, R., & Popova, L. (2007). Effect of exogenous treatment with salicylic acid on photosynthetic activity and antioxidant capacity of chilled wheat plants. *General and Applied Plant Physiology, 33*, 155–170.

Yoshida, S. (1981). Physiological analysis of rice yield. In *Fundamentals of rice crop science* (pp. 231–251). Los Banos: International Rice Research Institute.

Yoshioka, M., Uchida, S., Mori, H., Komayama, K., Ohira, S., Morita, N., Nakanishi, T., & Yamamoto, Y. (2006). Quality control of photosystem II: Cleavage of reaction center D1 protein in spinach thylakoids by FtsH protease under moderate heat stress. *The Journal of Biological Chemistry, 281*, 21660–21669.

You, L., Rosegrant, M. W., Wood, S., & Sun, D. (2009). Impact of growing season temperature on wheat productivity in China. *Agricultural and Forest Meteorology, 149*, 1009–1014.

Yruela, I. (2005). Copper in plants. *Brazilian Journal of Plant Physiology, 17*, 145–256.

Yue, M., Li, Y., & Wang, X. (1998). Effects of enhanced ultraviolet-B radiation on plant nutrients and decomposition of spring wheat under field conditions. *Environmental and Experimental Botany, 40*, 187–196.

Zabihi-e-mahmoodabad, R., Jamaati-e-somarin, S., Khayatnezhad, M., & Gholamin, R. (2011). Effect of cold stress on germination and growth of wheat cultivars. *Advances in Environmental Biology, 5*, 94–97.

Zahedi, H., Noormohammadi, G., Rad, A. H. S., Habibi, D., & Boojar, M. M. A. (2009). Effect of zeolite and foliar application of selenium on growth, yield and yield component of three canola cultivar under conditions of late season drought stress. *Notulae Scientia Biologicae, 1*, 73–80.

Zeng, F. R., Zhao, F. S., Qiu, B. Y., Ouyang, Y. N., Wu, F. B., & Zhang, G. P. (2011). Alleviation of chromium toxicity by silicon addition in rice plants. *Agricultural Sciences in China, 10*, 1188–1196.

Zhang, B., Liu, W., Chang, S. X., & Anyia, A. O. (2010). Water deficit and high temperature affected water use efficiency and arabinoxylan concentration in spring wheat. *Journal of Cereal Science, 52*, 263–269.

Zhang, J. H., Huang, W. D., Liu, Y. P., & Pan, Q. H. (2005). Effects of temperature acclimation pretreatment on the ultrastructure of mesophyll cells in young grape plants (*Vitis vinifera* L. cv. Jingxiu) under cross-temperature stresses. *Journal of Integrative Plant Biology, 47*, 959–970.

Zhou, X., Yang, L., Qi, Y., Guo, P., & Chen, L. (2015). Mechanisms on boron-induced alleviation of aluminum-toxicity in *Citrus grandis* seedlings at a transcriptional level revealed by cDNA-AFLP analysis. *PloS One, 10*(3), e0115485. doi: 10.1371/journal.pone.0115485.

Zhu, J. K. (2007). Plant salt stress. In A. O'Daly (Ed.), *Encyclopedia of life sciences* (pp. 1–3). Chichester: Wiley.

Zlatev, Z., & Lidon, F. J. C. (2012). An overview on drought induced changes in plant growth, water relations and photosynthesis. *Emirates Journal of Food and Agriculture, 24*, 57–72.

Zlatev, Z. S., Lidon, F. J. C., & Kaimakanova, M. (2010). Plant physiological responses to UV-B radiation. *Emirates Journal of Food and Agriculture, 24*, 481–501.

Chapter 11
The Rhizosphere and Plant Nutrition Under Climate Change

Tarek Alshaal, Hassan El-Ramady, Abdullah H. Al-Saeedi, Tarek Shalaby, Tamer Elsakhawy, Alaa E.-D. Omara, Aliaa Gad, Eman Hamad, Ayman El-Ghamry, Ahmed Mosa, Megahed Amer, and Neama Abdalla

Abstract The plant root–soil interfaces could be considered the rhizosphere area, which is the most important active zone in the soils for different microbial activities, biodegradation of pollutants and plant nutrition. Polluted soils are characterized by low organic matter content, limiting their microbial activity, nutrient availability and degradation of pollutants. Soil phyto- and/or bioremediation is mainly based on the use of plant roots and their associated soil microorganisms, whereas conventional approaches are based on physico-chemical methods in soil remediation. Plant root exudates are the most important compounds in the rhizosphere, which play a crucial role in the interactions between plant roots and soil microbes. It is worthy to mention that several plant species and soil microbes have been used in soil remediation for

T. Alshaal • H. El-Ramady (✉)
Soil and Water Department, Faculty of Agriculture,
Kafrelsheikh University, Egypt
e-mail: alshaaltarek@gmail.com; ramady2000@gmail.com

A.H. Al-Saeedi
College of Agricultural and Food Sciences, King Faisal University,
Al-Hassa 31982, Saudi Arabia
e-mail: 2012ahs@gmail.com

T. Shalaby
Horticulture Department, Faculty of Agriculture, Kafrelsheikh University,
Kafr El-Shaikh, Egypt

College of Agricultural and Food Sciences, King Faisal University,
Al-Hassa 31982, Saudi Arabia
e-mail: tashalaby@yahoo.com

T. Elsakhawy • A.E.-D. Omara • A. Gad • E. Hamad
Agriculture Microbiology Department, Soil, Water and Environment Research Institute (SWERI), Sakha Agricultural Research Station, Agriculture Research Center (ARC),
Giza, Egypt
e-mail: drelsakhawyg@gmail.com; alaa.omara@yahoo.com; aliaa_ali_94@yahoo.com

M. Naeem et al. (eds.), *Essential Plant Nutrients*,
DOI 10.1007/978-3-319-58841-4_11

different pollutants. The role of rhizosphere and its significance in plant nutrition are mainly controlled by the change in climatic attributes including temperature, moisture content, precipitation, etc. Therefore, global warming and climate changes do have a great and serious effect on the agricultural production through their effects on the rhizosphere and in turn plant nutrition. Hence, the aim of this review is to evaluate the significance of rhizosphere in plant nutrition under the changing climate. Soil biological activity and its security will be also highlighted.

Keywords Plant nutrition • Rhizosphere • Climate change • Root exudates • Pollution

11.1 Introduction

It is well known that the rhizosphere is a very complex, important zone surrounding the plant roots (Otero et al. 2015; Kumar et al. 2016a; Jia et al. 2017), which is responsible for the diversity and structure of soil microbial community (Oyelami and Semple 2015; Schlich and Hund-Rinke 2015; Jia et al. 2017; Li et al. 2017) and represents the great and vital soil microbial activity, where different soil microbes are classified into three groups, namely, passive, negative and neutral microbes (Baetz 2016; Mohammadi et al. 2017). The most important and beneficial soil microbes include plant growth-promoting microbes such as *Azotobacter*, pseudomonads, phosphate solubilizers and AM fungi (Yazdani et al. 2009; Noori and Saud 2012; Sabannavar and Lakshman 2009; Kumar et al. 2016a; Verbon and Liberman 2016; Vergani et al. 2017). Several investigations have been conducted about the diversity of microbial community and its beneficial aspects as well as the economic importance of agricultural crops including maize, wheat, rice, oilseed rape, vegetables and medicinal plants (Solanki et al. 2011; Kumar et al. 2013; Kumar et al. 2016a; Čerevková et al. 2017; Chen et al. 2017b; Hernández et al. 2017).

Due to anthropogenic activities, environmental pollution has occurred, which includes soil (Imadi et al. 2016; Liu et al. 2017b; Huang et al. 2017b), water (Han

A. El-Ghamry • A. Mosa
Soil and Water Department, Faculty of Agriculture, Mansoura University, Egypt
e-mail: ame.egypt@gmail.com; ahmedmosa@mans.edu.eg

M. Amer
Soils, Water and Environment Research Institute (SWERI), Agricultural Research Center, Giza, Egypt
e-mail: dr_megahed120@yahoo.com

N. Abdalla
Plant Biotechnology Department, Genetic Engineering Division, National Research Center, Dokki, Giza, Egypt
e-mail: neama_ncr@yahoo.com

et al. 2016; Saha et al. 2017; Smith et al. 2017), sediment (Bertrand et al. 2015; Gao et al. 2016; Gubelit et al. 2016; Palanques et al. 2017), air (Achakzai et al. 2017; Fu and Gu 2017; Sammarco et al. 2017), etc. Large areas from agricultural lands around the universe have deteriorated due to soil pollution, which leads to a decline in soil biodiversity and decrease in crop productivity and poses several risks to human health (Lu et al. 2015; Balkhair and Ashraf 2016; Zhang et al. 2017d). It is believed that climate changes have direct and indirect effects on plant–soil–microbe–pollutant interactions and on the remediation of polluted soils (Abhilash et al. 2013a, b; Tripathi et al. 2015; Bojko and Kabala 2017), which means that elevated CO_2 may increase the photosynthetic rate of plants thereby enhancing its productivity, the exudation of roots and the rhizospheric soil microbial activity (Gömöryová et al. 2013; Abhilash and Dubey 2014; Abhilash et al. 2015; Duan et al. 2015; Xu et al. 2016; Lee and Kang 2016; Huang et al. 2017a; Xue et al. 2017) and also a decrease in soil pH as well as an increase in the soil content of dissolved organic matters, which will accelerate the secretion of plant root that enhances the availability of pollutants in the soil for uptake by plants as well as the degradation by soil microbes (Kim and Kang 2011; Rajkumar et al. 2013; Tripathi et al. 2015; Guarino and Sciarrillo 2017). Therefore, further investigations are essential to understand the nexus of plant and microbe pollutants in soils under different climatic change conditions (Rajkumar et al. 2013; Tripathi et al. 2015; Yadav et al. 2015).

Therefore, the main target of this review is to collect different links between climate changes and their impacts on the rhizosphere zone on one hand and different soil biological processes for plant nutrition on the other hand. The rhizosphere also will be highlighted including the effects of pollutants, root exudates and plant nutrition as well as the security of soils.

11.2 The Rhizosphere: More than a Treasure for Plant Nutrition

The rhizosphere is a place where the most important plant–soil interactions happen including beneficial, antagonistic or neutral interactions (Baetz 2016). It is a real treasure in plant nutrition due to its significance in fate, behaviour and uptake of nutrients by plants (Kayler et al. 2017; Rugova et al. 2017). Several biological and ecological processes or transformations take place in the rhizosphere area that control microbial activity and plant growth as well as nutrient uptake by plants (Callesen et al. 2016; Huang et al. 2016). Processes in the rhizosphere definitely differ in the bulk soil (Ibekwe et al. 2017; Rugova et al. 2017). The rhizosphere area has very dynamic interfaces between (1) soil and plant roots; (2) plant root systems, soil and their microorganisms; and (3) soil, plant roots and invertebrates (e.g. Bais et al. 2006; Hartmann et al. 2008; Singh et al. 2016; Rugova et al. 2017). The fate and behavior of organic compounds, which released by the plant roots and soil microorganisms is controlled by the rhizosphere conditions (Cai et al. 2017). The rhizosphere area is very rich in several compounds including low-molecular-weight root exudates and high-molecular-weight humic substances, such as mucilage or polysaccharides and proteins (Jha et al. 2015). Low-molecular-weight compounds are a water-soluble fraction that include various secondary metabolites like flavonoids, glucosinolates and terpenoids (Jha et al. 2015; Baetz 2016), gaseous molecules like CO_2 and H_2, inorganic ions (e.g. HCO_3^-, OH^-, H^+), amino acids (Zhang et al. 2017a), organic acid anions and carbohydrates (Haichar et al. 2014; Montiel-Rozas et al. 2016; Rugova et al. 2017).

The plant root exudates represent a wide range and complex mixture of hundreds of organic compounds, which could be classified into different categories based on their molecular weight and solubility in water (Strickland et al. 2012; Huang et al. 2014; Zhang et al. 2014; Mommer et al. 2016). Moreover, root exudates play great roles in the amelioration of plant stresses (Doornbos et al. 2012; Baetz and Martinoia 2014; Nalam and Nachappa 2014; Xu et al. 2015; Baetz 2016; Dubrovskaya et al. 2017), in the increase of soil microbial activity (Singh and Mukerji 2006; Nannipieri et al. 2008; Mommer et al. 2016; Swamy et al. 2016; Shcherbakova et al. 2017), in the biodegradation of pollutants (Liu et al. 2015; Jha et al. 2015; Hou et al. 2016; Jia et al. 2016; Verbon and Liberman 2016; Dubrovskaya et al. 2017; Vergani et al. 2017) and in plant nutrition (Neumann 2007; Cesco et al. 2010, 2012; Chen et al. 2016; Meier et al. 2017). Therefore, they have several functions in the beneficial ecological interactions with soil microbial communities in the rhizosphere, including interactions between plants and soil microorganisms (Chen et al. 2016; Lareen et al. 2016) and plant–plant communication (Mommer et al. 2016; Cai et al. 2017), as well as the tripartite interactions (Haichar et al. 2014; Hardoim et al. 2015; Baetz 2016; Vergani et al. 2017).

Therefore, rhizosphere is not only a real treasure for the nutrition of plants but also a very important area in the protection of plants through the action of plant root exudates. Day by day, a lot about the rhizosphere have been newly discovered due to promising "omics" tools that aid in further understanding different interactions among plants, soil and their microbes for sustainable production and productivity in agriculture.

11.2.1 The Plant Root Exudates and Plant Stresses

It is well documented that about 40% from photosynthetically fixed carbon may be released as a plant root exudate (Doornbos et al. 2012). Many biotic stressors on plant roots within the rhizosphere area have been reported including plant diseases or pathogens (e.g. virus, fungi, bacteria), insects and nematodes (Baetz and Martinoia 2014; Alexander et al. 2016). Furthermore, several plant stresses including abiotic (e.g. soil salinity, drought, flooding, freezing, pollutants like trace elements and/or petroleum, crude oil) and biotic stresses (attack by herbivores or pathogens) hugely damage cultivated plants. So plant root exudates represent one of the most important belowground plant defences (Bruce and Pickett 2007; Doornbos et al. 2012; Nalam and Nachappa 2014; Baetz 2016; Le Fevre and Schornack 2016; Johnson et al. 2016; Zuverza-Mena et al. 2017; Arpaia et al. 2017). These phytochemicals or plant root exudates have a direct or indirect potential to protect plants against pathogens. It is reported that some plant root exudates have the ability to mimic or behave like pesticides or antibacterial or antifungal agents and are released from the damaged roots to protect the plants (Baetz 2016).

Some plant root exudates also attract beneficial soil microorganisms, which release secondary metabolites like antibiotics that protect plants against soilborne diseases (Baetz 2016). Several studies have been published regarding root exudates and their roles in plant stresses or belowground plant defence (De Coninck et al. 2015; Alexander et al. 2016; Baetz 2016; Johnson et al. 2016; Le Fevre and Schornack 2016; Plouznikoff et al. 2016; Vos and Kazan 2016; Dubrovskaya et al. 2017). Conclusively, several strategies are employed by plants as belowground defence against biotic and abiotic stresses. There is an urgent need for a comprehensive understanding of the interactions between plant roots and pathogenic and beneficial microorganisms in order to increase crop productivity and minimize crop losses (Vos and Kazan 2016).

11.2.2 The Plant Root Exudates and Soil Microbial Activity

Soil microbial activity is the main source of the vital activities in soil besides plant root exudations. Hence, soil microbial populations in the rhizosphere have different and continuous flow of both low- and high-molecular-weight organic substances derived from plant roots, which control soil microbial activity and different community structures (Nannipieri et al. 2008). Soil microbial activity and its diversity in the rhizosphere have been extensively studied in numerous books, chapters and articles (e.g. Bashir et al. 2016; Hakeem and Akhtar 2016; Hakeem et al. 2016; Kumar et al. 2016a; Swamy et al. 2016; de Medeiros et al. 2017). These studies depend mainly on the interaction between cultivated plants and rhizosphere properties, which include:

1. Effects of applied pesticides (e.g. Álvarez-Martín et al. 2016; Franco-Andreu et al. 2016; Lv et al. 2017; Mauffret et al. 2017)
2. Effects of soil earthworm in presence of plants (Aghababaei et al. 2014; Aghababaei and Raiesi 2015; Lv et al. 2016; Zhang et al. 2016a; Kim et al. 2017; Liu et al. 2017b)
3. Effects of soil pollution including organic and inorganic pollutants (Parelho et al. 2016; Hansda et al. 2017; Le et al. 2017; Tong et al. 2017; Wang et al. 2017b)
4. Effects of different soil characterizations like soil organic matter, soil redox potential, etc. (Khan et al. 2016; Su et al. 2017; Xiao et al. 2017)
5. Effects of climate change (Bojko and Kabala 2017; Zhang et al. 2016b, 2017b)
6. Effects of plant characterization (Li et al. 2017; Mohammadi et al. 2017; Zhang et al. 2017c)
7. Effects of application of soil amendments and fertilizers (Meena et al. 2016; Abad-Valle et al. 2017; Wang et al. 2017c, d)
8. Effects of applied nanomaterials (Nogueira et al. 2012; Oyelami and Semple 2015; Schlich and Hund-Rinke 2015; Schlich et al. 2016; Liang et al. 2017)
9. Effects of different stresses (Cheng et al. 2016; Lee and Kang 2016; Xue et al. 2017; Wu et al. 2017)

10. Effects of tillage and other agricultural practices (Kabiri et al. 2016; Tautges et al. 2016; León et al. 2017; Yuan et al. 2017)
11. Effects of transgenic plants on microbial diversity in the rhizosphere soil (Chaudhry et al. 2012; Canfora et al. 2014; Sahoo et al. 2015; Turrini et al. 2015; Guan et al. 2016; Arpaia et al. 2017)

Therefore, there is a crucial understanding for the principles of interaction among microbes–plants and microbes–microbes. Hence, the communication among microbes and plants could provide the researchers with the most important and beneficial microbial populations for agricultural sectors. Moreover, the associated plants with microbial communities can be considered essential for both plant growth and its health (Berg et al. 2014; Mueller and Sachs 2015; Cai et al. 2017). A great progress has been achieved regarding different interactions between plant and soil microbes and their rhizosphere microbiomes (Berendsen et al. 2012; Ofek et al. 2014; Ai et al. 2015; Mueller et al. 2016; Cai et al. 2017). These interactions could be characterized by (1) the competition between plants and soil microorganisms for nutrients, (2) the dynamics of nutrients and their function in the rhizosphere, (3) the activity of enzymes in the rhizosphere, (4) soil respiration and its rates and (5) the biodiversity of microorganisms in the rhizosphere (Nannipieri et al. 2008; Duan et al. 2015; Cavagnaro et al. 2016; Zhu et al. 2017).

Therefore, the microbial activity in the rhizosphere mainly depends on cultivated plant species because of the different compositions in plant root exudates. Rhizosphere microbes could be classified into three groups in accordance to their interactions with plants: (1) microbes that have negative effects on plants, e.g. phytopathogens; (2) microbes that have positive effects on plants, i.e. promotion of plant growth and symbiosis; and (3) microbes that have neutral effects or no benefits. Symbiosis represents the most important interaction between plants and rhizobia, arbuscular mycorrhizal fungi (AMF), plant growth-promoting rhizobacteria (PGPR) and plant growth-promoting endophytic fungi (PGPF). These symbioses or beneficial microbial inoculums are mainly used in producing sustainable agri-biotechnological products including biopesticides, biofertilizers, bioremediators and phyto-stimulators. It is worthy to mention that plant roots can modify the soil microbial structure and its communities. The best approach to study these modifications can be achieved through using specific compounds occurring in root exudations. Several environmental factors including soil pH, light intensity, soil temperature, soil aeration, soil texture, nutrient status and soil microbes could be considered as the main qualitative and quantitative factors controlling the composition of root exudates.

11.2.3 The Plant Root Exudates and Biodegradation of Pollutants

It is estimated that about 25 and 44% from global soils are highly and moderately degraded, respectively, because of different pollutants including metals, metalloids, persistent organic pollutants, pesticides, radionuclides, etc. (Tripathi et al. 2015). Plant roots and their biochemistry, biology and genetic evolution have been intensively reported during the last decades, but some rhizosphere processes for these roots including root exudations and root border cells are not well understood (Bashir et al. 2016). The plant root system and its engineering were and are still one of the most important topics in plant sciences including the rhizosphere and its microbiomes as well as metabolomics for plant growth and its health (Berendsen et al. 2012; Chapman et al. 2012; Abhilash and Dubey 2015; Mueller and Sachs 2015; Dessaux et al. 2016; van Loon 2016; van Dam and Bouwmeester 2016; Freund and Hegeman 2017). Furthermore, the rhizosphere plays a great and significant role in enhancing both phytoremediation and biodegradation processes of different pollutants (Liu et al. 2017e) (Fig. 11.1).

Soil pollution was and is still one of the challenges facing the universe, and day by day, new approaches are needed for remediation processes. The major sources of pollutants or xenobiotic compounds include (1) agrochemicals (e.g. pesticides, mineral fertilizers), (2) industrial pollutants (pollutants from chemical, petrochemical and pharmaceutical industries), (3) mining wastes and (4) chlorinated organic compound (Jaiswal and Verma 2016). The degradation of different pollutants in the rhizosphere has been conducted using terrestrial plants and their surrounding soils (Alvarez et al. 2012; Bertrand et al. 2015; Kahlon 2016; Jaiswal and Verma 2016;

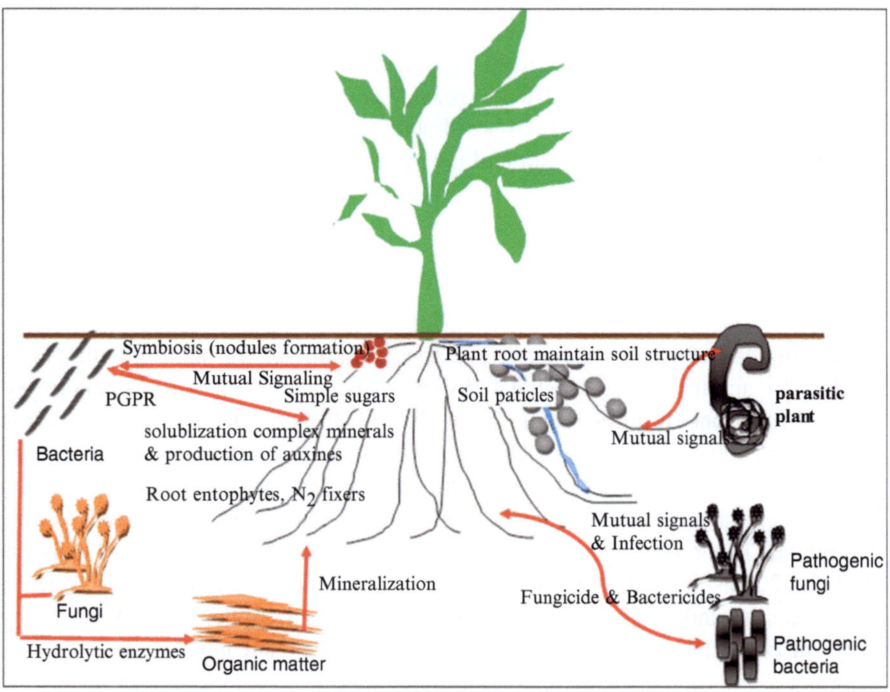

Fig. 11.1 The rhizosphere and its importance for plant nutrition starting from root exudates to the role of plant growth-promoting *Rhizobium* (PGPR), the role of different soil microbial communities, the potential of soil organic matter in this area and effects of plant–soil–microbe interactions on the uptake of nutrients by plant roots

Guarino and Sciarrillo 2017). Conventional approaches in soil remediation are based on the physico-chemical methods, whereas the bioremediation techniques are based on plant roots and their associated microorganisms. Furthermore, bioremediation is the most promising, cost-effective, efficient and more sustainable technology (Jaiswal and Verma 2016). Common soil bioremediation process includes removing organic pollutants (e.g. polycyclic aromatic hydrocarbons or pesticides) and heavy metals as presented in several studies, for example, by Lee et al. (2012), Edwards and Kjellerup (2013), Masciandaro et al. (2013), Winarso et al. (2016), Alvarez et al. (2017), Liu et al. (2017e) and Jia et al. (2017).

Remediation of pollutants by plant root exudates, or rhizodegradation, includes three aspects: (1) many enzymes released from roots and other materials, which participate as organic pollutants (Sun et al. 2013; Kvesitadze et al. 2015; Tripathi et al. 2015; Tiwari et al. 2017), (2) root exudates that increase the bioavailability of pollutants as surfactants by reducing their sorption (Mitton et al. 2012; Jia et al. 2016; Guo et al. 2017) and (3) plant root exudates that enhance the growth of soil microbes and then promote soil microbial communities (Meng and Zhu 2011; Liu et al. 2017b; Guo et al. 2017). Due to the difficulty of the remediation process in soil contaminated with multiple pollutants, more advanced studies should be conducted to properly understand the degradation and accumulation of these pollutants using the omics technology (e.g. genomic, metagenomic, proteomic, metabolomic approaches). Definitely, the use of suitable plants in soil remediation is the most important factor in controlling the phyto-uptake of pollutants from soils and enhancing soil microbial activity (Tripathi et al. 2015; Jia et al. 2016; Guo et al. 2017). Therefore, interactions between selected plants and associated microbes should be kept in mind in sustainable remediation for soils contaminated with multiple pollutants (Tripathi et al. 2015).

11.2.4 The Plant Root Exudates and Plant Nutrition

As mentioned before, the rhizosphere is an essential zone for plant growth and its production, and it also plays a vital and significant role in the nutritional and physiological functions of crops (Kumar et al. 2016a). In general, there is a very close relationship between plant roots and their exudations and plant nutrition. There are direct and indirect effects of plant root exudates on the nutrition of plants including the uptake, transformation, translocation and accumulation of different soluble compounds in the rhizosphere (Neumann 2007; Nannipieri et al. 2008; Doornbos et al. 2012; Bashir et al. 2016; Liu et al. 2017c; Meier et al. 2017). Moreover, the bioavailability and uptake of various nutrients in the rhizosphere are influenced by different characterizations of soil, plant root exudates and nutrient speciation as well as the microbial interactions with both soil and plants (Kumar et al. 2016a; Rashid et al. 2016; Nguyen et al. 2017). The nutrient transfer from plant cells to microbial cells and its dynamics are mainly impacted by the physiological aspects of the microbe–plant interactions (Munroe et al. 2015; Callesen et al. 2016; Das et al. 2017; Nietfeld et al. 2017). The diffusion of different nutrients and their supplements in the rhizosphere are mainly controlled by the gradient in diffusion of nutrients between soil surfaces and root surfaces (Kumar et al. 2016a).

Under the frame of plant nutrition, the complex interactions among plants, soils and microbes should be understood. These interactions could control the uptake, mobilization, transfer, translocation, accumulation of nutrients and their bioavailability (Lambers et al. 2009; Miki 2012; Kumar et al. 2016a) and biogeochemistry cycles and formation of soils in the rhizosphere or terrestrial ecosystem (Maheshwari et al. 2012; Perlatti et al. 2016). These interactions also include all botanical characterization, which—in turn—alters the growth and reproduction of plants (Miki 2012; Compant et al. 2016). On the other hand, several compounds called root exudates have been identified from the rhizosphere zone, which have many roles in plant nutrition:

1. Acquisition of both nutrients and water through rhizosphere modification and soil with mucilage, fetchers (i.e. seeking and fetching like phytosiderophores), modifiers (i.e. rhizosphere modification with protons and reductants) and ectoenzymes (i.e. converting unusable organic forms into usable ones like phosphatases)
2. Protection against physical stress and pathogens through amelioration of the rhizosphere, modification of soil interfaces and production of both phytoalexins and antibiotics
3. Protection against toxic elements and competition through complexation of toxic elements like aluminium or sequestering like sodium as well as the root exudates that could modify the rhizosphere through phyto-actives like allelochemicals
4. Establishment of symbiotic relationships with microbes such as *Rhizobium*, mycorrhiza (AMF), *Azotobacter*/*Azospirillum* through nitrogen fixation and phosphorus and mineral uptake (Maheshwari et al. 2012).

Therefore, plant root exudates are one of the most important sources for plant nutrition, which help plants directly or indirectly through promoting soil microbial activity and protecting them against pathogens. Plant nutrition processes should start from the rhizosphere and its plant root exudates. Hence, further researches are essential for investigating the role of plant root exudates in the nutrition and protection of plants.

11.3 Soil Enzymes in Rhizosphere and Their Role for Plant Nutrition

It is well known that soil enzymes are the most important soil quality indicator, which play a role in the transformation of elements in the soil and biochemical functions in the dynamics of organic matters in soils (Bakshi and Varma 2017; Maddela et al. 2017). The main sources of soil enzymes include soil animals, dead and living microbes, plant roots and plant and animal residues (Table 11.1).

These soil enzymes could be released through extracellular enzymes from microbial activities and plant roots in the soil. The main function of soil enzymes is catalysing many reactions including the cycling of nutrients, the decomposition of organic

Table 11.1 Some important soil enzymes and their functions in both nutrient cycling and soil organic matter decomposition

Enzyme	Substrate	Enzyme reaction	Significance of enzyme
I. Nutrient cycling			
Amidase	Carbon and N compounds	N-mineralization	Plant available NH_4^+
Phosphatase	Phosphorous	Release of PO_4^{3+}	Plant available P
Sulphatase	Sulphur	Release of SO_4^{2-}	Plant available S
Urease	Nitrogen	Release of NH_3 and CO_2	Plant available NH_4^+
II. Decomposition of organic matter			
β-glucosidase	Carbon compounds	Cellulose hydrolysis	Energy for microorganisms
Fluorescein diacetate (FDA) hydrolysis	Organic matter	Carbon and various nutrients	Energy for microorganisms as measure of microbial biomass

Source: Yang et al. (2012)

Table 11.2 Effects of climatic zone and soil characterization on soil biological, physical and chemical properties in three cities: Germany, Hungary and Egypt

Element or parameter	Germany (Braunschweig)	Hungary (Kolontar)	Egypt (Kafr El-Sheikh)
Chemical soil properties			
Soil EC (dS m^{-1})	1.09	1.25	3.47
Soil pH (1:2.5 suspension)	5.2	7.86	7.88
Physical soil analysis			
Soil texture	Loam	Loam	Silty loam
Biological soil assessments			
Microbial counts (CFU)			
Bacteria	4.1×10^6	5.40×10^6	5.1×10^7
Actinomycetes	1.3×10^5	3.37×10^6	3.3×10^6
Fungi	1.4×10^4	1.45×10^5	1.6×10^6
Dehydrogenase activity (μg TPF g^{-1} DM d^{-1})	53.5	55.21	6.91
Alkaline phosphatase activity (μg p-NP g^{-1})	191	0.53	35.5
Climatic zone	Temperate seasonal climate	Continental climate	Arid climate

EC soil salinity, *TPF* triphenylformazan, *p*-NP *p*-nitrophenol
Sources: Germany (El-Ramady 2008), Hungary (Alshaal et al. 2013) and Egypt (El-Ramady et al. 2014a)

wastes and the formation of organic matter (Burke et al. 2011; Mele 2011; Yang et al. 2012; Gul and Whalen 2016). Due to their important roles in nearly all biological activities in soils, soil enzymes need more better understanding of their roles in agro-ecosystems as well as the proper response under different global changes (Mele 2011; Yang et al. 2012; Burns et al. 2013; San Emeterio et al. 2016; Allen and Allen 2017).

It is worthy to mention that soil enzymes are the main driving force in the circulation of materials and the flow of energy in the agroecosystems (Shan et al. 2008; Yang et al. 2012). There are several factors influencing the activity of soil enzymes including soil texture, soil fertility, plant diversity and its community composition, temperature, soil pH, soil salinity (EC), soil organic matter, soil enzyme synthesis and secretion, etc. (Table 11.2). Therefore, it would be so helpful to achieve an appropriate strategy in order to conserve the biodiversity of soil in a sustainable manner, and change detection studies also should be performed (Jain and Pandey 2016).

Several investigations have been conducted concerning the relationship between soil enzymes and plant nutrition or soil fertility (Yang et al. 2012; Burns et al. 2013; Prasanna et al. 2016; Allen and Allen 2017; Guangming et al. 2017; Touceda-González et al. 2017; Wang et al. 2017a). Therefore, soil enzymes and their activities are close to different properties of soils and global environmental changes. These enzymes could be considered also very important indicators for both soil quality and its biological activities. Hence, it is a crucial issue to investigate different factors influencing the activities of soil enzymes.

11.4 The Biological Activities in Rhizosphere Under Climate Change

Many soil biological activities have been recorded as indicators for soil quality including soil enzymes, soil organic matter, soil microbial biomass, soil respiration and potentially mineralizable N (Allen et al. 2011; Bojko and Kabala 2017; Xue et al. 2017). They are mainly controlled by environmental conditions including temperature, water potential and pH and likely to be responsive to atmospheric warming and more frequent and extreme variations in precipitation patterns (Burns et al. 2013). Therefore, climate changes will have important effects to different functions of agroecosystem including nutrient cycling and different plant–microbe interactions, which will ultimately affect the productivity and net balance of carbon (Burns et al. 2013; Dutta and Dutta 2016; Zhang et al. 2016b; Bojko and Kabala 2017). Due to climate change and global warming, many global environmental problems arise such as high concentration of CO_2, elevated global temperature, sea level rise and some extreme weather events including droughts, rainfall and flooding (Coffey et al. 2016; Dutta and Dutta 2016; Sofi et al. 2016; Ennigrou et al. 2017). Thus, climate changes could be characterized mainly through fluctuations in both global temperature and moisture including elevated CO_2 and temperature as well as sea level rise (Drigo et al. 2008; French et al. 2009; Balser et al. 2010; Todd-Brown et al. 2012; Park et al. 2014; Saleem et al. 2015; Dutta and Dutta 2016;). Therefore, changing of climate could affect soil biological activities.

Many issues on soil biological activity and changing climate should be addressed including different impacts of global change on soil biota, environmental stress and its effects on soil enzymes under elevated temperature and CO_2. Different strategies should be mitigated for soil biological activities under global change. Facts and perspectives are listed as follows (Dutta and Dutta 2016):

1. Nearly 3.1×10^5 kg of carbon can be found in the soil representing more than the two-thirds from the total C in the terrestrial environments (Davidson et al. 2006).
2. About 10% of the atmospheric carbon dioxide (CO_2) annually passes through the soil.
3. Every year, 5–7.5 Mg of carbon could be emitted or released from the terrestrial surface flux as a carbon dioxide.
4. Every year, about 1.2×10^{14} kg of C is taken up by the soil autotrophic microbes, whereas the soil heterotrophic microbes cumulatively emit about 1.19×10^{14} kg of C (Singh et al. 2010).
5. The global CO_2 content in the atmosphere recorded 380 and 407.69 ppm in 2005 and 2017, respectively, as below:

Atmospheric CO_2 content	Dec. 1983	Dec. 1993	Dec. 2003	Dec. 2013	Dec. 2014	Dec. 2015	Dec. 2016	Jan. 2017
Part per million (ppm)	343.05	356.91	375.99	396.84	398.91	401.85	404.48	406.07

Source: https://www.co2.earth/annual-co2 accessed on 1 February 2017

6. The amount of nitrous oxide (N_2O) represents more than one-third from the greenhouse gases produced from the nitrification and denitrification from agricultural lands (IPCC 2007).
7. The emission rate ranges from 1 to 5% N_2O and deposition rate 10–50 kg N_2O from each 1000 kg fertilizer (Singh et al. 2010).
8. Soil biological processes may produce some greenhouse gases including CH_4, N_2O and CO_2 beside the soil represents the pool for C in the terrestrial lands.
9. The annual natural emission rate of methane (NH_4) is nearly 250 million tons produced by rumens, termite, oceans and paddy rice (Singh et al. 2010).
10. It is predicted that the sea surface temperature or the global warming will be increased from 1.18 to 6.48 °C under low and high CO_2 emission, respectively (Meehl et al. 2007; Dutta and Dutta 2016).
11. The global annual amount of C produced from the decomposition of soil microbes is 7.5–9 times the anthropogenic emissions (Crowther et al. 2015).
12. The annual production of CO_2 from decomposition of plant organic matter, microorganisms in soil, is 5.5×10^{12} kg eight times more than that by humans (Zimmer 2010; Dutta and Dutta 2016).
13. Climate changes could cause some modifications and complex changes in the soil microbial communities (Castro et al. 2010) and their diversity (Shade et al. 2012; Zhang et al. 2016b; Zhang et al. 2017c).
14. Climate change (i.e. elevated temperature under different levels of soil moisture) could increase soil respiration, which is very sensitive to changes in climate including soil temperature and moisture as well as precipitation (Aanderud et al. 2013; Brye et al. 2016; Feng et al. 2017).

15. Different fluctuations in soil moisture and its temperature can change soil biological activities including enzymatic and microbial activities (Schimel et al. 2007; Rezanezhad et al. 2014; Daou et al. 2016).
16. Climate changes including drought, flooding, freezing stresses as well as rainfall could cause the variability in soil moisture and its respiratory activity (Aanderud et al. 2011, 2013; Classen et al. 2015a, b; Zhang et al. 2017e).
17. The elevated atmospheric CO_2 could increase the photosynthetic rate of plants, hence accelerating soil biological activities, and carbon availability for soil microbes enhancing soil respiration which in turn releases carbon to the atmosphere (Bardgett et al. 2008; Tao et al. 2017).
18. Due to the importance of soil carbon, the Framework Convention on Climate Change (COP21) in Paris (2015) has adopted an initiative called "four per thousand". This initiative is already planned to enhance the content of soil organic carbon in world soils at the rate of 0.4% per year to a 40 cm soil depth (Lal 2016; Minasny et al. 2017).
19. The strategy of this initiative depends on promoting the soil organic C through adoption of some recommended agro-management practices including restoration of degraded soils, conservation agriculture, agroforestry, improved grazing, cover cropping, biochar, etc. (Lal 2016; Habtemariam et al. 2017).
20. The main target of the "four per thousand" proposal that includes the concept of both soils and agriculture are the main solutions in handling the global issues of changing climate, the insecurity of foods and the environmental pollution (Lal 2016; Lozano-García et al. 2017).
21. It is reported that the global temperature will increase by 1 °C in the next 25 years (by the end of 2040) and 2–4 °C degrees by the end of 2100 (IPCC 2014; McCarl 2015, 2017).

Therefore, the changing in climate could directly and indirectly impact on the soil biological activities through the effects of elevated CO_2, global warming, fluctuation in precipitation (flooding and drought), etc. Thus, different global and national strategies in dealing with this changing in climate should be mitigated to promote more soil biological activities.

11.5 Plant Nutrition Under Climate Change

No doubt that soil fertility and plant nutrition were and are still one of the most important fields in agricultural sciences (El-Ramady et al. 2014a, b). Therefore, due to the limitations in both natural water and land resources as well as the need for improving the quantity and quality of food supply worldwide, a new dimension for the importance of plant nutrition is needed. Under global climate change, there is a great and serious threat for the global agricultural production. Hence, several issues should be addressed under the umbrella of plant nutrition including environmental sustainability, cropping systems, climate changes and human health and well-being (El-Ramady et al. 2014b; Lal 2016; Tao et al. 2017). Furthermore, the relationship between changing of climate and plant nutrition is very complex, where all plant growth stages including development, biochemistry, physiology and yield of plants are influenced by climatic conditions (El-Ramady et al. 2015a, b; Lal et al. 2015; Wang et al. 2017b). In more details, indirect effects of climate changes on plants include many changes in botanical processes such as the structure, function of plant roots and its biodiversity as well as activity of rhizosphere microbes (El-Ramady et al. 2015b; Short et al. 2016; Ferrarini et al. 2017; Zhang et al. 2017e). It is worthy to mention that climate is the main factor controlling the biogeochemistry, transformation, uptake and behaviour of plant nutrients in the rhizosphere (El-Ramady et al. 2015b; Bilton et al. 2016; Huang et al. 2017a).

More than 1132 books have been published on Springer websites of which 52 are already published on January 2017. The most important published books by Springer are listed as follows:

1. Chen et al. (2017a): *Handbook of Climate Change Mitigation and Adaptation*
2. Kolokytha et al. (2017). *Sustainable Water Resources Planning and Management Under Climate Change*
3. Uitto et al. (2017). *Evaluating Climate Change Action for Sustainable Development*
4. Ahmed and Stockle (2017). *Quantification of Climate Variability, Adaptation and Mitigation for Agricultural Sustainability*
5. Salawitch et al. (2017). *Paris Climate Agreement: Beacon of Hope*

Several studies have been conducted regarding the effects of climate changes (e.g. elevated CO_2 and temperature) on plant nutrition such as the studies by Bielenberg and BassiriRad (2005), Abenavoli et al. 2012; Iversen and Norby (2014), Aljazairi and Nogués 2015; Wu and Malmström (2015), Gruwez et al. (2016), Wu et al. (2016), etc. These studies include different topics or issues like effects of climate change on root morphology and its architecture, changes in the kinetics and uptake of nutrients by roots and effects of elevated CO_2 or temperature on many plant physiological processes including nutrient availability by plant roots. The general and most important figures concerning the relationship between climate change and plant nutrition are as follows:

1. It is well known that the atmospheric CO_2 could be assimilated into the organic compounds through the photosynthesis and then allocated to several processes within the plants (Iversen and Norby 2014; Chen et al. 2015).
2. The four main global change factors controlling plant productivity are elevated CO_2, elevated temperature, changes in precipitation regimes and deposition of nitrogen (Short et al. 2016; Rao et al. 2017).
3. The response of plants to climate change represents very complex interactions including a change in the composition of plant communities, the competition and different interactions with insect herbivores (Iversen and Norby 2014; Bilton et al. 2016; Ferrarini et al. 2017).
4. There is a strong relation between CO_2 content in plant leaf and its interaction with temperature through the biochemistry of photosynthesis, where this interaction may effect on the whole plant as well as agroecosystem (Xu et al. 2016; Greer 2017).

5. It is very difficult to follow the effects of temperature on plant productivity and C allocation compared with the effects from elevated atmospheric CO_2 (Xu et al. 2016; Shi et al. 2017).
6. It is very difficult to study the response of plant metabolism to increased temperature (drought) alone without considering other factors such as the availability of nitrogen or nutrients and water (Liu et al. 2017a).
7. There is a difference between the response of plant production to elevated temperature (drought) and the response of the agroecosystem or the net C balance in the tropical areas because the warming action may increase the CO_2 efflux from the soil (soil respiration) more than this response in case of plant productivity (Xu et al. 2016; Anandhi and Blocksome 2017).
8. The changes in precipitation patterns have many effects on plant productivity mainly representing soil moisture levels depending on some processes like the rainfall rate, the soil-plant evapotranspiration rate, the soil water transport and the water redistribution in the soil profile by plant roots (Liu et al. 2017d).
9. The elevated precipitation rate leads to increase in plant above- and below-ground productivity (aboveground > belowground production), resulting in a decrease in the root/shoot ratio (Wu et al. 2011; Wang et al. 2015; Skarpaas et al. 2016).
10. The increase in both air and soil temperature leads to the acceleration of evaporation rates from soils. So it is important to investigate the effects of the elevated temperature and the soil water availability on the plant productivity alone and in different combinations (Iversen and Norby 2014; Kumar et al. 2016b).
11. It is expected that the changing in climate may shift crop zones, and some zones will become unsuitable for some crops, and others will be projected to reduce their productivity (McCarl 2017).

Therefore, plant nutrition like other plant biological processes totally depends on the climatic attributes including temperature, precipitation and so on. These climatic attributes have direct and indirect impacts on the whole plant growth stages starting from the germination of seeds up to the postharvest and dandling plant products. This impact definitely is different from plant to other and the growth conditions from soil to other growth media as well as the microclimate surrounding cultivated plants.

11.6 Soil Security and Plant Nutrition Under Climate Change

Because soil is the main source for food, feed, fibre and fuel production, soil and agriculture must be reconnected together in order to meet the needs of humanity. At the same time, there is an urgent need to enhance as well as sustain the agricultural productivity, mitigation of global climate changes, restoration of soil and water resource quality and improvement of the global biodiversity (Lal 2009; Koch 2017).

As reported by McBratney et al. (2014), soil security depends on the previous global environmental problems or challenges (security of water, food and energy, the abatement of global climate changes, global biodiversity protection and delivery of ecosystem service) through the main functions of soil including:

1. The production of biomass
2. Storing, filtration and transformation of substances and nutrients as well as water
3. Acting as a carbon pool
4. Acting as a biodiversity pool
5. Acting as a physical and cultural environments
6. Acting as a source of raw materials
7. Archiving of the geological and cultural heritage (McBratney et al. 2014)

It is well known that several environmental challenges faced by humanity nowadays include the security of water, energy and food, protection the global biodiversity and climate change (Bouma and McBratney 2013; Zhang and Vesselinov 2017). Due to the importance of security item, many global securities have been identified including water security, environment security, food security and energy security. The main target of all previous securities involves how to provide all population worldwide with enough, proper and safe food, water and energy (Koch et al. 2012; Bouma and McBratney 2013; Koch et al. 2013; McBratney et al. 2014; Carré et al. 2017; McBratney et al. 2017; Field 2017; Lewis et al. 2017; McCarl 2017; Murphy 2017). Soil security is defined in the great book, *Global Soil Security*, edited by Field et al. (2017), which is based on the rationale (McBratney et al. 2017), soil dimensions (Field 2017) including soil capability (Bouma et al. 2017), soil condition (Lewis et al. 2017), soil capital (McCarl 2017), connectivity (Carré et al. 2017) and finally codification (Koch 2017) as well as the securitization (McBratney and Jarrett 2017). Also, soil security could be defined as "the maintenance and improvement of the world's soil resource to produce food, fibre and fresh water, contribute to energy and climate sustainability and maintain the biodiversity and the overall protection of the ecosystem" (McBratney et al. 2017).

The relationship between soil security and plant nutrition can be distinguished from the seven functions for soil through the acting of this soil as a source for all plant growth needs, as well as the source for the proposal environment for almost plant nutrients. Hence, there is a need for integrating new perspectives to address the global soil security including the integral ecology (Grunwald et al. 2017a), the integration economics (McCarl 2017) and the integrative soil security (Grunwald et al. 2017b). Therefore, all previous approaches related to soil security will lead to the *security of plant nutrition* because plant nutrition is a part and parcel of the soil (El-Ramady et al. 2014a, b). Finally, there is no way to make any progress in agriculture through plant nutrition without integral soil security (Grunwald et al. 2017b; Koch 2017). Furthermore, any agriculture policy must include all soil security frameworks, and any achievement in the agriculture sector should be performed through a secure agricultural soil resource, and the connectivity between soil security and plant nutrition must be addressed (Koch 2017).

Several climate attributes including temperature, precipitation, extreme events and others have great effects on the security of soil (McCarl 2017). These effects include the mitigation and adaptation to climate change as follows:

1. Crop management should be changed to include new agenda for dates of planting, harvesting and the whole growing seasons, where the longer growing seasons may allow to add double cropping (Seifert and Lobell 2015; McCarl 2017).
2. The link between climate change and soil security is closed to the other types of securities such as the security of water (Misra 2014; Leong 2016; Ludwig and Roson 2016; Duran-Encalada et al. 2017; Francés et al. 2017; Tan et al. 2017), energy (Almeida Prado et al. 2016; De Llano-Paz et al. 2016; Matsumoto and Andriosopoulos 2016; Partridge et al. 2017), food (Misra 2014; Mârza et al. 2015; Adl 2016; Ali and Erenstein 2016; Pérez et al. 2016; Rigolot et al. 2017) and environment (Kang et al. 2017).
3. It is very common to notice that one or two securities can be interacted with the climate change, for example, water and food security (Misra 2014; Gohar and Cashman 2016), water and energy security (Markovska et al. 2016) or food, water and energy security (Taniguchi et al. 2015; Gondhalekar and Ramsauer 2016; Zhang and Vesselinov 2017). This reflects the significant role of climate change and its effects on agriculture, which in turn includes water, food and energy resources.
4. It could be concluded that climate change and energy, food and water securities do have consequences for soil and its security positive and negative relationships. That means the soil and its security differ from the long-term to short-term involving different agricultural practices considering the rhizosphere zone.
5. The relationship between climate change and soil security or plant nutrition may face great challenges including the ability of the society to save the healthier soil, feed, food, fuel and cleaner water as well as the climate mitigation (McCarl 2017).

11.7 Conclusion

The rhizosphere zone was and is still one of the hotspots in soil sciences, where great interactions between plant roots and soil as well as soil microbes occur. This area is the main place for plant root exudations, which are the most important source for nutrition and protection of plants and soil microbial activity as well and can also help the plant directly and/or indirectly in its nutrition through promoting soil microbial activity and protecting them against some pathogens. The study of plant nutrition processes should be started from the rhizosphere and its plant root exudates. Hence, further researches are essential for investigating the role of plant root exudates in the nutrition and protection of plants. Concerning climate changes, soil biological activity and its security should be addressed including different impacts of global change on soil biota, environmental stress and its effects on soil enzymes under elevated temperature and CO_2. Different strategies should be mitigated for soil biological activities under global change. Furthermore studies are needed to explore different impacts of climate changes on rhizosphere interactions and the agroecosystem processes including soil security.

Acknowledgements The authors thank the outstanding contribution of the research teams from the Science and Technology Development Fund (STDF), Egypt, and the Federal Ministry of Education and Research of the Federal Republic of Germany (BMBF/DLR), (Project ID 5310) for their help. Great support from this German Egyptian Research Fund (GERF) is gratefully acknowledged.

References

Aanderud, Z. T., Jones, S. E., Schoolmaster Jr., D. R., Fierer, N., & Lennon, J. T. (2013). Sensitivity of soil respiration and microbial communities to altered snowfall. *Soil Biology and Biochemistry, 57*, 217–227.

Aanderud, Z. T., Schoolmaster Jr., D. R., & Lennon, J. T. (2011). Plants mediate the sensitivity of soil respiration to rainfall variability. *Ecosystems, 14*, 156–167.

Abad-Valle, P., Iglesias-Jiménez, E., & Álvarez-Ayuso, E. (2017). A comparative study on the influence of different organic amendments on trace element mobility and microbial functionality of a polluted mine soil. *Journal of Environmental Management, 188*, 287–296.

Abenavoli, M. R., Panuccio, M. R., & Sorgonà, A. (2012). Root form and function in plant as an adaptation to changing climate. In P. Ahmad & M. N. V. Prasad (Eds.), *Environmental adaptations and stress tolerance of plants in the era of climate change*. New York: Springer. doi:10.1007/978-1-4614-0815-4_8.

Abhilash, P. C., & Dubey, R. K. (2014). Integrating aboveground–belowground responses to climate change. *Current Science, 106*(12), 1637–1638.

Abhilash, P. C., & Dubey, R. K. (2015). Root system engineering: prospects and promises. *Trends in Plant Science, 20*, 1360–1385.

Abhilash, P. C., Dubey, R. K., Tripathi, V., Srivastava, P., Verma, J. P., & Singh, H. B. (2013a). Remediation and management of POPs-contaminated soils in a warming climate: Challenges and perspectives. *Environmental Science and Pollution Research, 20*, 5879–5885.

Abhilash, P. C., Dubey, R. K., Tripathi, V., Srivastava, P., Verma, J. P., & Singh, H. B. (2013b). Adaptive soil management. *Current Science, 104*, 1275–1276.

Abhilash, P. C., Tripathi, V., Dubey, R. K., & Edrisi, S. A. (2015). Coping with changes: Adaptation of trees in a changing environment. *Trends in Plant Science, 20*, 137–138.

Achakzai, K., Khalid, S., Adrees, M., Bibi, A., Ali, S., Nawaz, R., & Rizwan, M. (2017). Air pollution tolerance index of plants around brick kilns in Rawalpindi, Pakistan. *Journal of Environmental Management, 190*, 252–258.

Adl, S. (2016). Rhizosphere, food security, and climate change: A critical role for plant-soil research. *Rhizosphere, 1*, 1–3.

Aghababaei, F., & Raiesi, F. (2015). Mycorrhizal fungi and earthworms reduce antioxidant enzyme activities in maize and sunflower plants grown in cd-polluted soils. *Soil Biology and Biochemistry, 86*, 87–97.

Aghababaei, F., Raiesi, F., & Hosseinpur, A. (2014). The combined effects of earthworms and arbuscular mycorrhizal fungi on microbial biomass and enzyme activities in a calcareous soil spiked with cadmium. *Applied Soil Ecology, 75*, 33–42.

Ahmed, M., & Stockle, C. O. (2017). *Quantification of climate variability, adaptation and mitigation for agricultural sustainability*. Cham, Switzerland: Springer. doi:10.1007/978-3-319-32059-5.

Ai, C., Liang, G., Sun, J., Wang, X., He, P., Zhou, W., & He, X. (2015). Reduced dependence of rhizosphere microbiome on plant-derived carbon in 32-year long-term inorganic and organic fertilized soils. *Soil Biology and Biochemistry, 80*, 70–78. doi:10.1016/j.soilbio.2014.09.028.

Alexander, H. D., Moczygemba, J., & Dick, K. (2016). Growth and survival of thornscrub forest seedlings in response to restoration strategies aimed at alleviating abiotic and biotic stressors. *Journal of Arid Environments, 124*, 180–188.

Ali, A. & Erenstein, O. (2016). Assessing farmer use of climate change adaptation practices and impacts on food security and poverty in Pakistan. *Climate Risk Management*. (in press).

Aljazairi, S., & Nogués, S. (2015). The effects of depleted, current and elevated growth [CO_2] in wheat are modulated by water availability. *Environmental and Experimental Botany, 112*, 55–66.

Allen, D. E., Singh, B. P., & Dalal, R. C. (2011). Soil health indicators under climate change: A review of current knowledge. In B. P. Singh et al. (Eds.), *Soil health and climate change, soil biology* (Vol. 29). Berlin: Springer-Verlag. doi:10.1007/978-3-642-20256-8_2.

Allen, M. F., & Allen, E. B. (2017). Mycorrhizal mediation of soil fertility amidst nitrogen eutrophication and climate change. In N. C. Johnson et al. (Eds.), *Mycorrhizal mediation of soil fertility, structure, and carbon storage*. Oxford: Elsevier. doi:10.1016/B978-0-12-804312-7.01001-9.

Almeida Prado Jr., F., Athayde, S., Mossa, J., Bohlman, S., Leite, F., & Oliver-Smith, A. (2016). How much is enough? An integrated examination of energy security, economic growth and climate change related to hydropower expansion in Brazil. *Renewable and Sustainable Energy Reviews, 53*, 1132–1136.

Alshaal, T., Domokos-Szabolcsy, E., Marton, L., Czako, M., Katai, J., Balogh, P., Elhawat, N., El-Ramady, H., & Fari, M. (2013). Phytoremediation of bauxite-derived red mud by giant reed. *Environmental Chemistry Letters, 11*, 295–302. doi:10.1007/s10311-013-0406-6.

Alvareza, A., Yanez, M. L., Benimeli, C. S., & Amoroso, M. J. (2012). Maize plants (*Zea mays*) root exudates enhance lindane removal by native *Streptomyces* strains. *International Biodeterioration and Biodegradation, 66*, 14–18.

Alvarez, A., Saez, J. M., Costa, J. S. D., Colin, V. L., Fuentes, M. S., Cuozzo, S. A., Benimeli, C. S., Polti, M. A., & Amoroso, M. J. (2017). Actinobacteria: Current research and perspectives for bioremediation of pesticides and heavy metals. *Chemosphere, 166*, 41–62.

Álvarez-Martín, A., Hilton, S. L., Bending, G. D., Rodríguez-Cruz, M. S., & Sánchez-Martín, M. J. (2016). Changes in activity and structure of the soil microbial community after application of azoxystrobin or pirimicarb and an organic amendment to an agricultural soil. *Applied Soil Ecology, 106*, 47–57.

Anandhi, A., & Blocksome, C. E. (2017). Developing adaptation strategies using an agroecosystem indicator: Variability in crop failure temperatures. *Ecological Indicators, 76*, 30–41.

Arpaia, S., Birch, A. N. E., Kiss, J., van Loon, J. J. A., Messéan, A., Nuti, M., Perry, J. N., Sweet, J. B., & Tebbe, C. C. (2017). Assessing environmental impacts of genetically modified plants on non-target organisms: The relevance of *in planta* studies. *Science of the Total Environment*. doi:10.1016/j.scitotenv.2017.01.039.

Baetz, U. (2016). Root exudates as integral part of belowground plant defence. In C. M. F. Vos & K. Kazan (Eds.), *Belowground defence strategies in plants, signaling and communication in plants*. Cham, Switzerland: Springer. doi:10.1007/978-3-319-42319-7_3.

Baetz, U., & Martinoia, E. (2014). Root exudates: The hidden part of plant defense. *Trends in Plant Science, 19*(2), 90–98.

Bais, H. P., Weir, T., Perry, L., Gilroy, S., & Vivanco, J. M. (2006). The role of root exudates in rhizosphere interactions with plants and other organisms. *Annual Review of Plant Biology, 57*, 233–266.

Bakshi, M., & Varma, A. (2017). Soil enzyme: The state-of-art. In G. Shukla & A. Varma (Eds.), *Soil enzymology, soil biology* (Vol. 22). Berlin: Springer-Verlag. doi:10.1007/978-3-642-14225-3_1.

Balkhair, K. S., & Ashraf, M. A. (2016). Field accumulation risks of heavy metals in soil and vegetable crop irrigated with sewage water in western region of Saudi Arabia. *Saudi Journal of Biological Sciences, 23*(Supplement 1), S32–S44.

Balser, T. C., Gutknecht, J. L. M., & Liang, C. (2010). How will climate change impact soil microbial communities? In G. R. Dixon & E. L. Tilston (Eds.), *Soil microbiology and sustainable crop production*. New York: Springer. doi:10.1007/978-90-481-9479-7_10.

Bardgett, R. D., Freeman, C., & Ostle, N. J. (2008). Microbial contributions to climate change through carbon cycle feedbacks. *The ISME Journal, 2*, 805–814.
Bashir, O., Khan, K., Hakeem, K. R., Mir, N. A., Rather, G. H., & Mohiuddin, R. (2016). Soil microbe diversity and root exudates as important aspects of rhizosphere ecosystem. In K. R. Hakeem & M. S. Akhtar (Eds.), *Plant, soil and microbes*. Cham, Switzerland: Springer. doi:10.1007/978-3-319-29573-2_15.
Berendsen, R. L., Pieterse, C. M. J., & Bakker, P. A. H. M. (2012). The rhizosphere microbiome and plant health. *Trends in Plant Science, 17*(8), 478–486.
Berg, G., Grube, M., Schloter, M., & Smalla, K. (2014). Unraveling the plant microbiome: Looking back and future perspectives. *Frontiers in Microbiology, 5*, 148. doi:10.3389/fmicb.2014.00148.
Bertrand, J.-C., Doumenq, P., Guyoneaud, R., Marrot, B., Martin-Laurent, F., Matheron, R., Moulin, P., & Soulas, G. (2015). Applied microbial ecology and bioremediation: Microorganisms as major actors of pollution elimination in the environment. In J.-C. Bertrand et al. (Eds.), *Environmental microbiology: Fundamentals and applications—Microbial ecology*. Dordrecht: Springer. doi:10.1007/978-94-017-9118-2_16.
Bielenberg, D. G., & BassiriRad, H. (2005). Nutrient acquisition of terrestrial plants in a changing climate. In H. BassiriRad (Ed.), *Ecological Studies, Nutrient acquisition by plants an ecological perspective* (Vol. 181). Berlin: Springer-Verlag.
Bilton, M. C., Metz, J., & Tielbörger, K. (2016). Climatic niche groups: A novel application of a common assumption predicting plant community response to climate change. *Perspectives in Plant Ecology, Evolution and Systematics, 19*, 61–69.
Bojko, O., & Kabala, C. (2017). Organic carbon pools in mountain soils: Sources of variability and predicted changes in relation to climate and land use changes. *Catena, 149*(Part 1), 209–220.
Bouma, J., & McBratney, A. (2013). Framing soils as an actor when dealing with wicked environmental problems. *Geoderma, 200–201*, 130–139.
Bouma, J., van Ittersum, M. K., Stoorvogel, J. J., Batjes, N. H., Droogers, P., & Pulleman, M. M. (2017). Soil capability: Exploring the functional potentials of soils. In D. J. Field et al. (Eds.), *Global soil security, progress in soil science* (pp. 27–44). Cham, Switzerland: Springer. doi:10.1007/978-3-319-43394-3_3.
Bruce, T. J. A., & Pickett, J. A. (2007). Plant defence signalling induced by biotic attacks. *Current Opinion in Plant Biology, 10*(4), 387–392.
Brye, K. R., McMullen, R. L., Silveira, M. L., Motschenbacher, J. M. D., Smith, S. F., Gbur, E. E., & Helton, M. L. (2016). Environmental controls on soil respiration across a southern US climate gradient: A meta-analysis. *Geoderma Regional, 7*(2), 110–119.
Burke, D. J., Weintraub, M. N., Hewins, C. R., & Kalisz, S. (2011). Relationship between soil enzyme activities, nutrient cycling and soil fungal communities in a northern hardwood forest. *Soil Biology and Biochemistry, 43*(4), 795–803.
Burns, R. G., DeForest, J. L., Marxsen, J., Sinsabaugh, R. L., Stromberger, M. E., Wallenstein, M. D., Weintraub, M. N., & Zoppini, A. (2013). Soil enzymes in a changing environment: Current knowledge and future directions. *Soil Biology and Biochemistry, 58*, 216–234.
Cai, F., Pang, G., Miao, Y., Li, R., Li, R., Shen, Q., & Chen, W. (2017). The nutrient preference of plants influences their rhizosphere microbiome. *Applied Soil Ecology, 110*, 146–150.
Callesen, I., Harrison, R., Stupak, I., Hatten, J., Raulund-Rasmussen, K., Boyle, J., Clarke, N., & Zabowski, D. (2016). Carbon storage and nutrient mobilization from soil minerals by deep roots and rhizospheres. *Forest Ecology and Management, 359*, 322–331.
Canfora, L., Sbrana, C., Avio, L., Felici, B., Scatà, M. C., Neri, U., & Benedetti, A. (2014). Risk management tools and the case study *Brassica napus*: Evaluating possible effects of genetically modified plants on soil microbial diversity. *Science of the Total Environment, 493*, 983–994.
Carré, F., Caudeville, J., Bonnard, R., Bert, V., Boucard, P., & Ramel, M. (2017). Soil contamination and human health: A major challenge for global soil security. In D. J. Field et al. (Eds.), *Global soil security, progress in soil science* (pp. 275–295). Cham, Switzerland: Springer. doi:10.1007/978-3-319-43394-3_2.

Castro, H. F., Classen, A. T., Austin, E. E., Norby, R. J., & Schadt, C. W. (2010). Soil microbial community responses to multiple experimental climate change drivers. *Applied and Environmental Microbiology, 76*(40), 999–1007.

Cavagnaro, T. R., Cunningham, S. C., & Fitzpatrick, S. (2016). Pastures to woodlands: Changes in soil microbial communities and carbon following reforestation. *Applied Soil Ecology, 107*, 24–32.

Čerevková, A., Miklisová, D., & Cagáň, Ľ. (2017). Effects of experimental insecticide applications and season on soil nematode communities in a maize field. *Crop Protection, 92*, 1–15.

Cesco, S., Mimmo, T., Tonon, G., Tomasi, N., Pinton, R., Terzano, R., Neumann, G., Weisskopf, L., Renella, G., Landi, L., & Nannipieri, P. (2012). Plant-borne flavonoids released into the rhizosphere: Impact on soil bio-activities related to plant nutrition. A review. *Biology and Fertility of Soils, 48*, 123–149. doi:10.1007/s00374-011-0653-2.

Cesco, S., Neumann, G., Tomasi, N., Pinton, R., & Weisskopf, L. (2010). Release of plant-borne flavonoids into the rhizosphere and their role in plant nutrition. *Plant and Soil, 329*, 1–25. doi:10.1007/s11104-009-0266-9.

Chapman, N., Miller, A. J., Lindsey, K., & Whalley, W. R. (2012). Roots, water, and nutrient acquisition: Let's get physical. *Trends in Plant Science, 17*(12), 701–710.

Chaudhry, V., Dang, H. Q., Tran, N. Q., Mishra, A., Chauhan, P. S., Gill, S. S., Nautiyal, C. S., & Tuteja, N. (2012). Impact of salinity-tolerant MCM6 transgenic tobacco on soil enzymatic activities and the functional diversity of rhizosphere microbial communities. *Research in Microbiology, 163*(8), 511–517.

Chen, B., Zou, D., & Jiang, H. (2015). Elevated CO_2 exacerbates competition for growth and photosynthesis between *Gracilaria lemaneiformis* and *Ulva lactuca*. *Aquaculture, 443*, 49–55.

Chen, W.-Y., Suzuki, T., & Lackner, M. (2017a). *Handbook of climate change mitigation and adaptation*. Cham, Switzerland: Springer. doi:10.1007/978-3-319-14409-2.

Chen, Z., Wang, H., Liu, X., Zhao, X., Lu, D., Zhou, J., & Li, C. (2017b). Changes in soil microbial community and organic carbon fractions under short-term straw return in a rice–wheat cropping system. *Soil and Tillage Research, 165*, 121–127.

Chen, Z.-J., Tian, Y.-H., Zhang, Y., Song, B.-R., Li, H.-C., & Chen, Z.-H. (2016). Effects of root organic exudates on rhizosphere microbes and nutrient removal in the constructed wetlands. *Ecological Engineering, 92*, 243–250.

Cheng, G., Wang, L., & Lan, H. (2016). Cloning of *PEPC-1* from a C4 halophyte *Suaeda aralocaspica* without Kranz anatomy and its recombinant enzymatic activity in responses to abiotic stresses. *Enzyme and Microbial Technology, 83*, 57–67.

Classen, A. T., Sundqvist, M. K., Henning, J. A., Newman, G. S., Moore, J. A. M., Cregger, M. A., Moorhead, L. C., & Patterson, C. M. (2015a). Direct and indirect effects of climate change on soil microbial and soil microbial plant interactions: What lies ahead? *Ecosphere, 6*(8), 1–21.

Classen, A. T., Sundqvist, M. K., Henning, J. A., Newman, G. S., Moore, J. A. M., Cregger, M. A., Moorhead, L. C., & Patterson, C. M. (2015b). Direct and indirect effects of climate change on soil microbial and soil microbial-plant interactions: What lies ahead? *Ecosphere, 6*(8), 130.

Coffey, R., Benham, B., Wolfe, M. L., Dorai-Raj, S., Bhreathnach, N., O'Flaherty, V., Cormican, M., & Cummins, E. (2016). Sensitivity of streamflow and microbial water quality to future climate and land use change in the west of Ireland. *Regional Environmental Change, 16*, 2111–2128. doi:10.1007/s10113-015-0912-0.

Compant, S., Saikkonen, K., Mitter, B., Campisano, A., & Mercado-Blanco, J. (2016). Editorial special issue: Soil, plants and endophytes. *Plant and Soil, 405*, 1–11. doi:10.1007/s11104-016-2927-9.

Crowther, T. W., Thomas, S. M., Maynard, D. S., Baldrian, P., Covey, K., Frey, S. D., van Diepen, L. T. A., & Bradford, M. A. (2015). Biotic interactions mediate soil microbial feedbacks to climate change. *Proceedings of the National Academy of Sciences, 112*(22), 7033–7038.

Daou, L., Périssol, C., Luglia, M., Calvert, V., & Criquet, S. (2016). Effects of drying–rewetting or freezing–thawing cycles on enzymatic activities of different Mediterranean soils. *Soil Biology and Biochemistry, 93*, 142–149.

Das, S., Chou, M.-L., Jean, J.-S., Yang, H.-J., & Kim, P. J. (2017). Arsenic-enrichment enhanced root exudates and altered rhizosphere microbial communities and activities in hyperaccumulator *Pteris vittata*. *Journal of Hazardous Materials, 325*, 279–287.

Davidson, E., Janssens, I., & Luo, Y. (2006). On the variability of respiration in terrestrial ecosystems: Moving beyond Q(10). *Global Change Biology, 12*, 154–164.

De Coninck, B., Timmermans, P., Vos, C., Cammue, B. P. A., & Kazan, K. (2015). What lies beneath: Belowground defense strategies in plants. *Trends in Plant Science, 20*(2), 91–101.

De Llano-Paz, F., Fernandez, P. M., & Soares, I. (2016). Addressing 2030 EU policy framework for energy and climate: Cost, risk and energy security issues. *Energy, 115*(Part 2), 1347–1360.

de Medeiros, E. V., Duda, G. P., dos Santos, L. A. R., de Sousa Lima, J. R., de Almeida-Cortêz, J. S., Hammecker, C., Lardy, L., & Cournac, L. (2017). Soil organic carbon, microbial biomass and enzyme activities responses to natural regeneration in a tropical dry region in Northeast Brazil. *Catena, 151*, 137–146.

Dessaux, Y., Grandclément, C., & Faure, D. (2016). Engineering the Rhizosphere. *Trends in Plant Science, 21*(3), 266–278.

Doornbos, R. F., van Loon, L. C., & Bakker, P. A. H. M. (2012). Impact of root exudates and plant defense signaling on bacterial communities in the rhizosphere. A review. *Agronomy for Sustainable Development, 32*, 227–243.

Drigo, B., Kowalchuk, G. A., & van Veen, J. A. (2008). Climate change goes underground: Effects of elevated atmospheric CO_2 on microbial community structure and activities in the rhizosphere. *Biology and Fertility of Soils, 44*, 667–679. doi:10.1007/s00374-008-0277-3.

Duan, B., Zhang, Y., Xu, G., Chen, J., Paquette, A., & Peng, S. (2015). Long-term responses of plant growth, soil microbial communities and soil enzyme activities to elevated CO_2 and neighbouring plants. *Agricultural and Forest Meteorology, 213*, 91–101.

Dubrovskaya, E., Pozdnyakova, N., Golubev, S., Muratova, A., Grinev, V., Bondarenkova, A., & Turkovskaya, O. (2017). Peroxidases from root exudates of *Medicago sativa* and *Sorghum bicolor*: Catalytic properties and involvement in PAH degradation. *Chemosphere, 169*, 224–232.

Duran-Encalada, J. A., Paucar-Caceres, A., Bandala, E. R., & Wright, G. H. (2017). The impact of global climate change on water quantity and quality: A system dynamics approach to the US–Mexican transborder region. *European Journal of Operational Research, 256*(2), 567–581.

Dutta, H., & Dutta, A. (2016). The microbial aspect of climate change. *Energy, Ecology and Environment, 1*(4), 209–232. doi:10.1007/s40974-016-0034-7.

Edwards, S. J., & Kjellerup, B. V. (2013). Applications of biofilms in bioremediation and biotransformation of persistent organic pollutants, pharmaceuticals/personal care products, and heavy metals. *Applied Microbiology and Biotechnology, 97*, 9909–9921. doi:10.1007/s00253-013-5216-z.

El-Ramady, H. (2008). *A contribution on the bio-actions of rare earth elements in the soil/plant environment*. Julius Kühn-Institut, Bundesforschungsinstitut für Kulturpfl anzen, Quedlinburg, Braunschweig, Germany.

El-Ramady, H., Abdalla, N., Alshaal, T., Domokos-Szabolcsy, É., Elhawat, N., Prokisch, J., Sztrik, A., Fári, M., El-Marsafawy, S., & Shams, M. S. (2015a). Selenium in soils under climate change, implication for human health. *Environmental Chemistry Letters, 13*(1), 1–19. doi:10.1007/s10311-014-0480-4.

El-Ramady, H., Abdalla, N., Alshaal, T., Elhenawy, A. S., Shams, M. S., Faizy, S. E.-D. A., Belal, E. B., Shehata, S. A., Ragab, M. I., Amer, M. M., Fari, M., Sztrik, A., Prokisch, J., Selmar, D., Schnug, E., Pilon-Smits, E. A. H., El-Marsafawy, S. M., & Domokos-Szabolcsy, E. (2015b). Giant reed for selenium phytoremediation under changing climate. *Environmental Chemistry Letters, 13*(4), 359–380. doi:10.1007/s10311-015-0523-5.

El-Ramady, H. R., Alshaal, T. A., Amer, M., Domokos-Szabolcsy, É., Elhawat, N., Prokisch, J., & Fári, M. (2014a). Soil quality and plant nutrition. In H. Ozier-Lafontaine & M. Lesueur-Jannoyer (Eds.), *Sustainable agriculture Reviews 14: Agroecology and global change, sustainable agriculture Reviews 14*. Cham, Switzerland: Springer. doi:10.1007/978-3-319-06016-3_11.

El-Ramady, H. R., Alshaal, T. A., Shehata, S. A., Domokos-Szabolcsy, É., Elhawat, N., Prokisch, J., Fári, M., & Marton, L. (2014b). Plant nutrition: From liquid medium to micro-farm. In

H. Ozier-Lafontaine & M. Lesueur-Jannoyer (Eds.), *Sustainable agriculture 449 Reviews 14: Agroecology and global change, sustainable agriculture Reviews 14*. Cham, Switzerland: Springer. doi:10.1007/978-3-319-06016-3_12.

Ennigrou, A., Casabianca, H., Laarif, A., Hanchi, B., & Hosni, K. (2017). Maturation-related changes in phytochemicals and biological activities of the Brazilian pepper tree (*Schinus terebinthifolius* Raddi) fruits. *South African Journal of Botany, 108*, 407–415.

Feng, J., Wang, J., Ding, L., Yao, P., Qiao, M., & Yao, S. (2017). Meta-analyses of the effects of major global change drivers on soil respiration across China. *Atmospheric Environment, 150*, 181–186.

Ferrarini, A., Alatalo, J. M., Gervasoni, D., & Foggi, B. (2017). Exploring the compass of potential changes induced by climate warming in plant communities. *Ecological Complexity, 29*, 1–9.

Field, D. J. (2017). Soil security: Dimensions. In D. J. Field et al. (Eds.), *Global soil security, progress in soil science* (pp. 4–23). Cham, Switzerland: Springer. doi:10.1007/978-3-319-43394-3_2.

Field, D. J., Morgan, C. L. S., & McBratney, A. B. (2017). *Global soil security, Progress in soil science series*. Cham, Switzerland: Springer. doi:10.1007/978-3-319-43394-3.

Francés, G. E., Quevauviller, P., González, E. S. M., & Amelin, E. V. (2017). Climate change policy and water resources in the EU and Spain. A closer look into the water framework directive. *Environmental Science & Policy, 69*, 1–12.

Franco-Andreu, L., Gómez, I., Parrado, J., García, C., Hernández, T., & Tejada, M. (2016). Behavior of two pesticides in a soil subjected to severe drought: Effects on soil biology. *Applied Soil Ecology, 105*, 17–24.

French, S., Levy-Booth, D., Samarajeewa, A., Shannon, K. E., Smith, J., & Trevors, J. T. (2009). Elevated temperatures and carbon dioxide concentrations: Effects on selected microbial activities in temperate agricultural soils. *World Journal of Microbiology and Biotechnology, 25*, 1887–1900. doi:10.1007/s11274-009-0107-2.

Freund, D. M., & Hegeman, A. D. (2017). Recent advances in stable isotope-enabled mass spectrometry-based plant metabolomics. *Current Opinion in Biotechnology, 43*, 41–48.

Fu, S., & Gu, Y. (2017). Highway toll and air pollution: Evidence from Chinese cities. *Journal of Environmental Economics and Management, 83*, 32–49.

Gao, B., Han, L., Hao, H., & Zhou, H. (2016). Pollution characteristics of mercury (hg) in surface sediments of major basins, China. *Ecological Indicators, 67*, 577–585.

Gohar, A. A., & Cashman, A. (2016). A methodology to assess the impact of climate variability and change on water resources, food security and economic welfare. *Agricultural Systems, 147*, 51–64.

Gömöryová, E., Ujházy, K., Martinák, M., & Gömöry, D. (2013). Soil microbial community response to variation in vegetation and abiotic environment in a temperate old-growth forest. *Applied Soil Ecology, 68*, 10–19.

Gondhalekar, D. & Ramsauer, T. (2016). Nexus City: Operationalizing the urban Water-Energy-Food Nexus for climate change adaptation in Munich, Germany. *Urban Climate*. (in press).

Greer, D. H. (2017). Temperature and CO_2 dependency of the photosynthetic photon flux density responses of leaves of *Vitis vinifera* cvs. Chardonnay and merlot grown in a hot climate. *Plant Physiology and Biochemistry, 111*, 295–303.

Grunwald, S., Clingensmith, C. M., Gavilan, C. P., Mizuta, K., Wilcox, R. K. K., Pinheiro, É. F. M., Ceddia, M. B., & Ross, C. W. (2017a). Integrating new perspectives to address global soil security: Ideas from integral ecology. In D. J. Field et al. (Eds.), *Global soil security, progress in soil science*. Cham, Switzerland: Springer. doi:10.1007/978-3-319-43394-3_28.

Grunwald, S., Mizuta, K., Ceddia, M. B., Pinheiro, É. F. M., Wilcox, R. K. K., Gavilan, C. P., Ross, C. W., & Clingensmith, C. M. (2017b). The meta soil model: An integrative multi- model framework for soil security. In D. J. Field et al. (Eds.), *Global soil security, progress in soil science*. Cham, Switzerland: Springer. doi:10.1007/978-3-319-43394-3_27.

Gruwez, R., De Frenne, P., De Schrijver, A., Vangansbeke, P., & Verheyen, K. (2016). Climate warming and atmospheric deposition affect seed viability of common juniper (*Juniperus communis*) via their impact on the nutrient status of the plant. *Ecological Research*. doi:10.1007/s11284-016-1422-3.

Guan, Z., Lu, S., Huo, Y., Guan, Z.-P., Liu, B., & Wei, W. (2016). Do genetically modified plants affect adversely on soil microbial communities? *Agriculture, Ecosystems & Environment, 235*, 289–305.

Guangming, L., Xuechen, Z., Xiuping, W., Hongbo, S., Jingsong, Y., & Xiangping, W. (2017). Soil enzymes as indicators of saline soil fertility under various soil amendments. *Agriculture, Ecosystems & Environment, 237*, 274–279.

Guarino, C., & Sciarrillo, R. (2017). Effectiveness of in situ application of an integrated phytoremediation system (IPS) by adding a selected blend of rhizosphere microbes to heavily multi-contaminated soils. *Ecological Engineering, 99*, 70–82.

Gubelit, Y., Polyak, Y., Dembska, G., Pazikowska-Sapota, G., Zegarowski, L., Kochura, D., Krivorotov, D., Podgornaya, E., Burova, O., & Maazouzi, C. (2016). Nutrient and metal pollution of the eastern gulf of Finland coastline: Sediments, macroalgae, microbiota. *Science of the Total Environment, 550*, 806–819.

Gul, S., & Whalen, J. K. (2016). Biochemical cycling of nitrogen and phosphorus in biochar-amended soils. *Soil Biology and Biochemistry, 103*, 1–15.

Guo, M., Gong, Z., Miao, R., Su, D., Li, X., Jia, C., & Zhuang, J. (2017). The influence of root exudates of maize and soybean on polycyclic aromatic hydrocarbons degradation and soil bacterial community structure. *Ecological Engineering, 99*, 22–30.

Habtemariam, L. T., Kassa, G. A., & Gandorfer, M. (2017). Impact of climate change on farms in smallholder farming systems: Yield impacts, economic implications and distributional effects. *Agricultural Systems, 152*, 58–66.

Haichar, F. Z., Santaella, C., Heulin, T., & Achouak, W. (2014). Root exudates mediated interactions belowground. *Soil Biology and Biochemistry, 77*, 69–80.

Hakeem, K. R., & Akhtar, M. S. (2016). *Plant, soil and microbes: Mechanisms and molecular interactions* (Vol. 2). Cham, Switzerland: Springer. doi:10.1007/978-3-319-29573-2.

Hakeem, K. R., Akhtar, M. S., & Abdullah, S. N. A. (2016). *Plant, soil and microbes: Implications in crop science* (Vol. 1). Cham, Switzerland: Springer. doi:10.1007/978-3-319-27455-3.

Han, D., Currell, M. J., & Cao, G. (2016). Deep challenges for China's war on water pollution. *Environmental Pollution, 218*, 1222–1233.

Hansda, A., Kumar, V., & Anshumali. (2017). Influence of cu fractions on soil microbial activities and risk assessment along cu contamination gradient. *Catena, 151*, 26–33.

Hardoim, P. R., van Overbeek, L. S., Berg, G., Pirttilä, A. M., Compant, S., Campisano, A., Döring, M., & Sessitsche, A. (2015). The hidden world within plants: Ecological and evolutionary considerations for defining functioning of microbial endophytes. *Microbiology and Molecular Biology Reviews, 79*(3). doi:10.1128/MMBR.00050-14.

Hartmann, A., Rothballer, M., Schmid, M., & Hiltner, L. (2008). A pioneer in rhizosphere microbial ecology and soil bacteriology research. *Plant and Soil, 312*, 7–14.

Hernández, M., Conrad, R., Klose, M., Ma, K., & Lu, Y. (2017). Structure and function of methanogenic microbial communities in soils from flooded rice and upland soybean fields from Sanjiang plain, NE China. *Soil Biology and Biochemistry, 105*, 81–91.

Hou, Y., Liu, X., Zhang, X., Chen, X., & Tao, K. (2016). Effects of key components of *S. triqueter* root exudates on fractions and bioavailability of pyrene–lead co-contaminated soils. *International journal of Environmental Science and Technology, 13*, 887–896. doi:10.1007/s13762-015-0927-6.

Huang, S., Jia, X., Zhao, Y., Bai, B., & Chang, Y. (2017a). Elevated CO_2 benefits the soil microenvironment in the rhizosphere of *Robinia pseudoacacia* L. seedlings in cd- and Pb-contaminated soils. *Chemosphere, 168*, 606–616.

Huang, H., Ouyang, W., Wu, H., Liu, H., & Andrea, C. (2017b). Long-term diffuse phosphorus pollution dynamics under the combined influence of land use and soil property variations. *Science of the Total Environment, 579*, 1894–1903.

Huang, M., Chen, J., Cao, F., Jiang, L., & Zou, Y. (2016). Rhizosphere processes associated with the poor nutrient uptake in no-tillage rice (*Oryza sativa* L.) at tillering stage. *Soil and Tillage Research, 163*, 10–13.

Huang, X. F., Chaparro, J. M., Reardon, K., Zhang, R., Shen, Q., & Vivanco, J. M. (2014). Rhizosphere interactions: Root exudates, microbes, and microbial communities. *Botany, 92*, 267–275.

Ibekwe, A. M., Ors, S., Ferreira, J. F. S., Liu, X., & Suarez, D. L. (2017). Seasonal induced changes in spinach rhizosphere microbial community structure with varying salinity and drought. *Science of the Total Environment, 579*, 1485–1495.

Imadi, S. R., Ali, Z., Hasan, H., & Gul, A. (2016). Soil pollution and remediation. In K. R. Hakeem & M. S. Akhtar (Eds.), *Plant, soil and microbes*. Cham, Switzerland: Springer. doi:10.1007/978-3-319-29573-2_18.

IPCC. (2014). *Climate change 2014: Impacts, adaptation, and vulnerability, Contribution of working group II to the fifth assessment report of the intergovernmental panel on climate change*. Cambridge, UK: Cambridge University Press.

IPCC, Intergovernmental Panel on Climate Change. (2007). *Climate change 2007: The physical science basis*. Cambridge: Cambridge University Press.

Iversen, C., & Norby, R. (2014). Terrestrial plant productivity and carbon allocation in a changing climate. In B. Freedman (Ed.), *Global environmental change*. Dordrecht: Springer. doi:10.1007/978-94-007-5784-4_2.

Jain, R., & Pandey, A. (2016). Soil enzymes and microbial endophytes as indicators of climate variation along an altitudinal gradient with respect to wheat rhizosphere under mountain ecosystem. *Rhizosphere, 2*, 75–84.

Jaiswal, D. K., & Verma, J. P. (2016). The significance of plant-associated microbial rhizosphere for the degradation of xenobiotic compounds. In A. Singh et al. (Eds.), *Plant responses to xenobiotics*. Singapore: Springer Nature Singapore Pte Ltd. doi:10.1007/978-981-10-2860-1_13.

Jha, P., Panwar, J., & Jha, P. N. (2015). Secondary plant metabolites and root exudates: Guiding tools for polychlorinated biphenyl biodegradation. *International journal of Environmental Science and Technology, 12*, 789–802. doi:10.1007/s13762-014-0515-1.

Jia, H., Lu, H., Dai, M., Hong, H., Liu, J., & Yan, C. (2016). Effect of root exudates on sorption, desorption, and transport of phenanthrene in mangrove sediments. *Marine Pollution Bulletin, 109*(1), 171–177.

Jia, Z., Deng, J., Chen, N., Shi, W., Tang, X., & Xu, H. (2017). Bioremediation of cadmium-dichlorophen co-contaminated soil by spent *Lentinus edodes* substrate and its effects on microbial activity and biochemical properties of soil. *Journal of Soils and Sediments, 17*, 315–325. doi:10.1007/s11368-016-1562-7.

Johnson, S. N., Benefer, C. M., Frew, A., Griffiths, B. S., Hartley, S. E., Karley, A. J., Rasmann, S., Schumann, M., Sonnemann, I., & Robert, C. A. M. (2016). New frontiers in belowground ecology for plant protection from root-feeding insects. *Applied Soil Ecology, 108*, 96–107.

Kabiri, V., Raiesi, F., & Ghazavi, M. A. (2016). Tillage effects on soil microbial biomass, SOM mineralization and enzyme activity in a semi-arid Calcixerepts. *Agriculture, Ecosystems & Environment, 232*, 73–84.

Kahlon, R. S. (2016). Biodegradation and bioremediation of organic chemical pollutants by *Pseudomonas*. In R. S. Kahlon (Ed.), *Pseudomonas: Molecular and applied biology*. Cham, Switzerland: Springer. doi:10.1007/978-3-319-31198-2_9.

Kang, S., X Hao, T. D., L Tong, X. S., Lu, H., Xi, L., Huo, Z., Li, S., & Ding, R. (2017). Improving agricultural water productivity to ensure food security in China under changing environment: From research to practice. *Agricultural Water Management, 179*, 5–17.

Kayler, Z., Keitel, C., Jansen, K., & Gessler, A. (2017). Experimental evidence of two mechanisms coupling leaf-level C assimilation to rhizosphere CO_2 release. *Environmental and Experimental Botany, 135*, 21–26.

Khan, K. S., Mack, R., Castillo, X., Kaiser, M., & Joergensen, R. G. (2016). Microbial biomass, fungal and bacterial residues, and their relationships to the soil organic matter C/N/P/S ratios. *Geoderma, 271*, 115–123.

Kim, S., & Kang, H. (2011). Effects of elevated CO_2 and Pb on phytoextraction and enzyme activity. *Water, Air, and Soil Pollution, 219*, 365–375.

Kim, Y.-N., Robinson, B., Lee, K.-A., Boyer, S., & Dickinson, N. (2017). Interactions between earthworm burrowing, growth of a leguminous shrub and nitrogen cycling in a former agricultural soil. *Applied Soil Ecology, 110*, 79–87.

Koch, A. (2017). Soil security for agricultural productivity: The policy disconnect and a promising future. In D. J. Field et al. (Eds.), *Global soil security, progress in soil science* (pp. 425–435). Cham, Switzerland: Springer. doi:10.1007/978-3-319-43394-3_39.

Koch, A., McBratney, A. B., Adams, M., Field, D. J., Hill, R., Lal, R., Abbott, L., Angers, D., Baldock, J., Barbier, E., Bird, M., Bouma, J., Chenu, C., Crawford, J., Flora, C. B., Goulding, K., Grunwald, S., Jastrow, J., Lehmann, J., Lorenz, K., Minasny, B., Morgan, C., O'Donnell, A., Parton, W., Rice, C. W., Wall, D. H., Whitehead, D., Young, I., & Zimmermann, M. (2013). Soil security: Solving the global soil crisis. *Global Policy, 4*(4), 434–441.

Koch, A., McBratney, A. B., & Lal, R. (2012). Global soil week: Put soil security on the global agenda. *Nature, 492*, 186.

Kolokytha, E., Oishi, S., & Teegavarapu, R. S. V. (2017). *Sustainable water resources planning and management under climate change*. Singapore: Springer. doi:10.1007/978-981-10-2051-3.

Kumar, V., Bisht, S., Teotia, P., Sharma, S., & Solanki, A. S. (2013). Interaction between *G. fasciculatum* and *A. chroococcum* for yield, nutrients uptake and cost economy of *Lepidium sativum* in Indian arid region. *Thai Journal of Agricultural Science, 46*(1), 21–28.

Kumar, V., Kumar, M., Shrivastava, N., Bisht, S., Sharma, S., & Varma, A. (2016a). Interaction among rhizospheric microbes, soil, and plant roots: Influence on micronutrient uptake and bioavailability. In K. R. Hakeem & M. S. Akhtar (Eds.), *Plant, soil and microbes*. Cham, Switzerland: Springer. doi:10.1007/978-3-319-29573-2_8.

Kumar, R., Mina, U., Gogoi, R., Bhatia, A., & Harit, R. C. (2016b). Effect of elevated temperature and carbon dioxide levels on maydis leaf blight disease tolerance attributes in maize. *Agriculture, Ecosystems & Environment, 231*, 98–104.

Kvesitadze, G., Khatisashvili, G., Sadunishvili, T., & Kvesitadze, E. (2015). Plants for remediation: Uptake, translocation and transformation of organic pollutants. In M. Öztürk et al. (Eds.), *Plants, pollutants and remediation*. Dordrecht: Springer. doi:10.1007/978-94-017-7194-8_12.

Lal, R. (2009). Soils and world food security. *Soil and Tillage Research, 102*(1), 1–4.

Lal, R. (2016). Beyond COP 21: Potential and challenges of the "4 per thousand" initiative. *Journal of Soil and Water Conservation, 71*, 20A–25A.

Lal, R., Singh, B. R., Mwaseba, D. L., Kraybill, D., Hansen, D. O., & Eik, L. O. (2015). *Sustainable intensification to advance food security and enhance climate resilience in Africa*. Heidelberg: Springer. doi:10.1007/978-3-319-09360-4.

Lambers, H., Mougel, C., Jaillard, B., & Hinsinger, P. (2009). Plant-microbe-soil interactions in the rhizosphere: An evolutionary perspective. *Plant and Soil, 321*, 83–115. doi:10.1007/s11104-009-0042-x.

Lareen, A., Burton, F., & Schaefer, P. (2016). Plant root-microbe communication in shaping root microbiomes. *Plant Molecular Biology, 90*, 575–587. doi:10.1007/s11103-015-0417-8.

Le Fevre, R., & Schornack, S. (2016). Belowground defence strategies in plants: Parallels between root responses to beneficial and detrimental microbes. In C. M. F. Vos & K. Kazan (Eds.), *Belowground defence strategies in plants, signaling and communication in plants*. Cham, Switzerland: Springer. doi:10.1007/978-3-319-42319-7_2.

Le, T. T., Son, M.-H., Nam, I.-H., Yoon, H., Kang, Y.-G., & Chang, Y.-S. (2017). Transformation of hexabromocyclododecane in contaminated soil in association with microbial diversity. *Journal of Hazardous Materials, 325*, 82–89.

Lee, K.-Y., Bosch, J., & Meckenstock, R. U. (2012). Use of metal-reducing bacteria for bioremediation of soil contaminated with mixed organic and inorganic pollutants. *Environmental Geochemistry and Health, 34*, 135–142. doi:10.1007/s10653-011-9406-2.

Lee, S.-H., & Kang, H. (2016). Elevated CO_2 causes a change in microbial communities of rhizosphere and bulk soil of salt marsh system. *Applied Soil Ecology, 108*, 307–314.

León, P., Espejo, R., Gómez-Paccard, C., Hontoria, C., Mariscal, I., Renella, G., & Benito, M. (2017). No tillage and sugar beet foam amendment enhanced microbial activity of degraded acidic soils in south West Spain. *Applied Soil Ecology, 109*, 69–74.

Leong, C. (2016). Resilience to climate change events: The paradox of water (in)-security. *Sustainable Cities and Society, 27*, 439–447.

Lewis, K. L., DeLaune, P., & Keeling, W. (2017). Securing our soil in intensive monoculture cropping systems. In D. J. Field et al. (Eds.), *Global soil security, progress in soil science* (pp. 145–151). Cham, Switzerland: Springer. doi:10.1007/978-3-319-43394-3_13.

Li, H., Yang, S., Xu, Z., Yan, Q., Li, X., van Nostrand, J. D., He, Z., Yao, F., Han, X., Zhou, J., Deng, Y., & Jiang, Y. (2017). Responses of soil microbial functional genes to global changes are indirectly influenced by aboveground plant biomass variation. *Soil Biology and Biochemistry, 104*, 18–29.

Liang, S., Jin, Y., Liu, W., Li, X., Shen, S., & Ding, L. (2017). Feasibility of Pb phytoextraction using nano-materials assisted ryegrass: Results of a one-year field-scale experiment. *Journal of Environmental Management, 190*, 170–175.

Liu, G., Du, Q., & Li, J. (2017a). Interactive effects of nitrate-ammonium ratios and temperatures on growth, photosynthesis, and nitrogen metabolism of tomato seedlings. *Scientia Horticulturae, 214*, 41–50.

Liu, G., Ling, S., Zhan, X., Lin, Z., Zhang, W., & Lin, K. (2017b). Interaction effects and mechanism of Pb pollution and soil microorganism in the presence of earthworm. *Chemosphere, 173*, 227–234.

Liu, B., Liu, X., Huo, S., Chen, X., Wu, L., Chen, M., Zhou, K., Li, Q., & Peng, L. (2017c). Properties of root exudates and rhizosphere sediment of *Bruguiera gymnorrhiza* (L.) *Journal of Soils and Sediments, 17*, 266–276. doi:10.1007/s11368-016-1541-z.

Liu, L., Monaco, T. A., Sun, F., Liu, W., Gan, Y., & Sun, G. (2017d). Altered precipitation patterns and simulated nitrogen deposition effects on phenology of common plant species in a Tibetan Plateau alpine meadow. *Agricultural and Forest Meteorology, 236*, 36–47.

Liu, S.-H., Zeng, G.-M., Niu, Q.-Y., Liu, Y., Zhou, L., Jiang, L.-H., Tan, X., Xu, P., Zhang, C., & Cheng, M. (2017e). Bioremediation mechanisms of combined pollution of PAHs and heavy metals by bacteria and fungi: A mini review. *Bioresource Technology, 224*, 25–33.

Liu, W., Hou, J., Wang, Q., Yang, H., Luo, Y., & Christie, P. (2015). Collection and analysis of root exudates of *Festuca arundinacea* L. and their role in facilitating the phytoremediation of petroleum-contaminated soil. *Plant and Soil, 389*, 109–119. doi:10.1007/s11104-014-2345-9.

Lozano-García, B., Muñoz-Rojas, M., & Parras-Alcántara, L. (2017). Climate and land use changes effects on soil organic carbon stocks in a Mediterranean semi-natural area. *Science of the Total Environment, 579*, 1249–1259.

Lu, Y., Song, S., Wang, R., Liu, Z., Meng, J., Sweetman, A. J., Jenkins, A., Ferrier, R. C., Li, H., Luo, W., & Wang, T. (2015). Impacts of soil and water pollution on food safety and health risks in China. *Environment International, 77*, 5–15.

Ludwig, R., & Roson, R. (2016). Climate change, water and security in the Mediterranean: Introduction to the special issue. *Science of the Total Environment, 543*(Part B), 847–850.

Lv, M., Shao, Y., Lin, Y., Liang, C., Dai, J., Liu, Y., Fan, P., Zhang, W., & Fu, S. (2016). Plants modify the effects of earthworms on the soil microbial community and its activity in a subtropical ecosystem. *Soil Biology and Biochemistry, 103*, 446–451.

Lv, T., Carvalho, P. N., Zhang, L., Zhang, Y., Button, M., Arias, C. A., Weber, K. P., & Brix, H. (2017). Functionality of microbial communities in constructed wetlands used for pesticide remediation: Influence of system design and sampling strategy. *Water Research, 110*, 241–251.

Maddela, N. R., Golla, N., & Vengatampalli, R. (2017). *Soil enzymes influence of sugar industry effluents on soil enzyme activities*. Cham, Switzerland: Springer. doi:10.1007/978-3-319-42655-6.

Maheshwari, D. K., Kumar, S., Maheshwari, N. K., Patel, D., & Saraf, M. (2012). Nutrient availability and management in the rhizosphere by microorganisms. In D. K. Maheshwari (Ed.), *Bacteria in agrobiology: Stress management*. Berlin: Springer-Verlag. doi:10.1007/978-3-642-23465-1_15.

Markovska, N., Duić, N., Mathiesen, B. V., Guzović, Z., Piacentino, A., Schlör, H., & Lund, H. (2016). Addressing the main challenges of energy security in the twenty-first century—Contributions of the conferences on sustainable development of energy, water and environment systems. *Energy, 115*(Part 3), 1504–1512.

Mârza, B., Angelescu, C., & Tindeche, C. (2015). Agricultural insurances and food security. The new climate change challenges. *Procedia Economics and Finance, 27*, 594–599.

Masciandaro, G., Macci, C., Peruzzi, E., Ceccanti, B., & Doni, S. (2013). Organic matter–microorganism–plant in soil bioremediation: A synergic approach. *Reviews in Environmental Science and Biotechnology, 12*, 399–419. doi:10.1007/s11157-013-9313-3.

Matsumoto, K., & Andriosopoulos, K. (2016). Energy security in East Asia under climate mitigation scenarios in the 21st century. *Omega, 59*(Part A), 60–71.

Mauffret, A., Baran, N., & Joulian, C. (2017). Effect of pesticides and metabolites on groundwater bacterial community. *Science of the Total Environment, 576*, 879–887.

McBratney, A., Field, D. J., & Koch, A. (2014). The dimensions of soil security. *Geoderma, 213*, 203–213. http://dx.doi.org/10.1016/j.geoderma.2013.08.013.

McBratney, A. B., Field, D. J., Morgan, C. L. S., & Jarrett, L. E. (2017). Soil security: A rationale. In D. J. Field et al. (Eds.), *Global soil security, progress in soil science* (pp. 3–14). Cham, Switzerland: Springer. doi:10.1007/978-3-319-43394-3_1.

McBratney, A. B., & Jarrett, L. E. (2017). Securitisation. In D. J. Field et al. (Eds.), *Global soil security, progress in soil science* (pp. 437–441). Cham, Switzerland: Springer. doi:10.1007/978-3-319-43394-3_40.

McCarl, B. A. (2015). Elaborations on climate adaptation in U.S. Agriculture. *Choices 2nd Quarter, 30*.

McCarl, B. A. (2017). Economics, energy, climate change, and soil security. In D. J. Field et al. (Eds.), *Global soil security, progress in soil science* (pp. 195–205). Cham, Switzerland: Springer. doi:10.1007/978-3-319-43394-3_17.

Meehl, G. A., et al. (2007). In S. Solomon, D. Qin, M. Z. Chen, M. Marquis, K. B. Averyt, M. Tignor, & H. H. Miller (Eds.), *Climate change 2007: The physical science basis, Contribution of working group I to the fourth assessment report of the intergovernmental panel on climate change* (pp. 748–845). Cambridge: Cambridge University Press.

Meena, M. D., Joshi, P. K., Jat, H. S., Chinchmalatpure, A. R., Narjary, B., Sheoran, P., & Sharma, D. K. (2016). Changes in biological and chemical properties of saline soil amended with municipal solid waste compost and chemical fertilizers in a mustard–pearl millet cropping system. *Catena, 140*, 1–8.

Meier, I. C., Finzi, A. C., & Phillips, R. P. (2017). Root exudates increase N availability by stimulating microbial turnover of fast-cycling N pools. *Soil Biology and Biochemistry, 106*, 119–128.

Mele, P. M. (2011). Soil biota, soil health and global change. In B. P. Singh et al. (Eds.), *Soil health and climate change, Soil biology* (Vol. 29). Berlin: Springer-Verlag. doi:10.1007/978-3-642-20256-8_8.

Meng, L., & Zhu, Y. G. (2011). Pyrene biodegradation in an industrial soil exposed to simulated rhizodeposition: How does it affect functional microbial abundance? *Environmental Science & Technology, 45*, 1579–1585.

Miki, T. (2012). Microbe-mediated plant–soil feedback and its roles in a changing world. *Ecological Research, 27*, 509–520. doi:10.1007/s11284-012-0937-5.

Minasny, B., Malone, B. P., McBratney, A. B., Angers, D. A., Arrouays, D., Chambers, A., Chaplot, V., Chen, Z.-S., Cheng, K., Das, B. S., Field, D. J., Gimona, A., Hedley, C. B., Hong, S. Y., Mandal, B., Marchant, B. P., Martin, M., McConkey, B. G., Mulder, V. L., O'Rourke, S., Richer-de-Forges, A. C., Odeh, I., Padarian, J., Paustian, K., Pan, G., Poggio, L., Savin, I., Stolbovoy, V., Stockmann, U., Sulaeman, Y., Tsui, C.-C., Vågen, T.-G., van Wesemael, B., & Winowiecki, L. (2017). Soil carbon 4 per mille. *Geoderma, 292*, 59–86.

Misra, A. K. (2014). Climate change and challenges of water and food security. *International Journal of Sustainable Built Environment, 3*(1), 153–165.

Mitton, F. M., Gonzalez, M., Pena, A., & Miglioranza, K. S. B. (2012). Effects of amendments on soil availability and phytoremediation potential of aged p,p′-DDT, p,p′-DDE and p,p′-DDD residues by willow plants (Salix sp.) *Journal of Hazardous Materials, 203*, 62–68.

Mohammadi, M. F., Jalali, S. G., Kooch, Y., & Said-Pullicino, D. (2017). The effect of landform on soil microbial activity and biomass in a Hyrcanian oriental beech stand. *Catena, 149*(Part 1), 309–317.

Mommer, L., Kirkegaard, J., & van Ruijven, J. (2016). Root–root interactions: Towards a rhizosphere framework. *Trends in Plant Science, 21*(3), 209–217.

Montiel-Rozas, M. M., Madejón, E., & Madejón, P. (2016). Effect of heavy metals and organic matter on root exudates (low molecular weight organic acids) of herbaceous species: An assessment in sand and soil conditions under different levels of contamination. *Environmental Pollution, 216*, 273–281.

Mueller, C. A., Obermeier, M. M., & Berg, G. (2016). Bioprospecting plant-associated microbiomes. *Journal of Biotechnology, 235*, 171–180.

Mueller, U. G., & Sachs, J. L. (2015). Engineering microbiomes to improve plant and animal health. *Trends in Microbiology, 23*, 606–617. doi:10.1016/j. tim.2015.07.009.

Munroe, J. W., Soto, G., Filho, E. d. M. V., Fulthorpe, R., & Isaac, M. E. (2015). Soil microbial and nutrient properties in the rhizosphere of coffee under agroforestry management. *Applied Soil Ecology, 93*, 40–46.

Murphy, B. (2017). Testing the links between soil security, sustainable land management practices and land evaluation. In D. J. Field et al. (Eds.), *Global soil security, progress in soil science* (pp. 87–97). Cham, Switzerland: Springer. doi:10.1007/978-3-319-43394-3_8.

Nalam, V. J., & Nachappa, P. (2014). The role of roots in plant defense responses to aboveground herbivores. In A. Morte & A. Varma (Eds.), *Root engineering*, *Soil biology* (Vol. 40). Berlin: Springer-Verlag. doi:10.1007/978-3-642-54276-3_17.

Nannipieri, P., Ascher, J., Ceccherini, M. T., Landi, L., Pietramellara, G., Renella, G., & Valor, F. (2008). Effects of root exudates in microbial diversity and activity in rhizosphere soils. In C. S. Nautiyal & P. Dion (Eds.), *Molecular mechanisms of plant and microbe coexistence, Soil biology* (Vol. 15). Berlin: Springer-Verlag. doi:10.1007/978-3-540-75575-3.

Neumann, G. (2007). Root exudates and nutrient cycling. In P. Marschner & Z. Rengel (Eds.), *Soil biology, Nutrient cycling in terrestrial ecosystems* (Vol. 10). Berlin: Springer-Verlag.

Nguyen, T. X. T., Amyot, M., & Labrecque, M. (2017). Differential effects of plant root systems on nickel, copper and silver bioavailability in contaminated soil. *Chemosphere, 168*, 131–138.

Nietfeld, H., Prenzel, J., Helmisaari, H.-S., Polle, A., & Beese, F. (2017). Modeling of mineral nutrient uptake of spruce tree roots as affected by the ion dynamics in the rhizosphere: Upscaling of model results to field plot scale. *Ecological Modelling, 345*, 150–164.

Nogueira, V., Lopes, I., Rocha-Santos, T., Santos, A. L., Rasteiro, G. M., Antunes, F., Gonçalves, F., Soares, A. M. V. M., Cunha, A., Almeida, A., Gomes, N. N. C. M., & Pereira, R. (2012). Impact of organic and inorganic nanomaterials in the soil microbial community structure. *Science of the Total Environment, 424*, 344–350.

Noori, M. S. S., & Saud, H. M. (2012). Potential plant growth promoting activity of Pseudomonas sp isolated from paddy soil in Malaysia as biocontrol agent. *Plant Pathology and Microbiology, 3*(2), 1–4.

Ofek, M., Voronov-Goldman, M., Hadar, Y., & Minz, D. (2014). Host signature effect on plant root-associated microbiomes revealed through analyses of resident vs. active communities. *Environmental Microbiology, 16*, 2157–2167. doi:10.1111/1462-2920.12228.

Otero, X. L., González-Guzman, A., Souza-Junior, V. S., Pérez-Alberti, A., & Macías, F. (2015). Soil processes and nutrient bioavailability in the rhizosphere of *Bolax gummifera* in a subantarctic environment (Martial Mountains, Ushuaia—Argentina). *Catena, 133*, 432–440.

Oyelami, A. O., & Semple, K. T. (2015). Impact of carbon nanomaterials on microbial activity in soil. *Soil Biology and Biochemistry, 86*, 172–180.

Palanques, A., Lopez, L., Guillén, J., Puig, P., & Masqué, P. (2017). Decline of trace metal pollution in the bottom sediments of the Barcelona City continental shelf (NW Mediterranean). *Science of the Total Environment, 579*, 755–767.

Parelho, C., Rodrigues, A. S., Barreto, M. C., Ferreira, N. G. C., & Garcia, P. (2016). Assessing microbial activities in metal contaminated agricultural volcanic soils—An integrative approach. *Ecotoxicology and Environmental Safety, 129*, 242–249.

Park, S., Y-S, S., & Hegeman, A. D. (2014). Plant metabolomics for plant chemical responses to Belowground community change by climate change. *Journal of Plant Biology, 57*, 137–149. doi:10.1007/s12374-014-0110-5.

Partridge, T., Thomas, M., Harthorn, B. H., Pidgeon, N., Hasell, A., Stevenson, L., & Enders, C. (2017). Seeing futures now: Emergent US and UK views on shale development, climate change and energy systems. *Global Environmental Change, 42*, 1–12.

Pérez, I., Janssen, M. A., & Anderies, J. M. (2016). Food security in the face of climate change: Adaptive capacity of small-scale social-ecological systems to environmental variability. *Global Environmental Change, 40*, 82–91.

Perlatti, F., Ferreira, T. O., Sartor, L. R., & Otero, X. L. (2016). Copper biogeochemistry in response to Rhizosphere soil processes under four native plant species growing spontaneously in an abandoned mine site in NE Brazil. *Water, Air, and Soil Pollution, 227*, 142. doi:10.1007/s11270-016-2840-0.

Plouznikoff, K., Declerck, S., & Calonne-Salmon, M. (2016). Mitigating abiotic stresses in crop plants by arbuscular mycorrhizal fungi. In C. M. F. Vos & K. Kazan (Eds.), *Belowground defence strategies in plants, signaling and communication in plants*. Cham, Switzerland: Springer. doi:10.1007/978-3-319-42319-7_15.

Prasanna, R., Kanchan, A., Kaur, S., Ramakrishnan, B., Ranjan, K., Singh, M. C., Hasan, M., Saxena, A. K., & Shivay, Y. S. (2016). Chrysanthemum growth gains from beneficial microbial interactions and fertility improvements in soil under protected cultivation. *Horticultural Plant Journal*. (in press).

Rajkumar, M., Prasad, M. N. V., Swaminathan, S., & Freitas, H. (2013). Climate change driven plant–metal–microbe interactions. *Environment International, 53*, 74–86.

Rao, Z., Guo, W., Cao, J., Shi, F., Jiang, H., & Li, C. (2017). Relationship between the stable carbon isotopic composition of modern plants and surface soils and climate: A global review. *Earth-Science Reviews, 165*, 110–119.

Rashid, M. I., Mujawar, L. H., Shahzad, T., Almeelbi, T., Ismail, I. M. I., & Oves, M. (2016). Bacteria and fungi can contribute to nutrients bioavailability and aggregate formation in degraded soils. *Microbiological Research, 183*, 26–41.

Rezanezhad, F., Couture, R.-M., Kovac, R., O'Connell, D., & Van Cappellen, P. (2014). Water table fluctuations and soil biogeochemistry: An experimental approach using an automated soil column system. *Journal of Hydrology, 509*, 245–256.

Rigolot, C., de Voil, P., Douxchamps, S., Prestwidge, D., Van Wijk, M., Thornton, P. K., Rodriguez, D., Henderson, B., Medina, D., & Herrero, M. (2017). Interactions between intervention packages, climatic risk, climate change and food security in mixed crop–livestock systems in Burkina Faso. *Agricultural Systems, 151*, 217–224.

Rugova, A., Puschenreiter, M., Koellensperger, G., & Hann, S. (2017). Elucidating rhizosphere processes by mass spectrometry: A review. *Analytica Chimica Acta*. doi:10.1016/j.aca.2016.12.044.

Sabannavar, S. J., & Lakshman, H. C. (2009). Effect of rock phosphate solubilization using mycorrhizal fungi and phosphobacteria on two high yielding varieties of *Sesamum indicum* L. *World Journal of Agricultural Sciences, 5*(4), 470–479.

Saha, N., Rahman, M. S., Ahmed, M. B., Zhou, J. L., Ngo, H. H., & Guo, W. (2017). Industrial metal pollution in water and probabilistic assessment of human health risk. *Journal of Environmental Management, 185*, 70–78.

Sahoo, R. K., Ansari, M. W., Tuteja, R., & Tuteja, N. (2015). Salt tolerant *SUV3* overexpressing transgenic rice plants conserve physicochemical properties and microbial communities of rhizosphere. *Chemosphere, 119*, 1040–1047.

Salawitch, R. J., Canty, T. P., Hope, A. P., Tribett, W. R., & Bennett, B. F. (2017). *Paris climate agreement: Beacon of hope*. Cham, Switzerland: Springer. doi:10.1007/978-3-319-46939-3.

Saleem, M., Pervaiz, Z. H., & Traw, M. B. (2015). Theories, mechanisms and patterns of microbiome species coexistence in an era of climate change. In M. Saleem et al. (Eds.), *Microbiome community ecology, springer briefs in ecology*. Cham, Switzerland: Springer. doi:10.1007/978-3-319-11665-5_2.

Sammarco, M., Tse, R., Pau, G., & Marfia, G. (2017). Using geosocial search for urban air pollution monitoring. *Pervasive and Mobile Computing, 35*, 15–31.

San Emeterio, L., Múgica, L., Ugarte, M. D., Goicoa, T., & Canals, R. M. (2016). Sustainability of traditional pastoral fires in highlands under global change: Effects on soil function and nutrient cycling. *Agriculture, Ecosystems & Environment, 235*, 155–163.

Schimel, J., Balser, T. C., & Wallenstein, M. (2007). Microbial stress response physiology and its implications for ecosystem function. *Ecology, 88*, 1386–1394.

Schlich, K., Beule, L., & Hund-Rinke, K. (2016). Single versus repeated applications of CuO and Ag nanomaterials and their effect on soil microflora. *Environmental Pollution, 215*, 322–330.

Schlich, K., & Hund-Rinke, K. (2015). Influence of soil properties on the effect of silver nanomaterials on microbial activity in five soils. *Environmental Pollution, 196*, 321–330.

Seifert, C. A., & Lobell, D. B. (2015). Response of double cropping suitability to climate change in the United States. *Environmental Research Letters, 10*, 2.

Shade, A., Peter, H., Allison, S. D., et al. (2012). Fundamentals of microbial community resistance and resilience. *Frontiers of Aquatic Microbiology, 3*, 417.

Shan, Q., Yu, Y., Yu, J., & Zhang, J. (2008). Soil enzyme activities and their indication for fertility of urban forest soil. *Frontiers of Environmental Science & Engineering China, 2*(2), 218–223. doi:10.1007/s11783-008-0037-5.

Shcherbakova, E. N., Shcherbakov, A. V., Andronov, E. E., Gonchar, L. N., Kalenskaya, S. M., & Chebotar, V. K. (2017). Combined pre-seed treatment with microbial inoculants and Mo nanoparticles changes composition of root exudates and rhizosphere microbiome structure of chickpea (*Cicer arietinum* L.) plants. *Symbiosis*. doi:10.1007/s13199-016-0472-1.

Shi, J., Li, J., Zhang, D. D., Zheng, J., Shi, S., Ge, Q., Lee, H. F., Zhao, Y., Zhang, J., & Lu, H. (2017). Two centuries of April-July temperature change in southeastern China and its influence on grain productivity. *Science Bulletin, 62*(1), 40–45.

Short, F. T., Kosten, S., Morgan, P. A., Malone, S., & Moore, G. E. (2016). Impacts of climate change on submerged and emergent wetland plants. *Aquatic Botany, 135*, 3–17.

Singh, B. K., Bardgett, R. D., Smith, P., & Reay, D. S. (2010). Microorganisms and climate change: Terrestrial feedbacks and mitigation options. *Nature Reviews. Microbiology, 8*, 779–790.

Singh, G., & Mukerji, K. G. (2006). Root exudates as determinant of rhizospheric microbial biodiversity. In K. G. Mukerji, C. Manoharachary, & J. Singh (Eds.), *Soil biology, Microbial activity in the rhizosphere* (Vol. 7). Berlin: Springer-Verlag.

Singh, N., Srivastava, S., Rathaur, S., & Singh, N. (2016). Assessing the bioremediation potential of arsenic tolerant bacterial strains in rice rhizosphere interface. *Journal of Environmental Sciences, 48*, 112–119.

Skarpaas, O., Meineri, E., Bargmann, T., Pötsch, C., Töpper, J., & Vandvik, V. (2016). Biomass partitioning in grassland plants along independent gradients in temperature and precipitation. *Perspectives in Plant Ecology, Evolution and Systematics, 19*, 1–11.

Smith, L., Inman, A., Lai, X., Zhang, H., Fanqiao, M., Jianbin, Z., Burke, S., Rahn, C., Siciliano, G., Haygarth, P. M., Bellarby, J., & Surridge, B. (2017). Mitigation of diffuse water pollution from agriculture in England and China, and the scope for policy transfer. *Land Use Policy, 61*, 208–219.

Sofi, J. A., Lone, A. H., Ganie, M. A., Dar, N. A., Bhat, S. A., Mukhtar, M., Dar, M. A., & Ramzan, S. (2016). Soil microbiological activity and carbon dynamics in the current climate change scenarios: A review. *Pedosphere, 26*(5), 577–591.

Solanki, A. S., Kumar, V., & Sharma, S. (2011). AM fungi, *A. chroococcum*, yield, nutrient uptake and economics of *Chlorophytum borivillianum* in Indian arid region. *Journal of Agricultural Technology, 7*(4), 983–991.

Strickland, M. S., Wickings, K., & Bradford, M. A. (2012). The fate of glucose, a low molecular weight compound of root exudates, in the belowground food web of forests and pastures. *Soil Biology and Biochemistry, 49*, 23–29.

Su, J.-Q., Xia, Y., Yao, H.-Y., Li, Y.-Y., An, X.-L., Singh, B. K., Zhang, T., & Zhu, Y.-G. (2017). Metagenomic assembly unravel microbial response to redox fluctuation in acid sulfate soil. *Soil Biology and Biochemistry, 105*, 244–252.

Sun, J., Liu, J., Yu, M., Wang, C., Sun, Y., Zhang, A., Wang, T., Lei, Z., & Jiang, G. (2013). *In vivo* metabolism of 2,2′,4,4′-tetrabromodiphenyl ether (BDE-47) in young whole pumpkin plant. *Environmental Science & Technology, 47*, 3701–3707.

Swamy, M. K., Akhtar, M. S., & Sinniah, U. R. (2016). Root exudates and their molecular interactions with rhizospheric microbes. In K. R. Hakeem & M. S. Akhtar (Eds.), *Plant, soil and microbes*. Cham, Switzerland: Springer. doi:10.1007/978-3-319-29573-2_4.

Tan, M. L., Ibrahim, A., Yusop, Z., Chua, V. P., & Chan, N. W. (2017). Climate change impacts under CMIP5 RCP scenarios on water resources of the Kelantan River basin, Malaysia. *Atmospheric Research, 189*, 1–10.

Taniguchi, M., Masuhara, N. & Burnett, K. (2015). Water, energy, and food security in the Asia Pacific region. *Journal of Hydrology: Regional Studies*. (in press).

Tao, F., Feng, Z., Tang, H., Chen, Y., & Kobayashi, K. (2017). Effects of climate change, CO_2 and O_3 on wheat productivity in eastern China, singly and in combination. *Atmospheric Environment, 153*, 182–193.

Tautges, N. E., Sullivan, T. S., Reardon, C. L., & Burke, I. C. (2016). Soil microbial diversity and activity linked to crop yield and quality in a dry land organic wheat production system. *Applied Soil Ecology, 108*, 258–268.

Tiwari, B., Sellamuthu, B., Ouarda, Y., Drogui, P., Tyagi, R. D., & Buelna, G. (2017). Review on fate and mechanism of removal of pharmaceutical pollutants from wastewater using biological approach. *Bioresource Technology, 224*, 1–12.

Todd-Brown, K. E. O., Hopkins, F. M., Kivlin, S. N., Talbot, J. M., & Allison, S. D. (2012). A framework for representing microbial decomposition in coupled climate models. *Biogeochemistry, 109*, 19–33. doi:10.1007/s10533-011-9635-6.

Tong, J., Miaowen, C., Juhui, J., Jinxian, L., & Baofeng, C. (2017). Endophytic fungi and soil microbial community characteristics over different years of phytoremediation in a copper tailings dam of Shanxi, China. *Science of the Total Environment, 574*, 881–888.

Touceda-González, M., Álvarez-López, V., Prieto-Fernández, Á., Rodríguez-Garrido, B., Trasar-Cepeda, C., Mench, M., Puschenreiter, M., Quintela-Sabarís, C., Macías-García, F., & Kidd, P. S. (2017). Aided phytostabilisation reduces metal toxicity, improves soil fertility and enhances microbial activity in cu-rich mine tailings. *Journal of Environmental Management, 186*(Part 2), 301–313.

Tripathi, V., Fraceto, L. F., & Abhilash, P. C. (2015). Sustainable clean-up technologies for soils contaminated with multiple pollutants: Plant-microbe-pollutant and climate nexus. *Ecological Engineering, 82*, 330–335.

Turrini, A., Sbrana, C., & Giovannetti, M. (2015). Belowground environmental effects of transgenic crops: A soil microbial perspective. *Research in Microbiology, 166*(3), 121–131.

Uitto, J. I., Puri, J., & van den Berg, R. D. (2017). *Evaluating climate change action for sustainable development*. Cham, Switzerland: Springer. doi:10.1007/978-3-319-43702-6.

van Dam, N. M., & Bouwmeester, H. J. (2016). Metabolomics in the Rhizosphere: Tapping into Belowground chemical communication. *Trends in Plant Science, 21*(3), 256–265.

van Loon, L. C. (2016). The intelligent behavior of plants. *Trends in Plant Science, 21*(4), 286–294.

Verbon, E. H., & Liberman, L. M. (2016). Beneficial microbes affect endogenous mechanisms controlling root development. *Trends in Plant Science, 21*(3), 218–229.

Vergani, L., Mapelli, F., Zanardini, E., Terzaghi, E., Di Guardo, A., Morosini, C., Raspa, G., & Borin, S. (2017). Phyto-rhizoremediation of polychlorinated biphenyl contaminated soils: An outlook onplant-microbe beneficial interactions. *Science of the Total Environment, 575*, 1395–1406.

Vos, C. M. F., & Kazan, K. (2016). *Belowground defence strategies in plants, Signaling and communication in plants series*. Cham, Switzerland: Springer. doi:10.1007/978-3-319-42319-7.

Wang, X., Ciais, P., Li, L., Ruget, F., Vuichard, N., Viovy, N., Zhou, F., Chang, J., Wu, X., Zhao, H., & Piao, S. (2017a). Management outweighs climate change on affecting length of rice growing period for early rice and single rice in China during 1991–2012. *Agricultural and Forest Meteorology, 233*, 1–11.

Wang, M., Faber, J. H., & Chen, W. (2017b). Application of stress index in evaluating toxicological response of soil microbial community to contaminants in soils. *Ecological Indicators, 75*, 118–125.

Wang, Y., Hu, N., Ge, T., Kuzyakov, Y., Z-L Wang, Z., Li, Z., Tang, Y., Chen, C. W., & Lou, Y. (2017c). Soil aggregation regulates distributions of carbon, microbial community and enzyme activities after 23-year manure amendment. *Applied Soil Ecology, 111*, 65–72.

Wang, H., Nie, Y., Butterly, C. R., Wang, L., Chen, Q., Tian, W., Song, B., Xi, Y., & Wang, Y. (2017d). Fertilization alters microbial community composition and functional patterns by changing the chemical nature of soil organic carbon: A field study in a Halosol. *Geoderma, 292*, 17–24.

Wang, Y., Gong, J.-R., Liu, M., Luo, Q., Xu, S., Pan, Y., & Zhai, Z. (2015). Effects of land use and precipitation on above- and below-ground litter decomposition in a semi-arid temperate steppe in Inner Mongolia, China. *Applied Soil Ecology, 96*, 183–191.

Winarso, S., Pandutama, M. H., & Purwanto, L. D. (2016). Effectively of humic substance extracted from palm oil compost as liquid fertilizer and heavy metal bioremediation. *Agriculture and Agricultural Science Procedia, 9*, 146–157.

Wu, J., Franzén, D., & Malmström, M. E. (2016). Nutrient flows following changes in source strengths, land use and climate in an urban catchment, Råcksta Träsk in Stockholm, Sweden. *Ecological Modelling, 338*, 69–77.

Wu, J., & Malmström, M. E. (2015). Nutrient loadings from urban catchments under climate change scenarios: Case studies in Stockholm, Sweden. *Science of the Total Environment, 518–519*, 393–406.

Wu, M., Li, W., Dick, W. A., Ye, X., Chen, K., Kost, D., & Chen, L. (2017). Bioremediation of hydrocarbon degradation in a petroleum-contaminated soil and microbial population and activity determination. *Chemosphere, 169*, 124–130.

Wu, Z. T., Dijkstra, P., Koch, G. W., Peñuelas, J., & Hungate, B. A. (2011). Responses of terrestrial ecosystems to temperature and precipitation change: A meta-analysis of experimental manipulation. *Global Change Biology, 17*, 927–942.

Xiao, W., Feng, S., Liu, Z., Su, Y., Zhang, Y., & He, X. (2017). Interactions of soil particulate organic matter chemistry and microbial community composition mediating carbon mineralization in karst soils. *Soil Biology and Biochemistry, 107*, 85–93.

Xu, G., Singh, S. K., Reddy, V. R., Barnaby, J. Y., Sicher, R. C., & Li, T. (2016). Soybean grown under elevated CO_2 benefits more under low temperature than high temperature stress: Varying response of photosynthetic limitations, leaf metabolites, growth, and seed yield. *Journal of Plant Physiology, 205*, 20–32.

Xu, W., Wang, Z., & Wu, F. (2015). Companion cropping with wheat increases resistance to *Fusarium* wilt in watermelon and the roles of root exudates in watermelon root growth. *Physiological and Molecular Plant Pathology, 90*, 12–20.

Xue, S., Yang, X., Liu, G., Gai, L., Zhang, C., Ritsema, C. J., & Geissen, V. (2017). Effects of elevated CO_2 and drought on the microbial biomass and enzymatic activities in the rhizospheres of two grass species in Chinese loess soil. *Geoderma, 286*, 25–34.

Yadav, B. K., Akhtar, M. S., & Panwar, J. (2015). Rhizospheric plant-microbe interactions: Key factors to soil fertility and plant nutrition. In N. K. Arora (Ed.), *Plant microbes Symbiosis: Applied facets*. New Delhi, India: Springer. doi:10.1007/978-81-322-2068-8_6.

Yang, L., Zhang, Y., & Li, F. (2012). Soil enzyme activities and soil fertility dynamics. In A. K. Srivastava (Ed.), *Advances in citrus nutrition*. New York: Springer. doi:10.1007/978-94-007-4171-3_11.

Yazdani, M., Bahmanyar, M. A., Pirdashti, H., & Esmaili, M. A. (2009). Effect of phosphate solubilization microorganisms (PSM) and plant growth promoting rhizobacteria (PGPR) on yield and yield components of corn (*Zea mays* L.) *Proceedings of World Academy of Science, Engineering and Technology, 37*, 90–92.

Yuan, Z., Liu, H., Han, J., Sun, J., Wu, X., & Yao, J. (2017). Monitoring soil microbial activities in different cropping systems using combined methods. *Pedosphere, 27*(1), 138–146.

Zhang, Y., Dong, S., Gao, Q., Liu, S., Zhou, H., Ganjurjav, H., & Wang, X. (2016a). Climate change and human activities altered the diversity and composition of soil microbial community in alpine grasslands of the Qinghai-Tibetan Plateau. *Science of the Total Environment, 562*, 353–363.

Zhang, C., Mora, P., Dai, J., Chen, X., Giusti-Miller, S., Ruiz-Camacho, N., Velasquez, E., & Lavelle, P. (2016b). Earthworm and organic amendment effects on microbial activities and metal availability in a contaminated soil from China. *Applied Soil Ecology, 104*, 54–66.

Zhang, H., Huang, B., Dong, L., Hu, W., Akhtar, M. S., & Qu, M. (2017a). Accumulation, sources and health risks of trace metals in elevated geochemical background soils used for greenhouse vegetable production in southwestern China. *Ecotoxicology and Environmental Safety, 137*, 233–239.

Zhang, J., Li, M., & Zheng, G. (2017b). Effect of stand age on soil microbial community structure in wolfberry (*Lycium barbarum* L.) fields. *Acta Ecologica Sinica, 37*(1), 10–17.

Zhang, Q., Yang, J., Koide, R. T., Li, T., Yang, H., & Chu, J. (2017c). A meta-analysis of soil microbial biomass levels from established tree plantations over various land uses, climates and plant communities. *Catena, 150*, 256–260.

Zhang, Z., Yuan, Y., Zhao, W., He, H., Li, D., He, W., Liu, Q., & Yin, H. (2017d). Seasonal variations in the soil amino acid pool and flux following the conversion of a natural forest to a pine plantation on the eastern Tibetan Plateau, China. *Soil Biology and Biochemistry, 105*, 1–11.

Zhang, P., Zhang, J., & Chen, M. (2017e). Economic impacts of climate change on agriculture: The importance of additional climatic variables other than temperature and precipitation. *Journal of Environmental Economics and Management, 83*, 8–31.

Zhang, N., Wang, D., Liu, Y., Li, S., Shen, Q., & Zhang, R. (2014). Effects of different plant root exudates and their organic acid components on chemotaxis, biofilm formation and colonization by beneficial rhizosphere-associated bacterial strains. *Plant and Soil, 374*, 689–700. doi:10.1007/s11104-013-1915-6.

Zhang, X., & Vesselinov, V. V. (2017). Integrated modeling approach for optimal management of water, energy and food security nexus. *Advances in Water Resources, 101*, 1–10.

Zhu, S., Huang, X., Ho, S.-H., Wang, L., & Yang, J. (2017). Effect of plant species compositions on performance of lab-scale constructed wetland through investigating photosynthesis and microbial communities. *Bioresource Technology, 229*, 196–203.

Zimmer, C. (2010). *The microbe factor and its role in our climate future*. Retrieved December 15, 2015, from http://e360.yale.edu/feature/the_microbe_factor_and_its_role_in_our_climate_future/2279/.

Zuverza-Mena, N., Martínez-Fernandez, D., Du, W., Hernandez-Viezcas, J. A., Bonilla-Bird, N., Lopez-Moreno, M. L., Komarek, M., Peralta-Videa, J. R., & Gardea-Torresdey, J. L. (2017). Exposure of engineered nanomaterials to plants: Insights into the physiological and biochemical responses: A review. *Plant Physiology and Biochemistry, 110*, 236–264.

Chapter 12
Regulatory Role of Mineral Nutrients in Nurturing of Medicinal Legumes Under Salt Stress

M. Naeem, Abid A. Ansari, Sarvajeet Singh Gill, Tariq Aftab, Mohd. Idrees, Akbar Ali, and M. Masroor A. Khan

Abstract Medicinal legumes (Fabaceae family) bear primary and secondary metabolites and other important compounds such as nutraceuticals, pharmaceuticals, and other useful products. The medicinal legumes are potential sources of glycosides (aloe-emodin, chrysophenol, emodin, rhein, etc.), antibiotics, flavonoids, alkaloids and phytochemicals. Hence, it is the need of the hour to escalate the yield and quality of these legumes. In fact, balanced nutrition of crop plants plays a vital role in sustaining the yield and quality of medicinal plants together with maintaining the fertility status of soils on long-term basis. The role of mineral nutrition is of vital importance in the cultivation of these plants. The yield of most crop plants increases linearly with the amount of fertilizers absorbed. With a balanced mineral nutrients supply, the maximum genetic potential of plants can be realized successfully. The productivity as well as the quality of a crop is affected by environmental factors; the uptake and utilization of mineral nutrients from the soil, or from the

M. Naeem (✉) • T. Aftab • M.M.A. Khan
Plant Physiology Section, Department of Botany, Aligarh Muslim University, Aligarh 202002, India
e-mail: naeemgaur@gmail.com

A. Ali
Department of Botany, Dayalbagh Educational Institute, Agra 282005, India
Plant Physiology Section, Department of Botany, Aligarh Muslim University, Aligarh 202002, India
e-mail: akbar.agra.alig@gmail.com

A.A. Ansari
Department of Biology, Faculty of Science, University of Tabuk, Tabuk 71491, Saudi Arabia

S.S. Gill
Stress Physiology and Molecular Biology Lab, Centre for Biotechnology, MD University, Rohtak 124001, Haryana, India

M. Idrees
Department of Mathematics and Sciences, College of Arts and Applied Sciences, Dhofar University, P. O. Box 2509, Salalah 211, Sultanate of Oman

M. Naeem et al. (eds.), *Essential Plant Nutrients*,
DOI 10.1007/978-3-319-58841-4_12

fertilizers applied to the plants is of prime importance. Among the macro-nutrients, nitrogen (N), phosphorus (P), potassium (K) and calcium (Ca) are major components of metabolic molecules that affect the growth and metabolism of plants significantly. They also play important structural and physiological roles in the overall development of plants. On the other hand, out of many crop production strategies, foliar application of mineral fertilizers in the form of aqueous sprays is also a successful method of administering it. Salinity is a major cause of decrease in agricultural productivity and the increasing level of salt accumulation in soils has become an exigent issue for Indian soils. However, as in the majority of cultivated plants, growth and yield of medicinal plants can be affected by environmental constraints such as salinity and drought. The above technique could be applied to ameliorate the productivity and quality of these medicinal legumes under salinity stress. This review covers the possible role of mineral nutrients application on selected medicinal legumes as well as their ameliorative effects on these legumes under salt stress.

Keywords Medicinal legumes • Nutrients • Phytochemicals • Salt stress

12.1 Introduction

Various environmental stresses such as high winds, extreme temperatures, soil salinity, drought and flood have affected the production and cultivation of agricultural crops. Among these factors, soil salinity is one of the devastating environmental stresses, which causes major reductions in cultivated land area, crop productivity and quality (Yamaguchi and Blumwald 2005; Shahbaz and Ashraf 2013). It has been stated that an approximate area of seven million hectares of land is covered by saline soil in India (Patel et al. 2011). Salt stressed soils are known to restrain the growth of plants (Paul 2012). According to Jamil et al. (2011) study, more than 50% of the arable land would be salinized by the year 2050. An excess of soluble salts in the soil causes osmotic stress, specific ion toxicity, and ionic imbalances that lead to plant death or considerable yield losses in the crop plants. A number of studies on salt stress response and tolerance in many crop plants have progressed rapidly in recent years. Salinity has highly adverse impacts on growth and productivity of agricultural plants worldwide. Soil salinity, resulting from natural processes or from crop irrigation with saline water, occurs in many arid and semi-arid regions of the world (Läuchli and Epstein 1990). It is reported that 20% of the cultivated lands are salt-affected worldwide. Alleviation of salinity stress is the main issue in salt-affected regions to ensure agricultural sustainability. An excess of soluble salts in the soil leads to osmotic stress, specific ion toxicity, and ionic imbalances (Munns et al. 1995 (Fig. 12.1); Ashraf 2004; Munns et al. 2006), leading to plant death or yield losses both in conventional crop species and medicinal plants (Rout and Shaw 2001). All these factors cause adverse effects on plant growth and development at physiological and biochemical levels (Munns and James 2003), and at the molecular level (Tester and Davenport 2003). Salinity stress generates oxidative stress in

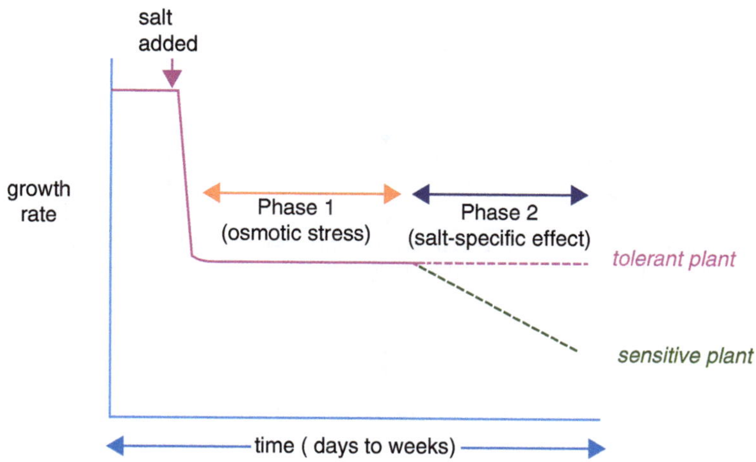

Fig. 12.1 Scheme of the two-phase growth response to salinity (Source: Munns et al. 1995)

plant tissues (Abel et al. 2003), which is manifested by ROS such as singlet oxygen, superoxide anion, hydrogen peroxide, and hydroxyl radicals (Gosset et al. 1994). Plants have evolved various protective mechanisms to eliminate or reduce ROS. Free radical reactions, in participation with oxidative radicals, have been shown to be involved in many biological reactions, causing damage to lipids, proteins, membranes, and nucleic acids, thus giving rise to a variety of metabolic disorders (Cavalcanti et al. 2006).

Salinity affects almost all aspects of plant development such as germination, vegetative growth and reproductive development hindering seed germination, seedling growth, enzyme activity (Seckin et al. 2009; Horie et al. 2012), DNA, RNA, protein synthesis and mitosis (Tabur and Demir 2010; Javid et al. 2011). Salinity imposes ion toxicity, osmotic stress, nutrients (N, Ca, K, P, Fe, Zn) deficiency and oxidative stress on plants, and thus limits water uptake from soil (Fig. 12.2). Moreover, soil salinity significantly reduces P uptake due to phosphate ions precipitate with Ca ions (Bano and Fatima 2009). Some elements, such as sodium, chlorine and boron, have specific toxic effects on plants. Accumulation of surplus sodium in cell wall can rapidly lead to osmotic stress, and consequently cell death (Munns 2002). Also salt stress disturbs photosynthesis mainly through a reduction in leaf area, chlorophyll pigment and stomatal conductance, and to a lesser extent through a decrease in photosystem II efficiency (Netondo et al. 2004).

Furthermore, morphological, physiological, biochemical, genetic and genomic analyses have made important discoveries of salt-responsive genes, proteins and metabolites involved in different pathways in the cells (Khan and Basha 2016). Plant salinity tolerance is a multifaceted physiological aspect that involves adaptations to signalling and metabolic networks. However, continuously growing environmental stresses such as salt stress cause retardation in growth and yield. Plants use various tolerance mechanisms against environmental challenges. Noticeably, salinity has been reported to affect growth, mineral nutrition, and the yield and composition of medicinal legumes (Khan and Basha 2016).

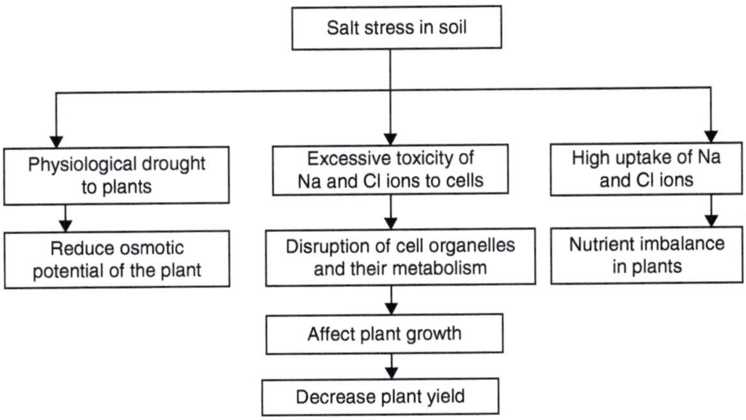

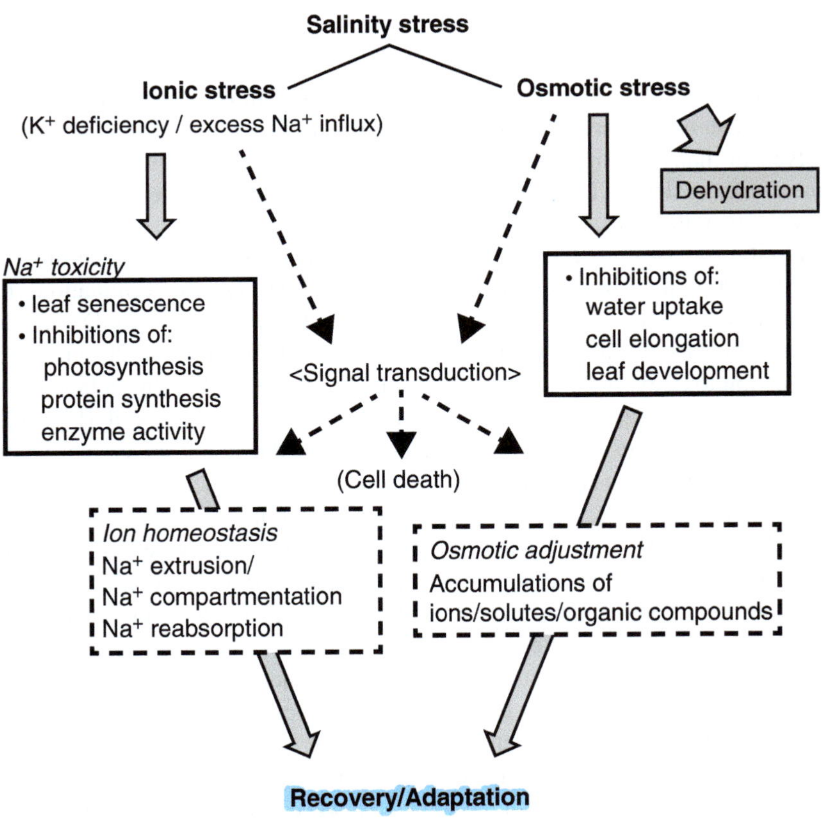

Fig. 12.2 Adaptive responses of plants to salt stress (Source: Horie et al. 2012)

Generally, legumes used as food crops, play an imperative nutritional role in the diet of millions of people living in the developing countries. They provide a significant part of the diet of vegetarians, being vital sources of protein, calcium, iron, phosphorus and other minerals. Although, legumes are good sources of fodders, forages and green manures. Legumes cultivation continuously enhances fertility of soils through their involvement in the processes of nitrogen fixation. In fact, legumes require high amount of P and Ca for their growth, nodule formation, and N_2-fixation. Noticeably, various environmental stresses can severely affected productivity and quality of leguminous crops. Development of stress-tolerant cultivars of leguminous crops is an urgent need due to alarming threat of the global warming and changing climatic patterns explosion in the near future. Hence, exhaustive research efforts should be expanded for maximizing the yield of potentially useful leguminous medicinal plants particularly through scientific strategies. Therefore, proper agricultural and management practices (using mineral nutrients) are the best alternative that can help agriculturalists to boost crop productivity worldwide. However, foliar application (the application of nutrients through aerial parts of the plants specially leaves, and their absorption) is an alternative method for the administration of fertilizers (Wittwer and Teubner 1967).

Majority of the medicinal plants belong to angiospermic families, of which legume family (Fabaceae) is the third largest one, distributed in approximately 650 genera and 20,000 species. A handsome number of medicinal legumes are potential sources of glycosides (e.g. aloe-emodin, chrysophenol, emodin, and rhein, etc.), antibiotics, flavonoids, alkaloids, and phytochemicals, which are used in drug manufacturing by various pharmaceutical industries (Tyler et al. 1976; Morris 2003). Medicinal legumes produce primary and secondary metabolites and other phytochemicals such as nutraceuticals, pharmaceuticals, pesticides and other industrial products. Thus, escalation of yield and quality of the medicinal legumes is of paramount importance. The diverse ways to enhance yield and quality of medicinal legumes essentially include the supply of mineral nutrients as per the soil demand. In fact, balanced nutrition of crops plays a vital role in sustaining the productivity of medicinal plants together with maintaining the fertility status of soils on long-term basis.

For the cultivation of these medicinal legumes, the role of mineral nutrition is of primary importance. Using optimal quantity of fertilizer application, the yield and quality of these medicinally important leguminous plants could be improved in order to meet their increasing demands. Among the factors responsible for achieving higher yield, adequate nutrient supply is considered one of the most effective tools (Munsi 1992). Therefore, optimum concentrations of mineral nutrients should be set up to achieve the desired yield of these plants. Appropriate mineral nutrition is a basic requirement of every crop plant and deficiency of any of the minerals can result in great yield losses.

Of the essential mineral nutrients, N, P, K, and Ca are considered to be of prime importance as they are required by plants in large quantities. These nutrients, by virtue of their function in the generation of energy, producing the building molecules, participating in the repair of protoplasm and regulation of metabolic process, and maintaining the physical organization and function of livings cells, play several important roles in enhancing crop production and maintaining the soil fertility.

This chapter provides an overview of the responses of various medicinal leguminous plants under non stress and salt stress conditions and its mitigation through successful strategies (basal and foliar application of mineral nutrients) with regard to morphological, physiological and biochemical aspects including yield attributes as well as active constituents of the various medicinally important leguminous plants namely Black gram (*Phaseolus mungo* L.), *Cassia tora* (*Cassia obtusifolia* L.), Cluster bean (*Cyamopsis tetragonoloba* L. **Taub.**), Coffee senna (*Senna occidentalis* L.), Fenugreek (*Trigonella foenum-graecum* L.), French bean (*Phaseolus vulgaris* L.), Hyacinth bean (*Lablab purpureus* L.), Lucern (*Medicago sativa* L.) Mung bean (*Vigna radiata* L), Senna (*Cassia angustifolia* Vahl.), *Senna sophera* (*Cassia sophera* L.), and Soybean (*Glycine max* L.). However, general description and therapeutic uses of the above mentioned plants have earlier been reported (Naeem et al. 2012).

12.2 Responses of Medicinal Legumes Towards Salt Stress: Amelioration Through Mineral Nutrients

The effects of mineral nutrients on the growth, physiological and biochemical modifications as well as quality characteristics of the numerous medicinal legumes under normal and stress conditions are reviewed in the following manner.

12.2.1 Black gram (*Phaseolus mungo* L.)

Five different levels of salt (NaCl) on the germination of *P. vulgaris* L. seed was investigated by Cokkizgin (2012). He performed experiment on Petri dishes to determine the salt effects on germination of common bean. The germination of the cultivar (Tegmen) was studied using distilled water (control) and under osmotic potential of 0.3, 0.6, 0.9, 1.2, and 1.5 MPa NaCl. The results indicated that the Mean germination time (MGT), Germination index (GI), Coefficient of velocity of germination (CVG), Germination percentage (GP), and Seed vigor index (SVI) varied between 3.13 and 3.78 days, 6.88 and 3.93, 0.156 and 0.153, 90 and 60%, and 867.0 and 290.3, respectively. All the examined parameters were decreased with increasing NaCl concentration, except MGT. The maximum and minimum GI, GP, CVG, and SVI were observed at the control condition (0.0 MPa) and highest osmotic potential (1.5 MPa) of NaCl, respectively.

Devi et al. (2012) applied different concentrations of sodium chloride (NaCl) (i.e., 25, 50, 100 and 200 mM) that induced salinity in black gram caused significant reduction in germination, seedling growth, root and shoot lengths, fresh weight and leaf area. A reduction in protein content, and activity of nitrate reductase was observed in all the NaCl treated plants. However, the proline level, free amino acid, peroxidase and catalase activities were increased with the increase in the concentration of NaCl. Foliar application of 15 ppm IAA to the salt stressed plants caused an alleviating effect on the salt stressed plants and increased crop yield.

Surendar et al. (2013) studied the effect of basal N in combination with foliar spray of bioregulators and micronutrients on growth and productivity of black gram. They estimated photosynthetic pigments and foliage soluble protein content at different phenological phases of black gram. Seed yield were assessed at the time of harvest. There was significant increase in the leaf area index and specific leaf weight due to basal application of N (25 kg ha^{-1}) with foliar spray of urea 2% and 0.1 ppm brassinolide. The specific leaf area was also greatly increased by the basal application of N along with foliar spray of urea and brassinolide.

According to a study of Sangeetha and Subramani (2014), the morphological and biochemical changes of black gram (*V. mungo* L.) under NaCl stress were affected. The black gram seeds were treated with various concentrations viz., control, 10, 25, 50, 75, 100 and 150 ppm of NaCl. The growth parameters such as germination percentage, seedling growth and photosynthetic pigments were decreased with increasing concentration of NaCl treatment, while metabolites such as sugar, protein, and proline content were increased at increasing concentrations of NaCl.

Salinity reduces the ability of plant utilize water and causes a reduction in growth rates, as well as changes in plant metabolic processes (Velmani et al. 2015). Furthermore it decreased plant growth and yield including qualitative and quantitative changes in photosynthetic pigments, proteins, reducing sugar, starch content depending on the plant species.

A field experiment was also carried out the effect of growth regulator, organic and inorganic foliar nutrition on yield attributes of black gram (Mishra 2016). He studied the effect of basal application of N in combination with foliar spray of urea and plant growth regulators. Among the treatments, 25 kg ha^{-1} of N with foliar spray of urea 2% and 0.1 ppm brassinolide significantly expressed the higher values in growth attributes. Among the treatments T7 performed its superiority and had higher leaf area (102.7, 246.0, 567.0 and 494.3 mg g^{-1}) at vegetative stage, flowering stage, pod filling stage and harvest stage, respectively.

12.2.2 Cassia tora L. (*Cassia obtusifolia* L.)

You-Sheng et al. (2004) investigated the changes in concentrations of H_2O_2 and O^{2-}, some antioxidative enzyme activities and several physiological parameters involved in oxidative damage to plasma membrane in the root tips of *C. tora* L. Their results indicated that 20 mmol L^{-1} aluminum (Al) caused increases in electrolytes leakage, malondlaldehyde (MDA) content and intense staining with Evans blue in root tips, while treatment with 5 mmol L^{-1} SA suppressed the Al-induced increase in MDA.

12.2.3 Cluster bean (*Cyamopsis tetragonoloba* L.) Taub

Two year's field experiment was conducted by Garg et al. (2006) to study the effects of seed treatment (500 ppm) or foliar application of 1000 ppm of thiourea at 25 and 40 DAS or their combination, on growth, yield, net photosynthesis and nitrogen metabolism of cluster bean. Thiourea application either as pre-sowing seed treatment or as foliar spray significantly increased plant height, leaf area, dry matter production and seed yield as compared to the untreated control plants during both the years. However, maximum favourable effects were obtained with combined application of seed treatment and foliar spray. The beneficial effects of thiourea were noticed to increase the net photosynthetic rates and the concentrations of total chlorophyll and starch in the leaves. Thiourea also reflected a positive role in enhancing nitrogen metabolism as it significantly improved activity of nitrate reductase and soluble protein.

Further, 2-year field study on clusterbean (*C. tetragonoloba* L. Taub. cv. RGC-936) was conducted by Burman et al. (2007). The experiments were set up in a split-split-plot design with three levels of P (0, 20, and 40 kg ha^{-1}) and two levels of N (0 and 20 kg ha^{-1}) with and without thiourea application (seed treatment with 500 mg kg^{-1} followed by two foliar sprays of 1000 mg kg^{-1} each at 25 and 40 DAS). Phosphorous (P) and nitrogen (N) application either alone or in combination with thiourea resulted in significantly higher net photosynthetic rates and concentrations of chlorophyll, starch, soluble protein, and total free amino acids as well as nitrate reductase activity compared to control plants at both vegetative and flowering stages. Seed yield, dry-matter production, harvest index, and water-use efficiency were significantly enhanced by the above mentioned treatments. They concluded that the improvement of P and N status of arid-zone soils coupled with thiourea application could significantly improve the yield of clusterbean under rainfed conditions.

Manivasagaperumal et al. (2011) applied various concentrations of zinc sulphate (0, 10, 25, 50, 100,150 and 200 mg L^{-1}) on the seeds of cluster bean in order to assess the effect of metal on germination, growth and biochemical parameters. The results indicated that low level of zinc (10 and 25 mg L^{-1}) showed significant changes in the all the parameters; whereas the higher concentrations (50–200 mg L^{-1}) reduced the above studied attributes except proline content. Vijayarengan (2013) raised cluster bean plants in pots containing the soil amended with various levels of zinc (control, 50, 100, 150, 200 and 250 mg kg^{-1} soil). Zinc treatment at all levels tested (except 50 and 100 mg kg^{-1}) decreased the various growth and yield parameters, biochemical constituents of cluster bean plants. However the 50 and 100 mg kg^{-1} Zn level in the soil showed a positive effect on the overall growth and dry matter yield and biochemical constituents of cluster bean plants. But the proline content and antioxidant enzymes (catalase, peroxidase and polyphenol oxidase) of cluster bean plants decreased at low levels (50 and 100 mg kg^{-1}) and increased at high levels (150–250 mg kg^{-1}) of Zn in the soil.

Deepika and Dhingra (2014) assessed salinity stress tolerance status of the variety (HG 2–20) of cluster bean. Physiological parameters such as membrane injury of leaves and roots, and leaf succulence increased under salinity stress, while chlorophyll fluorescence, relative water content, aerial biomass and harvest index were decreased. Soluble sugars, amino acids and proline content increased significantly, while protein and starch content decreased with the increasing level of salinity. Salinity reduced the number of flowers per plant, stamen length, number of pollen per flower, pollen size and their fecundity, pistil length and number of ovules. Similarly, guar gum content of seeds, seed yield and number of pods per plant also decreased under salinity stress. Fathima and Sudha (2016) concluded that the combined application of major and micro nutrient @ 20:50:50 kg ha^{-1} N, P_2O_5, K_2O and 10:10 kg ha^{-1} $ZnSO_4$: $FeSO_4$ recorded higher seed yield of clusterbean, the crude gum, crude protein content, net returns and B:C ratio over other treatment combinations.

12.2.4 Coffee senna (*Senna occidentalis* L.)

Coffee senna (*Senna occidentalis* L.) plants were subjected to five P levels: 0, 25, 50, 75 and 100 mg P kg^{-1} soil (P_0, P_1, P_2, P_3 and P_4, respectively). Naeem and Khan (2009) conducted a pot experiment at AMU, Aligarh, India, under P-deficient soil. Their data indicated that soil-applied P significantly improved most of the attributes studied. Out of five P levels, 75 mg kg^{-1} soil (P_3) proved the best and enhanced fresh and dry weights, total chlorophyll and carotenoid content, nitrate reductase activity and leaf-NPK and Ca content, analyzed at 120, 270 and 300 DAS. The number of pods, seed-yield per plant and seed-protein content (330 DAS) were also significantly enhanced by the P_3 level, except the number of seeds per pod, 100-seed weight and total anthraquinone glycosides content, respectively. Physiological parameters (transpiration rate, stomatal conductance and net photosynthetic rate) were also enhanced by this treatment.

Plant biological yield was comparatively low in calcium-deficient soil of Aligarh, western Uttar Pradesh, India (Naeem et al. 2010). Calcium deficiency poses a serious yield and quality limitation for several crops, including medicinal herbs, in this region of India (Naeem 2007). Calcium application through soil enhanced crop productivity, photosynthetic efficiency, enzymatic activities, and contents of nutraceuticals in coffee senna, a medicinal legume. Plants were grown in pots containing soil supplied with five levels of Ca, viz. 0, 40, 80, 120 and 160 mg Ca kg^{-1} soil applied as $CaCl_2$. Calcium application (120 mg Ca kg^{-1} soil) significantly increased the growth, physiological, biochemical, yield, and quality attributes studied at three growth stages (120, 270, and 300 DAS). Further, it increased the seed-yield and seed-protein content by 27.6 and 10.6%, respectively, compared to the control.

12.2.5 Fenugreek (*Trigonella foenum-graecum* L.)

A study was carried out to assess the effect of applied Fe on yield and Fe enrichment of fenugreek grown in the Fe-deficient alkaline field (Chhibba et al. 2007). Both green and seed yield as well as Fe content increased significantly with Fe fertilizer in the Fe deficient soils. The crop was responded significantly to soil as well as the foliar application of Fe. They concluded that the foliar mode proved significantly superior as compared to soil application as reflected by the magnitude of increase in crop yield and the Fe concentration in plants.

Gour et al. (2009) conducted a field experiment to find out the effect of P and PGRs on growth and yield of fenugreek. They observed that 60 kg P ha^{-1} gave higher values of growth and yield (17.62 q ha^{-1}) of the crop. Pariari et al. (2009) carried out a field experiment to study the effect of boron (B) and zinc (Zn) on growth and seed yield of fenugreek. Zinc was applied in four concentrations (0.1, 0.2, 0.3 and 0.4%) and B was applied in two concentrations (0.1 and 0.2%). They reported that foliar application of B at 0.1% and Zn at 0.2% twice had been found to be effective in enhancing most of the yield attributes and seed yield of fenugreek. It is also observed that lower concentration is more effective than higher concentration.

Babar et al. (2014) conducted a study to alleviate the salinity-induced harmful effect on biomass production and physiochemical attributes of fenugreek by foliar application of salicylic acid. Two varieties (Deli Kabul and Kasuri) were grown in salt treated (100 mM NaCl) and untreated (0 mM NaCl) growth medium. Salinity stress significantly reduced the growth biomass in both varieties. Their results indicated that growth medium salinity induced reduction in biomass production, gas exchange attributes, and also chlorophyll contents.

Verma et al. (2014) studied the effect of vermicompost and sulphur on growth, yield and nutrient uptake of fenugreek. They applied 16 treatment combinations with four levels of vermicompost (0, 2, 4, and 6 t ha^{-1}) and four levels of sulphur (0, 20, 40, and 60 kg S ha^{-1}). They reported that application of S up to 40 kg ha^{-1} resulted in significantly higher plant height(34.06, 50.2 and 59.0 cm) and branches per plant (3.53, 5.67 and 8.00) at 60, 90 DAS and at harvest, total (27.06) and effective root nodules (15.00), leghaemoglobin content in root nodules (1.91 mg g^{-1}), seed (15.05 q ha^{-1}) and straw yields (38.80 q ha^{-1}), net returns (35,311 Rs ha^{-1}), total uptake of N (4.48 kg ha^{-1}), P (.683 kg ha^{-1}), K (3.29 kg ha^{-1}) and S (0.497 Kg ha^{-1}) and available N (140.97 kg ha^{-1}) content in soil after crop harvest over lover levels of S application. The application of 40 kg ha^{-1} S increased seed yield by 29.10% and 11.55% and net returns by 36.55% and 12.40%, respectively, over control and 60 kg S ha^{-1}.

A pot experiment was conducted in the natural conditions of net house of the Department of Botany, Aligarh Muslim University, Aligarh (UP), India, to evaluate the effect of four concentrations of GA_3 (0, 10^{-7}, 10^{-6} and 10^{-5} M), alone and in combination with P (40 kg P ha^{-1}), on growth, biochemical and yield attributes of fenugreek (Dar et al. 2015a). Compared to the control, the combination of GA_3 and P (P_{40} + 10^{-6} M GA_3) significantly increased the activities of nitrate reductase (30.8%) and carbonic anhydrase (30.7%) enzymes; it also enhanced the seed yield

(140.6%) and the content of total chlorophyll (28.5%) and carotenoids (26%). There was also noticed significant increase (19.51%) in the content of seed trigonelline.

Considering the medicinal importance of *T. foenum-graecum* L., another pot experiment was conducted to investigate the effect of foliar application of Co-60 gamma irradiated chitosan (IC) with soil applied P on growth, biochemical and quality attributes of fenugreek (Dar et al. 2015b). Four concentrations of irradiated chitosan were used (0, 40, 80 and 120 mg L^{-1}) individually as well as in combination with single dose of P 40 kg ha^{-1}. Un-irradiated chitosan (UN) 40 mg L^{-1} and de-ionized water were used as control. IC significantly affected almost all the parameters including seed yield, trigonelline content and trigonelline yield. Among all treatments, P40 + IC 40 mg L^{-1} proved the best, which increased total alkaloid content by 34.9%, seed yield by 125.4%, and trigonelline content by 17.8%. The photosynthetic pigments and activities of NR and CA enzymes were also significantly improved by this combination.

Nancy et al. (2015) investigated the effects of selenium on growth, selenium accumulation, photosynthetic pigments, antioxidant activity and the micronutrients variations. They grown plants under greenhouse condition applied with different doses of sodium selenate (20, 40 and 50 μg mL^{-1}). Application of selenium enhanced the growth parameters and reduced the levels of chlorophyll at higher concentrations.

12.2.6 French bean (*Phaseolus vulgaris* L.)

Bayram et al. (2014) investigated the response of common bean (*P. vulgaris* L. Volare) to salinity in different growth mediums (soil and hydroponic culture). The relative water content and total chlorophyll content decreased with 300 mM NaCl in both cultures but did not change with 150 mM treatment in soil culture. Similarly, the malondialdehyde content did not change with 150 mM treatment in soil culture, whereas it increased in all other treatments. The highest increase in H_2O_2 and SOD activity was observed with 300 mM treatment in hydroponic culture. Catalase activity did not change with 150 mM treatment in soil culture but decreased with 300 mM treatment in both cultures. However, APX activity decreased in all treatments, except in the roots in the hydroponic culture. They concluded that salt stress induced more serious oxidative damage in the hydroponic culture compared to the soil culture.

The growth and yield response of *P. vulgaris* to different basal and foliar nutrients application was studied (Rahman et al. 2014). Their experiment was based on seven treatments including: Control (Water spray), DAP (solid from) 123.5 kg ha^{-1}, P (K_2SO_4) 100 kg ha^{-1}, B 60 mg L^{-1}, Mo 30 mg L^{-1}, Zn 300 mg L^{-1} and B + Mo + Zn (60 + 30 + 300 mg L^{-1}, respectively). Their results showed that foliar feeding of combined application of micronutrients (B + Mo + Zn) significantly increased the plant height, number of branches per plant, number of pods per plant, number of seeds per pod, biological and seed yields.

12.2.7 Hyacinth bean (*Lablab purpureus* L.)

The objective of Abdel-Wahab et al. (2002) was to study the nitrogen-fixing potentialities of *L. purpureus* under the effects of the extreme environmental conditions prevailing in the southern part of Egypt aiming at the recommendation of its propagation in areas of the National giant projects. *L. purpureus* inoculated with *Rhizobium* sp. Strain I4 (21 days after planting) grown on Nile valley and Wadi Allaqi soils was relatively tolerant to mild levels of salinity, but the nodule number was reduced to about 35% of the control plants when subjected to a high salt level (120 mM NaCl). Lablab plants were similarly affected by different rates of water deficits. This legume was tolerant to moderate levels of drought. The nodule number and weight at 50% of field capacity was about 70% of the control. These values were reduced to 45–55% at a field capacity of 16.5%. Absolute nitrogenase activity, leghaemoglobin content of nodules and protein content of bacteroids and cytosol were moderately affected by mild levels of NaCl and drought but significantly reduced to about 25–35% of the control treatments. Their results also indicated that plants grown on the soil of Nile Valley exhibited slightly higher values (nitrogenase, protein, etc.) than those grown on Wadi Allaqi soil.

Elegeil (2003) conducted five pot experiments to study the effect of salt- stress and *Bradyrhizobium* inoculation (strain TAL 209) on growth and symbiotic properties of lablab bean cultivars in different soil textures amended with N, P and chicken manure. His results indicated that lablab bean cultivars varied in their response to inoculation and /or salinity. Nodulation was more affected by salinity than the plant growth. *Bradyrhizobium* strain TAL 209 significantly improved nodulation under normal conditions and salt stress. Salinity significantly reduced shoot and root fresh and dry weights, nodules number and nodules dry weight of lablab bean. Plants grown in silt soil showed the highest results in all parameters measured followed by clay and sandy soils, respectively, under salt stress. Under normal and saline conditions cultivar Sh. Black gave the highest results in all parameters measured followed by Sh. White and Sh. Brown, respectively.

Karmegam and Daniel (2008) carried out a field trial using 14 different treatments, using vermicomposts of *Polyalthia longifolia* leaf litter + cowdung (1:1, VC-PL), pearl millet cobs + cowdung (1:1, VC-PT) and a weed, *Rottboellia exaltata* + cowdung (1:1, VC-RE), chemical fertilizer and combination of vermicompost and chemical fertilizer with *L. purpureus* (L.) Sweet. Their results showed that all the growth and yield parameters i.e. leaf area index, total chlorophyll content of leaves, dry matter production, number of primary branches per plant, day of first flower appearance, length of fruits, dry weight of 100 seeds, yield (fruit) per plant, yield per plot and yield per hectare were significantly higher in the plots which received vermicompost, chemical fertilizer and vermicompost + chemical fertilizer mixture than in the control plots ($P < 0.05$). The highest fruit yield (fresh weight) of 109 tonnes ha^{-1} was recorded in the treatment which received 2.5 tonnes of vermicompost (VC-RE prepared using *Perionyx ceylanensis*) + ½ dose of recommended NPK ha^{-1}, while it was 61.9 tonnes ha^{-1} in control plots without vermicompost and/

or chemical fertilizer. The available soil nutrients (N, P and K) were higher in the plots which received vermicompost than in the plots that received chemical fertilizer and those that received chemical fertilizer + vermicompost mixture. The uptake of nutrients by the plants was high in the plots which received chemical fertilizer + vermicompost mixture over the plots which received chemical fertilizer and vermicompost. The available NPK after harvest in soil collected from the plots treated with chemical fertilizers were 244, 32.9 and 100 kg ha^{-1} which were higher than in the control plots and lower than in the plots that received chemical fertilizer + vermicompost mixture. The available NPK values in soil which received VC-RE prepared using *P. ceylanensis* + ½ dose of recommended NPK ha^{-1} were 262, 45.2 and 110 kg ha^{-1}, respectively.

Khan et al. (2005) studied the effect of Ca application on vegetative growth, physiological attributes, seed yield, and quality of hyacinth bean in response to application of Ca at various rates (0, 0.2, 0.4, 0.6, and 0.8 g per pot). Calcium application at 0.6 g per pot improved the vegetative growth as well as most of the physiological parameters and yield and quality attributes. It also enhanced the seed-yield and seed-protein content by 30.3 and 11.6%, respectively, over the control. Ca application also increased the pH of the soil that presumably enhanced the root-nodulation capability. The addition of Ca to soil also significantly enhanced the nodulation in the roots and activity of nitrate reductase in the leaves.

In another pot experiment, Naeem et al. (2009a) studied the effect of soil-Applied Ca on nitrogen fixation, photosynthesis, enzymatic activities, contents of nutraceuticals, and yield and quality attributes of hyacinth bean. Calcium was applied to the soil as $CaCl_2$ at five levels, viz. 0, 40, 80, 120, and 160 mg Ca kg^{-1} soil. The performance of the crop was assessed at 60, 90, 120, and 150 days after sowing (DAS). Calcium application proved to be significantly effective on most of the parameters studied. Of the five Ca levels, 120 mg kg^{-1} soil showed the best results, improving most of the attributes studied significantly at all the sampling dates. It increased the seed yield, seed-protein content, and seed tyrosinase activity by 30.3, 16.6, and 20.3%, respectively, compared to the control. This study depicted that an optimum Ca dose should be included in the fertilizer recommendations for hyacinth and other beans in this region.

According to another study of Naeem et al. (2010), P deficiency causes a serious yield and quality constraint of beans at Aligarh, Western Uttar Pradesh, India. To address the problem, a pot experiment was conducted to study the effect of basal P application on the agricultural performance of this medicinal legume. The plants were grown in pots containing soil supplied with five levels of P viz. 0, 25, 50, 75 and 100 mg P kg^{-1} soil as KH_2PO_4. The growth and other physiological attributes, leaf nutrient contents, nodule-nitrogen and leghemoglobin content were studied at 60, 90 and 120 days after sowing (DAS), photosynthesis and other related parameters were measured at 90 DAS and yield and quality attributes were recorded at harvest (150 DAS). Nitrate reductase and carbonic anhydrate activities, leaf-N, -P, -K and -Ca contents and nodule-nitrogen and leghemoglobin contents reached the maximum extent at 60 DAS. At 90 and 120 DAS, the values decreased significantly. Chlorophyll content, carotenoids content, and photosynthesis were at maximum level at 90 DAS. At various growth stages, P application at 75 mg P kg^{-1} soil resulted in maximum amelioration of most of the parameters studied. It increased the seed-yield by 38.3%, seed-protein content by 14.9% and seed-carbohydrate content by 5.0%, relative to the control. It was concluded that there was a hidden hunger of hyacinth bean for P owing to soil-P deficiency that was ameliorated effectively by its basal dressing at 75 mg P kg^{-1} soil.

D'souza and Devaraj (2010) evaluated the effect of salinity on 10-day old seedlings of hyacinth bean (HA-4 cultivar). Seventy two hours of exposure of salt stress (100–500 mM) reduced dry and fresh weights, leaf surface area, root and shoot lengths, total chlorophyll, and RWC. Oxidative stress markers, H_2O_2, glutathione, TBARS, proline, ascorbic acid, total phenols, and total soluble sugar contents were significantly elevated. Salinity enhanced antioxidant enzymes, POX, and GR activities and reduced that of CAT in concentration and time dependent manner in leaves. Metabolic enzyme ß-amylase activity increased in both leaves and roots. Acid phosphatase decreased in leaves and elevated in roots. Further, they studied intensity of constitutive isozymes correlated with in vitro levels under stress, but the protein band patterns differed from controls. Further, Lablab showed reasonable tolerance up to 300 mM NaCl, but leaves and roots differed in their response.

D'souza and Devaraj (2013) assessed the ameliorating effect of $CaCl_2$ on NaCl-stressed seedlings of hyacinth bean. Ten day old seedlings were stressed either solely with NaCl (100–500 mM NaCl) or with NaCl (100–500 mM NaCl) + $CaCl_2$ (10 mM) and compared with control and $CaCl_2$ (10 mM) treated seedlings. NaCl-stressed seedlings showed reduced growth as indicated by growth index and relative water content (RWC) while a comparatively less decline in these parameters was seen in seedlings stressed with supplemental $CaCl_2$. An enhancement in levels of H_2O_2, MDA, GSH, ASC, TSS and photosynthetic pigments noted in stressed seedling supplemented with $CaCl_2$ was evident of its role in ameliorating salinity stress. Supplementation was also found to increase the activity of metabolic enzyme AMY paving the way for partial amelioration of stress caused by salinity. Further, in a different study, D'souza and Devaraj (2015) investigated the effects of short-term salinity stress and spermidine pre-treatment on polyamine metabolism, mineral composition and growth of Hyacinth bean. The levels of diaminoputrescine (DAP), putrescine (Put), spermidine (Spd) and spermine (Spm) were evaluated as possible biomarkers for salinity stress. Ten-day old seedlings were subjected to stress at five different concentrations of NaCl (100–500 mM) and pre-treatment with 0.5 mM Spd in the presence of 300 mM NaCl under strictly controlled growth conditions. Salinity stress resulted in significant reduction in growth index, fresh and dry weight and total chlorophyll content of Hyacinth bean, but pretreatment with Spd significantly enhanced these growth parameters. The results also showed that exogenous application of Spd maintained higher values of $[K^+]/[Na^+]$ as compared with NaCl-stressed plants. Their study showed that exogenous Spd can be applied as short-term pre-treatment prior to introduction of salt stress to help elevate salt tolerance in Hyacinth bean seedlings.

12.2.8 Lucern (*Medicago sativa* L.)

Rogers et al. (1998) examined the effect of Na_2SO_4 and NaCl, at electrical conductivity levels (2–17 dS m^{-1}) on the growth and tissue ion concentrations of 16 lines of lucerne (*M. sativa* L.) under greenhouse condition. Dry matter production was negatively correlated with shoot concentrations of Na^+, Cl^- and S^{2-} and generally lines that were more tolerant to salinity had lower concentrations of those ions in the shoots. They concluded that lucerne is moderately tolerant to Na_2SO_4 -predominated salinity.

Mohammadi et al. (2008) evaluated the responses of two alfalfa cultivars differing in salt tolerance in terms of root nitrogen remobilization rates (RNRR) and their relationship with the ionic status of the plants. Three levels of salinity stress with electrical conductivities (ECs) of 1.2, 7 and 12 ds m^{-1} were established in irrigation water by using tap water with and without NaCl. It was found that alfalfa shoot growth was highly dependent on RNRR under salinity stress. Unlike in some other species, the shoot K^+ concentration and contents of alfalfa plants were significantly reduced by increasing salt stress. The salt tolerance recognized in the Bami cultivar may be attributed to the 339% increase in its selectivity rates of K^+ over Na^+ in ion transport from the soil to the shoots, as the shoot Na^+ content did not increase with increasing salt levels.

12.2.9 Mung bean (*Vigna radiata* L.)

Chakrabarti and Mukherji (2002) investigated the metabolic alterations in *V. radiata* (L.) Wilczek under NaCl salinity (E.C. value 4.0 m mhos/cm). Application of NaCl resulted in about 7 and 9% decrease in phenol content in mung bean leaf and root respectively. In leaf, NaCl caused 40% increase in polyphenol oxidase enzyme activity over the control set. This effect was accentuated in root, where salinity caused 200% increase in the enzyme activity. In leaf and root of mung bean plant, ascorbic acid content decreased about 29 and 31% respectively under salinity stress as compared with control. Ascorbic acid oxidase enzyme activity increased under stress by about 55 and 23% respectively in leaf and root.

Naeem et al. (2005) conducted a pot experiment to study the effect of four levels of Ca (0, 15, 30 and 45 kg Ca ha^{-1}) on mung bean (*V. radiata* L. Wilczek). Observations were recorded on fresh and dry weights per plant, number and dry weight of nodules per plant, nodule-nitrogen content and total chlorophyll and carotenoids contents. Among the four levels of Ca, 45 kg Ca ha^{-1} proved the best for all the parameters studied. The application of Ca (45 kg Ca ha^{-1}) significantly enhanced fresh and dry weights, number and dry weight of nodules per plant, nodule-nitrogen content, total chlorophyll content and total carotenoids content over the respective controls.

Ali et al. (2010) conducted field experiments to evaluate the influence of three levels of phosphatic fertilizer on mung bean for two consecutive kharif seasons i.e.

2007 and 2008. The experiment comprised of four treatments viz., control, phosphatic fertilizer @ 30 kg ha^{-1} with started dose of nitrogen, phosphatic fertilizer @ 57 kg ha^{-1} and phosphatic fertilizer @ 84 kg ha^{-1}. Their results revealed that all the levels of phosphatic fertilizer showed significant impact on mung bean compared to that of control plots, However, treatment of phosphatic fertilizer @ 84 kg ha^{-1} out yielded rest of the treatments giving the maximum yield components and grain yield during both years.

Sadeghipour et al. (2010) conducted a field experiment to investigate the effect of different N and P levels on seed yield and yield components of mung bean variety Partowa. Five levels of nitrogen (0, 30, 60, 90 and 120 kg N ha^{-1}) and six levels of P (0, 30, 60, 90, 120 and 150 kg P_2O_5 ha^{-1}) were the treatment variables. Application of N and P fertilizers significantly increased the seed yield. The maximum seed yield (224.2 g m^{-2}) was obtained when 90 kg N and 120 kg P_2O_5 ha^{-1} was applied. It was reported that the sustained increase in seed yield was mainly due to higher number of pods per plant, number of seeds per pod and 1000-seeds weight. Thus, application of 90 kg N and 120 kg P_2O_5 ha^{-1} appeared to be the optimum N and P levels for harvesting the highest yield of mung bean.

Mondal et al. (2011) investigate the effect of N was prayed alone and with micronutrients (0.1% w/v of B, Mo, Zn, Ca and Fe) at reproductive stage on morphological, growth and yield attributes in two mung bean genotypes, one bold (MB-16) and another small seeded (MB-35). Foliar application of N or N plus micronutrients increased leaf area, specific leaf weight, chlorophyll content, total dry mass, flower number and reproductive efficiency, yield attributes and yield over the control. Foliar application of N and N plus micronutrients had no significant influence on harvest index and grain protein content but had significant influence on yield attributes and yield.

Samreen et al. (2013) grown four varieties of mung bean (Ramazan, Swat mung I, NM92 and KMI) containing sand nutrient solutions with and without Zn under hydroponic condition. Each variety was applied with Zn solutions at three levels i.e. 0, 1 and 2 lM concentrations. Plant growth, chlorophyll contents, crude proteins and Zn contents were noted to be higher when greater supply of zinc doses was applied. Plant phosphorous contents declined with supply of Zn from 1 lM to 2 lM compared to the control signifying a Zn/P complex foundation possibly in roots of plant, preventing the movement of P to plant. Plant copper and Mg contents increased whereas Fe showed competitive behavior with Zn while K, Na and Mn plant contents were non-significantly reduced with Zn increase from control to 2 lM.

Saviour and Stalin (2013) conducted a field experiment to study the effect of applied Zn and B to the residual black gram in maize-black gram cropping system. Results revealed that soil application of $ZnSO_4$ and Borax @ 50 kg and 10 kg ha^{-1} respectively combined with recommended 100% NPK ha^{-1} to the proceeding crop, significantly recorded the highest dry pod (690.3 kg ha^{-1}) and haulm yield (803.2 kg ha^{-1}) with increase being 52.16 and 50.7% over control for the successive residual black gram. The same treatment also greatly influenced the yield attributes viz., the highest number of pods $plant^{-1}$ and 100 seed weight. Foliar application of micronutrients (Mn and Zn) on quality and quantity characteristics of mung bean under water stress was also conducted (Jafar Dokht et al. 2015). Four levels of micronutrients included: no foliar application, 3000 and 3000 ppm foliar application of Zn and Mn, and foliar application of Zn + Mn. The results showed that the highest plant height, number of capsule in plant, number of seed in capsule, 1000 seeds-weight and biological yield and harvest index obtained from optimum irrigation treatment. The highest grain yield was achieved from optimum irrigation treatment and foliar application of Zn + Mn. The highest percentage of carbohydrate, zinc and manganese was achieved from micronutrients foliar application. However, micronutrients application led to decreasing phosphorous uptake in plant. Their results showed that interaction between irrigation treatments and micronutrients foliar significantly increased plant height, grain yield and biological yield. The highest grain yield (451 kg ha^{-1}) was achieved from optimum irrigation and foliar application of Zn + Mn. Mir et al. (2015) studied the response of *V. radiata* (L.) Wilczek towards the salt stress (100 mM) and Zn toxicity (zinc sulphate at 200 mg kg^{-1} soil). The salinity and Zn stress significantly reduced the plant growth, gas exchange parameters but increased antioxidant enzyme activity, proline content and electrolyte leakage in the plant.

Marimuthu and Surendran (2015) studied the effect of P sources (mono- and diammonium phosphate) with brassinolide and salicylic acid on growth and yield of black gram in sandy loam soils. The treatments include 100% recommended dose of NPK along with foliar application of monoammonium phosphate (MAP), diammonium phosphate (DAP), brassinolide (0.25 ppm), and salicylic acid (100 ppm) along with the combination of these treatments. TNAU pulse wonder at 5.0 kg ha^{-1} and TNAU micronutrient mixture (MN) at 5 kg ha^{-1} were also applied. Their results revealed that application of 100% recommended dose of NPK + DAP 2% + TNAU pulse wonder 5.0 kg ha^{-1} was statistically significant and recorded higher plant growth (37.62 cm), number of pods/ plant (37.15), yield (1162 kg ha^{-1}), and benefit cost ratio (2.98) over the other treatments. The lowest black gram yield (730 kg ha^{-1}) was recorded for control.

A field experiment was conducted to evaluate the effect of P, Mo and Co nutrition on crop growth, yield and quality of mungbean grown in a acidic soil (pH 4.5) of northeast India by Awomi et al. (2016). Nutrients at higher doses significantly increased the growth, yield attributes, yield and quality parameters of mungbean. Application of 60 kg P_2O_5 ha^{-1} resulted in maximum growth attributes and number of nodules, leading to 36.5, 19 and 7.5% higher grain yield (426 kg ha^{-1}) over control (272 kg ha^{-1}), 20 (345 kg ha^{-1}) and 40 kg P_2O_5 (394 kg ha^{-1}) treatments, respectively. According to their study, crop showed maximum growth and yield with application of 1.5 kg Mo ha^{-1}; the yield was 31.9 and 5.4% higher than the control (272 kg ha^{-1}) and 0.75 kg Mo ha^{-1} (378 kg ha^{-1}) treatments, respectively. The grain yield at 1.0 kg Co ha^{-1} (395 kg ha^{-1}) was 31 and 3% higher than that in control and 0.5 kg Co ha^{-1} treatment, respectively. Protein content, protein yield and NPK uptake increased with increasing levels of P, Mo and Co application.

HanumanthaRao et al. (2016) reviewed on salinity and high temperatures (HT) stresses on mung bean grown as a fallow crop (mung bean-rice-wheat to replace fallow-rice-wheat) and/or a relay crop in cereal cropping systems. Salinity tolerance comprises multifaceted responses at the molecular, physiological and plant canopy levels. In HTs, adaptation of physiological and biochemical processes gradually may lead to improvement of heat tolerance in plants. At the field level, managing or manipulating cultural practices can mitigate adverse effects of salinity and HT. Greater understanding of physiological and biochemical mechanisms regulating these two stresses will contribute to an evolving profile of the genes, proteins, and metabolites responsible for mung bean survival. We focus on abiotic stresses in legumes in general and mung bean in particular, and highlight gaps that need to be bridged through future mung bean research. Recent findings largely from physiological and biochemical fronts are examined, along with a few agronomic and farm-based management strategies to mitigate stress under field conditions.

Sivakumar and Nandhitha (2017) conducted a study the impact of seed soaking with PGRs and nutrients (potassium chloride and calcium chloride) on mung bean (*V. radiata* L.) variety Co (Gg) eight under salt stress condition. The seed and seedling growth under normal and saline (150 mM NaCl) conditions were studied to determine their usefulness in increasing relative salt tolerance by using nutrients. Under salt stress, seedling characters were significantly affected compared to normal condition. All pre-soaking seed treatments increased the germination percentage, shoot and root length, vigour index and stress tolerance index over control. Na^+/K^+ ratio was decreased by $CaCl_2$ (0.5%) and KCl (1%) treatments. It was concluded that treatments with hormones and nutrients have reduced the severity of the salinity effect and the amelioration was observed in all the treatments compare to control.

12.2.10 Senna (*Cassia angustifolia* Vahl.)

Arshi et al. (2002) studied the effect of NaCl concentrations to assess their impact on the growth and metabolic changes in senna (*C. angustifolia* Vahl.). According to their study, five treatments (0, 40, 80, 120, and 160 mM NaCl) were given to the plants at three phenological stages, i.e. at pre-flowering, (45 days after sowing, DAS); flowering (75 DAS) and post-flowering (90 DAS) stages. A significant reduction in the biomass and length of the roots and shoots, photosynthetic rate, stomatal conductance, the total chlorophyll content, protein content, nitrate reductase activity, and reduced nitrogen content of the leaves was observed at each phenological stage with each salt concentration applied. Contrary to this, proline and nitrate contents of the leaves increased markedly. The post-flowering stage was most sensitive to NaCl treatment. Further, Arshi et al. (2005) investigated the effect of NaCl and $CaCl_2$ on growth parameters, ionic relations, and proline level in senna (*C. angustifolia*) plant under a pot culture experiment using NaCl (80 and 160 mM), $CaCl_2$ (5 and 10 mM), and the combined salt of $NaCl^+$ $CaCl_2$ (80 + 10 and 160 + 10 mM). These treatments were given to the growing plant separately at pre-flowering (A1), flowering (A2) and post-flowering (A3) stages. Compared to the controls, biomass of the root, shoot, and leaf decreased significantly with each NaCl treatment, whereas the reverse was observed with $CaCl_2$ treatments. Combined treatments of NaCl + $CaCl_2$ applied at different stages reduced the biomass, but this reduction was less than the one observed with NaCl treatments alone. However, proline accumulation in the leaves was 8 times higher than in the controls with treatment of 160 mM NaCl + 10 mM $CaCl_2$, whereas it was 5 times higher with NaCl (160 mM) alone. The concentration of K^+ and Ca^{2+} was inhibited with NaCl treatments, while Na + and Cl^- levels increased in the different plant parts. $CaCl_2$ treatment enhanced the K^+ and Ca^{2+} concentration, while the combined treatments mitigated the adverse effect caused by NaCl. Thus Ca could alleviate the NaCl induced inhibition of plant growth via the maintenance of net K^+ to Na^+ selectivity and the enhancement of proline accumulation in the leaves. Shitole and Dhumal (2011) examined the response of *C. angustifolia* under different NaCl stress. The first treatment of NaCl (25 mM NaCl concentration) was given on 45 days after sowing (DAS) and the subsequent treatments were given in three stages at the interval of 15 days. A significance reduction in growth and yield parameters with the increase of salinity levels was observed. However, the content of sennoside content in leaves and pods increased with 25 mM concentration of NaCl.

12.2.11 Senna sophera (*Cassia sophera* L.)

Naeem and Khan (2009) conducted a pot experiment according to simple randomised design, to find out the performance of *Senna sophera* (a medicinal herb) under five basal levels of P (0, 25, 50, 75 and 100 mg P kg^{-1} soil). The basal dressing of P proved beneficial for most of the parameters studied at 120, 150, 180 and 210 days after sowing (DAS), with 75 mg P kg^{-1} soil proving best. Application of this treatment, for example, gave 20.4% higher photosynthetic rate at 150 DAS, 16.7% higher carbonic anhydrase activity and 15.6% higher nitrate reductase activity at 120 DAS, and 24.5% higher seed yield and 13.6% higher seed protein content at 210 DAS than the no P control.

In view of the medicinal importance of *Cassia sophera* L., a hypothesis was designed to determine whether Ca application through soil could enhance the photosynthetic efficiency, enzymatic activities, nitrogen assimilation, yield and quality attributes (Naeem et al. 2009b). The plants were grown in pots containing soil supplied with five levels of Ca, viz. 0, 40, 80, 120 and 160 mg Ca kg^{-1} soil (Ca_0, Ca_1, Ca_2, Ca_3 and Ca_4, respectively) applied as $CaCl_2$. The performance of the crop was assessed in terms of various growth, physiological, biochemical, yield and quality attributes at 120, 150, 180 and 210 days after sowing. Calcium application proved significantly effective on most of the attributes studied. Of the five Ca levels, Ca_3 showed the best results that significantly stimulated most of the attributes studied at the three growth stages.

12.2.12 Soybean (*Glycine max* L.)

The salinity sensitivity of soybean was studied to determine the effect of salinity on seed germination, shoot and root dry weights, and leaf mineral contents (Essa 2002). Three soybean cultivars, Lee, Coquitt, and Clark 63, were planted in soils of different salinity levels. The electrical conductivity (EC) of the soils used in this experiment was 0.5 dS m^{-1}). The soil salinity treatments were 0.5, 2.5 4.5, 6.5 and 8.5 dS m^{-1}). Germination percentages were recorded 10 days after planting. Shoot and root dry weights of 45-day-old plants were measured. Nutrient concentrations for Na^+, K^+, Ca^{2+}, Mg^{2+} and Cl) were determined. Germination percentages were significantly reduced with increasing salinity levels. The cultivar Lee was less affected by salinity stress than Coquitt and Clark 63. At 8.5 dS m^{-1} a significant reduction in plant height was found in all three cultivars. However, Lee plants were taller than plants of the other two cultivars. Salinity stress induced a significant increase in leaf sodium (Na+) and chloride (Cl) in all cultivars. However, the cultivar Lee maintained lower Na^+ and Cl^+ concentrations, a higher potassium (K^+) concentration and a higher K^+/Na^+ ratio at higher salinity levels than Coquitt and Clark 63. Saline stress reduced the accumulation of K^+, Ca^{2+} and Mg^{2+} in the leaves of the cultivars studied. Concluding, they suggested that Lee is the most tolerant cultivar, and that there is a relationship between the salt tolerance of the cultivar and macronutrient accumulation in the leaves.

Torres-García et al. (2009) studied the effect of different concentrations of salinity (0, 1000 and 2000 ppm of NaCl) on snap bean (*P. vulgaris* L.) at the vegetative stage and beginning of flowering. NaCl at 2000 ppm reduced biomass and pod production in 35.5 and 45%, respectively. Amirjani in the year (2010) explored the effect of salinity stress on growth, mineral composition, proline content, antioxidant enzymes of soybean. The plants were exposed to 0, 50, 100 and 200 mM NaCl. Increasing salinity level to 50, 100 and 200 mM resulted in a reduction of plant height of 30%, 47% and 76% and a reduction of fresh weight of 32%, 54% and 76%, respectively. Nitrogenase activity had a decrease of 60% and ammonium content a significant increase (100%) at 200 mM salt concentration. Proline accumulation and diamine oxidase (DAO), were studied by reading the absorption of chromophore at 520 nm using spectrophotometer. Seedlings subjected to salt stress in the presence. Both DAO activity and proline content were increased in soybean under 50 to 200 mM NaCl. Using atomic absorption spectrophotometer, ion uptake of for Na^+, K^+, Ca^{2+} and Mg^{2+} were determined. The Na^+ content significantly increased but the contents of K^+, Ca^{2+} and Mg^{2+} decreased significantly as salinity treatment concentrations increased. Further, a significant decrease in superoxide dismutase, catalase and peroxidase activities under 100 and 200 mM salt were noticed.

Arshi et al. (2010) and to analyze the changes in growth, reactive oxygen metabolism in terms of H_2O_2 content, lipid peroxidation (TBARS), free radical quenching systems (nonenzymatic and enzymatic antioxidants) and ion accumulation in different plant parts of cultivars (JS-335 and Bragg) of soybean (*Glycine max*). NaCl and $CaCl_2$ stress 15-day-old seedlings were treated with solutions of 25, 50 and 100 mM NaCl alone and in combination of 10 mM $CaCl_2$ i.e., 25 + 10 mM, 50 + 10 mM and 100 + 10 mM. Observations recorded at 30 days after sowing displayed significant decreases in plant biomass, leaf water potential, leaf area, chlorophyll content and the contents of glutathione (GSH) and ascorbate (AsC) on application of NaCl alone. However, H_2O_2 content and lipid peroxidation (TBARS) in leaves were enhanced, consequently raising the activities of SOD, APX, GR and CAT. Application of NaCl + $CaCl_2$ alleviated adverse effects of NaCl stress. The Na^+ and Cl^- contents in different plant parts increased with NaCl as well as with NaCl + $CaCl_2$ treatments. The maximum accumulation occurred in roots, followed by the stem and the leaves. The K^+ and Ca^{2+} contents decreased under NaCl stress; but NaCl + $CaCl_2$ treatment reduced the extent of decrease caused by NaCl.

Kondetti et al. (2012) studied the effect of salinity stress on eleven (Co-1, CoSoy-2, DS-40, GujratSoy-1, JS-80-21, MACS-13, MAUS-2, NRC-2, PalamSoy, Pusa-16 and Shilageet) Indian soybean varieties under salinity (NaCl) levels (0, 120, 180, 240 and 300 mM). Salinity had adverse effects on germination and all the physiological parameters (root and shoot lengths, root/shoot ratio, dry matter production in root and shoot, moisture content in root and shoot) for early seedling growth. The varietal difference was pronounced at high (240 and 300 mM) concentrations of NaCl. Co-1, GujratSoy-1 and NRC-2 varieties were salt sensitive and CoSoy-2, DS-40, PalamSoy, Pusa-16 varieties were salt tolerant, and rest varieties were moderate in their response towards salt.

Hamayun et al. (2010) evaluated the effect of silicon (Si) in alleviating salinity and drought induced physio-hormonal changes in soybean grown in perlite. The plant growth attributes viz. shoot length, plant fresh and dry weights were improved with elevated Si nutrition, while they decreased with NaCl and polyethylene glycol (PEG) application. The adverse effects of NaCl and PEG on plant growth were alleviated by adding 100 mg L^{-1} and 200 mg L^{-1} Si to salt and drought stressed treatments. It was observed that Si effectively mitigated the adverse effects of NaCl on soybean than that of PEG. Bioactive GA1 and GA4 contents of soybean leaves increased, when Si was added to control or stressed plants. Jasmonic acid (JA) contents sharply increased under salinity and drought stress but declined when the plants were supplemented with Si. Similarly, free SA level also increased with NaCl and PEG application. However, free SA level further increased with the addition of Si to salt treated plants, but decreased when Si was given to PEG treated plants.

Goli et al. (2015) applied micro-nutrients-chelating agent citric acid on soybean cultivar, Bolivar (maturity group V) at V3 and R3 (pod initiation) stage. Their results showed that applications of Cu, Zn, B and Mo increased three unrolled trifoliate leaves Cu, Zn, B by 26.5%, 13.8%, 113% and Mo increased to 179 mg kg^{-1}, respectively in the leaves. Also, the application of Cu, Zn, B and Mo increased Cu, Zn, B by 55.5%, 8.2%, 28.6% and Mo increased to 202 mg kg^{-1} respectively in soybean seeds. Application of Mn had no direct effect on increasing Mn either in leaves or in seeds, however, Mn and Mn + CA treatment affected other mineral contents. Further, application of Cu, Zn, Mo, B and CA increased macro-nutrients K, N, P, Mg, and S. Irrespective of the applications, the nutrient increase trend in seed was Na > Fe > Zn > Mn > B > Cu > Mo. However, Mo application resulted in the following seed nutrient accumulation pattern: Na > Mo > Fe > Zn > Mn > B > Cu.

Barbosa et al. (2016) estimated the agronomic efficiency of a powder fertilizer (a source of micronutrient; 6.8% Mn, 3.9% Zn, 2.1% Fe, 1.2% Cu and 1.1% B), applied on soybean crop. The treatments consisted of five doses of the fertilizer: 0, 33.33, 66.66, 133.32 and 66.66 kg ha^{-1} + 1.4 ton ha^{-1} of calcium oxide. The contents of Mn, Zn and Fe increased in the soil in all treatments, but no effect was observed in leaf contents with the application. The levels of Zn and Mn are adequate. The dose of 133.32 kg ha^{-1} leads to an increase in the productivity of soybean plants in the Cerrado biome.

12.3 Conclusions

Of the 19 elements, macro nutrients (N, P and K) are considered and treated separately as they are removed by most of the crops in relatively much larger quantities. These nutrients play several crucial roles in metabolic and regulatory process in plants. Nitrogen has a pronounced influence on plant growth and development in general, and particularly for leguminous crops in the process of nitrogen-fixation. Applying excess of plant requirements can increase chances of loss to the environment; therefore, proper nutrients fertilizer application is important for proper nutrient management. Ameliorative effect, pertaining to growth, yield and quality (active constituents) may also be possible as the threshold quantities of these macronutrients could have activated the enzymes required in different metabolic processes. These nutrients have ameliorative effect on the growth and productivity of enormous medicinal legumes and the recommended fertilizer dose may be used for maximum production. Soil fertilization with micronutrients also increases micronutrient levels in cultivated areas providing better yield of different cultures in different production systems. The most feasible and successfully adopted technique is the basal or foliar application of mineral nutrients for exploiting the full genetic potential of a crop under normal as well as adverse conditions. According to Hochmuth (2003) the type, level, and mode (soil or foliar application) of nutrient fertilizers play important role for the maximum production of desired active constituents of the plants.

Abiotic stresses are major limiting factors of crop yields and cause losses of billions of dollars annually around the world. Understanding as to how the plants respond to adverse conditions and adapt to a changing environment at the molecular level might help the plants to cope better with such kind of stresses. There are various physiological, cellular and molecular strategies in the plants that have capacity to tolerate any unfavourable conditions. Among various salinity responses, controlling ion uptake, transport and balance, osmotic regulation, hormone metabolism, antioxidant metabolism, and stress signalling play critical roles in plant adaptation to salinity stress. It is the need of the hour to focus on the study of intercellular and intracellular molecular interaction involved in salinity stress response towards these leguminous plants. Plant scientists are exploring different ways to improve yield limits of existing demand of these plants.

References

Abel, A. J., Sutherland, M. W., Guest, D. I. (2003). Production of reactive oxygen species during nonspecific elicitation, non-host resistance and field resistance expression in cultures of tobacco cells. *Functional Plant Biology,* 30, 91–99.

Abdel-Wahab, A. M., Shabebw, M. S. A., & Younis, M. A. M. (2002). Studies on the effect of salinity, drought stress and soil type on nodule activities of *Lablab purpureus* (L.) sweet (Kashrangeeg). *Journal of Arid Environments, 51*, 587–602.

Ali, M. A., Abbas, G., Mohy-ud-Din, Q., Ullah, K., Abbas, G., & Aslam, M. (2010). Response of mungbean (Vigna radiata) to phosphatic fertilizer under arid climate. *The Journal of Animal Plant Sciences, 20*, 83–86.

Amirjani, M. R. (2010). Effect of salinity stress on growth, mineral composition, proline content, antioxidant enzymes of soybean. *American Journal of Plant Physiology, 5*, 350–360.

Arshi, A., Abdin, M. Z., & Iqbal, M. (2002). Growth and metabolism of senna as affected by salt stress. *Biologia Plantarum, 45*, 295–298.

Arshi, A., Abdin, M. Z., & Iqbal, M. (2005). Ameliorative effects of $CaCl_2$ on growth, ionic relations, and proline content of senna under salinity stress. *Journal of Plant Nutrition, 28*, 101–125.

Arshi, A., Ahmad, A., Aref, I. M., & Iqbal, M. (2010). Calcium interaction with salinity-induced effects on growth and metabolism of soybean (*Glycine max* L.) cultivars. *Journal of Environmental Biology, 31*, 795–801.

Ashraf, M. (2004). Some important physiological selection criteria for salt tolerance in plants. *Flora, 199*, 361–376.

Awomi, T. A., Singh, A. K., Kumar, M., & Bordoloi, L. J. (2016). Effect of phosphorus, molybdenum and cobalt nutrition on yield and quality of mungbean (*Vigna radiata* L.) in acidic soil of northeast India. *Indian Journal of Hill Farming, 25*, 22–26.

Babar, S., Siddiqi, E. H., Hussain, I., Bhatti, K. H., & Rasheed, R. (2014). Mitigating the effects of salinity by foliar application of salicylic acid in fenugreek. *Physiology Journal*, 869058. 6 pages.

Bano, A., & Fatima, M. (2009). Salt tolerance in *Zea mays* (L.) following inoculation with *Rhizobium* and *Pseudomonas*. *Biology and Fertility of Soils, 45*, 405–413.

Barbosa, J. M., Rezende, C. F. A., Leandro, W. M., Ratke, R. F., Flores, R. F., & da Silva, A. R. (2016). Effects of micronutrients application on soybean yield. *AJCS, 10*, 1092–1097.

Bayram, D., Dinler, B. S., & Tasci, E. (2014). Differential response of bean (*Phaseolus vulgaris* L.) roots and leaves to salinity in soil and hydroponic culture. *Notulae Botanicae Horti Agrobotanici Cluj-Napoca, 42*, 219–226.

Burman, U., Garg, B. K., & Kathju, S. (2007). Interactive effects of phosphorus, nitrogen, and thiourea on clusterbean (*Cyamopsis tetragonoloba* L.) under rainfed conditions of the Indian arid zone. *Journal of Plant Nutrition and Soil Science, 170*, 803–810.

Cavalcanti, B. C., Costa-Lotufo, L. V., Moraes, M. O., Burbano, R. R., Silveira, E. R., Cunha, K. M., Rao, V. S., Moura, D. J., Rosa, R. M., Henriques, J. A., & Pessoa, C. (2006). Genotoxicity evaluation of kaurenoic acid, a bioactive diterpenoid present in Copaiba oil. *Food and Chemical Toxicology, 44*, 388–392.

Chakrabarti, N., & Mukherji, S. (2002). Effect of phytohormone pretreatment on metabolic changes in *Vigna radiata* under salt stress. *Journal of Environmental Biology, 23*, 295–300.

Chhibba, I. M., Nayyar, V. K., & Kanwar, J. S. (2007). Influence of mode and source of applied iron on fenugreek (*Trigonella corniculata* L.) in a typic ustochrept in Punjab, India. *International Journal of Agriculture and Biology, 9*, 254–256.

Cokkizgin, A. (2012). Salinity stress in common bean (*Phaseolus vulgaris* L.) seed germination. *Notulae Botanicae Horti Agrobotanici Cluj-Napoca, 40*, 177–182.

D'souza, M. R., & Devaraj, V. R. (2010). Biochemical responses of hyacinth bean (*Lablab purpureus*) to salinity stress. *Acta Physiologiae Plantarum, 32*, 341–353.

D'souza, M. R., & Devaraj, V. R. (2013). Role of calcium in increasing tolerance of hyacinth bean to salinity. *Journal of Applied Biology & Biotechnology, 1*, 011–020.

D'souza, M. R., & Devaraj, V. R. (2015). Pre-treatment with spermidine reverses inhibitory effects of salt stress in hyacinth bean (*Lablab purpureus*). *Journal of Chemical and Pharmaceutical Research, 7*, 504–509.

Dar, T. A., Uddin, M., Khan, M. M. A., Ali, A., Hashmi, N., & Idrees, M. (2015a). Cumulative effect of gibberellic acid and phosphorus on crop productivity, biochemical activities and trigonelline production in *Trigonella foenum-graecum* L. *Cogent Food & Agriculture, 1*, 995950.

Dar, T. A., Uddin, M., Khan, M. M. A., Ali, A., Mir, S. R., & Varshney, L. (2015b). Effect of Co-60 gamma irradiated chitosan and phosphorus fertilizer on growth, yield and trigonelline content of *Trigonella foenum-graecum* L. *Journal of Radiation Research and Applied Science, 8*, 446–458.

Deepika, & Dhingra, H. R. (2014). Effect of salinity stress on morpho-physiological, biochemical and yield characters of cluster bean [*Cyamopsis tetragonoloba* (L.) Taub.] *Indian Journal of Plant Physiology, 19*, 393–398.

Devi, R. G., Gurusaravanan, V. P., & Gurusaravanan, P. (2012). Alleviating effect of IAA on salt stressed *Phaseolus mungo* (L.) with reference to growth and biochemical characteristics. *Recent Research in Science and Technology, 4*, 22–24.

Elegeil, G. A. G. B. (2003). *Effect of salt-stress on nodulation and growth of Lablab bean (Lablab purpureus) in different soil textures*. M. Sc. Thesis, University of Khartoum.
Essa, T. A. (2002). Effect of salinity stress on growth and nutrient composition of three soybean (*Glycine max* L. Merrill) cultivars. *Journal of Agronomy and Crop Science, 188*, 86–93.
Fathima, K. J., & Sudha, T. (2016). Effect of major and micro nutrients on seed yield, quality and economics of clusterbean (*Cyamopsis tetragonoloba* L. Taub). *Journal of Farm Sciences, 29*(2), 273–275.
Garg, B. K., Burman, U., & Kathju, S. (2006). Influence of thiourea on photosynthesis, nitrogen metabolism and yield of clusterbean (*Cyamopsis tetragonoloba* (L.) Taub.) under rainfed conditions of Indian arid zone. *Plant Growth Regulation, 48*, 237–245.
Goli, M. B., Pande, M., Bellaloui, N., & De Wrachien, D. (2015). Effects of soil applications of micro-nutrients and chelating agent citric acid on mineral nutrients in soybean seeds. *Agricultural Sciences, 6*, 1404–1411.
Gosset, D. R., Millhollon, E. P., & Lucas, M. C. (1994). Antioxidant response to NaCl stress in salt-tolerant and salt sensitive cultivars of cotton. *Crop Science, 34*, 706–714.
Gour, R., Naruka, I. S., Singh, P. P., Rathore, S. S., & Shaktawat, R. P. S. (2009). Effect of phosphorus and plant growth regulators on growth and yield of Fenugreek (*Trigonella foenum-graecum* L.) *Journal of Spices and Aromatic Crops, 18*, 33–36.
Hamayun, M., Sohn, E. Y., Khan, S. A., Shinwari, Z. H., Khan, A. L., & Lee, E. J. (2010). Silicon alleviates the adverse effects of salinity and drought stress on growth and endogenous plant growth hormones of soybean (*Glycine max* L.) *Pakistan Journal of Botany, 42*, 1713–1722.
HanumanthaRao, B., Nair, R. M., & Nayyar, N. (2016). Salinity and high temperature tolerance in mungbean [*Vigna radiata* (L.) Wilczek] from a physiological perspective. *Frontiers in Plant Science, 7*, 957.
Horie, T., Karahara, I., & Katsuhara, M. (2012). Salinity tolerance mechanisms in glycophytes: An overview with the central focus on rice plants. *Rice, 5*, 1–18.
Hochmuth, G. J. (2003). Progress in mineral nutrition and nutrient management for vegetable crops in the last 25 years. *HortScience, 38*, 999–1003.
Jafar Dokht, R., Mosavi Nik, S. M., Mehraban, A., & Basiri, M. (2015). Effect of water stress and foliar micronutrient application on physiological characteristics and nutrient uptake in mung bean. *Electronic Journal of Crop Production, 8*, 121–141.
Jamil, A., Riaz, S., Ashraf, M., & Foolad, M. R. (2011). Gene expression profiling of plants under salt stress. *Critical Reviews in Plant Sciences, 30*, 435–458.
Javid, M. G., Sorooshzadeh, A., Moradi, F., Sanavy Seyed, A. M. M., & Allahdadi, I. (2011). The role of phytohormones in alleviating salt stress in crop plants. *Aust J Crop Sci, 5*, 726–734.
Karmegam, N., & Daniel, T. (2008). Effect of vermicompost and chemical fertilizer on growth and yield of hyacinth bean, *Lablab purpureus* (L.) sweet. *Dynamic Soil, Dynamic Plant, 2*, 77–81.
Khan, M. M. A., Naeem, M., & Siddiqui, M. H. (2005). Calcium fertilization ameliorates growth, yield and quality of hyacinth bean (*Lablab purpureus* L.). In *Proceedings of 1st International edible legume conference & 4th world cowpea congress*, Durban, South Africa, April 17–21.
Khan, P. S. S. V., & Basha, P. O. (2016). Salt stress and leguminous crops: Present status and prospects. In M. M. Azooz & P. Ahmad (Eds.), *Legumes under environmental stress: Yield, improvement and adaptations*. Hoboken, NJ: John Wiley & Sons, Inc.
Kondetti, P., Jawali, N., Apte, S. K., & Shitole, M. G. (2012). Salt tolerance in Indian soybean (*Glycine max* (L.) Merill) varieties at germination and early seedling growth. *Annals of Biological Research, 3*, 1489–1498.
Läuchli, A., & Epstein, E. (1990). Plant responses to saline and sodic conditions. In K. K. Tanji (Ed.), *Agricultural Salinity Assessment and Management. ASCE manuals and reports on engineering practice No. 71* (pp. 113–137). New York: ASCE.
Manivasagaperumal, R., Balamurugan, S., Thiyagarajan, G., & Sekar, J. (2011). Effect of zinc on germination, seedling growth and biochemical content of cluster bean (*Cyamopsis tetragonoloba* (L.) Taub). *Current Botany, 2*, 11–15.
Marimuthu, S., & Surendran, U. (2015). Effect of nutrients and plant growth regulators on growth and yield of black gram in sandy loam soils of Cauvery new delta zone, India. *Cogent Food and Agriculture, 1*, 1010415.

Mir, B. A., Khan, T. A., & Fariduddin, Q. (2015). 24-epibrassinolide and spermidine modulate photosynthesis and antioxidant systems in *Vigna radiata* under salt and zinc stress. *International Journal of Advanced Research (IJAR), 3*, 592–608.

Mishra, B. P. (2016). Effects of nitrogen and growth regulators on yield *Phaseolus mungo* L. *International Journal of Advanced Research and Development, 1*, 39–42.

Mohammadi, H., Poustini, K., & Ahmadi, A. (2008). Root nitrogen remobilization and ion status of two alfalfa (*Medicago sativa* L.) cultivars in response to salinity stress. *Journal of Agronomy and Crop Science, 194*, 126–134.

Mondal, M. M. A., Rahman, M. A., Akter, M. B., & Fakir, M. S. A. (2011). Effect of foliar application of nitrogen and micronutrients on growth and yield in mungbean. *Legume Research, 34*, 166–171.

Morris, J. B. (2003). Bio-functional legumes with nutraceutical, pharmaceutical, and industrial uses. *Economic Botany, 57*, 254–261.

Munns, R., Schachtman, D., & Condon, A. (1995). The significance of a two-phase growth response to salinity in wheat and barley. *Functional Plant Biology, 22*(561), 569.

Munns, R. (2002). Comparative physiology of salt and water stress. *Plant, Cell & Environment, 25*, 239–250.

Munns, R., & James, R. A. (2003). Screening methods for salt tolerance: A case study with tetraploid wheat. *Plant and Soil, 253*, 201–218.

Munns, R., James, R. A., & Lauchli, A. (2006). Approaches to increasing the salt tolerance of wheat and other cereals. *Journal of Experimental Botany, 57*, 1025–1043.

Munsi, P. S. (1992). Nitrogen and phosphorus nutrition response in Japan's mint cultivation. *Acta Horticulturae, 306*, 436–443.

Naeem, M., Khan, M. N., & Singh, M. (2005). Effect of calcium fertilization on growth, photosynthesis pigments and nodulation of mung bean (*Vigna radiata* L. Wilczek). *Indian Journal of Applied & Pure Biology, 20*, 253–254.

Naeem, M. (2007). Effect of phosphorus and calcium on selected medicinally important leguminous plants. PhD Thesis, Aligarh Muslim University, Aligarh, India.

Naeem, M., Khan, M. M. A., Moinuddin, & Khan, M. N. (2009a). Role of calcium in ameliorating photosynthetic capacity, nitrogen-fixation, enzyme activities, nutraceuticals and crop productivity of hyacinth bean (*Lablab purpureus* L.) under calcium deficient soil. *Medicinal and Aromatic Plant Science and Biotechnology, 3*, 64–73.

Naeem, M., Idrees, M., & Khan, M. M. A. (2009b). Calcium ameliorates photosynthetic capacity, nitrate reductase and carbonic anhydrase activities, nitrogen assimilation, yield and quality attributes of *Cassia sophera* L. - **a medicinal legume**. *Physiology and Molecular Biology of Plants, 15*, 237–247.

Naeem, M., & Khan, M. M. A. (2009). Phosphorus ameliorates crop productivity, photosynthesis, nitrate reductase activity and nutrient accumulation in coffee senna (*Senna occidentalis* L.) under phosphorus-deficient soil. *Journal of Plant Interactions, 4*, 145–153.

Naeem, M., Khan, M. M. A., Moinuddin, Idrees, M., & Aftab, T. (2010). Phosphorus ameliorates crop productivity, photosynthetic efficiency, nitrogen-fixation, activities of the enzymes and content of nutraceuticals of *Lablab purpureus* L. *Scientia Horticulturae, 126*, 205–214.

Naeem, M., Khan, M. M. A., & Moinuddin. (2010). Calcium chloride stimulates crop yield, photosynthetic efficiency, enzyme activities and nutraceuticals of coffee senna (*Senna occidentalis* L.) under calcium deficient soil. *Asian Australian Journal of Plant Science & Biotechnology, 4*, 30–37.

Naeem M., Khan M. M. A., & Moinuddin. (2012). Role of mineral nutrients in cultivation of medicinal legumes. In: Naeem M., Khan MMA, Moinuddin (Guest Editors) Mineral Nutrition of Medicinal and Aromatic Plants, MAPSB 6 (Special Issue 1) 24-38, Global Science Books, Japan

Nancy, D., Thribuvana, P., & Arulselvi, P. I. (2015). Impact of selenium fortification in Fenugreek (*Trigonella foenum-graecum*). *Indo American Journal of Pharmaceutical Research, 5*, 635–640.

Netondo, G. W., Onyango, J. C., & Beck, E. (2004). Sorghum and salinity: II. Gas exchange and chlorophyll fluorescence of sorghum under salt stress. *Crop Science, 44*, 806–811.

Pariari, A., Khan, S., & Imam, M. N. (2009). Influence of boron and zinc on increasing productivity of Fenugreek seed (*Trigonella foenum graecum* L.) *Journal of Crop and Weed, 5*, 57–58.

Patel, B. B., Patel, B. B., & Dave, R. S. (2011). Studies on infiltration of saline–alkali soils of several parts of Mehsana and Patan districts of north Gujarat. *Journal of Applied Technology in Environmental Sanitation, 1*, 87–92.

Paul, D. (2012). Osmotic stress adaptations in rhizobacteria. *Journal of Basic Microbiology, 52*, 1–10.
Rahman, I. U., Afzal, A., Iqbal, Z., Ijaz, F., Sohail, S. S., Manan, S., & Afzal, M. (2014). Response of common bean (*Phaseolus vulgaris*) to basal applied and foliar feeding of different nutrients application. *American-Eurasian Journal of Agricultural & Environmental Sciences, 14*, 851–854.
Rout, N. P., Shaw, B. P. (2001). Salt tolerance in aquatic macrophytes: Ionic relation and interaction. *Biologia Plantarum, 55*, 91–95.
Rogers, M. E., Grieve, C. M., & Shannon, M. C. (1998). The response of lucerne (*Medicago sativa* L.) to sodium sulphate and chloride salinity. *Plant and Soil, 202*, 271–280.
Sadeghipour, O., Monem, R., & Tajali, A.A. (2010). Production of mungbean (*Vigna radiata* L.) as affected by nitrogen and phosphorus fertilizer application. *Journal of Applied Sciences, 10*, 843–847.
Samreen, T., Shah, H. U., Ullah, S., & Javid, M. (2013). Zinc effect on growth rate, chlorophyll, protein and mineral contents of hydroponically grown mungbeans plant (*Vigna radiata*). *Arabian Journal of Chemistry*. doi:10.1016/j.arabjc.2013.07.00.
Sangeetha, S., & Subramani, A. (2014). Sodium chloride stress induced alterations in germination, growth and biomolecules of black gram (*Vigna mungo* L.) *International Journal of Environment and Bioenergy, 9*, 17–28.
Saviour, M. N., & Stalin, P. (2013). Influence of zinc and boron in residual blackgram productivity. *Indian Journal of Science and Technology, 6*, 5105–5108.
Seckin, B., Sekmen, A. H., & Turkan, I. (2009). An enhancing effect of exogenous mannitol on the antioxidant enzyme activities in roots of wheat under salt stress. *Journal of Plant Growth Regulation, 28*, 12–20.
Shahbaz, M., & Ashraf, M. (2013). Improving salinity tolerance in cereals. *Critical Reviews in Plant Sciences, 32*, 237–249.
Shitole, S. M., & Dhumal, K. N. (2011). Effect of NaCl stress on growth, yield and sennoside content of medicinal plant *Cassia angustifolia* Vahl. *International Journal of Current Research, 3*, 181–186.
Sivakumar, R., & Nandhitha, G. K. (2017). Impact of PGRs and nutrients pre-soaking on seed germination and seedling characters of mung bean under salt stress. *Legume Research, 40*(1), 125–131.
Surendar, K. K., Vincent, S., Vanagamudi, M., & Vijayaraghavan, H. (2013). Influence of plant growth regulators and nitrogen on leaf area index, specific leaf area, specific leaf weight and yield of black gram (*Vigna mungo* L.) *Plant Gene and Trait, 4*(7), 37–42.
Tabur, S., & Demir, K. (2010). Role of some growth regulators on cytogenetic activity of barley under salt stress. *Plant Growth Regulation, 60*, 99–104.
Tester, M., & Davenport, R. (2003). Na^+ tolerance and Na^+ transport in higher plants. *Annals of Botany, 91*, 503–527.
Torres-García, J. R., Escalante-Estrada, J. A., Rodríguez-González, M. T., Ramírez-Ayala, C., & Martínez-Moreno, D. (2009). Exogenous application of growth regulators in snap bean under water and salinity stress. *Journal of Stress Physiology & Biochemistry, 5*, 13–21.
Tyler, V. E., Brady, L. R., & Robbers, J. E. (1976). Glycosides. In *Pharmacognosy* (pp. 76–103). Philadelphia: Lea & Febiger.
Velmani, S., Murugesan, S., & Arulbalachandran, D. (2015). Growth and biochemical characteristics of black gram (*Vigna mungo* (L.) Hepper) under NaCl salinity. *International Journal of Current Trends in Research, 4*, 13–17.
Verma, S. R., Shivran, A. C., Bhanwaria, R., & Singh, M. (2014). Effect of vermicompost and sulphur on growth, yield and nutrient uptake of fenugreek (*Trigonella foenum-graecum* L.) *The Bioscan, 9*, 667–670.
Vijayarengan, P. (2013). Changes in growth, biochemical constituents and antioxidant potentials in cluster bean *Cyamopsis tetragonoloba* L. Taub under zinc stress. *International Journal of Current Science, 5*, 37–49.
Wittwer, S. H. & Teubner, F. G. (1967). Foliar absorption of mineral nutrients. *Annual Review of Plant Physiology, 10*, 13–32.
Yamaguchi, T., & Blumwald, E. (2005). Developing salt-tolerant crop plants: Challenges and opportunities. *Trends in Plant Science, 10*, 615–620.
You-Sheng, W., Jin, W., Zhi-Min, Y., Qing-Ya, W., Bo, L., Shao-Qiong, L., Ya-Ping, L., Song-Hua, W., & Xin, S. (2004). Salicylic acid modulates aluminum-induced oxidative stress in roots of *Cassia tora*. *Acta Botanica Sinica, 46*, 819–828.

Chapter 13
Role of Iron in Alleviating Heavy Metal Stress

Zaid ul Hassan, Shafaqat Ali, Muhammad Rizwan, Qasim Ali, Muhammad Zulqarnain Haider, Muhammad Adrees, and Afzal Hussain

Abstract Heavy metals naturally present in soils usually result from human activities such as agricultural practices, mining, automobile, sewage processing, and metal industries. Higher concentrations of these metals in surrounding environment showed toxic effects on plants and animals. Heavy metals entered in soil-plant environment through various anthropogenic activities are taken up and accumulated in various plant parts. Higher concentrations of these metals showed toxic symptoms in plants. Heavy metals at higher dosage negatively affect plants physiological, morphological, and biochemical traits. On the other hand, plants used different strategies to cope with damaging effects induced by metal toxicity. There are some metals such as macro and micro nutrients, which are essentially required by plants for their growth and development processes. Micronutrient such as iron plays a key role in minimizing toxic effects of heavy metals and limits their entry in food chain. It has been thoroughly documented by many researchers that Fe has potential to alleviate metal toxicity by limiting metals uptake in different plants. Reports suggested that Fe improves plant physiological, morphological, and biochemical parameters by neutralizing metals toxicity. However, Fe deficiency resulted in malnutrition that affects human population worldwide. Various strategies have been used to enhance food quality, improve Fe uptake from soil and increased Fe shortage through a process known as biofortification. Fe uptake can be enhanced by overexpressing genes. Micronutrients level in plants could also be enhanced through agricultural practices, plant breeding, and biotechnology techniques.

Keywords Heavy metals • Fe • Anthropogenic activities • Physiological • Morphological • Biochemical • Micronutrient • Biofortification

Z. ul Hassan • S. Ali (✉) • M. Rizwan • M. Adrees • A. Hussain
Department of Environmental Sciences and Engineering, Government College University, Allama Iqbal Road, 38000 Faisalabad, Pakistan
e-mail: shafaqataligill@yahoo.com

Q. Ali • M.Z. Haider
Department of Botany, Government College University, Allama Iqbal Road, 38000 Faisalabad, Pakistan

M. Naeem et al. (eds.), *Essential Plant Nutrients*,
DOI 10.1007/978-3-319-58841-4_13

13.1 Introduction

Heavy metals are key environmental pollutants, and their toxicity is a major threat to ecological, nutritional, and environmental concerns (Lenntech Water Treatment and Air Purification 2004). Major sources of these metals toxicity in agricultural soils are usage of fertilizers, sludge application, industrial waste, and other anthropogenic activities (Passariello et al. 2002). Heavy metal exposure results in the production of reactive oxidative stress in plants which is the indirect cause of heavy metals toxicity (Mithofer et al. 2004). It has been well documented by many researchers that heavy metals negatively affect plant morphological, physiological, and biochemical parameters (Ali et al. 2011, 2013; Qadir et al. 2014; Chandra et al. 2009; Fozia et al. 2008). Other than metals toxic effects, there are some metals (Fe, Zn, Ni, Mn, Cu, and Co) which at lower concentration are essential for plants growth and development (Reeves and Baker 2000). These essential heavy metals play a significant role in plant physiological and biochemical functions. They participate in redox reactions and central part of many enzymatic activities (Nagajyoti et al. 2010).

Iron (Fe) is an essential micronutrient for all living organisms including plants. It plays a significant role in plant's physiological and biological function (Römheld and Nikolic 2006). Fe is required by plants for their normal growth parameters. It carried out various activities from photosynthesis to respiration (Welch and Graham 2004). Iron has been found beneficial regarding its role as reducing heavy metals toxicity in various plants. It was observed that Fe declined Cd toxicity by enhancing plant growth, photosynthetic pigments, and chloroplast quality in almond seedlings as documented by (Nada et al. 2007). Furthermore, Fe application reduced oxidative stress induced by Cd stress and maintains stability in chloroplast, chlorophyll contents, and thylakoid complexes in Indian mustard (Qureshi et al. 2010). Human survival also needs Fe content. However, its deficiency caused nutritional problems worldwide (Welch and Graham 2004). Iron deficiency could be filled by various agronomic practices, plant breeding, and Fe biofortification using Fe fertilizers. Biofortification completes cereal-based dietary needs for human population (Lucca et al. 2002). Transgenic approaches that overexpress various genes have also been used which are imperative for Fe uptake and transportation (Masuda et al. 2013).

Fe at higher concentration showed toxic effect on plants. Iron toxicity in plants is associated to Fe^{+2} uptakes from roots and translocation to leaves via transpiration flow. It is the reduction of Fe^{+2} and Fe^{+3} through microbial activities which give indications of iron toxicity (Becker and Asch 2005). Excess concentration of Fe^{+2} result in the formation of free radicals that disturb cellular structures and membrane damage (de Dorlodot et al. 2005).

13.2 Essential Heavy Metals

Heavy metals (Zn, Cu, and Fe) are essential micronutrients required by plants and animals for their survival (Wintz et al. 2002). Essential heavy metals are called trace elements due to their small concentrations i.e. 10 mg kg^{-1} in environmental atmosphere. Essential heavy metals play a key role in plants physiological and biochemical functions. Major functions of essential heavy metals are contribution in reduction, oxidation reactions, and central component of enzymes. Copper plays a significant role in plant photosynthesis (Chatterjee et al. 2006). Copper can easily gain or lose electron and electron donor in the photosystem of plants. It is a cofactor of oxidase, monooxygenase, and dioxygenase and enzymes that take part in the removal of superoxide radicals. Zinc is necessary for the reliability of ribosome. It contributes to various oxidation processes in plants and formation of carbohydrates. Zn also plays an important role in transcript parts and cofactor of RNA polymerase. Nickel is another essential micronutrients required by living organisms including plants and animals. Magnesium is also required in enzyme reactions. Furthermore, iron is an essential micronutrient required by various metabolic processes and vital to all living organisms. Fe is also an essential constituent of heme-based protein like hemoglobin and cytochrome. It also performs vital functions in numerous nonheme iron contained protein and other metabolic processes. Fe is found as component of protein and activates many oxidation-reduction reactions (Nagajyoti et al. 2010). However, higher concentrations of these essential heavy metals in plants result in toxic effects (Monni et al. 2000).

13.3 Iron Uptake by Plants

Plants acquire Fe through rhizosphere. Fe availability to plants is very low. Fe availability depends upon redox potential and pH of the soil. At higher soil pH, Fe is easily oxidized and exists in insoluble ferric oxides. Fe is readily available to roots for uptake at lower pH value. Some crops such as rice are mostly under consideration at Fe shortage. Fe is essential for plant growth and development processes (Takahashi et al. 2001; Walker and Connolly 2008). Different strategies are used by plants concerning primary iron uptake and bioavailability. Acidification-reduction transport strategy is being used by non-graminaceous plants to acquire Fe solubility before Fe uptake (Walker and Connolly 2008). Plants enhanced Fe (III) solubility by lower soil pH. Ferric chelate reductase also improves Fe solubility by reducing Fe (III) as Fe (II) has more solubility power than Fe (III). Reduction strategy maintains Fe uptake to regulate iron transporter (IRT1) which influx or efflux Fe across root epidermal plasma membrane. Iron diffusion between root cells are tested by exodermis and endodermis cell layers (Chen et al. 2011). These strategies enhanced under Fe shortage conditions and mostly used by plants facilitating by gene-based chelate ferric reductase (Robinson et al. 1999). Genes (*FRO2* and *IRTI*) are essential for iron uptake and plant growth. Both genes are particularly expressed in root epidermis (Kobayashi and Nishizawa 2012). Metal transporter such as AtIRT1 and NRAMP1 takes part in iron uptake from the soil (Cailliatte et al. 2010). Iron transporters are engaged in iron take up from soils (Walker and Connolly 2008). Grass plants used distinct mechanism in the form of chelate strategy to take up Fe from soil. Chelation is attained by strong Fe (III) chelator from mugineic acid (MA) class named phytosiderophores (Kobayashi and

Nishizawa 2012). Rice is different from other crops because it takes up iron by involving both above described strategies by using Fe (III)-phytosiderophores transporters (Inoue et al. 2009; Lee et al. 2009). Iron uptake from root to shoot is done through xylem. Fe shortage improved organic acids such as malate and citrate in xylem (Lopez-Millan et al. 2000). Reports suggested that citrate contributes largely in Fe complex in xylem (Rellan-Alvarez et al. 2008). Xylem citrate levels are necessary for xylem iron distribution in cells (Green and Rogers 2004). MATE family transporters FRD3 (ferric chelate defective3) are involved for carrying Fe and citrate in xylem (Durrett et al. 2007). Once again Fe is taken up into cells through apoplastic space as Fe arrives in leaves. As discussed earlier, citrate is essentially required for proper transportation of Fe (Green and Rogers 2004). However, exact role of citrate in xylem is still not fully described.

13.4 Metabolic Alterations and Plants Fe Status

Fe is essential for all living organisms and plants contained Fe as dietary source for human nutrition across the world. Scientists have been exploring during past decades on Fe uptake by roots; its transportation throughout the organs, tissues, and cellular partitions; synthesis of heme-based groups; and its accumulation (Balk and Pilon 2011). Fe is essential for metabolic functions of many organelles for instance: photosynthetic electron transport and respiratory functions. However, its deficiency negatively affects entire cellular metabolic activities. Plants effectively respond to alteration in Fe availability in the surrounding environment by following their metabolism and adaptable Fe homeostasis to maintain Fe gaining, transportation, and exploitation. The mechanism of Fe as signal sensing is required to be further evaluated. Up till now, only some transcriptional study regarding Fe homeostasis and signaling pathways have been identified (Hindt and Guerinot 2012). To cope with Fe deficiency, plants undergo different mechanisms to fulfill the Fe requirement as documented by many researchers (Zocchi 2006; Briat et al. 2003). Fe deficiency disturbs photosynthetic apparatus and subcellular compartments such as mitochondria and chloroplast. Fe supply among subcellular compartments is critical and managed by Fe transporter (Conte and Walker 2011; Kobayashi and Nishizawa 2012).

13.5 Fe and Plant Productivity

Sustainable growth is the major area of concern regarding agricultural production. It requires enhanced crop production and quality under changing climate conditions in the form of increased CO_2 and temperature (Pretty 2008). Mineral nutrients are essential as linked to crop productivity and quality enhancement. Minerals like Fe play a key role in photosynthetic CO_2 fixation. Fe homeostasis is required for photosynthetic efficiency (Boyd et al. 2007). Fe shortage disturbs photosynthetic rate and chloroplast structure in higher plants (Eberhard et al. 2008). Fe deficiency also results in leaves chlorosis and alters electron transport chain (Sharma 2007; Msilini et al. 2011). However, date regarding Fe homeostasis on photosynthetic efficiency and biomass production is limited. Photosynthetic efficiency and its functioning are largely Fe dependent (Layer et al. 2010; Yadavalli et al. 2012). Chloroplasts Iron (Fe), in heavy metal stress:and plant productivity are independent for Fe-S cluster biogenesis (Couturier et al. 2013; Balk and Schaedler 2014). These artificial groups are involved in electron transfer chain in thylakoid membranes where they operate as cofactor of protein complexes (Eberhard et al. 2008). The synthesis of chlorophyll b also involved a Rieske Fe-S cluster (Eggink et al. 2004). Increased CO_2 concentration also effect plants. Enhanced CO_2 increased net photosynthetic rate in plants by restricting photorespiration and improved c-assimilation to enhance plant growth and development. Decreased stomatal conductance also enhanced assimilation rate (Lammertsma et al. 2011).

13.6 Partitioning of Iron

Partitioning of iron through plant organs and tissues is done by a different transporter after iron uptake from soil. Iron usage is key concerned to mitochondria and chloroplast within the plant cells (Nouet et al. 2011). About 70–90% chloroplast iron take part in cellular Fe (Nouet et al. 2011). Iron partitioning changes with respect to cell types and related organs. Major portion of total iron is localized in the vacuoles of endodermal cells as in the nucleolus in arabidopsis-grown embryo (Kim et al. 2006; Roschzttardtz et al. 2009). Current studies have shown that iron is also found in arabidopsis leaves and in the embryo of pea plant (Roschzttardtz et al. 2011).

13.7 Role of Vacuole in Iron Storage

Role of vacuole in iron storage is effectively described in arabidopsis-grown embryos. Iron imaging techniques have shown that iron is accumulated in globoids which are embedded in protein storage vacuoles in endodermal cells of arabidopsis-grown embryos (Kim et al. 2006; Roschzttardtz et al. 2009). Iron is also linked to globoids in wheat crop. Main transporter (AtVIT1) permitted iron uptake into vacuoles in arabidopsis-grown embryos and expressed in seed development (Kim et al. 2006). Iron mobilizes from vacuole to other plant parts are done through AtNRAMP3 and AtNRAMP4. Both AtNRAMP3 and AtNRAMP4 recover iron from embryo cells during germination (Lanquar et al. 2005).

13.8 Significance of Mitochondria and Chloroplast as Fe Transporters

Fe not only transports and exports from vacuole but also enters and exits from other organelles. Mitochondria and chloroplast required Fe in excess to perform their metabolic functions. Iron is essentially required for the synthesis of Fe-S gather up in mitochondria and appropriate functioning of respiratory electron transport. However, Fe shortage declined oxygen usage in plant roots and respiration activities (Vigani et al. 2009). Chloroplast largely required Fe for their metabolic functions. Plastids hold about 80% Fe in their leaf cells. The production of reactive oxygen species by photosynthetic electron transport chain reacts with Fe and results in oxidative damage. Hence, Fe content in chloroplast must be tightly controlled in cellular partitioning. Fatefully, a little knowledge has been acquired about Fe transportation in or out of the chloroplast (Jeong and Guerinot 2009). Mitochondria and chloroplast are key sites for iron utilization and biosynthesis of heme. Fe transportation to plastids is done through PIC1 (Duy et al. 2011). Fe reduction is essential before its entry to plastids as recommended by ferric reductase AtFRO7 localization in chloroplast (Jeong et al. 2008). A homologue of IREG2/FPN2 which is MAR1 involved in Fe transportation in plastids on the basis of MAR1 expression under Fe shortage (Conte et al. 2009). Mitochondria iron introduces MRS3/4 or mitoferrin transporters in yeast and animals (Shaw et al. 2006). Mutant study confirmed that mitoferrin plays a key role in providing Fe to mitochondria. Export of Fe-S in mitochondria is catalyzed by ABC transporter which is similar in properties to yeast (Bernard et al. 2009).

13.9 Fe Transporters and their Possible Functions

Transporter	Function	References
AtATM3	Transfer Fe-S cluster from mitochondria	Bernard et al. (2009)
AtFPN1	Transport Fe into xylem and Fe outward transportation in plasma membrane	Morrissey et al. (2009)
AtFPN2	Transition metals entry into vacuole and declined metals toxicity during Fe shortage	Morrissey et al. (2009)
AtIRT1	Chief iron uptake from soil	Vert et al. (2002)
AtIRT2	Fe entry to cortical vesicles	Vert et al. (2009)
AtNAP14	Fe access to plastids	Shimoni-Shor et al. (2010)
AtNRAMP3	Fe transportation in vacuoles	Lanquar et al. (2005)
PIC1	Fe entry into chloroplast	Duy et al. (2007)
VIT1	Fe localization in Arabidopsis embryos	Roschzttardtz et al. (2009)
OsYSL2	Fe transportation from root to shoot and entry in seeds	Koike et al. (2004)
OsYSL18	Fe carrying from xylem to phloem	Aoyama et al. (2009)

13.10 Enhancing Fe Content through Plant Breeding

Plant breeding technique has been a pretty good option to enhance Fe content in target plant species. Fe concentration in various crops varies from 6 to 22 mg kg^{-1} in rice, 15–360 mg kg^{-1} in wheat, and 10–160 mg kg^{-1} in maize (White and Broadley 2005). Cultivated varieties have lower content of Fe to fulfill Fe intake needs despite genetic variation. However, Fe improving programs have been developed in the form of quantitative trait loci (QTLs) to enhance grain Fe content (Blair and Izquierdo 2012; Lung'aho et al. 2011). Bioavailability of Fe has been recorded in maize (Tako et al. 2013), cereal (Xu et al. 2012), and legume crops (Blair and Izquierdo 2012). This leads to Fe biofortification breeding strategy (Blair and Izquierdo 2012).

13.11 Iron Biofortification and Agronomic Practices

Fe deficiency and reduced bioavailability disturb plant growth traits. Fe shortage also results in nutritional risks in human worldwide. Especially infants and pregnant women are severely affected. Ultimately, it disturbs immune system. Plant biofortification is an effective strategy through biotechnology to cope with Fe deficiency (Grotz and Guerinot 2002). Application of Fe fertilizers fulfilled plants Fe deficiency. Fe fertilizers belong to three groups are inorganic compounds, Fe chelates, and natural Fe complexes (Abadia et al. 2011). Synthetic Fe chelates are usually expensive. However, it is largely applicable from soilless horticulture to additional significant field-grown crops (Alvarez-Fernandez et al. 2007). Foliar Fe fertilizer is the best alternative to expensive Fe chelates because it is cheaper and environmentally safe as compared to Fe chelates. Foliar Fe application depends upon various factors including stomata penetration, plasma membrane of leaf cells, and ultimately reached to chloroplast (El-Jendoubi et al. 2014). Another alternative to Fe chelates is intercropping system which helps to combat with Fe deficiency in non-grass plants. This intercropping system enhanced Fe concentration in the shoots and seeds of non-grass plants (Zuo and Zhang 2009). Plant breeding is another technique to enhance Fe content. Increased Fe content was found in rice, wheat, and maize crops (White and Broadley 2005). Cultivated varieties have lower Fe content which is managed through Fe biofortification breeding (Blair and Izquierdo 2012). Transgenic approaches have also been used to improve Fe deficiency. GMO (genetically modified organism) technology is being used nowadays to alter cellular and molecular parameters regarding Fe homeostasis (Kobayashi and Nishizawa 2012). Recently, transgenic approaches are being used to overexpress various genes that are important for Fe uptake, accumulation, and transportation (Masuda et al. 2013).

13.12 Role of Fe in Alleviating Heavy Metal Stress in Various Plants

Inorganic nutrients are required by plants for their growth and development processes. Nutrients deficiency imbalance plant growth traits. However, sufficient supply of mineral nutrients may also reduce heavy metals toxic effects in plants. Therefore, beneficial role of Fe in ameliorating heavy metals stress has been observed in many crops. It was investigated that Fe reduced Cd toxic effects by alleviating complex protein, oxidative stress, chlorophyll content, and retention of chloroplast in *Indian mustard* (Qureshi et al. 2010) and provided shield to photosynthetic tissues in *Lupinus albus* (Zornoza et al. 2010).

Reports suggested that Fe improved plant growth, photosynthetic pigments, and energetic photosynthesis by reducing Cd uptake and transportation in plant parts. Fe maintains the value of chloroplast by facilitating plants to neutralize Cd toxic effects as observed in almond seedlings (Nada et al. 2007). Sheng et al. (2008) documented that soil application of Fe fertilizer effectively enhanced rice grain, shoots, and roots by limiting Cd concentration. It was also observed that Fe declined oxidative stress induced by Cd in rice. Fe improved plant growth, chlorophyll content, and MDA in both leaves and roots of rice by alleviating Cd toxic effects. High Fe nutrition value declined Cd concentration in leaves and roots of rice. It is the high affinity of iron transport that reduces Cd uptake and translocation in target plant (Shao et al. 2007). Liu et al. (2008) investigated that Fe application improves rice production by limiting Cd uptake and accumulation in target specie. Furthermore, iron is a key cofactor of antioxidant enzymes. It was reported that Fe enhanced antioxidant enzymes and secure oxidative damage by ameliorating toxic effects of Cd in rice seedlings (Sharma et al. 2004). Higher dosage of Fe with some phosphates addition was found to reduce Pb and As dissolving power in water and reduced their uptake in plants (Xenidis et al. 2010). Feng et al. (2013) investigated that Pb accumulation in *T. latifolia*roots, and result showed that Fe alleviate Pb toxic effects. Pb toxicity was also limited in *T. latifolia* roots by low-cost phytostabilization method. Similarly, Rodriguez-Hernandez et al. (2015) reported enhanced accumulation of Cd and Pb in *Typha latifolia* under Fe and Ca treatment. It was noted that increased Fe application declined Cd and Pb uptake and transportation in the roots and shoots of *Typha latifolia* plant.

Shao et al. (2008) documented that use of iron fertilizer enhanced Fe concentration in rice grain. Application of Fe fertilizer with EDTA declined Cd accumulation and uptake in rice shoots and roots. Similarly, another study was conducted in which interaction between Fe-EDTA and $Cdcl_2$ was studied. Results showed that addition of Cd reduced growth of leaf area, water, and protein contents. However, Fe-EDTA application neutralized toxic effects of Cd and improved rice growth parameters (Ali et al. 2014). In addition, another study was conducted regarding role of Fe fertilizer in rice seedlings under Cd stress. Reports have shown that significant correlation was found between Fe and Cd. It was the enhanced Fe nutrition that declined adverse of Cd in rice (Liu et al. 2010). Hossain et al. (2009) also investigated that

As reduced rice growth and straw yield. Fe application enhanced root biomass, grain, and straw yield in rice by alleviating As toxic effects. Other than this, Fe is a cofactor of many enzymes and results illustrated that Fe nutrition enhanced enzymatic activities that may improve defense mechanism against oxidative stress induced by heavy metals (Gratao et al. 2005). Sharma et al. (2004) investigated the effects of Fe in barley seedlings under Cd stress. An increased in enzymes activities were observed. Report suggested that enhanced enzyme activities may alleviate the toxic effects of Cd by preventing membrane damage induced by reactive oxygen species.

Fe plaque usually formed in rice root surfaces under flooded conditions. It was observed that Fe plaque also ameliorated toxic effects of heavy metals and reduce their uptake such as Zn (Greipsson 1995), As (Syu et al. 2013; Li et al. 2015), Pb (Liu et al. 2011; Cheng et al. 2014), Cd (Cheng et al. 2014), Se (Zhou and Shi 2007; Huang et al. 2015), Cr (Hu et al. 2014), Cd, and Zn (Xu and Yu 2013) in rice crop.

On the other hand, Fe deficiency enhanced heavy metals toxicity. Effects of Cd under Fe shortage were observed on physiological responses of corn (*Zea mays L.*) and wheat (*Triticum aestivum L.*) crops. Results have shown that Cd reduced chlorophyll content, phytosiderophores (PS) release, and root and shoot biomass under Fe-deficient conditions as compared to Fe-sufficient treatment. Fe shortage improved Cd accumulation in both corn and wheat. These findings suggested that Cd might decline plants ability to acquire Fe by limiting phytosiderophores (PS) release (Bao et al. 2012). Similarly, effects of Fe deficiency were observed in *Solanum nigrum* L under Cd stress. It was found that Fe deficiency enhanced Cd accumulation in plants by variable soil pH and redox potential which is required for the phytoremediation of Cd from soil. Fe shortage enhanced Cd uptake and decreased plant biomass (Bao et al. 2009). In addition, it was also reported that Cd disturbs Fe uptake and accumulation in cucumber plants. Increased Cd concentration declined Fe content and vice versa (Kovacs et al. 2010). Nakanishi et al. (2006) also reported that Fe deficiency improved Cd uptake and translocation in rice grains (Table 13.1).

13.13 Concluding Remarks

Heavy metals are renowned regarding their toxic effects worldwide. Plants and animals are largely affected by toxic effects of these metals. Heavy metals emerge from anthropogenic activities in the surrounding environment. Higher concentrations of heavy metals largely affect plant health as documented by many researchers. Plant physiological, morphological, and biochemical parameters are negatively affected by metals toxicity. On the other hand, there are some metals which are essentially required by plants for their growth traits known as micro- and macronutrients. Fe is one of the key essential micronutrient beneficial to plants health. Beside Fe importance in plant growth attributes, it also plays a significant role in alleviating heavy metals stress in many plants. Fe application improved plant growth, photosynthetic

Table 13.1 Role of Fe in alleviating heavy metals stress in different plants

Fe dosage	HM conc.	Plant species	Effect	References
Fe + EDTA at 5, 10, 20 ppm	Cd at 0, 50, 100 μM	Rice (*Oryza sativa* L.)	Improve plant growth, leaf area, and leaf water content; reduce Cd toxic effects, decrease proline, MDA content, antioxidant enzyme activities	Ali et al. (2014)
Fe at 2.77, 5.54, and 8.31 μM	Cd and Pb at 10 μM	*Typha latifolia*	Decrease Cd and Pb uptake and translocation in plant shoots and roots, absorb Pb on roots at maximum Fe	Rodriguez-Hernandez et al. (2015)
Fe at 1.89 and 16.8 mg L^{-1}	Cd at 5 μM	Rice (*Oryza sativa* L.)	Increased MDA content, improve plant growth and SPAD value, enhanced antioxidant enzyme activities	Shao et al. (2007)
Fe at 0.54–2.6 mg kg^{-1}	Pb at 45–199 mg kg^{-1} Cd at 1.1–3.5 mg kg^{-1}	Rice plant tissues	Promote metal deposition on root surfaces, limit Pb and Cd translocation, and distribution in plant tissues	Cheng et al. (2014)
Fe at 10, 30,50, 80, and 100 mg L^{-1}	Cd at 0.1 and 1 mg L^{-1}	Rice (*Oryza sativa* L.)	Decrease Cd supply in shoots and roots, inhibit Cd uptake and translocation within rice plant, decrease radioactivity of ^{109}Cd in shoots of seedlings	Liu et al. (2007)
Fe at 10 and 250 μM	Cd at 25 μM	Barley	Enhance antioxidant enzyme activities, improve plant growth and biochemical parameters, reduce Cd toxic effects	Sharma et al. (2004)
Fe at 40 μM	Cd as $Cdcl_2$	Indian mustard	Reduce oxidative stress and metal toxicity, stabilize thylakoid complex, retention of chloroplast and chlorophyll contents	Qureshi et al. (2010)

pigments, transpiration rate, and enzymatic activities by limiting metal toxic effects. Fe also stabilizes chloroplast value in plants by neutralizing metal toxicity. Increased Fe concentration in flooded condition forms Fe plaque. Fe plaque form in plant roots has also been used to alleviate many heavy metal toxicities in plants such as rice. However, Fe deficiency in cereal plants caused dietary problems such as malnutrition in human population worldwide. Fe deficiency limits plant growth and ultimately leads to plant death. Understanding the Fe shortage response in plants, it is vital to enhance plant health and human diet which depends upon Fe plant sources. Biofortification could be the best choice globally to achieve sufficient levels of minerals in human diet. Fe shortage could be filled by various agronomic practices, plant breeding techniques, and biotechnology in addition to transgenic approaches involving genes which are strongly recommended.

By combining above reports, it could be illustrated that Fe inhibits metal uptake and transportation in plant parts by enhancing plant physiological, morphological, and biochemical parameters. Conversely, excessive Fe concentration disturbs plant functions by limiting their growth. Higher Fe dosage inhibits plant growth and development processes. Excessive Fe concentration caused physical injury, loss in chlorophyll, photooxidative damage, and root cell death in plants.

References

Abadia, J., et al. (2011). Towards a knowledge-based correction of iron chlorosis. *Plant Physiology and Biochemistry, 49*, 471–482.

Ali, S., Zeng, F., Qiu, L., & Zhang, G. (2011). The effect of chromium and aluminum on growth, root morphology, photosynthetic parameters and transpiration of the two barley cultivars. *Biologia Plantarum, 55*, 291–296.

Ali, S., Farooq, M. A., Jahangir, M. M., Abbas, F., Bharwana, S. A., & Zhang, G. P. (2013). Effect of chromium and nitrogen form on photosynthesis and anti-oxidative system in barley. *Biologia Plantarum, 57*, 758–763.

Ali, R. K., Najafi, F., & Rezaei, M. (2014). The influence of cadmium toxicity on some physiological parameters as affected by iron in rice (*Oryza Sativa* L.) plant. *Journal of Plant Nutrition, 37*, 1202–1213.

Alvarez-Fernandez, A., et al. (2007). Determination of synthetic ferric chelates used as fertilizers by liquid chromatography-electrospray/mass spectrometry in agricultural matrices. *Journal of the American Society for Mass Spectrometry, 18*, 37–47.

Aoyama, T., et al. (2009). OsYSL18 is a rice iron (III)–deoxymugineic acid transporter specifically expressed in reproductive organs and phloem of lamina joints. *Plant Molecular Biology, 70*, 681–692.

Balk, J., & Pilon, M. (2011). Ancient and essential: The assembly of Fe-S cluster in plants. *Trends in Plant Science, 16*, 218–226.

Balk, J., & Schaedler, T. A. (2014). Iron cofactor assembly in plants. *Annual Review of Plant Biology, 65*, 125–153.

Bao, T., Sun, L., Sun, T., Zhang, P., & Niu, Z. (2009). Iron-deficiency induces cadmium uptake and accumulation in *Solanum nigrum* L. *Bulletin of Environmental Contamination and Toxicology, 82*, 338–342.

Bao, T., Sun, T. H., & Sun, L. N. (2012). Effect of cadmium on physiological responses of wheat and corn to iron deficiency. *Journal of Plant Nutrition, 35*, 1937–1948.

Becker, M., & Asch, F. (2005). Iron toxicity in rice-conditions and management concepts. *Journal of Plant Nutrition and Soil Science, 168*, 558–573.

Bernard, D. G., Cheng, Y., Zhao, Y., & Balk, J. (2009). An allelic mutant series of ATM3 reveals its key role in the biogenesis of cytosolic iron-sulfur proteins in *Arabidopsis*. *Plant Physiology, 151*, 590–602.

Blair, M. W., & Izquierdo, P. (2012). Use of the advanced backcross- QTL method to transfer seed mineral accumulation nutrition traits from wild to Andean cultivated common beans. *Theoretical and Applied Genetics, 125*, 1015–1031.

Boyd, P. W., et al. (2007). Mesoscale iron enrichment experiments 1993–2005: Synthesis and future directions. *Science, 315*, 612–617.

Briat, J. F., et al. (2003). Iron utilization and metabolism in plants. *Current Opinion in Plant Biology, 10*, 276–282.

Cailliatte, R., Schikora, A., Briat, J. F., Mari, S., & Curie, C. (2010). High-affinity manganese uptake by the metal transporter NRAMP1 is essential for *Arabidopsis* growth in low manganese conditions. *Plant Cell, 22*, 904–917.

Chandra, R., Bharagava, R. N., Yadav, S., & Mohan, D. (2009). Accumulation and distribution of toxic metals in wheat (*Triticum aestivum* L.) and Indian mustard (*Brassica campestris* L.) irrigated with distillery and tannery effluents. *Journal of Hazardous Materials, 162*, 1514–1521.

Chatterjee, C., Gopal, R., & Dube, B. K. (2006). Physiological and biochemical responses of French bean to excess cobalt. *Journal of Plant Nutrition, 29*, 127–136.

Chen, T., Cai, X., Wu, X., Karahara, I., Schreiber, L., & Lin, J. (2011). Casparian strip development and its potential function in salt tolerance. *Plant Signaling & Behavior, 6*, 1499–1502.

Cheng, H., Wang, M., Wong, M. H., & Ye, Z. (2014). Does radial oxygen loss and iron plaque formation on roots alters Cd and Pb uptake and distribution in rice plant tissues? *Plant and Soil, 375*, 137–148. doi:10.1007/s11104-013-1945-0.

Conte, S. S., & Walker, E. L. (2011). Transporters contributing to iron trafficking in plants. *Molecular Plant, 4*, 464–476.

Conte, S., Stevenson, D., Furner, I., & Lloyd, A. (2009). Multiple antibiotic resistance in Arabidopsis is conferred by mutations in a chloroplast-localized transport protein. *Plant Physiology, 151*, 559–573.

Couturier, J., et al. (2013). The iron–sulfur cluster assembly machineries in plants: Current knowledge and open questions. *Frontiers in Plant Science, 4*, 259.

de Dorlodot, S., Lutts, S., & Bertin, P. (2005). Effects of ferrous iron toxicity on the growth and mineral composition of an inter specific rice. *Journal of Plant Nutrition, 28*, 1–20.

Durrett, T. P., Gassmann, W., & Rogers, E. E. (2007). The FRD3- mediated efflux of citrate into the root vasculature is necessary for efficient iron translocation. *Plant Physiology, 144*, 197–205.

Duy, D., Wanner, G., Meda, A., von Wiren, N., Soll, J., & Philippar, K. (2007). PIC1, an ancient permease in Arabidopsis chloroplasts, mediates iron transport. *Plant Cell, 19*, 986–1006.

Duy, D., Stube, R., Wanner, G., & Philippar, K. (2011). The chloroplast permease PIC1 regulates plant growth and development by directing homeostasis and transport of iron. *Plant Physiology, 155*, 1709–1722.

Eberhard, S., et al. (2008). The dynamics of photosynthesis. *Annual Review of Genetics, 42*, 463–515.

Eggink, L. L., et al. (2004). Synthesis of chlorophyll b: Localization of chlorophyllide a oxygenase and discovery of a stable radical in the catalytic subunit. *BMC Plant Biology, 4*, 5.

El-Jendoubi, H., et al. (2014). The effects of foliar fertilization with iron sulfate in chlorotic leaves are limited to the treated area. A study with peach trees (*Prunus persica* L. Batsch) grown in the field and sugar beet (*Beta vulgaris* L.) grown in hydroponics. *Frontiers in Plant Science, 5*, 2.

Feng, H., Qian, Y., Gallagher, F. J., Wu, M., Zhang, W., Yu, L., Zhu, Q., Zhang, K., Liu, C. J., & Tappero, R. (2013). Lead accumulation and association with Fe on *Typha latiofolia* root from an urban brownfield site. *Environmental Science and Pollution Research, 20*, 3743–3750. doi:10.1007/s11356-012-1298-x.

Fozia, A., Muhammad, A. Z., Muhammad, A., & Zafar, M. K. (2008). Effect of chromium on growth attributes in sunflower (*Helianthus annuus* L.) *Journal of Environmental Sciences, 20*, 1475–1480.

Gratao, P. L., Polle, A., Lea, P. J., & Azevedo, R. A. (2005). Making the life of heavy metal stressed plants a little easier. *Functional Plant Biology, 32*, 32481–32494.

Green, L. S., & Rogers, E. E. (2004). FRD3 controls iron localization in *Arabidopsis*. *Plant Physiology, 136*, 2523–2531.

Greipsson, S. (1995). Effect of iron plaque on roots of rice on growth of plants in excess zinc and accumulation of phosphorus in plants in excess copper or nickel. *Journal of Plant Nutrition, 18*, 1659–1665.

Grotz, N., & Guerinot, M. L. (2002). Limiting nutrients: An old problem with new solutions? *Current Opinion in Plant Biology, 5*, 158–163.

Hindt, M. N., & Guerinot, M. L. (2012). Getting a sense for signals: Regulation of the plant iron deficiency response. *Biochimica et Biophysica Acta, 1823*, 1521–1530.

Hossain, M. B., Jahiruddin, M., Leoppert, R. H., Panaullah, G. M., Islam, M. R., & Duxbury, J. M. (2009). The effects of iron plaque and phosphorus on yield and arsenic accumulation in rice. *Plant and Soil, 317*, 167–176.

Hu, Y., Huang, Y. Z., & Liu, Y. X. (2014). Influence of iron plaque on chromium accumulation and translocation in three rice (*Oryza sativa* L.) cultivars grown in solution culture. *Chemistry and Ecology, 30*, 29–38.

Huang, Q., Qi, W., Luo, Z., Yu, Y., Jiang, R., & Li, H. (2015). Effects of root iron plaque on selenite and selenate dynamics in rhizosphere and uptake by rice (*Oryza sativa*). *Plant and Soil, 388*, 255–266.

Inoue, H., Kobayashi, T., Nozoye, T., Takahashi, M., Kakei, Y., Suzuki, K., Nakazono, M., Nakanishi, H., Mori, S., & Nishizawa, N. K. (2009). Rice OsYSL15 is an iron-regulated iron (III) deoxymugineic acid transporter expressed in the roots and is essential for iron uptake in early growth of the seedlings. *The Journal of Biological Chemistry, 284*, 3470–3479.

Jeong, J., & Guerinot, M. L. (2009). Homing in on iron homeostasis in plants. *Trends in Plant Science, 14*, 280–285.

Jeong, J., Cohu, C., Kerkeb, L., Pilon, M., Connolly, E. L., & Guerinot, M. L. (2008). Chloroplast Fe(III) chelate reductase activity is essential for seedling viability under iron limiting conditions. *Proceedings of the National Academy of Sciences of the United States of America, 105*, 10619–10624.

Kim, S. A., Punshon, T., Lanzirotti, A., Li, L., Alonso, J. M., Ecker, J. R., Kaplan, J., & Guerinot, M. L. (2006). Localization of iron in Arabidopsis seed requires the vacuolar membrane transporter VIT1. *Science, 314*, 1295–1298.

Kobayashi, T., & Nishizawa, N. (2012). Iron uptake, translocation and regulation in higher plants. *Annual Review of Plant Biology, 63*, 131–152.

Koike, S., et al. (2004). OsYSL2 is a rice metal–nicotianamine transporter that is regulated by iron and expressed in the phloem. *The Plant Journal, 39*, 415–424.

Kovacs, K., Kuzmann, E., Vertes, A., Levai, L., Cseh, E., & Fodor, F. (2010). Effect of cadmium on iron uptake in cucumber roots: A Mössbauer-spectroscopic study. *Plant and Soil, 327*, 49–56.

Lammertsma, E. I., et al. (2011). Global CO_2 rise leads to reduced maximum stomatal conductance in Florida vegetation. *Proceedings of the National Academy of Sciences of the United States of America, 108*, 4035–4040.

Lanquar, V., Lelievre, F., Bolte, S., Hames, C., Alcon, C., Neumann, D., Vansuyt, G., Curie, C., Schroder, A., Kramer, U., et al. (2005). Mobilization of vacuolar iron by AtNRAMP3 and AtNRAMP4 is essential for seed germination on low iron. *The EMBO Journal, 24*, 4041–4051.

Layer, G., et al. (2010). Structure and function of enzymes in heme biosynthesis. *Protein Science, 19*, 1137–1161.

Lee, S., Chiecko, J. C., Kim, S. A., Walker, E. L., Lee, Y., Guerinot, M. L., & An, G. (2009). Disruption of OsYSL15 leads to iron inefficiency in rice plants. *Plant Physiology, 150*, 786–800.

Lenntech Water Treatment and Air Purification. (2004). *Water treatment*. Rotterdamseweg, Netherlands: Lenntech.

Li, R., Zhou, Z., Zhang, Y., Xie, X., Li, Y., & Shen, X. (2015). Uptake and accumulation characteristics of arsenic and iron plaque in rice at different growth stages. *Communications in Soil Science and Plant Analysis, 46*, 2509–2522.

Liu, H. J., Zhang, J. L., & Zhang, F. S. (2007). Role of iron plaque in Cd uptake by and translocation within rice (*Oryza sativa* L.) seedlings grown in solution culture. *Environmental and Experimental Botany, 59*, 314–320.

Liu, H. J., Zhang, J. L., Christie, P., & Zhang, F. S. (2008). Influence of iron plaque on uptake and accumulation of Cd by rice (*Oryza sativa* L.) seedlings grown in soil. *The Science of the Total Environment, 394*, 361–368.

Liu, H. J., Zhang, J. L., Christie, P., & Zhang, F. S. (2010). Influence of iron fertilization on cadmium uptake by rice seedlings irrigated with cadmium solution. *Communications in Soil Science and Plant Analysis, 41*, 584–594.

Liu, J., Leng, X., Wang, M., Zhu, Z., & Dai, Q. (2011). Iron plaque formation on roots of different rice cultivars and the relation with lead uptake. *Ecotoxicology and Environmental Safety, 54*, 1304–1309.

Lopez-Millan, A. F., Morales, F., Abadia, A., & Abadia, J. (2000). Effects of iron deficiency on the composition of the leaf apoplastic fluid and xylem sap in sugar beet: Implications for iron and carbon transport. *Plant Physiology, 124*, 873–884.

Lucca, P., Hurrell, R., & Potrykus, I. (2002). Fighting iron deficiency anemia with iron-rich rice. *Journal of the American College of Nutrition, 21*, 184–190.

Lung'aho, M. G., et al. (2011). Genetic and physiological analysis of iron biofortification in maize kernels. *PloS One, 6*, 20429.

Masuda, H., et al. (2013). Iron biofortification of rice using different transgenic approaches. *Rice, 6*, 40.

Mithofer, A., Schulze, B., & Boland, W. (2004). Biotic and heavy metal stress response in plants: Evidence for common signals. *FEBS Letters, 566*, 1–5.

Monni, S., Salemma, M., & Millar, N. (2000). The tolerance of Empetrumnigrum to copper and nickel. *Environmental Pollution, 109*, 221–229.

Morrissey, J., et al. (2009). The ferroportin metal efflux proteins function in iron and cobalt homeostasis in Arabidopsis. *Plant Cell, 21*, 3326–3338.

Msilini, N., et al. (2011). Inhibition of photosynthetic oxygen evolution and electron transfer from the quinone acceptor QA to QB by iron deficiency. *Photosynthesis Research, 107*, 247–256.

Nada, E., Ferjani, B. A., Ali, R., Imed, B. M., & Makki, B. (2007). Cadmium-induced growth inhibition and alteration of biochemical parameters in almond seedlings grown in solution culture. *Acta Physiologia Plantarum, 29*, 57–62. doi:10.1007/s11738-006-0009-y.

Nagajyoti, P. C., Lee, K. D., & Sreekanth, T. V. M. (2010). Heavy metals, occurrence and toxicity for plants: A review. *Environmental Chemistry Letters, 8*, 199–216. doi:10.1007/s10311-010-0297-8.

Nakanishi, H., Ogawa, I., Ishimaru, Y., Mori, S., & Nishizawa, N. K. (2006). Iron deficiency enhances cadmium uptake and translocation mediated by the Fe^{2+}transporters OsIRT1 and OsIRT2 in rice. *Soil Science & Plant Nutrition, 52*, 464–469.

Nouet, C., Motte, P., & Hanikenne, M. (2011). Chloroplastic and mitochondrial metal homeostasis. *Trends in Plant Science, 16*, 395–404.

Passariello, B., Giuliano, V., Quaresima, S., Barbaro, M., Caroli, S., Forte, G., Garelli, G., & Iavicoli, I. (2002). Evaluation of the environmental contamination at an abandoned mining site. *Microchemical Journal, 73*, 245–250.

Pretty, J. (2008). Agricultural sustainability: Concepts, principles and evidence. *Philosophical Transactions of the Royal Society of London. Series B, Biological Sciences, 363*, 447–465.

Qadir, S., Hameed, A., Nisa, N. T., Azooz, M. M., Wani, M. R., Hasannuzaman, M., Kazi, A. G., & Ahmad, P. (2014). Chapter 1 Brassicas: Responses and tolerance to heavy metal stress. In *Improvement of crops in the era of climatic changes* (10.1007/978–1–4614-8824-8_1 ed.). New York: Springer. http://www.researchgate.net/publication/265251513.

Qureshi, M. I., D'Amici, G. M., Fagioni, M., Rinalducci, S., & Zolla, L. (2010). Iron stabilizes thylakoid protein-pigment complexes in indian mustard during Cd-phytoremediation as revealed by BN-SDS-PAGE and ESI-MS/MS. *Journal of Plant Physiology, 167*, 761–770. doi:10.1016/j.jplph.2010.01.017.

Reeves, R. D., & Baker, A. J. M. (2000). Metal-accumulating plants. In I. Raskin & B. D. Ensley (Eds.), *Phytoremediation of toxic metals: Using plants to clean up the environment* (pp. 193–229). New York: Wiley.

Rellan-Alvarez, R., Abadia, J., & Alvarez-Fernandez, A. (2008). Formation of metal–nicotianamine complexes as affected by pH, ligand exchange with citrate and metal exchange: A study by electrospray ionization time-of-flight mass spectrometry. *Rapid Communications in Mass Spectrometry, 22*, 1553–1562.

Robinson, N. J., Procter, C. M., Connolly, E. L., & Guerinot, M. L. (1999). A ferric–chelate reductase for iron uptake from soils. *Nature, 397*, 694–697.

Rodriguez-Hernandez, M. C., Bonifas, I., Alfaro-De la Torre, M. C., Flores-Flores, J. L., Banuelos-Hernandez, B., & Patino-Rodriguez, O. (2015). Increased accumulation of cadmium and lead under Ca and Fe deficiency in *Typha latifolia*: A study of two pore channel (TPC1) gene responses. *Environmental and Experimental Botany, 115*, 38–48.

Römheld, V., & Nikolic, M. (2006). Iron. In A. V. Barker & D. J. Pilbeam (Eds.), *Handbook of plant nutrition* (pp. 329–350). Boca Raton: CRC Press.

Roschzttardtz, H., Conejero, G., Curie, C., & Mari, S. (2009). Identification of the endodermal vacuole as the iron storage compartment in the *Arabidopsis embryo. Plant Physiology, 151*, 1329–1338.

Roschzttardtz, H., Grillet, L., Isaure, M. P., Conejero, G., Ortega, R., Curie, C., & Mari, S. (2011). Plant cell nucleolus as a hot spot for iron. *The Journal of Biological Chemistry, 286*, 27863–27866.

Shao, G., Chen, M., Wang, W., Mou, R., & Zhang, G. (2007). Iron nutrition affects cadmium accumulation and toxicity in rice plants. *Plant Growth Regulation, 53*, 33–42.

Shao, G. S., Chen, M. X., Wang, D. Y., CM, X., Mou, R. X., Cao, Z. Y., & Zhang, X. F. (2008). Using iron fertilizer to control Cd accumulation in rice plants: A new promising technology. *Science in China Series C: Life Sciences, 51*, 245–253.

Sharma, S. (2007). Adaptation of photosynthesis under iron deficiency in maize. *Journal of Plant Physiology, 164*, 1261–1267.

Sharma, S. S., Kaul, S., Metwally, A., Goyal, K. C., Finkemeier, I., & Dietz, K. J. (2004). Cadmium toxicity to barley (*Hordeum vulgare*) as affected by varying Fe nutritional status. *Plant Science, 166*, 1287–1295.

Shaw, G. C., Cope, J. J., Li, L., Corson, K., Hersey, C., Ackermann, G. E., Gwynn, B., Lambert, A. J., Wingert, R. A., Traver, D., et al. (2006). Mitoferrin is essential for erythroid iron assimilation. *Nature, 440*, 96–100.

Sheng, S. G., Xue, C. M., Ying, W. D., Mei, X. C., Yun, C. Z., & ZX, F. (2008). Using iron fertilizer to control Cd accumulation in rice plants: A new promising technology. *Science in China. Series C, Life Sciences, 51*, 245–253.

Shimoni-Shor, E., Hassidim, M., Yuval-Naeh, N., & Keren, N. (2010). Disruption of Nap14, a plastid-localized non-intrinsic ABC protein in *Arabidopsis thaliana* results in the over accumulation of transition metals and in aberrant chloroplast structures. *Plant, Cell & Environment, 33*, 1029–1038.

Syu, C. H., Jiang, P. Y., Huang, H. H., Chen, W. T., Lin, T. H., & Lee, D. Y. (2013). Arsenic sequestration in iron plaque and its effect on As uptake by rice plants grown in paddy soils with high contents of As, iron oxides, and organic matter. *Soil Science & Plant Nutrition, 59*, 463–471.

Takahashi, M. T., Nakanishi, H., Kawasaki, S., Nishizawa, N. K., & Mori, S. (2001). Enhanced tolerance of rice to low iron availability in alkaline soils using barley nicotianamine aminotransferase genes. *Nature Biotechnology, 19*, 466.

Tako, E., et al. (2013). High bioavailability iron maize (*Zea mays* L.) developed through molecular breeding provides more absorbable iron in vitro (Caco-2 model) and in vivo (*Gallus gallus*). *Nutrition Journal, 12*, 3.

Vert, G., et al. (2002). IRT1, an Arabidopsis transporter essential for iron uptake from the soil and for plant growth. *Plant Cell, 14*, 1223–1233.

Vert, G., Barberon, M., Zelazny, E., Seguela, M., Briat, J. F., & Curie, C. (2009). Arabidopsis IRT2 cooperates with the high affinity iron uptake system to maintain iron homeostasis in root epidermal. Cells. *Planta, 229*, 1171–1179.

Vigani, G., Maffi, D., & Zocchi, G. (2009). Iron availability affects the function of mitochondria in cucumber roots. *The New Phytologist, 182*, 127–136.

Walker, E. L., & Connolly, E. L. (2008). Time to pump iron: Iron-deficiency signaling mechanisms of higher plants. *Current Opinion in Plant Biology, 11*, 530–535.

Welch, R. M., & Graham, R. D. (2004). Breeding of micronutrients in staple food crops from a human nutrition perspective. *Journal of Experimental Botany, 55*, 353–364.

White, P. J., & Broadley, M. R. (2005). Biofortifying crops with essential mineral elements. *Trends in Plant Science, 10*, 586–593.

Wintz, H., Fox, T., & Vulpe, C. (2002). Responses of plants to iron, zinc and copper deficiencies. *Biochemical Society Transactions, 30*, 766–768.

Xenidis, A., Stouraiti, C., & Papassiooi, N. (2010). Stabilization of Pb and As in soils by applying combined treatment with phosphates and ferrous iron. *Journal of Hazardous Materials, 177*, 929–937.

Xu, B., & Yu, S. (2013). Root iron plaque formation and characteristics under N_2 flushing and its effects on translocation of Zn and Cd in paddy rice seedlings (*Oryza sativa*). *Annals of Botany, 111*, 1189–1195.

Xu, Y., et al. (2012). Molecular mapping of QTLs for grain zinc, iron, and protein concentration of across two environments. *Field Crops Research, 138*, 57–62.

Yadavalli, V., et al. (2012). Alteration of proteins and pigments influence the function of photosystem I under iron deficiency from *Chlamydomonas reinhardtii. PloS One, 7*, 35084.

Zhou, X. B., & Shi, W. M. (2007). Effect of root surface iron plaque on se translocation and uptake by Fe-deficient rice. *Pedosphere, 17*(5), 580–587.

Zocchi, G. (2006). Metabolic changes in iron-stressed dicotyledonous plants. In L. L. Barton & J. Abadia (Eds.), *Iron nutrition in plants and rhizospheric microorganisms* (pp. 359–370). Dordrecht, Netherlands: Springer.

Zornoza, P., Sánchez-Pardo, B., & Carpena, R. R. O. (2010). Interaction and accumulation of manganese and cadmium in the manganese accumulator *Lupinus albus. Journal of Plant Physiology, 167*, 1027–1032. doi:10.1016/j.jplph.2010.02.011.

Zuo, Y., & Zhang, F. (2009). Iron and zinc biofortification strategies in dicot plants by intercropping with graminaceous species. A review. *Agronomy for Sustainable Development, 29*, 63–71.

Chapter 14
Role of Zinc in Alleviating Heavy Metal Stress

Zaid ul Hassan, Shafaqat Ali, Muhammad Rizwan, Afzal Hussain, Zaheer Akbar, Nasir Rasool, and Faraht Abbas

Abstract Heavy metals pollution is continuously increasing and posing threats to our environment. This heavy metals' presence in soils accumulates in plants which become part of food supply. Heavy metals at their higher concentrations negatively affect plant growth parameters. Zn is one of the essential micronutrient which takes part in plant physiological functions and showed beneficial effects in plant growth, development processes and yield. Zn plays important role in cellular functions in all living organisms. Nevertheless, heavy metal concentrations are increasing in our agricultural soils through different anthropogenic activities which need to be discussed. Considering present scenario, this review was conducted to illustrate the promotive and beneficial role of Zn in alleviating heavy metals stress in various plants. It was observed that Zn not only ameliorate different metals' toxic levels but also improve plant growth attributes by inhibiting heavy metals uptake in plant parts. Heavy metals induced oxidative stress and decline plant growth, biomass, chlorophyll contents, photosynthetic traits and many metabolic functions. Zn combat heavy metals toxicity by generating antioxidant defence system against oxidative damage and improved plant growth parameters by alleviating metals toxicity in different plants. However, Zn inadequate availability declined crop yield. Additionally, Zn deficiency disturbs plant growth and leaf chlorosis. Various techniques in the form of bio-fortification have been used to overcome Zn deficiency.

Keywords Heavy metals • Zn • Anthropogenic activities • Micronutrient • Oxidative stress • Physiological • Photosynthetic • Metabolic • Bio-fortification

Z. ul Hassan • S. Ali (✉) • M. Rizwan • A. Hussain • Z. Akbar • F. Abbas
Department of Environmental Sciences and Engineering, Government College University, Allama Iqbal Road, 38000 Faisalabad, Pakistan
e-mail: shafaqataligill@yahoo.com

N. Rasool
Department of Chemistry, Government College University, Allama Iqbal Road, 38000 Faisalabad, Pakistan

M. Naeem et al. (eds.), *Essential Plant Nutrients*,
DOI 10.1007/978-3-319-58841-4_14

14.1 Introduction

Heavy metal concentrations have been enhanced in agricultural soils and posing threats due to various human activities such as industrial and mining operations (Ikenaka et al. 2010). A few of the lower molecular weight heavy metals are essential nutrients like Zn, Ni, Cu, Si, Fe, etc. but mostly are toxic and so are non-essential. Their toxicity depends upon the bioavailability and sensitivity of the target organism (Rascio and Navari-Izzo 2011). Heavy metals not only showed toxicity in humans and animals but also in other organisms like plants that are severely affected. Plants are constantly attacked by abiotic stresses (Gratao et al. 2008), and many heavy metals are present in our agricultural soils affecting soil fertility (Przedpelska and Wierzbicka 2007). A reduction in plant biomass, root growth, leaf chlorosis, and morphological parameters were noted under heavy metals stress. Plants need to evolve methods to cope with heavy metals. They used different strategies to handle heavy metals stress. In first strategy, plants tried to avoid heavy metal entrance through roots by reducing bioavailability of soil metals or reducing metal uptake protein transport. Metal solubility in roots is affected by factors like pH, minerals properties, microorganism activities and cation exchange capacity (Ghosh and Singh 2005). The exudation of carbohydrates from roots acidifies rhizosphere, and metal solubility enhanced their toxicity (Martinez and Motto 2000). In second strategy, plants tolerate and restrict heavy metal accumulation from plant roots to above parts by ensuring nutrients homeostasis (Pollard et al. 2002). Plant strategies need to be regulated to cope with heavy metals stress. Agricultural soils contained many heavy metals like zinc (Zn), cadmium (Cd), copper (Cu), chromium (Cr), nickel (Ni), lead (Pb), etc. (Yadav 2009).

Zinc (Zn), a transition heavy metal, is 24th most abundantly found in earth crust (Alloway 2013). Major sources of Zn occurrence are anthropogenic activities, agrochemicals, sewer sludge and high fertilizer usage (Alloway 2008c). Knowing the importance of Zn in plants, Zn is an essential micronutrient required by plants (Alloway 2008b). Among essential nutrients, Zn plays an important role in maintaining heavy metals homeostasis in plant cells. In this way, it reduces metal toxicity by limiting their availability (Appenroth 2010). Zn like other micronutrients enhanced plant growth and development processes (Hansch and Mendel 2009). It plays key role in many metabolic functions. Zn activates many enzymes which are involved in protein synthesis, lipid and nucleic acid metabolic activities (Bonnet et al. 2000). Zn plays a major role in abiotic resistance in various plants. It has been investigated that Zn reduces Cu toxicity and improved Zn uptake in plants (Hafeez et al. 2013). Zn also ameliorates Cd toxicity and induced oxidative stress in rice (Hassan et al. 2005) and wheat (Zhao et al. 2011). Addition of Zn also reduced Cu stress in rice (*Oryza sativa* L.) seedlings by generating antioxidant mechanism against oxidative stress and restricts Cu toxic effects (Thounaojam et al. 2014). These findings suggested that Zn has the ability to reduce heavy metals stress by enhancing plant growth attributes.

14.2 Why Zinc?

Cereals which are lower in Zn level have been used by developing and developed countries as an essential food item. Zn deficiency in cereals is the key reason for worldwide malnutrition (Chandel et al. 2010). Micronutrient as Zn is essential for all living organisms worldwide (Andreini et al. 2006). Zn deficiency largely affects world's population (Maret and Sandstead 2006). Its shortage is also a health risk factor across the developed and developing countries. Zn deficiency not only affects individuals but also the economy of developing countries. It has been a great challenge for economist to cope with problems like malnutrition-caused disorders (Chandel et al. 2010). Food fortification has been found as effective strategy to handle mineral nutrients deficiency across the world (Horton 2006).

14.3 Biochemical Functions of Zinc

The tendency of Zn to form complexes with ligands determined its metabolic functions. Zn is predominantly found as lower molecular weight complexes, metal ions and cell walls insoluble structures. Zn is inactivated by complex formation at intercellular level (Alloway 2004). Zn affects the capacity of water uptake and its translocation to various plant parts (Disante et al. 2010). It also involved in signal transduction through protein kinase (Hansch and Mendel 2009). Moreover, Zn reduces heat stress (Peck and McDonald 2010) or salt stress (Tavallali et al. 2010). The synthesis of tryptophan which is precursor of important growth hormone requires Zn (Brennan 2005). The structural arrangement of macromolecules, ion transport system and membrane maintenance is also done by Zn (Disante et al. 2010). Zn is required in protein synthesis and energy (Hansch and Mendel 2009). Zn is also involved in other metabolic functions like lipids and carbohydrate while forming complexes with DNA and RNA (Marschner 2012). Zn is tightly bound in cellular membrane and restricts the production of toxic radicals (Alloway 2004; Brennan 2005; Disante et al. 2010). Zn is also involved in antioxidant enzymes defence system (Cakmak 2000).

14.4 Zn Role in Human Health

WHO, food and agricultural organization in mid-1990s investigated the required amount of Zn needed by humans in their dietary intake (WHO, FAO, IAEA 2002). Human adults required 15 mg Zn per day for their dietary intake. Zn plays a key role in the cellular performance of all living creatures and enhanced immune system in humans. Zn also acts as structural part of human body enzymes. Zn deficiency disturbed human growth periods. Most affected are immune, central nervous, skeletal, epidermal and reproductive systems. However, long-term intake of Zn may affect inclusion of dietary zinc (Lonnerdal 2000). The major cause of Zn deficiency is the dietary intake shortage of absorbable Zn. Insufficient dietary of Zn was observed that affected 25% of the world's population (Maret and Sandstead 2006: King and Cousins 2006). Zn deficiency enhanced threats for malnutrition which cause children death across the world (Bhutia 2014). The immune system is largely affected by malnutrition. It was reported by the World Health Organization (WHO 2002) that Zn shortage ranks 5th among other health threat factors in emergent countries and seventh overall in the world (Chandel et al. 2010). Zn deficiency disturb the immune system by showing infection symptoms including weight loss, late wound healing, growth problems and dermatitis along other related symptoms. Zn importance was observed as host immunity in patients after 60s, and its shortage leads to chronicity (Foster and Samman 2012). Different diseases were noted in Zn-deficient patients in the form of viral, bacteria and parasitic diseases (Prasad 2009). Zn shortage in humans reduces testosterone levels that largely affect immune functions mainly body weight loss, hyperammonaemia and neurosensory organs (Prasad 2008). This problem was sorted out by growing staple food grain to overcome Zn shortage disorders. It was observed that sufficient amount of Zn also enhanced crop yield (Cakmak 2008). Zn deficiency can be overcome by using various techniques like dietary modification and bio-fortification (Roohani et al. 2013).

14.5 Zn and the Immune System

Zn affects the immune system in different ways (Overbeck et al. 2008). Zn functions as signal molecule in the immune system. Zn is necessary for the normal development and working of cell-induced innate immunity. Zn deficiency affects intracellular killing and the production of cytokines which is the basic messenger of the immune system. Zn is required for the synthesis of DNA and RNA transcription. Zn is also required for cell division, and its deficiency leads to cell death. Zn functions as antioxidant and maintains cellular membrane. Zn deficiency also deteriorated thymus and lymphoid tissues in target animals (Shankar and Prasad 1998).

14.6 Zinc Deficiency in Plants

Zn is a key micronutrient which takes part in plant physiological functions (Broadley et al. 2007). One of the extensively wide range abiotic stresses arises from Zn shortage in agricultural calcareous soils as documented by Singh et al. (2005). Zn is one of the most prevalent disorders among various crops (Naik and Das 2008). Zn deficiency depends upon plant species and type of experiment. Different mechanisms are involved in the deficiency of Zn. It is quite significant mechanism that Zn may be used in plant tissues and Zn uptake in various plant parts (Genc et al. 2006). Zn deficiency inhibits protein synthesis and ultimately reduces plant root development (Fageria 2004). A decrease in nutrient and water absorption from soil under Zn shortage leads to plant growth reduction (Epstein and Bloom 2005). Cereal crops like rice are affected by Zn shortage. About 70% rice is produced in rich water conditions increased in bicarbonate and phosphorus that reduce Zn content (Cakmak 2008). Factors that caused Zn shortage in rice staple food are high pH, calcite, phosphorus and bicarbonates in agricultural soils (Alloway 2009). Soil factors which affect Zn availability are those that maintain Zn content in soil solution and its sorption into soil solution. These soil factors are clay, moisture, microbial activity, macronutrients, calcium carbonate and total Zn content (Alloway 2008a).

14.7 Zinc Enrichment Techniques

Plant-derived food has been key source of dietary minerals chiefly for developing countries where plant food is an important supply of diet. Zn concentration in plants is low as compared to animal food supply-caused malnutrition problem (Cakmak 2008). Zn deficiency in edible parts of staple foods can be enhanced by various techniques like fertilizer management, genetic engineering, cultural management practices, bio-fortification, etc. (Cakmak 2009). Zn gathered up in rice and wheat grains can be enhanced by managing fertilizer and irrigation practices (Hao et al. 2007). A well-judged usage of fertilizer Zn improves crop productivity and plant grains (Phattarakul et al. 2012). Recently, most focus has been found on bio-fortification. This technique effectively alleviated malnutrition of micronutrient and protein. It involves enhanced staple food grain plants that are selectively bred to improve nutritional value (Chandel et al. 2010). In bio-fortification, required micronutrients move from roots to edible parts of cereal crops (Palmgren et al. 2008). Bio-fortification has been known to be cost-effective as it enhanced micronutrients through breeding and biotechnology (Mayer et al. 2008).

14.8 Zinc Status in Soil, Significance, Uptake and Translocation in Plants

Zn is the 24th most abundant element found in earth crust and beneficial to human, plants, animals and microorganisms (Alloway 2013). Zn occurs in soil, air, water, different anthropogenic sources, usage of fertilizers, sewage sludge and enhanced agrochemical addition in the environment. Weathering of older rocks also contributes to Zn in soils. Total amount of Zn in soils have been estimated to be 55 mg Zn kg^{-1} (Alloway 2008c). Amount of Zn in soil depends upon soil texture. Distribution of Zn in soils is determined by various factors like absorption and precipitation (Alloway 2008c). Soil pH extensively determines Zn distribution in soils (Alloway 2008c). Zn is an essential micronutrient required by plants. It regulates protein metabolism and carbohydrate, control membrane functions and restrict pathogenic attack (Alloway 2008c). Zn also regulates gene expression, synthesis of protein and cofactor of many enzymes (Alloway 2008c, 2009). About 36% involvement of Zn protein in gene expression was documented by Andreini et al. (2006). Recently, molecular genetics and molecular mechanisms have been evolved to determine the mechanism of Zn ions uptake and translocation in plants (Hall and Williams 2003). A particular gene sequence has been identified at first in *Arabidopsis thaliana* and was not affected by Zn fertilization (Burleigh et al. 2003).

Plants develop different strategies to assimilate and toxicity prevention of metal micronutrients from the soils by chelation, sequestration and transport (Colangelo and Guerinot 2006). Zn mainly occurs in +2 oxidation states in soil which is taken by plant roots (Alloway 2008c). Zn uptake needs to be solubilizing in rhizosphere through acidification and secretion of chelator of lower molecular weight (Sinclair and Kramer 2012). Soil pH, organic matter, clay content, moisture content, micronutrients level and microbial activity are key factors which control Zn transport from soils to plants (Alloway 2008c). Zn translocation occurs from soils to plant shoots. Translocation take place from older plant leaves to growing parts (Alloway 2008c). Zn in the form of free ion passes through cell plasma membrane chelated by ligands in cytosol (Freisinger 2008). Zn is immobilized in root and transported into vacuole (Deinlein et al. 2012). Zn uptake in plants is supported by various protein families (Bouain et al. 2014). Zn uptake also occurs through xylem from roots to plant different shoot parts (Broadley et al. 2007). Higher Zn level in phloem indicates that Zn translocation occurs through both xylem and phloem tissues (Haslett et al. 2001). The composition and concentration of growth media determined the level of Zinc uptake in plants. Zinc forms complexes with organic ligands or moves as divalent cation in nutrient medium or in the soils (Kabata-Pendias and Pendias 2001).

14.9 Plants Response to Heavy Metals and HMs-Plants Interaction

Heavy metals toxicity depends upon bioavailability of metals, time of exposure and usual conditions of target plants. Heavy metals are distributed into two categories, essential micronutrients and non-essential elements. Essential micronutrients mostly at lower concentration enhanced plant growth and development processes (Hansch and Mendel 2009). Heavy metals concentration should be up to a certain limit to avoid their toxic effects. Plants adopt regulatory mechanism to maintain heavy metal micronutrient homeostasis. When this mechanism reaches a certain threshold, heavy metals toxicity will be prominent. Non-essential heavy metals showed no biological function in plants and disturb plant growth cells and metabolic functions. Even lower concentrations of these non-essential heavy metals extensively pose damaging effects (Appenroth 2010). Plants uptake and translocation of heavy metals occur from outer environment. Higher concentrations of metals pose toxic effects by affecting plant growth and metabolic functions. Heavy metals can be detoxified by mechanism-involved chelation. Chelation of metal ions is done by ligand. Different ligands include amino acid, peptide and organic acids (Cobbett and Goldsbrough 2002: Trampczynska et al. 2010).

In context to HMs-plants interaction, heavy metals accumulation in plants showed toxic effects on the biosynthesis of photosynthetic pigments (Vesely et al. 2011), impair photosynthesis and alter chloroplast membrane (Ventrella et al. 2011). It is well known that heavy metals negatively affect photosynthetic rate, plant growth, stomatal conductance, transpiration rate and crop yield (Yusuf et al. 2012). These effects of heavy metals on growth, photosynthesis and enzymatic activities are dose dependent (Babu et al. 2010). Plants have developed various mechanisms to avoid heavy metals toxic effects (Latowski et al. 2011). Plants' survival under heavy metals stress relies on their ability to resist oxidative damage and antioxidant activity. Furthermore, heavy metals have been found to enhance CAT, SOD, APX and DHAR under metal stress such as Cd in rice and under Pb stress in wheat seedling (Iqbal et al. 2010). These findings suggested that enzymes could be used as a biomarker for heavy metals pollution (Dazy et al. 2009). Heavy metals have been categorized into two groups, essential and non-essential heavy metals. Essential metals such as Ni, Zn, Co, Fe and Mo are required by plants for their normal growth and development processes. It has been found that Zn improves plant physiological parameters (Hansch and Mendel 2009). Zn regulates many enzymatic activities and engaged in signal transduction (Broadley et al. 2007). On the other hand, non-essential heavy metals such as Cd, Hg and Pb are toxic to plants (Jarup 2003), and even lower metal concentration caused strong response from plants (Appenroth 2010). Therefore, detoxification of these heavy metals is essential for plants survival.

14.10 Zinc Dual Behaviour

Excessive Zn concentration inhibits plant biomass, cell division (Cakmak 2000; Khudsar et al. 2004; Tsonev and Lidon 2012), nutrient uptake (Kaya et al. 2000) and enzyme activities (Khudsar et al. 2004). Zn toxicity depends upon many factors like time of exposure, growth medium and target specie. Zn regulates gene expression required for heavy metals tolerance. However, Zn deficiency inhibits plant growth and development in the form of chlorosis and smaller leaves formation. Moreover, quality of harvested crop is also affected by Zn deficiency (Cakmak 2000). Zn also has an effect on water uptake capacity and transportation in different plant parts (Peck and McDonald 2010). Zn is mandatory for the structural arrangement of macromolecules and interrelates protein membranes (Dang et al. 2010). Zn is also required in large amount by protein unlike other heavy metals (Cakmak 2008b). It was observed that 10% proteome of eukaryotic cells were from Zn-containing protein and 36% involved in gene expression (Andreini et al. 2006). Zn application as fertilizer improved plants productivity. However, its deficiency negatively affects plant growth parameters (Mousavi et al. 2007).

On the other hand, Zn is an essential micronutrient which at higher dosage disturbs plant metabolic functions by affecting hydrogenase activities, balanced ribosomal fraction and cytochrome production (Tisdale et al. 1984). Higher Zn concentration stunted plant growth, alter mitotic activity and induce structural disorders (Tewari et al. 2008). Excessive Zn concentration induces oxidative stress that negatively affects many plant physiological parameters (Maksymiec et al. 2008) that ultimately affect photosynthetic electron transport (Cakmak 2000). In addition, membrane permeability, nutrients uptake and their translocation were also affected by Zn exposure (Wang et al. 2009). One of the chief benefits of Zn usage in plants is that it detoxifies abiotic stress like heavy metal as documented by Zhao et al. (2011).

14.11 Zinc Ameliorates Heavy Metal Stress

Zinc interacts with heavy metals in various ways. As documented earlier, Zn plays a significant role in detoxifying heavy metals stress in plants. It was observed that Zn ameliorates toxic effects of Cu by reducing its uptake and improved Zn translocation in different plants (Hafeez et al. 2013). Many observations have shown that Zn maintain cell membrane reliability and restrict the formation of free radical that causes cellular damage in plant roots (Cakmak 2000). Hassan et al. (2005) documented the role of Zn in alleviating Cd induced growth inhibition and oxidative stress in rice cultivars. It was noted that Cd declines plant growth, biomass and photosynthetic rate, while enzyme activities and malondialdehyde content were enhanced under Cd dosage. Zn treatment reduced Cd toxicity as shown by improved plant growth, biomass, photosynthetic rate and chlorophyll content, while decreasing in malondialdehyde content and antioxidant enzyme activities. Zhao et al. (2011) investigated that Zn also ameliorates Cd stress in winter wheat. Cd concentrations inhibit plant growth and superoxide dismutase activities, while enhancing proline content, catalase and peroxide activities. Results have shown that Zn at lower concentration increased Cd toxicity in wheat plants. On the other hand, higher level of Zn decreased Cd toxicity in wheat growth parameters. Higher concentration of Zn also ameliorates Cu-induced toxic effects in rice seedlings. Cu induces phytotoxic symptoms and oxidative damage in rice seedlings which is counteracted by Zn treatment (Arvind and Prasad 2005a, b). Cu induces reactive oxidative stress in the form of peroxides, etc. that impose threat to plants. Zn ameliorates Cu-induced stress and declines ROS (Arvind and Prasad 2005a, b; Upadhyay and Panda 2009). The enhanced production of ROS under Cu stress increases lipid peroxidation which is directly related to growth inhibition traits.

Zn application declined the rate of lipid peroxidation by reducing Cu stress. Moreover, Cu caused changes in antioxidant enzyme activities. Cu-alone treatment enhanced the activities of SOD, CAT and GPX. Enzymatic activities also detoxify ROS production (Singh et al. 2012). Zn application enhanced antioxidant efficiency of rice seedlings under Cu stress (Arvind and Prasad 2005a, b; Upadhyay and Panda 2009). These findings suggested that Zn plays a significant role in detoxifying heavy metals by modulating them and antioxidant enzyme mechanism to decline oxidative stress in plants. Cherif et al. (2011) noted interaction between Zn and Cd in tomato plants (*Solanum lycopersicum*). Cd induced oxidative stress and decline in chlorophyll content. Cd stress enhanced enzyme activities like SOD, while CAT, APX and GR were declined. Zn treatment at lower level enhanced the functional activities of CAT, APX, SOD and GR. Zn clearly declined Cd accumulation in tomato and reduced Cd toxicity. Several other researchers have also investigated the ameliorating role of Zn against Cd toxicity as documented in bread and durum wheat (Hart et al. 2002), angiosperm (Clemens 2009) and *Ceratophyllum demersum* (Arvind and Prasad 2003). Moreover, Zn also alleviated Cd-induced oxidative stress in *Ceratophyllum demersum* L (coontail). Results showed that Cd induced membrane damage, oxidative stress as indicated by lipid peroxidation and lipoxygenase activity. Zn treatment ameliorated membrane damage and oxidative stress by counteracting with antioxidant enzyme activities that direct membrane peroxidation (Arvind and Prasad 2003). Zn treatment under Cd stress enhanced glutathione content by suppressing Cd toxicity. It was also investigated that Zn alleviated Cd uptake and accumulation in *Chara australis* (R. Br.) (Clabeaux et al. 2013) and marigold (*Calendula officinalis* L.) plants (Moustakas et al. 2011). Moreover, Arvind and Prasad (2005a, b) extensively reviewed the role of Zn in alleviating Cd stress in *Ceratophyllum demersum* L. Cd-induced chlorosis and necrosis were efficiently ameliorated by Zn treatment. A strong competition between Zn and Cd-reduced Cd uptake was found due to enhanced Zn accumulation. Zn application also improved ROS defence system and antioxidant capacity in plant system. Zn stimulated CAT, SOD and APX activity by lessening Cd-induced toxicity. An increased ascorbate-glutathione cycle under Zn treatment detoxifies Cd. Zn plays an important role in improving GSH activity by limiting Cd toxicity. Additionally, Zn ameliorated Cd toxicity in barley plant by enhancing growth parameters and declining oxidative stress as investigated. Zn also alleviated Cd uptake and its concentration in target plant parts (Akay and Koleli 2007). Zhao et al. (2011) documented that Cd declined winter wheat growth and enzyme activities. Application of Zn at lower concentration enhanced Cd concentration and inhibits plant growth. However, higher dosage of Zn alleviated Cd toxicity and improved plant growth parameters. Zn also ameliorated toxic effects of Cd on human health. Cd induced vascular disorder, cancer and diabetes and ultimately caused mortality. Zn counteracts the toxicity of Cd by limiting its effects (McCarty 2012). Koleli et al. (2004) documented the role of Zn against Cd stress in bread and durum wheat cultivars. Increased Cd treatment enhanced necrotic patches in older leaves and decreased shoot dry weight in target cultivars. Zn treatment ameliorated Cd toxic effects by improving plant defence system against oxidative damage and challenging Cd for binding with enzymes, lipids and membrane protein. Similarly, Cd declined rice growth, biomass, chlorophyll content and photosynthetic rate in the shoots and roots of target plant species. An increase in malondialdehyde content and antioxidant enzyme activities was also noted under Cd stress. However, Zn application alleviated Cd toxicity and enhanced plant growth, chlorophyll content and photosynthetic rate. Zn also decreased malondialdehyde content and antioxidant enzyme activities by limiting Cd stress (Hassan et al. 2005).

Above findings showed that Zn usually at lower concentration ameliorated various heavy metals toxicity in different plant species (Table 14.1).

14.12 Conclusions and Future Perspectives

Heavy metals toxicity in plants is well known. They damage plant growth processes negatively. Heavy metals induced toxicity by generating oxidative stress which declines plant physiological parameters. Reports suggested that heavy metals provoke oxidative damage that leads to reduced plant growth, biomass, photosynthetic parameters, gas exchange attributes, enzyme activities and other metabolic functions. Need of the time is to overcome heavy metals toxic effects. Zn is one of the key micronutrient required by plants. Zn not only growth promotive but also alleviate heavy metals toxicity and their uptake in plants. It enhanced plant physiological functions by alleviating heavy metals toxic effects in various plant species. Zn counteracts heavy metals toxicity by modulating heavy metals uptake and antioxidant defence system against oxidative stress. Reports suggested that these promotive effects were mostly observed at lower Zn concentration. Zn enhanced cell functioning at higher level. However, Zn at higher concentration destabilized metabolic functions and not attributed to protective role in plants. Severe toxic effects were observed at higher concentration of Zn like heavy metals. Metals toxicity and higher Zn concentration could be managed by phytoremediation. Zn addition in some crops improves their physiological traits and fulfils food supply worldwide. However, Zn shortage causes reduction in yield and plant development defects. Biofortification is recommended as a useful technique to cope with Zn deficiency. Zn fertilizers could be used to enhanced crop growth and production in land areas of Zn deficiency.

Until now, Zn interaction and its alleviating role were observed in a few heavy metals. However, further studies are required in future regarding Zn interaction with other heavy metals and its alleviating role. Zn biochemistry and physiological intact with plants has also been less studied that needs to be further investigated.

Table 14.1 Role of Zn in alleviating heavy metals stress

Zn dosage	HM conc.	Plant species	Effect	References
10, 50, 100, and 150 µmol/L	Cd at 10 µmol/L	Tomato (*Solanum lycopersicum*)	Stimulate growth, photosynthetic pigments, POD, CAT, APX and GR activity at 10 and 50 µmol/L, reduce these parameters at100, and 150 µmol/L	Cherif et al. (2011)
50 µM	Cu at 200 and 500 µM	Rice (*Oryza sativa* L.)	Root length increased by 20 and 28%,13 and 8% in shoots, decline Cu uptake by 25 and 30%, decrease H_2O_2 by 18 and 30%, reduce MDA in root and shoot, increased GSH content, stimulate SOD, CAT and GPX activity	Thounaojam et al. (2014)
10,50,100 and 200 µM	Cd at 10 µM	*Ceratophyllum demersum* L. (coontail)	Enhance dry weight, increase lipid peroxidation by 48%, decrease Cd uptake and accumulation, little increase in lipoxygenase activity, increase in SOD by 263% and CAT by 435%,	Arvind and Prasad (2003)
10,50,100 and 200 µM	Cd at 200 µM	*Ceratophyllum demersum* L	Reduce Cd uptake by 26%, alleviate oxidative stress, lipid peroxidation and electrical conductivity, improve metabolic functions, inhibit free radical formation	Arvind and Prasad (2005a, b)
0, 1, 2, 4 and 6 kg/da	Cd at 1.5 and 3 kg/da	Barley (*Hordeum vulgare* L. cv. Karatay-94)	Improve grain yield, change N and K concentrations, decrease Cd uptake and concentration, reduce oxidative stress	Akay and Koleli (2007)
0.5, 5 and 50 mg L^{-1}	Cd at 5 and 50 mg L^{-1}	Winter wheat	Reduce Cd toxicity, improve plant growth traits, enzyme activities and proline content at higher Zn dosage	Zhao et al. (2011)
0, 1.5, 20, 100,150 mg (kg soil)$^{-1}$	Cd at 8 mg (kg soil)$^{-1}$	*Chara australis* (R. Br.)	Alleviate Cd stress, enhance glutathione concentration, no effect on growth and Cd uptake, Zn protect strength of plants allowing Cd tolerance	Clabeaux et al. (2013)
0, 10 mg kg^{-1} soil	Cd at 10, 25 $mgkg^{-1}$ soil	Bread and durum wheat cultivars	Disturb growth traits of both cultivars, improve antioxidant defence system against oxidative stress, enhance membrane protein content and lipids	Koleli et al. (2004)
0.2 and 1 µM	Cd at 0.1 and 5 µM	Rice	Increased plant growth, chlorophyll content, photosynthetic rate by alleviating Cd stress, decreased malondialdehyde content and antioxidant enzyme activities	Hassan et al. (2005)

References

Akay, A., & Koleli, N. (2007). Interaction between cadmium and zinc in barley (*hordeum vulgare* l.) grown under field conditions. *Bangladesh Journal of Botany, 36*, 13–19.

Alloway, B. J. (2004). *Zinc in soils and crop nutrition*. Publication of International Zinc Association. Retrieved October 4, 2010, from http://www.iza.com/Documents/Communications/Publications/ALLOWAY_PRINT.pdf.\.

Alloway, B. J. (2008a). *Zinc in soils and crop nutrition*. Brussels, Belgium: International Zinc Association.

Alloway, B. J. (2008b). Copper and Zinc in soils: Too little or too much. *New Zealand Trace Elements Group Conference 13–15th February 2008*, University of the Waikato, Hamilton, New Zealand.

Alloway, B. J. (2008c). *Zinc in soils and crop nutrition* (2nd ed.). Brussels Belgium/Paris, France: International Zinc Association/International Fertilizer Industry Association.

Alloway, B. J. (2009). Soil factors associated with zinc deficiency in crops and humans. *Environmental Geochemistry and Health, 31*, 537–548.

Alloway, B. J. (2013). Heavy metals and metalloids as micronutrients for plants and animals. In B. J. Alloway (Ed.), *Heavy metals in soils* (pp. 195–209). Dordrecht, The Netherlands: Springer.

Andreini, C., Banci, L., Bertini, I., & Rosato, A. (2006). Zinc through the three domains of life. *Journal of Proteome Research, 5*, 3173–3178.

Appenroth, K. J. (2010). Definition of "heavy metals" and their role in biological systems. *Soil Heavy Metals, 19*, 19–29.

Arvind, P., & Prasad, M. N. V. (2003). Zinc alleviates cadmium-induced oxidative stressin *Ceratophyllum demersum* L.: A free floating freshwater macrophyte. *Plant Physiology and Biochemistry, 41*, 391–397.

Arvind, P., & Prasad, M. N. V. (2005a). Modulation of cadmium-induced oxidative stress in Ceratophyllum demersum by zinc involves ascorbate-glutathione cycle and glutathione metabolism. *Plant Physiology and Biochemistry, 43*, 107–116.

Arvind, P., & Prasad, M. N. V. (2005b). Cadmium-Zinc interactions in a hydroponic system using Ceratophyllum demersum L.: Adaptive Ecophysiology, biochemistry and molecular toxicology. *Brazilian Journal of Plant Physiology, 17*, 3–20.

Babu, N. G., Sarma, P. A., Attitalla, I. H., & Murthy, S. D. S. (2010). Effect of selected heavy metal ions on the photosynthetic electron transport and energy transfer in the thylakoid membrane of the cyanobacterium, *Spirulina platensis*. *Academic Journal of Plant Sciences, 3*, 46–49.

Bhutia, D. T. (2014). Protein energy malnutrition in India: The plight of our under five children. *Journal of Family Medicine and Primary Care, 3*, 63–70.

Bonnet, M., Camares, O., & Veisseire, P. (2000). Effects of zinc and influence of Acremonium lolii on growth parameters, chlorophyll a fluorescence and antioxidant enzyme activities of ryegrass (*Lolium perene* L. cv Apollo). *Journal of Experimental Botany, 51*, 945–953.

Bouain, N., Shahzad, Z., Rouached, A., Khan, G. A., Berthomieu, P., et al. (2014). Phosphate and zinc transport and signalling in plants: Toward a better understanding of their homeostasis interaction. *Journal of Experimental Botany, 65*(20), 5725–5741. doi:10.1093/jxb/eru314.

Brennan, R. F. (2005). *Zinc application and its availability to plants*. PhD dissertation, Division of Science and Engineering, School of Environmental Science, Murdoch University.

Broadley, M. R., White, P. J., Hammond, J. P., Zelko, I., & Lux, A. (2007). Zinc in plants. *The New Phytologist, 173*, 677–702.

Burleigh, S. H., Kristensen, B. K., & Bechmann, I. E. (2003). A plasma membrane zinc transporter from *Medicago truncatula* is upregulated in roots by Zn fertilization, yet down-regulated by arbuscular mycorrhizal colonization. *Plant Molecular Biology, 52*, 1077–1088.

Cakmak, I. (2000). Role of zinc in protecting plant cells from reactive oxygen species. *The New Phytologist, 146*, 185–205.

Cakmak, I. (2008). Enrichment of cereal grains with zinc: Agronomic or genetic bio-fortification? *Plant and Soil, 302*, 1–17.
Cakmak, I. (2009). Bio-fortification of cereal grains with zinc by applying zinc fertilizers. *Biozoom, 1*, 2–7.
Chandel, G., Datta, K., & Datta, S. K. (2010). Detection of genomic changes in transgenic Bt rice populations through genetic fingerprinting using amplified fragment length polymorphism (AFLP). *GM Crops, 1*, 327–336.
Cherif, J., Mediouni, C., Ammar, W. B., & Jemal, F. (2011). Interactions of zinc and cadmium toxicity in their effects on growth and in antioxidative systems in tomato plants (*Solanum lycopersicum*). *Journal of Environmental Sciences, 23*, 837–844.
Clabeaux, B. L., Navarro, D. A., Aga, D. S., & Bisson, M. A. (2013). Combined effects of cadmium and zinc on growth, tolerance, and metal accumulation in Chara australis and enhanced phytoextraction using EDTA. *Ecotoxicology and Environmental Safety, 98*, 236–243.
Clemens, S. (2009). Toxic metal accumulation, responses to exposure and mechanisms of tolerance in plants. *Biochimie, 88*, 1707–1719.
Cobbett, C., & Goldsbrough, P. (2002). Phytochelatins and metallothioneins: Roles in heavy metal detoxification and homeostasis. *Annual Review of Plant Biology, 53*, 159–182.
Colangelo, E. P., & Guerinot, M. L. (2006). Put the metal to the petal: Metal uptake and transport throughout plants. *Current Opinion in Plant Biology, 9*, 322–330.
Dang, H. R., Li, Y., Sun, X., & Zhang, Y. L. (2010). Absorption, accumulation and distribution of zinc in highly-yielding winter wheat. *Agricultural Sciences in China, 9*, 965–973.
Dazy, M., Masfaraud, J. F., & Ferard, J. F. (2009). Induction of oxidative stress biomarkers associated with heavy metal stress in *Fontinalis antipyretica* Hedw. *Chemosphere, 75*, 297–302.
Deinlein, U., Weber, M., Schmidt, H., Rensch, S., Trampczynska, A., Hansen, T. H., Husted, S., Schjoerring, J. K., Talke, I. N., Kramer, U., & Clemens, S. (2012). Elevated nicotianamine levels in Arabidopsis halleri roots play a key role in Zn hyperaccumulation. *Plant Cell, 24*, 708–723.
Disante, K. B., Fuentes, D., & Cortina, J. (2010). Response to drought of Zn-stressed *Quercus suber* L. seedlings. *Environmental and Experimental Botany, 70*, 96–103.
Epstein, E., & Bloom, A. (2005). *Mineral nutrition of plants: Principles and perspectives*. Sunderland: Sinauer Associates.
Fageria, N. K. (2004). Dry matter yield and nutrient uptake by lowland rice at different growth stages. *Journal of Plant Nutrition, 27*, 947–958.
Foster, M., & Samman, S. (2012). Zinc and regulation of inflammatory cytokines: Implications for cardio-metabolic disease. *Nutrients, 4*, 676–694.
Freisinger, E. (2008). Plant MTs—Long neglected members of the metallothioneins superfamily. *Dalton Transactions, 47*, 6663.
Genc, Y., McDonald, G. K., & Graham, R. D. (2006). Contribution of different mechanisms to zinc efficiency in bread wheat during early vegetative stage. *Plant and Soil, 281*, 353–367.
Ghosh, M., & Singh, S. P. (2005). A review on phytoremediation of heavy metals and utilization of its byproducts. *Applied Ecology and Environmental Research, 3*, 1–18.
Gratao, P. L., Monteiro, C. C., Antunes, A. M., Peres, L. E. P., & Azevedo, R. A. (2008). Acquired tolerance of tomato (*Lycopersicon esculentum* cv. Micro-Tom) plants to cadmium-induced stress. *Annals of Applied Biology, 153*, 321–333.
Hafeez, B., Khanif, Y. M., & Saleem, M. (2013). Role of zinc in plant nutrition—A review. *American Journal of Experimental Agriculture, 3*, 374–391.
Hall, J. L., & Williams, L. E. (2003). Transition metal transporters in plants. *Journal of Experimental Botany, 54*, 2601–2613.
Hansch, R., & Mendel, R. R. (2009). Physiological functions of mineral micronutrients (Cu, Zn, Mn, Fe, Ni, Mo, B, Cl). *Current Opinion in Plant Biology, 12*, 259–266.
Hao, H., Wei, Y., Yang, X., Feng, Y., & Wu, C. (2007). Effects of different nitrogen fertilizer levels on Fe, Mn, Cu and Zn concentrations in shoot and grain quality in rice (Oryza sativa). *Rice Science, 14*, 289–294.

Hart, J. J., Welch, R. M., Norvell, W. A., & Kochian, L. V. (2002). Transport interactions between cadmium and zinc in roots of bread and durum wheat seedlings. *Physiologia Plantarum, 116*, 73–78.

Haslett, B. S., Reid, R. J., & Rengel, Z. (2001). Zinc mobility in wheat: Uptake and distribution of zinc applied to leaves or roots. *Annals of Botany, 87*, 379–386.

Hassan, M. J., Zhang, G., Wu, F., Wei, K., & Chen, Z. (2005). Zinc alleviates growth inhibition and oxidative stress caused by cadmium in rice. *Journal of Plant Nutrition and Soil Science, 168*, 255–261.

Horton, S. (2006). The economics of food fortification. *The Journal of Nutrition, 136*, 1068–1071.

Ikenaka, Y., Nakayama, S. M. M., Muzandu, K., Choongo, K., Teraoka, H., Mizuno, N., & Ishizuka, M. (2010). Heavy metal contamination of soil and sediment in Zambia. *African Journal of Environmental Science and Technology, 4*, 729–739.

Iqbal, N., Masood, A., Nazar, R., Syeed, S., & Khan, N. A. (2010). Photosynthesis, growth and antioxidant metabolism in mustard (*Brassica juncea* L.) cultivars differing in Cd tolerance. *Agricultural Sciences in China, 9*, 519–527.

Jarup, L. (2003). Hazards of heavy metal contamination. *British Medical Bulletin, 68*, 167–182.

Kabata-Pendias, A., & Pendias, H. (2001). *Trace elements in soils and plants*. Boca Raton: CRC Press.

Kaya, C., Higgs, D., & Burton, A. (2000). Plant growth, phosphorus nutrition and acidphosphatase enzyme activity in three tomato cultivers grown hydroponically at different zinc concentrations. *Journal of Plant Nutrition, 23*, 569–579.

Khudsar, T., Mahmooduzzafar Iqbal, M., & Sairam, R. K. (2004). Zinc-induced changes inmorpho-physiological and biochemical parameters in *Artemisia annua*. *Biologia Plantarum, 48*, 255–260.

King, J. C., & Cousins, R. J. (2006). Zinc. In M. E. Shils, M. Shike, A. C. Ross, B. Caballero, & R. J. Cousins (Eds.), *Modern nutrition in health and disease* (10th ed., pp. 271–285). Baltimore: Lippincot Williams and Wilkins.

Koleli, N., Eker, S., & Cakmak, I. (2004). Effect of zinc fertilization on cadmium toxicity in durum and bread wheat grown in zinc-deficient soil. *Environmental Pollution, 131*, 453–459.

Latowski, D., Kuczyńska, P., & Strzałka, K. (2011). Xanthophyll cycle-a mechanism protecting plants against oxidative stress. *Redox Report, 16*, 78–90.

Lonnerdal, B. (2000). Dietary factors influencing zinc absorption. *The Journal of Nutrition, 130*, 1378–1383.

Maksymiec, W., Drazkiewicz, M., & Skorzynska-Polit, E. (2008). *Responses of higher plants to heavy metal stress. Abiotic stress and plant responses* (pp. 139–163). New Delhi: IK International Publishing House.

Maret, W., & Sandstead, H. H. (2006). Zinc requirements and the risks and benefits of zinc supplementation. *Journal of Trace Elements in Medicine and Biology, 20*, 3–18.

Marschner, H. (2012). *Mineral nutrition of higher plants* (3rd ed.). London: Academic Press.

Martinez, C., & Motto, H. (2000). Solubility of lead, zinc and copper added to mineral soils. *Environmental Pollution, 107*, 153–158.

Mayer, J. E., Pfeiffer, W. H., & Beyer, P. (2008). Biofortified crops to alleviate micronutrient malnutrition. *Current Opinion in Plant Biology, 11*, 166–170.

McCarty, M. F. (2012). Zinc and multi-mineral supplementation should mitigate the pathogenic impact of cadmium exposure. *Medical Hypotheses, 79*, 642–648.

Mousavi, S. R., Mohammad, G., & Goudarz, A. (2007). Effect of zinc and manganese foliar application on yield, quality and enrichment on potato (*Solanum tuberosum* L.) *Asian Journal of Plant Sciences, 8*, 1256–1260.

Moustakas, N. K., Akoumianaki-Ioannidou, A., & Barouchas, P. E. (2011). The effects of cadmium and zinc interactions on the concentration of cadmium and zinc in pot marigold (Calendula officinalis L.) *Australian Journal of Crop Science, 5*(3), 277.

Naik, S. K., & Das, D. K. (2008). Relative performance of chelated zinc and zinc sulphate for lowland rice (*Oryza sativa* L). *Nutrition Cycle in Agroecosystem, 81*, 219–227.

Overbeck, S., Rink, L., & Haase, H. (2008). Modulating the immune response by oral zinc supplementation: A single approach for multiple diseases. *Archivum Immunologiae et Therapiae Experimentalis, 56*, 15–30.

Palmgren, M. G., Clemens, S., Williams, L. E., Kramer, U., Borg, S., Schjorring, J. K., & Sanders, D. (2008). Zinc biofortification of cereals: Problems and solutions. *Trends in Plant Science, 13*, 464–473.

Peck, A. W., & McDonald, G. K. (2010). Adequate zinc nutrition alleviates the adverse effects of heat stress in bread wheat. *Plant and Soil, 337*, 355–374.

Phattarakul, N., Rarkasem, B., Li, L. J., LH, W., Zou, C. Q., Ram, H., Sohu, V. S., Kang, B. S., Surek, H., Kalayci, M., Yazici, A., Zhang, F. S., & Cakmak, I. (2012). Biofortificaiton of rice grain with zinc through zinc fertilization in different countries. *Plant and Soil, 361*, 131–141.

Pollard, A. J., Powell, K. D., Harper, F. A., & Smith, J. A. C. (2002). The genetic basis of metal hyperaccumulation in plants. *Critical Reviews in Plant Sciences, 21*, 539–566.

Prasad, A. S. (2008). Zinc in human health: Effect of zinc on immune cells. *Molecular Medicine, 14*, 353–357.

Prasad, A. S. (2009). Zinc: Role in immunity, oxidative stress and chronic inflammation. *Current Opinion in Clinical Nutrition and Metabolic Care, 12*, 646–652.

Przedpelska, E., & Wierzbicka, M. (2007). *Arabidopsis arenosa* (Brassicaceae) from a lead-zinc waste heap in southern Poland- a plant with high tolerance to heavy metals. *Plant and Soil, 299*, 43–53.

Rascio, N., & Navari-Izzo, F. (2011). Heavy metal hyperaccumulating plants: How and why do they do it? And what makes them so interesting? *Plant Science, 180*, 169–181.

Roohani, N., Hurrell, R., Kelishadi, R., & Schullin, R. (2013). Zinc and its importance for human health: An integrative review. *Journal of Research in Medical Sciences, 18*, 144–157.

Shankar, A. H., & Prasad, A. S. (1998). Zinc and immune function: The biological basis of altered resistance to infection. *The American Journal of Clinical Nutrition, 68*(suppl), 447–463.

Sinclair, S. A., & Kramer, U. (2012). The zinc homeostasis network of land plants. *Biochimica et Biophysica Acta, 1823*(9), 1553–1567.

Singh, B., Natesan, S. K. A., Singh, B. K., & Usha, K. (2005). Improving zinc efficiency of cereals under zinc deficiency. *Current Science, 88*, 36–44.

Singh, V. P., Srivastava, P. K., & Prasad, S. M. (2012). Differential effect of UV-B radiation on growth, oxidative stress and ascorbate-glutathione in two cyanobacteria under copper stress. *Plant Physiology and Biochemistry, 61*, 61–70.

Tavallali, V., Rahemi, M., Eshghi, S., Kholdebarin, B., & Ramezanian, A. (2010). Zinc alleviates salt stress and increases antioxidant enzyme activity in the leaves of pistachio (*Pistacia vera* L. 'Badami') seedlings. *Turkish Journal of Agriculture and Forestry, 34*, 349–359.

Tewari, R. K., Kumar, P., & Sharma, P. N. (2008). Morphology and physiology of zinc-stressed mulberry plants. *Journal of Plant Nutrition and Soil Science, 171*, 286–294.

Thounaojam TC, Panda P, Choudhury S , Patra HK, Panda SK (2014) Zinc ameliorates copper-induced oxidative stress in developing rice (*Oryza sativa* L.) seedlings. Protoplasma (251):61–69.

Tisdale, S. L., Nelson, W. L., & Beaten, J. D. (1984). *Zinc in soil fertility and fertilizers* (4th ed.pp. 382–391). New York: Macmillan.

Trampczynska, A., Kupper, H., Meyer-Klaucke, W., Schmidt, H., & Clemens, S. (2010). Nicotianamine forms complexes with Zn (II) in vivo. *Metallomics, 2*, 57–66.

Tsonev, T., & Lidon, F. J. C. (2012). Zinc in plants-An overview. *Emirates Journal of Food & Agriculture, 24*, 322–333.

Upadhyay, R., & Panda, S. K. (2009). Zinc reduced copper toxicity induced oxidative stress by promoting antioxidant defense in freshly grown duckweed *Spirodela polyrhiza*. *Journal of Hazardous Materials, 175*, 1081–1084.

Ventrella, A., Catucci, L., Piletska, E., Piletsky, S., & Agostiano, A. (2011). Interactions between heavy metals and photosynthetic materials studied by optical techniques. *Bioelectrochemistry, 77*, 19–25.

Vesely, T., Neuberg, M., Trakal, L., Szakova, J., & Tlustoa, P. (2011). Water lettuce *Pistia stratiotes* L. response to lead toxicity. *Water, Air, and Soil Pollution, 223*, 1847–1859.

Wang, C., Zhang, S. H., Wang, P. F., Qian, J., Hou, J., Zhang, W. J., & Lu, J. (2009). Excess Zn alters the nutrient uptake and induces the antioxidative responses in submerged plant *Hydrilla verticillata* (L.f) Royle. *Chemosphere, 76*, 938–945.

World Health Organization (WHO). (2002). *The World Health Report 2002*. Geneva: WHO.

WHO, FAO, & IAEA. (2002). *Trace elements in human health and nutrition* (pp. 230–245). Geneva: WHO.

Yadav, S. K. (2009). Heavy metals toxicity in plants: An overview on the role of glutathione and phytochelatins in heavy metal stress tolerance of plants. South Afr. *Journal of Botany, 76*, 167–179.

Yusuf, M., Fariduddin, Q., Varshney, P., & Ahmad, A. (2012). Salicylic acid minimizes nickel and/or salinity-induced toxicity in Indian mustard (*Brassica juncea*) through an improved antioxidant system. *Environmental Science and Pollution Research, 19*, 8–18.

Zhao, A. Q., Tian, X. H., WH, L., Gale, W. J., XC, L., & Cao, Y. X. (2011). Effect of zinc on cadmium toxicity in winter wheat. *Journal of Plant Nutrition, 34*, 1372–1385.

Chapter 15
Calcium Application Enhances Plant Salt Tolerance: A Review

Bouzid Nedjimi

Abstract Environmental stressors such as drought and soil salinity limit crop production especially in arid and semiarid areas. Improvement of crop production with chemical applications was largely used to enhance salt tolerance of plant species. Calcium (Ca) is an essential element for plant nutrition, and it's indispensable to regulating many biochemical functions in plants subjected to salt stress. Exogenous application of this element can play an important role in enhancing plant stress tolerance. In spite of its acclaimed protective role, unfortunately, limited research has been focused to review the effectiveness of Ca to improve plant salt tolerance. In this review chapter, some knowledge of successful application of Ca to improve plant stress tolerance was presented.

Keywords Salinity tolerance • Calcium amendment • Growth rate • Membranes integrity • Nutrient acquisition • Photosynthetic efficiency • Root hydraulic conductance

Abbreviations

APX	Ascorbate peroxidase
CAT	Catalase
DHAR	Dehydroascorbate reductase
EL	Electrolyte leakage
GA	Gibberellic acid
GB	Glycine betaine
GR	Glutathione reductase
L_0	Root hydraulic conductivity
POX	Peroxidase

B. Nedjimi (✉)
Laboratory of Exploration and Valorization of Steppe Ecosystem, Faculty of Science of Nature and Life, University of Djelfa, Cité Aîn Chih, P.O. Box 3117, Djelfa 17000, Algeria
e-mail: bnedjimi@yahoo.fr

M. Naeem et al. (eds.), *Essential Plant Nutrients*,
DOI 10.1007/978-3-319-58841-4_15

P5CS	Pyrroline-5-carboxylate synthetase
SOD	Superoxide dismutase
RWC	Relative water content
$\Psi\pi$	Osmotic potential
$\Psi\omega$	Water potential

15.1 Introduction

In recent decades, the extension of soil salinity in the world constitutes a serious economic problem. Above 50% of irrigated soils were affected by secondary salinization especially in arid areas. In these areas, the inappropriate methods of irrigation and unsuitable agricultural practices were the origin of this phenomenon. Therefore the preventive applications against salinization were needed (Rengasamy 2006).

Calcium (Ca) is an essential major element to plant growth and was considered as second messenger and exerts a crucial role in transduction signal in many biological activities (Xiong et al. 2006). It is required to maintain integrity of plasma membrane and in osmotic adjustment (Hepler 2005). Plants absorbed Ca from the soil solution by root hairs and transport them through the xylem sap to leaves using evapotranspiration forces. Habitually, the old leaves contain considerable amount of Ca comparably to young leaves, due to the less mobility of this element in the phloem sap (White 1998). Calcium is known to protect the membranes integrity and prevent electrolyte leakage (EL) caused by salt stress. It was considered as secondary messenger for signal transduction of salt stress stimuli (Martínez-Ballesta et al. 2008).

Saline soils' reclamation involves the application of chemical amendments like gypsum or others (sulphur, calcium chloride, calcium nitrate, etc.). Also Ca supplementations were necessary especially in acid soils when the solution pH was lowered (Qadir et al. 2001).

In many plant species, improvement of salinity tolerance has been achieved by Ca supplementation (Rengel 1992). However, this approach, which may considerably contribute to enhance crop production in salt soils, has not received great attention in the literature. The present chapter is focused on the role of Ca applications to improve plant salt tolerance.

15.2 Amelioration of Plant Salt Tolerance by Calcium Addition

To remove excess soluble salts, leaching with water and drainage were the simple methods applied to reclaiming saline lands especially in soils with high permeability. However, application of chemical amendments was also useful, particularly in the case of soils with heavy textures (clay soils) (Szabolcs 1989).

Calcium supplementation to plants may help reduce adverse effects of salinity stress (Epstein 1998). External application of Ca can penetrate through cell wall of the roots and be transported to shoots, where it would participate to improved stress tolerance in different development stages (Nedjimi et al. 2010).

15.2.1 Germination

A considerable number of researches have been conducted to overcome the detrimental effects of salinity in seeds germination by Ca implication. For example, in *Phragmites karka*, it was determined that addition of 5, 10 or 25 mM of $CaCl_2$ to seeds mitigated the adverse effects of NaCl stress on both final and rate of germination (Zehra et al. 2012). Furthermore, alleviation of salt stress on germination seeds of *Urochondra setulosa*, a halophytic grass, was achieved when 10 mM of $CaCl_2$ was applied at the same time with various soluble salts ($MgSO_4$, NaCl or Na_2SO_4) (Shaikh et al. 2007). Gul and Khan (2006) studied seed germination in five halophytic coastal species (*Arthrocnemum indicum*, *A. macrostachyum*, *Desmostachya bipinnata*, *Halopyrum mucronatum* and *Urochondra setulosa*) grown with NaCl or in combination with 10 mM of $CaCl_2$. According to them, high salinity decreased the germination percentage in all the species studied. However, application of Ca reduced inhibition of final seed germination. Likewise, Tobe et al. (2002) indicated that Ca efficaciously alleviated the toxicity of various soluble salts on the germination of *Kalidium capsicum.* In another study, the same authors reported that germination of *Haloxylon ammodendron* seeds declined with higher NaCl or $MgCl_2$ levels, while this adverse effect was reversed by treatment of seeds with $CaCl_2$ (Tobe et al. 2004). However Bonilla et al. (2004) showed that addition of extra Ca to medium prevented the reduction in final germination of pea plants (*Pisum sativum*) caused by 75 mM NaCl. Working with *Chenopodium album*, a salt-tolerant species, Yao et al. (2010) found that the percentage of seed germination exposed to 300 mM NaCl was significantly improved by the addition of low concentrations of $CaCl_2$.

15.2.2 Growth

Many researches indicate the beneficial effects of Ca amendments on plant growth and crop production under salinity stress, for example, in tomato, barley, pistachio, rice, pea, pepper, muskmelon, etc. (see Table 15.1). Exogenous application of Ca on strawberry (*Fragaria* × *ananassa*) has been shown to improve growth, fruit yield, and water use of NaCl-stressed plants compared to unstressed plants (Kaya et al. 2002). Ameliorating effects of Ca application on plants treated with 400 mM NaCl was studied in *Atriplex halimus subsp. schweinfurthii* seedlings (Nedjimi and Daoud

Table 15.1 Effects of exogenous application of calcium on some plant species

Plant species	Effects of Ca application	References
Amaranthus tricolor	Enhancement of antioxidant enzymes (SOD, CAT and APX), lipid peroxidation reduction	Sunukumar et al. (2011)
Arabidopsis thaliana	Increase in plant biomass and decline in EL and anthocyanins	Kaddour et al. (2012)
Arachis hypogaea	Ca addition increases GB production but decreases proline content	Girija et al. (2002)
Arthrocnemum indicum	Enhancement of germination by Ca application	Gul and Khan (2006)
Atriplex canescens	Growth, L_o and chlorophyll content improvement	Nedjimi and Daoud (2009a)
Atriplex halimus	Growth and RCW improvement, EL decreases	Nedjimi and Daoud (2009b)
Cakile maritima	Enhancement of antioxidant enzymes (SOD, POX, GR, DHAR)	Ben Amor et al. (2010)
Capsicum annuum	Plant-water relations, aquaporin functionality improvement	Cabañero et al. (2004), Martínez-Ballesta et al. (2008)
Cassia angustifolia	Biomass, leaf area, stomatal conductance and photosynthetic rate improvement	Arshi et al. (2006a)
Centella asiatica	Enhancement of proline content and P5CS activity	Murugan and Sathish (2005)
Chenopodium album	Improved seed germination and seedling survival	Yao et al. (2010)
Cichorium intybus	Growth, photosynthetic efficiency and nitrogen assimilation improvement	Arshi et al. (2006b)
Cornus stolonifera	Shoot height and cell wall nutrient improvement	Renault (2005)
Cucumis melo	Hydraulic conductance improvement	Carvajal et al. (2000)
Cucumis sativus	Enhancement of K and Ca contents and reduced the uptake of Na	Lei et al. (2014)
Fragaria × ananassa	Growth, fruit yield and water use improvement, EL decreases	Kaya et al. (2002)
Haloxylon ammodendron	Amelioration of seed germination, seedling growth and reduced K outflux	Tobe et al. (2004)
Hordeum vulgare	Root growth and nutrient acquisition improvement	Shabala et al. (2003)
Helianthus tuberosus	Up-regulation of antioxidant enzymes (CAT, POD, SOD)	Xue et al. (2008)
Linum usitatissimum	Enhancement of antioxidative defence system and proline and GB accumulation	Khan et al. (2010)
Lycopersicon esculentum	Growth, RWC and fruit yield improvement	Tuna et al. (2007)

(continued)

Table 15.1 (continued)

Plant species	Effects of Ca application	References
Lygeum spartum	Growth, L_0, $\Psi\pi$ and $\Psi\omega$ improvement	Nedjimi et al. (2010)
Oryza sativa	Growth and photosynthetic rate improvement, EL decreases	Cha-um et al. (2012)
Phragmites karka	Amelioration of final and rate of germination	Zehra et al. (2012)
Pistacia vera	Amelioration of shoots and roots DW, leaf area and nutrient acquisition	Hojjatnooghi et al. (2014)
Pisum sativum	Amelioration of seed germination, seedling growth	Bonilla et al. (2004)
Psidium guajava	Amelioration of shoot K and Ca contents	Ebert et al. (2002)
Sorghum bicolor	Improvement of ionic relations and proline accumulation	Colmer et al. (1996)
Triticum aestivum	Enhancement of antioxidant enzymes (APX, CAT, POD, SOD)	Al-Whaibi et al. (2012)
Vigna unguiculata	Chlorophyll fluorescence parameters improvement	Murillo-Amador et al. (2006)
Vigna unguiculata	Glycolipid and membrane integrity amelioration	Guimarães et al. (2011)
Urochondra setulosa	Alleviation of the inhibitory effects of salts on final germination percentage	Shaikh et al. (2007)

2009b). Briefly, addition of 40 mM Ca improves significantly shoot and root growth, RWC and lowered root membrane permeability, compared to untreated plants. Besides, in tomato plants (*Lycopersicon esculentum*), additional application of 5 mM $CaSO_4$ improved growth, leaf RCW and fruit yield of the salt-stressed plants (Tuna et al. 2007). It was also determined that application of 0.5 or 1 mM of ($Ca(NO_3)_2{\bullet}4H_2O$) to nutrient solution was effective to improving shoot and root growth and leaf area and reducing Na accumulation in pistachio seedlings (*Pistacia vera*) grown under salt stress (Hojjatnooghi et al. 2014). Furthermore, under salt stress conditions, supplementation of 20 mM $CaCl_2$ increased shoot growth of *Lygeum spartum* by 25% compared to non-treated plants (Nedjimi et al. 2010).

15.2.3 Mineral Nutrients

In general, addition of Ca to plants subjected to salinity was associated with a reduction in the concentration of potential toxic ions such as Na and Cl and an enhancement of K and Ca. For example, in *Cornus stolonifera* plants stressed with sodium sulphate, the supply of Ca significantly decreased the salt-induced efflux K which caused improvement of shoot height and growth rate (Renault 2005). In a similar case, exogenous application of Ca to barley (*Hordeum vulgare*) plants growing under salt stress resulted in reduced Na accumulation and maintained K concentration in the shoot (Shabala et al. 2003). Tuna et al. (2007) showed that application of $CaSO_4$ to the nutrient solution containing high salt concentration increased the concentrations of K and Ca and reduced the concentration of Na in the shoots of tomato (*Lycopersicon esculentum*). While working with guava seedlings (*Psidium guajava*) subjected to saline conditions, supplementary of 10 mM $Ca(NO_3)_2$ resulted in a decrease in shoot Na accumulations and an increase in K and Ca contents (Ebert et al. 2002). Likewise, combined treatments of NaCl with Ca applied at different doses could alleviate the salt inhibition of *Sorghum bicolor* plant growth via the maintenance of K and Ca and the enhancement of proline accumulation (Colmer et al. 1996). Tobe et al. (2004) showed that the addition of low concentrations of $CaCl_2$ reduced K outflux from seedlings of *Haloxylon ammodendron* but caused no appreciable decrease in the influx of Na into seedlings. The ameliorative effect of supplemental Ca in mineral acquisition was also showed recently in cucumber plants (*Cucumis sativus*) through enhanced K and Ca contents and reduced the uptake of Na (Lei et al. 2014).

15.2.4 Photosynthesis

The inhibitory effect of Ca on salt toxicity has also been reported in many studies. Supplementation of Ca to NaCl-stressed plants of rice (*Oryza sativa*) increased their growth due to photosynthetic rate improvement and halted increased electrolyte leakage (EL) (Cha-um et al. 2012). In addition, foliar application of Ca induced a significant improvement of salt tolerance of cowpea plants (*Vigna unguiculata*) by enhancement of growth and chlorophyll fluorescence parameters (Murillo-Amador et al. 2006). The improvement in photosynthetic rate of NaCl-stressed *Cassia angustifolia* plants by Ca application was suggested to be associated with improvement of stomatal conductance and PSII efficiency (Arshi et al. 2006a). The same authors showed that $CaCl_2$ application ameliorated the negative effects of salinity growth performance, photosynthetic efficiency and nitrogen assimilation of *Cichorium intybus* grown under increase NaCl treatments (Arshi et al. 2006b). Recently, Zrig et al. (2016) observed that Ca supplementation did not improved photosynthetic rate in two almond species (*Prunus amygdalus*) due to degradation of Chl *a* and non-stomatal limitations.

15.2.5 Water Relations and Membrane Integrity

In addition to its direct improvement role, either through positive effects on growth and photosynthesis apparatus, Ca may also protect plants from salt stress indirectly via its role in membrane integrity and water relations. In albardine (*Lygeum spartum*), for example, Ca-treated plants showed an increase in leaf water potential ($\Psi\omega$)

and osmotic potential (Ψπ) compared to untreated plants (Nedjimi et al. 2010). Furthermore, Ca treatment alleviated the adverse effects of salinity on growth, root hydraulic conductivity (L_0) and electrolyte leakage (EL) in two saltbush species, *Atriplex halimus* and *A. canescens* (Nedjimi and Daoud 2009a) (Fig. 15.1). Similarly, tolerance to salinity was improved in two *Arabidopsis thaliana* accessions by additional application of 5 mM $Ca(NO_3)_2$; in addition, Ca-treated plants showed a low electrolyte leakage (EL) and anthocyanins (Kaddour et al. 2012). Carvajal et al.

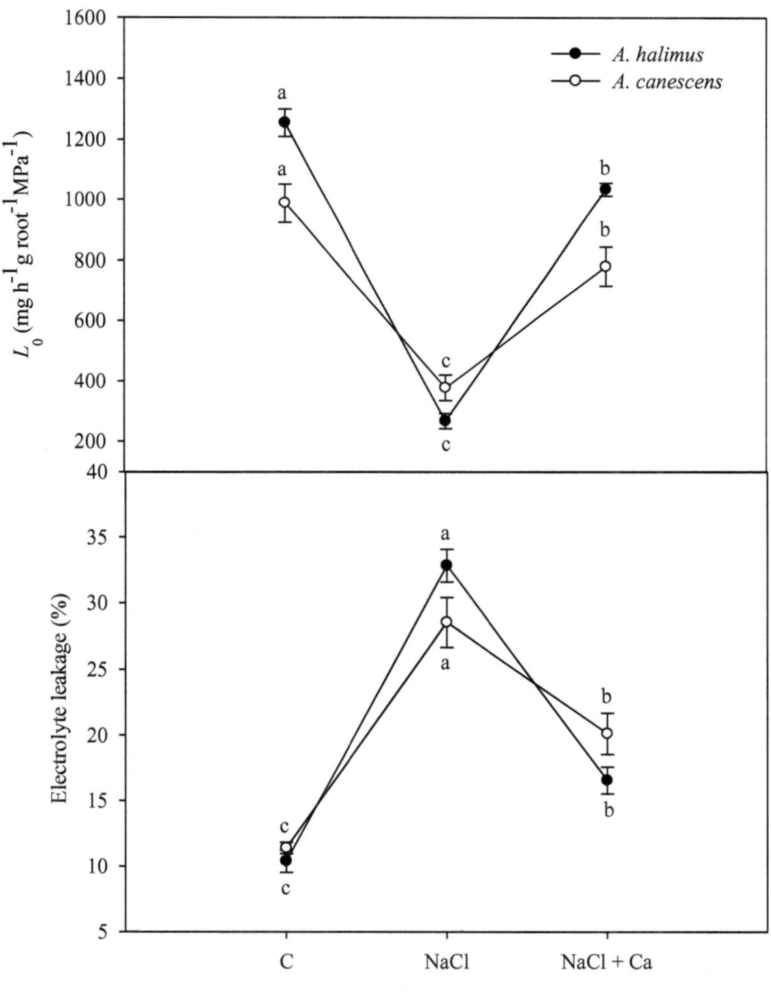

Fig. 15.1 Root hydraulic conductivity (L_0) and electrolyte leakage (EL) of two *Atriplex* species grown in hydroponic solution containing 400 mM NaCl with or without application of 40 mM $CaCl_2$. Values with different letters are significantly different using LSD at $P < 0.05$ (C: control) (Nedjimi and Daoud 2009a)

(2000) found that co-existence of $CaCl_2$ and NaCl together in growth medium is necessary for alleviation of salt stress in muskmelon plants (*Cucumis melo*) by improvement of hydraulic conductance. Likewise, exogenous application of Ca has also been shown to alleviate salt stress in paper plants (*Capsicum annuum*) through promising plant water relations and aquaporin functionality (Cabañero et al. 2004; Martínez-Ballesta et al. 2008). In another culture experiment, growth of cowpea plants (*Vigna unguiculata*) under salt stress was promoted by exogenous application of 10 mM $CaSO_4$, due to glycolipid production and amelioration of membrane integrity (Guimarães et al. 2011).

15.2.6 Oxidative Stress Protection

Several works showed that applied Ca alleviated the NCl-induced oxidative stress by reducing membrane damage coupled with enhanced activities of antioxidative enzymes. For example, addition of 20 mM $Ca(NO_3)_2$ counteracted the adverse effects of salinity on early seedling growth of *Cakile maritima*, through enhancement of antioxidant enzyme activities (SOD, POX, GR and DHAR) (Ben Amor et al. 2010). In *Amaranthus tricolor*, supplemental Ca induced salt tolerance in stressed plants. Thus, specifically, application of 5 mM of external Ca to plants treated with 150 mM NaCl increased their salt tolerance by enhancement of antioxidant enzymes (SOD, CAT and APX) and reduction of lipid peroxidation (Sunukumar et al. 2011). In wheat (*Triticum aestivum* cv. *Samma*) cultivated hydroponically under 90 mM of NaCl, application of 40 mM of Ca increased activities of peroxidase (POD), catalase (CAT), superoxide dismutase (SOD), glutathione reductase (GR) and ascorbate peroxidase (APX) which habitually contribute to increased salt tolerance (Al-Whaibi et al. 2012). Similarly, in a hydroponic culture experiment with Jerusalem artichoke plants (*Helianthus tuberosus*), the activities of superoxide dismutase (SOD), peroxidase (POD) and catalase (CAT) were decreased by NaCl. However, addition of 10 mM $CaCl_2$ to the salt solutions significantly restored the activities of these antioxidant enzymes (Xue et al. 2008). While working with *Linum usitatissimum*, Khan et al. (2010) found that although the activities of antioxidant enzymes (SOD, CAT and POX) were increased by salt stress, the combined addition of $CaCl_2$ and GA_3 to salt-stressed plants further enhanced the activities of these enzymes by 25, 7 and 48%, respectively, in comparison to plants stressed with NaCl alone.

15.2.7 Osmolyte Production

Numerous studies have indicated that exogenous application of Ca provided significant protection against salt-induced damages in plants (see Table 15.1). This element improved salt tolerance by enhancing their growth and osmolyte production. For example, Girija et al. (2002) reported the enhancement of osmoprotant production, especially the glycine betaine (GB) content, in response to Ca supplementation to *Arachis hypogaea* plants stressed with NaCl salinity. Also, addition of $CaCl_2$ was effective in alleviating the negative effects of salt stress on *Pennisetum typoidies* plants through enhancement of proline contents (Gobinathan et al. 2009). The ameliorative effect of supplemental Ca was also demonstrated in *Centella asiatica*. When the salt-stressed calli were supplemented with 25 mM $CaCl_2$, the proline content significantly increased compared to those grown on 200 mM NaCl alone, simultaneous with increase in activity of proline biosynthetic enzyme, pyrroline-5-carboxylate synthetase (P5CS) (Murugan and Sathish 2005). However, application of 10 mg $CaCl_2$ kg^{-1} in combination with 10^{-6} M GA to salt-stressed plants of *Linum usitatissimum* appears to confer greater osmoprotection by the additive role with NaCl in proline and GB accumulation which habitually contribute to increased cell osmoregulation (Khan et al. 2010).

15.3 Concluding Remarks

Calcium is a vital element for plant physiology and has a key role in many biochemical reactions. It is an essential constituent in cell structure and signal transduction as a second messenger. Exogenously applied Ca to medium culture was suggested to mitigate the negative effects of salt stress on growth, antioxidant systems and photosynthetic activity. Thus growth of many plant species was enhanced by Ca addition subjected to salinity. This element is thought to play adaptive roles to maintain membrane integrity, nutrient acquisition and protecting subcellular structures in stressed plants. Application of supplementary Ca to plants growing under salinity was considered as promising and low costly method to enhance their tolerance. However optimal dose and appropriate growth stage must be exactly determined.

Acknowledgments This work has been supported by internal funds of the Algerian Ministry of Higher Education and Scientific Research (CNEPRU Project no. D04N01UN170120140017).

References

Al-Whaibi, M. H., Siddiqui, M. H., & Basalah, M. O. (2012). Salicylic acid and calcium-induced protection of wheat against salinity. *Protoplasma, 249*, 769–778.

Arshi, A., Abdin, M. Z., & Iqbal, M. (2006a). Sennoside content and yield attributes of *Cassia angustifolia* Vahl. as affected by NaCl and $CaCl_2$. *Scientia Horticulturae, 111*, 84–90.

Arshi, A., Abdin, M. Z., & Iqbal, M. (2006b). Effect of $CaCl_2$ on growth performance, photosynthetic efficiency and nitrogen assimilation of *Cichorium intybus* L. grown under NaCl stress. *Acta Physiolia Plantarum, 28*, 137–147.

Ben Amor, N., Megdiche, W., Jiménez, A., Sevilla, F., & Abdelly, C. (2010). The effect of calcium on the antioxidant systems in the halophyte *Cakile maritima* under salt stress. *Acta Physiolia Plantarum, 32*(3), 453–461.

Bonilla, I., El-Hamdaoui, A., & Bolaños, L. (2004). Boron and calcium increase *Pisum sativum* seed germination and seedling development under salt stress. *Plant and Soil, 267*, 97–107.

Cabañero, F. J., Martínez, V., & Carvajal, M. (2004). Does calcium determine water uptake under saline conditions in pepper plants, or is it water flux which determines calcium uptake?. *Plant Science, 166*, 443–450.

Carvajal, M., Cerda, A., & Martínez, V. (2000). Does calcium ameliorate the negative effect of NaCl on melon root water transport by regulating aquaporin activity?. *New Phytologist, 145*, 439–447.

Cha-um, S., Singh, H. P., Samphumphuang, T., & Kirdmanee, C. (2012). Calcium-alleviated salt tolerance in indica rice (*Oryza sativa* L. spp. *indica*): Physiological and morphological changes. *Australian Journal of Crop Science, 6*(1), 176–182.

Colmer, T. D., Fan, T. W. M., Higashi, R. M., & Läuchli, A. (1996). Interactive effects of Ca^{2+} and NaCl salinity on the ionic relations and proline accumulation in the primary root tip of *Sorghum bicolor*. *Physiologia Plantarum, 97*, 421–424.

Ebert, G., Eberle, J., Ali-Dinar, H., & Ludders, P. (2002). Ameliorating effects of $Ca(NO_3)_2$ on growth, mineral uptake and photosynthesis of NaCl-stressed guava seedlings (*Psidium guajava* L.). *Scientia Horticulturea, 93*, 125–135.

Epstein, E. (1998). How calcium enhances plant salt tolerance. *Science, 40*, 1906–1907.

Girija, C., Smith, B. N., & Swamy, P. M. (2002). Interactive effects of sodium chloride and calcium chloride on the accumulation of proline and glycinebetaine in peanut (*Arachis hypogaea* L.). *Environmental and Experimental Botany, 47*, 1–10.

Gobinathan, P., Murali, P. V., & Panneerselvam, R. (2009). Interactive effects of calcium chloride on salinity-induced proline metabolism in *Pennisetum typoidies*. *Advance Biological Research, 3*, 168–173.

Guimarães, F. V. A., de Lacerda, C. F., Marques, E. C., de Miranda, M. R. A., de Abreu, C. E. B., Prisco, J. T., & Gomes-Filho, E. (2011). Calcium can moderate changes on membrane structure and lipid composition in cowpea plants under salt stress. *Plant Growth Regulation, 65*, 55–63.

Gul, B., & Khan, M. A. (2006). Role of calcium in alleviating salinity effects in coastal halophytes. In M. A. Khan & D. J. Weber (Eds.), *Ecophysiology of high salinity tolerant plants* (pp. 107–114). Dordrecht, The Netherlands: Springer.

Hepler, P. K. (2005). Calcium: A central regulator of plant growth and development. *Plant Cell, 17*(8), 2142–2155.

Hojjatnooghi, F., Mozafari, V., Tajabadipour, A., & Hokmabadi, H. (2014). Effects of salinity and calcium on the growth and chemical composition of pistachio seedlings. *Journal of Plant Nutrition, 37*, 928–941.

Kaddour, R., Mahmoudi, M., Baâtour, O., Tarchoun, I., Nasri, N., Ben, S. I., Berthomieu, P., Gruber, M., & Lachaâl, M. (2012). Physiological and molecular responses of two *Arabidopsis* accessions to calcium amendment and salt constraint. *Acta Physiologiae Plantarum, 34*, 439–450.

Kaya, C., Kirnak, H., Higgs, D., & Saltali, K. (2002). Supplementary calcium enhances plant growth and fruit yield in strawberry cultivars grown at high (NaCl) salinity. *Scientia Horticulturae, 93*, 65–74.

Khan, M. N., Siddiqui, M. H., Mohammad, F., Naeem, M., Masroor, M., & Khan, A. (2010). Calcium chloride and gibberellic acid protect linseed (*Linum usitatissimum* L.) from NaCl stress by inducing antioxidative defense system and osmoprotectant accumulation. *Acta Physiolia Plantarum, 32*, 121–132.

Lei, B., Huang, Y., Xie, J. J., Liu, Z. X., Zhen, A., Fan, M. L., & Bie, Z. L. (2014). Increased cucumber salt tolerance by grafting on pumpkin rootstock and after application of calcium. *Biologia Plantarum, 58*(1), 179–184.

Martínez-Ballesta, C. M., Cabañero, F., Olmos, E., Periago, P. M., Maurel, C., & Carvajal, M. (2008). Two different effects of calcium on aquaporins in salinity-stressed pepper plants. *Planta, 228*, 15–25.

Murillo-Amador, B., Jones, H. G., Kaya, C., Aguilar, R. L., Garcia-Hernandez, J. L., Troyo-Dieguez, E., Avila-Serrano, N. Y., & Rueda-Puente, E. (2006). Effects of foliar application of calcium nitrate on growth and physiological attributes of cowpea (*Vigna unguiculata* L. Walp.) grown under salt stress. *Environmental and Experimental Botany, 58*, 188–196.

Murugan, K., & Sathish, D. K. (2005). Ameliorative effect by calcium on NaCl salinity stress related to proline metabolism in the callus of *Centella asiatica* L. *Journal of Plant Biochemistry and Biotechnology, 14*, 205–207.

Nedjimi, B., & Daoud, Y. (2009a). Effects of calcium chloride on growth, membrane permeability and root hydraulic conductivity in two *Atriplex* species grown at high (sodium chloride) salinity. *Journal of Plant Nutrition, 32*, 1818–1830.

Nedjimi, B., & Daoud, Y. (2009b). Ameliorative effect of $CaCl_2$ on growth, membrane permeability and nutrient uptake in *Atriplex halimus* subsp. *schweinfurthii* grown at high (NaCl) salinity. *Desalination, 249*, 163–166.

Nedjimi, B., Daoud, Y., Carvajal, M., & Martínez-Ballesta, M. C. (2010). Improvement of the adaptation of *Lygeum spartum* L. to salinity under the presence of calcium. *Communications in Soil Science and Plant Analysis, 41*(19), 2301–2317.

Qadir, M., Schubert, S., Ghafoor, A., & Murtaza, G. (2001). Amelioration strategies for sodic soil: A review. *Land Degradation & Development, 12*, 357–386.

Renault, S. (2005). Response of red-osier dogwood (*Cornus stolonifera*) seedlings to sodium sulphate salinity: Effects of supplemental calcium. *Physiologia Plantarum, 123*, 75–81.

Rengasamy, P. (2006). World salinization with emphasis on Australia. *Journal of Experimental Botany, 57*, 1017–1023.

Rengel, Z. (1992). The role of calcium in salt toxicity. *Plant, Cell & Environment, 15*, 625–632.

Shabala, S., Shabala, L., & van Volkenburgh, E. (2003). Effect of calcium on root development and root ion fluxes in salinised barley seedlings. *Functional Plant Biology, 30*, 507–514.

Shaikh, F., Gul, B., Li, W., Liu, X., & Khan, M. A. (2007). Effect of calcium and light on the germination of *Urochondra setulosa* under different salts. *Journal of Zhejiang University Science. B, 8*(1), 20–26.

Sunukumar, S. S., Harish, S. R., Manoj, G. S., Sreelekshmi, S. G., Krishnan, R., Lubaina, A. S., & Murugan, K. (2011). Ameliorative effect by calcium on NaCl salinity stress related to reactive oxygen species metabolism in *Amaranthus tricolor* L. *Journal of Research in Biology, 6*, 411–418.

Szabolcs, I. (1989). Amelioration of soils in salt affected areas. *Soil Technology, 2*, 331–344.

Tobe, K., Li, X., & Omasa, K. (2002). Effect of sodium magnesium and calcium salts on seed germination and radicle survival of a halophyte, *Kalidium caspicum* (Chenopodiaceae). *Australian Journal of Botany, 50*, 163–169.

Tobe, K., Li, X., & Omasa, K. (2004). Effects of five different salts on seed germination and seedling growth of *Haloxylon ammodendron* (Chenopodiaceae). *Seed Science Research, 14*, 345–353.

Tuna, A. L., Kaya, C., Ashraf, M., Altunlu, H., Yokas, I., & Yagmur, B. (2007). The effects of calcium sulphate on growth, membrane stability and nutrient uptake of tomato plants grown under salt stress. *Environmental and Experimental Botany, 59*, 173–178.

White, P. J. (1998). Calcium channels in the plasma membrane of root cells. *Annals of Botany, 81*, 173–183.

Xiong, T. C., Bourque, S., Lecourieux, D., Amelot, N., Grat, S., Brière, C., Mazars, C., Pugin, A., & Ranjeva, R. (2006). Calcium signaling in plant cell organelles delimited by a double membrane. *Biochimica et Biophysica Acta, 1763*, 1209–1215.

Xue, Y. F., Liu, L., Liu, Z. P., Mehta, S. K., & Zhao, G. M. (2008). Protective role of Ca against NaCl toxicity in Jerusalem artichoke by up-regulation of antioxidant enzymes. *Pedosphere, 18*, 766–774.

Yao, S., Chen, S., Zhao, J., Xu, D., Lan, H., & Zhang, F. (2010). Effect of three salts on germination and seedling survival of dimorphic seeds of *Chenopodium album*. *Botany, 88*, 821–828.

Zehra, A., Gul, B., Ansari, R., & Khan, M. A. (2012). Role of calcium in alleviating effect of salinity on germination of *Phragmites karka* seeds. *South African Journal of Botany, 78*, 122–128.

Zrig, A., Tounekti, T., Ben Mohamed, H., Abdelgawad, H., Vadel, A. M., Valero, D., & Khemira, H. (2016). Differential response of two almond rootstocks to chloride salt mixtures in the growing medium. *Russian Journal of Plant Physiology, 63*, 143–151.

Chapter 16
Short-Term Transformation and Dynamics of Main Nutrients in Soil

Katalin Sárdi

Abstract It is known that nutrient availability plays a significant role in soil fertility; organic and inorganic sources of essential nutrients are affected by several factors including transformation processes. These processes are controlled by the physical, chemical, and biological soil properties. Therefore, transformation of available forms into unavailable ones is always closely related to the given soil type.

For ensuring the balanced nutrient supply in crop production, interrelations in the soil-crop-nutrient system should be taken into consideration, both naturally available sources and the determinant factors affecting specific plant-available forms.

This chapter reviews the importance of short-term nutrient transformation processes in soils, providing the basic and essential information on the determinant factors on them, and gives an outlook on the methodologies of experiments in studying.

Keywords Short-term nutrient transformation • Nutrient availability soil properties • Mobilization and immobilization • Experimental approaches

16.1 Introduction

Sustainable crop production depends on the efficiency of nutrient management; consequently, high yield levels can be maintained when each part of the production technology is adapted for the agro-ecological conditions with special respect to characteristics determinant for soil fertility. It is known that nutrient availability plays a significant role in soil fertility; organic and inorganic sources of essential

K. Sárdi (✉)
Department of Crop Production and Soil Science, University of Pannonia, Hungary 16 Deák F. st., H-8360 Keszthely, Hungary
e-mail: sardi@georgikon.hu

M. Naeem et al. (eds.), *Essential Plant Nutrients*,
DOI 10.1007/978-3-319-58841-4_16

nutrients are affected by several factors including transformation processes. These processes are controlled by the physical, chemical, and biological soil properties. Therefore, transformation of available forms into unavailable ones is always closely related to the given soil type.

Recently, the "4R" concept in fertilization (i.e., the right source, the right rate, the right time, and the right place) has been commonly agreed and applied as it is essential in any advanced cropping technologies.

16.2 Importance of Short-Term Nutrient Transformation Processes in Soil Nutrient Dynamics and Nutrient Management

Soils serve as the principal source of crop nutrients; required amounts for root uptake ensuring yields may originate either from the available forms or from the large amounts of reserves (generally expressed in kg or g per hectare), referred as *nutrient pools*.

The main pools including labile and non-labile ones or primary soil nutrient reservoirs typically present in a soil:

- Ions dissolved in the soil solution
- Adsorbed/exchangeable cations and anions
- Primary and secondary minerals
- Organic matter

Among these pools, soil solution is considered the critical one as all the nutrients are taken up by plant roots from the dissolved phase held here. Therefore, replenishment is required from the solid phase for ensuring the continuous uptake to sustain plant growth and development. The amounts of nutrient ions dissolved in soil solution (referred the intensity factor) and that of entering the solution phase (quantity factor) are important measures of soil nutrient behavior and fertility. The concept of Q/I describing the mineral nutrient status of a soil was first proposed by Schofield (1955a, b).

According to our present understanding, it is evident that for most soil nutrients, there are no specific and discrete fractions of nutrient forms, they create equilibrium with one another thus can be well-defined parts of nutrient dynamics. From the aspect of root availability, the two main directions of the transformations are *mobilization and immobilization*. Mobilization and immobilization of nutrients in soils are determinant factors of nutrient dynamics and cycling; they are influenced by several factors and have an impact on the effectiveness of fertilization.

Soil properties may show a great spatial variability within a site, therefore a large number of replicate samples are required to estimate nutrient pools on a landscape scale (Fig. 16.1).

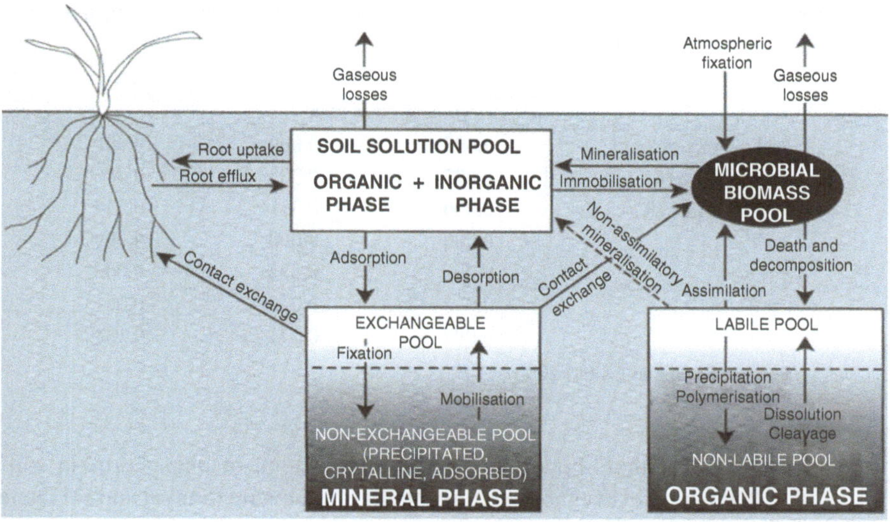

Fig. 16.1 Conceptual diagram of major nutrient pools and pathways in soil. Reproduced from McLaughlin et al. (1999), with permission from CSIRO Publishing

Several elements entering the surface of soil through atmospheric deposition may also contribute to the amounts of soil nutrients. Plant nutrients are exposed to various transformation processes in soils, both in the short-term and the long-term scale (global nutrient cycling). According to the main types of these transformations, they may be biological and/or chemical.

16.3 Role of Abiotic and Biotic Factors Influencing Nutrient Transformation and Bioavailability of Major Essential Nutrients

16.3.1 Abiotic Factors

16.3.1.1 Soil Physical Characteristics

16.3.1.1.1 Environmental Conditions

Environmental conditions—especially temperature and moisture—have a strong impact on soil nutrient dynamics. It was generally observed that there is a strong interaction between soil temperature, moisture conditions, and nutrient transformation processes. Soil organic matter (SOM) decomposition has primary importance among these.

Among the abiotic factors, temperature and moisture (namely, freezing-thawing and drying-rewetting) are determinant both on soil P and K availability. It has been

Table 16.1 Statistical significance of temperature, fertilizer rates, and their interaction on wet and dry potassium fixation

Factor	K_{ex}	K_{wf}	K_{df}
	Level of significance		
Soil characteristics	**0.000**	**0.000**	**0.000**
Temperature (t)	**0.007**	0.305	0.446
Fertilizer rate (R)	**0.000**	0.351	0.085
S × t	**0.000**	**0.000**	**0.000**
S × R	0.061	**0.032**	**0.002**
t × R	0.090	**0.032**	0.061
S × t × R	0.097	0.241	0.131

(Source: Sardi and Csitari 2000)

studied by several authors how microbial immobilization of phosphorus in soils exposed to fluctuation in drying-rewetting and freeze-thawing (Sanyal and De Datta 1991). Influence of a relatively high and a low temperature (25 °C and 0 °C) on potassium fixation capacities was studied by Sardi and Csitari (2000). Soil samples represented seven sites of multilocation long-term fertilization trials. The role of main soil characteristics, fertilizer rates and temperature on exchangeable K content (K_{ex}), and wet and dry K fixation (K_{wf} and K_{df}) capacities as well as their interaction were analyzed by applying stepwise regression analyses (Table 16.1).

From the results summarized above, it was established that soil characteristics (S), fertilizer rates (R), and temperature significantly determined exchangeable K content (K_{ex}) in soils, while wet and dry K fixation were only dependent on soil characteristics.

After evaluating the effect of temperature on exchangeable potassium content, it was concluded that at 25 °C, 40.7% of the value of K_{ex} could be attributed to soil K_2O content, total quantity of kaolinite plus chlorites in soils. At 0 °C, 63.5% of the value of exchangeable K was attributed to soil K_2O, humus, and illite content. Comparing the correlation coefficients obtained for the variables studied, it was concluded that the correlation between K_{ex} and K_{wf} was the closest.

16.3.1.2 Soil Chemical Characteristics

16.3.1.2.1 Soil Reaction (Soil pH, Acidity, and Alkalinity)

Soil reaction and exchangeable acidity are important parameters in standard soil analyses controlling numerous chemical and biological processes during soil nutrient transformation, consequently, nutrient dynamics.

Soil reaction (soil pH, measured in a suspension with distilled water or a salt solution, e.g., KCl or $CaCl_2$) is considered the most important chemical property of a soil, one of the determinant characteristics for fertility (Schofield 1955a, b). Actual information on soil pH is vital for estimating nutrient availability and—as it can be critical for productivity—for crop yield expectations of farmers. It is important to understand that in intensive agricultural production, a decline in soil pH occurs over time. High inputs of several acid-forming fertilizer sources (such as ammonium-N), leaching, soil respiration by microbes and plant roots, root excretion of organic acids, dry and wet atmospheric deposition, and acid rain are the main factors (Bloom 2000).

Numerous methods are available either for routine soil testing or for research. These methods were developed for the application under field conditions for the rapid estimation of acidity/alkalinity and in laboratory for the analytical measurements. Portable pH meters, color kits, or pH strip papers offer low-cost approaches for immediate in-field information, and they can be effective when decisions are required on which samples are to be sent to a lab for analytical determinations. Recently, automated instruments have become more popular for measuring soil pH. These instruments save labor costs and improve accuracy of standard measurements; protocols are internationally accepted (Kalra 1995; Motsara and Roy 2008). Potential (hydrolitic) acidity of soils (Y1 value) determined in a 1n Ca-acetate solution serves as a good indicator of the rapid acidification process which cannot be demonstrated from other soil parameters (Sárdi and Máté 2012). Soil pH has a strong influence on buffering capacity, oxidation-reduction, precipitation-dissolution equilibrium, consequently, on chemical and biological nutrient transformations in soils. Natural soil acidification may be considerably accelerated under intensive farming. Unfavorable impacts on soil may occur in a very short period when acid-producing fertilizers are used especially in soils having lower buffering capacities.

Another example for the effect of soil pH is as follows: short-term chemical phosphorus transformation/immobilization is significantly influenced by soil reaction; the main steps are shown in the following:

- In neutral pH range:

$$Ca(H_2PO_4)_2 + Ca(HCO_3)_2 \Leftrightarrow 2CaHPO_4^{*} H_2O + CO_2$$

- In alkaline pH range:

$$Ca(H_2PO_4)_2 + 2Ca(HCO_3)_2 \Leftrightarrow Ca_3(PO_4)_2^{*} 4H_2O + 4CO_2$$
$$Ca(H_2PO_4)_2 \Rightarrow CaHPO_4 \Rightarrow Ca_3(PO_4)_2 \Rightarrow \text{apatit}$$

- In acidic pH range:

$$Ca(H_2PO_4)_2 + 2Al(OH)_3 \Rightarrow 2AlPO_4 + Ca(OH)_2 + 4H_2O$$
$$Ca(H_2PO_4)_2 + Fe(OH)_3 \Rightarrow FePO_4 + CaHPO_4 + 3H_2O$$

Although there are some differences among elements in this aspect, it is widely demonstrated that in terms of pH, the (bio)availability of most plant nutrients is ranging between 4.0 and 9.0, with the optimum interval 6.0–6.5. Both strong acidity and alkalinity will result in a significant decline in available forms of essential nutrients. However, with *adaptive* cropping technologies, especially *nutrient management*, sufficient yield levels are possible both in acidic and alkali soils (see the recommendations included in the concept of 4R).

16.3.1.2.2 Ion Exchange

Ion exchange and cation exchange capacity (CEC) shows the amount of cations that can be absorbed on the negative charge sites of clay minerals.

Ion exchange is of primary importance in nutrient dynamics; the measurement of the exchange capacity of a given soil involves the replacement of readily exchangeable ions by a standard or "index" ion under controlled environment. Whereas among anions, chlorine is the most commonly used index ion (Sposito 2000). Ion exchange reactions describing the kinetic and thermodynamic nature of these processes are expressed by chemical equations developed by several scientists (Sparks and Suarez 1991; Sumner and Miller 1996 etc.). The availability of soil potassium is controlled by numerous soil and environmental factors such as moisture, temperature, and others (Grimme et al. 1971; Mc Lean and Watson 1985). Potassium content of the soil solution together with easily exchangeable potassium plays the major role in plant uptake, and they are recognized as the plant available amounts (Quémener 1979).

On the other hand, it was observed by several authors that the release from the non-exchangeable potassium reserves may contribute to the potassium supplying capacity of soils (Németh 1975; Mengel and Weichens 1979). This varies with the site characteristics, and the amounts of potassium released tend to become available to meet crop requirements.

16.3.1.2.3 Clay Mineralogy

Soils show considerable differences in clay content and clay mineralogy. Therefore, different levels of available potassium develop even at the same K fertilizer rates (Debreczeni and Sárdi 1990).

Potassium dynamics in soils is dominated by clay mineralogy, i.e., quality and quantity of layer silicates characterized by isomorphic substitution of lower valence cations in either or both the tetrahedral or octahedral sheets. Therefore, negative charges are typical and these can be balanced by other cations adsorbed by the external surface or fixed between the layers (McBride 1994; Goldberg et al. 2000).

Ion exchange in clays can be characterized from two viewpoints: either a thermodynamic or a kinetic. Amounts of potassium in adsorption-desorption are reversible reactions having their specific thermodynamical parameters (Sparks and Jardine 1981). Studying the fixation in soils described already in 1894 (Dyer 1894) is of primary interest both from theoretical and practical aspect.

Table 16.2 Mineral type and cation exchange capacity of clay minerals

Name	Mineral type	CEC mmol kg^{-1}
Kaolinite	1:1	10–100
Smectites	2:1	800–1200
Montmorillonite		
Vermiculite	2:1	1200–1500
Mica	2:1	200–400
Chlorite	2:1:1	200–400

Exchange capacity is defined as the total amount of cations that can be attracted and adsorbed at the solid surface of the soil particles especially the layered clay minerals. Cation exchange capacity (CEC) is usually expressed in moles per unit mass: CEC mmol $kg^{-}1$.

Cation exchange capacities typical for representative clay minerals are summarized in Table 16.2 (after Bohn et al. 2001).

16.3.1.2.4 Chemisorption and Precipitation

Chemisorption means the adsorption of inorganic ions to the solid phase which involves a chemical reaction between the surface and the adsorbate. With chemisorption, new chemical bonds are generated at the adsorbent surface. The strong attraction to the soil particles may be either electrical (e.g., K^+ ions) or chemical which is typical for phosphates.

The majority of Phosphorus fertilizers are applied in water-soluble form to ensure P requirements of crops; however, it should be taken into consideration that only a very low percentage of this amount will remain immediately available in the soil solution. A higher part of phosphate anions will be chemically adsorbed on the surface of metal-containing compounds (typically Al- and Fe-hydroxides). These metal-phosphate complexes remain strongly fixed for longer periods, especially in acidic environments (Sanyal and De Datta 1991).

Phosphorus becomes unavailable through precipitation when plant-available inorganic P reacts with dissolved Fe, Al, Mn (in acid soils), or Ca (in alkaline soils) to form phosphate minerals (McBride 2000).

It is commonly believed that phosphorus precipitation and strong adsorption may occur especially in calcareous soils generating P retention, and it is difficult to distinguish between the two mechanisms. Measurement of P is often carried out by single or sequential extractions, which often involve the solvent systems of Na-citrate, Na-acetate, Na-bicarbonate, and others.

On the other hand, soils contain various P-containing minerals; weathering on the long-term scale results slowly available forms to plants.

16.3.1.3 Biotic Factors

16.3.1.3.1 Interactions in the Soil-Plant-Nutrient System

Soil-plant-nutrient interactions are determinant, and it is worth noting that they are variable among the growing seasons; therefore, it has an outstanding importance to study the average of these interactions providing the reliable estimates of the field/site. Conclusions from only a selected year's specific results may be misleading.

For sustainable crop production, one must consider the dynamics of crop uptake as well as the dynamics of nutrient transformation in the soil and recognize dominant weather factors during the cropping year.

Among the most important factors controlling the availability of nutrients to plant roots are the following:

(a) The concentration of the ion in the soil solution.
(b) The buffer capacity of the soil, i.e., the rate of replenishment/desorption of the ion from the solid phase.
(c) The intensity of root uptake, i.e., ion absorption capacity which is genetically determined by plant species and their varieties/hybrids. Crop nutrient requirement is closely related to the extent/area of the root system as well as the efficiency of nutrient use.
(d) External factors such as environmental conditions, including soil moisture, crop-weed competition for water, nutrients, light, etc.

16.3.1.4 Biological Transformation of Nutrients

Biological transformation of nutrient elements and their various forms are controlled by several reactions including mineralization of organic forms subject to the activity of soil microorganisms. Soil organic matter (SOM) is the great reservoir of nutrients, containing organic materials in all and variable stages of decomposition/mineralization.

Distribution of soil nutrient forms or various fractions receives significant attention both from the viewpoint of research and practice.

Based on the type of bound, organic and inorganic pools are dominant for nitrogen, phosphorus, and sulfur:

Organic pool—occurs in several complex forms such as proteins, amino acids, and their derivatives.

For example, their proportion of total soil N varies between 20 and 40%.

Amino sugars, ranging between 5 and 10%. Information on the main properties of the remaining 50% is rather limited. Proteins are bound with clays, lignin, and other compounds in the solid phase and rather resistant to decomposition. Most organic soil P are inositol phosphates, their proportion of total soil P is between 10 and 50%, and the remaining part consists of P lipids, proteids, and nucleic and phytic acids, which are products of microbial degradation of plant residues.

Similarly to N, sulfur and phosphorus mineralization depends on climate (temperature, moisture and aeration, soil pH), influencing the activity of heterotrophic and autotrophic soil microorganisms (e.g. *Streptomyces sp.* and *Thiobacillus* in case of sulfur and *Penicillium* in case of phosphorus). Phosphorus mineralization is governed by phosphatase enzymes capable to catalyze the reaction of organic P.

Inorganic pool—easily or completely soluble and available compounds.

Inorganic N forms are mainly occurring as ammonium, nitrate, and nitrite, in lesser amounts as nitrous oxide and nitric oxide. Apart from the amounts applied to soil in mineral fertilizers, ammonium and nitrate are produced by the aerobic decomposition of soil organic matter. Nitrification is considered the aerobic biological oxidation of ammonium-N to nitrate-N, via nitrite-N controlled by bacterial activity.

From the aspect of root availability, they are divided in two main groups:

Bioavailable and unavailable forms.
Based on their extent of transformation: mineralizable amounts.

These forms are known to play a significant role in fertility. Quantities of organic compounds decrease with crop growth and increase after the harvest which is due to the enhanced activity of plant growth-promoting rhizobacteria (PGRP).

The capacity in supply to crops strongly depends on quantities of SOM as an ultimate natural nutrient reserve maintaining soil productivity.

16.4 Experimental Approaches

(a) Chemical methods (extraction)
(b) Biological methods
(c) Simulation and modeling

There are numerous experimental approaches and methodologies internationally recognized and accepted to study the short-term nutrient transformation processes in soils. The main characteristics are based on their main type: in vivo, in vitro, or in situ.

16.4.1 The Importance of Soil Sampling

For agricultural purposes, i.e., for any cropping system, for any laboratory methods selected upon the principal goal of soil analyses, sampling is the first and determinant step.

Possible errors during sampling were investigated from the middle of the last century (Hemingway 1955); it was concluded that the sampling errors were considerably exceeding those of analytical errors. Based on these assessments, sampling error may even reach 80% of the total error.

Sampling methods should provide a representative quantity of soil for the given area. Random sampling may not provide sufficient information on soil variability within the field. Systematic sampling can be either applying the conventional "W" or "X" pattern for determining the average soil test levels for the field or plot. Recently, sampling protocols were developed and introduced to avoid systematic errors.

Recently, using a GPS and a grid pattern sampler for site-specific nutrient management based on the spatial variability of soil, digital mapping is an important tool to improve the efficiency of fertilization.

On the other hand, similar systematic errors can occur during laboratory analyses. When combined with soil sampling errors, these may generate misleading data and false data interpretation.

Following carefully the instructions of sampling as well as sample preparation and those of appropriate laboratory determinations, they are among the key questions for efficient fertilization practice.

16.4.2 Chemical Methods (Extraction)

Soil testing is considered as the rapid *chemical laboratory determination* of nutrient status, availability conditions, furthermore, includes laboratory data interpretation and the fertilizer recommendation program based on the results of chemical analyses. Soil testing represents the most comprehensive practical application of our knowledge in soil science and should not be simplified as a certain series of laboratory analyses (Kamprath 2000).

Soil testing is considered as the main tool in evaluating soil nutrient status, availability conditions for cropping, thus serves as a basis in any fertilizer recommendation system.

Results of soil tests determined for various soil types were correlated and validated worldwide to study the relationships between test values and crop responses to amounts of nutrients in numerous extracts.

Standardization for analytical procedures including accuracy requirements, quality control, and validation is often connected with international sample and data exchange programs, e.g., the International Soil Analytical Exchange (ISE) Program or the Association of Official Analytical Chemists (AOAC) of the USA.

The main questions related to accuracy and reliability are the following:

(a) Which methods can be considered the best soil testing methodologies to minimize or avoid inaccuracy or uncertainty in a recommendation system?
(b) Are test values able to predict yield levels?

16.4.2.1 Extraction of Available Nutrients

During the last few decades, several methodologies have been developed and introduced for rapid soil tests to determine the quantities of easily soluble or extractable elements (Olsen et al. 1954; Egner et al. 1960; Peck 1990; Jones 1998).

Laboratory determination of available soil nutrients using extractants is the primary tool of soil testing. According to the extent of availability, several methodologies have been developed in the past decades. Terms used for these amounts providing estimates of availability:

(a) Readily available (dissolved ions in the soil solution)
(b) Easily available (ions easily exchangeable from the solid phase by plant roots)
(c) Forms of low availability, strongly bonded, or adsorbed/fixed in soil minerals

Already from the end of the nineteenth century, several methods have been developed using water as an extractant in order to introduce rapid routine soil testing for fertilizer recommendations (Dyer 1894). Hot water extraction was successfully used for the determination of several elements such as boron and phosphorus (Berger and Truog 1944, van der Paauw 1969 etc). The electro-ultrafiltration (EUF) method developed by Németh (1976) and the continuous water extraction were suitable for routine P determinations. The EUF method was used also for determining the release of nutrients into the soil solution (Füleky 1978). Hot water percolation (HWP) is a rapid soil extraction; results are in close correlation with those of conventional soil testing (Füleky and Czinkota 1993).

Availability of phosphorus in different extracts was studied by numerous authors (Blake et al. 2003). Soil K status determined by various methodologies such as EUF can be correlated with K removal of plants (Sinclair 1982).

For the determination of these forms, i.e., the quantification in availability, laboratory incubation experiments were carried out at 10 °C and 40 °C temperature. Changes of soil phosphorus content in a long-term fertilization field trial conducted on a brown forest soil (USA taxonomy, Orthic Eutrochrept; FAO taxonomy, Eutric Cambisol) were studied in these experiments. Soil samples showed three increasing levels of P resulted by a 10-year intensive P fertilization (referred as P0, P1, and P2, respectively); samples were collected after 30 years of stopping fertilization. Available phosphorus contents of soil samples were determined using three approaches: in water (modified Murphy-Riley method), ammonium hydrocarbonate (Olsen, pH = 8.5), and ammonium lactate (AL, pH = 3.7) extract. Changes in the amounts of P were determined after 2 and 60 days of incubation in four freshly applied "new" treatments with increasing additions of phosphorus: 0, 100, 500, and 1000 mg P_2O_5 per kg of soil, representing agronomic and extreme P rates to be studied for environmental considerations. From the results of the experiments, it was established that after 2 days of incubation, at 10 °C temperature, both agronomic and extreme P rates resulted in significant increases in phosphorus content in each extract. On the other hand, after 60 days, even higher values were obtained in

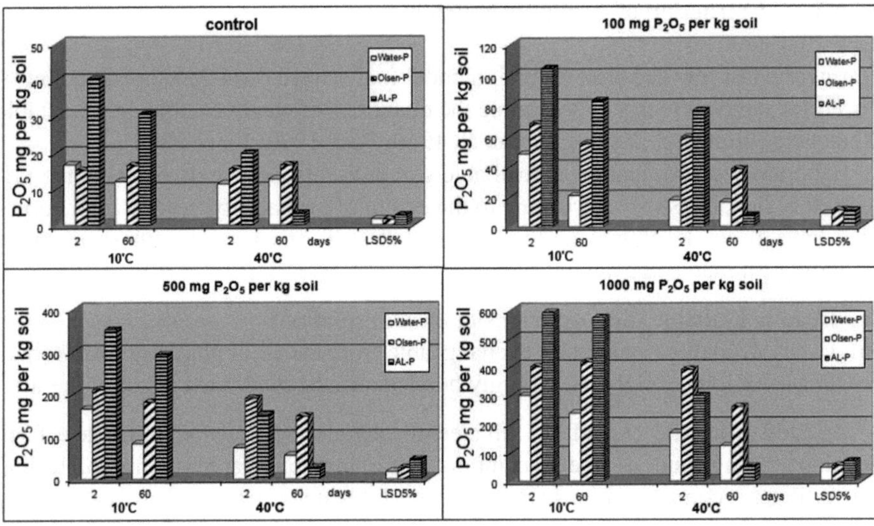

Fig. 16.2 Changes in the available P contents of an Eutric Cambisol after 2 and 60 days of incubation

several treatments. Correlation between water-P, Olsen-P, and AL-P values was evaluated at both temperatures.

From the results shown in Fig. 16.2, it was established that high temperature at the constant moisture favored the short-term immobilization of available phosphorus forms (Sárdi et al. 2006).

16.4.3 Biological Methods

Biological methods are considered to provide good estimation for the dynamics of nutrient transformation processes. Differences in nutrient absorption capacity of plant species may be also important in experimentation from this aspect.

Perennial ryegrass (*Lolium perenne* L.) is often used as a test plant known for its ability to utilize nutrients from soil reserves (Chaminade 1960). It was reported by several authors that perennial ryegrass showed better ability in taking up both phosphorus and potassium from fertilizers (Barrow and Campbell 1972, Barrow 1980; Mengel and Kirkby 1987).

Among several experimental approaches, chemical and biological methodologies are most common (i.e., with plants grown in pot experiments or in field trials).

16.4.3.1 Microbial Activity

Agricultural soils represent a special ecosystem containing a certain variety of soil microorganisms.

The process of organic decomposition is basically oxidation, exposed to microbial activity in the soil supposed that aeration is sufficient. Though the overall reaction has several steps, it can be characterized basically with oxygen consumption and carbon dioxide release. From the point when microbial activity reaches its maximum, the microbial biomass increases and may be increased up to the one third of the organic fraction of a soil (Brady 1990).

16.4.4 Simulation and Modeling

Research on model experiments developed for describing and understanding a great variety of processes existing in soils, characterizing and managing, or prognosting them receives international recognition.

Physically based mathematical models introduced from the early 1970s and their current state were reviewed recently (Shein 2015).

For the bioavailability assessment of major soil nutrients, several simulation methodologies have been developed and introduced in the past decades. Ion exchange resins may serve as a good approach for assessing availability (van Raij et al. 1986) and predicting root uptake. Greenhouse studies indicated strong correlation of PST (phytoavailability soil tests) resin capsules with nutrient uptake of soybean (Sardi et al. 1996) when several micronutrients were included in their research, extracting P, K, Ca, S, Al, Mn, Fe, Zn, and Cu.

Model experiments are carried out in order to obtain advanced concepts of organic C and N transformation both on the short and the long term, for the better understanding of the influence of soil aggregates on surface soil physical properties and others. The results contribute to the development in the knowledge related to C and N dynamics and improve models predicting biogeochemical processes within the agroecosystems.

It is commonly agreed that long-term fertilization trials may serve as model experiments in studying these questions. These experiments provide useful information, e.g., for the quantification of soil phosphorus transformation processes.

It should be considered that the main problem is the validation and adaptation of these rather diverse models, and software packages are required in the future.

16.5 Nitrogen Transformation and Dynamics

16.5.1 Mineralization

Most of the soil nitrogen pool exists in organic forms and structures. It is generally agreed that soil organic N pool can be found in five fractions based on their behavior during hydrolysis from acid insoluble N to hydrolyzable unidentified N (Baldock and Nelson 2000).

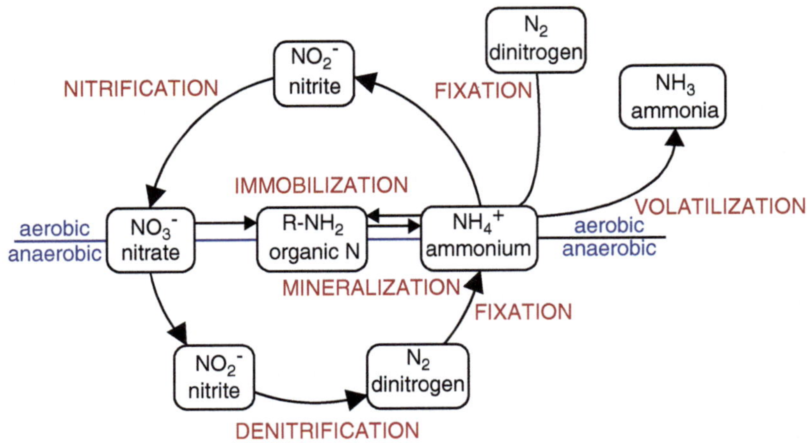

Fig. 16.3 Main transformation processes of nitrogen during cycling in the soil-plant-atmosphere system

The short-term transformation of N from SOM is influenced by several factors including microorganisms, plant roots, and soil fauna such as earthworms, physical soil disturbance, etc. (Fig. 16.3).

Isotope dilution methodology provides a good estimation of gross N mineralization and nitrification rates. Rates of nitrogen mineralization can be studied under controlled environments (e.g., in greenhouse experiments) using 15 N pool dilution techniques (Davidson et al. 1991; Whalen et al. 2001) where gross N transformations could be estimated. The balance between gross rates of N mineralization and N consumption in soils influenced both by the presence or absence of plants and by physical soil disturbance could be quantified. From the results, it was established that the rates of gross N transformations differ between bulk and ryegrass root-associated soils and that clipping of plants and the presence of earthworms exert marked effects on short-term N transformations.

16.6 Phosphorus Transformation and Dynamics

16.6.1 Immobilization

Among physicochemical processes controlling soil P concentrations and availability characteristics, sorption and desorption are of primary importance. According to Beckett and White (1964), there are four distinguished forms of soil phosphates:

1. Immediately labile P held on net-exchange sites
2. P held at occluded net-exchange sites
3. P held at surface isotopic-exchange sites
4. P held within more perfect crystal lattices from which it can be mobilized very slowly

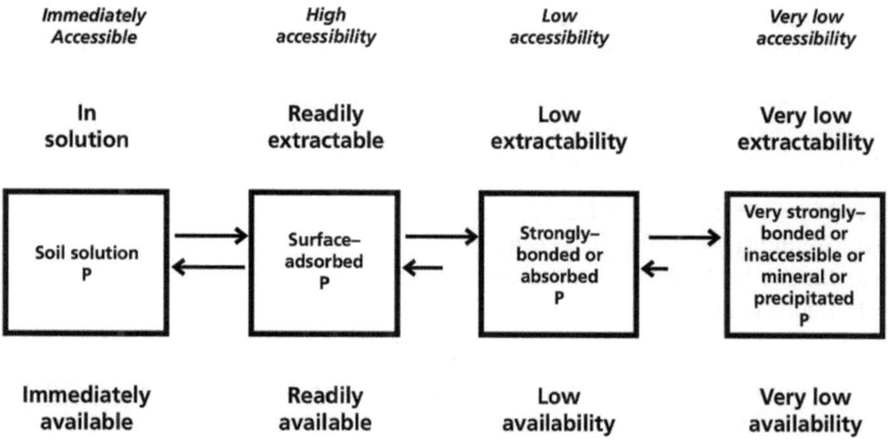

Fig. 16.4 Conceptual diagram for the forms of inorganic P in soils categorized in terms of extractability, with permission of FAO

The relationship between soluble and labile P in acidic, neutral, and calcareous soils was reviewed by Mattingly (1975). Labile P exists in equilibrium with other phosphates:

Labile P ↔ Solution P ↔ Crystalline P

Several authors (Tiessen et al. 1984; Sanyal and De Datta 1991) reviewed the chemistry of phosphorus transformations in soil. Rates of desorption play an important role in P supplying capacity of soils affected by the main soil properties (Sharpley 1983).

Figure 16.4 summarizes the concept of the behavior of soil P with the equilibrium, including the main fractions of extractability and availability.

As it has a primary importance for the achievable yield levels, an assessment of plant-available P status of European croplands were performed by Tóth et al. (2014). It was concluded that a repeated topsoil survey, i.e., Land Use/Land Cover Area Frame Survey (LUCAS), can be useful to monitor future changes of soil nutrient status in the EU, including phosphorus.

In the regions where agricultural soils received high or excessive P rates for several decades, the enrichment of soil P and consequently strongly positive balances were calculated. After the period of intensive fertilization, it was supposed that residual phosphorus in these soils may still have an influence on the phosphorus dynamics.

From the results of experiments reported by Sárdi et al. (2006), it was evident that:

(a) Effects of both freshly applied and residual P were affected by soil characteristics.
(b) Residual effects of intensive P fertilization still showed themselves in water-P values.
(c) Decline of water-soluble P fraction was significant after 60 days of incubation especially in the higher P additions.

(d) Permanent high temperature favored the immobilization of soil P in both soils.
(e) Fertilization history has markedly influenced the recovery of freshly added P.

Phosphorus transformation in soils is also influenced by soil microorganisms in several ways. Bacterial activity may result in various changes related to short-term P transformation processes included in the P cycle:

(a) Mineralization rate of P may be altered.
(b) Solubility/availability is affected.
(c) Immobilization may be enhanced.
(d) Redox conditions may be influenced.

Soil microorganisms (e.g., *Aspergillus*, *Bacillus*, *Fusarium*, *Pseudomonas*, *Micrococcus*, *Mycobacterium*, *Flavobacterium*, *Penicillium*) are known for the secretion of various organic acids and enzymes favoring the solubility of phosphorus compounds in soils. The main acids are acetic, formic, fumaric, and lactic. The rate and intensity of the P solubilization process are dependent of pH, temperature, moisture, and other factors. Commercially used species of phosphate-solubilizing bacteria and fungi are *Bacillus polymyxa, Bacillus megaterium, Pseudomonas striata, Aspergillus, Penicillium awamori, and Mycorrhiza.*

On the other hand, microorganisms require soluble/directly available phosphates for their cell synthesis, resulting decreases in the plant-available P pool, and when their life cycle terminates, these phosphates will become available to plants again.

16.6.2 Phosphorus Retention in Soils

Phosphorus retention—an important phenomenon—is the ability of a soil controlling the release of phosphates from the solid phase to the soil solution. P retention capacity of a soil is controlled by the main soil characteristics; it is commonly determined under laboratory conditions. According to the results of a study carried out with more than one hundred soil samples from Canada (Ige et al. 2007), Ca, Mg, and Al content in the Mehlich 3 extract had significant direct effect on soils' P retention capacity.

Studying the short-term processes of soil P transformation, e.g., characterizing the decline of available forms, may serve as useful information in understanding the phosphorus cycling in soils. Among weak extractants used for agronomic purposes, water is suitable for the evaluation of intensity factors. Amounts of water soluble P, i.e., the immediately available P, estimate a significant part of "labile" P fraction (Kamprath and Watson 1980).

Changes in the availability of phosphorus in different extracts (including acidic, neutral, and alkaline extractants) were studied by using different approaches (Hedley et al. 1982, Tiessen and Moir 1993 etc., Blake et al. 2003). Results of experiments comparing various extractants were reported by several authors (Sárdi et al. 2006).

From the results, it was evident that correlations between the soil P test values were highly significant even at the $P = 0.1\%$ level in each cases (R^2 values ranged

Table 16.3 Correlation between results of soil test determined in different extracts

10 °C	2 days Equation ($n = 48$)	R^2	60 days Equation ($n = 48$)	R^2
Water P–Olsen P	$y = -0.0016x^2 + 1.8052x + 0.2096$	**0.9608**	$y = -0.0057x^2 + 3.2652x - 23.196$	**0.971**
Water P–AL P	$y = 1.5617x + 18.019$	**0.8902**	$y = -0.0077x^2 + 4.4979x - 20.569$	**0.9761**
Olsen P–AL P	$y = 1.1141x + 10.357$	**0.8511**	$y = -0.0007x^2 + 1.6864x - 1.2249$	**0.9851**
40 °C				
Water P–Olsen P	$y = -0.0064x^2 + 3.2054x - 24.008$	**0.949**	$y = -0.0154x^2 + 4.3431x - 27.044$	**0.9114**
Water P–AL P	$y = -0.0044x^2 + 2.4854x + 0.3707$	**0.9777**	$y = -0.0217x^2 + 5.163x - 43.598$	**0.4984**
Olsen P–AL P	$y = -0.0008x^2 + 1.0941x + 8.3578$	**0.9519**	$y = -4E - 05x^2 + 1.0281x - 5.207$	**0.5371**

between 0.4984 and 0.9851, $n = 48$). Closest correlation described by a quadratic equation was found between water P and Olsen P values at both temperatures and incubation time intervals (see Table 16.3).

Amounts of water soluble P, i.e., the immediately available P, estimate a significant part of "labile" P fraction (Fig. 16.5).

It is very important and commonly agreed that soil properties may play a significant role in the effectiveness of residual and freshly applied phosphorus fertilizers. However, the quantitative relationships are not completely understood.

Relationships between important soil parameters and the P supplying or retention characteristics of several soil types were reported by Sárdi and Csathó 2002a, b. Changes in organic and inorganic phosphorus fractions of soils of long-term fertilization field experiments in Hungary were related to characteristics and fertilization history of soils (Füleky 1975; Sisák et al. 2002).

The objective of the study was to find a simple way for the comparison and evaluation of the two approaches in nutrient dynamics of soils. In order to obtain this, results of a biological method (i.e., pot experiments) and stepwise regression analyses were correlated. From these results, a rank correlation was also made from the experimental and calculated results for the evaluation.

For the unfertilized control (N0P0K0), the equation represented the amount of P retention of soils:

$$P_{ret} = 24.9\ \text{pH} - 0.353\ P_f + 5.11\ CaCO_3\% + 24.7\ \text{Humus}\ \% - 213.9$$

$R = 0.913 \qquad n = 86$

Where P_{ret} denotes P retention and P_f = freshly added P.

From these results, expectable average P supply power or retention capacity of soils per unit area (hectare) can be calculated. In the upper soil layer of 20 cm, the average mass is approximately 2.6 million kg soil. Estimated P supply and retention capacity of soils involved in the study showed a wide range: the difference between maximum and minimum values assumed almost 270 kg P per hectare. The supply ability amounted +72.02 kg ha^{-1} in a Calcaric Fluvisol. On the other hand, the P retention of soil could be expected to reach −194.8 kg ha^{-1} in the strongly acidic site, an Ochric Luvisol.

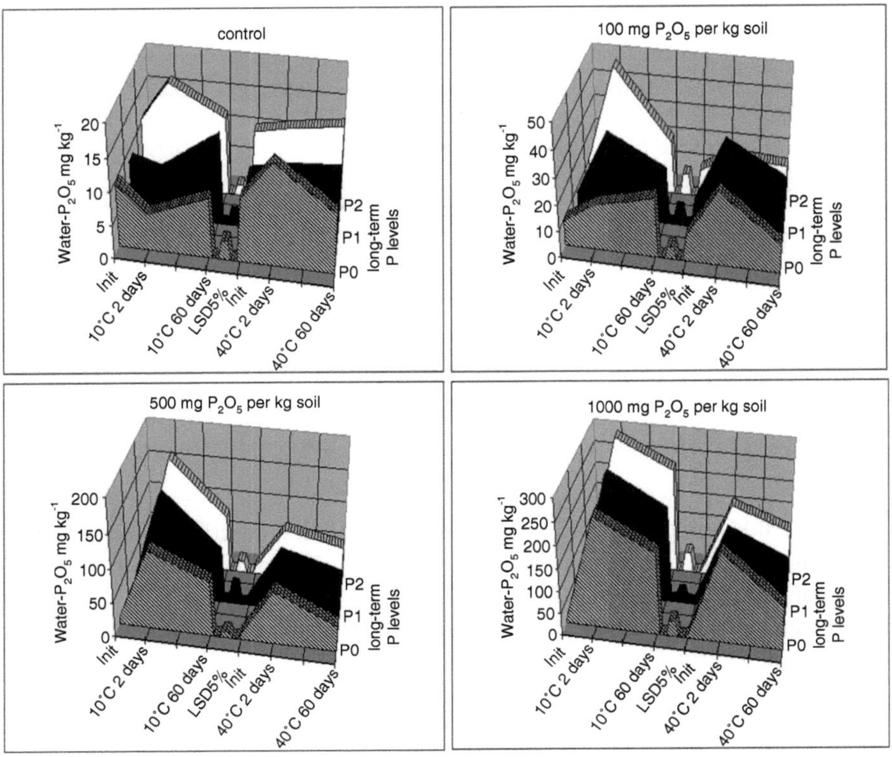

Fig. 16.5 Changes in water P content of the Eutric Cambisol induced by laboratory incubation

The P supplying power of 26 European benchmark soils was investigated by means of desorption/sorption isotherms (Q/I plots) and by desorption graphs determined at increasing water-to-soil ratios. From the results, it was concluded that European agricultural soils vary widely in their capacity to supply P to surface runoff waters. They also indicate that the traditional determination of P reserves and Q/I curves with a fixed solution-to-soil ratio is likely of limited use for predicting the P loading by eroded soil particles transported to watercourses (Hartikainen and Simojoki 2010).

16.7 Potassium Transformation and Dynamics

16.7.1 Adsorption and Fixation

In a cropping system, potassium behavior in soils affects the efficiency of nutrients; consequently, it may potentially become one of the limiting factors of yield levels related to nutrient management.

16.7.1.1 Dynamic Equilibrium Processes of Soil Potassium

The most widely accepted concept distinguishes the following forms of soil in four pools:

(a) Soil solution K (K_{sol})
(b) Exchangeable K (K_{ex})
(c) Fixed or non-exchangeable K in clay minerals (K_{fix})
(d) K in the lattice of primary clay minerals (K_{struct})

The fractions K_{sol}, K_{ex}, and K_{fix} are interrelated through reversible exchange processes (Grimme & Németh 1979). Plant removal of soil potassium and/or leached into the subsoil is replenished by K ions released either from the K_{ex} and K_{fix} pools. This release can be simultaneous from both pools to the soil solution or can show a linear exchange process from the non-exchangeable to the exchangeable pool and from the exchangeable pool to the soil solution.

The amount of K in each pool varies and depends on past cropping history, past fertilizer, and manure use, i.e., the K balance, soil pH, and soil water content.

Potassium fixation means the transformation of available K into unavailable forms, significant for both the farmers' practice and scientific aspect. Dynamics of soil potassium, i.e., the extent and magnitude of soil fixing capacity and the quantities of K release, is controlled by clay content and clay mineralogy (Beckett 1970).

Fixation characteristics and the role of clay mineralogy were widely studied by several authors from the 1970s (Barshad and Kishk 1970, Amberger et al. 1974, Quémener 1979, Sparks and Huang 1985, Bouabid and Bloom 1991 etc.) (Fig. 16.6).

Quantities of K fixation including wet and dry fixation related to soil K levels resulted by fertilization history were studied by Sardi and Csitari (1998). Wet fixation was lower in soils where higher levels of exchangeable K were determined as

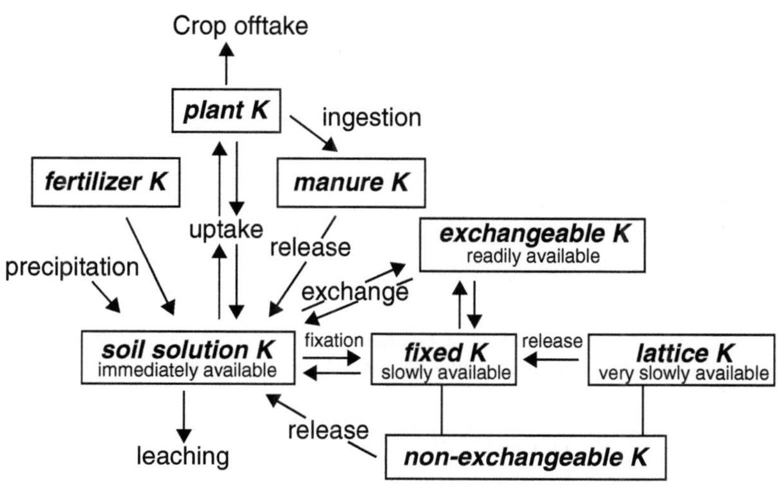

Fig. 16.6 Conceptual diagram of potassium dynamics and transformation processes (after Syers 1998)

the result of higher rates of K fertilizers. On the other hand, higher values in dry fixation of K were observed with higher K fertilizer rates. In soils having more than 20% clay, a better correlation was observed between exchangeable K content and wet fixation. Similarly, in soils above 20% clay, a close positive correlation was obtained between exchangeable K and dry fixation capacity of soils ($R^2 = 0.9464$).

16.7.1.2 Summary

These rather diverse aspects of soil nutrient transformation processes existing in the short-term scale discussed above, it can be concluded that their significance is the impact on soil productivity and crop yield levels including efficiency of nutrient management. Therefore, much attention is needed both from the theoretical and practical point of view.

The main challenges of agricultural activities are to develop adaptive technologies in order to maximize yield levels and quality based on integrated cropping systems and, at the same time, to minimize the nutrient losses and environmental risks. All these require the functional understanding and good estimates of both favorable and unfavorable nutrient transformation of soil nutrients.

References

Amberger, A., Gutser, R., & Teicher, H. (1974). Potassium nutrition of plants and K dynamics on potassium fixing soil. *Plant and Soil, 40*, 269–284.

Baldock, J. A., & Nelson, P. N. (2000). Soil organic matter. In M. E. Sumner (Ed.), *Handbook of soil science* (pp. B25–B84). Boca Raton, FL: CRC Press.

Barrow, N. J. (1980). Evaluation and utilization of residual phosphorus in soils (Chapter 13). In *The role of phosphorus in agriculture* (pp. 333–359). Madison, WI: American Society of Agronomy.

Barrow, N. J., & Campbell, N. A. (1972). Methods of measuring residual value of fertilizers. *Australian Journal of Experimental Agriculture and Animal Husbandry, 12*, 502–510.

Barshad, I., & Kishk, F. M. (1970). Factors affecting potasium fixation and cation exchange capacities of soil vermiculite clays. *Clay Minerals, 18*, 127–131.

Beckett, P. H. T., & White, R. E. (1964). Studies on the phosphate potential of soils: III. The pool of labile inorganic phosphate. *Plant and Soil, 21*, 253–282.

Beckett, P. H. T. (1970). 'Fixed' potassium and the residual effects of potassium fertilizers. *Potash Review,* Subject 16, Suite 52, 1–12.

Berger, K. C., & Truog, E. (1944). Boron tests and determination for soils and plants. *Soil Science, 57*, 25–36.

Blake, L., Johnston, A. E., Poulton, P. R., & Goulding, K. W. T. (2003). Changes in soil phosphorus fractions following positive and negative phosphorus balances for long periods. *Plant and Soil, 254*, 245–261.

Bloom, P. R. (2000). Soil pH and pH buffering. In M. E. Sumner (Ed.), *Handbook of soil science* (pp. B-333–B-350). Boca Raton, FL: CRC Press.

Bohn, H., McNeal, B., & O'Connor, G. (2001). *Soil chemistry* (3rd ed.). New York: Wiley. Chapters 5–9.

Bouabid, R. B., & Bloom, P. R. (1991). Potassium fixation and charge characteristics of soil clays. *Soil Science Society of America Journal, 55*, 1493–1498.

Brady, N. C. (1990). *The nature and properties of soils*. New York: MacMillan. 621 p.

Chaminade, R. (1960). Experimentation en petit vases de végétation types d'essais pour tester l'efficacité des engrais humiques. *Annals of Agronomy, 2*, 121–133.

Davidson, E. A., Hart, S. C., Shanks, C. A., & Firestone, M. K. (1991). Measuring gross nitrogen mineralization, immobilization, and nitrification by ^{15}N isotopic pool dilution in intact soil cores. *Journal of Soil Science, 42*(3), 335–349.

Debreczeni, K., & Sárdi, K. (1990). Evaluation of soil potassium supply using a method of biological testing. *Acta Agronomica, 39*(3–4), 227–240.

Dyer, B. (1894). On the analytical determination of probably available "mineral" plant food in soils. *Journal of the Chemical Society, Transactions, 65*, 115–167.

Egner, H., Riem, H., & Domingo, W. (1960). Untersuchungen über die chemische Bodenanalyse als Grundlage für die Beurteilung des Nährstoff-zustandes der Böden. II. Chemische Extractionsmethoden zur Phosphor und Kaliumbestimmung. *Kungliga Lantbrukshögskolans Annaler, 26*, 199–215.

Füleky, G. (1975). Changes of the soil phosphorus status in long-term field experiments. II. *Agrokémia és Talajtan, 24*(3–4), 291–302.

Füleky, G. (1978). Available phosphorus content of soil affected by P fertilization and its change in time. *Communications in Soil Science and Plant Analysis, 9*, 851–863.

Füleky, G., & Czinkota, I. (1993). Hot water percolation (HWP): A new rapid soil extraction method. *Plant and Soil, 157*, 131–135.

Goldberg, S., Lebron, I., & Suarez, D. I. (2000). Soil colloidal behaviour, Chapter 6. In M. E. Sumner (Ed.), *Handbook of soil science* (pp. B-195–B-240). Boca Raton, FL: CRC Press.

Grimme, H., & Németh, K. (1979). *The evaluation of soil K status by means of soil testing* (pp. 61–70). Bern: IPI Research Topics.

Grimme, H., Németh, K., & Braunschweig, v. L. C. (1971). Some factors controlling potassium availability in soils. *Proceedings of International Symposium on Soil Fertility Evaluation., 1*, 33–43.

Hartikainen, H., & Simojoki, A. (2010). Phosphorus supplying potential of European soils. *Proceedings of the 19th World Congress of Soil Science* (pp. 102–105). http://www.iuss.org. ISBN 978–0–646-53783-2.

Hedley, M. J., Stewart, J. W. B., & Chauhan, B. S. (1982). Changes in inorganic and organic soil phosphorus fractions induced by soil cultivation practices and by laboratory incubations. *Soil Science Society of America Journal, 46*, 970–976.

Hemingway, R. G. (1955). Soil-sampling errors and advisory analyses. *The Journal of Agricultural Science, 46*, 1–8. doi:10.1017/S0021859600039563.

Ige, D. V., Akinremi, O. O., & Flaten, D. N. (2007). Direct and indirect effect of soil properties on phosphorus retention capacity. *Soil Science Society of America, 71*, 95–100.

Jones Jr., J. B. (1998). Soil test methods: Past, present and future use of soil extractants. *Communications in Soil Science and Plant Analysis, 29*(11–14), 1543–1552.

Kalra, Y. P. (1995). Determination of pH of soils by different methods: Collaborative study. *Journal of AOAC International, 78*, 310–321.

Kamprath, E. J., & Watson, M. E. (1980). Conventional soil and tissue tests for assessing the phosphorus status of soils, Chapter 16. In *The role of phosphorus in agriculture* (pp. 433–469). Madison, WI: American Society of Agronomy.

Kamprath, E. J. (2000). Soil fertility and plant nutrition. In M. E. Sumner (Ed.), *Handbook of soil science*. Boca Raton, FL: CRC Press.

Mattingly, G. E. G. (1975). Labile phosphate in soils. *Soil Science, 119*(5), 369–375.

Mc Lean, E. O., & Watson, M. E. (1985). Soil measurements of plant available potaasium. In R. D. Munson (Ed.), *Potassium in agriculture* (pp. 277–308). Madison, WI: ASA, CSSA, SSSA.

McBride, M. B. (1994). *Environmental chemistry of soils*. New York: Oxford University Press. 406 p.

McBride, M. B. (2000). Chemisorption and precipitation reactions, Chapter 8. In M. E. Sumner (Ed.), *Handbook of soil science* (pp. B-265–B-302). Boca Raton, FL: CRC Press.
McLaughlin, M. J., Reuter, D. J., & Rayment, G. E. (1999). Soil testing—Principles and concepts. In K. I. Peverill, L. A. Sparrow, & D. J. Reuter (Eds.), *Soil analysis. An interpretation manual* (pp. 1–21). CSIRO Publishing: Melbourne.
Mengel, K., & Weichens, B. (1979). Die Bedeutung der nich austauschbaren Kaliumfraction des Bodens für die Ertragsbildung von Wiedelgras. *Zeitschrift für Pflanzenernährung und Bodenkunde, 142*, 836–847.
Mengel, K., & Kirkby, E. A. (1987). *Principles of plant nutrition*. Worblaufen, Bern: International Potash Institute.
Motsara, M. R., & Roy, R. N. (2008). Chapter 3. Soil analysis. In *Guide to laboratory establishment for plant nutrient analysis* (pp. 17–77). Rome, Italy: FAO Fertilizer and Plant Nutrition Bulletin 19.
Németh, K. (1975). The effect of K fertilization and K removal by ryegrass in pot experiments on the concentration of the soil solution of various soils. *Plant and Soil, 42*, 97–107.
Németh, K. (1976). The determination of effective and potential availability of nutrients in the soil by electroultrafiltration (EUF). *Applied Sciences and Development, 8*, 89–111.
Olsen, S. R., Cole, C. V., Watanabe, F. S., & Dean, L. A. (1954). *Estimation of available phosphorus in soils by extraction with sodium bicarbonate*. Washington: United States Department of Agriculture; Circular 939.
van der Paauw, F. (1969). Entwicklung und Verwertung einer neuen Wasserextraktionsmethode für die Bestimmung der pflanzenaufnehmbaren Phosphorsaure. Sonderh. *Landwirtschaftliche Forschung, 23/II*, 102–109.
Peck, T. R. (1990). Soil testing: Past, present and future. *Communications in Soil Science and Plant Analysis, 21*, 1165–1186.
Quémener, J. (1979). *The measurement of soil potassium* (pp. 5–48). Bern: IPI Research Topics. No. 4.
van Raij, B., Quaggio, J. A., & de Silva, N. M. (1986). Extraction of phosphorus, potassium, calcium and magnesium from soils by an ion-exchange resin procedure. *Communications in Soil Science and Plant Analysis, 17*, 547–566.
Sanyal, S. K., & De Datta, S. K. (1991). Chemistry of phosphorus transformations in soil. In *Advances in soil science* (Vol. 16, pp. 1–94). New York: Springer-Verlag.
Sárdi, K., & Csathó, P. (2002a). Studies on the phosphorus dynamics in pot experiments with different soil types. *Communications in Soil Science and Plant Analysis, 33*(15–18), 3045–3058.
Sárdi, K., & Csathó, P. (2002b). Studies on the phosphorus retention of different soil types in a pot experiment with prerennial ryegrass. *Agrokémia és Talajtan, 51*(1–2), 176–184.
Sardi, K., Sabbe, W. E., & Wolf, N. A. (1996). Nutrient adsorption characteristics of resin capsules used for the PST (Phytoavailability Soil Test) in four Arkansas soils. *Communications in Soil Science and Plant Analysis, 27*(3–4), 713–726.
Sárdi, K., & Csitári, G. (1998). Potassium fixation of different soil types and nutrient levels. *Communications in Soil Science and Plant Analysis, 29*(11–14), 1843–1850.
Sárdi, K., & Csitári, G. (2000). Studies on the potassium fixation of soils at different temperatures. *Communications in Soil Science and Plant Analysis, 31*(11–14), 2359–2365.
Sárdi, K., Csathó, P., & Osztoics, E. (2006). Changes in the soil phosphorus content of a long-term fertilization field trial studied in laboratory incubations. *Communications in Soil Science and Plant Analysis, 37*(1–24), 2833–2840.
Sardi, K., & Máté, F. (2012). A short-term study on the influences of fertilizer sources on soils. In *Proceedings of the 12th ESA Congress, Helsinki*, 20–24 August 2012 (pp. 396–397).
Schofield, R. K. (1955a). The measurement of soil pH. *Soil Science Society of America Journal, 19*(2), 164–167.
Schofield, R. K. (1955b). Can a precise meaning be given to "available" soil phosphorus. *Soils and Fertilizers, 18*, 373–375.

Sharpley, A. N. (1983). Effect of soil properties on the kinetics of phosphorus desorption. *Soil Science Society of America Journal, 47*, 462–467.
Shein, E. V. (2015). Physically based mathematical models in soil science: History, current state, problems and outlook (analytical review). *Eurasian Soil Science, 48*(7), 712–718.
Sinclair, A. H. (1982). A comparison of electro-ultrafiltration and quantity/intensity measurements of soil potassium with its uptake by ryegrass in Scottish soils. *Plant and Soil, 64*, 85–94.
Sisák, I., Sárdi, K., & Palkovics, M. (2002). Water-soluble P as affected by freshly applied and residual P and P fractions of soil. *Communications in Soil Science and Plant Analysis, 33*(15–18), 2813–2823.
Sparks, D. L., & Jardine, P. M. (1981). Comparison of kinetic equations to describe K-Ca exchange in pure and in mixed systems. *Soil Science Society of America, 138*, 115–122.
Sparks, D. L., & Huang, P. M. (1985). Physical chemistry of soil potassium. In R. D. Munson (Ed.), *Potassium in agriculture* (pp. 201–276). Madison, WI: American Society of Agronomy.
Sparks, D. L., & Suarez, D. L. (Eds.). (1991). *Rates of soil chemical processes*. Madison, WI: Soil Science Society of America Special Publication 27.
Sposito, G. (2000). Ion exchange phenomena. In M. E. Sumner (Ed.), *Handbook of soil science* (pp. B-241–B-264). Boca Raton, FL: CRC Press.
Sumner, M. E., & Miller, W. P. (1996). Cation exchange capacity and exchange coefficients. In D. L. Sparks (Ed.), *Methods of soil analysis. Part 3: Chemical methods* (pp. 1201–1229). Madison, WI: Soil Science Society of America.
Syers, J. K. (1998). Soil and plant potassium in agriculture. *Proceedings No. 411, The International Fertiliser Society, York, UK*. 32 p.
Tiessen, H., & Moir, J. O. (1993). Characterisation of available P by sequential extraction. In M. R. Carter (Ed.), *Soil sampling and methods of analysis* (pp. 75–86). London: Lewis Publishers.
Tiessen, H., Stewart, J. W. B., & Cole, C. V. (1984). Pathways of phosphorus transformations in soils of differing pedogenesis. *Soil Science Society of America Journal, 48*, 853–858.
Tóth, G., Guicharnaud, R. A., Tóth, B., & Hermann, T. (2014). Phosphorus levels in croplands of the European Union with implications for P fertilizer use. *European Journal of Agronomy, 55*, 52–52.
Whalen, J. K., Bottomley, P. J., & Myrold, D. D. (2001). Short-term nitrogen transformations in bulk and root-associated soils under ryegrass. *Soil Biology and Biochemistry, 33*(14), 1937–1945.

Part IV
Molecular Mechanisms in Plant Nutrition

Chapter 17
Genetic Engineering and Molecular Strategies for Nutrient Manipulation in Plants

Claudio Inostroza-Blancheteau, Felipe Aquea, Felipe Moraga, Cristian Ibañez, Zed Rengel, and Marjorie Reyes-Díaz

Abstract Plants require an adequate balance of mineral nutrients in each stage of development to achieve maximum yield. Deficiencies of mineral nutrients are common in crops worldwide. To solve this problem in modern agriculture, the fertilizer applications are necessary, but this practice may be associated with undesirable environmental impacts as well as the high cost of fertilizers. However, improving nutrient use efficiency (NUE) via genetic manipulation may result in increased plant capacity to capture and utilize nutrients. In this chapter, we presented the advances made through genetic engineering and molecular strategies in a range of plant species aimed at enhancing uptake, translocation, and remobilization of nutrients as a sustainable way to increase crop productivity and quality.

C. Inostroza-Blancheteau (✉)
Núcleo de Investigación en Producción Alimentaria (NIPA-UCT), Escuela de Agronomía, Facultad de Recursos Naturales, Universidad Católica de Temuco, P.O. Box 56-D, Temuco, Chile
e-mail: claudio.inostroza@uct.cl

F. Aquea • F. Moraga
Laboratorio de Bioingeniería, Facultad de Ingeniería y Ciencias, Universidad Adolfo Ibáñez, Santiago, Chile

C. Ibañez
Laboratorio de Biología Molecular y Bioquímica, Departamento de Biología, Universidad de La Serena, P.O. Box 554, La Serena, Chile

Z. Rengel
Soil Science and Plant Nutrition, School of Earth and Environment, The University of Western Australia, Crawley, WA 6009, Australia

M. Reyes-Díaz
Departamento de Ciencias Químicas y Recursos Naturales, Facultad de Ingeniería y Ciencias, Universidad de La Frontera, P.O. Box 54-D, Temuco, Chile

Center of Plant, Soil Interaction and Natural Resources Biotechnology, Scientific and Technological Bioresource Nucleus (BIOREN), Universidad de La Frontera, Temuco, Chile

M. Naeem et al. (eds.), *Essential Plant Nutrients*,
DOI 10.1007/978-3-319-58841-4_17

Keywords Genetic manipulation • Nutrient use efficiency • Molecular strategies • Fertilizers

17.1 Introduction

Nutrients are essential for functioning of all living organisms. An adequate balance of nutrients is needed for each stage of development and to achieve maximum yield potential (López-Arredondo et al. 2013). Each plant species and genotype within species requires different amounts of nutrients for their biological processes depending on how nutrients are used and remobilized within the plant (Rengel 2001; Rengel and Damon 2008; Shukla et al. 2014). The availability of mineral nutrients in arable soils is relatively low in general and depends on the soil properties and seasonal variation (Mitra 2015) as well as plant species (genotype) characteristics (Rengel and Marschner 2005).

Macronutrients such as nitrogen (N), phosphorus (P), and potassium (K) are supplied regularly as fertilizers in most cropping systems (Edmeades 2003), whereas micronutrients such as iron (Fe), zinc (Zn), and manganese (Mn) are added in specific situations (Fageria et al. 2002), particularly when biofortification of edible plant parts is required (Velu et al. 2014; Saha et al. 2015). However, for all nutrients it is necessary to ensure sufficient supply in the soil solution for maintaining the productivity in a particular agricultural system regarding the quantity and the quality of agricultural produce. Hence, fertilizer applications are crucial, even though fertilization increases the cost of production and may result in negative environmental impacts (Hawkesford 2012, 2014).

There is currently an intensive world-wide research effort focused on improving nutrient use efficiency to at least partly offset the economic and environmental costs of fertilization. Numerous reports indicated that nutrient uptake, transport, and redistribution in plants are regulated genetically (e.g., Orsel et al. 2002; Hammond et al. 2004; Ning et al. 2015). Hence, conventional breeding as well as transgenic approaches can be used to genetically modify plants to improve uptake, translocation, utilization, and remobilization of mineral nutrients. In particular, genetic engineering is a power tool that can be used in several ways, for example, introgression of novel genes, overexpression or silencing of genes to alter abundance and/or activity of specific proteins, manipulation of regulatory pathways involved in nutrient uptake and utilization, and disruption of inhibitors of nutrient absorption/utilization (e.g., Lonnerdal 2003; Johnson et al. 2011; Abhishek et al. 2015). Therefore, in this chapter, we presented the advances made through genetic engineering and molecular strategies in a range of plant species aimed at enhancing uptake, translocation, and remobilization of nutrients as a sustainable way to increase crop productivity and quality.

17.2 Genetic Approaches for Improving Macronutrient Uptake

Macronutrients are required in large amounts and their use in fertilization is required regularly. Hence, there is strong interest in improving transport of macronutrients from soil into crop and pasture plants. Some of these molecular strategies are specified for each macronutrient, specie, target gene, and improved result (see Table 17.1).

17.2.1 Nitrogen

Nitrogen is one of the main factors that limit plant growth, yield, and quality (Correia et al. 2005; Riquelme et al. 2007). Plants acquire N from the soil solution mainly as nitrate (NO_3^-) and ammonium (NH_4^+), with NO_3^- being the predominant form (Larcher 2003; Marschner 2012). It is estimated that one third of the applied nitrogen is not taken up by crops due to NO_3^- leaching (Podgornik and Pintar 2007).

Plants can take up NH_4^+ or NO_3^- from soil through root-specific transporters (Masclaux-Daubresse et al. 2010). Two gene families of nitrate transports (NRT1 and NRT2) encoding NO_3^- uptake proteins have been identified in plants (Marschner 2012). The NRT1 gene family mediates the root low-affinity transporters (LATs), and NRT2 gene family encodes the high-affinity transporters (HATs) when the external nitrate concentration is low (Williams and Miller 2001). The NRT1 family includes genes that are either induced or constitutively expressed, whereas NRT2 genes are inducible in response to NO_3^- (Miller et al. 2007). So far, five different nitrate transporters have been identified: NRT2, the chloride channel family (CLC), slow anion channel-associated homologues (SLAC/SLAH), aluminum-activated malate transporter 4 (ALMT4), and nitrate transporter 1/peptide transporter family (NPF), with 35 genes well characterized and remaining 38 still needing detailed characterization (Krapp et al. 2014; Léran et al. 2014).

Arabidopsis NO_3^- uptake mutant, *chlorate resistant 1* (*CHL1*), was first designated as the NRT1.1 transporter (Tsay et al. 1993). This transporter has dual affinity (Wang et al. 1998; Liu et al. 1999). Most of NRT1/PTR transporters, with the exception of *AtNRT1.1*, have been reported as proton-coupled and low-affinity nitrate transporters (Chiu et al. 2004; Almagro et al. 2008; Wang and Tsay 2011). Huang et al. (1999) reported that *NRT1.2* was constitutively expressed and located in root epidermis of *Arabidopsis*, whereas *NRT1.4* was expressed in *Arabidopsis* petioles, with a mutation of this gene resulting in a reduced petiole nitrate content and altered leaf development, indicating a critical role of *AtNRT1.4* in regulating leaf nitrate homeostasis (Chiu et al. 2004; David et al. 2014). Similarly, in rice (*Oryza sativa*) *OsNRT1* was constitutively expressed in the root epidermis and root hairs as a low-affinity nitrate transporter (Lin et al. 2000).

Table 17.1 Genetic strategies for improving macronutrient uptake in some plant species

Macronutrients	Species	Gene target	Results	References
Nitrogen	*Arabidopsis*	*ZmAMT1;1a* *ZmAMT1;3*	Ectopic expression of both genes in *Arabidopsis* mutant (ammonium uptake-defective) roots conferred high-affinity ammonium uptake	Gu et al. (2013)
	Arabidopsis	*AtCLCa*	GFP fluorescence showed specific localization of *AtCLCa* as a NO_3^-/H^+ exchanger in vacuole	De Angeli et al. (2006)
	Oocytes of *Xenopus laevis*	*NRT2.4*	Under nutrient-deficient conditions, this nitrate transporter gene functions as high-affinity transporter	Kiba et al. (2012)
	Arabidopsis	*NRT2.5*	Reduction of *NRT2.5* expression decreased high-affinity nitrate uptake without impacting low-affinity uptake. Growth analyses of multiple mutants revealed that *NRT2.5* is required for ensuring the efficient nitrate uptake of nitrogen-starved adult plants	Lezhneva et al. (2014)
	Oocytes of *Xenopus laevis*	*NRT1.5*	Functional analysis of cRNA-injected *Xenopus laevis* oocytes showed that *NRT1.5* is a low-affinity nitrate transporter and pH dependent	Lin et al. (2008)
Phosphorous	*Arabidopsis*	*Pht1;4*	Its expression in lateral roots provokes phosphate absorption and translocation during P-limiting	Misson et al. (2004)
	Arabidopsis	*Pht1;9*	*Pht1;9* encodes a functional plasma membrane transporter that mediates high-affinity Pi/H^+ symport in yeast and is induced in *Arabidopsis* roots under P starvation	Remy et al. (2012)
	Arabidopsis	*OsPTF1*	Overexpression of *OsPTF1* enhanced tolerance to Pi starvation in transgenic rice plants	Yi et al. (2005)
	Arabidopsis	*AtPHT1;8* *AtPHT1;9*	*AtPHT1;8* and *AtPHT1;9* are involved in the translocation of Pi from the root to the shoot as demonstrated by the use of knockout mutants	Lapis-Gaza et al. (2014)
	Arabidopsis	*PHT5*	*PHT5* overexpression leads to massive Pi sequestration into vacuoles, altering regulation of Pi starvation-responsive genes	Liu et al. (2016)
Potassium	*Arabidopsis*	*NHX1* *NHX2*	These genes are involved for active K^+ uptake at the tonoplast. *nhx1 nhx2* mutants showed similar sensitivity to salinity stress	Barragán et al. (2012)
	Arabidopsis	*AtHAK5* *AKT1*	*athak5* mutant plants exhibited growth defects at 10 μM K^+, but at K^+ concentrations above 20 μM, *athak5* mutants were visibly indistinguishable from the wild type	Pyo et al. (2010)
	Arabidopsis	*NHX1* *NHX2*	The roles of *NHX1* and *NHX2* in regulating intravacuolar K^+ and pH are essential to cell expansion and flower development as demonstrated in double knockout *nhx1 nhx2* of *Arabidopsis*	Bassil et al. (2011)
	Arabidopsis	*AtHAK5*	AtHAK5 promoter-β-glucuronidase and green fluorescent protein fusions showed *AtHAK5* promoter activity in the epidermis and vasculature of K^+-deprived roots. The results demonstrate an in vivo function for *AtHAK5* in the inducible high-affinity K^+ uptake system in *Arabidopsis* roots	Gierth et al. (2005)
	Arabidopsis	*TPK1*	Characterization of *TPK1* using electrophysiological, reverse genetics, and homologous expression approaches provides evidence that it encodes the previously characterized VK	Gobert et al. (2007)

Macronutrients	Species	Gene target	Results	References
Calcium	Carrot	*AtCAX1*	Carrots transgenic with AtCAX1 genes increase calcium concentration.	Connoly (2008)
	Tomato	*AtCAX4*	Tomatoes transgenic prolonged its useful life, suggesting that AtCAX4 can serve to obtain a better condition of the fruit.	Park et al. (2005)
Magnesium	Rice	*OsMGT1*	A short-term uptake experiment with stable isotope Mg showed that knockout of *OsMGT1* resulted in decreased Mg uptake, suggesting an important role of this Mg transporter gene	Chen et al. (2012)
	Nicotiana benthamiana	*AtMGT1*	Overexpression of *ATMGT1* encoding a Mg transport protein can improve Al tolerance in plants	Deng et al. (2006)
	Saccharomyces cerevisiae	*AtMRS2-11*	Overexpression of *AtMRS2-11* gene increases accumulation of protein, but the plants showed no detectable phenotype, suggesting this gene is involved in Mg transport into chloroplasts/ plastids	Drummond et al. (2006)
	Arabidopsis	*MRS2-7*	Studies of ectopic overexpression of *MRS2-7* result in complementation and increased biomass accumulation confirming its role in Mg absorption	Gebert et al. (2009)
	Arabidopsis	*MGT2* *MGT3*	Knockout of *MGT2*and *MGT3* genes, these vacuolar Mg transporters decreased the osmotic potential in their tissues rose and growth in plant	Gilliham et al. (2011)
Sulfur	Oilseed rape	*BnSULTR4;1* *BnSULTR4;2*	High expression tonoplast transporters, *BnSULTR4;1* and *BnSULTR4;2*, would be involved in S remobilization in leaves	Dubousset et al. (2009)
	Arabidopsis	*PHR1*	*phr1* mutant showed that *SULTR1;3* was upregulated following phosphate deficiency and regulated by PHR1 transcription factor, suggesting an up- or downregulation on the expression of sulfate transporters	Rouached et al. (2011)
	Tobacco	*miR395*	Overexpression of *OsamiR395h* impaired its sulfate homeostasis and sulfate distribution in leaves. *NtaSULTR2* was identified to be the target of miR395 in tobacco which belongs to low-affinity sulfate transporter group. Both miR395 and NtaSULTR2 respond to sulfate starvation in tobacco	Yuan et al. (2016)
	Arabidopsis	*SULTR3*	A dissection of the seed proteome of the *sultr3* mutants revealed protein changes characteristic of a sulfur stress response, supporting a role for these transporters in providing sulfate to the embryo	Zuber et al. (2010)

Various other transporters involved in NO_3^- transport and accumulation such as *NRT1.5* (root-to-shoot transport), *NRT1.6* (seed development), *NAXT1* (NO_3^- efflux), *CLCa*, and *CLCb* (into vacuole) have been reported in *Arabidopsis* (De Angeli et al. 2006; Segonzac et al. 2007; Almagro et al. 2008; Lin et al. 2008; von der Fecht-Bartenbach et al. 2010). Furthermore, the members of NRT2 gene family (HATs), such as *NRT2.1*, *NRT2.2*, and *NRT2.4* are actively involved in nitrate transport at low NO_3^- availability. Under N deprivation, a quadruple mutant of *Arabidopsis* roots, expressing *AtNRT2.1*, *AtNRT2.2*, *AtNRT2.4*, and *AtNRT2.5*, revealed that these four NRT2 transporters have high-affinity nitrate influx activity of approximately 95% of the total HATs under N limitation, with *AtNRT2.1* the major contributor (Lezhneva et al. 2014). The spatio-temporal distribution of these four AtNRT2 transporters is critical for efficient nitrate uptake to maintain growth under N starvation (Kiba et al. 2012; Lezhneva et al. 2014; Kiba and Krapp 2016).

Otherwise, through functional complementation of yeast mutant defective in high-affinity ammonium uptake, the first NH_4^+ transporter genes were identified in yeast and *Arabidopsis* (Marini et al. 1994; Ninnemann et al. 1994). Ammonium transport is mediated by high-affinity transporters of the AMT/MEP/Rh (AMT 1.1, 1.2, 1.3, and 1.5) superfamily (Ludewig et al. 2007), including uptake of NH_4^+ from soil (Nath and Tuteja 2016). It is reported that all AtAMTs have different ammonium affinity levels and transport capacities, indicating their different roles in plants (Kiba and Krapp 2016). The NH_4^+ transporters are preferentially expressed in root hairs, which is consistent with their role in NH_4^+ nutrition (Lauter et al. 1996). In fact, *Arabidopsis* contains six AMT genes, which encode high-affinity ammonium transporters (*AtAMT1;1*, *AtAMT1;2*, *AtAMT1;3*, *AtAMT1;5*, and *AtAMT2;1*) and are expressed in roots, being upregulated under N limitation (Loqué et al. 2006; Yuan et al. 2007; Kiba and Krapp 2016). Studies of ammonium influx in *Arabidopsis* have used triple and quadruple mutants, showing that *AtAMT1;1*, *AtAMT1;2*, and *AtAMT1;3* jointly are responsible for 90% of high-affinity uptake capacity under N starvation, with AtAMT1;5 probably explaining the remaining 10% capacity (Loqué et al. 2006; Yuan et al. 2007; Kiba and Krapp 2016).

In *Arabidopsis*, *AtAMT1;1*, *AtAMT1;3*, and *AtAMT1;5* are mainly expressed at the root tip and in epidermal cells, absorbing ammonium directly from soil, whereas *AtAMT1;2* is localized in the endodermis and cortex, transporting apoplastic ammonium into the cell symplast (Loqué et al. 2006; Yuan et al. 2007). Furthermore, rice and poplar contain 10 and 14 putative AMT genes, respectively (Sonoda et al. 2004; Couturier et al. 2007), suggesting a coordinated regulation of AMT genes depending on plant species and habitat (Kiba and Krapp 2016). The AMT genes in rice, tomato, and maize grown under N limitation were inducible by ammonium (Sonoda et al. 2004; Gu et al. 2013).

Through research currently performed with molecular strategies, it is possible to suggest that to enhance NO_3^- and NH_4^+ uptake by genetically manipulating genes of the ANR1 and AMT pathways is key to better N nutrition because these cellspecific transporter genes are highly regulated. Creating plants with better ability to absorb NO_3^- and NH_4^+, depending on nitrogen availability, will be a useful tool for farmers. On the other hand, such plants may lead to lesser consumption of nitrogenous fertilizers as well as reducing NO_3^- leaching and thus contamination of ground water.

17.2.2 Phosphorus (P)

Phosphorus (P) is an important nutrient required for plant growth (Richardson et al. 2009), making up 0.2–0.5% of plant dry weight (Vance et al. 2003). The P plays an important role in photosynthesis, respiration, storage and energy transfer reactions, cell division, development of reproductive structures, crop maturity, root growth, protein synthesis, and other processes (Byrne et al. 2011; Yao et al. 2011). In soils, phosphorus is often the most limiting plant nutrient. And soils contain a significant amount of total P, but it has low availability due to high fixation in soils (Shen et al. 2011). The total soil P content can be in the order of 200–3000 mg P kg^{-1} soil (Richardson et al. 2009). Organic P generally constitutes the majority of P in soils; however, it is unavailable for plant uptake (Wang et al. 2004; Richardson et al. 2009). Instead, plants require soluble orthophosphate (Pi) as their main source of P, which represents <1% of the total P pool in soils (Shen et al. 2011). Plants are only able to absorb P from the soil solution as orthophosphate anions (HPO_4^{2-} and H_2PO_4); however, this P form is extremely reactive and can be immobilized by precipitations with cations such as Ca^{2+}, Mg^{2+}, Fe^{3+}, and Al^{3+}, reducing its availability to plants (Gyaneshwar et al. 2002).

Plants have low- and high-affinity plasma membrane Pi transporters, being important for orthophosphate uptake into roots (Remy et al. 2012). Four Pi transporter gene families [phosphate transporter 1 (PHT1), PHT2, PHT3, and PHT4] have been identified in *A. thaliana*, including their respective orthologs in other plant species (López-Arredondo et al. 2013). The members of PHT1 family are localized in the plasma membranes and are responsible for the external Pi acquisition and/or Pi translocation between cells or tissues (Nussaume et al. 2011), whereas the members of PHT2, PHT3, and PHT4 families belong to Pi transporters in mitochondria, plastids, or Golgi and are involved in energy metabolism and stress responses (Versaw and Harrison 2002; Irigoyen et al. 2011; Zhu et al. 2012; Hassler et al. 2012). In the *Arabidopsis* genome, these Pi transporter gene family seems to be at least three members (*AtPT1*, *AtPT2*, and *AtPT4*), mapping to a specific region of chromosome 5 (Smith et al. 1997; Lu et al. 1997). Interestingly, some of these Pi transporter genes such as *StPT1* and *StPT2* are strongly upregulated in response to Pi starvation, while for others the expression is constitutive (Leggewie et al. 1997).

Phosphate transporters are critical for orthophosphate distribution throughout the plant, including remobilization between source and sink tissues. A phosphate transporter essential for orthophosphate transfer to the shoot is the phosphate efflux transporter (PHO1) (Dong et al. 1998). It is a main participant in orthophosphate homeostasis regulation and may underpin strategies for optimizing orthophosphate distribution in plants (Arpat et al. 2012).

In *Arabidopsis* nine transporter genes have been cloned encoding high-affinity Pi carriers expressed mainly in root epidermal cells and highly responsive to Pi deficiency (López-Arredondo et al. 2013). The contribution of these transporters to Pi acquisition has been determined with null mutants and gene overexpression in *Arabidopsis* (Lin et al. 2009a) and rice (Yi et al. 2005). In *Lycopersicon esculentum*, the expression of two *LePT1* and their protein localization also support their involvement in Pi acquisition at the root-soil interface (Rausch and Bucher 2002).

All cloned genes belong to the pht1 family of plant P transporters, which is highly conserved (Rausch and Bucher 2002). The *Pht1* genes code for the plasma membrane proteins; even though these genes are expressed in different organs including root and shoot tissues and reproductive organs, the highest expression is commonly found in root hairs (Mudge et al. 2002). The *Pht1* genes are involved in Pi uptake from soil, Pi translocation among plant tissues, and Pi remobilization from senescent organs (Fan et al. 2013). Homologous *Pht1* genes have been identified in several species, and they share conserved functions in Pi uptake and Pi affinity (Ai et al. 2009). In *Arabidopsis*, PHT1 members (*AtPHT1;1*, *AtPHT1;4*, *AtPHT1;5*, *AtPHT1;8*, and *AtPHT1;9*) have been functionally characterized using heterologous expression systems, mutant lines, and/or transgenic plants (Muchhal et al. 1996; Misson et al. 2004: Shin et al. 2004; Nagarajan et al. 2011; Remy et al. 2012; Lapis-Gaza et al. 2014; Gu et al. 2016). Using microRNA (miRNA) silencing to create single and multiple mutants in *Arabidopsis*, Ayadi et al. (2015) demonstrated that PHT1;1, PHT1;2, and PHT1;3 transporters are involved in P uptake. On the other hand, in rice, Yang et al. (2012) found that *OsPHT1;11* and *1;13* are required for the development of the symbiosis between AM fungi and plant roots, similarly *PHT1;4* in *Medicago truncatula* (Javot et al. 2007). Secco et al. (2013) showed that *OsPHT1;3* is one of the most strongly induced genes in Pi-starved rice plants, whereas *OsPHT1;5* showed higher expression in shoots than roots during vegetative stage.

The study of members of PHT2, PHT3, and PHT4 families has received less attention than PHT1 family. The *PHT2;1*, localized in chloroplast, is responsible for Pi/H^+ symport into the chloroplast (Zhao et al. 2003), affecting Pi translocation in leaves (Versaw and Harrison 2002). In *Arabidopsis*, *Pht2;1* expression does not respond to P deficiency (Daram et al. 1999; Versaw and Harrison 2002). Using a Pht2;1::GFP fusion protein, this Pi transporter was localized to the chloroplast, varying its expression during the photoperiod, suggesting a role in photosynthesis (Versaw and Harrison 2002).

Despite PHT3-mediated high-affinity Pi transport in mitochondria being crucial for plant respiration under prolonged Pi starvation (Gu et al. 2016), the physiological and molecular roles of PHT3s are rarely studied and thus poorly understood. One report shows that PHT3s affect salt susceptibility in *Arabidopsis*, regulating ATP accumulation (Zhu et al. 2012).

The PHT4s are involved in a wide range of biological processes, such as maintenance of Pi homeostasis, carbon metabolism, pathogen resistance, and tolerance to salt and high light intensity (Cubero et al. 2009; Irigoyen et al. 2011; Wang et al. 2011; Hassler et al. 2012; Karlsson et al. 2015; Miyaji et al. 2015). Hassler et al. (2012) demonstrated that *Arabidopsis* mutants with loss of function exhibited dwarf growth and a lack of protein glycosylation and cell wall hemicellulose synthesis. These features typically are associated with Golgi functions. Although the total phosphate content in *pht4;6* mutant plants was similar to wild type, the mutants exhibited clear symptoms of cellular Pi deficiency (Hassler et al. 2012). The *PHT4;6* is also critical for senescence-associated processes in *Arabidopsis* (Hassler et al. 2016).

After examining the nomenclature of PHT1-PHT4 Pi transporters in *Arabidopsis*, another phosphate transporter family (PHT5) was identified that includes SYG1/PHO81/XPR1-Major Facilitator Superfamily (SPX-MFS) proteins associated with vacuolar transport (Liu et al. 2016). Wang et al. (2015) showed that rice OsSPX-MFS3 is responsible for the vacuolar Pi export; however, the results of Liu et al. (2016) suggested that its paralogue (OsSPX-MFS1) and orthologs (*Arabidopsis* PHT5 proteins) contribute to Pi import into the vacuoles. The identification of both Pi influx and efflux vacuolar systems in the same plant species could offer clear and insightful information on how Pi is translocated in and out of the vacuoles for metabolic needs and during Pi remobilization.

Several studies have been postulated that phosphatases may participate in Pi transport, given their capacity to bind Pi (Lee 1988; Duff et al. 1991). Under P starvation, intracellular phosphatases may play a role in phosphate recycling (Smyth and Chevalier 1984). Phosphate starvation-induced acid phosphatase (*AtACP5*) has been involved in phosphate mobilization in *Arabidopsis* (Poirier and Bucher 2002). In tomato, a novel phosphate starvation-induced gene (*LePS2*) coding for an acid phosphatase has been characterized (Baldwin et al. 2008). In *Cucumis melo*, Fita et al. (2012) reported the expression of nine Pi starvation-responsive genes as well as their differential response among nine diverse accessions. The accessions with higher phosphorus use efficiency (PUE) also had higher expression of Cm-PAP10.1 and Cm-PAP10.2 (purple acid phosphatases) and Cm-RNS1 (ribonuclease, RNS), whereas the accessions with lower PUE had low expression of these genes. Hence, higher mobilization and remobilization of P may be a preferential source of diversity among melon accessions regarding tolerance to Pi starvation.

The knowledge and tools related to Pi transporters may ultimately contribute to a reduction in the use of P fertilizers. In addition, the use of cultivars with improved uptake and/or use efficiency of P is crucial for enhancing the effectiveness of P fertilization.

17.2.3 Potassium (K)

The potassium ion (K^+) is the most important and abundant cation in the cytoplasm and chloroplast (Wang and Wu 2015). The K^+ in plant cells constitutes 2–10% of plant dry weight, with cytoplasmic K^+ concentration at approximately 100 mM as the optimal K^+ concentration for cytoplasmic enzyme activities (Leigh and Wyn Jones 1984). In contrast, the typical K^+ concentration at the surfaces of roots in soil varies from 0.1 to 1 mM (Maathuis 2009). Therefore, plant root cells absorb K^+ from the soil against K^+ concentration gradient, a process conducted by K^+ transporters and channels (Wang and Wu 2015). Thus, early researches point out that K^+ uptake in higher plants is accomplished by high- and low-affinity mechanisms that alternate depending on the external K^+ supply (Epstein et al. 1963). At low external K^+

concentration (below 0.2 mM), the high-affinity K^+ transporters are induced (Maathuis and Sanders 1994), whereas at high external K^+ concentration (above 0.3 mM), the low-affinity K^+ channels operate (Coskun et al. 2013). To date, a large number of genes encoding K^+ transporters and channels have been cloned and identified in many higher-plant species (Mäser et al. 2001; Véry and Sentenac 2003; Ward et al. 2009; Li et al. 2016a). These genes belong to three families that encode K^+ channel proteins such as Shaker, TPK (tandem-pore K^+), and Kir (K^+ inward rectifier) (Véry and Sentenac 2003; Gambale and Uozumi 2006; Lebaudy et al. 2007) or K^+ transporter families, including K^+ uptake permeases/high-affinity K^+ transporters/ K^+ transporter (KUP/HAK/KT), high-affinity K^+ transporter (HKT), Na^+/H^+ exchanger (NHX), and cation/H^+ exchangers (CHX) (Gierth and Mäser 2007).

Plant Shaker channels show notable voltage sensitivity and K^+ selectivity. These channels are divided into three categories based on their voltage dependence (1) inward-rectifying channels activated by hyperpolarization potential and mainly mediating K^+ uptake, (2) outward-rectifying channels activated by depolarization and functioning in K^+ efflux, and (3) weakly rectifying channels activated by hyperpolarization and mediating both K^+ uptake and K^+ release depending on the membrane potential (Wang and Wu 2015). Plant TPK channels and Kir-like channels appear not to have a voltage sensor, having weak sensitivity to membrane potential (Lebaudy et al. 2007).

Transporters encoded by the KUP/HAK/KT family have been verified as performing K^+ transport activities in plant cells (Rigas et al. 2001; Elumalai et al. 2002; Gierth et al. 2005). Most studies in plants have indicated that HKT transporters function as Na^+ transporters (Gierth and Mäser 2007), and only a few studies suggested that these transporters are Na^+/K^+ symporters (Rubio et al. 1995; Gassmann et al. 1996). Similarly, the most members of the NHX and CHX families have been identified as Na^+/H^+ antiporters, and a few are K^+/H^+ antiporters (Cellier et al. 2004; Song et al. 2004; Padmanaban et al. 2007; Zhao et al. 2008; Bassil et al. 2011; Barragán et al. 2012).

It is reported that Shaker K^+ channel AKT1 and KUP/HAK/KT transporter HAK5 of *Arabidopsis* have been expressed primarily in roots and function in K^+ uptake from the external environment, indicating that these two transport proteins mediate almost all K^+ absorption in *Arabidopsis* roots (Lagarde et al. 1996; Hirsch et al. 1998; Gierth et al. 2005; Pyo et al. 2010). In *Arabidopsis* and barley, there is Shaker channel for K^+ translocation and secretion from root cortex cells into the xylem, which is an outward-rectifying channel (Wegner and Raschke 1994; Gaymard et al. 1998). The K^+ loading and unloading in phloem tissues of *Arabidopsis* plants are facilitated mainly by the weakly rectifying K^+ channels, such as AKT2 (Marten et al. 1999; Lacombe et al. 2000).

The stomatal guard cells and pollen cells are two special types of cells, and they possess unique K^+ transport, with both inward and outward K^+ channel components (Holdaway-Clarke and Hepler 2003; Pandey et al. 2007; Sirichandra et al. 2009). They were initially identified in stomatal guard cells, mediating K^+ uptake and release (Schroeder et al. 1984; Hosoi et al. 1988). Functional expression of K^+ channel from *A. thaliana* in *Saccharomyces cerevisiae* showed that the inward K^+ channels KAT1 and KAT2 control the K^+ influx across the plasma membrane during stomatal opening (Anderson et al. 1992; Kwak et al. 2001). Hosy et al. (2003) reported that the GORK, an outward K^+ channel, conducts the K^+ efflux during stomatal closure in *A. thaliana*. The pollen-specific Shaker channel SPIK mediates K^+ influx across the plasma membrane in pollen and pollen tube cells, which regulates its growth and development (Mouline et al. 2002). The TPK1 was functionally characterized, showing functions in K^+ transport across the vacuolar membrane, and plays a role in intracellular K^+ homeostasis (Gobert et al. 2007), whereas the physiological function of TPK2, TPK3, and TPK5, targeted to the tonoplast, remains unknown.

17.2.4 Calcium (Ca)

Calcium is an important constituent of the cell walls (Lamport and Várnai 2013) as well as a crucial intracellular messenger (White and Broadley 2003; Dodd et al. 2010). Calcium deficiency is rare in nature, but if plants are deficient in Ca, mechanical damage provoked by strong winds, insects, animals, or fungi infections may be exacerbated due to weakened cell walls (Dodd et al. 2010). Ca is not a mobile cation in plants, meaning it cannot be translocated via phloem. Hence, young leaves cannot benefit from remobilization and instead have to obtain Ca from soil via the xylem-mediated transpiration pathway (Baas et al. 2003; Gilliham et al. 2011). However, transpiration rates are low in new tissues (especially in the apices), and therefore, calcium deficiency in these new tissues might be critical for the future plant development.

Identification of plasma membranes Ca^{2+} channel genes as well as non-specific cation transporter genes in root cells allowed improvements in calcium absorption by genetic manipulation. For instance, overexpression of *Arabidopsis* H^+/cation exchanger (*CAX*) genes in *Solanum lycopersicon* cells resulted in plants with more root biomass and nearly 20% increase in Ca^{2+} concentration through the entire plant. In addition, transgenic tomatoes had fruits with a prolonged shelf life, suggesting that enhanced *CAX4* expression can serve to biofortify fruits (Park et al. 2005). Using a different approach (combination of tilling and comparative genomics) in an inbred mapping population of *Brassica rapa*, Graham et al. (2014) enhanced the knowledge of calcium homeostasis. Using two quantitative trait loci (eQTLs) previously identified to respond to altered Ca supply in *B. rapa* (O'Lochlainn et al. 2011), Graham et al. (2014) screened the *Arabidopsis thaliana* leaf ionomic database to identify putative genes associated with these eQTLs. They identified genes coding for four potential Ca^{2+} transporters (*ACA8*, *ECA4*, *CAX1*, and *CAX3*) and one cation efflux protein (*MTP5*) involved in Ca homeostasis in plants. Further characterization of *B. rapa cax1* mutants showed a phenotype with a paler/yellow leaves with chlorophyll content significantly lower than in the WT plants (Graham et al. 2014). Although the relationship between impaired chlorophyll biosynthesis and leaf Ca homeostasis remains to be fully understood, this research demonstrated that manipulating Ca transporters might be useful in biofortifying edible crops. Moreover, CAX proteins play a role in avoiding salt toxicity in halophytes, as well as transporting cadmium in metal hyperaccumulator plants or being pH regulators in others, demonstrating the versatility of these cation/proton exchangers in enhancing abiotic stress tolerance in plants (Pittman and Hirschi 2016).

To reduce the negative effects of Ca deficiency, horticulturists and florists in the 1990s realized that manipulating the boron (B) concentration in the soil might affect development of the young root tips, which are the only root part capable of taking up calcium (Clarkson 1985). Ganmore-Neumann and Davidov (1993) and Shams et al. (2010) demonstrated in rose plantlets that a relatively high concentration of Ca (as well as B) in the nutrient solution improved significantly the young root development. However, in *Solanum tuberosum* (Abdulnour et al. 2000) and *Brassica napus* (Asad et al. 1997) grown in the nutrient solution with high B concentration, Ca absorption was hampered, demonstrating a narrow range of B supply between deficiency and toxicity, especially in plants adapted to arid conditions (Nable et al. 1997).

Several studies have shown that sufficient Ca supply may enhance plant performance during environmental stresses such as heat (Knight 1999), drought (Weng et al. 2012; Wang et al. 2014), and salinity (Bose et al. 2015; Pehlivan et al. 2016) by activation and/or regulation of antioxidant pathways (Bose et al. 2014) that contributes to stabilizing and extending plant cell life. Using *Arabidopsis* mutants hypersensitive to NaCl but not to other salts such as lithium and cesium chlorides, Guan et al. (2013) discovered that a calcium sensor protein localized in the cell nucleus (RSA1) interacted with a bHLH transcription factor called RITF1, with RSA1-RITF1 regulating transcription of several genes involved in detoxification of reactive oxygen species produced by the salt stress. Interestingly, these two proteins also regulate the expression of the plasma membrane-localized Na^+/H^+ antiporter (*SOS1*) gene, which together with *SOS2* and *SOS3* is part of the salt overly sensitive (SOS) pathway involved in the ion homeostasis in plants that includes efflux and influx regulation of Ca, Na, and H cations through the plasma and vacuolar membranes (Guo et al. 2009; Guan et al. 2013).

Our database search of RSA1 homologs revealed that except grapes (*Vitis vinifera*), many monocot and dicot crops have a single copy of the *RSA1* gene. These results suggest that relatively simple manipulation of a calcium sensor protein might be a useful approach to enhancing the salt stress tolerance in plants.

17.2.5 Magnesium (Mg)

Since the Green Revolution started in the 1960s, an emphasis on high-input management techniques has provided new high-yielding crop cultivars that have reduced famine and improved living standards for the world population (Borlaug 1983). However, modern fertilization practices have changed the equilibrium of minerals in soils due to the utilization of excessive amounts of fertilizers that contain nitrogen, phosphorus, and K (NPK) minerals. As a consequence, magnesium (Mg) has been one of the most affected due to competition with K and Ca, reducing the availability of Mg in soil (Poschenrieder et al. 1995).

Magnesium is an essential nutrient involved in cell development (Castiglioni et al. 2013), stabilization of macromolecules (Sreedhara and Cowan 2002; Politi et al. 2010), enzymatic activities (Hermans et al. 2010), sucrose phloem loading (Hermans and Verbruggen 2005), and chlorophyll formation (Fiedor et al. 2008) and is necessary for maintaining the homeostasis of reactive oxygen species (ROS) under aluminum stress (Bose et al. 2013), among other important roles (Hermans and Verbruggen 2005; Cakmak and Yazici 2010; Rengel et al. 2015; Guo et al. 2016). Magnesium deficiency can affect activity of crucial enzymes and proteins involved in cell division, photosynthetic carbon fixation, metabolism, and transcription of stress-tolerance genes (Chen et al. 2012).

Cereals cultivated in soils with low Mg availability produce grain with depleted Mg content (Murphy et al. 2008; Cakmak 2013; Rosanoff et al. 2013). Magnesium deficiency interferes with sugar export via phloem, reducing growth (e.g., in birch, *Betula pendula*, Ericsson and Kahr 1995). For instance, in maize, Mg deficiency impaired N and carbon assimilations, mainly inhibiting activity of nitrate reductase, sucrose-phosphate synthase, and phosphoenolpyruvate carboxylase, which significantly reduced growth as well as chlorophyll and protein synthesis (Zhao et al. 2012). Interestingly, cerium chloride added to the Mg-deficient solutions promoted N and carbon assimilations in maize, increasing chloroplast activity and improving growth; these findings suggested that cerium can partly substitute for Mg^{2+}, but the mechanisms are unknown.

In *Arabidopsis*, *Mitochondrial RNA Splicing 2/Magnesium Transporter (MRS2/MGT)* gene family has been characterized as Mg transporters (Gregan et al. 2001; Kolisek et al. 2003). This family contains 10 gene members, all characterized in *Arabidopsis* (6 homolog genes were identified in rice; Schock et al. 2000; Li et al. 2001). The Mg transporter genes are classified in terms of their high-, medium-, and low-affinity for Mg^{2+} (Cong et al. 2012). All *MRS2/MGT* genes showed complementary functioning and high homology in the conserved GMN motif, but were highly diverse outside of this region, characterized as leucine-rich (Bui et al. 1999). Ectopic overexpression of *Arabidopsis MGT1* gene in *Nicotiana benthamiana* made these plants tolerant to Mg^{2+} deficiency and, interestingly, alleviated root inhibition under Al toxicity (Deng et al. 2006; Chen and Ma 2013) because Al^{3+} may compete with Mg^{2+} for active sites on membrane transporters (Pandey et al. 2013).

In *Arabidopsis*, knockout of Mg transporter 7 (*AtMGT7*) gene, which is highly expressed in roots, resulted in reduced growth and impaired Mg uptake, confirming its role in Mg absorption (Gebert et al. 2009). In contrast, when MRS2 family members (*mrs2-1*, *mrs2-5*, *mrs2-7*, *mrs2-10*, *mrs2-5*, *mrs2-1*, and *mrs25 mrs2-10* genes) were knocked out in *Arabidopsis*, membrane transport of Mg^{2+} was not affected (Gebert et al. 2009). However, when Mg concentration in growth media was lowered to 50 μM, some mutants showed decreased growth; an impairment related to membrane stabilization or calcium signaling was considered as possible explanation affecting physiological, biochemical, structural, and molecular aspects.

A non-related member of MRS2/MGT family located in the vascular system, Mg^{2+}/H^{+} exchanger (MHX) gene, was the first Mg^{2+} transporter discovered in plants (Shaul et al. 1999). Its function is related to adjustment of osmotic potential in the phloem cells (David-Assael et al. 2006). Otherwise, ectopic overexpression of *Arabidopsis* MHX (*AtMHX*) gene in tobacco vacuole plants showed that this transporter is efficient in taking up Mg^{2+}, Zn^{2+}, Cd^{2+}, and Fe^{2+} but not Ca^{2+}, Co^{2+}, Ni^{2+}, and Cu^{2+}, demonstrating that *AtMHX* is involved in selective proton homeostasis mediated by the vacuolar H^{+} ATPase (Berezin et al. 2008; Gaash et al. 2013). Likewise, *AtMHX* as well as *AtMGT10* and *AtMRS2-11* genes may be involved in translocation of Mg^{2+} into the chloroplasts in *Arabidopsis* plants (Drummond et al. 2006). In *Arabidopsis*, MRS2-2, MRS2-3, and MRS2-11 are three genes related to accumulation of Mg^{2+} and other minerals in seeds. Their utilization as QTL marker to potentially manipulate mineral accumulation in seeds has been proposed (Waters and Grusak 2008).

Two-pore channel 1 (TPC1), a two-pore Ca^{2+} channel located in the tonoplast, can release vacuolar Mg into the cytoplasm to adjust osmotic potential when Ca^{2+} levels are low (Larisch et al. 2016). When *MGT2* and *MGT3* genes, two vacuolar Mg^{2+} transporters, were knocked out in *Arabidopsis*, the osmotic potential in their tissues rose and plant growth was decreased. Also, chlorophyll content was importantly reduced in the transgenic plants. A hyper-accumulation of K in the vacuoles and impairment capacity to use Mg for remobilization and synthesis of chlorophyll were suggested as plausible explanations for these results (Gilliham et al. 2011).

In summary, despite advances in deciphering the mechanisms involved in uptake and transport of Mg inside the plant and the consequences of a lack of Mg in the physiology and plant adaptation to their environment, many questions still remain unsolved. We expect that advances in genomics, transcriptomics, proteomics, metabolomics, ionomics, and other new molecular approaches will contribute enormously to unraveling the complex and dynamic molecular pathways between plants and the rhizosphere, especially regarding magnesium, a vital element for crop cultivation that has been neglected for too long.

17.2.6 Sulfur (S)

Sulfur is a crucial component of proteins through the amino acids cysteine and methionine and an active constituent of numerous coenzymes and prosthetic groups, iron sulfur centers, coenzyme-A, thiamine, lipoic acid, S-adenosylmethionine, and glutathione, among others. Plant sulfur is also an important component of two major classes of natural products, the glucosinolates in Brassicaceae and alliins in Allium species (Kopriva et al. 2015). Sulfur is assimilated in the inorganic form into cysteine through cysteine biosynthesis pathways (Noji and Saito 2007). The sulfate uptake and assimilation are regulated by the demand for the reduced sulfur, availability of nutrients, and environmental conditions.

Sulfate transport across the internal barriers and plasma membranes is very complex. The transporters can be divided into four main groups: high-affinity SULTRs and low-affinity transporters, the transporters in the plastid and symbiosome membranes, and the vacuolar sulfate exporter (Gigolashvili and Kopriva 2014).

The main high-affinity sulfate transporters in root are SULTR1;1 and SULTR1;2. The SULTR1;1 is necessary during S deprivation (Rouached et al. 2008). Similarly, SULTR1;2 is located in the root epidermal and cortical plasma membrane, consistent with its role in uptake of sulfate from the soil solution (Davidian and Kopriva 2010). Low-affinity transporter genes such as *SULTR2;1* and *SULTR2;2* are expressed in xylem parenchyma cells and are implicated in the long-distance transport (Takahashi et al. 2000).

The SULTR3 family is involved in the translocation of sulfate at different stages of seed development (Zuber et al. 2009). The *sultr3;2 Arabidopsis* mutant showed increased accumulation of sulfate in mature seeds. Moreover, the proportion of sulfate in total sulfur was significantly increased, accompanied by a reduction in free cysteine content. Hence, *sultr3;2* mutant showed different sulfate partitioning in seeds compared with the wild type (Zuber et al. 2010).

The SULTR4 family is likely localized in the tonoplast, with *SULTR4;1* and *SULTR4,2* involved in vacuolar SO_4^{2-} remobilization to the cytosol (Kataoka et al. 2004). In *Brassica napus*, SULTR4 transporters were differentially expressed in leaves (*BnSULTR4;2* more highly than *BnSULTR4;1*) in response to sulfate limitation (Parmar et al. 2007; Dubousset et al. 2009).

In order to increase efficiency of sulfate uptake, translocation, storage, and remobilization, transgenic plants have been studied (Buchner et al. 2004; Davidian and Kopriva 2010). For example, using the *phr1* mutant, it was shown that the transcription factor PHR1 (crucial in regulating translocation of sulfate from shoot to roots under phosphate starvation in *Arabidopsis*) acted to overexpress (upregulate) the *SULTR1;3* gene (Rouached et al. 2011). On the other hand, under sulfate deprivation, the microRNA395 was significantly upregulated. This microRNA targets APS and *SULTR2;1* that are involved in the sulfate metabolism pathway resulting in overaccumulation of sulfate in *Arabidopsis* shoots but not roots (Liang et al. 2010). Recently, Yuan et al. (2016) cloned a microRNA395 gene from *O. sativa*, overexpression of *OsMicroRNA395* in tobacco-induced endogenous *NtaMicroRNA395* (*Nicotiana tabacum* miR395) and *NtaSULTR2* in tobacco leaves under low sulfate, thus altering sulfate homeostasis (Yuan et al. 2016). These significant advances allowed accelerated progress in understanding regulation of sulfate transporters and generally sulfur metabolism in plants.

17.3 Genetic Approach for Improving Micronutrients Uptake

Micronutrients are essential elements for plant growth and play an important role in balanced crop nutrition. Deficiency of any of these elements can reduce growth, even when all other nutrients are present in adequate amounts. Several genetic approaches have been described recently with an aim to improve uptake of micronutrients from soil. Some of these molecular strategies are specified for each micronutrient, specie, target gene and improved result (see Table 17.2).

Table 17.2 Genetic strategies for improving micronutrient uptake in some plant species

Micronutrients	Species	Gene target	Results	References
Boron	*Arabidopsis*	*AtNIP5;1*	Improved root elongation under B-limiting conditions	Kato et al. (2009)
	Tomato	*AtBOR1*	Strong expression of *AtBOR1* improved growth in tomato under B-deficient conditions	Uraguchi et al. (2014)
	Arabidopsis	*AtBOR1*	Improved seed yields compared with wild type under nutrient-deficient conditions	Miwa et al. (2006)
	Tobacco	*S6PDH*	Increase in plant growth and yield when growing with limited or interrupted soil B supply	Brown et al. (1999)
	Arabidopsis	*CmBOR1*	Improved growth of the aerial portions of the transgenic plants grown under deficient conditions	Cañon et al. (2013)
Manganese	*Arabidopsis*	*NRAMP1*	The overexpression provoked a hypertolerance of the plants to Mn deficiency	Cailliatte et al. (2010)
Copper	Tobacco	*TCU-1*	The expression of a heterologous copper transporter in tobacco enhanced acquisition of copper	Singh et al. (2011)
Zinc	Barley	*AtZIP1*	Overexpression increased short-term zinc uptake and seed zinc content after zinc deprivation	Ramesh et al. (2004)
	Rice	*OsIRT1*	Overexpression leads to increased iron and zinc accumulation in rice	Lee and An (2009)
	Arabidopsis	*ZmZIP7*	Zn concentration was elevated in roots and shoots of ZmZIP7-overexpressing plants	Li et al. (2016b)
	Tomato	*AtHMA4*	Higher Zn concentration was observed in leaves of AtHMA4-expressing lines compared to wild type	Kendziorek et al. (2014)
	Rice	*MxIRT1*	Zn concentration in seeds increased threefold in sense lines when compared to the wild type	Tan et al. (2015)

Micronutrients	Species	Gene target	Results	References
Iron	*Arabidopsis*	*FRO2*	Transgenic plants grow better on low iron as compared with wild-type plants	Connolly et al. (2003)
	Tobacco	flavodoxin	Transgenic lines expressing a cyanobacterial Fld in chloroplasts were able to grow in Fe-deficient media that severely compromised survival of WT plants	Tognetti et al. (2007)
	Rice	*OsIRT1*	Overexpression led to increased iron and zinc accumulation in rice	Lee and An (2009)
	Arabidopsis	*ZmZIP7*	Fe concentration was elevated in roots, shoots, and seeds of ZmZIP7-overexpressing plants	Li et al. (2016a, b)
	Rice	*Fer-NAS-NAAT-IDS3*	Transgenic rice showed increased Fe concentration when the plants were cultivated in both commercially supplied soil and calcareous soil	Masuda et al. (2013a, b)
	Apple	*MdbHLH104*	Overexpression increased the tolerance to Fe deficiency in transgenic apple plants and calli	Zhao et al. (2016)
	Cassava	*FEA1*	Fe concentration in mature cassava storage roots was increased from 10 to 36 mg/kg in the highest iron accumulating transgenic lines	Ihemere et al. (2012)
	Cassava	*AtVIT1*	Transgenic plants showed 2–4 times higher values of iron content when compared with wild-type plants	Narayanan et al. (2015)
	Rice	*AhIRT1*	Induced expression of *AhIRT1* in rice plants resulted in high tolerance to low Fe availability in calcareous soils	Xiong et al. (2014)
	Tobacco	*AhIRT1*	Induced expression of *AhIRT1* in tobacco plants resulted in accumulation of Fe in young leaves under Fe-deficient conditions	Xiong et al. (2014)
	Rice	*MxIRT1*	Fe concentration in seeds increased threefold in sense lines when compared to the wild type	Tan et al. (2015)

17.3.1 Boron (B)

Boron (B) is taken up from soil in the form of boric acid. To date, the primary B function is undoubtedly its structural role in the cell wall (Brown et al. 2002). More than 90% of B in plants is found in cell walls, forming borate ester cross-linked with rhamnogalacturonan II dimers, essential for the structure and function of the extracellular matrix (O'Neill et al. 2001). In addition, González-Fontes et al. (2008) suggested a possible role of B as a cellular signal through transcription factors, which could explain why many physiological processes are affected when vascular plants are subjected to B deficiency. Boron is important for root elongation (Kouchi and Kumazawa 1975), leaf expansion (Dell and Huang 1997), viability of pollen grains, and elongation of pollen tubes (Cheng and Rerkasem 1993).

Application of B fertilizer is one approach to alleviate B deficiency in crops (Schon and Blevins 1990), but B excess is also toxic for plants (Nable et al. 1997). A narrow B concentration range exists between deficient and toxic levels for plants, which complicate B fertilizer application (Francois 1984; Gupta et al. 1985; Schon and Blevins 1990).

Two types of proteins are involved in B homeostasis in plants. The first group is related to the major intrinsic protein (MIP) superfamily, which is classified into subfamilies including NOD-26-like intrinsic proteins (NIPs), tonoplast intrinsic proteins (TIPs), and plasma membrane intrinsic proteins (PIPs). In *Arabidopsis*, NIP5;1 facilitates B uptake in epidermal, cortical, and endodermal cells under B deprivation conditions (0.1 μM H_3BO_3). Under low B conditions, *NIP5;1* is upregulated at the mRNA level, and NIP5;1 mutants exhibited severe growth reduction of shoots and roots (Takano et al. 2006). The overexpression of the coding region of *NIP5;1* had no effect on plant growth under low B conditions (Kato et al. 2009). However, inserting the CaMV 35S enhancer into the promoter region of *NIP5;1* enhanced tolerance to low B levels (Kato et al. 2009).

The second group corresponds to the B transporters named BORs, which are supposedly borate transporters. BOR proteins were first identified in the *Arabidopsis* mutant *bor1-1*. This mutant needs high B concentration to complete the growth cycle, showing smaller rosette leaves and inability to produce seeds under low B conditions (3 μM H_3BO_3), unlike its wild-type relative. The *bor1-1* mutant phenotype is reversed in a growth medium supplemented with B (Noguchi et al. 1997). Subsequent studies using molecular markers identified the gene responsible for this phenotype, encoding for a boron transporter (boric acid/borate) known as *AtBOR1*, the first B carrier described in biological systems (Takano et al. 2002). In *Arabidopsis*, the overexpression of *BOR1* enhanced the root-to-shoot translocation of B, especially to the shoot apex, resulted in greater shoot growth, and allowed normal seed production under B-limiting conditions compared with wild-type plants that failed to produce seed (Miwa et al. 2006). In tomato, the overexpression of *Arabidopsis BOR1* resulted in normal leaf development, even under B-deficient conditions, and higher B accumulation in shoots/fruits compared with non-transgenic tomato plants (Uraguchi et al. 2014). Moreover, the expression of *Citrus macrophylla BOR1* (*CmBOR1*) increased tolerance to B deficiency in *A. thaliana* (Cañon et al. 2013). All the results mentioned above suggest that BOR1 is a xylem-loading B transporter.

Another strategy for improving micronutrient uptake is genetic manipulation of nutrient movement and remobilization within plants to enhance plant growth under conditions of variable soil nutrient availability. In the case of B, the manipulation of sorbitol synthesis has an effect on phloem mobility of B. Transgenic tobacco plants expressing the gene *S6PDH* exhibited a marked increase in within-plant B mobility, resulting in an increase in plant growth and yield when the plants grew with limited or interrupted soil B supply (Brown et al. 1999).

17.3.2 Manganese (Mn)

Manganese is an essential metal nutrient required for normal plant growth and development (Rengel 2003). It is of particular importance for most photosynthetic organisms as an indispensable component in the oxygen-evolving complex of the PSII, where a cluster of Mn atoms acts as the catalytic center for light-induced water oxidation. In addition, Mn is required as a cofactor of many enzymes, such as the Mn-dependent superoxide dismutase (MnSOD), a principal antioxidant enzyme in mitochondria (Goussias et al. 2002; Morgan et al. 2008; Holley et al. 2011). Manganese deficiency occurs in plants grown on calcareous or alkaline soils that favor Mn oxidation and immobilization of Mn^{2+}. In addition, excess iron in the culture media can compete with Mn and trigger Mn deficiency. Yellowing of the young leaves in dicotyledonous plants, or development of gray specks on the mature leaves of cereals, denotes Mn deficiency (Marschner 2012). Moreover, the patterning and development of root hairs in *Arabidopsis* is altered by Mn deficiency (Yang et al. 2008).

The transporter family NRAMP/divalent cation transporter 1/divalent metal transporter 1 (DCT1/DMT1) has been shown to play a major role in metal homeostasis in different species (Nevo and Nelson 2006). In *Arabidopsis*, six members of the NRAMP family have been identified, yet only NRAMP1, NRAMP3, and NRAMP4 have been functionally characterized and assigned a biological function. The *nramp1-1* loss-of-function mutant grows poorly, contains less Mn than the wild type, and fails to take up Mn under limited Mn supply, demonstrating that NRAMP1 is the major high-affinity Mn transporter in *Arabidopsis* (Cailliatte et al. 2010). The overexpression of *NRAMP1* provoked hypertolerance to Mn deficiency, as shown by the size of these plants and their Mn content, both of which were higher than in wild-type plants (Cailliatte et al. 2010).

17.3.3 Copper (Cu)

Copper (Cu) is a cofactor of proteins that are involved in electron transfer reactions and is an essential micronutrient for plants. Copper plays a role in photosynthesis, respiration, ethylene perception, reactive oxygen metabolism, and cell wall remodeling (Burkhead et al. 2009). Symptoms of Cu deficiency include decreased growth rate, distortion or whitening (chlorosis) of young leaves, curling of leaf margins, and damage to the apical meristem, as well as an impairment of fruit formation. Copper deficiency in forests severely affects wood production (Ruiter 1969).

Plants take Cu from soil as Cu^{+} by high-affinity transporters belonging to the Cu transporter (Ctr) family (COPT1, COPT2, COPT3, and COPT5) (Grotz and Guerinot 2006; Puig et al. 2007) and as Cu^{2+} by ZIP transporters (ZIP2 and ZIP4) (Grotz et al. 1998; Wintz et al. 2003). With the objective of developing plants with improved Cu acquisition, a high-affinity Cu transporter gene (*tcu-1*) was cloned from fungus *Neurospora crassa* and introduced into a model plant (*Nicotiana tabacum*). All the transgenic lines tested showed enhanced Cu acquisition compared with control plants, and Cu concentration in root and shoot biomass was higher in transgenic lines than wild-type plants. Although the shoots of transgenic plants accumulated higher levels of Cu compared with control plants, roots retained a high proportion of Cu (Singh et al. 2011).

17.3.4 Zinc (Zn)

Zinc is an essential micronutrient for all living organisms because it acts as a catalytic or structural cofactor in many enzymes and regulatory proteins. However, Zn can be toxic when it is in excess. Therefore, plants have developed a Zn homeostasis system comprising coordination of Zn uptake, transport, traffic, and sequestration in order to ensure a sufficient Zn supply.

The identification of the first plant family of Zn transporter (ZRT) genes (zinc iron permeases (ZIP), ZIP1, ZIP2, ZIP3, and ZIP4) and iron-regulated transporter 3 (IRT3) that respond to Zn deficiency was carried out by screening an *A. thaliana* cDNA expression library for clones that restored growth and Zn uptake across the plasma membrane under Zn-limited conditions when expressed in a *Saccharomyces cerevisiae zrt1 zrt2* double mutant lacking both high- and low-affinity Zn uptake system and thus being extremely sensitive to Zn deficiency (Grotz et al. 1998; Lin et al. 2009b; Assuncao et al. 2010). The cDNA sequences of clones that suppressed the zrt1 zrt2 phenotype revealed protein products similar to the low- and high-affinity Zn transporters *ZRT1* and *ZRT2* in *S. cerevisiae* and the iron uptake transporter IRT1 in *Arabidopsis* (Eide et al. 1996; Grotz et al. 1998). The overexpression of *Arabidopsis* ZIP1 in barley grown in soil resulted in increased Zn and iron accumulation (Ramesh et al. 2004). The identification of these transporters has been described in different species (e.g., Darbani et al. 2015; Ricachenevsky et al. 2015).

The overexpression of the maize ZIP7 elevated Zn and Fe contents in *Arabidopsis* (Li et al. 2016b). When *OsZIP4* was overexpressed in rice plants under the control of the CaMV 35S promoter, Zn concentration in roots was 10 times higher than in the vector controls. However, Zn concentration in seeds of 35S-OsZIP4 plants was four times lower than controls (Ishimaru et al. 2007).

Iron-regulated transporter 1 encodes the major Fe transporter at the root surface in *Arabidopsis* (Eide et al. 1996; Vert et al. 2002). Transgenic rice plants overexpressing *OsIRT1* under the control of the maize ubiquitin promoter showed an increase in Fe and Zn contents in shoots, roots, and mature seeds (Lee and An 2009). Those plants showed enhanced tolerance to Fe deficiency at the seedling stage (Lee and An 2009). The same phenomena were observed when the *Malus xiaojinensis IRT1* was overexpressed in rice (Tan et al. 2015).

Another step in the control of metal translocation to shoots is the rate of xylem loading. Heavy metal P-type ATPase (HMA4, type of P1B-ATPase) is a key transporter responsible for loading of Zn and Cd into the xylem vessels in *A. thaliana*, *Arabidopsis halleri*, and *Noccaea caerulescens* (Mills et al. 2003, 2005; Hussain et al. 2004; Verret et al. 2004; Courbot et al. 2007; Hanikenne et al. 2008). HMA4 drives the efflux of metals from the xylem parenchyma cell (Mills et al. 2003, 2005). The ectopic expression of *35S::AtHMA4* in tobacco increased Zn and decreased Cd concentration in shoots, and this modification pattern could be useful for biofortification purposes (Siemianowski et al. 2011; see also Slamet-Loedin et al. 2015). The ectopic expression of *AtHMA4* in tomato modified Zn accumulation depending on the Zn supply (Kendziorek et al. 2014). Even though genetic engineering and molecular approaches have allowed elucidation of some structures involved in Zn homeostasis, there is still little known about the regulation of Zn homeostasis in plants. Also, continual exploration of the complex genetic traits related to uptake, transport, and remobilization of Zn in grain crops (e.g., Hussain et al. 2016) is needed.

17.3.5 Iron (Fe)

Iron (Fe) is essential for cellular respiration, synthesis of chlorophyll, and photosynthetic electron transport. In addition, Fe serves as a cofactor for a wide range of enzymes (e.g., cytochromes, catalase, and peroxidase isozymes contain Fe as a prosthetic group) (Rout and Sahoo 2015). Given that Fe exists in soil solution mostly as sparingly soluble ferric complexes under aerobic conditions, plants had to evolve specific strategies for enhancing Fe acquisition (Rout and Sahoo 2015). In recent years, various components of Fe uptake and transport have been elucidated in detail, including metal transporters, synthesis of Fe chelators, and transcription factors regulating Fe uptake. Genetic strategies for improving Fe uptake have been summarized recently (Bashir et al. 2013; López-Arredondo et al. 2013; Masuda et al. 2013a, b). Some examples are summarized in Table 17.2.

The transcription factors belonging to the IVc subgroup of the basic helix-loop-helix bHLH family influence the activity of Fe chelate reductase and acidification of the rhizosphere to regulate plant growth and development under Fe-deficient conditions (Zhang et al. 2015). In *Malus domestica*, the overexpression of a *BHLH104* homologous gene resulted in greater Fe accumulation than in the wild-type control under Fe deficiency (Zhao et al. 2016). In addition, IDE-binding factor 1 (IDEF1), a transcription factor, regulated the Fe homeostasis under Fe deficiency and allowed a higher tolerance to Fe deficiency when *OsIRO2* was overexpressed in rice (Kobayashi et al. 2012).

Vacuolar sequestration is a mechanism of regulating Fe homeostasis in plants (Kim et al. 2006). To enhance Fe concentration in cassava (*Manihot esculenta*) storage roots, a codon-optimized *Arabidopsis*vacuolar iron transporter 1 (VIT1) version under transcriptional control of the type I patatin promoter was expressed (Narayanan et al. 2015). Transgenic plants showed around 4 times higher Fe concentration in storage roots than wild-type plants, as well as 4 and 16 times higher Fe concentration in young stem and stem base tissues, respectively (Narayanan et al. 2015). Ferric reductase defective 3 (FRD3) is the best characterized protein involved in iron transport and belongs to multidrug and toxic compound extrusion (MATE) family associated to Al tolerance (Schmidt 2006; Kochian et al. 2015). A FRD3 gene tissue-specific named *OsFRDL1* was expressed in root pericycle cells of rice; this gene is similar to *FRD3* of *Arabidopsis* and coded for a citrate transporter necessary for Fe translocation (Yokosho et al. 2009). Despite significant progress in understanding Fe transport in plants, translocation and chelation of Fe have not been elucidated fully.

17.4 Future Directions

In recent years, a wide range of studies described the identification and functional characterization of many gene families coding for mineral transporters in a number of plant species. With the recent advances in genetic engineering, there is an opportunity to understand the mineral transport at the whole plant level, as well as determine the physiological, biochemical, and molecular mechanisms underpinning uptake, utilization, and remobilization of minerals in higher plants. The overexpression of appropriate endogenous genes or introgression of foreign genes is a key tool for improving nutrient-use efficiency. The molecular basis of the transporter genes induced by mineral nutrients, transcriptions factor, or regulatory elements is now beginning to be understood. However, more research is still indispensable at the molecular level to integrate and elucidate important structural, functional, and regulatory points in the complex network of nutrient transport in plants. Finally, it is necessary for breeders and researchers to develop new strategies and protocols of genetic engineering to improve nutrient-use efficiency using biotechnological tools.

References

Abdulnour, J. E., Donnelly, D. J., & Barthakur, N. N. (2000). The effect of boron on calcium uptake and growth in micropropagated potato plantlets. *Potato Research, 43*, 287–295.

Abhishek, B., Sahrawat, K. L., Shiv, K., Rohit, J., Parihar, A. K., Ummed, S., Deepak, S., & Singh, N. P. (2015). Genetics- and genomics-based interventions for nutritional enhancement of grain legume crops: Status and outlook. *Journal of Applied Genetics, 56*, 151–161.

Ai, P., Sun, S., Zhao, J., Fan, X., Xin, W., Guo, Q., Yu, L., Shen, Q., Wu, P., Miller, A. J., & Xu, G. (2009). Two rice phosphate transporters, *OsPht1;2* and *OsPht1;6*, have different functions and kinetic properties in uptake and translocation. *The Plant Journal, 57*, 798–780.

Almagro, A., Lin, S. H., & Tsay, Y. F. (2008). Characterization of the Arabidopsis nitrate transporter *NRT1.6* reveals a role of nitrate in early embryo development. *Plant Cell, 20*, 3289–3299.

Anderson, J. A., Huprikar, S. S., Kochian, L. V., Lucas, W. J., & Gaber, R. F. (1992). Functional expression of a probable *Arabidopsis thaliana* potassium channel in *Saccharomyces cerevisiae*. *Proceedings of the National Academy of Sciences of the United States of America, 89*, 3736–3740.

Arpat, A. B., Magliano, P., Wege, S., Rouached, H., Stefanovic, A., & Poirier, Y. (2012). Functional expression of *PHO1* to the Golgi and trans-golgi network and its role in export of inorganic phosphate. *The Plant Journal, 71*, 479–491.

Asad, A., Bell, R., Dell, B., & Huang, L. (1997). Development of a boron buffered solution culture system for studying plant boron nutrition. *Plant and Soil, 188*, 21–32.

Assuncao, A. G., Herrero, E., Lin, Y. F., Huettel, B., Talukdar, S., Smaczniak, C., Immink, R. G. H., van Eldike, M., Fierse, M., Schat, H., & Aarts, M. G. M. (2010). *Arabidopsis thaliana* transcription factors *bZIP19* and *bZIP23* regulate the adaptation to zinc deficiency. *Proceedings of the National Academy of Sciences of the United States of America, 107*, 10296–10301.

Ayadi, A., David, P., Arrighi, J. F., Chiarenza, S., Thibaud, M. C., Nussaume, L., & Marin, E. (2015). Reducing the genetic redundancy of Arabidopsis PHOSPHATE TRANSPORTER1 transporters to study phosphate uptake and signaling. *Plant Physiology, 167*, 1511–1526.

Baas, R., Van Oers, S., Silber, A., Bernstein, N., Ioffe, M., Keinan, M., & Bar-Tal, A. (2003). Calcium distribution in cut roses as related to transpiration. *The Journal of Horticultural Science and Biotechnology, 78*, 1–9.

Baldwin, J. C., Karthikeyan, A. S., Cao, A., & Raghothama, K. G. (2008). Biochemical and molecular analysis of *LePS2;1*: A phosphate starvation induced protein phosphatase gene from tomato. *Planta, 228*, 273–280.

Barragán, V., Leidi, E. O., Andrés, Z., Rubio, L., De Luca, A., Fernández, J. A., Cubero, B., & Pardo, J. M. (2012). Ion exchangers NHX1 and NHX2 mediate active potassium uptake into vacuoles to regulate cell turgor and stomatal function in Arabidopsis. *Plant Cell, 24*, 1127–1142.

Bashir, K., Nozoye, T., Ishimaru, Y., Nakanishi, H., & Nishizawa, N. K. (2013). Exploiting new tools for iron bio-fortification of rice. *Biotechnology Advances, 31*, 1624–1633.

Bassil, E., Tajima, H., Liang, Y. C., Ohto, M. A., Ushijima, K., Nakano, R., Esumi, T., Coku, A., Belmonte, M., & Blumwald, E. (2011). The Arabidopsis Na^+/H^+ antiporters NHX1 and NHX2 control vacuolar pH and K^+ homeostasis to regulate growth, flower development, and reproduction. *Plant Cell, 23*, 3482–3497.

Berezin, I., Mizrachy-Dagry, T., Brook, E., Mizrahi, K., Elazar, M., Zhuo, S., Saul-Tcherkas, V., & Shaul, O. (2008). Overexpression of *AtMHX* in tobacco causes increased sensitivity to Mg^{2+}, Zn^{2+}, and Cd^{2+} ions, induction of V-ATPase expression, and a reduction in plant size. *Plant Cell Reports, 27*, 939–949.

Borlaug, N. (1983). Contributions of conventional plant breeding to food production. *Science, 219*, 689–693.

Bose, J., Babourina, O., Shabala, S., & Rengel, Z. (2013). Low-pH and aluminum resistance in Arabidopsis correlates with high cytosolic magnesium content and increased magnesium uptake by plant roots. *Plant & Cell Physiology, 54*, 1093–1104.

Bose, J., Rodrigo-Moreno, A., & Shabala, S. (2014). ROS homeostasis in halophytes in the context of salinity stress tolerance. *Journal of Experimental Botany, 65*, 1241–1257.

Bose, J., Rodrigo-Moreno, A., Lai, D., Xie, Y., Shen, W., & Shabala, S. (2015). Rapid regulation of the plasma membrane H1-ATPase activity is essential to salinity tolerance in two halophyte species, *Atriplex lentiformis* and *Chenopodium quinoa*. *Annals of Botany, 115*, 481–494.

Brown, P. H., Bellaloui, N., Hu, H., & Dandekar, A. (1999). Transgenically enhanced sorbitol synthesis facilitates phloem boron transport and increases tolerance of tobacco to boron deficiency. *Plant Physiology, 119*, 17–20.

Brown, P. H., Bellaloui, N., Wimmer, M. A., Bassil, E. S., Ruiz, J., Hu, H., Pfeffer, H., Dannel, F., & Romheld, V. (2002). Boron in plant biology. *Plant Biology, 4*, 205–223.

Buchner, P., Stuiver, C. E. E., Westerman, S., Wirtz, M., Hell, R., Hawkesford, M. J., & De Kok, L. J. (2004). Regulation of sulfate uptake and expression of sulfate transporter genes in *Brassica oleracea* as affected by atmospheric H_2S and pedospheric sulfate nutrition. *Plant Physiology, 136*, 3396–3408.

Bui, D. M., Gregan, J., Jarosch, E., Ragnini, A., & Schweyen, R. J. (1999). The bacterial magnesium transporter CorA can functionally substitute for its putative homologue Mrs2p in the yeast inner mitochondrial membrane. *The Journal of Biological Chemistry, 274*, 20438–20443.

Burkhead, J. L., Reynolds, K. A., Abdel-Ghany, S. E., Cohu, C. M., & Pilon, M. (2009). Copper homeostasis. *The New Phytologist, 182*, 799–816.

Byrne, S. L., Foito, A., Hedley, P. E., Morris, J. A., Stewart, D., & Barth, S. (2011). Early response mechanisms of perennial ryegrass (*Lolium perenne*) to phosphorus deficiency. *Annals of Botany, 107*, 243–254.

Cailliatte, R., Schikora, A., Briat, J. F., Mari, S., & Curie, C. (2010). High-affinity manganese uptake by the metal transporter *NRAMP1* is essential for Arabidopsis growth in low manganese conditions. *Plant Cell, 22*, 904–917.

Cakmak, I. (2013). Magnesium in crop production, food quality and human health. *Plant and Soil, 368*, 1–4.

Cakmak, I., & Yazici, A. (2010). Magnesium: A forgotten element in crop production. *Better Crops, 94*, 1–3.

Cañon, P., Aquea, F., Rodríguez-Hoces de la Guardia, A., & Arce-Johnson, P. (2013). Functional characterization of *Citrus macrophylla BOR1* as a boron transporter. *Physiologia Plantarum, 149*, 329–339.

Castiglioni, S., Cazzaniga, A., Albisetti, W., & Maier, J. (2013). Magnesium and osteoporosis: Current state of knowledge and future research directions. *Nutrients, 5*, 3022–3033.

Cellier, F., Conéjéro, G., Ricaud, L., Luu, D. T., Lepetit, M., Gosti, F., & Casse, F. (2004). Characterization of *AtCHX17*, a member of the cation/H^+ exchangers, CHX family, from *Arabidopsis thaliana* suggests a role in K^+ homeostasis. *The Plant Journal, 39*, 834–846.

Chen, Z., & Ma, J. (2013). Magnesium transporters and their role in Al tolerance in plants. *Plant and Soil, 368*, 51–56.

Chen, Z., Yamaji, N., Motoyama, R., Nagamura, Y., & Ma, J. (2012). Up-regulation of a magnesium transporter gene *OsMGT1* is required for conferring aluminum tolerance in rice. *Plant Physiology, 159*, 1624–1633.

Cheng, C., & Rerkasem, B. (1993). Effects of boron on pollen viability in wheat. *Plant and Soil, 155*, 313–315.

Chiu, C. C., Lin, C. S., Hsia, A. P., RC, S., Lin, H. L., & Tsay, Y. F. (2004). Mutation of a nitrate transporter, *AtNRT1:4*, results in a reduced petiole nitrate content and altered leaf development. *Plant & Cell Physiology, 45*, 1139–1148.

Clarkson, D. T. (1985). Factors affecting mineral nutrient acquisition by plants. *Annual Review of Plant Physiology, 36*, 77–115.

Cong, Y. X., Luo, D. F., Chen, K. M., Jiang, L. X., & Guo, W. L. (2012). The development of magnesium transport systems in organisms. *Journal of Agricultural Biotechnology, 20*, 837–848.

Connolly, E. L., Campbell, N. H., Grotz, N., Prichard, C. L., & Guerinot, M. L. (2003). Overexpression of the *FRO2* ferric chelate reductase confers tolerance to growth on low iron and uncovers posttranscriptional control. *Plant Physiology, 133*, 1102–1110.

Connolly, E. L. (2008). Raising the bar for biofortification: Enhanced levels of bioavailable calcium in carrots. *Trends in Biotechnology, 26*, 401–403.

Correia, C. M., Moutinho Pereira, J. M., Coutinho, J. F., Björn, L. O., & Torres-Pereira, J. M. G. (2005). Ultraviolet-B radiation and nitrogen affect the photosynthesis of maize: A Mediterranean field study. *European Journal of Agronomy, 22*, 337–347.

Coskun, D., Britto, D. T., Li, M., Oh, S., & Kronzucker, H. J. (2013). Capacity and plasticity of potassium channels and high-affinity transporters in roots of barley and Arabidopsis. *Plant Physiology, 162*, 496–511.

Courbot, M., Willems, G., Motte, P., Arvidsson, S., Roosens, N., Saumitou-Laprade, P., & Verbruggen, N. (2007). A major quantitative trait locus for cadmium tolerance in *Arabidopsis halleri* colocalizes with *HMA4*, a gene encoding a heavy metal ATPase. *Plant Physiology, 144*, 1052–1065.

Couturier, J., Montanini, B., Martin, F., Brun, A., Blaudez, D., & Chalot, M. (2007). The expanded family of ammonium transporters in the perennial poplar plant. *The New Phytologist, 174*, 137–150.

Cubero, B., Nakagawa, Y., Jiang, X. Y., Miura, K. J., Li, F., Raghothama, K. G., Bressan, R. A., Hasegawa, P. M., & Pardo, J. M. (2009). The phosphate transporter *PHT4;6* is a determinant of salt tolerance that is localized to the Golgi apparatus of Arabidopsis. *Molecular Plant, 2*, 535–552.

Daram, P., Brunner, S., Rausch, C., Steiner, C., Amrhein, N., & Bucher, M. (1999). *Pht2;1* encodes a low-affinity phosphate transporter from Arabidopsis. *Plant Cell, 11*, 2153–2166.

Darbani, B., Noeparvar, S., & Borg, S. (2015). Deciphering mineral homeostasis in barley seed transfer cells at transcriptional level. *PloS One, 10*, e0141398.

David, L. C., Dechorgnat, J., Berquin, P., Routaboul, J. M., Debeaujon, I., Daniel-Vedele, F., & Ferrario-Méry, S. (2014). Proanthocyanidin oxidation of Arabidopsis seeds is altered in mutant of the high-affinity nitrate transporter *NRT2.7*. *Journal of Experimental Botany, 65*, 885–893.

David-Assael, O., Berezin, I., Shoshani-Knaani, N., Saul, H., Mizrachy-Dagri, T., Chen, J. X., Brook, E., & Shaul, O. (2006). *AtMHX* is an auxin and ABA-regulated transporter whose expression pattern suggests a role in metal homeostasis in tissues with photosynthetic potential. *Functional Plant Biology, 33*, 661–672.

Davidian, J. C., & Kopriva, S. (2010). Regulation of sulfate uptake and assimilation—The same or not the same? *Molecular Plant, 3*, 314–325.

De Angeli, A., Monachello, D., Ephritikhine, G., Frachisse, J. M., Thomine, S., Gambale, F., & Barbier-Brygoo, H. (2006). The nitrate/proton antiporter *AtCLCa* mediates nitrate accumulation in plant vacuoles. *Nature, 442*, 939–942.

Dell, B., & Huang, L. B. (1997). Physiological response of plants to low boron. *Plant and Soil, 193*, 103–120.

Deng, W., Luo, K. M., Li, D. M., Zheng, X. L., Wei, X. Y., Smith, W., Thammina, C., LT, L., Li, Y., & Pei, Y. (2006). Overexpression of an Arabidopsis magnesium transport gene, *AtMGT1*, in *Nicotiana benthamiana* confers Al tolerance. *Journal of Experimental Botany, 57*, 4235–4243.

Dodd, A. N., Kudla, J., & Sanders, D. (2010). The language of calcium signaling. *Annual Review of Plant Biology, 61*, 593–620.

Dong, B., Rengel, Z., & Delhaize, E. (1998). Uptake and translocation of phosphate by *pho2* mutant and wild-type seedlings of *Arabidopsis thaliana. Planta, 205*, 251–256.

Drummond, R. S. M., Tutone, A., Li, Y. C., & Gardner, R. C. (2006). A putative magnesium transporter *AtMRS2-11* is localized to the plant chloroplast envelope membrane system. *Plant Science, 170*, 78–89.

Dubousset, L., Abdallah, M., Desfeux, A. S., Etienne, P., Meuriot, F., Hawkesford, M. J., Gombert, J., Ségura, R., Bataillé, M. P., Rezé, S., Bonnefoy, J., Ameline, A. F., Ourry, A., Le Dily, F., & Avice, J. C. (2009). Remobilization of leaf S compounds and senescence in response to restricted sulphate supply during the vegetative stage of oilseed rape are affected by mineral N availability. *Journal of Experimental Botany, 60*, 3239–3253.

Duff, S. M. G., Plaxton, W. C., & Lefebvre, D. D. (1991). Phosphate starvation responses in plant cells: De novo synthesis and degradation of acid phosphatase. *Proceedings of the National Academy of Sciences of the United States of America, 88*, 9538–9542.

Edmeades, D. C. (2003). The long-term effects of manures and fertilisers on soil productivity and quality: A review. *Nutrient Cycling in Agroecosystems, 66*, 165–180.

Eide, D., Broderius, M., Fett, J., & Guerinot, M. L. (1996). A novel iron-regulated metal transporter from plants identified by functional expression in yeast. *Proceedings of the National Academy of Sciences of the United States of America, 93*, 5624–5628.

Elumalai, R. P., Nagpal, P., & Reed, J. W. (2002). A mutation in the Arabidopsis *KT2/KUP2* potassium transporter gene affects shoot cell expansion. *Plant Cell, 14*, 119–131.

Epstein, E., Rains, D., & Elzam, O. (1963). Resolution of dual mechanisms of potassium absorption by barley roots. *Proceedings of the National Academy of Sciences of the United States of America, 49*, 684–692.

Ericsson, T., & Kahr, M. (1995). Growth and nutrition of birch seedlings at varied relative addition rates of magnesium. *Tree Physiology, 15*, 85–93.

Fageria, N. K., Baligar, V. C., & Clark, R. B. (2002). Micronutrients in crop production. *Advances in Agronomy, 77*, 185–268.

Fan, C., Wang, X., Hu, R., Wang, Y., Xiao, C., Jiang, Y., Zhang, X., Zheng, C., & YF, F. (2013). The pattern of phosphate transporter 1 genes evolutionary divergence in *Glycine max* L. *BMC Plant Biology, 13*, 48.

von der Fecht-Bartenbach, J., Bogner, M., Dynowski, M., & Ludewig, U. (2010). CLC-b-mediated NO_3^-/H^+ exchange across the tonoplast of Arabidopsis vacuoles. *Plant & Cell Physiology, 51*, 960–968.

Fiedor, L., Kania, A., Myśliwa-Kurdziel, B., Orzeł, L., & Stochel, G. (2008). Understanding chlorophylls: Central magnesium ion and phytyl as structural determinants. *Biochimica et Biophysica Acta, 1777*, 1491–1500.

Fita, A., Bowen, H. C., Hayden, R. M., Nuez, F., Pico, B., & Hammond, J. P. (2012). Diversity in expression of phosphorus (P) responsive genes in *Cucumis melo* L. *PloS One, 7*, e35387.

Francois, L. E. (1984). Effect of excess boron on tomato yield, fruit size, and vegetative growth. *Journal of the American Societyfor Horticultural Science, 109*, 322–324.

Gaash, R., Elazar, M., Mizrahi, K., Avramov-Mor, M., Berezin, I., & Shaul, O. (2013). Phylogeny and a structural model of plant MHX transporters. *BMC Plant Biology, 13*, 75.

Gambale, F., & Uozumi, N. (2006). Properties of Shaker-type potassium channels in higher plants. *The Journal of Membrane Biology, 210*, 1–19.

Ganmore-Neumann, R., & Davidov, S. (1993). Uptake and distribution of calcium in rose plantlets as affected by calcium and boron concentration in culture medium. *Plant and Soil, 155*, 151–154.

Gassmann, W., Rubio, F., & Schroeder, J. I. (1996). Alkali cation selectivity of the wheat root high-affinity potassium transporter HKT1. *The Plant Journal, 10*, 869–882.

Gaymard, F., Pilot, G., Lacombe, B., Bouchez, D., Bruneau, D., Boucherez, J., Michaux-Ferrière, N., Thibaud, J. B., & Sentenac, H. (1998). Identification and disruption of a plant Shaker-like outward channel involved in K^+ release into the xylem sap. *Cell, 94*, 647–655.

Gebert, M., Meschenmoser, K., Svidova, S., Weghuber, J., Schweyen, R., Eifler, K., Lenz, H., Weyand, K., & Knoop, V. (2009). A root-expressed magnesium transporter of the *MRS2/MGT* gene family in *Arabidopsis thaliana* allows for growth in low-Mg^{2+} environments. *Plant Cell, 21*, 4018–4030.

Gierth, M., & Mäser, P. (2007). Potassium transporters in plants-involvement in K^+ acquisition, redistribution and homeostasis. *FEBS Letters, 581*, 2348–2356.

Gierth, M., Mäser, P., & Schroeder, J. I. (2005). The potassium transporter *AtHAK5* functions in K+ deprivation-induced high-affinity K^+ uptake and AKT1 K^+ channel contribution to K^+ uptake kinetics in Arabidopsis roots. *Plant Physiology, 137*, 1105–1114.

Gigolashvili, T., & Kopriva, S. (2014). Transporters in plant sulfur metabolism. *Frontiers in Plant Science, 5*, 442.

Gilliham, M., Dayod, M., Hocking, B., Xu, B., Conn, S., Kaiser, B., Leigh, R., & Tyerman, S. (2011). Calcium delivery and storage in plant leaves: Exploring the link with water flow. *Journal of Experimental Botany, 62*, 2233–2250.

Gobert, A., Isayenkov, S., Voelker, C., Czempinski, K., & Maathuis, F. J. M. (2007). The two-pore channel *TPK1* gene encodes the vacuolar K^+ conductance and plays a role in K^+ homeostasis. *Proceedings of the National Academy of Sciences of the United States of America, 104*, 10726–10731.

González-Fontes, A., Rexach, J., Navarro-Gochicoa, M. T., Herrera-Rodríguez, M. B., Beato, V. M., Maldonado, J. M., & Camacho-Cristóbal, J. J. (2008). Is boron involved solely in structural roles in vascular plants? *Plant Signaling & Behavior, 3*, 24–26.

Goussias, C., Boussac, A., & Rutherford, A. W. (2002). Photosystem II and photosynthetic oxidation of water: An overview. *Philosophical Transactions of the Royal Society of London. Series B, Biological Sciences, 357*, 1369–1420.

Graham, N. S., Hammond, J. P., Lysenko, A., Mayes, S., O'Lochlainn, S., Blasco, B., Bowen, H. C., Rawlings, C. J., Rios, J. J., Welham, S., Carion, P. W., Dupuy, L. X., King, G. J., White, P. J., & Broadley, M. R. (2014). Genetical and comparative genomics of Brassica under altered Ca supply identifies Arabidopsis Ca-transporter orthologs. *Plant Cell, 26*, 2818–2830.

Gregan, J., Kolisek, M., & Schweyen, R. J. (2001). Mitochondrial Mg2þ homeostasis is critical for group II intron splicing in vivo. *Genes & Development, 15*, 2229–2237.

Grotz, N., & Guerinot, M. L. (2006). Molecular aspects of Cu, Fe and Zn homeostasis in plants. *Biochimica et Biophysica Acta, 1763*, 595–608.

Grotz, N., Fox, T., Connolly, E., Park, W., Guerinot, M. L., & Eide, D. (1998). Identification of a family of zinc transporter genes from Arabidopsis that respond to zinc deficiency. *Proceedings of the National Academy of Sciences of the United States of America, 95*, 7220–7224.

Gu, R., Duan, F., An, X., Zhang, F., von Wiren, N., & Yuan, L. (2013). Characterization of AMT-mediated high-affinity ammonium uptake in roots of maize (*Zea mays* L.) *Plant & Cell Physiology, 54*, 1515–1524.

Gu, M., Chen, A., Sun, S., & Xu, G. (2016). Complex regulation of plant phosphate transporters and the gap between molecular mechanisms and practical application: What is missing? *Molecular Plant, 9*, 396–416.

Guan, Q., Wu, J., Yue, X., Zhang, Y., & Zhu, J. (2013). A nuclear calcium-sensing pathway is critical for gene regulation and salt stress tolerance in Arabidopsis. *PloS One, 9*, e1003755.

Guo, K. M., Babourina, O., & Rengel, Z. (2009). Na^+/H^+ antiporter activity of the *SOS1* gene: Lifetime imaging analysis and electrophysiological studies on Arabidopsis seedlings. *Physiologia Plantarum, 137*, 155–165.

Guo, W., Nazim, H., Liang, Z., & Yang, D. (2016). Magnesium deficiency in plants: An urgent problem. *The Crop Journal, 4*, 83–91.

Gupta, U. C., Jame, Y. W., Campbell, C. A., Leyshon, A. J., & Nicholaichuk, W. (1985). Boron toxicity and deficiency: A review. *Canadian Journal of Soil Science, 65*, 381–409.

Gyaneshwar, P., Kumar, G. N., Parekh, L. J., & Poole, P. S. (2002). Role of soil microorganisms in improving P nutrition of plants. *Plant and Soil, 245*, 83–93.

Hammond, J. P., Broadley, M. R., & White, J. P. (2004). Genetic responses to phosphorus deficiency. *Annals of Botany, 94*, 323–332.

Hanikenne, M., Talke, I. N., Haydon, M. J., Lanz, C., Nolte, A., Motte, P., Kroymann, J., Weigel, D., & Krämer, U. (2008). Evolution of metal hyperaccumulation required cis-regulatory changes and triplication of HMA4. *Nature, 453*, 391–395.

Hassler, S., Lemke, L., Jung, B., Möhlmann, T., Krüger, F., Schumacher, K., Espen, L., Martinoia, E., & Neuhaus, H. E. (2012). Lack of the Golgi phosphate transporter PHT4;6 causes strong developmental defects, constitutively activated disease resistance mechanisms and altered intracellular phosphate compartmentation in Arabidopsis. *The Plant Journal, 72*, 732–744.

Hassler, S., Jung, B., Lemke, L., Novák, O., Strnad, M., Martinoia, E., & Neuhaus, H. E. (2016). Function of the Golgi-located phosphate transporter *PHT4;6* is critical for senescence-associated processes in Arabidopsis. *Journal of Experimental Botany, 67*, 4671–4684.

Hawkesford, M. J. (2012). Improving nutrient use efficiency in crop. In *eLS*. Chichester, UK: Wiley.

Hawkesford, M. J. (2014). Reducing the reliance on nitrogen fertilizer for wheat production. *Journal of Cereal Science, 59*, 276–283.
Hermans, C., & Verbruggen, N. (2005). Physiological characterization of Mg deficiency in *Arabidopsis thaliana. Journal of Experimental Botany, 56*, 2153–2161.
Hermans, C., Vuylsteke, M., Coppens, F., Cristescu, S., Harren, F., Inzé, D., & Verbruggen, N. (2010). Systems analysis of the responses to long-term magnesium deficiency and restoration in *Arabidopsis thaliana. The New Phytologist, 187*, 132–144.
Hirsch, R. E., Lewis, B. D., Spalding, E. P., & Sussman, M. R. (1998). A role for the AKT1 potassium channel in plant nutrition. *Science, 280*, 918–921.
Holdaway-Clarke, T. L., & Hepler, P. K. (2003). Control of pollen tube growth: Role of ion gradients and fluxes. *The New Phytologist, 159*, 539–563.
Holley, A. K., Bakthavatchalu, V., Velez-Roman, J. M., & St. Clair, D. K. (2011). Manganese superoxide dismutase: Guardian of the powerhouse. *International Journal of Molecular Sciences, 12*, 7114–7162.
Hosoi, S., Lino, M., & Shimazaki, K. (1988). Outward-rectifying K^+ channels in stomatal guard cell protoplasts. *Plant & Cell Physiology, 29*, 907–911.
Hosy, E., Vavasseur, A., Mouline, K., Dreyer, I., Gaymard, F., Porée, F., Boucherez, J., Lebaudy, A., Bouchez, D., Very, A. A., Simonneau, T., Thibaud, J. B., & Sentenac, H. (2003). The Arabidopsis outward K^+ channel GORK is involved in regulation of stomatal movements and plant transpiration. *Proceedings of the National Academy of Sciences of the United States of America, 100*, 5549–5554.
Huang, N. C., Liu, K. H., Lo, H. J., & Tsay, Y. F. (1999). Cloning and functional characterization of an Arabidopsis nitrate transporter gene that encodes a constitutive component of low affinity uptake. *Plant Cell, 11*, 1381–1392.
Hussain, D., Haydon, M. J., Wang, Y., Wong, E., Sherson, S. M., Young, J., Camakaris, J., Harper, J. F., & Cobbett, C. S. (2004). P-type ATPase heavy metal transporters with roles in essential zinc homeostasis in Arabidopsis. *Plant Cell, 16*, 1327–1339.
Hussain, S., Rengel, Z., Mohammadi, S. A., Ebadi-Segherloo, A., & Maqsood, M. A. (2016). Mapping QTL associated with remobilization of zinc from vegetative tissues into grains of barley (*Hordeum vulgare*). *Plant and Soil, 399*, 193–208.
Ihemere, U. E., Narayanan, N. N., & Sayre, R. T. (2012). Iron biofortification and homeostasis in transgenic cassava roots expressing the algal iron assimilatory gene, *FEA1*. *Frontiers in Plant Science, 3*, 171.
Irigoyen, S., Karlsson, P. M., Kuruvilla, J., Spetea, C., & Versaw, W. K. (2011). The sink specific plastidic phosphate transporter *PHT4;2* influences starch accumulation and leaf size in Arabidopsis. *Plant Physiology, 154*, 1765–1777.
Ishimaru, Y., Masuda, H., Suzuki, M., Bashir, K., Takahashi, M., Nakanishi, H., Mori, S., & Nishizawa, N. K. (2007). Overexpression of the *OsZIP4* zinc transporter confers disarrangement of zinc distribution in rice plants. *Journal of Experimental Botany, 58*, 2909–2915.
Javot, H., Penmetsa, R. V., Terzaghi, N., Cook, D. R., & Harrison, M. J. (2007). A *Medicago truncatula* phosphate transporter indispensable for the arbuscular mycorrhizal symbiosis. *Proceedings of the National Academy of Sciences of the United States of America, 104*, 1720–1725.
Johnson, A. A. T., Kyriacou, B., Callahan, D. L., Carruthers, L., Stangoulis, J., Lombi, E., & Tester, M. (2011). Constitutive overexpression of the *OsNAS* gene family reveals single-gene strategies for effective iron- and zinc-biofortification of rice endosperm. *PloS One, 6*, e24476.
Karlsson, P. M., Herdean, A., Adolfsson, L., Beebo, A., Nziengui, H., Irigoyen, S., Ünnep, R., Zsiros, O., Nagy, G., Garab, G., Aronsson, H., Versaw, W. K., & Spetea, C. (2015). The Arabidopsis thylakoid transporter *PHT4;1* influences phosphate availability for ATP synthesis and plant growth. *The Plant Journal, 84*, 99–110.
Kataoka, T., Watanabe-Takahashi, A., Hayashi, N., Ohnishi, M., Mimura, T., Buchner, P., Hawkesford, M. J., Yamaya, T., & Takahashi, H. (2004). Vacuolar sulfate transporters are essential determinants controlling internal distribution of sulfate in Arabidopsis. *Plant Cell, 16*, 2693–2704.

Kato, Y., Miwa, K., Takano, J., Wada, M., & Fujiwara, T. (2009). Highly boron deficiency-tolerant plants generated by enhanced expression of *NIP5;1*, a boric acid channel. *Plant & Cell Physiology, 50*, 58–66.

Kendziorek, M., Barabasz, A., Rudzka, J., Tracz, K., Mills, R. F., Williams, L. E., & Antosiewicz, D. M. (2014). Approach to engineer tomato by expression of *AtHMA4* to enhance Zn in the aerial parts. *Journal of Plant Physiology, 171*, 1413–1422.

Kiba, T., & Krapp, A. (2016). Plant nitrogen acquisition under low availability: Regulation of uptake and root architecture. *Plant & Cell Physiology, 57*, 707–714.

Kiba, T., Feria-Bourrellier, A. B., Lafouge, F., Lezhneva, L., Boutet-Mercey, S., Orsel, M., Bréhaut, V., Miller, A., Daniel-Vedele, F., Sakakibara, H., & Krapp, A. (2012). The Arabidopsis nitrate transporter *NRT2.4* plays a double role in roots and shoots of nitrogen-starved plants. *Plant Cell, 24*, 245–258.

Kim, T., Punshon, A., Lanzirotti, L., Li, J. M., Alonso, J. R., Ecker, J., Kaplan, M. L., & Guerinot, M. L. (2006). Localization of iron in Arabidopsis seed requires the vacuolar membrane transporter VIT1. *Science, 314*, 1295–1298.

Knight, H. (1999). Calcium signaling during abiotic stress in plants. *International Review of Cytology, 195*, 269–324.

Kobayashi, T., Itai, R. N., Aung, M. S., Senoura, T., Nakanishi, H., & Nishizawa, N. K. (2012). The rice transcription factor *IDEF1* directly binds to iron and other divalent metals for sensing cellular iron status. *The Plant Journal, 69*, 81–91.

Kochian, L. V., Piñeros, M., Liu, J., & Magalhaes, J. V. (2015). Plant adaptation to acid soils: The molecular basis for crop aluminum resistance. *Annual Review of Plant Biology, 66*, 571–598.

Kolisek, M., Zsurka, G., Samaj, J., Weghuber, J., Schweyen, R. J., & Schweigel, M. (2003). Mrs2p is an essential component of the major electrophoretic Mg2þ influx system in mitochondria. *The EMBO Journal, 22*, 1235–1244.

Kopriva, S., Calderwood, A., Weckopp, S. C., & Koprivova, A. (2015). Plant sulfur and big data. *Plant Science, 241*, 1–10.

Kouchi, H., & Kumazawa, K. (1975). Anatomical responses of root tips to boron deficiency II. Effect of boron deficiency on the cellular growth and development in root tips. *Soil Science & Plant Nutrition, 21*, 137–150.

Krapp, A., David, L. C., Chardin, C., Girin, T., Marmagne, A., Leprince, A. S., Chaillou, S., Ferrario-Méry, S., Meyer, C., & Daniel-Vedele, F. (2014). Nitrate transport and signaling in Arabidopsis. *Journal of Experimental Botany, 65*, 789–798.

Kwak, J. M., Murata, Y., Baizabal-Aguirre, V. M., Merrill, J., Wang, M., Kemper, A., Hawke, S. D., Tallman, G., & Schroeder, J. I. (2001). Dominant negative guard cell K^+ channel mutants reduce inward-rectifying K^+ currents and light-induced stomatal opening in Arabidopsis. *Plant Physiology, 127*, 473–485.

Lacombe, B., Pilot, G., Michard, E., Gaymard, F., Sentenac, H., & Thibaud, J. B. (2000). A Shaker-like K^+ channel with weak rectification is expressed in both source and sink phloem tissues of Arabidopsis. *Plant Cell, 12*, 837–851.

Lagarde, D., Basset, M., Lepetit, M., Conejero, G., Gaymard, F., Astruc, S., & Grignon, C. (1996). Tissue-specific expression of Arabidopsis *AKT1* gene is consistent with a role in K^+ nutrition. *The Plant Journal, 9*, 195–203.

Lamport, D., & Várnai, P. (2013). Periplasmic arabinogalactan glycoproteins act as a calcium capacitor that regulates plant growth and development. *The New Phytologist, 197*, 58–64.

Lapis-Gaza, H. R., Jost, R., & Finnegan, P. M. (2014). Arabidopsis PHOSPHATE TRANSPORTER1 genes *PHT1;8* and *PHT1;9* are involved in root-to-shoot translocation of orthophosphate. *BMC Plant Biology, 14*, 334.

Larcher, W. (2003). Plants under stress. In W. Larcher (Ed.), *Physiological ecology* (pp. 321–448). Berlin: Springer.

Larisch, N., Kirsch, S. A., Schambony, A., Studtrucker, T., Böckmann, R. A., & Dietrich, P. (2016). The function of the two-pore channel TPC1 depends on dimerization of its carboxy-terminal helix. *Cellular and Molecular Life Sciences, 73*, 2565–2581.

Lauter, F. R., Ninnemann, O., Bucher, M., Riesmeier, J., & Frommer, W. B. (1996). Preferential expression of an ammonium transporter and two putative nitrate transporters in root hairs of tomato. *Proceedings of the National Academy of Sciences of the United States of America, 93*, 8139–8144.

Lebaudy, A., Véry, A. A., & Sentenac, H. (2007). K^+ channel activity in plants: Genes, regulations and functions. *FEBS Letters, 581*, 2357–2366.

Lee, R. B. (1988). Phosphate influx and extracellular phosphatase activity in barley roots and rose cells. *The New Phytologist, 109*, 141–148.

Lee, S., & An, G. (2009). Overexpression of *OsIRT1* leads to increased iron and zinc accumulations in rice. *Plant, Cell & Environment, 32*, 408–416.

Leggewie, G., Willmitzer, L., & Riesmeier, J. W. (1997). Two cDNAs from potato are able to complement a phosphate uptake-deficient yeast mutant: Identification of phosphate transporters from higher plants. *Plant Cell, 9*, 381–392.

Leigh, R. A., & Wyn Jones, R. G. (1984). A hypothesis relating critical potassium concentrations for growth to the distribution and functions of this ion in the plant cell. *The New Phytologist, 97*, 1–13.

Léran, S., Varala, K., Boyer, J. C., Chiurazzi, M., Crawford, N., Daniel-Vedele, F., David, L., Dickstein, R., Fernandez, E., Forde, B., Gassmann, W., Geiger, D., Gojon, A., Gong, J. M., Halkier, B. A., Harris, J. M., Hedrich, R., Limami, A. M., Rentsch, D., Seo, M., Tsay, Y. F., Zhang, M., Coruzzi, G., & Lacombe, B. (2014). A unified nomenclature of NITRATE TRANSPORTER 1/PEPTIDE TRANSPORTER family members in plants. *Trends in Plant Science, 19*, 5–9.

Lezhneva, L., Kiba, T., Feria-Bourrellier, A. B., Lafouge, F., Boutet-Mercey, S., Zoufan, P., Sakakibara, H., Daniel-Vedele, F., & Krapp, A. (2014). The Arabidopsis nitrate transporter *NRT2.5* plays a role in nitrate acquisition and remobilization in nitrogen-starved plants. *The Plant Journal, 80*, 230–241.

Li, L., Tutone, A. F., Drummond, R. S., Gardner, R. C., & Luan, S. (2001). A novel family of magnesium transport genes in Arabidopsis. *Plant Cell, 13*, 2761–2775.

Li, J., Zhang, H., Lei, H., Jin, M., Yue, G., & Su, Y. (2016a). Functional identification of a GORK potassium channel from the ancient desert shrub *Ammopiptanthus mongolicus* (Maxim.) Cheng f. *Plant Cell Reports, 35*, 803–815.

Li, S., Zhou, X., Zhao, Y., Li, H., Liu, Y., Zhu, L., Guo, J., Huang, Y., Yang, W., Fan, Y., Chen, J., & Chen, R. (2016b). Constitutive expression of the *ZmZIP7* in Arabidopsis alters metal homeostasis and increases Fe and Zn content. *Plant Physiology and Biochemistry, 106*, 1–10.

Liang, G., Yang, F., & Yu, D. (2010). MicroRNA395 mediates regulation of sulfate accumulation and allocation in *Arabidopsis thaliana*. *The Plant Journal, 62*, 1046–1057.

Lin, C. M., Koh, S., Stacey, G., SM, Y., Lin, T. Y., & Tsay, Y. F. (2000). Cloning and functional characterization of a constitutively expressed nitrate transporter gene, *OsNRT1*, from rice. *Plant Physiology, 122*, 379–388.

Lin, S. H., Kuo, H. F., Canivenc, G., Lin, C. S., Lepetit, M., Hsu, P. K., et al. (2008). Mutation of the Arabidopsis NRT1.5 nitrate transporter causes defective root-to-shoot nitrate transport. *Plant Cell, 20*, 2514–2528.

Lin, W. Y., Lin, S. I., & Chiou, T. J. (2009a). Molecular regulators of phosphate homeostasis in plants. *Journal of Experimental Botany, 60*, 1427–1438.

Lin, Y. F., Liang, H. M., Yang, S. Y., Boch, A., Clemens, S., Chen, C. C., JF, W., Huang, J. L., & Yeh, K. C. (2009b). Arabidopsis *IRT3* is a zinc-regulated and plasma membrane localized zinc/iron transporter. *The New Phytologist, 182*, 392–404.

Liu, K. H., Huang, C. Y., & Tsay, Y. F. (1999). CHL1 is a dual-affinity nitrate transporter of Arabidopsis involved in multiple phases of nitrate uptake. *Plant Cell, 11*, 865–874.

Liu, T. Y., Huang, T. K., Yang, S. Y., Hong, Y. T., Huang, S. M., Wang, F. N., Chiang, S. F., Tsai, S. Y., WC, L., & Chiou, T. J. (2016). Identification of plant vacuolar transporters mediating phosphate storage. *Nature Communications, 7*, 11095.

Lonnerdal, B. (2003). Genetically modified plants for improve trace element nutrition. *The Journal of Nutrition, 133*, 1490–1493.

López-Arredondo, D. L., Leyva-González, M. A., Alatorre-Cobos, F., & Herrera-Estrella, L. (2013). Biotechnology of nutrient uptake and assimilation in plants. *The International Journal of Developmental Biology, 57*, 595–610.

Loqué, D., Yuan, L., Kojima, S., Gojon, A., Wirth, J., Gazzarrini, S., Ishiyama, K., Takahashi, H., & von Wirén, N. (2006). Additive contribution of *AMT1;1* and *AMT1;3* to high-affinity ammonium uptake across the plasma membrane of nitrogen-deficient Arabidopsis roots. *The Plant Journal, 48*, 522–534.

Lu, Y. P., Zhen, R. G., & Rea, P. A. (1997). *AtPT4*: A fourth member of the Arabidopsis phosphate transporter gene family (Accession no. U97546) Plant Gene Register PGR 97-082. *Plant Physiology, 114*, 747.

Ludewig, U., Neuhduser, B., & Dynowski, M. (2007). Molecular mechanisms of ammonium transport and accumulation in plants. *FEBS Letters, 581*, 2301–2308.

Maathuis, F. J. M. (2009). Physiological functions of mineral macronutrients. *Current Opinion in Plant Biology, 12*, 250–258.

Maathuis, F. J. M., & Sanders, D. (1994). Mechanism of high-affinity potassium uptake in roots of *Arabidopsis thaliana. Proceedings of the National Academy of Sciences of the United States of America, 91*, 9272–9276.

Marini, A. M., Vissers, S., Urrestarazu, A., & André, B. (1994). Cloning and expression of the *MEP1* gene encoding an ammonium transporter in *Saccharomyces cerevisiae. The EMBO Journal, 13*, 3456–3463.

Marschner, P. (2012). *Mineral nutrition of higher plants* (3rd ed.). San Diego: Academic Press.

Marten, I., Hoth, S., Deeken, R., Ache, P., Ketchum, K. A., Hoshi, T., & Hedrich, R. (1999). AKT3, a phloem-localized K^+ channel, is blocked by protons. *Proceedings of the National Academy of Sciences of the United States of America, 96*, 7581–7586.

Masclaux-Daubresse, C., Daniel-Vedele, F., Dechorgnat, J., Chardon, F., Gaufichon, L., & Suzuki, A. (2010). Nitrogen uptake, assimilation and remobilization in plants: Challenges for sustainable and productive agriculture. *Annals of Botany, 105*, 1141–1157.

Mäser, P., Thomine, S., Schroeder, J. I., Ward, J. M., Hirschi, K., Sze, H., Talke, I. N., Amtmann, A., Maathuis, F. J., Sanders, D., Harper, J. F., Tchieu, J., Gribskov, M., Persans, M. W., Salt, D. E., Kim, S. A., & Guerinot, M. L. (2001). Phylogenetic relationships within cation transporter families of Arabidopsis. *Plant Physiology, 126*, 1646–1667.

Masuda, H., Aung, M. S., & Nishizawa, N. K. (2013a). Iron biofortification of rice using different transgenic approaches. *Rice, 6*, 40.

Masuda, H., Kobayashi, T., Ishimaru, Y., Takahashi, M., Aung, M. S., Nakanishi, H., Mori, S., & Nishizawa, N. K. (2013b). Iron-biofortification in rice by the introduction of three barley genes participated in mugineic acid biosynthesis with soybean ferritin gene. *Frontiers in Plant Science, 4*, 132.

Miller, A. J., Fan, X., Orsel, M., Smith, S. J., & Wells, D. M. (2007). Nitrate transport and signalling. *Journal of Experimental Botany, 58*, 2297–2306.

Mills, R. F., Krijger, G. C., Baccarini, B. J., Hall, J. L., & Williams, L. E. (2003). Functional expression of *AtHMA4*, a P_{-1B}-type ATPase of the Zn/Co/Cd/Pb subclass. *The Plant Journal, 35*, 164–176.

Mills, R. F., Francini, A., Ferreira da Rocha, P. S. C., Bacarini, P. J., Aylett, M., Krijger, G. C., & Williams, L. E. (2005). The plant P_{-1B}-type ATPase *AtHMA4* transports Zn and Cd and plays a role in detoxification of transition metals supplied at elevated levels. *FEBS Letters, 579*, 783–791.

Misson, J., Thibaud, M. C., Bechtold, N., Raghothama, K., & Nussaume, L. (2004). Transcriptional regulation and functional properties of Arabidopsis *Pht1;4*, a high affinity transporter contributing greatly to phosphate uptake in phosphate deprived plants. *Molecular Biology, 55*, 727.

Mitra, G. N. (2015). *Regulation of nutrient uptake by plants: A biochemical and molecular approach* (pp. 1–195). New Delhi, India: Springer.

Miwa, K., Takano, J., & Fujiwara, T. (2006). Improvement of seed yields under boron-limiting conditions through overexpression of *BOR1*, a boron transporter for xylem loading, in *Arabidopsis thaliana. The Plant Journal, 46*, 1084–1091.

Miyaji, T., Kuromori, T., Takeuchi, Y., Yamaji, N., Yokosho, K., Shimazawa, A., Sugimoto, E., Omote, H., Ma, J. F., Shinozaki, K., & Moriyama, Y. (2015). *AtPHT4;4* is a chloroplast-localized ascorbate transporter in Arabidopsis. *Nature Communications, 6*, 5928.

Morgan, M. J., Lehmann, M., Schwarzlnder, M., Baxter, C. J., Sienkiewicz-Porzucek, A., Williams, T. C. R., & Finkemeier, I. (2008). Decrease in manganese superoxide dismutase leads to reduced root growth and affects tricarboxylic acid cycle flux and mitochondrial redox homeostasis. *Plant Physiology, 147*, 101–114.

Mouline, K., Véry, A. A., Gaymard, F., Boucherez, J., Pilot, G., Devic, M., Bouchez, D., Thibaud, J. B., & Sentenac, H. (2002). Pollen tube development and competitive ability are impaired by disruption of a Shaker K^+ channel in Arabidopsis. *Genes & Development, 16*, 339–350.

Muchhal, U. S., Pardo, J. M., & Raghothama, K. G. (1996). Phosphate transporters from the higher plant *Arabidopsis thaliana*. *Proceedings of the National Academy of Sciences of the United States of America, 93*, 10519–10523.

Mudge, S. R., Rae, A. L., Diatloff, E., & Smith, F. W. (2002). Expression analysis suggests novel roles for members of the Pht1 family of phosphate transporters in Arabidopsis. *The Plant Journal, 31*, 341–353.

Murphy, K., Reeves, P., & Jones, S. (2008). Relationship between yield and mineral nutrient concentrations in historical and modern spring wheat cultivars. *Euphytica, 163*, 381–390.

Nable, R. O., Banuelos, G. S., & Paull, J. G. (1997). Boron toxicity. *Plant and Soil, 193*, 181–198.

Nagarajan, V. K., Jain, A., Poling, M. D., Lewis, A. J., Raghothama, K. G., & Smith, A. P. (2011). Arabidopsis *Pht1;5* mobilizes phosphate between source and sink organs and influences the interaction between phosphate homeostasis and ethylene signaling. *Plant Physiology, 156*, 1149–1163.

Narayanan, N., Beyene, G., Chauhan, R. D., Gaitán-Solis, E., Grusak, M. A., Taylor, N., & Anderson, P. (2015). Overexpression of Arabidopsis *VIT1* increases accumulation of iron in cassava roots and stems. *Plant Science, 240*, 170–181.

Nath, M., & Tuteja, N. (2016). NPKS uptake, sensing, and signaling and miRNAs in plant nutrient stress. *Protoplasma, 253*, 767–786.

Nevo, Y., & Nelson, N. (2006). The NRAMP family of metal-ion transporters. *Biochimica et Biophysica Acta, 1763*, 609–620.

Ning, L., Sun, P., Wang, Q., Ma, D., Hu, Z., Zhang, D., Zhang, G., Cheng, H., & Yu, D. (2015). Genetic architecture of biofortification traits in soybean (*Glycine max* L. Merr.) revealed through association analysis and linkage mapping. *Euphytica, 204*, 353–369.

Ninnemann, O., Jauniaux, J. C., & Frommer, W. B. (1994). Identification of a high affinity NH^{4+} transporter from plants. *The EMBO Journal, 13*, 3464–3471.

Noguchi, K., Yasumori, M., Imai, T., Naito, S., Matsunaga, T., Oda, H., Hayashi, H., Chino, M., & Fujiwara, T. (1997). *bor1-1*, an *Arabidopsis thaliana* mutant that requires a high level of boron. *Plant Physiology, 115*, 901–906.

Noji, M., & Saito, K. (2007). Metabolic engineering of sulfur assimilation in plants. In R. T. Verpoorte, A. W. Alfermann, & T. S. Johnson (Eds.), *Applications of plant metabolic engineering* (pp. 297–309). Dordrecht, The Netherlands: Springer.

Nussaume, L., Kanno, S., Javot, H., Marin, E., Pochon, N., Ayadi, A., Nakanishi, T. M., & Thibaud, M. C. (2011). Phosphate import in plants: Focus on the PHT1 transporters. *Frontiers in Plant Science, 2*, 83.

O'Lochlainn, S. O., Amoah, S., Graham, N. S., Alamer, K., Rios, J. J., Kurup, S., Stoute, A., Hammond, J. P., Ostergaard, L., King, G. J., White, P. J., & Broadley, M. R. (2011). High Resolution Melt (HRM) analysis is an efficient tool to genotype EMS mutants in complex crop genomes. *Plant Methods, 7*, 43.

O'Neill, M. A., Eberhard, S., Albersheim, P., & Darvill, A. G. (2001). Requirement of borate cross-linking of cell wall rhamnogalacturonan II for Arabidopsis growth. *Science, 294*, 846–849.

Orsel, M., Filleur, S., Fraisier, V., & Daniel-Vedele, F. (2002). Nitrate transport in plant: Which gene and which control? *Journal of Experimental Botany, 53*, 825–833.

Padmanaban, S., Chanroj, S., Kwak, J. M., Li, X., Ward, J. M., & Sze, H. (2007). Participation of endomembrane cation/H^+ exchanger *AtCHX20* in osmoregulation of guard cells. *Plant Physiology, 44*, 82–93.

Pandey, S., Zhang, W., & Assmann, S. M. (2007). Roles of ion channels and transporters in guard cell signal transduction. *FEBS Letters, 581*, 2325–2336.

Pandey, P., Srivastava, R. K., & Dubey, R. S. (2013). Salicylic acid alleviates aluminum toxicity in rice seedlings better than magnesium and calcium by reducing aluminum uptake, suppressing oxidative damage and increasing antioxidative defense. *Ecotoxicology, 22*, 656–670.

Park, S., Cheng, N. H., Pittman, J. K., Yoo, K. S., Park, J., Smith, R. H., & Hirschi, K. D. (2005). Increased calcium levels and prolonged shelf life in tomatoes expressing Arabidopsis H^+/Ca^{2+} transporters. *Plant Physiology, 139*, 1194–1206.

Parmar, S., Buchner, P., & Hawkesford, M. J. (2007). Leaf developmental stage affects sulfate depletion and specific sulfate transporter expression during sulfur deprivation and specific sulfate transporter expression during sulfur deprivation in *Brassica napus* L. *Plant Biology, 9*, 647–653.

Pehlivan, N., Sun, L., Jarrett, P., Yang, X., Mishra, N., Chen, L., Kadioglu, S., Shen, G., & Zhang, H. (2016). Co-overexpressing a plasma membrane and a vacuolar membrane sodium/proton antiporter significantly improves salt tolerance in transgenic Arabidopsis plants. *Plant & Cell Physiology, 57*(5), 1069–1084. doi:10.1093/pcp/pcw055.

Pittman, J., & Hirschi, K. (2016). CAX-ing a wide net: Cation/H^+ transporters in metal remediation and abiotic stress signaling. *Plant Biology, 18*(5), 741–749. doi:10.1111/plb.12460.

Podgornik, M., & Pintar, M. (2007). Causes of nitrate leaching from agriculture land in Slovenia. *Acta Agriculturae Slovenica, 89*, 207–220.

Poirier, Y., & Bucher, M. (2002). Phosphate transport and homeostasis in Arabidopsis. *The Arabidopsis Book, 1*, e0024. doi:10.1199/tab.0024.

Politi, Y., Batchelor, D., Zaslansky, P., Chmelka, B., Weaver, J., Sagi, I., Weiner, S., & Addadi, L. (2010). Role of magnesium ion in the stabilization of biogenic amorphous calcium carbonate: A structure–function investigation. *Chemistry of Materials, 22*, 161–166.

Poschenrieder, C., Llugany, M., & Barcelo, J. (1995). Short-term effects of pH and aluminum on mineral-nutrition in maize varieties differing in proton and aluminum tolerance. *Journal of Plant Nutrition, 18*, 1495–1507.

Puig, S., Andrés-Colás, N., García-Molina, A., & Peñarrubia, L. (2007). Copper and iron homeostasis in Arabidopsis: Responses to metal deficiencies, interactions and biotechnological applications. *Plant, Cell & Environment, 30*, 271–290.

Pyo, Y. J., Gierth, M., Schroeder, J. I., & Cho, M. H. (2010). High-affinity K^+ transport in Arabidopsis: *AtHAK5* and *AKT1* are vital for seedling establishment and postgermination growth under low-potassium conditions. *Plant Physiology, 153*, 863–875.

Ramesh, S. A., Choimes, S., & Schachtman, D. P. (2004). Overexpression of an Arabidopsis zinc transporter in *Hordeum vulgare* increases short-term zinc uptake after zinc deprivation and seed zinc content. *Plant Molecular Biology, 54*, 373–385.

Rausch, C., & Bucher, M. (2002). Molecular mechanisms of phosphate transport in plants. *Planta, 216*, 23–37.

Remy, E., Cabrito, T. R., Batista, R. A., Teixeira, M. C., Sá-Correia, I., & Duque, P. (2012). The *Pht1;9* and *Pht1;8* transporters mediate inorganic phosphate acquisition by the *Arabidopsis thaliana* root during phosphorus starvation. *The New Phytologist, 195*, 356–371.

Rengel, Z. (2001). Genotypic differences in micronutrient use efficiency in crops. *Communications in Soil Science and Plant Analysis, 32*, 1163–1186.

Rengel, Z. (2003). Heavy metals as essential nutrients. In M. N. V. Prasad & J. Hagemeyer (Eds.), *Heavy metal stress in plants: Molecules to ecosystems* (2nd ed., pp. 271–294). Berlin: Springer-Verlag.

Rengel, Z., & Damon, P. M. (2008). Crops and genotypes differ in efficiency of potassium uptake and use. *Physiologia Plantarum, 133*, 624–636.

Rengel, Z., & Marschner, P. (2005). Nutrient availability and management in the rhizosphere: Exploiting genotypic differences. *The New Phytologist, 168*, 305–312.

Rengel, Z., Bose, J., Chen, Q., & Tripathi, B. N. (2015). Magnesium alleviates plant toxicity of aluminium and heavy metals. *Crop & Pasture Science, 66*, 1298–1307.

Ricachenevsky, F. K., Menguer, P. K., Sperotto, R. A., & Fett, J. P. (2015). Got to hide your Zn away: Molecular control of Zn accumulation and biotechnological applications. *Plant Science, 236*, 1–17.

Richardson, A. E., Barea, J. M., McNeill, A. M., & Prigent-Combaret, C. (2009). Acquisition of phosphorus and nitrogen in the rhizosphere and plant growth promotion by microorganisms. *Plant and Soil, 321*, 305–339.

Rigas, S., Debrosses, G., Haralampidis, K., Vicente-Agullo, F., Feldmann, K. A., Grabov, A., Dolan, L., & Hatzopoulos, P. (2001). *TRH1* encodes a potassium transporter required for tip growth in Arabidopsis root hairs. *Plant Cell, 13*, 139–151.

Riquelme, A., Wellmann, E., & Pinto, M. (2007). Effects of ultraviolet-B radiation on common bean (*Phaseolus vulgaris* L.) plants grown under nitrogen deficiency. *Environmental and Experimental Botany, 60*, 360–367.

Rosanoff, A., Weaver, C., & Rude, R. (2013). Suboptimal magnesium status in the United States: Are the health consequences underestimated? *Nutrition Reviews, 70*, 153–164.

Rouached, H., Wirtz, M., Alary, R., Hell, R., Arpat, A. B., Davidian, J. C., Fourcroy, P., & Berthomieu, P. (2008). Differential regulation of the expression of two high-affinity sulfate transporters, *SULTR1.1* and *SULTR1.2*, in Arabidopsis. *Plant Physiology, 147*, 897–911.

Rouached, H., Secco, D., Arpat, B., & Poirier, Y. (2011). The transcription factor *PHR1* plays a key role in the regulation of sulfate shoot-to-root flux upon phosphate starvation in Arabidopsis. *BMC Plant Biology, 11*, 19.

Rout, G. R., & Sahoo, S. (2015). Role of iron in plant growth and metabolism. *Reviews in Agricultural Science, 3*, 1–24.

Rubio, F., Gassmann, W., & Schroeder, J. I. (1995). Sodium-driven potassium uptake by the plant potassium transporter *HKT1* and mutations conferring salt tolerance. *Science, 270*, 1660–1663.

Ruiter, H. J. (1969). Suspected copper deficiency in radiata pine. *Plant and Soil, 31*, 197–200.

Saha, S., Mandal, B., Hazra, G. C., Dey, A., Chakraborty, M., Adhikari, B., Mukhopadhyay, S. K., & Sadhukhan, R. (2015). Can agronomic biofortification of zinc be benign for iron in cereals? *Journal of Cereal Science, 65*, 186–191.

Schmidt, W. (2006). Iron stress responses in roots of strategy I plants, chapter 11. In L. L. Barton & J. Abadia (Eds.), *Iron nutrition in plants and rhizospheric microorganisms* (pp. 1–435). Dordrecht, Netherlands: Springer.

Schock, I., Gregan, J., Steinhauser, S., Schweyen, R., Brennicke, A., & Knoop, V. (2000). A member of a novel *Arabidopsis thaliana* gene family of candidate Mg^{2+} ion transporters complements a yeast mitochondrial group II intron-splicing mutant. *The Plant Journal, 24*, 489–501.

Schon, M. K., & Blevins, D. G. (1990). Foliar boron applications increase the final number of branches and pods on branches of field-grown soybeans. *Plant Physiology, 92*, 602–607.

Schroeder, J. I., Hedrich, R., & Fernandez, J. M. (1984). Potassium-selective single channels in guard cell protoplasts of *Vicia faba*. *Nature, 312*, 361–362.

Secco, D., Jabnoune, M., Walker, H., Shou, H., Wu, P., Poirier, Y., & Whelan, J. (2013). Spatio-temporal transcript profiling of rice roots and shoots in response to phosphate starvation and recovery. *Plant Cell, 25*, 4285–4304.

Segonzac, C., Boyer, J. C., Ipotesi, E., Szponarski, W., Tillard, P., Touraine, B., Sommerer, N., Rossignol, M., & Gibrat, R. (2007). Nitrate efflux at the root plasma membrane: Identification of an Arabidopsis excretion transporter. *Plant Cell, 19*, 3760–3777.

Shams, M., Etemadi, N., Baninasab, B., Ramin, A. A., & Khoshgoftarmanesh, A. H. (2010). Effect of boron and calcium on growth and quality of 'easy lover' cut rose. *Journal of Plant Nutrition, 35*, 1303–1313.

Shaul, O., Hilgemann, D., de-Almeida-Engler, J., Van Montagu, M., Inze, D., & Galili, G. (1999). Cloning and characterization of a novel Mg^{2+}/H^+ exchanger. *The EMBO Journal, 18*, 3973–3980.

Shen, J., Yuan, L., Zhang, J., Li, H., Bai, Z., Chen, X., Zhang, W., & Zhang, F. (2011). Phosphorus dynamics: From soil to plant. *Plant Physiology, 156*, 997–1005.

Shin, H., Shin, H. S., Dewbre, G. R., & Harrison, M. J. (2004). Phosphate transport in Arabidopsis: *Pht1;1* and *Pht1;4* play a major role in phosphate acquisition from both low- and high-phosphate environments. *The Plant Journal, 39*, 629–642.

Shukla, R., Sharma, Y. K., & Shukla, A. K. (2014). Molecular mechanism of nutrient uptake in plant. *International Journal of Current Research and Academic Review, 2*, 142–154.

Siemianowski, O., Mills, R. F., Williams, L. E., & Antosiewicz, D. M. (2011). Expression of the P1B-type ATPase AtHMA4 in tobacco modifies Zn and Cd root to shoot partitioning and metal tolerance. *Plant Biotechnology Journal, 9*, 64–74.

Singh, S., Korripally, P., Vancheeswaran, R., & Eapen, S. (2011). Transgenic *Nicotiana tabacum* plants expressing a fungal copper transporter gene show enhanced acquisition of copper. *Plant Cell Reports, 30*, 1929–1938.

Sirichandra, C., Wasilewska, A., Vlad, F., Valon, C., & Leung, J. (2009). The guard cell as a single-cell model towards understanding drought tolerance and abscisic acid action. *Journal of Experimental Botany, 60*, 1439–1463.

Slamet-Loedin, I. H., Johnson-Beebout, S. E., Impa, S., & Tsakirpalogloul, N. (2015). Enriching rice with Zn and Fe while minimizing Cd risk. *Frontiers in Plant Science, 6*, 121.

Smith, F. W., Ealing, P. M., Dong, B., & Delhaize, E. (1997). The cloning of two Arabidopsis genes belonging to a phosphate transporter family. *The Plant Journal, 11*, 83–92.

Smyth, D. A., & Chevalier, P. (1984). Increases in phosphatase and b-glucosidase activities in wheat seedlings in response to phosphorus-deficient growth. *Journal of Plant Nutrition, 7*, 1221–1231.

Song, C. P., Guo, Y., Qiu, Q., Lambert, G., Galbraith, D. W., Jagendorf, A., & Zhu, J. K. (2004). A probable $Na^+(K^+)/H^+$ exchanger on the chloroplast envelope functions in pH homeostasis and chloroplast development in *Arabidopsis thaliana*. *Proceedings of the National Academy of Sciences of the United States of America, 101*, 10211–10216.

Sonoda, Y., Ikeda, A., Yamaya, T., & Yamaguchi, J. (2004). Feedback regulation of the ammonium transporter gene family *AMT1* by glutamine in rice. *Plant & Cell Physiology, 45*, S98–S98.

Sreedhara, A., & Cowan, J. (2002). Structural and catalytic roles for divalent magnesium in nucleic acid biochemistry. *Biometals, 15*, 211–223.

Takahashi, H., Watanabe-Takahashi, A., Smith, F. W., Blake-Kalff, M., Hawkesford, M. J., & Saito, K. (2000). The roles of three functional sulphate transporters involved in uptake and translocation of sulphate in *Arabidopsis thaliana*. *The Plant Journal, 23*, 171–182.

Takano, J., Noguchi, K., Yasumori, M., Kobayashi, M., Gajdos, Z., Miwa, K., Hayashi, H., Yoneyama, T., & Fujiwara, T. (2002). Arabidopsis boron transporter for xylem loading. *Nature, 420*, 337–340.

Takano, J., Wada, M., Ludewig, U., Schaaf, G., von Wirén, N., & Fujiwara, T. (2006). The Arabidopsis major intrinsic protein NIP5;1 is essential for efficient boron uptake and plant development under boron limitation. *Plant Cell, 18*, 1498–1509.

Tan, S., Han, R., Li, P., Yang, G., Li, S., Zhang, P., Wang, W. B., Zhao, W. Z., & Yin, L. P. (2015). Overexpression of the *MxIRT1* gene increases iron and zinc content in rice seeds. *Transgenic Research, 24*, 109–122.

Tognetti, V. B., Zurbriggen, M. D., Morandi, E. N., Fillat, M. F., Valle, E. M., Hajirezaei, M. R., & Carrillo, N. (2007). Enhanced plant tolerance to iron starvation by functional substitution of chloroplast ferredoxin with a bacterial flavodoxin. *Proceedings of the National Academy of Sciences of the United States of America, 104*, 11495–11500.

Tsay, Y. F., Schroeder, J. I., Feldmann, K. A., & Crawford, N. M. (1993). The herbicide sensitivity gene *CHL1* of Arabidopsis encodes a nitrate-inducible nitrate transporter. *Cell, 72*, 705–713.

Uraguchi, S., Kato, Y., Hanaoka, H., Miwa, K., & Fujiwara, T. (2014). Generation of boron-deficiency-tolerant tomato by overexpressing an *Arabidopsis thaliana* borate transporter *AtBOR1*. *Frontiers in Plant Science, 5*, 125.

Vance, C. P., Uhde-Stone, C., & Allan, D. (2003). Phosphorus acquisition and use: Critical adaptation by plants for securing non-renewable resources. *The New Phytologist, 15*, 423–447.

Velu, G., Ortiz-Monasterio, I., Cakmak, I., Hao, Y., & Singh, R. P. (2014). Biofortification strategies to increase grain zinc and iron concentrations in wheat. *Journal of Cereal Science, 59*, 365–372.

Verret, F., Gravot, A., Auroy, P., Preveral, S., Forestier, C., Vavasseur, A., & Richaud, P. (2004). Overexpression of *AtHMA4* enhances root-to-shoot translocation od zinc and cadmium and plant metal tolerance. *FEBS Letters, 576*, 306–312.

Versaw, W. K., & Harrison, M. J. (2002). A chloroplast phosphate transporter, *PHT2;1*, influences allocation of phosphate within the plant and phosphate-starvation responses. *Plant Cell, 14*, 1751–1766.

Vert, G., Grotz, N., Dedaldechamp, F., Gaymard, F., Guerinot, M. L., Briat, J. F., & Curie, C. (2002). IRT1, an Arabidopsis transporter essential for iron uptake from the soil and for plant growth. *Plant Cell, 14*, 1223–1233.

Véry, A. A., & Sentenac, H. (2003). Molecular mechanisms and regulation of K^+ transport in higher plants. *Annual Review of Plant Biology, 54*, 575–603.

Wang, Y. Y., & Tsay, Y. F. (2011). Arabidopsis nitrate transporter *NRT1.9* is important in phloem nitrate transport. *Plant Cell, 23*, 1945–1957.

Wang, Y., & Wu, W. H. (2015). Genetic approaches for improvement of the crop potassium acquisition and utilization efficiency. *Current Opinion in Plant Biology, 25*, 46–52.

Wang, R., Liu, D., & Crawford, N. M. (1998). The Arabidopsis CHL1 protein plays a major role in high-affinity nitrate uptake. *Proceedings of the National Academy of Sciences of the United States of America, 95*, 15134–15139.

Wang, Y., Ribot, C., Rezzonico, E., & Poirier, Y. (2004). Structure and expression profile of the Arabidopsis *PHO1* gene family indicates a broad role in inorganic phosphate homeostasis. *Plant Physiology, 135*, 400–411.

Wang, G. Y., Shi, J. L., Ng, G., Battle, S. L., Zhang, C., & Lu, H. (2011). Circadian clock-regulated phosphate transporter *PHT4;1* plays an important role in Arabidopsis defense. *Molecular Plant, 4*, 516–526.

Wang, W. H., Chen, J., Liu, T. W., Chen, J., Han, A. D., Simon, M., Dong, X. J., He, J. X., & Zheng, H. L. (2014). Regulation of the calcium-sensing receptor in both stomatal movement and photosynthetic electron transport is crucial for water use efficiency and drought tolerance in Arabidopsis. *Journal of Experimental Botany, 65*, 223–234.

Wang, C., Yue, W., Ying, Y., Wang, S., Secco, D., Liu, Y., Whelan, J., Tyerman, S. D., & Shou, H. (2015). Rice SPX-Major Facility Super family 3, a vacuolar phosphate efflux transporter, is involved in maintaining phosphate homeostasis in rice. *Plant Physiology, 169*, 2822–2831.

Ward, J. M., Mäser, P., & Schroeder, J. I. (2009). Plant ion channels: Gene families, physiology, and functional genomics analyses. *Annual Review of Physiology, 71*, 59–82.

Waters, B., & Grusak, M. (2008). Quantitative trait locus mapping for seed mineral concentrations in two *Arabidopsis thaliana* recombinant inbred populations. *The New Phytologist, 179*, 1033–1047.

Wegner, L. H., & Raschke, K. (1994). Ion channels in the xylem parenchyma of barley roots. *Plant Physiology, 105*, 799–813.

Weng, H., Yoo, C., Gosney, M., Hasegawa, P., & Mickelbart, M. (2012). Poplar *GTL1* is a Ca^{2+}/calmodulin-binding transcription factor that functions in plant water use efficiency and drought tolerance. *PloS One, 7*, e32925.

White, P. J., & Broadley, M. R. (2003). Calcium in plants. *Annals of Botany, 92*, 487–511.

Williams, L., & Miller, A. (2001). Transporters responsible for the uptake and partitioning of nitrogenous solutes. *Annual Review of Plant Physiology and Plant Molecular Biology, 52*, 659–688.

Wintz, H., Fox, T., YY, W., Feng, V., Chen, W., Chang, H. S., Zhu, T., & Vulpe, C. (2003). Expression profiles of *Arabidopsis thaliana* in mineral deficiencies reveal novel transporters involved in metal homeostasis. *The Journal of Biological Chemistry, 278*, 47644–47653.

Xiong, H., Guo, X., Kobayashi, T., Kakei, Y., Nakanishi, H., Nozoye, T., Zhang, L., Shen, H., Qiu, W., Nishizawa, N. K., & Zuo, Y. (2014). Expression of peanut iron regulated transporter 1 in tobacco and rice plants confers improved iron nutrition. *Plant Physiology and Biochemistry, 80*, 83–89.

Yang, T. J., Perry, P. J., Ciani, S., Pandian, S., & Schmidt, W. (2008). Manganese deficiency alters the patterning and development of root hairs in Arabidopsis. *Journal of Experimental Botany, 59*, 3453–3464.

Yang, S. Y., Grønlund, M., Jakobsen, I., Grotemeyer, M. S., Rentsch, D., Miyao, A., Hirochika, H., Kumar, C. S., Sundaresan, V., Salamin, N., Catausan, S., Mattes, N., Heuer, S., & Paszkowski, U. (2012). Nonredundant regulation of rice arbuscular mycorrhizal symbiosis by two members of the *PHOSPHATE TRANSPORTER1* gene family. *Plant Cell, 24*, 4236–4251.

Yao, Y. A., Sun, H. Y., FS, X., Zhang, X. J., & Liu, S. Y. (2011). Comparative proteome analysis of metabolic changes by low phosphorus stress in two *Brassica napus* genotypes. *Planta, 233*, 523–537.

Yi, K., Wu, Z., Zhou, J., Du, L., Guo, L., Wu, Y., & Wu, P. (2005). *OsPTF1*, a novel transcription factor involved in tolerance to phosphate starvation in rice. *Plant Physiology, 138*, 2087–2096.

Yokosho, K., Yamaji, N., Ueno, D., Mitani, N., & Ma, J. F. (2009). OsFRDL1 is a citrate transporter required for efficient translocation of iron in rice. *Plant Physiology, 149*, 297–305.

Yuan, L., Loqué, D., Kojima, S., Rauch, S., Ishiyama, K., Inoue, E., Takahashi, H., & von Wirén, N. (2007). The organization of high-affinity ammonium uptake in Arabidopsis roots depends on the spatial arrangement and biochemical properties of AMT1-type transporters. *Plant Cell, 19*, 2636–2652.

Yuan, N., Yuan, S., Li, Z., Li, D., & Hu Q Luo, H. (2016). Heterologous expression of a rice miR395 gene in *Nicotiana tabacum* impairs sulfate homeostasis. *Scientific Reports, 6*, 28791.

Zhang, J., Liu, B., Li, M., Feng, D., Jin, H., Wang, P., Liu, J., Xiong, F., Wang, J., & Wang, H. B. (2015). The bHLH transcription factor bHLH104 interacts with IAA-LEUCINE RESISTANT3 and modulates iron homeostasis in Arabidopsis. *Plant Cell, 27*, 787–805.

Zhao, L., Versaw, W. K., Liu, J., & Harrison, M. J. (2003). A phosphate transporter from *Medicago truncatula* is expressed in the photosynthetic tissues of the plant and located in the chloroplast envelope. *The New Phytologist, 157*, 291–302.

Zhao, J., Cheng, N. H., Motes, C. M., Blancaflor, E. B., Moore, M., Gonzales, N., Padmanaban, S., Sze, H., Ward, J. M., & Hirschi, K. D. (2008). *AtCHX13* is a plasma membrane K^+ transporter. *Plant Physiology, 148*, 796–807.

Zhao, H., Zhou, Q., Zhou, M., Li, C., Gong, X., Liu, C., Qu, C., Wang, L., Si, W., & Hong, F. (2012). Magnesium deficiency results in damage of nitrogen and carbon cross-talk of maize and improvement by cerium addition. *Biological Trace Element Research, 148*, 102–109.

Zhao, Q., Ren, Y. R., Wang, Q. J., Yao, Y. X., You, C. X., & Hao, Y. J. (2016). Overexpression of *MdbHLH104* gene enhances the tolerance to iron deficiency in apple. *Plant Biotechnology Journal, 14*(7), 1633–1645. doi:10.1111/pbi.12526.

Zhu, W., Miao, Q., Sun, D., Guodong, Y., Wu, C., Huang, J., & Zheng, C. (2012). The mitochondrial phosphate transporters modulate plant responses to salt stress via affecting ATP and gibberellin metabolism in *Arabidopsis thaliana*. *PloS One, 7*, e43530.

Zuber, H., Aubert, G., Davidian, J. C., Thompson, R., & Gallardo, K. (2009). Sulfur metabolism and transport in developing seeds. In A. Sirko, L. J. De Kok, S. Haneklaus, M. J. Hawkesford, H. Rennenberg, K. Saito, E. Schnug, & I. Stulen (Eds.), *Sulfur metabolism in plants* (pp. 113–118). Leiden: Backhuys Publishers.

Zuber, H., Davidian, J. C., Aubert, G., Aimé, D., Belghazi, M., Lugan, R., Heintz, D., Wirtz, M., Hell, R., Thompson, R., & Gallardo, K. (2010). The seed composition of Arabidopsis mutants for the group 3 sulfate transporters indicates a role in sulfate translocation within developing seeds. *Plant Physiology, 154*, 913–926.

Part V
Management of Plant Nutrients

Chapter 18
Nutrient Management for Improving Crop, Soil, and Environmental Quality

Kaushik Batabyal

Abstract Agricultural intensification and mechanization has resulted overall deterioration in soil-based ecosystem making it poor reserve of nutrients and organic matter and contributes to loss of biodiversity, thereby damaging sustainability of agricultural production, soil resilience capacity, and environmental quality. All these necessitate adopting strategies to improve resource use efficiency to meet burgeoning demand for food from shrinking land areas. Efficient nutrient management strategies pave the way to combat the challenges by tackling the over and under use of nutrients, checking different kinds of losses from the system, and improving use efficiency of the crops. Application of targeted, sufficient, and balanced quantities of inorganic fertilizers will be necessary to make nutrients available for high yields without polluting the environment. At the same time, every effort should be made to improve the availability and use of secondary nutrients and micronutrients, organic fertilizers, and soil conservation practices to augment crop yield and quality in an efficient and environmentally benign manner, without sacrificing soil health and/or productivity of future generations.

Keywords Agricultural intensification • Soil-based ecosystem • Nutrient management • Soil resilience • Environmental quality • Micronutrients • Soil health

K. Batabyal (✉)
Department of Agricultural Chemistry and Soil Science, Bidhan Chandra Krishi Viswavidyalaya, Mohanpur, Nadia, West Bengal 741252, India

Laboratory of Plant Stress Responses, Faculty of Agriculture, Kagawa University, Miki-cho, Kita-gun, Kagawa 761-0795, Japan
e-mail: kbatabyal@rediffmail.com

M. Naeem et al. (eds.), *Essential Plant Nutrients*,
DOI 10.1007/978-3-319-58841-4_18

18.1 Introduction

The present era of intensive high-yield agriculture is much dependent on chemical fertilizers. Agricultural intensification over the last few decades has strongly increased global food production, but at high environmental cost (Conway 1997; Tilman et al. 2001), and generally leads to imbalances in nutrient budgets and increased loss of potential pollutants to the atmosphere, particularly ammonia and greenhouse gases like nitrous oxide and methane, and to water, particularly nitrate and phosphate. Moreover, in large areas of the world, future production potential was compromised by soil erosion and soil salinization. Agricultural intensification can have negative local consequences, such as increased erosion, lower soil fertility, and reduced biodiversity; negative regional consequences, such as pollution of groundwater and eutrophication of rivers and lakes; and negative global consequences, including impacts on atmospheric constituents and climate.

In the later part of the twentieth century, scientists expressed concern about nutrient enrichment of surface water bodies and groundwater. Fertilizer recommendation began incorporation of stronger references to managing fertilizers for crop productivity in the context of protecting natural resources (soil and water) and the environment. Due to environmental, economic (e.g., rising cost of fossil fuels), and land size constraints, further increases in food supplies projected for the coming 50 years must be attained through improved resource use efficiency rather than more agricultural inputs, especially N and P fertilizer applications (Matson et al. 1997; Cassman 1999; Tilman et al. 2002).

The use of mineral fertilizers is among key factors driving the increased global agricultural production required to feed the rising human population. Synthetic N compounds produce roughly half of today's world food (Erisman et al. 2008). The global use of fertilizers is highly unbalanced: overfertilization in North America, Western Europe, China, and India causes environmental pollution, while underfertilization in Africa, Eurasia, and parts of Latin America causes soil mining (National Geographic 2013). In addition, there is often an imbalance in the ratio of nutrients. This is largely because the recovery rates of fertilizer N, P, and K by crops are as low as 50%, 25%, and 40%, respectively (Prasad 2009). Imbalanced and excess use of fertilizer nutrients to crops may lead to NO_3^- enrichment in groundwater (Kohler et al. 2006) and also in fleshy vegetables (Ju et al. 2007), high N_2O emission from soils (Eichner 1990; Bouwman et al. 2002), and preponderance of unwelcome guests (pests and diseases) onto our crop plants (Huber 1980; Krauss 1999; Altieri and Nicholls 2003). Further, it is established that imbalanced fertilization or application of only N causes net depletion of soil organic carbon (SOC) stock (Mandal et al. 2007). Again, inappropriate P fertilizer use or mismanagement vis-a-vis significant losses in various forms would lead not only to exhaust the present stock within a few decades but also might cause serious environmental problems. Major losses in the form of runoff and erosion and leaching through soil are known to cause global epidemic of eutrophication in freshwater, estuarine, and near-shore ocean environments, loss in potable water resources, aquatic biodiversity, and formation of large ocean "dead zone." Micronutrient deficiencies are also widespread in well-developed industrialized countries. More than three billion people globally suffer from Fe and Zn deficiencies (Graham et al. 2001). Heavy and monotonous consumption of cereal-based foods with low concentrations and reduced bioavailability of Fe and Zn has been considered a major reason for the widespread deficiencies of Zn and Fe in developing countries (Welch and Graham 1999; Graham et al. 2001).

In achieving the aim of fertilizers, the desire is that the nutrients end up only in the target plant. In reality, however, nutrients in fertilizers do not all end up in the plant; up to 20–80% of nutrients in fertilizers may be lost to the environment or temporarily accumulate in the soil due to several complex soil chemistries that preclude their immediate availability to plant. Presently, emphasis has been given on improving the use efficiency of fertilizers through the 4R nutrient stewardship principle, i.e., the use of fertilizer from the right source, at the right rate and at the right time, and with the right placement (IPNI 2014). A range of agronomic practices are pursued to implement the 4R approach, including precision application, deep placement, row application, coating of fertilizers for slow release to check nutrient losses, and improved availability of nutrients to plant (Chien et al. 2009). Unfortunately, the overall progress achieved through these practices has been insufficient to address the flaws of current fertilizers. In order to avoid the accompanying acceleration of environmental degradation, the efficiency of fertilizer nutrient use must be increased greatly. The capacity of the soil system to supply nutrients and retain applied nutrients is undermined by practices that diminish the role of soil organisms and lead to depletion in soil organic matter.

The real challenge is to keep the space of production under condition of decreasing per capita arable land without losing land productivity. In spite of continued growth of inputs, there has been no matching growth in agricultural production, indicating a decrease in total factor productivity due to (1) wide nutrient gap between nutrient demand and supply, (2) soil nutrient imbalances under intensive cropping system, (3) high nutrient turnover in soil-plant system coupled with low and imbalanced fertilizer use, (4) emerging deficiencies of secondary nutrients and micronutrients in soils, and, of course, (5) soil degradation. Formulation of efficient nutrient management strategies may help in bridging the gap.

18.2 Efficient Nutrient Management Strategy

Nutrient management is the internationally accepted strategy to address farm non-point source or field nutrient losses. A tactical nutrient management plan developed from this process must be based on a firm set of strategic objectives agreed upon by the farmers and society. Traditionally nutrient management has been concerned with optimizing the economic return from nutrients used for crop production. Today, the agronomic and economic requirements of nutrient management remain central, but in addition, the process must consider the potential impact of long-term use of these nutrients on crop, soil, and environmental quality.

The consumption of nitrogen and phosphorus fertilizers has increased manifold since green revolution compared to potash fertilizer. Further increases in nitrogen and phosphorus application are unlikely to be as effective at increasing yields because of diminishing returns which means that the efficiency is highest with the first increments of added fertilizer nutrient and it declines at higher levels of addition. The use efficiency of N, P, K, and micronutrients varies from 25 to 35, 15 to 20, 40 to 50, and below 5%, respectively. The reasons behind low use efficiency of applied nutrient elements are attributed due to the fact that the fertilizer use is inadequate (around 8–10 mt of negative nutrient balance in Indian soil), imbalanced, nonintegrated, and poorly managed. These cause significant amount to be lost from agricultural fields and reach to nonpoint sources like river, pond, lake, etc. Such nonpoint nutrient losses harm off-site ecosystems, water quality, and aquatic ecosystems and contribute to changes in atmospheric composition. Nitrogen loading to estuaries and coastal waters and phosphorus loading to lakes, rivers, and streams are responsible for over-enrichment, eutrophication, and low oxygen conditions that endanger fisheries. Again, cumulative negative nutrient balances heighten the impact of climatic factors, insecure tenure arrangements, and land and demographic pressures on soil fertility. To address the above problems, it is urgent to increase nutrient use efficiency through wise management of costly fertilizer nutrient which is the most challenging issue in sustainable food production.

Efficient nutrient management (ENM) practices can reduce trace gas and total losses of fertilizer and maintain yields. These ENM practices, which require greater knowledge about efficient use of nutrients, can substitute for higher levels of those inputs and might ultimately allow farmers to remain competitive in an era of economic liberalization and expanding free trade. At the same time, they reduce the environmental costs of agriculture, some of which are directly felt in the local regions, and others of which are globally important. An integration of agronomic knowledge of practical alternatives, economic analysis of on-farm costs and benefits, and biogeochemical analysis of their consequences in soils and the atmosphere can provide the basis for the identification or development of win-win solutions.

Management strategies for efficient management and utilization of plant nutrients for enhancing crop, soil, and environmental quality can be broadly grouped as follows.

18.3 Balanced Fertilization

The requirement of nutrients such as nitrogen, phosphate, and potash are soil and crop specific. The use of right ratio of nutrients as per soil or crop requirement is known as "balanced fertilization." Balanced fertilization should not only be confined with three major nutrients (N, P, K), and therefore, fertilization done based on the insufficiency of any plant nutrients should be the guiding principle of balanced fertilization. So, balanced fertilization should also include secondary nutrients and micronutrients, both of which are often most readily available from organic fertilizers such as animal and green manures. This is because application only of NPK for decades has induced micronutrient deficiency and soil deterioration (Bindraban et al. 2012), with yield responses of between 20 and 70% increases found upon application of secondary nutrients and micronutrients, depending on the crop (Voortman and Bindraban 2015). Balanced fertilization is nothing but application of nutrients either through fertilizer or organic manure or combination of both in right quantity and right proportion for optimum crop productivity without deteriorating soil health.

Global consumption of fertilizer is increasing at an alarming rate and at imbalanced proportion causing surplus of some nutrients while others become deficit in soil (Fig. 18.1; IFA 2013). The big three grain producers, i.e., China, India, and the United States, account for more than half of world fertilizer consumption. In the United States, the growth in fertilizer use came to an end in 1980. China's fertilizer use climbed rapidly in recent decades but has leveled off since 2007. In contrast, India's fertilizer consumption is still on the rise, growing 5% annually. Intensive farming practices, particularly with wheat (*Triticum aestivum* L.) and rice (*Oryza sativa* L.) in India, have virtually mined nutrients from the soil. The already imbalanced consumption ratio of 6.2:4:1 (N:P:K) in 1990–1991 has widened to 7:2.7:1 in 2000–2001 and 5:2:1 in 2009–2010 compared with a target ratio of 4:2:1 (Indian Fertilizer Scenario 2013). As food grain production increased with time, the number of elements deficient in Indian soils increased from one (N) in 1950 to nine (N, P, K, S, B, Cu, Fe, Mn, and Zn) in 2005–2006 (Shukla et al. 2014).

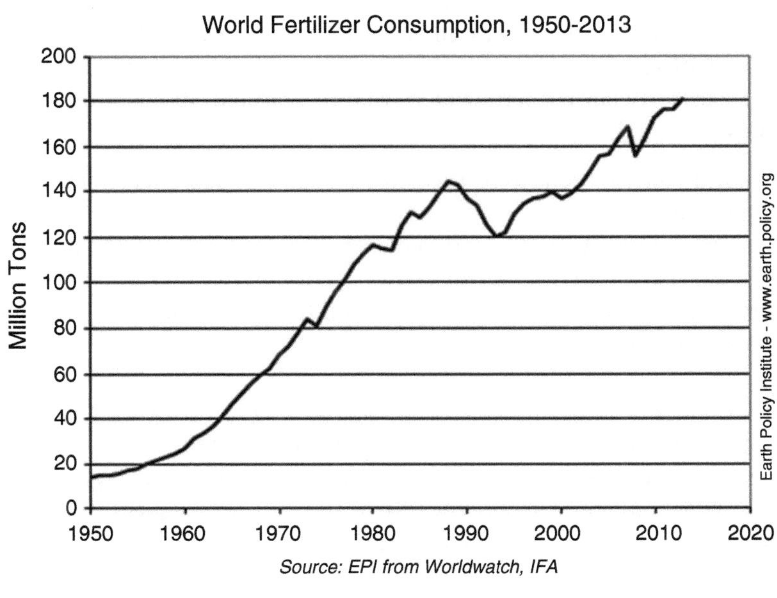

Fig. 18.1 Global fertilizer consumption (Source: IFA 2013)

Adequate and balanced application of fertilizer nutrients is one of the most common practices for improving the efficiency of N fertilizer and is equally effective in both developing and developed countries. For example, in many regions of India, crop production is still constrained much by imbalanced use of fertilizers than their too little application. In a review based on 241 sites of experiments in China, India, and North America, balanced fertilization with N, P, and K increased first-year recoveries an average of 54% compared to recoveries of only 21% where N was applied alone (Fixen et al. 2005). Multi-location, on-farm field experiments in India demonstrated the importance of balanced fertilization in increasing yield of crops and improving N use efficiency (Table 18.1). Based on several balanced nutrient management experiments, agronomic efficiency of applied N was improved by applying P and K fertilizers, by 6.7 kg sorghum grain kg^{-1} N, 10.3 kg pearl millet grain kg^{-1} N, and 19.5 kg maize grain kg^{-1} N. Nitrogen use efficiency improved from a deplorably low 6–20% in rain-fed pearl millet, maize, and sorghum (Prasad 2009).

The soil content of N and P that may increase as a result of long-term positive balances shows little relationship to N and P emissions. Simple N balances (N input minus crop off take) and especially cumulative balances over a number of years are of limited value in predicting the nitrate concentrations in soil water draining to rivers, because of the overriding impact of transient balances between soil and crop processes (Withers and Lord 2002). In particular, soil nitrate concentrations are not so much dependent on accumulated soil N, but on the short-term temporal balance between N mineralization, immobilization, and crop demand. Fundamental differences between N and P behavior in soil and in their mobility through the landscape, therefore, govern the time scales over which surpluses of N and P become important.

A long-term pragmatic policy for all fertilizer nutrients including secondary nutrients and micronutrients is essential to achieve balanced nutrition to overcome degradation of soil fertility and soil health and improve crop productivity on a sustainable basis. Balanced fertilization can be ensured through the following measures:

- Chemical nutrient needs to be used only on soil test-based recommendations in optimum quantities.
- Customized fertilizer and or fortified/coated fertilizers need to be promoted on large scale similar to other fertilizers.
- Micronutrients should be the part of any soil fertility or integrated plant nutrient supply (IPNS) program.

Table 18.1 Effect of balanced application of fertilizer N, P, and K on yield and agronomic efficiency of applied N in rain-fed crops in India (Source: Prasad 2009)

Crop	Yield (t/ha)			Agronomic efficiency (kg grain kg^{-1} N)	
	Control	N	NPK	N	NPK
Pearl millet	1.05	1.24	1.65	4.7	15.0
Maize	1.67	2.45	3.24	19.5	39.0
Sorghum	1.27	1.48	1.75	5.3	12.0

18.4 Balanced Fertilization and C Sequestration

Soil can act as a source or sink for atmospheric carbon dioxide. There is a critical need for the development of best management practices that enhance SOC sequestration as it not only helps mitigate climate change effect by offsetting the rising levels of carbon dioxide into the atmosphere but also improves the quality of soil that is important for sustainable agricultural production. Optimum level of soil organic matter can be maintained through crop rotation, fertility maintenance including use of inorganic fertilizers and organic manures, tillage methods, and other cropping system components. Among these, proper cropping system and balanced fertilization are believed to offer the greatest potential for increasing SOC storage in agricultural soils (Lal 2006).

Long-term field experiment (LTFE) results have demonstrated that judicious application of fertilizers with appropriate crop rotation and conservation tillage produced optimum biomass yield and lead to C sequestration in agricultural soils. The balanced application of NPK (100% or 150% NPK) showed higher accumulation of soil organic C over imbalanced use of fertilizers (100% N and 100% NP) in different cropping systems (maize-wheat-cowpea, rice-wheat-jute, maize-wheat, soybean-wheat) over three decades under dissimilar climate and soil (Table 18.2) (Bhattacharyya et al. 2004; Manna et al. 2006). Total organic C content in the entire 0–45 cm soil profile in maize-wheat-cowpea cropping system in Delhi, India, fol-

Table 18.2 Long-term manuring and fertilization effect on total organic C contents in soil

Soil	Inceptisol, Delhi, 31 years	Inceptisol, Barrackpore, West Bengal, 30 years	Alfisol, Palampur, Himachal Pradesh, 29 years	Inceptisol, Almora, Uttaranchal, 30 years
Cropping	Maize-wheat-cowpe[a]	Rice-wheat-jute	Maize-wheat	Soybean-wheat
Depth	0–45 cm	0–15 cm	0–15 cm	0–45 cm
Treatment				
50% NPK	51.5[cd]	–	9.4	–
100% NPK	54.1[c]	7.4[b]	8.9	22.81
150% NPK	63.5[b]	–	–	–
100% NP	53.0[c]	6.3[c]	7.3	21.05
100% N	51.1[d]	5.7[d]	8.6	–
100% NPK + FYM	71.1[a]	7.9[a]	12.7	31.42
Control	48.7[e]	5.1[e]	7.1	18.08
CD ($p = 0.05$)	–	–	0.4	–
Source	Manna et al. (2006)		Bhattacharyya et al. (2004)	

Table 18.3 Average annual crop residue C returned to soil through above- and below-ground biomass under balanced and imbalanced fertilization for a rice-wheat-jute cropping system (Source: Majumder et al. 2008)

Treatment/Cropping systems	Leaf C	Stubble biomass C	Root biomass C	Rhizode position C	Aquatic biomass C	Annual crop residue C return to soil
	Mg ha^{-1}					
Control	0.13	0.06	0.75	1.06	0.70	2.69
N	0.14	0.14	1.54	2.32	0.70	4.83
NP	0.16	0.17	1.85	2.74	0.70	5.62
NPK	0.20	0.17	2.11	2.97	0.70	6.15
NPK + FYM	0.25	0.19	2.37	3.18	0.70	6.68

lowed the order: 150% NPK + FYM > 150% NPK > 100% NPK > 100% NP = 100% *N* = 50% NPK > control. From the same experiment, it was reported that continuous adoption of 100% NPK + FYM treatment in maize-wheat-cowpea cropping system in semiarid subtropical India might sequester 1.83 Tg C year-1 which corresponds to about 1% of the fossil fuel emissions by India. Again, intensive agricultural practices with liberal use of fertilizers and manures have shown decline in organic C content in soils of Indo-Gangetic plains in India. What is the possible reason for such decline? Is it related to imbalanced fertilization of crops? Under imbalanced fertilization practices, quantity and quality of crop residues left over in the field/soil have been found to be inadequate to cause a net buildup in SOC. But soil test-based balanced fertilization may cause a net enrichment of SOC (Table 18.3). For sustenance of SOC level (zero change due to cropping) under rice-based cropping systems of the Indo-Gangetic plains of India, Mandal et al. (2007) reported that rice-rice system with balanced fertilization causes both a higher enrichment of SOC and also stabilization of applied C through amendments as compared to other rice-based cropping systems. Such higher enrichment of C in the rice-rice system has been ascribed to a number of factors, of which retardation of oxidation, inclusion of extra C inputs through photosynthetic organisms, and recalcitrant character to SOC under rice ecology are important.

18.5 Use of Customized Fertilizers

Customized fertilizers are multi-nutrient carrier designed to contain macro- and/or micronutrient forms, both from inorganic and organic sources, manufactured through a systematic process of granulation, satisfying the crop nutritional needs, specific to its site, soil, and stage, validated by a scientific crop model capability developed by an accredited fertilizer manufacturing or marketing company (Goel et al. 2011). It aims at a balanced distribution of plant nutrients in the fields and provides the best nutritional package for premium quality plant growth and yield (Shivey 2011). Moreover, customized fertilizers are farmers' system compatible. It maximizes crop yield without harming environment, increases fertilizer use efficiency, promotes nutrient balance, prevents multi-nutrient deficiencies, enhances crop productivity, improves cost-benefit ratio, restores soil health, and prevents pollution.

When fertilizer is applied in crop field, it segregates and causes uneven distribution near root zone and results unbalanced application and low response. On the other hand, urea prills, DAP granules, MOP crystals with or without sulfur pustules, and Zn/B powders upon bulk blending increase the specific gravity. Fusion blending of different fertilizers particles in customized fertilizer gives balanced application and efficient crop response. Application of 100% RDF through customized fertilizer registered an increase in rice yield of 22.2% over and above under application of conventional fertilizers. Addition of Zn through customized fertilizer could have attributed to increase in number of tillers and reduced the spikelet sterility (Kaleswari 2013).

18.6 Integrated Plant Nutrient Management

The basic concept underlying the principle of integrated plant nutrient management (IPNM) system is maintaining or adjusting plant nutrient supply to achieve a given level of crop production by optimizing the benefits from all possible sources of plant nutrients. It reduces inorganic fertilizer requirement, improves crop yield and quality, restores organic matter in soil, enhances nutrient use efficiency, and maintains soil quality in terms of physical, chemical, and biological properties. Therefore IPNM relies on a number of factors, including appropriate nutrient application and conservation and the transfer of knowledge about IPNM practices to farmers and researchers. Balanced application of appropriate fertilizers is a major component of IPNM. Development and formulation of efficient IPNM strategy can result in agronomically feasible, economically viable, and environmentally sound sustainable crop production systems by enhancing soil fertility, improving soil quality and C sequestration, reducing N losses and emission of greenhouse gases, and increasing energy conserving efficiency of the crop production system (Batabyal et al. 2016a). Classical field experiments at the Rothamsted Experimental Station in England have provided a wealth of integrated nutrient management related information on crops grown continuously and in rotation under a variety of soil fertility amendments. A number of lessons can be learned about appropriate and balanced fertilization from these experiments. Continuously cropped wheat, without the benefit of organic and inorganic fertilizers, typically has low yields, on the order of 1.2 tons per hectare. Short fallow rotations of 1–3 years have little effect on yields. The application of organic and inorganic fertilizers can increase average wheat yields to 6–7 tons per hectare. Wheat yields are highest (9.4 tons per hectare) when farmyard manure is applied, wheat is grown in rotation, and inorganic fertilizers are used to top up nitrogen availability (Rothamsted Experimental Station 1991; IFA 1995).

Besides, IPNM technology proved to be an economically sound and environment-friendly practice. It helped to produce better-quality crops with higher value-added products such as crude protein, dietary fiber, and vitamin C (Batabyal et al. 2016b).

Further, it concomitantly maintained better soil quality by improving soil organic carbon stock, microbial biomass carbon, bulk density, and extractable plant available nutrients. Mandal (2005) made a soil quality assessment study using a few long-term experiments with different management practices and found that cultivation without any fertilization (control) or only with N caused a net degradation of soil quality. Cultivation even with application of balanced NPK could hardly maintain such quality at the level where no cultivation was practiced. Only integrated use of organic and inorganic sources of nutrients has aggraded the system (Table 18.4). Batabyal et al. (2016b) also reported that INM involving conjoint application of organic and inorganic sources of nutrients for 6 years could stabilize, on average, 33.4% of the added C into soil organic C (SOC) stock under subtropical condition, and this was more so when FYM was the preferred organic (Fig. 18.2). The magnitude of stabilization of applied C into SOC was increased with increasing the level of organic C added through different sources as evidenced by the strong linear relationship between the changes in SOC stock (C sequestration) and the cumulative C inputs to the soils over the years, suggesting that even after 6 years of C additions at a reasonable rate (1.46 Mg ha^{-1} $year^{-1}$) the soils of the experimental site are still unsaturated in their capacity for storing C and, therefore, have great potential for further C sequestration. Mandal et al. (2007) studied the long-term effect of manuring and fertilization under various rice-based cropping systems on C buildup in some Inceptisols and reported that total soil organic C (0–0.20 m) was highest in the NPK + FYM (38.8 Mg C ha^{-1}) followed by NPK (35.2 Mg C ha^{-1}) > fallow (33.1 Mg C ha^{-1}) and control (28.2 Mg C ha^{-1}) treatments (Table 18.5). Therefore, sufficient and balanced application of organic and inorganic fertilizers is a major component of integrated nutrient management.

Formulation and adoption of careful strategies to propagate the long-term usefulness of IPNS in providing nutrients and improving the soil health, educative extension efforts about the economic and environmental benefits of IPNS, regulations for prohibiting the burning of crop residues, and some incentives for encouraging the crop residue incorporation as a means of disposal could lead to the adoption of such eco-friendly practices. For example, few proven INM technologies under Indian subtropical condition include:

(a) Green manuring in rice-wheat, rice-mustard, and rice-rapeseed is cost effective and economically viable.
(b) Under constrained water resources, GM produced during the mild-rainy season and applied to rapeseed is more beneficial than rice-applied GM.

Table 18.4 Soil quality change (as % over no cultivation fallow) under different management practices and cropping systems (Source: Mandal 2005)

Treatment/cropping systems	Rice-Rice	Rice-Wheat	Rice-Field pea	Rice-Lentil	Jute-Rice-Wheat
Control	−28.6	−56.0	−19.7	−8.0	−49.0
N only	−14.0	−9.1	−7.1	−11.7	−35.0
NPK only	0.9	−10.8	31.7	−9.7	19.0
NPK + FYM	35.0	18.7	51.2	8.6	45.1

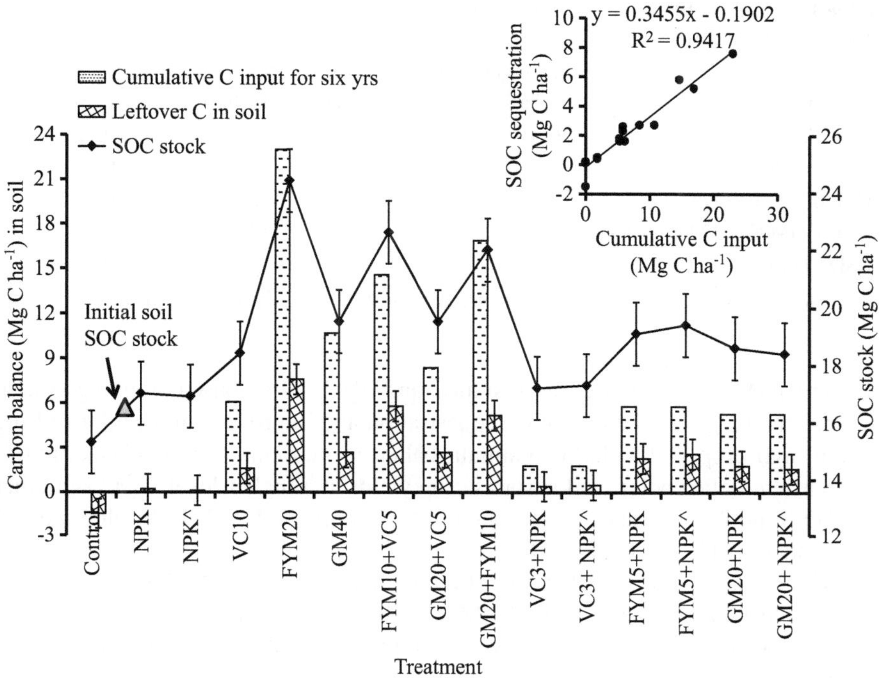

Fig. 18.2 Soil organic C (SOC) stock and C balance in soil after 6 years of cauliflower production. NPK, N-P-K at 200-44-82 kg ha^{-1}; NPK^, 125% of NPK; *VC* vermicompost, *FYM* farmyard manure, *GM* green manure; numbers followed by VC, FYM, and GM indicate dose in Mg ha^{-1} (Source: Batabyal et al. 2016b)

Table 18.5 Variation in total soil organic C content (0–0.2 m) across treatments and cropping systems (Source: Mandal et al. 2007)

Cropping system	Age (years)	Treatment					
		Initial soil	Control	NPK	NPK + FYM/Compost	Fallow	Mean
R-M-S	7	37.29	25.81^{Bc}	39.17^{Aab}	40.18^{Ba}	38.33^{Ab}	35.87^{AB}
R-W-F	19	33.99	30.13^{Ac}	35.18^{Bb}	37.19^{Ca}	34.02^{Bb}	34.13^{B}
R-F-B	20	30.45	28.99^{Ac}	31.68^{Cb}	36.14^{CDa}	31.37^{Cb}	32.05^{C}
R-W-J	34	27.91	26.16^{Bd}	30.06^{Db}	34.61^{Da}	28.49^{Dc}	29.83^{D}
R-F-R	38	31.60	29.77^{Ad}	39.74^{Ab}	46.09^{Aa}	33.17^{Bc}	37.19^{A}
Mean		32.25	28.17^{d}	35.17^{b}	38.84^{a}	33.08^{c}	33.81^{A}

R-M-S rice-mustard-sesame, *R-W-F* rice-wheat-fallow, *R-F-B* rice-fallow-berseem, *R-W-J* rice-wheat-jute, *R-F-R* rice-fallow-rice, where compost was used. Different capital letters within columns and different small letters within rows are significantly different at $P = 0.05$ according to Duncan multiple range test (DMRT) for separation of means

(c) Supply of nutrients through the integrated use of 20 t GM and 60 kg FN/ha provides advantages over the use of 120 kg FN/ha alone, producing greater yields of rice and wheat while reducing the use of FN by >50% in rice and 25% in wheat. It significantly reduces denitrification losses and diminishes the accumulation of residual NO_3^- in the soil profile and hence reduces the chances of NO3 leaching to groundwater providing environmental benefits (Table 18.6).

Table 18.6 Effect of INM on rice yield, denitrification losses, N_2O emissions, and soil organic C (Source: Aulakh 2010)

Treatment	Rice yield (q/ha)	Denitrification losses (kg/ha)	N_2O emissions (kg/ha)	Soil organic C (%)
Control	34	18	6.9	0.37
120 kg FN/ha	56	58	12.4	0.37
GM20 + 32 kg FN/ha	59	50	11.8	0.41
CR6 + GM20 + 32 kg FN/ha	59	52	11.8	0.49
LSD (0.05)	2	6	3.4	0.04

FN = 88 kg N/ha through 20 t/ha *Sesbania* green manure; CR = 6 t/ha crop residues

(d) The INM through GM, crop residues, and FN in a rice-based cropping system has the long-term benefit of C sequestration and improved soil health resulting in high crop yields, helps maintain balanced nutrients supply, checks multi-nutrient deficiencies, and sustains crop yields at a higher level. SOC and other labile pools of C and N were significantly improved with INM plots after 4 years of study (Aulakh 2010).

(e) The INM involving application of recommended dose of N-P-K at the rate of 200-44-82 kg ha^{-1} plus either FYM at 5 Mg ha^{-1} or vermicompost at 3 Mg ha^{-1} was the best technology for harvesting higher yield of cauliflower with its quality produce and maintaining ecological health (Batabyal et al. 2016b).

Nutrient conservation in the soil is another critical component of INM. Soil conservation technologies prevent the physical loss of soil and nutrients through leaching and erosion and fall into three general categories. First, practices such as terracing, alley cropping, and low-till farming alter the local physical environment of the field and thereby prevent soil and nutrients from being carried away. Second, mulch application, cover crops, intercropping, and biological nitrogen fixation act as physical barriers to wind and water erosion and help to improve soil characteristics and structure. Lastly, organic manures such as animal and green manures also aid soil conservation by improving soil structure and replenishing secondary nutrients and micronutrients (Kumwenda et al. 1996).

18.7 Site-Specific Nutrient Management

Site-specific nutrient management (SSNM) approach has been developed as an alternative to blanket recommendations for NPK fertilizers (Dobermann et al. 2004), and it is formed from the QUEFTS (quantitative evaluation of the fertility of tropical soils) model developed by Janssen et al. (1990). The SSNM approach entails a set of nutrient management principles (Fig. 18.3) which aims to develop fertilizer recommendations on a field-specific basis to account for the variability in indigenous nutrient supply and yield target. Site-specific nutrient management, whether based on nutrient status of soil or plant in a given field, ensures that

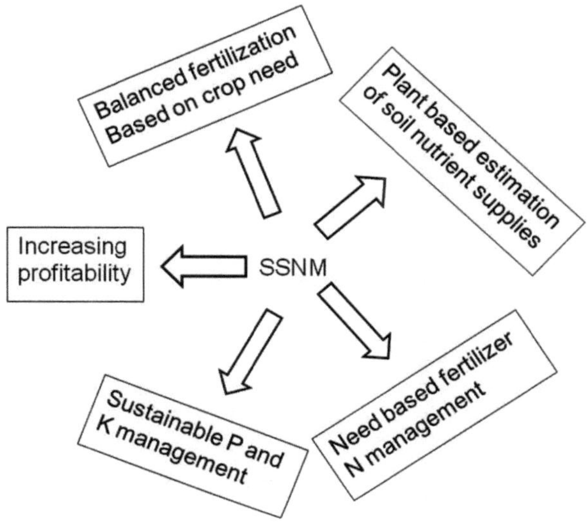

Fig. 18.3 Principles of SSNM

nutrients applied via fertilizers are managed according to the needs of the soil-plant system. Thus, as compared to blanket fertilizer recommendations for different crops, which are still prevalent in several developing countries, site-specific nutrient management ensures that soil health is maintained on a long-term basis.

Several field experiments have confirmed the positive implications of SSNM approach in Indian condition to account for large variability in the indigenous soil nutrient supply, yield targets, and nutrient demand of crops. By conducting field experiments across 56 irrigated wheat sites representing six irrigated rice-wheat regions in Punjab, India, Khurana et al. (2008) found at least 0.5 Mg ha^{-1} (12%) yield gain of wheat through SSNM approach compared to farmers' fertilizer practice (FFP) with 12–20% higher N, P, and K accumulations in plant under SSNM approach (Table 18.7). The gross return above fertilizer cost (GRF) was about 13% greater with SSNM than with FFP. Improved timing and/or splitting of fertilizer N increased N recovery efficiency from 0.17 kg kg^{-1} in FFP plots to 0.27 kg kg^{-1} in SSNM plots with 63% greater agronomic N use efficiency compared to FFP. Although the SSNM strategies have produced tangible yield gains, along with higher efficiency and improved soil health, the process is quite intensive and feasible in small domains only. Integration of SSNM with GIS-based spatial variability mapping has the potential to become a useful technique for use in large domains (Singh et al. 2012).

18.8 Rhizosphere Nutrient Management

The overall principle of INM is to maximize biological potential for improving crop productivity and resource use efficiency through root zone/rhizosphere management. In this particular environment, exchange of energy, nutrients, and molecular signals takes place, rendering the chemistry and biology of this

Table 18.7 Effect of site-specific nutrient management (SSNM) on wheat grain yield and plant nutrient accumulation in six regions of Punjab, India, during 2003–2004 and 2004–2005 (Source: Khurana et al. 2008)

	SSNM	FFP[a]	Δ[b]	$P >$ [T][c]	SSNM	FFP	Δ	$P >$ [T]	SSNM	FFP	Δ	$P >$ [T]	SSNM	FFP	Δ	$P >$ [T]
Regions	Grain yield[d] (Mg ha^{-1})				PNA[d] (kg ha^{-1})				PPA[d] (kg ha^{-1})				PKA[d] (kg ha^{-1})			
Gurdaspur	5.1	4.6	0.5	0.006	105	94	11	0.009	18	16	1.7	0.008	80	59	21	0.002
Hoshiarpur	4.9	4.4	0.5	0.008	97	87	10	0.011	17	15	1.9	0.012	72	51	21	<0.001
Ludhiana	5.4	5.1	0.3	0.025	121	113	8	0.023	21	20	1.3	0.017	116	100	16	0.009
Patiala	4.8	4.4	0.4	0.011	102	94	8	0.017	16	14	1.8	0.007	105	94	11	0.010
Faridkot	4.5	3.8	0.7	0.004	94	81	13	0.003	15	13	2.3	0.006	97	90	7	0.008
Firozpur	3.9	3.0	0.9	0.007	81	67	14	0.005	11	9	2.1	<0.002	84	69	15	0.002
All[e]	4.7	4.2	0.5	0.006	100	89	11	0.006	16	14	2.0	0.003	92	77	15	0.002

[a]Farmers' fertilizer practice in which all crop and fertilizer management was done by the farmer on a single field (0.1–0.4 ha) with no interference by the researcher. Average fertilizer nutrients applied under FFP to irrigated wheat crop across all regions were 143 and 14 kg ha^{-1} N and P, respectively. During the experimental period, no farmer applied fertilizer K in any of the 56 sites selected for the study

[b]Δ = SSNM−FFP

[c]Probability of a significant mean difference between SSNM and FFP

[d]Grain yield and plant nutrient (N [PNA], P [PPA], and K [PKA]) accumulations for each region are averages of 2 years (2003–2004 and 2004–2005)

[e]Combined data averaged for all six regions and 2 years (2003–2004 and 2004–2005). Values shown are based on measurements from 8 to 11 sites in each region

environment different from the bulk soil. Plant roots take up nutrients from soils via the rhizosphere, a narrow zone of the soil that is directly influenced by root growth, root secretions, and associated soil microorganisms. In cropping systems, a rhizosphere continuum in the root zone can be formed due to root/rhizosphere interactions among individual plants. Rhizosphere is the important interface where interactions among plants, soils, and microorganisms occur and is a "bottleneck" controlling nutrient transformations, availability, and flow from soils to plants. Therefore, the chemical and biological processes occurring in the rhizosphere determine the mobilization and acquisition of soil nutrients together with microbial dynamics, and also control NUE by crops, and thus profoundly influence cropping system productivity and sustainability.

Based on a better understanding of rhizosphere processes, the key steps of INM are (1) optimizing nutrient inputs and taking all possible sources of nutrients into consideration, (2) dynamically matching soil nutrient supply with crop requirement spatially and temporally, (3) effectively reducing N losses in intensively managed cropping systems, and (4) taking all possible yield increase measures into consideration.

18.9 Use of Nanofertilizers

Since fertilizers, particularly synthetic fertilizers, have a major potential to pollute soil, water, and air, in recent years, many efforts were done to minimize these problems by agricultural practices and the design of the new improved fertilizers. The appearances of nanotechnology open up potential novel applications to solve these problems. Nanotechnology is a novel, innovative, interdisciplinary scientific approach that involves designing, development, and application of material and devices at molecular level in nanometer scale. Nanostructure fertilizer exhibits novel physicochemical properties, which determines their interaction with biological substances and process. Nanofertilizer mainly delays the release of the nutrients and extends the fertilizers' effect period. Encapsulation of fertilizers within a nanoparticle is one of these new facilities which are done in three ways: a) the nutrient can be encapsulated inside nanoporous materials, b) coated with thin polymer film, or c) delivered as particle or emulsions of nanoscale dimensions (Rai et al. 2012).

The use of nanofertilizers can facilitate nutrient transport to the rhizosphere when needed, and in more suitable amounts and composition, thereby improving use efficiency (Mura et al. 2013). Nanotechnology is showing promise and may help improve the nutrient efficiency of not only phosphate fertilizer but also nitrogen and potassium, besides micronutrients like zinc and boron. The preliminary results of increased agricultural use of nanotechnology by densely populated countries such as China and India indicate that this technology might have a great impact on reducing hunger, malnutrition, and child mortality (Gogos et al. 2012).

It has been well established that nanoparticles can penetrate (and be uptaken) and translocate in plants (Ma et al. 2010; Wang et al. 2013), thus suggesting a new nutrient delivery system using nanoscale porous domains for ultimate plant growth and productivity.

It is claimed that controlled nutrient release and increase water retention in the soil are responsible for better yield under nanofertilizer application. Nano-sized TiO_2 promoted photosynthesis and nitrogen metabolism (Zheng et al. 2005); carbon nanotubes penetrate tomato seeds and affect their germination and growth rates. Nanofertilizers combined with nanodevices synchronize the release of fertilizer N and P with their uptake by crop, preventing undesirable nutrient losses to soil, water, and air via direct internalization by crops, which comes by avoiding the interaction of nutrients with soil, microorganisms, water, and air (DeRosa et al. 2010). Fertilizer incorporation into cochleate nanotubes had improved crop yield. Nanofertilizers release the nutrients in a controlled manner in response to reaction to different signals such as heat, moisture, etc. It is known that crops secrete carbonaceous compounds into the rhizosphere under nutrient stress which can be considered as environmental signals for incorporation into novel nanofertilizers (Sultan et al. 2009). These fertilizers can also improve the performance of fertilizers in other ways. For instance, due to its photocatalytic characteristic, nano-sized titanium dioxide added into fertilizers as a bactericidal, nanosilica particle that absorbed by roots can improve the plant's resistance to stress and thus increases the crop yield (DeRosa et al. 2010). Therefore, such nanofertilizers have a significant influence on energy, the economy, and the environment.

Delivery pathways of nanomaterials also play an important role in nanoparticle uptake by plant leaves. Wang et al. (2013) investigated aerosol-based nanoparticle delivery and transport thorough watermelon leaves. They also noted that aerosolized nanoparticles can be easily applied to leaf surfaces which enter the stomata via gas uptake, avoiding direct interaction with soil systems and reducing potential ecological risks. However, selecting the proper concentration of nanoparticles is important for realizing higher benefits for a target agro-economic trait. In an experiment Raliya et al. (2015) used nano-sized nutrients, viz., titanium dioxide (TiO_2) and zinc oxide (ZnO), to boost yield and quality of tomato without straining natural resources. Titanium dioxide (TiO_2) increased the light absorption and chlorophyll content in the plant, while zinc oxide nanoparticles had a twin role of being an essential nutrient and a cofactor for nutrient-mobilizing enzymes. With these the tomato plants were better able to absorb light and minerals producing nearly 82% (by weight) more fruit than untreated plants, and the fruit had higher antioxidant (lycopene) content (Fig. 18.4). They also reported that the aerosol technique used resulted in much greater uptake of nutrients by the plant in comparison to their soil application.

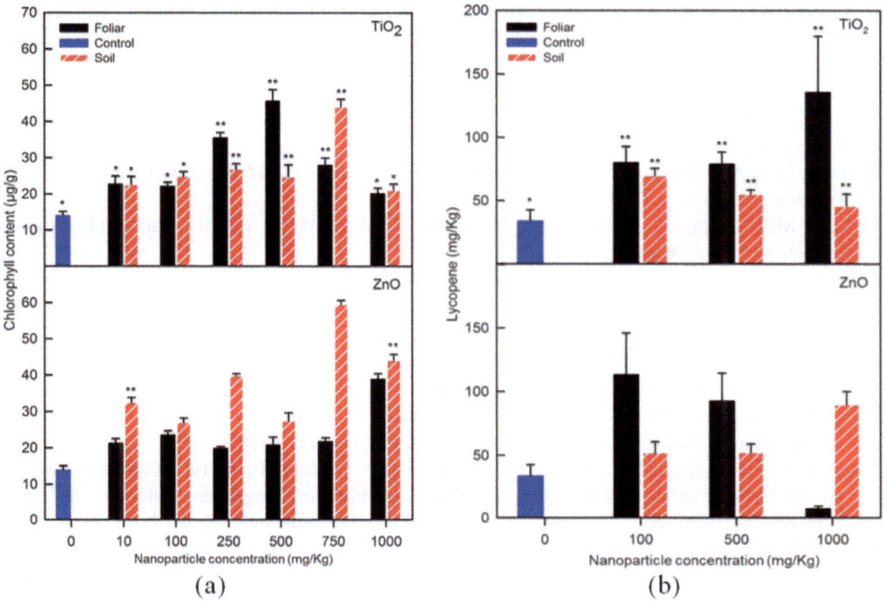

Fig. 18.4 Effects of TiO_2 and ZnO nanoparticles on (**a**) chlorophyll contents in the leaves of 28-day-old tomato plants and (**b**) lycopene content in tomato fruit. Error bar represents the standard deviation $n = 4$. *Asterisk*(s) above bar demonstrate significant difference ($p = 0.05$) (Source: Raliya et al. 2015)

References

Altieri, M. A., & Nicholls, C. I. (2003). Soil fertility management and insect pests: Harmonizing soil and plant health in agroecosystems. *Soil and Tillage Research, 72*, 203–211.

Aulakh, M. S. (2010). Integrated nutrient management for sustainable crop production, improving crop quality and soil health, and minimizing environmental pollution. In *19th World Congress of Soil Science, Soil Solutions for a Changing World*, 1–6 August 2010, Brisbane, Australia.

Batabyal, K., Mandal, B., & Hazra, G. C. (2016a). Nutrient management, energy input-output and economic analyses of eggplant production under subtropical conditions. *International Journal of Vegetable Science, 22*, 409–419. doi:10.1080/19315260.2016.1141825.

Batabyal, K., Mandal, B., Sarkar, D., Murmu, S., Tamang, A., Das, I., Hazra, G. C., & Chattopadhyay, P. S. (2016b). Comprehensive assessment of nutrient management technologies for cauliflower production under subtropical conditions. *European Journal of Agronomy, 79*, 1–13.

Bhattacharyya, T., Pal, D. K., Chandran, P., Mandal, C., Ray, S. K., Gupta, R. K., & Gajbhiye, K. S. (2004). *Managing soil carbon stocks in the Indo-Gangetic Plains, India* (p. 44). New Delhi, India: Rice-Wheat Consortium for the Indo-Gangetic Plains.

Bindraban, P. S., van der Velde, M., Ye, L., van den Berg, M., Materechera, S., Kiba, D. I., Tamene, L., Ragnarsdottir, K. V., Jongschaap, R., Hoogmoed, M., Hoogmoed, W., van Beek, C., & van Lynden, G. (2012). Assessing the impact of soil degradation on food production. *Current Opinion in Environmental Sustainability, 4*, 478–488.

Bouwman, A. F., Boumans, L. J., & Batjes, N. H. (2002). Emissions of N_2O and NO from fertilized fields: Summary of available measurement data. *Global Biogeochemical Cycles, 16*, 1–4.

Cassman, K. G. (1999). Ecological intensification of cereal production systems: Yield potential, soil quality, and precision agriculture. *Proceedings of the National Academy of Sciences of the United States of America, 96*, 5952–5959.

Chien, S. H., Prochnow, L. I., & Cantarella, H. (2009). Recent developments of fertilizer production and use to improve nutrient use efficiency and minimize environmental impacts. *Advances in Agronomy, 102*, 267–322.

Conway, G. (1997). *The doubly-green revolution: Food for all in the 21st century* (p. 335). London: Penguin.

DeRosa, M. C., Monreal, C., Schnitzer, M., Walsh, R., & Sultan, Y. (2010). Nano-technology in fertilizers. *Nature Nanotechnology, 5*, 91.

Dobermann, A., Witt, C., & Dawe, D. (Eds.). (2004). *Increasing the productivity of intensive rice systems through site-specific nutrient management*. Enfield, NH: Science Publishers, Inc./ International Rice Research Institute (IRRI).

Eichner, M. J. (1990). Nitrous oxide emissions from fertilized soils: Summary of available data. *Journal of Environmental Quality, 19*, 272–280.

Erisman, J. W., Sutton, M. A., Galloway, J., Klimont, Z., & Winiwarter, W. (2008). How a century of ammonia synthesis changed the world. *Nature Geoscience, 10*, 636–639.

Fixen, P. E., Jin, J., Tiwari, K. N., & Stauffer, M. D. (2005). Capitalizing on multi-element interactions through balanced nutrition—A pathway to improve nitrogen use efficiency in China, India and North America. *Science in China. Series C, Life Sciences, 48*, 1–11.

Goel, M. C., Singh, K. J. B., & Bhende, S. N. (2011). Response of application of customized fertilizer grade (CFG) on yield and quality of pomegranate. In M. K. Sheikh et al. (Eds.), *Proceedings IInd IS on Pomegranate and Minor, including Mediterranean Fruits (ISPMMF-2009)*, Acta Horticulturae, 890, ISHS 2011, pp. 333–340.

Gogos, A., Knauer, K., & Bucheli, T. D. (2012). Nanomaterials in plant protection and fertilization: Current state, foreseen applications, and research priorities. *Journal of Agricultural and Food Chemistry, 60*, 9781–9792.

Graham, R. D., Welch, R. M., & Bouis, H. E. (2001). Addressing micronutrient malnutrition through enhancing the nutritional quality of staple foods: Principles, perspectives and knowledge gaps. *Advances in Agronomy, 70*, 77–142.

Huber, D. M. (1980). The role of mineral nutrition in defense. In J. G. Horsfall & E. B. Cowling (Eds.), *Plant pathology: An advanced treatise* (pp. 381–406). New York: Academic Press.

IFA (International Fertilizer Industry Association). (1995). The efficient use of plant nutrients in agriculture. In *Fertilizers and agriculture*, special edition, Paris, France.

IFA (International Fertilizer Industry Association). (2013). *Assessment of fertilizer use by crop at the global level 2010–2010/11*. Paris: IFA.

Indian Fertilizer Scenario. (2013). Economics & Statistics (E&S) Wing, Department of Fertilizers, Ministry of Chemicals & Fertilizers 222 A, Shastri Bhawan, Government of India, New Delhi.

IPNI. (2014). *International Plant Nutrition Institute*. http://www.nutrientstewardship.com/4r-news/newsletter/ipni-issues-4r-plant-nutrition-manual.

Janssen, B. H., Guiking, F. C. T., Van der Eijk, D., Smaling, E. M. A., Wolf, J., & van Reuler, H. (1990). A system for quantitative evaluation of the fertility of tropical soils (QUEFTS). *Geoderma, 46*, 299–318.

Ju, X. T., Kou, C. L., Christie, P., Dou, Z. X., & Zhang, F. S. (2007). Changes in the soil environment from excessive application of fertilizers and manures to two contrasting intensive cropping systems on the North China Plain. *Environmental Pollution, 145*, 497–506.

Kaleswari, R. K. (2013). Impact of customized fertilizers on yield and soil properties of lowland rice ecosystem. *The Madras Agricultural Journal, 100*, 150–152.

Khurana, H. S., Phillips, S. B., Bijay-Singh, Alley, M. M., Dobermann, A., Sidhu, A. S., Yadvinder-Singh, & Peng, S. (2008). Agronomic and economic evaluation of site-specific nutrient management for irrigated wheat in northwest India. *Nutrient Cycling in Agroecosystems, 82*, 15–31.

Kohler, K., Duynisveld, W. H. M., & Bottcher, J. (2006). Nitrogen fertilization and nitrate leaching into groundwater on arable sandy soils. *Journal of Plant Nutrition and Soil Science, 169*, 185–195.

Krauss A. (1999). Balanced nutrition and biotic stress. In *IFA Agricultural Conference on Managing Plant Nutrition*. 29 June–2 July 1999, Barcelona, Spain.

Kumwenda, J. D. T., Waddington, S. R., Snapp, S. S., Jones, R. B., & Blackie, M. J. (1996). *Soil fertility management research for the maize cropping systems of smallholders in southern Africa: A review* (Natural Resources Group Paper 96-02). Mexico City: International Maize and Wheat Improvement Center (CIMMYT).

Lal, R. (2006). Carbon management in agricultural soils. *Mitigation and Adaptation Strategies for Global Change, 12*, 303–322.

Ma, X., Geiser-Lee, J., Deng, Y., & Kolmakov, A. (2010). Interactions between engineered nanoparticles (ENPs) and plants: Phytotoxicity, uptake and accumulation. *Science of the Total Environment, 408*, 3053–3061.

Majumder, B., Mandal, B., & Bandyopadhyay, P. K. (2008). Organic amendments influence soil organic carbon pools and crop productivity in a 19 years old rice-wheat agroecosystem. *Soil Science Society of America Journal, 72*, 775–785.

Mandal, B. (2005). *Assessment and improvement of soil quality and resilience for rainfed production system*. Completion Report. New Delhi: National Agricultural Technology Project, Indian Council of Agricultural Research, 30 pp.

Mandal, B., Majumder, B., & Bandyopadhyay, P. K. (2007). The potential of cropping systems and soil amendments for carbon sequestration in soils under long-term experiments in subtropical India. *Global Change Biology, 13*, 357–369.

Manna, M. C., Swarup, A., Wanjari, R. H., Singh, Y. V., Ghosh, P. K., Singh, K. N., Tripathi, A. K., & Saha, M. N. (2006). Soil organic matter in a West Bengal inceptisol after 30 years of multiple cropping and fertilization. *Soil Science Society of America Journal, 70*, 121–129.

Matson, P. A., Parton, W. J., Power, A. G., & Swift, M. J. (1997). Agricultural intensification and ecosystem processes. *Science, 277*, 504–509.

Mura, S., Seddaiu, G., Bacchini, F., Roggero, P. P., & Greppi, G. F. (2013). Advances of nanotechnology in agro-environmental studies. *Italian Journal of Agronomy, 8*, 127–140.

National Geographic. (2013). http://ngm.nationalgeographic.com/2013/05/fertilized-world/nitrogen-flow-graphic.

Prasad, R. (2009). *Enhancing nutrient use efficiency—Environmental benign strategies. Souvenir* (pp. 67–74). New Delhi: The Indian Society of Soil Science.

Rai, V., Acharya, S., & Dey, N. (2012). Implications of nanobiosensors in agriculture. *Journal of Biomaterials and Nanobiotechnology, 3*, 315–324.

Raliya, R., Nair, R., Chavalmane, S., Wang, W., & Biswas, P. (2015). Mechanistic evaluation of translocation and physiological impact of titanium dioxide and zinc oxide nanoparticles on the tomato (*Solanum lycopersicum* L.) plant. *Metallomics, 7*, 1584–1594.

Rothamsted Experimental Station. (1991). *Guide to the classical field experiments*. Harpenden, UK: AFRC Institute of Arable Crops Research.

Shivey, Y. S. (2011). Customized fertilizers: increased crop productivity and nutrient use efficiency. In *Proceedings International Conference on Issues for climate change, land use diversification and biotechnological tools for livelihood security, held in Meerut, 8–10 October, 2011*. Hi-tech Horticultural Society, Meerut, pp. 128–133.

Shukla, A. K., Tiwari, P. K., & Chandra, P. (2014). Micronutrients deficiencies vis-a-vis food and nutritional security of India. *Indian Journal of Fertilisers, 10*(12), 94–112.

Singh, V. K., Govil, V., Singh, S. K., Diwedi, B. S., Meena, M. C., Gupta, V. K., Majumdar, K., & Gangwar, B. (2012). Precision nutrient management strategies using GIS-based mapping in western Uttar Pradesh. *Better Crops-South Asia, 6*(1), 15–18.

Sultan, Y., Walsh, R., Monreal, C. M., & DeRosa, M. C. (2009). Preparation of functional aptamer films using layer-by-layer self-assembly. *Biomacromolecules, 10*, 1149–1154.

Tilman, D., Cassman, K. G., Matson, P. A., Naylor, R., & Polasky, S. (2002). Agricultural sustainability and intensive production practices. *Nature, 418*, 671–677.

Tilman, D., Fargione, J., Wolff, B., D'Antonio, C., Dobson, A., Howarth, R., Schindler, D., Schlesinger, W. H., Simberloff, D., & Swackhamer, D. (2001). Forecasting agriculturally driven global environmental change. *Science, 292*, 281–284.

Voortman, R., & Bindraban, P. S. (2015). *Beyond N and P: Toward a land resource ecology perspective and impactful fertilizer interventions in Sub-Saharan Africa. VFRC Report 2015/1* (p. 49). Washington DC: Virtual Fertilizer Research Center.

Wang, P., Menzies, N. W., Lombi, E., McKenna, B. A., Johannessen, B., Glover, C. J., Kappen, P., & Kopittke, P. M. (2013). Fate of ZnO nanoparticles in soils and cowpea (*Vigna unguiculata*). *Environmental Science & Technology, 47*, 13822–13830.

Welch, R. M., & Graham, R. D. (1999). A new paradigm for world agriculture: Meeting human needs- productive, sustainable, nutritious. *Field Crops Research, 60*, 1–10.

Withers, P. J. A., & Lord, E. I. (2002). Agricultural nutrient inputs to rivers and groundwaters in the UK: Policy, environmental management and research needs. *The Science of the Total Environment, 282*, 9–24.

Zheng, L., Hong, F., Eu, S., & Hu, C. (2005). Effect of nano Tio_2 on strength of naturally aged seeds and growth of spinach. *Biological Trace Element Research, 104*, 83–91.

Chapter 19
Leaching of Plant Nutrients from Agricultural Lands

I. Rashmi, Abhay Shirale, K.S. Kartikha, K.C. Shinogi, B.P. Meena, and S. Kala

Abstract Crop production is improved by application of fertilizer inputs, but improper management and overusage of fertilizers often lead to low use efficiency, accumulation and losses of nutrients from the soil. Leaching is the result of the release of ions in the soluble form and of the movement of water in the soil profile in removing these ions. The magnitude of loss is proportional to the element concentration in the soil solution and the amount of drained water. Nutrient leaching is becoming increasingly important because the movement of nutrient out of root zone represents an immediate loss to crops and an economic loss to farmer. The important inputs in agriculture in the form of nitrogen, phosphorus and potassium fertilizers are considered as nonpoint source of pollution. These nutrients undergo series of transformations by various physical, chemical and biological processes in the soil which make them available to crops as well as vulnerable to leaching losses. Soil nutrient leaching is governed by various factors like soil type, available nutrient content, amount and intensity of rainfall or irrigation water, the nature of the crop plant or extent of soil surface covered by it, etc. Nutrients like nitrogen and phosphorus leaching are of environmental concern and a potential risk to human health, and therefore efficient and judicious management should be adapted. To minimize

I. Rashmi (✉) • S. Kala
Soil Chemistry and Fertility, IISWC, Research Centre, Kota, Rajasthan, India
e-mail: rashmisac@rediffmail.com

A. Shirale
Soil Chemistry and Fertility, IISS, Bhopal, Madhya Pradesh, India
e-mail: abhayshirale@gmail.com

K.S. Kartikha
Soil Chemistry and Fertility, CPCRI, Regional Station, Vittal, Karnataka, India
e-mail: kskavukattu@gmail.com

K.C. Shinogi
Agriculture extension, IISS, Bhopal, Madhya Pradesh, India
e-mail: shinojikallely@gmail.com

B.P. Meena
Agronomy, IISS, Bhopal, Madhya Pradesh, India
e-mail: Bharatmeena24@gmail.com

M. Naeem et al. (eds.), *Essential Plant Nutrients*,
DOI 10.1007/978-3-319-58841-4_19

nutrient leaching, important management strategies like (a) reducing the nutrient input in soils with high soil test values; (b) using countermeasures like cover crop, crop rotation and conservation tillage and (c) environmental indexing of fields and consideration of spatial variability within fields in relation to their contribution to leaching losses within a catchment should be adapted at field level. The adoption of management practices depends upon how sensitive is the cropping system; soil and climate are at risk of this nutrient leaching.

Keywords Soil nutrient • Leaching • Nitrogen • Phosphorus • Potassium • Micronutrients • Management

19.1 Introduction

Soil nutrient depletion is both natural and human-induced process with respect to loss either by crop harvest and other anthropogenic activities either by leaching or erosion. Dynamically, it is the process by which the soil nutrient stock is shrinking because of continuous nutrient mining without sufficient replenishment of nutrients harvested in agricultural products and of nutrient losses by soil erosion and leaching. Leaching is a process in which soil nutrients are released into contacting water phase. The leaching of the essential plant nutrients is influenced by pedoclimatic and soil crop management practices. With increasing population and limited land resources, more emphasis should be given to improve nutrient use efficiency by minimizing its loss from the soil system. The dynamics of leaching of soil nutrients is explained in Fig. 19.1. Leaching involves various chemical, physical and biological transformation of soil nutrients. Leaching is an environmental concern when it contributes to groundwater contamination. A large portion of water percolation and nutrient leaching occurs through large soil pores. As water from rain, flooding or other sources seeps into the ground, it can dissolve chemicals and carry them into the underground water supply. Among the macronutrients, nitrogen (N) and phosphorus (P) are the two main growth-limiting nutrients of many agricultural crops, and they are consequently applied as fertilizers in quantities often exceeding crop requirements. These nutrients that are not fully absorbed by crops may be lost from the field through pathways such as volatilization, surface run-off or leaching to groundwater, causing economic losses and environmental pollution (Lundell et al. 2001). Leaching and nutrient removal are accelerated by human actions for exploitation of vegetation and soil resources. Understanding nutrient leaching from fertilized agricultural soils is important for several reasons. First, nutrients like NO_3^- can affect human and animal health. Second, the higher concentration of these essential nutrients loss to water bodies affect water quality and net phytoplankton productivity and causes eutrophication. Third, the loss of nutrients is a huge economical loss for farmers and the country.

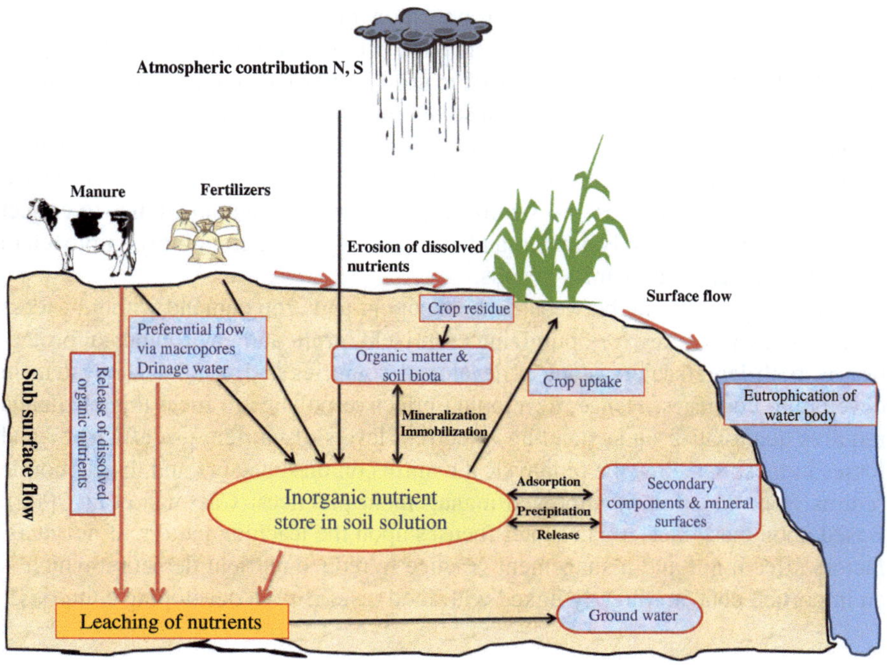

Fig. 19.1 Dynamics of nutrient leaching in soil

Problems of nutrient leaching vary substantially with both rainfall intensity and soil properties (Laird et al. 2010). The magnitude of leaching nutrients via rainfall or irrigation depends on the nature of the soil, the amount and intensity of rainfall or irrigation water, and the nature of the crop plant or extent of soil surface covered by it. The soil nutrient leaching is generally controlled by soil hydraulic properties like porosity, infiltration capacity, hydraulic conductivity, etc. and sorption of constituents on soil matrix (Mahmood-ul-Hassan et al. 2008; Ersahin et al. 2002; Clay et al. 2004). Nutrient leaching from agricultural soils depletes soil fertility, accelerates soil acidification, increases fertilizer costs for farmers, reduces crop yields and adversely affects the quality of surface and groundwater. Therefore, reducing the leaching loss of plant nutrient will improve nutrient use efficiency of cropping systems endowed with good soils and climate and can increase crop yields.

In general, leaching takes place especially when the mineralization and absorption by the plant are not synchronized and the water flow is enough to transport the solute in depth (McNeill et al. 2005; Arregui and Quemada 2006). The leaching loss of nitrate, phosphate and potassium below root zone represents the loss of valuable plant nutrient and can contribute to pollution of aquifers. Besides N, other cations like Ca, Mg and K leaching may be considerable under acidifying conditions when ammonia fertilizer is used (Hartemink 2008; Oliveria et al. 2002). Until recently, P leaching was seldom considered a significant pathway for transporting soil P to surface water as it was believed that most soils have considerable P adsorption capacity. Phosphates can cause eutrophication of water bodies and as little as 20–30 mg P L^{-1} can stimulate phytoplankton production (Daniel et al. 1998; Jiao et al. 2004). This calls the attention for the need of management practices to attenuate leaching losses not only of N but also of other nutrients. Quantification of nutrient leaching from agricultural soils is important for both fertilizer and environmental management. Different techniques like extraction of soil solution by capsules used in field studies, resin cores and suction cups; laboratory column studies to collect leachate, from cultivated lysimeter; and isotope analysis are commonly employed to understand leaching dynamics in soils.

Achieving synchrony between nutrient supply and crop demand without excess or deficiency is the key for optimizing crop yield, profit and environmental protection in both large-scale system in developed countries and small-scale system in developing countries. Hence, developing effective policies to meet this challenge requires quantitative understanding of current levels of nutrient use efficiency and losses in these systems, the biophysical controls on these factors and the economic returns from adoption of improved management practices (Cassman et al. 2002). Based upon these facts, this chapter focuses upon the leaching pattern of nutrients, factors affecting it and management practice to reduce nutrient depletion which is an important concern directly linked with food insecurity in developing countries.

19.2 Leaching Pattern of Macro- and Micronutrients in Soils

Leaching of soil nutrients is correlated with soil fertility, fertilizer application, crop nutrient uptake, soil clay content and precipitation (Stoorvogel and Smaling 1998) and may be insignificant in crop production systems compared with total nutrient depletion. Among the various nutrients, vertical downward movement of N and P through the profile with percolating water causes loading of nutrients in water bodies. Nutrient leaching, the movement of plant nutrients from soil to water, can have negative effects on aquatic ecosystems due to eutrophication, which reduces the oxygen available in water, causing species and habitat loss. Therefore in this section the leaching pattern of various nutrients is discussed in relation to soil properties.

19.2.1 Nitrogen Leaching

The role of N in ensuring food security is a success story in Indian grain production in the last three decades. Among the nutrient anions, nitrate is the most easily leached because of its negligible reaction with soil matrix and therefore very mobile in soil. Agriculture is considered as the major source of loss of N to environment. With N use efficiency below 50%, a major portion of applied N is lost by various pathways and serves as a major nonpoint source of pollution. The efficient use of N depends upon its judicious management which otherwise results in negative balance in agriculture system (Lehmann and Schroth 2003). The N use efficiency generally ranges from 20 to 40% (Goswami et al. 1988), and the major losses of N is through volatilization, denitrification, run-off and leaching. These losses not only cause wastage of N fertilizers but also have detrimental effect on environment. The dominant pathway of N loss is through leaching and can result in more than 40% of N loss (Chhabra et al. 2010). Nitrogen is absorbed by crops in the form of NO_3^- and NH_4^+ depending upon the crop species, soil type and management practices like submerged or upland conditions. The degradation of groundwater quality by N fertilizer use in intensive agriculture is essentially a nitrate leaching problem. High nitrate content of >10 ppm in drinking water is reported to cause methemoglobinemia in infants (USEPA 1985). Nitrate leaching is dominant in soils with high NO_3^- and water content. The NO_3^- anions are very soluble in water and leach down rapidly until intercepted by crop roots or carried through water by run-off. Estimation of different N losses under field conditions is of vital importance in understanding and managing various N losses. Nitrate leaching is not only common in coarse texture soil but is also reported in fine texture soils.

There are evidences of high nitrate content in groundwater in many states of Punjab, Andhra Pradesh, Delhi and Maharashtra which practice intensive agriculture (Rao and Puttanna 2000). Even in semiarid areas like Rajasthan with low intensive agriculture reports show high NO_3^- leaching (Ozha et al. 1993). The high nitrate leaching is also caused by dumping of organic manures and industrial waste on agricultural land. There is an alarming trend in nitrate pollution in India where three-fourth of the world population lives. In N cycle as shown in Fig. 19.2 nitrate

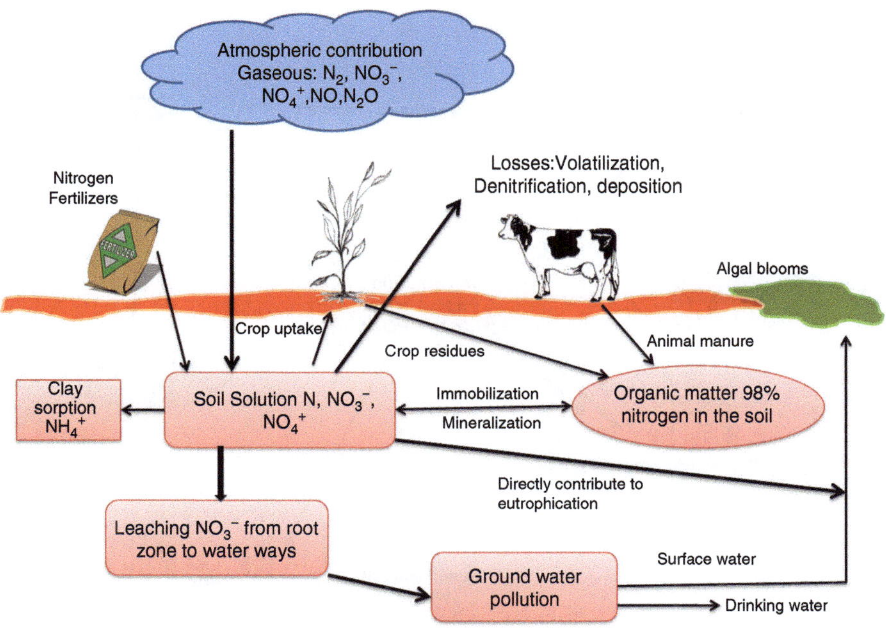

Fig. 19.2 Nitrogen mineralization and leaching loss in soil system

moves down with percolating water because of the high solubility of the ion and lower affinity of the ion to adsorption sites on the soil.

19.2.2 Phosphorus Leaching

Compared to N, phosphorus is considered to be relatively immobile in soil. Phosphorus is firmly bound in soils due to precipitation of P with calcium ions in calcareous soil and due to adsorption of P by Fe and Al oxides in acidic soil (Hinsinger 2001). Phosphorus is highly dynamic in soil as shown in Fig. 19.3 and undergoes a number of chemical transformations in soil. Levels of different pools of soil P have been affected not only by soil properties and climatic condition but also by rate and type of P applied. Only a relatively small fraction of soil P is available to plants. In response, massive amounts of fertilizer P is applied by farmers to increase available soil P which leads to accumulation and loss of P from soil.

In agriculture, P fertilizers are often applied in excess of immediate plant uptake leading to accumulation of P in top soil layer, which leads to P transport from soil system (Sims et al. 1998). The GIS-based fertility mapping of Indian soils revealed (Muralidharudu et al. 2011) high fertilizer consumption in Punjab followed by Andhra Pradesh and Tamil Nadu and Kerala. In Kerala, due to overuse of P fertilizers in excess of crop demand, buildup of soil P in pockets had led to its inefficient use and consequent environmental pollution. In Punjab, tremendous movement of residual P to deeper layers in one subtropical coarse-textured irrigated soil under

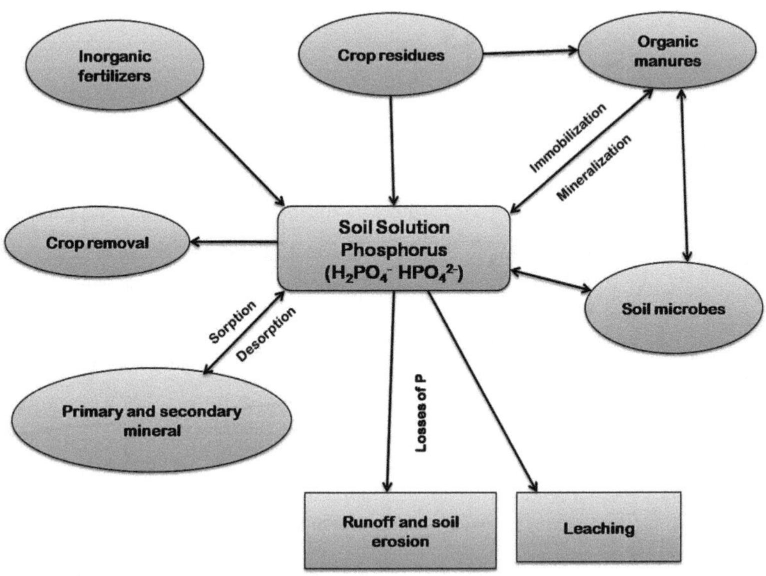

Fig. 19.3 Phosphorus dynamics in soil

long-term fertilizer applications, pointing to the potential for extensive leaching of fertilizer P, was reported by Aulakh et al. (2007). In Tamil Nadu Rajmohan and Elango (2005) reported possible contamination of groundwater quality due to usage of P fertilizers. Significant quantities of P leaching have been reported by many workers worldwide. In Rothamsted, >100 years of long-term fertilizer experiments revealed high content of total P of 3 mg L^{-1} in drainage waters from 65 cm depth (Heckrath et al. 1995). In another long-term fertilizer experiment, Whalen and Chang (2001) demonstrated that 7–15% of P applied for 16 years was not accounted for and moved through soil layers, eventually reaching groundwater. Phosphorus lost from agricultural soils can increase the fertility status of natural waters, commonly known as eutrophication. Eutrophication restricts water use for fisheries, recreation, industry and drinking because of the increased growth of undesirable algal and aquatic weeds and resulting oxygen shortages caused by their death and decomposition.

With declining phosphate rock reserve, it is the need of the hour to utilize this nonrenewable resource more judiciously for mineral P supply as global agricultural production depends on it. In India most of the soils are low to medium in available P status; however, high soil P had been reported in 8% states (Table 19.1), indicating that crop-based application of P should be implemented instead of blanket doses. Recently, Aulakh et al. (2007) and Garg and Aulakh (2010) reported tremendous movement of residual P to deeper layers in one subtropical irrigated soil under long-term fertilizer applications, pinpointing the potential for extensive leaching of fertilizer P. More information is therefore needed particularly under irrigated conditions, different cropping sequences, rate and frequency of applied P, and varying soil

Table 19.1 List of districts under high soil P status and area covered

State	Status	District	Area (%)
Bihar	High	Samastipur	3.2
Himachal Pradesh	High	Kinnaur	11.6
Jharkand	High	Lohardaga	1.9
Karnataka	High	Bangalore (R), Chamrajnagar, Davanagere, Mandya	4.2
Kerala	High	Ernakulam, Kollam, Kottayam, Thiruvananthapuram	24.1
Madhya Pradesh	High	Dindori, Guna, Hoshangabad, Rajgarh, Sehore	13
Orissa	High	Bolangir, Boudh, Keonjhar, Mayurbhanj, Phulbani, Sambalpur, Sonepur	
Punjab	High	Bathinda, Pategar Sahib, Gurdaspur, Hoshiarpur, Jalandhar, Kapurthala, Ludhiana, Ropar	52.8
Tamil Nadu	High	Coimbatore, Pudukkottai, Madurai, Nagapattinam, Nilgiri, Perambalur, Ramanathapuram, Thanjavur, Thiruallur, Trivvanamalai, Tiruvarur, Vellore	37.6
West Bengal	High	Hoogly	5.8

Subba Rao et al. (2015)

texture to find out the potential for P leaching and to develop strategies and policies for sustainable high productivity without detriment to the environment.

19.2.3 Potassium Leaching

Among the macronutrients K is abundant in soils. Quantities of mineral K, exchangeable, nonexchangeable and soil solution K is shown in Fig. 19.4. The amount of K in the soil depends upon parent material, soil type, weathering rate, fertilizer and manure addition, losses by crop removal, soil erosion and leaching. Sandy soils and organic soils are poor in K-bearing minerals and have nonexchangeable K and release little K and have low adsorption capacity. The cycling and availability of K is quite dynamic and affected by management practices. Potassium leaching in soil is generally dependent on the amount of exchangeable K in the soil, which to a great extent reflects the level of K input and resulting K surpluses (Askegaard et al. 2003; Alfaro et al. 2004; Kayser et al. 2007). Under conditions of intensive animal production and frequent application of organic sources resulting in constant K surpluses, K levels might further increase with the consequence of even larger K losses. Potassium leaching losses are related to the soil type, but the chemistry of the soil is not as important as its hydrology which controls the dynamics of K leaching.

The K fertilizer application in sandy soils with low clay content, less interaction of K with soil matrix and low buffering capacity of soil will result in increased K concentration in soil solution and subsequently leaching down the soil profile. The major factors which affect K leaching in soils are weathering of K feldspar, muscovite, biotite minerals, effect of field stones on packing density of soil, bypass flow, inputs from rain and organic matter addition. Thus soil mineralogical composition and its potential weathering capacity and leaching risk need to be taken into account when deciding on the K fertilizer requirements of a given field (Andersson et al. 2007).

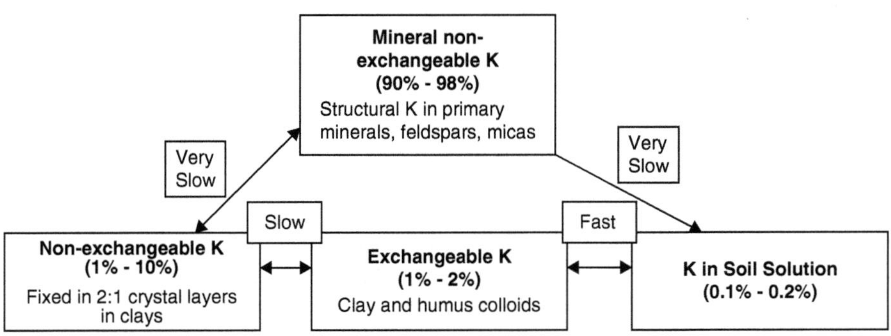

Fig. 19.4 Potassium dynamics in soils (© State of Victoria, Department of Economic Development, Jobs, Transport and Resources. Reproduced with permission)

19.2.4 Sulphur, Calcium and Magnesium Leaching

Sulphur transformation in soil is similar to nitrogen but does not leach readily as nitrate. Crops take sulphur (S) in the form of SO_4^{2-} and this form of sulphur is highly susceptible for rapid leaching from the soil especially if the soil is dominated by monovalent cations like sodium and potassium (Lehmann and Schroth 2003). High calcium (Ca) leaching is observed in soils which are treated with soil amendments for treating soil acidity. In a study conducted by Wong et al. (1992) where 1025 kg ha^{-1} of calcium hydroxide was applied for ameliorating soil acidity in corn crop, significant amount of Ca (312 kg ha^{-1}) was lost through leaching. The amount of Ca leaching also depends upon the volume of drained water. Both Ca and Mg are vulnerable to leaching because of their high solubility in their hydroxide forms. There are reports indicating 64 and 34% of Ca and Mg loss from cultivable areas where Ca and Mg amendments were applied to treat soil acidity (Wong et al. 1992). On comparing the percolation rate, it was observed that Ca leaches more compared to Mg. Results of experiments showed a decrease in relationship of exchangeable Ca:Mg ratio demonstrating high leaching losses of Ca compared to Mg (Oliveira et al. 1999). A portion of Ca and Mg along with K can leach from soils resulting in more H^+ charges on soil surface leading to Al saturation and acidity of soils.

19.2.5 Micronutrient Leaching

The use of micronutrients for both field and horticultural crops has increased in the last decades in India. As the micronutrients are mostly cation like copper (Cu), zinc (Zn), iron (Fe) and manganese (Mn), they remain adsorbed to soil surface and do not move significantly in soils. Therefore they usually do not leach beyond root zone and reach groundwater sources. These micronutrients can leach down under some particular conditions like those of high rainfall, higher application rate and coarse-textured soils. Molybdenum is an anion and is easily leachable from soil. However boron (B) fertilizers form boric acid in soils which are highly water soluble and is subjected to leaching losses. While mobility of B is less compared to NO_3^-, many experiments have shown a considerable amount of B loss via leaching pathway (Mortvedt 1994). Boron, like nitrates, is not readily held by soil particles and moves down through coarse-textured soils, often leaching below the root zones of many plants. Heavy leaching of boron leads to temporary deficiency of the nutrient in the soil. Most of the available boron from root zone temporarily leaches out as a result of heavy rain (Woods 1994). This occurs most often in sandy and rocky soils. During rainy or wet periods, plants may encounter long periods of low boron levels, resulting in boron-deficiency symptoms.

19.3 Factors Affecting Leaching of Soil Nutrients

The leaching of nutrients is a complex process, affected by many factors such as dose and solubility of the fertilizer, crop nutrient use efficiency, soil water storage capacity and soil water movement and the amount of mineralized nutrient in soil. Leaching process is indiscriminant in which all the constituents like organic, inorganic and other contaminants are released under a common set of chemical phenomena which may include mineral dissolution, desorption, complexation and mass transport process (Fig. 19.5). These phenomena are in turn affected by four major factors, viz., (a) internal chemical and physical reactions; (b) external stress from surrounding environment; (c) physical degradation of soil matrix due to erosion and cracking and (d) loss of matrix constituents due to leaching process itself. The major factors which affect the leaching of nutrients include:

19.3.1 Soil Texture

Soil texture is a major factor which controls nutrient leaching. Nitrogen loss can be significant in light textured sandy soils and can increase with higher levels of N application. The loss of N is attributed to the higher pore space of soil easing percolation losses of N. In sandy, loamy and sandy loam soil, urea leaches down rapidly through profile reaching groundwater as compared to clay loam soils. In fine textured clay and clay loam soil, urea transformation into NH_4^+-N and NO_3^--N forms

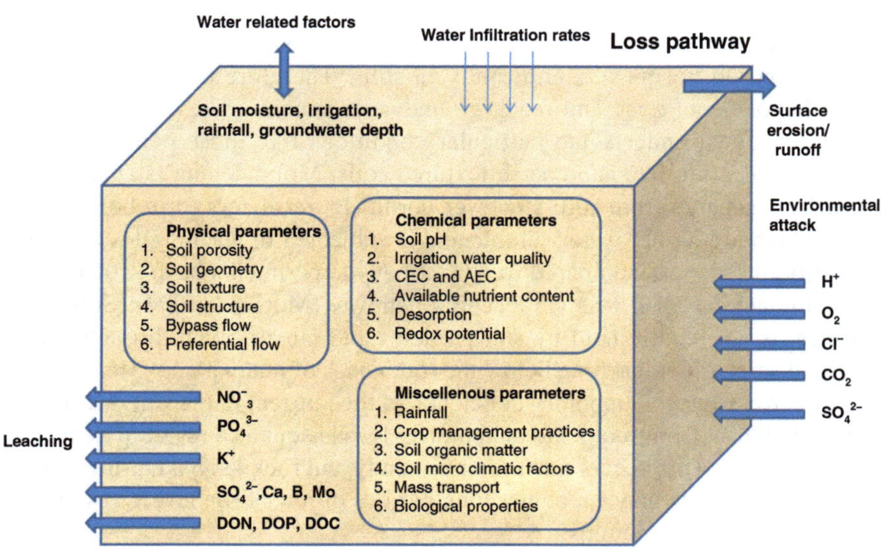

Fig. 19.5 Factors affecting leaching of nutrients

is rather slow, and with reduced percolation, the possibility of nitrate leaching is limited (Singh et al. 2006). In lowland rice fields of India less N was leached due to clayey texture of the soil, whereas in Indo-Gangetic soils with sandy loam texture, N losses were 10% of applied N levels (Ladha et al. 2005; Pathak et al. 2006). In a study reported by Sharma (1999), N loss varied from 80 to 133% under terrace cultivation, whereas it varied from 56 to 72% under bun method of cultivation. Soil texture along with water can affect N movement more rapidly leading to significant loss from agricultural soils. However water can infiltrate in soils with higher clay content that develop deep cracks which can carry water along with nutrients to subsoil layers. This can also be due to root-growing channels and faunal activity. Macropore or bypass flow can also increase leaching of nutrients following increased level of concentration resulting in rapid leaching giving less time to associate with the soil matrix.

The amount of K leaching depends upon the soil available K content which is related to soil texture. The soluble K is negatively related to the amount of sand content and positively related to the amounts of clay soils. Thus more K leaching is observed in sandy soils compared to clayey soils (Dassannayake 1990). In clay soil exchangeable K will be on greater side compared to sandy soils. However, high K losses have even been reported from clay soil due to preferential flow where K source is both organic and inorganic in a grassland system. The loss of K from clay soils could be twice that of sandy soil from grassland with application of farmyard manure. The loss of total K in a soil type is affected by hydrological properties where macropore flow is an important pathway and can cause large differences in K losses (Alfaro et al. 2003). In a study conducted in Kenya (Roy et al. 2003), N and K leaching was calculated based on mineral N and K percentage and amount of inorganic and organic N and K sources applied. This mineral nutrient content in soils was based on rainfall and clay content, and their leaching percentages are shown in Table 19.2.

19.3.2 Ion Exchange Capacity

The cation with higher affinity replaces the ions on soil surface and exposes them to leaching losses. Leaching of K is not considered a problem in soils with high clay content, but losses of K in coarse sandy soils with little clay content

Table 19.2 Nitrogen and potassium leaching percentages for different rainfall and clay content (Roy et al. 2003)

	Rainfall (mm)									
	1350		1500		1700		1900		2050	
Clay content (%)	N	K	N	K	N	K	N	K	N	K
<35	25	0.80	29	0.85	32.5	0.90	36	0.95	40	1
35–55	20	0.65	22.5	0.70	25	0.75	27.5	0.80	30	0.85
>55	15	0.50	16.5	0.55	17.5	0.60	18.5	0.65	20	0.70

and low K inputs can become a limiting factor in the development of negative K balances due to their low cation exchange capacity (Zhang et al. 2013). The leaching of K is also governed by the presence of other cations like Ca^{2+} in the soil solution, weathering of minerals like gypsum and calcite. The irrigation water containing high concentration of Ca^{2+}, Mg^{2+} and Na^{+} replaces K from exchange sites resulting in K desorption and accentuates K leaching (Kolahchi and Jalali 2007).

19.3.3 Climatic Factors

Among different climatic factors, rainfall and temperature are the most dominating ones influencing nutrient leaching from soil. Leaching occurs in a condition where rate of water input exceeds the requirement, which means less evapotranspiration and soil water level is above field capacity. Therefore nutrient losses are more in humid compared to arid climates. Precipitation causes heavy losses of N through leaching affecting crop productivity. The rainfall affects the soil hydric conditions and the volume of drained soil solution (Salcedo and Sampaio 1991). In areas with high rainfall, a significant amount of N will be lost. With annual rainfall of 800–1200 mm, nearly 50% of applied N is lost by leaching (Sharma 1999). With higher rainfall the loss of N increased due to easy movement of anion. Rainfall along with texture can enhance nutrient leaching. States like Punjab and Haryana with sandy soil receive 60–80 cm rainfall which results in nitrate leaching to groundwater due to high percolation of coarse texture soils. The concentration of mineral fertilizer and the quantity of the water passing through the soil profile are the major controlling factors of leaching of nutrient like NO_3^-. In arid and semiarid regions, K leaching is enhanced by the presence of gypsum and calcite in the soil (Jalali and Rowell 2003).

19.3.4 Soil Hydrology

Soil hydrology is determined by the way water moves in the soil. It is the major determinant of vulnerability of land to leaching or run-off of contaminants to water bodies. The transport of nutrients from soil to water is governed by the rate of water entry at the soil surface and the nature of water transmission within the soil. The storage and transport of water through soil is governed by soil fissures, worm burrows, pores and the continuity of the fissure or pore network within the soil profile. The nature of the pore network affects infiltration, drainage, soil aeration, surface run-off, soil erosion and chemical leaching (Webb et al. 2010). Soil hydraulic properties like water infiltration, conductance and retention have direct and indirect influence on soil factors and will effect nutrient movement. Soil nutrients like NO_3^- being nonreactive are considered to leach by preferential flow, whereas reactive ions like PO_4^{3-} are considered to accumulate on the surface. Later, reports illustrated PO_4^{3-} leaching to sub-surface water occurred through soil matrix (Schoumans and Groenendijk 2000) or by preferential flow (Djodjic et al. 2004) associated with paedological features such as macropores, continuous inter-aggregated voids, earthworm burrows, decayed root channels and other geometric anomalies. These preferential flow pathways have entirely different hydraulic properties than the soil matrix and represent a small fraction of the total cross-sectional area through which solutes travel rapidly, bypassing the soil matrix (Radulovich et al. 1992) and causing a rapid and accelerated breakthrough of surface-applied fertilizer nutrients (Buchter et al. 1995; Gaber et al. 1995). Water movement through cracks or macropores (preferential flow) can be as much as twenty times higher than in the same soils without cracks. This movement allows nutrient to flush through soil more rapidly than might be expected.

19.3.5 Source of Nutrient

Different sources of nutrients like organic and inorganic fertilizers affect nutrient leaching depending upon their transformation in soil. Application of organic matter and their mineralization enhance nutrient mobilization. Nutrient losses from organic sources are rather difficult to predict compared to mineral sources. Nutrients are often released from organic sources when there is no crop uptake, and this provides an opportunity for nutrient leaching (Camargo 1989). Organic sources like farmyard manure, green manure and animal waste are known to increase exchangeable form of nutrients compared to inorganic fertilizers, thus making them more susceptible for leaching losses. The application of N fertilizers with organic manure increased N leaching even by 200–300% over control in a study reported by Sharma (1999). The K leaching is highly influenced by the type of nutrient source. Reports indicate high K leaching in plots with dairy manure slurry were 25.5 ± 6.89 and 36.3 ± 0.66 mg L^{-1} in clay soil compared to sandy soil 2.5 ± 0.44 and 3.0 ± 0.43 mg L^{-1}. Application of pig slurry enhanced K leaching losses during storms if they occurred within 24 h of application (Edwards and Daniel 1993). The type of organic manure effects the types of K like soluble and exchangeable K. The nutrients associated with organic matter are only available for leaching after an equilibrium has been reached between the soil and the fertilizer (Williams 1988) which will take longer time than the dissolution of mineral fertilizer at surface. Even during FYM preparations, nutrient losses of 20% of N, 30% of P and 50% of K by leaching have been quantified by Reddy et al. (2010) which can have deleterious effect on nearby water sources. Organic fertilizers improve soil fertility and quality, but long-term application at high rates can also lead to more nitrate leaching and accumulation of P, if not managed well (Miao et al. 2011). The N present in organic manure undergoes ammonification and producing NH_4^+ ions. The acid produced during nitrification is expected from areas where inorganic fertilizer is used than from areas with slurry or FYM application, other conditions and management being the same. FYM is a major source of plant nutrients. Another study reported by Goulding et al. (2000) observed larger N leaching loss from the plots treated with FYM in the Broadbalk experiment. Fall application is part of the reason, because N mineralized in fall and early winter was not used by the crops, and subject to leaching. Organic manures are also rich source of organic P. A 15-year long-term application of manure and fertilizers study by Zhang et al. (2009) reported very high buildup of available P in the soil at the end of the study. Enriched P in the soil may be lost through run-off, or leached in soils with low P retention or in situations of organic P leaching, thus leading to water pollution (Edmeades 2003).

Leaching is less severe under crop compared to fallow but enhances with application of fertilizers. Acidifying fertilizers like urea and ammonium sulphate will have adverse effect with more release of protons during nitrification which will release cations from soil solution (Rowell 1988). Application of P fertilizers like triple superphosphate enhanced more P leaching than chicken manure in soil with a low P sorption capacity at P rates of 56 and 224 kg P ha^{-1}, and for the soil with a high P sorption capacity (Elliott et al. 2002), no difference was found between triple superphosphate and chicken manure applied at 56 kg P ha^{-1} but a greater P load was emitted from soils receiving triple superphosphate at a rate of 224 kg P ha^{-1}. Differences in P leaching from inorganic and organic fertilizer sources may be related to soil properties and the P fertilizer rate applied.

19.3.6 Quality of Irrigation Water and Affinity of Ions for Adsorption Sites

Soil irrigated with water containing high Ca content can displace K by competing for exchange sites resulting in K leaching. With higher Ca concentration the loss of K can be >1 kg ha^{-1} with every 100 mm rainfall. Irrigation with water in which the concentrations of Ca^{2+}, Mg^{2+} and Na^{+} are higher than those in high-quality water leads to an increase in K^{+} desorption and leaching (Meiri et al. 1984; Feigenbaum and Meiri 1988). According to WHO (1993) guidelines, 12 ppm K is critical with respect to water quality. The soluble K form will be less in sandy soils because of its low buffering capacity (Marrs et al. 1991). The secondary nutrients like Ca and Mg react with water to form $Ca(OH)_2$ and $Mg(OH)_2$, which increases solubility resulting in leaching of Ca and Mg. In poor quality irrigation, Na^{+} ion is the dominant ion which participates in ion-exchange processes and results in the displacement of base cations into solution and subjected to leaching to groundwater sources. Exchange occurs between solution Na^{+} and exchangeable cations (Ca^{2+}, Mg^{2+} and K^{+}), resulting in the displacement of these cations and anions into solution. Leaching of Ca^{2+}, Mg^{2+}, K^{+} and P from the soil with the application of solutions represents a significant loss of valuable nutrients. Thus sandy soil with high Na^{+} levels is at high risk for nutrient transfer into groundwater in high concentrations (Jalali and Merrikhpour 2008).

19.3.7 Presence of Accompanying Ions

In soils, leaching of nutrients depends upon the presence of other ions. Potassium leaching in soil is different from nitrate leaching. However, movement of exchangeable cations depends upon the concentration of free anions, as cations do not move independently. When anions such as nitrate and chloride are leached, equivalent amounts of cations will also be translocated as counter ions (Tinker and Nye 2000; Oborn et al. 2005). Nitrogen fertilization and high indexes of leaching of NO_3^- speeded up the losses of Ca and Mg as reported by Cahn et al. (1993) in Typic Acrudox of Amazon, due to acidification in the ploughable layer. Similarly, application of K fertilizers can increase exchangeable K pools in soil solution. This excess K accompanied by high rainfall results in K leaching along with nitrate leaching in field where both are applied indiscriminately. The loss of K in soil depends upon the concentration of competing ion (Ca^{2+}) and the amount of water flowing through soil. Under field conditions, enhanced CO_2 production due to root and microbial respiration reacts with water to form HCO_3 and enhances the dissolution of $CaCO_3$. The H^+ ions released at the root zone also solubilizes the calcite. The field with high flow rates of water, along with dissolution of calcite, increases Ca^{2+} content and induces K^+ leaching (Kolahchi and Jalali 2007). The movement of exchangeable cations depends upon the concentration of free anions, as cations do not move independently. When anions such as nitrate and chloride are leached, equivalent amounts of cations will also be translocated as counter ions (Tinker and Nye 2000; Oborn et al. 2005).

19.3.8 Type of Nutrients and Their Susceptibility to Leaching

Nitrate is easily leached down the soil profile due to low activity with the soil matrix. Even the organic matter mineralization release large amount of N prior to crop sowing often results in N leaching which usually occurs during rewetting of dry soils soon after first rainfall. Similarly, S is another anion susceptible to leaching. The SO_4^{2-} anion is also lost by mineralization of organic matter. Sulphate losses are more in soils dominated by potassium and sodium; the losses are limited in soil with high aluminium content (Havlin et al. 1999). Even reports suggest 2–86% of loss of organic S by leaching in forest areas of Amazonia (Lehmann and Schroth 2003). The susceptibility of P to leaching is less compared to N and S due to high fixation and precipitation of the former. However leaching of both organic and inorganic P forms is reported from soils cultivated under long-term manure application. Potassium is leached much smaller compared to Ca and Mg even when applied as fertilizer. However significant amount of K can leach under organic soils and sandy texture with high rainfall conditions. Among micronutrients, Mn and B are susceptible to leaching compared to other nutrients (Lehmann and Schroth 2003).

In soils the percolating water moves down both cations and anions that are electrically neutral, and therefore equivalent amount of anions are leached out along with cations. Both secondary nutrients cations like Ca and Mg are leached easily in soils. There exists a close relationship between cation and anion leaching in soils. In soils significant amount of Ca and Mg leaching was observed which was accentuated by fluxes of nitrate concentration. In some sandy soils, Mg was leached to application of KCl and K_2SO_4 fertilizers to crops (Havlin et al. 1999).

19.3.9 Soil Microclimatic Factors

Microclimatic factors of soil include soil temperature, moisture and aeration. In northern countries the major loss of N, P is leaching over winter. Following harvest of the crop till spring season, when no crop cover and high infiltration rate due to snowmelt result in leaching f nutrients. The intensity with which the nutrients are removed from soil solution, taken up by the plant roots or immobilized by microorganisms, also influences percolation rates and leaching (Ng Kee Kwong and Deville 1984; Orlando Filhol et al. 1995). The soil K dynamics is governed by the equilibrium in soil solution. Both soluble and exchangeable forms of K constitute nearly 10% of total K, which demonstrates a strong equilibrium between these two soil K pools, and leaching is strongly affected by these changes (Sparks and Huang 1985). The soluble K increases immediately in soil solution after application of fertilizers or organics and because of the exchange between soluble to exchangeable K pools. Later this soluble K leaches downwards with time.

19.3.10 Crop Management Practices

Crop management practices like application of FYM along with N fertilizers can increase N leaching by 200–400% over control in bun method of cultivation in a potato cropping system under rain-fed system (Sharma 1999). Soils under rice cropping system are highly vulnerable to N leaching. The loss of N applied through urea can result in 35 and 40% of N leaching in kharif and rabi each under Indo-Gangetic plain considered as food bowl of India (Chhabra et al. 2010). In pastures loss of nutrients can differ between cut and grazed crops. In grazed plots significant amount of K can leach with time and depend upon soil K content. Whereas in cut grassland, the most important way to reduce K losses is to avoid or reduce K inputs during the drainage season, especially in autumn, when the soil has not reached field capacity (Alfaro et al. 2004). Potassium leaching losses in grazed areas can occur by preferential flow or by percolation soon after rainfall. Manures are usually applied based on N content and therefore lead to P accumulation in soils. Organic P in compost (or manure) is known to be less susceptible to P sorption (Guo et al. 2009), so substantial leaching of P from topsoil amended with compost to deeper soil can occur (Eghball 2003).

The soil with deep ploughing for the incorporation of vegetal residues may induce the process of mineralization, thereby increasing the release of ions potentially leachable. Deep ploughing of organic sandy soil will increase mineralizable N under regular tillage practice resulting in major loss of N (Kayser et al. 2012). Tillage has twofold effect on nutrient leaching: first on soil nutrient concentration and second on water flow pattern. In general, tillage is expected to hasten decomposition of residues and more mineralization of organic amendments (Jiao et al. 2004). Nitrate-N loss from tilled soil was more compared to no till soils as reported by Power et al. (2001). However, greater percolation under no till led to more NO_3^--N movement compared to ploughed soils in soil profile (Eisenhauer et al. 1993). Higher P leaching was observed by Gaynor and Findlay (1995) who reported that 3-year average concentrations of dissolved reactive P in the tile drainage waters were 0.24 mg L^{-1} for conventional tillage and 0.54 mg L^{-1} for zero tillage. Leaching has also been reported in some of the typical soils like Terra Preta de Indio found in Amazon basin which are highly valued by farmers for their sustainable fertility and production potential. In these soils large leaching of NO_3^-, Ca and Mg was observed in Ferralsol by Cahn et al. (1992). Commercial crops like sugarcane where K fertilizers are applied annually are also susceptible for K leaching. In a study, sugarcane fertilized by 285 kg K_2O ha^{-1} contributed to K losses by leaching at the rate of 64–136 kg ha^{-1} (Ng Kee Kwong and Deville 1984).

19.4 Strategies to Reduce Nutrient Leaching

19.4.1 Soil Amendments

Use of slow release fertilizer is an important approach to reduce nutrient leaching by using organic fertilizer and increasing adsorption sites, thereby retaining the applied mineral fertilizers (Lehmann et al. 2002). In recent years, many amendments ($CaCl_2$, Al $(OH)_3$, etc.) and industry by-products (such as mill mud) have been suggested to reduce P and metal leaching in sandy soils (O'Connor et al. 2005). And liming acid soils improves crop yield and quality due to improved soil environment for root development. Liming also decreases losses of nutrients and metals in surface run-off by increasing soil holding capacity for these elements. Reactive P leaching loss was reduced by 36%, 17.5% and 40.4%, respectively, by amendment of $CaCl_2$, $CaCO_3$ or $CaCl_2$ + $CaCO_3$, as compared to chemical fertilizer application alone. Low PO_4-P concentrations occurred in the leachates of soils that received $CaCl_2$ or $CaCl_2$ + $CaCO_3$ amendment, and PO_4-P concentrations were reduced by 70.8% and 71.9%, respectively, as compared with chemical fertilizer alone. In addition, application of $CaCl_2$ and $CaCl_2$ + $CaCO_3$ significantly decreased the concentration of K, Mg, Cu and Fe in leachate.

Another amendment, coated calcium carbide (CCC) along with urea reduced the leaching loss of N in all the three top dressing events. The reduced leaching loss of N in urea CCC treatment is attributed to the fact that calcium carbide being a nitrification inhibitor caused most of the mineral nitrogen to remain in NH_4^+-N form, which was adsorbed on soil colloids. This finding is corroborated by greater retention of fertilizer N in soil in urea CCC treatment than that observed in the case of sole urea application. With the advancement of the crop-growth period, the leaching loss of fertilizer N decreased. This is attributed to higher N uptake by the plant at the later stage of crop growth. Charcoal amendments can reduce nutrient leaching by retaining them with water. Charcoal from both hardwood and softwood increases soil cation exchange capacity and retains nutrients and prevents them from leaching (Lehmann et al. 2002). Despite the success of synthetic nitrification and urease inhibitors in research farms, they have poor acceptability among farmers because of high cost. However, the use of product plants like neem for coating urea can be popularized among the farmers to affect N economy and minimize long-term environmental consequences of denitrification and nitrate leaching.

19.4.2 Crop Rotation

Crop rotation practices like including legume crops like soybean can not only reduce NO_3^--N leaching but can also mobilize residual N by symbiotic N fixation and meet crop requirements (Grant et al. 2002). In many cropping systems dissolved N and P lost by leaching are major concerns and influenced by many factors like application rate, soil type, rainfall, residual content and P and N mineralization status of the soil. Crop rotation should be adopted in irrigated and high rainfall areas. Crop rotation like soybean–corn are good examples for improving nitrogen use efficiency but reduces residual N from leaching when compared to monoculture (Huang et al. 1996). Unfortunately, in many areas farmers are accustomed to monoculture production system, which requires new crop management practices to be adopted.

19.4.3 Conservation Tillage

Conservational tillage is known to improve soil health and not directly increase the crop productivity. Conservational tillage compared to conventional tillage more emphasizes on nutrient conservation, minimizing losses to environment and operation costs rather than on yield potential where possible advantages of improving nutrient use efficiency can be seen (Raun and Johnson 1999). Minimum disturbance of soil and crop residue cover can decrease the losses of nutrients.

19.4.4 Best Management Practices (BMPs)

Best management practices were developed in order to conserve nutrients from losses to environment, but information is limited. Applying nutrients at the right rate, right time, and in the right place are important in BMP for achieving optimum nutrient efficiency. Water content of soil and irrigation are the important factors which control nutrient leaching to groundwater table. Growing N fixing crops on the hedges of the field will improve N fixation of crops and limit N leaching losses. Their vigorous root system mobilizes P, K and other nutrients (Saha et al. 2012). For nutrient conservation BMPs include all factors like crop health, insect and weed management, moisture and temperature regimes, supplies of all nutrients, and use of the best adapted cultivar or hybrid, all contributing to more efficient uptake of available nutrients and greater conversion of plant nutrient to grain yield.

Leaching loss of NO_3^- can be minimized by balanced fertilization, split application of urea synchronizing with crop demand, manipulation of water application and rooting depth, appropriate crop rotations and use of slow release fertilizers and nitrification inhibitors like N-serve, DCD, AM, CCC and neem-coated urea. Integrated use of urea and FYM could save 50% of urea N (Upadhyay et al. 2011) in terms of rice grain yield and improve physiological N use efficiency and highest fertilizer N recovery by rice plant, maximum retention of fertilizer N in soil, and hence minimum unaccounted loss for fertilizer N. Application of organic fertilizers in conjunction with inorganic is important to reduce leaching of nutrients. Most organic fertilizers are insoluble, so they tend to solubilize and leach slowly. Their rate of leaching is largely determined by the rate at which they are digested and converted to water-soluble forms by soil microbes. Bone meal, blood meal, feather meal and fish meal are examples of organic fertilizers that are not soluble in water, so they leach slowly.

Incorporation of vegetative residue into the soil instead of burning them is also important to reduce nutrient leaching. According to Franco et al. (2007), 29.8 Mg ha^{-1} of sugarcane crop residue was incorporated into the soil during field preparation. The residue contained 196.7 kg ha^{-1} of N, 150 kg ha^{-1} of K, 20 kg ha^{-1} of P, 60 kg ha^{-1} of Ca, 24.7 kg ha^{-1} of Mg and 29.3 kg ha^{-1} of S, a C/N relation approximately 100 and a relation C/S higher than 200, increasing the immobilization of N and S and decreasing the elements potentially leachable (Oliveira et al. 1999; McNeill et al. 2005). Thus nutrients added as organic residues will be released slowly, having little tendency to leaching, fixation, etc. (Woods 1994).

Best management practices can also be applied to the field based upon their leaching index values. Leaching index (LI) values are determined by the degree to which water percolates below the root zone in certain soils. Percolating water containing dissolved nitrates or other soluble nutrients can be a hazard to groundwater. The LI uses annual precipitation, hydrologic soil group and rainfall distribution data. The leaching index values for different depth are shown in Table 19.3.

Table 19.3 Leaching index values for soils

Hydrologic group	Soils	Leaching index (in.)	Rating
A	All soil	<2	Low
B	All soils	2–10	Moderate
C	All soils	>10	High
D	Soils with less than 20 in. bedrock	>10	High
E	Glacial outwash and well drained alluvial soils	>10	High

Williams and Kissel (1991)

19.4.5 Agroforestry Systems

In agroforestry, presence of continuous and intermittent trees can more efficiently use nutrients and will reduce leaching (Young 1997). In such system where crops like *Leucaena leucocephala* are intercropping with other field crops reduced nitrate leaching. In agroforestry systems, tree growth, with increased litter production, improves soil organic matter content and increases cation exchange capacity and will retain nutrient against leaching. Trees create channels through root growth and can absorb water and reduce infiltration and therefore leaching of nutrients. But in this system, when both tree and crops are present together, competition for light, water and nutrients is usual resulting in yield depression of crops. Therefore this system will maintain fertility and will reduce crop productivity. However selection of trees based upon their rooting pattern especially with those having few superficial roots, and deep roots spread laterally. These roots in subsoil will reduce leaching of nutrients without interfering crop nutrient requirements.

Tree-based intercropping (TBI) systems that combine agricultural alley crops with rows of hardwood trees have been implemented in many countries like Europe and the USA. The roots of 5–8-year-old hybrid poplars, growing in two TBI systems in southern Quebec, played a "safety-net" role of capturing nutrients leaching below the rooting zone of alley crops. Tree roots in the TBI system established on clay loam soil decreased subsoil NO_3^- leaching by 227 kg N ha^{-1} and 30 kg N ha^{-1} over two consecutive years, and decreased dissolved organic N (DON) leaching by 156 kg N ha^{-1} $year^{-1}$ in the second year of the study (Bergeron et al. 2011). The poplar trees have a greater affinity to NO_3^- compared to NH_4^+ ions and can reduce nitrate leaching from soils (Choi et al. 2005).

Losses equivalent to 29–387 kg ha^{-1} was reported by Jalali and Rowell (2003) when 780 mm of distilled water was applied to calcite- and gypsum-bearing soil columns in the laboratory. The amount of K leached was 1.8 times higher in control plot with no N application in a study conducted on sugarcane where the major contributor of K to leaching was through fast liberation of K from straw (Oliveira et al. 1999). The application of animal manure based upon N significantly increases loss of K by leaching. High level of K like 80 kg ha^{-1} was leached from soil in a maize-cultivated field supplied with cattle manure and pig slurry applied on the basis of N (Kayser et al. 2012). This illustrates that under conditions of high N leaching, K leaching can be large and can cause negative K balance.

19.5 Conclusion

The process of soil nutrient depletion via leaching is a potentially serious threat to world food security and sustainable agriculture. Leaching rate of any nutrient is determined by three major factors like (a) the degree of water solubility of a fertilizer combined with; (b) the amount of soil water at any time and (c) soil type, porosity and compaction. Leaching should be less severe under a crop than under bare fallow but will increase when fertilizers are added. Therefore good measurements are essential since the leaching of nutrient ions may be a key to understanding the chemical degradation of the soil during cultivation. Research is needed to better predict nutrient dynamics, not only with regard to fertilizer use efficiency but also with the aim of better environmental protection.

References

Alfaro, M., Jarvis, S. C., & Gregory, P. J. (2003). Potassium budgets in grassland systems as affected by nitrogen and drainage. *Soil Use and Management, 19*, 89–95.

Alfaro, M. A., Jarvis, S. C., & Gregory, P. J. (2004). Factors affecting potassium leaching in different soils. *Soil Use and Management, 20*, 182–189.

Andersson, S., Simonsson, M., Mattsson, L., Edwards, A. C., & Oborn, I. (2007). Response of soil exchangeable and crop potassium concentrations to variable fertilizer and cropping regimes in long-term field experiments on different soil types. *Soil Use and Management, 23*, 10–19.

Arregui, L. M., & Quemada, M. (2006). Drainage and nitrate leaching in a crop rotation under different N-fertilizer strategies: Application of capacitance probes. *Plant and Soil, 288*, 57–69.

Askegaard, M., Eriksen, J., & Olesen, J. E. (2003). Exchangeable potassium and potassium balances in organic crop rotations on a coarse sand. *Soil Use and Management, 19*, 96–103.

Aulakh, M. S., Garg, A. K., & Kabba, B. S. (2007). Phosphorus accumulation, leaching and residual effects on crop yields from long term applications in the subtropics. *Soil Use and Management, 23*, 417–427.

Bergeron, M., Simon, L., Robert, L. B., Joann, W., Cogliastro, A., Marie, F. J., & Paul, A. (2011). Reduced soil nutrient leaching following the establishment of tree-based intercropping systems in eastern Canada. *Agroforestry Systems, 83*, 321–330.

Buchter, B., Hinz, C., Flury, M., & Fluhler, H. (1995). Heterogeneous flow and solute transport in a natural grassland and arable soils on three contrasting parent materials. *Biodiversity and Conservation, 57*, 257–271.

Cahn, M. D., Bouldin, D. R., & Cravo, M. S. (1992). Nitrate sorption in the profile of an acid soil. *Plant and Soil, 143*, 179–183.

Cahn, M. D., Bouldin, D. R., Cravo, M. S., & Bowen, W. T. (1993). Cation and nitrate leaching in an Oxisol of the Brazilian Amazon. *Agronomy Journal, 85*, 334–340.

Camargo, P. B. (1989). *Nitrogen dynamics of fertilizer urea (15 N) and aqua ammonia (15 N) incorporated into the soil in the cultivation of sugarcane*. Masters dissertation, Piracicaba: ESALQ.

Cassman, K. G, Dobermann, A. R., & Walters, D. T. (2002). *Agroecosystems, nitrogen-use efficiency, and nitrogen management*. Agronomy & Horticulture—Faculty Publications. Paper 356.

Chhabra, A., Manjunath, K. R., & Panigrahy, S. (2010). Non point source pollution in Indian agriculture: Estimation of nitrogen losses from rice crop using remote sensing and GIS. *International Journal of Applied Earth Observation and Geoinformation, 12*, 190–200.

Choi, W. J., Chang, S. X., & Hao, X. (2005). Soil retention, tree uptake, and tree resorption of $^{15}NH_4NO_3$ and $NH_4\ ^{15}NO_3$ applied to trembling and hybrid aspens at planting. *Canadian Journal of Forest Research, 35*, 823–831.

Clay, D. E., Zheng, Z., Liu, Z., Clay, S. A., & Trooien, T. P. (2004). Bromide and nitrate movement through undisturbed soil columns. *Journal of Environmental Quality, 33*, 338–342.

Dassannayake, A. R. (1990). *Potassium leaching in stony and sandy soils*. M.Sc. Thesis, University of Reading Central Library, UK.

Djodjic, F., Borling, K., & Bergstrom, L. (2004). Phosphorous leachin g in relation to soil type and soil phosphorus content. *Journal of Environmental Quality, 33*, 678–684.

Edmeades, D. C. (2003). The long-term effects of manures and fertilizers on soil productivity and quality: A review. *Nutrient Cycling in Agroecosystems, 66*, 165–180.

Edwards, D. R., & Daniel, T. C. (1993). Runoff quality impacts of swine manure applied to fescue plots. *Transactions of the American Society of Agricultural Engineers, 36*, 8–86.

Eghball, B. (2003). Leaching of phosphorus fractions following manure or compost application. *Communications in Soil Science and Plant Analysis, 34*(19), 2803–2815.

Eisenhauer, D. E., Norton, M. J., Martin, D. L., & Spalding, R. F. (1993). Impact of tillage and crop rotation on water movement with furrow irrigation. In *Proceedings of Agricultural Research to Protect Water Quality, Minneapolis, MN, 21–24 February 1993*. Soil and Water Conservation Society, Ankeny. pp. 533–536.

Elliott, H. A., Oconnor, G. A., & Brinton, S. (2002). Phosphorus leaching from biosolids-amended sandy soils. *Journal of Environmental Quality, 31*, 681–689.

Ersahin, S., Papendick, I. R., Smith, J. L., Keller, C. K., & Manoranjan, V. M. (2002). Macropore transport of bromide as influenced by soil structure differences. *Geoderma, 108*, 207–223.

Feigenbaum, S., & Meiri, A. (1988). *The effect of potassium fertilization on cotton response and potassium distribution under irrigation with saline water* (BARD Report I-630-83). pp. 88–110.

Franco, H. C. J., Vitti, A. C., Faroni, E. C., Cantarella, H., & Trivelin, P. C. O. (2007). Nutrients stock in crops residues incorporated into the soil in reform areas with sugarcane. *STAB, 25*, 32–36.

Gaber, H. M., Inskeep, P. W., Comfort, S. D., & Mraith, J. M. (1995). Non-equilibrium transport of atrazine through large intact soil cores. *Soil Science Society of America Journal, 59*, 60–67.

Garg, A. K., & Aulakh, M. S. (2010). Effect of long-term fertilizer management and crop rotations on accumulation and downward movement of phosphorus in semi-arid subtropical irrigated soils. *Communications in Soil Science and Plant Analysis, 41*, 848–864.

Gaynor, J. D., & Findlay, W. I. (1995). Soil and phosphorus loss from conservation and conventional tillage in corn production. *Journal of Environmental Quality, 24*, 734–741.

Goswami, N. N., Prasad, R., Sarkar, M. C., & Singh, S. (1988). Studies on the effect of green manuring on N economy in a rice-wheat rotation using a ^{15}N technique. *Journal of Agriculture Science (Cambridge), 11*, 413–417.

Goulding, K. W. T., Poulton, P. R., Webster, C. P., & Howe, M. T. (2000). Nitrate leaching from the Broadbalk wheat experiment, Rothamsted, UK, as influenced by fertilizer and manure inputs and the weather. *Soil Use and Management, 16*, 244–250.

Grant, C. A., Brown, K. R., Racz, G. J., & Bailey, L. D. (2002). Influence of source, timing and placement of nitrogen fertilizer on seed yield and nitrogen accumulation in the seed of canola under reduced- and conventional-tillage management. *Canadian Journal of Plant Science, 82*, 629–638.

Guo, B., Liang, Y., Li, Z., & Han, F. (2009). Phosphorus adsorption and bioavailability in a paddy soils amended with pig manure compost and decaying rice straw. *Communications in Soil Science and Plant Analysis, 40*(13), 2185–2199.

Hartemink, A. E. (2008). Sugarcane for bioethanol: Soil and environmental issues. *Advances in Agronomy, 99*, 125–182.

Havlin, J. L., Beaton, J. D., Tisdale, S. L., & Nelson, W. L. (1999). *Soil fertility and fertilizers—An introduction to nutrient management* (6th ed.). New Jersey: Prentice Hall.

Heckrath, G., Brookes, P. C., Poulton, P. R., & Goulding, K. W. T. (1995). Phosphorus leaching from soils containing different phosphorus concentrations in the Broadbalk experiment. *Journal of Environmental Quality, 24*, 904–910.

Hinsinger, P. (2001). Bioavailability of soil inorganic P in the rhizosphere as affected by root-induced chemical changes: A review. *Plant and Soil, 237*, 173–195.

Huang, W. Y., Shank, D., & Hewitt, T. I. (1996). On-farm costs of reducing residual nitrogen on cropland vulnerable to nitrate leaching. *Review of Agricultural Economics, 18*, 325–339.

Jalali, M., & Merrikhpour, H. (2008). Effects of poor quality irrigation waters on the nutrient leaching and groundwater quality from sandy soil. *Environmental Geology, 53*(6), 1289–1298.

Jalali, M., & Rowell, D. L. (2003). The role of calcite and gypsum in the leaching of potassium in a sandy soil. *Experimental Agriculture, 39*, 379–394.

Jiao, Y., Hendershot, W. H., & Whalen, J. K. (2004). Agricultural practices influence dissolved nutrients leaching through intact soil cores. *Soil Science Society of America Journal, 68*, 2058–2068.

Kayser, M., Muller, J., & Isselstein, J. (2007). Potassium leaching from cut grassland and from urine patches. *Soil Use and Management, 23*, 384–392.

Kayser, M., Benke, M., & Isselstein, J. (2012). Potassium leaching following silage maize on a productive sandy soil. *Plant, Soil and Environment, 58*(12), 545–550.

Kolahchi, Z., & Jalali, M. (2007). Effect of water quality on the leaching of potassium from sandy soil. *Journal of Arid Environments, 68*, 624–639.

Ladha, J. K., Pathak, H., Krupnik, T. J., Six, J., & van Kessel, C. (2005). Efficiency of fertilizer nitrogen in cereal production: Retrospects and prospects. *Advances in Agronomy, 87*, 85–156.

Laird, D., Fleming, P., Wang, B., Horton, R., & Laird, D. K. (2010). Biochar impact on nutrient leaching from a Midwestern agricultural soil. *Geoderma, 158*, 436–442.

Lehmann, J., & Schroth, G. (2003). Nutrient leaching. In G. Schroth & S. N. Finclair (Eds.), *Trees, crops and soil fertility*. Wallingford: CAB International.

Lehmann, J., Silva, J. P. Jr, Rondon, M., Cravo, M. S., Greenwood, J., Nehls, T., Steiner, C., & Glaser, B. (2002). Slash-and-char—A feasible alternative for soil fertility management in the central Amazon? In *Proceedings of the World Congress of Soil Science, Thailand.*

Lundell, Y., Johannisson, C., & Hogberg, P. (2001). Ion leakage after liming or acidifying fertilization of Swedish forests: A study of lysimeters with and without active tree roots. *Forest Ecology and Management, 147*, 151–170.

Mahmood-ul-Hassan, M., Akhtar, M. S., & Nabi, G. (2008). Boron and zinc transport through intact columns of calcareous soils. *Pedosphere, 18*, 524–532.

Marrs, R. H., Gough, M. W., & Grifrths, M. (1991). Soil chemistry and leaching losses of nutrients from semi-natural grassland and arable soils on three contrasting parent materials. *Biodiversity and Conservation, 57*, 257–271.

McNeill, A. M., Eriksen, J., Bergstrom, L., Smith, K. A., Marstorp, H., Kirchmann, H., & Nilsson, I. (2005). Nitrogen and sulphur management: Challenges for organic sources in temperate agricultural systems. *Soil Use and Management, 21*, 82–93.

Meiri, A., Feigenbaum, S., & Sagiv, B. (1984). *Potassium fertilization under irrigation with saline and sodic water.* Report to Dead Sea Works, 301-00-81.

Miao, Y., Bobby, A. S., & Zhang, F. (2011). Long-term experiments for sustainable nutrient managementin China. A review. *Agronomy for Sustainable Development, 31*(2), 397–414.

Mortvedt, J. J. (1994). Needs for controlled-availability micronutrient fertilizers. *Fertilizer Research, 38*(3), 213–221.

Muralidharudu, Y., Reddy, K. S., Mandal, B. N., Rao, A. S., Singh, K. N., & Sonekar, S. (2011). *GIS based soil fertility maps of different states of India.* Bhopal: All India Coordinated Research Project on Soil Test Crop Response Correlation. IISS.

Ng Kee Kwong, K. F., & Deville, J. (1984). Nitrogen leaching from soils cropped with sugarcane under the humidtropical climate of Mauritius. Indian Ocean. *Journal of Environmental Quality, 13*(3), 471–474.

O'Connor, G. A., Brinton, S., & Silveira, M. L. (2005). Evaluation and selection of soil amendments for field testing to reduce P losses. *Soil and Crop Science Society of Florida, Proceedings, 64*, 22–34.

Oborn, I., Andrist-Rangel, Y., Askekaard, M., Grant, C. A., Watson, C. A., & Edwards, A. C. (2005). Critical aspects of potassium management in agricultural systems. *Soil Use and Management, 21*, 102–112.

Oliveira, M. W., Trivelin, P. C. O., Penatti, C. P., & Piccolo, M. C. (1999). Decomposition of nutrients from the straw of sugarcane in the field. *Pesquisa Agropecuária Brasileira, Brasília, 34*, 2359–2362.

Oliveria, M. W., Trivelin, P. C. O., Boaretto, A. E., Muraoka, T., & Moratatti, J. (2002). Leaching of nitrogen, potassium and magnesium in a sandy soil cultivated with sugarcane. *Pesquisa Agropecuária Brasileira, Brasília, 37*(6), 861–868.

Orlando Filhol, J., Bittencourt, V. C., & Alves, M. C. (1995). Vinasse application in a Brazilian sandy soil and nitrogen water table pollution. *STAB, 13*(6), 9–13.

Ozha, D. D., Varshey, C. P., Bohra, T. L., & Sharma, D. C. (1993). *Vignana Parishad Anusandhan Patrika, 36*, 147–157.

Pathak, H., Li, C., Wasserman, R., & Ladha, J. (2006). Simulation of nitrogen balance in the rice-wheat systems of the indo-Gangetic plains. *Soil Science Society of America Journal, 70*, 1612–1622.

Power, J. F., Wiese, R., & Flowerday, D. (2001). Managing farming systems for nitrate control: A research review from management systems evaluation areas. *Journal of Environmental Quality, 30*, 1866–1880.

Radulovich, R., Sollin, P., Baveye, P., & Solorzano, E. (1992). Bypass water flow through unsaturated microaggregated tropical soils. *Soil Science Society of America Journal, 56*, 721–726.

Rajmohan, N., & Elango, L. (2005). Nutrient chemistry of groundwater in an intensively irrigated region of southern India. *Environmental Geology, 47*, 820–830.

Rao, E. V. S. P., & Puttanna, K. (2000). Nitrates, agriculture and environment. *Current Science, 79*(9), 1163–1167.

Raun, W. R., & Johnson, G. V. (1999). Improving nitrogen use efficiency for cereal production. *Agronomy Journal, 91*, 357–363.

Reddy, K. S., Pax, A. F. C., Blamey, B., Ram, C., Dalal, C., Mohanty, M., et al. (2010). Leaching losses of nutrients from farmyard manure pits in Central India. *Proceedings of 19th World Congress of Soil Science, Soil Solutions for a Changing World, 1–6 August 2010, Brisbane, Australia*. Published on DVD.

Rowell, D. L. (1988). Soil acidity and alkalinity. In A. Wiid (Ed.), *Russell's soil conditions and plant growth* (pp. 844–898). Essex: Longman.

Roy, R. N., Mishra, R. V., Lesschen, J. P., & Smaling, E. M. A. (2003). *Assessment of soil nutrient balance: Approaches and methodology* (p. 14). Rome, Italy: FAO Fertilizer and Plant Nutrition Bulletin.

Saha, R., Chaudhary, R. S., & Somasundram, J. (2012). Soil health management under hill agroecosystem of North East India. *Applied and Environmental Soil Science*. doi:10.1155/2012/696174.

Salcedo, I. H., & Sampaio, E. V. S. B. (1991). Dynamics nutrients in sugarcane. V. Balance K in four crop cycles. *Pesquisa Agropecuária Brasileira, Brasília, 26*(9), 1323–1335.

Schoumans, O. F., & Groenendijk, P. (2000). Modeling soil phosphorus levels and phosphorus leaching from agricultural land in the Netherlands. *Journal of Environmental Quality, 29*, 111–116.

Sharma, U. C. (1999). Loss of nitrogen through leaching and runoff from two potato land-use systems on different soils: Impact of land-use change on nutrient loads from diffuse sources. *Proceedings of IUGG 99 Symposium HS3, Birmingham, Publ. no. 257, 1999.*

Sims, J. T., Simard, R. R., & Joern, B. C. (1998). Phosphorus loss in agricultural drainage: Historical perspective and current research. *Journal of Environmental Quality, 27*(2), 277–293.

Singh, V., Singh, Y., Singh, B., & Meelu, O. P. (2006). Distribution and leaching losses of applied urea-N in Sandy loam and clay loam soils under wetland and upland moisture regimes. *Journal of the Indian Society of Soil Science, 54*(2), 185–192.

Sparks, D. L., & Huang, P. M. (1985). Physical chemistry of soil potassium. In R. Munson (Ed.), *Potassium in Agriculture* (pp. 201–276). Madison WI: ASA CSSA & SSSA.

Stoorvogel, J. J., & Smaling, E. M. A. (1998). Research on soil fertility decline in tropical environments: Integration of spatial scales. *Nutrient Cycling in Agroecosystems, 50*, 151–158.

Subba Rao, A., Srivastava, S., & Ganeshamurty, A. N. (2015). Phosphorus supply may dictate food security prospects in India. *Current Science, 108*(7).

Tinker, P. B., & Nye, P. H. (2000). *Solute movement in the rhizosphere*. Oxford: Oxford University Press.

Upadhyay, R. K., Patra, D. D., & Tewari, S. K. (2011). Natural nitrification inhibitors for higher nitrogen use efficiency, crop yield, and for curtailing global warming. *Tropical Agriculture, 49*(1–2), 19–24.

USEPA (United State of Environmental Protection Agency). (1985). *Nitrate/Nitrite Health Advisory (draft)*. Washington, DC: USEPA, Office of Drinking Water.

Webb, T., Allan, H., Linda, L., McLeod, M., & Murray, C. (2010). *Environmental Canterbury Regional Council Report on Mapping of vulnerability of nitrate and phosphorus leaching, microbial bypass flow, and soil runoff potential for two areas of Canterbury*.

Whalen, K. J., & Chang, C. (2001). Phosphorus accumulation in cultivated soils from long-term annual application of cattle feedlot manure. *Journal of Environmental Quality, 30*, 229–237.

Williams, P. H. (1988). *The fate of potassium in grazed dairy pastures*. Ph.D. Thesis, Massey University Central Library, New Zealand.

Williams, J. R., & Kissel, D. E. (1991). Water percolation: An indicator of N leaching potential. In R. F. Follet (Ed.), *Managing nitrogen for groundwater quality and farm profitability*. Madison: Soil Science Society of America.

Wong, M. T. F., Kruijs, A. C. B., & Juo, A. S. R. (1992). Leaching loss of calcium, magnesium, potassium and nitrate derived from soil, lime and fertilizers as influenced by urea applied to undisturbed lysimeters in southeast Nigeria. *Fertilizer Research, 31*, 281–289.

Woods, W. G. (1994). An introduction to boron: History, sources, uses, chemistry. *Environmental Health Perspectives, 102*(Suppl 7), 5–11.

Young, A. (1997). *Agroforestry for soil management* (2nd ed.). Wallingford: CAB International.

Zhang, H., Xu, M., & Zhang, F. (2009). Long-term effects of manure application on grain yield under different cropping systems and ecological conditions in China. *The Journal of Agricultural Science, 14*, 31–42.

Zhang, S., Zhang, X., Liu, X., Liu, W., & Liu, Z. (2013). Spatial distribution of soil nutrient at depth in black soil of Northeast China: A case study of soil available potassium. *Nutrient Cycling in Agroecosystems, 95*, 319–331.

Chapter 20
Plant-Microbe Interactions for Phosphate Management in Tropical Soils

Nelson W. Osorio, Laura Osorno, Juan D. Leon, and Claudia Álvarez

Abstract Low soil phosphate (P) availability is a major limiting factor for tropical agriculture and forestry because this severely limits the plant productivity. A viable alternative consists of using arbuscular mycorrhizal fungi (AMF) that form a symbiotic association with plant roots, the AMF hyphae are by far most efficient than roots alone in taking up water and low-mobile nutrient from soils, particularly P. Also, there are phosphate-solubilizing microorganisms (PSM) capable of dissolving native and applied P compounds. With both microbial groups, plants enhance its capability of taking up P and thus overcome this limiting factor. The objective in this chapter was to discuss the effectiveness of both AMF and PSM to enhance plant P uptake in tropical soils.

Keywords Phosphate fixation • Phosphate cycling • Arbuscular mycorrhizal fungi • Phosphate- solubilizing microorganisms

20.1 Introduction

The low soil phosphate (P) availability is a major stressor for tropical agriculture and forestry (Wakelin et al. 2004; Oberson et al. 2006). This affects globally 2016 million ha, about half of them are in the tropical zone, which constraints plant productivity, particularly in volcanic ash soils and those highly weathered soils of the humid tropics (Fairhurst et al. 1999; Osorio and Habte 2009). Phosphate ions are either *adsorbed* on the surface of soil clays and oxides where they are retained in unavailable forms or *precipitated* with iron and aluminum ions leaving soil solution with scarce amount of plant available P (Smith 2002; Khan et al. 2007). This intensively occurs in volcanic ash soils (Andisols) and in highly weathered soils of the tropic (particularly in Oxisols and Ultisols). Consequently, the efficiency of soluble P fertilizers in these soils is quite low (5–10%), and high amounts of these fertilizers are required to achieve satisfactory crop yields, which it is not always possible by

N.W. Osorio (✉) • L. Osorno • J.D. Leon • C. Álvarez
Universidad Nacional de Colombia, Calle 59A No. 63-20, 050034 Medellin, Colombia
e-mail: nwosorio@unal.edu.co; nwosorio@gmail.com

M. Naeem et al. (eds.), *Essential Plant Nutrients*,
DOI 10.1007/978-3-319-58841-4_20

poor farmers of underdevelop tropical countries (Reddy et al. 2002; Osorio and Habte 2009; Batti and Yamar 2010).

Fortunately, there are fungi that form symbiotic associations with plant roots called *arbuscular mycorrhizal fungi* (AMF) and thus are capable of increasing water and nutrient uptake by plant increasing thus the effectiveness in taking up native and applied P (Osorio and Habte 2013). On the other hand, some soil microorganisms are able of solubilizing native and applied P and are called *phosphate-solubilizing microorganisms* (PSM). The role of these two microbial groups AMF and PSM to enhance plant P uptake in tropical soils will be discussed.

20.2 Phosphate Deficiency in Tropical Soils

Plant roots commonly uptake P ion ($H_2PO_4^-$) from the soil solution; nevertheless, its concentration is quite low (0.001–0.1 mg L^{-1}) (Scervino et al. 2010), which limits plant productivity in agricultural crops, grassland, and forestry. The low availability of P in tropical soils is due to a series of reactions that remove soluble P into the soil solid phase, which has been called P fixation (Barber 1995; Collavino et al. 2010).

In the tropics the soils that exhibit high P fixation capacity occupy 1018 million ha (Sanchez and Logan 1992): 659 million ha tropical America, 210 million ha in Africa, and 199 million ha in Asia. The term P fixation is used to describe *adsorption* and *precipitation* reactions (Collavino et al. 2010). In the first place, P ions are adsorbed on the surface of soil clay and oxides (it is particularly strong on iron and aluminum hydrous oxides, e.g., goethite, gibbsite);

In acidic soils (pH < 5.5), P *precipitation* occurs with active forms of aluminum [Al^{3+}, $Al(OH)^{2+}$, $Al(OH)_2^+$] and iron [(Fe^{3+}), $Fe(OH)^{2+}$, $Fe(OH)_2^+$] (Eqs. (20.1 and (20.2)) (Smith 2002). In neutral and alkaline soils (pH > 6.5), it also occurs mostly with calcium (Ca^{2+}) (Bohn et al. 1985). Initially, P ions precipitate to form amorphous (noncrystalline) compounds, which became crystalline over time (Brady and Weil 1999; Trolove et al. 2003; Havlin et al. 2004; Do Carmo Harta and Torrent 2007).

$$\mathrm{H}_2PO_4^- + Al\left(OH\right)_2^+ \leftrightarrow AlPO_4 \cdot 2H_2O \tag{20.1}$$

$$\mathrm{H}_2PO_4^- + Fe\left(OH\right)_2^+ \leftrightarrow \mathrm{FePO}_4 \cdot 2H_2O \tag{20.2}$$

20.3 Isotherm of Soil Phosphate Fixation

The use of isotherm of P sorption is a simple way to measure the soil capacity to fix P (Do Carmo Harta and Torrent 2007). In our laboratory we used the method developed by professors Fox and Kamprath (1970) at the North Carolina State

University and the University of Hawaii. Briefly, this consists of applying separately grading amounts of soluble P (e.g., KH_2PO_4; 0–2000 mg P kg^{-1}) dissolved in 30 mL of 0.01 M $CaCl_2{\bullet}2H_2O$ to aliquots of soils (3 g, dry basis) in plastic centrifuge tubes. Then, the tubes are shaken 30 min each 12 h for 6 days. After this incubation period, the tubes are centrifuged (15 min, 4000 rpm) and the supernatant filtered with filter paper (and membrane filters). The concentration of solution P at equilibrium is measured using the phosphomolybdate blue method (Murphy and Riley 1962); the remaining non-soluble P is considered fixed into the soil particles. A graph is constructed to show the relationship between adsorbed and solution P.

The isotherm of soil P fixation is also used to determine the amount of P fertilizers required (Hue and Fox 2010). In this case, it is necessary to know the critical soil solution P level associated to a desired yield for a given crop. Examples for this use are illustrated in the Fig. 20.1 with three Colombian soils (Mollisol, Oxisol, and Andisol). Whereas the Palmira soil (Valle del Cauca, Colombia) exhibited a low very low capacity to fix P ($P_{0.2}$ = 64 mg kg^{-1}); the Carimagua soil (Vichada, Colombia) and the Chinchina soil (Caldas, Colombia) exhibited medium and very high capacity to fix P ($P_{0.2}$ = 352 and 1658 mg kg^{-1}, respectively). These amounts of P coincide with the P requirements of soybean and tomato crops. In the case of corn, the P requirements for 95% of the maximum yield would be 3, 174 and 977 mg of P kg^{-1} for the Mollisol, Oxisol, and Andisol, respectively. If the corn yield desired is lower (75%), the soil solution P level should be 0.008 mg L^{-1}; consequently, the P required will be lower (0, 76, 604 mg of P kg^{-1}, respectively). In any case, the soluble P fertilizers must be applied in the root zone.

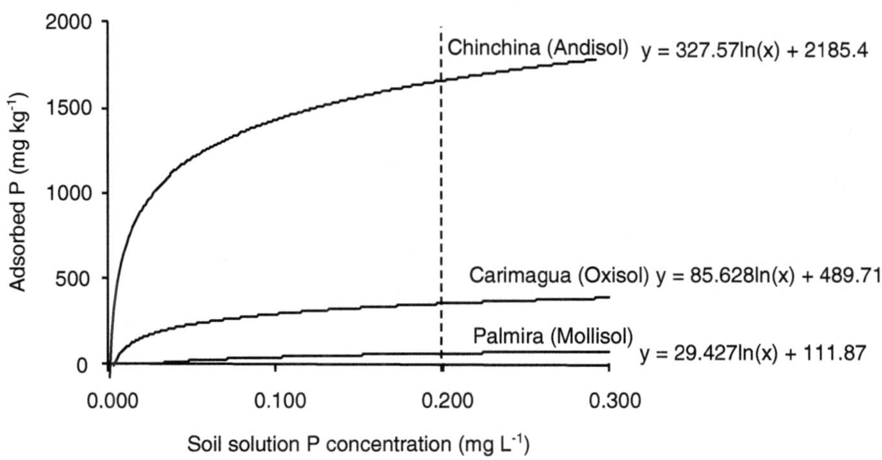

Fig. 20.1 Isotherms of P fixation for three soils of Colombia. The projection of the *dashed line* on the *Y*-axis shows the $P_{0.2}$ value, which measures the soil P fixation capacity (Source: Osorio (2014). Used with permission)

20.4 Litter P Cycling

It has been widely recognized that the low soil P availability in tropical ecosystems represents a limiting factor for plant productivity (Turner et al. 2006; Osorio et al. 2015). This means that even under natural conditions, structure and functioning of ecosystems may be affected in different ways. Numerous studies of the biogeochemical cycle in planted and natural forests in the tropics have demonstrated the strong character of P as a limiting nutrient. This has been determined from, among other indicators, low foliar P concentrations both in green leaves and leaf litter, low P returns through leaf litter, slow decomposition of fine litter and P nutrient release from leaf litter, and intense P reabsorption process from previous leaf abscission, and with this a high P use efficiency. These aspects are presented below with the aim to illustrate how the low availability of this nutrient in such ecosystems is limiting both functionality and productivity.

The major source of soil organic matter in terrestrial ecosystems is the fine litter production; the leaf litter constitutes the most abundant (~70%) and easily decomposable fraction (Meentemeyer 1978). These materials are accumulated on the soil surface and must be decomposed to release nutrients. In many tropical soils, nutrients released from the litter are the most relevant source of plant nutrients (Parzych and Trojanowski 2006), and the rate of litter decay controls nutrient input (León and Osorio 2014).

The rate of litter decomposition will determine the thickness of the litter layer accumulated on the forest floor. Overall, it has been found both a higher litterfall and a higher rate of litter decomposition in tropical lowland forests than in tropical highland forests (Songwe et al. 1988; McDonald and Healey 2000; Del Valle 2003). The release of P in leaf litter is the product of the decomposition process of such detritus, in which the nutrient is forming biomolecules of different structural complexity. In the case of P, it has been reported a general pattern of very low circulation through leaf litter and consequently a low P accumulation into the litter layer (León and Osorio 2014).

Although it has been noted that P is limiting especially in lowland forests and N in upland forests (Fisher et al. 2013), in many mountain ecosystems of the tropics, there is a low soil P availability, which leads to the limitation of the latter, particularly in volcanic ash soils (Nanzyo et al. 1993). This determines that P concentrations in leaf litter of these ecosystems are very low (0.02–0.1%: Lundgren 1978, Medina et al. 1981, Proctor et al. 1983, Vitousek et al. 1995, Cuevas and Lugo 1998, Vera et al. 1999).

When observed P return values through leaf litter (kg P ha^{-1} $year^{-1}$) for different tropical forest ecosystems (Table 20.1), it can be established that the overall cycling amounts of the element are very low. Although the nutrient use efficiency differs among species, the low soil P availability in most of tropical soils appears to be decisive in the low P recycling, and it probably constrains the productivity of both natural forests and forest plantations established with different objectives.

Table 20.1 Reported values of P flows through the leaf litterfall for lowland and highland tropical forests

Ecosystem	Kind of forest or species	P (kg ha^{-1} $year^{-1}$)	Source
Lowland forests	Mixed forests; Ghana	6.1	Nye (1961)
	Humid forests; Panama	9.4	Golley et al. (1975)
	Humid forests; Brazil	3.6	Scott et al. (1992)
	Forests dominated by *Peltogyne gracilipes*; Brazil	0.9	Villela and Proctor (1999)
	Plantations of *Pinus pinnata*; India	1.3	Singh et al. (1999)
	Humid forests; Indonesia	2.0–5.0	Hermansah et al. (2002)
	Flooding humid forests; Pacific, Colombia	3.5	Del Valle (2003)
	Humid forests; Mexico	2.1–5.8	Campo-Alves (2003)
	Mature mixed humid forest of *Tectona grandis*; India	6.0	Pande (2004)
	Mature humid forests, Amazonas; Colombia	1.1–4.4	Tobón et al. (2004)
	Rain forests; Nigeria	3.4	Muoghalu (2003)
	Forest plantations of *Pinus caribaea*, *Swietenia mahagoni*, *Swietenia macrophylla*; Puerto Rico	1–2	Lugo (1992)
	Forest plantations of *Pinus caribaea*, *Eucalyptus saligna*, *Eucalyptus* cf. *Patentinervis*, *Swietenia macrophylla* *Terminalia ivorensis*; Puerto Rico	2–4	Cuevas and Lugo (1998)
	Forest plantations of *Casuarina equisetifolia*, *Leucaena leucocephala*, *Eucalyptus robusta*; Puerto Rico	1–5	Parrotta (1999)
	Forest plantations of *Dalbergia sissoo*, *Azadirachta indica*, *Pongamia pinnata*, *Shorea robusta*; India	1–2	Singh et al. (1999)
	Forest plantations of *Eucalyptus tereticornis*, *Eucalyptus grandis*; Congo	4	Nzila et al. (2002)
	Humid forest, dominant species: *Acmena resa*, *Syzygium wesa*, *Cardwellia sublimis*; Australia	1	Herbohn and Congdon (1998)
	Dry forest, dominant species: *Shorea robusta*, *Madhuca indica*, *Boswellia serrata*, *Prosopis juliflora*; India	0.1	Singh et al. (2004)
	Rain forest, dominant species: *Palaquium formosanum*, *Bischofia javanica*, *Aglaia formosanum*, *Pouteria obovata*, *Macaranga tanarius*, *Pisonia umbellifera*; Taiwan	1–2	Liao et al. (2006)
	Forest plantations of Azadirachta indica; Colombia	0.06	Flórez-Flórez et al. (2013)
	Successional dry forest; Colombia	0.02	Restrepo et al. (2013)

(continued)

Table 20.1 (continued)

Ecosystem	Kind of forest or species	P (kg ha^{-1} year^{-1})	Source
Montane forests	Native forest; Tanzania	5.2	Lundgren (1978)
	Forest plantations of *Pinus patula*; Tanzania	1.3	Lundgren (1978)
	Forest plantations of *Cupressus lusitanica*; Tanzania	1.7	Lundgren (1978)
	Humid forest; New Guinea	2.5	Edwards and Grubb (1982)
	Humid forest; Venezuela	2.1	Fassbender and Grimm (1981)
	Humid forests; Colombia	1.1–3.4	Veneklaas (1991)
	Humid forest; Australia	1.2	Herbohn and Congdon (1998)
	Humid forest; Hawaii	1.1–4.6	Vitousek et al. (1995)
	Subtropical native forest; China	2.9–3.2	Liu et al. (2002)
	Rain forest, dominant taxa *Shorea, Syzygium, Tristaniopsis, Schima*; Borneo	1.2	Kitayama and Aiba (2002)
	Rain forest, dominant species: *Allophylus varians, Cinnamonum ovalifolium, Eugenia mabaeoides, Memecylon parvifolium, Michelia nilagirica*; Sri Lanka	4.4	Weerakkody and Parkinson (2006)
	Humid oak forest, *Quercus humboldtii*; Colombia	1.6	León et al. (2011)
	Forest plantations of *Pinus patula*; Colombia	1.7	León et al. (2011)
	Forest plantations of Cupressus lusitanica; Colombia	0.8	León et al. (2011)

20.5 P Release from Litter Decomposition

P constraints to tropical terrestrial ecosystems can be very critical because little P returned annually by the litter during the decomposition process in nutrient recycling process. An approach to determine the rate of nutrient release from the litter and their respective residence times was proposed by Jenny et al. (1949) to calculate the decomposition rate of organic debris. The original equation proposes to calculate the decomposition constant k_j (Eq. (20.1):

$$k_j = A / (F + A)$$

where A is the annual rate of litterfall (t ha^{-1} year^{-1}), k_j is the constant that represents the litter fraction that decays, and F is the litter amount accumulated before the measure of litterfall (t ha^{-1} year^{-1}). At equilibrium, the litter loss (decomposition) is compensated by the litter additions (litter production).

From this proposal, for each nutrient, a k_j can be calculated. High values of $_j$ indicate a fast return of the nutrient from litter. The mean residence time (MRT) of the nutrients in litter can be estimated from the inverse value of $_j$:

$$MRT_L = 1 / k_j = (F + A) / A$$

From the decay constant k_j, and the nutrient potential return value through litter (NPR), it can be calculated the nutrient real return (NRR).

$$NRR = k_j^* NPR$$

Plant productivity in ecosystem problems arising from the low soil P availability can occur in both lowlands and uplands, even under contrasting weather conditions. For example, in the highlands of Colombia (2490 m), León et al. (2011) studied coniferous forest plantations (*Pinus patula*) and native oak forests (*Quercus humboldtii*) growing on Andisols (Fulvudands and Hapludands) and found a strong P fixation and therefore a very low availability of the nutrient for plant nutrition.

These authors found very low P returns through leaf litter (NPR) in both types of forests (1.6–1.7 kg P ha^{-1} $year^{-1}$, Table 20.1). However, the release of P into the soil was even lower because of the very low decomposition rate of the leaf remains. Decomposition constants of materials composing the leaf litter showed very different values for the two types of ecosystems, as a result of the nature of the materials. The constants k_j found for the oak and pine forest were 0.6 and 0.4, respectively; what determines that the nutrient potential return values for P (NPRP) was lower than those listed above and then represents a very low P return (NRRP) with values of 0.9 and 0.7 kg P ha^{-1} $year^{-1}$, respectively. These values are extremely low if compared with common doses of P fertilization used in tropical lowland forests and montane tropical forests (ca. 50 kg ha^{-1} $year^{-1}$: Tanner et al. 1992, Vitousek et al. 1995, Kaspari et al. 2008); these values correspond to those commonly reported for forestry activities (50–65 kg ha^{-1} $year^{-1}$; Binkley 1986).

In a contrasting environment, such as the dry tropical forest, Restrepo et al. (2013) studied early successional forests dominated by species of *Croton leptostachyus* and plantations of *Azadirachta indica* and found very low P cycling values through litterfall. The P return values via leaf litter were only 0.22 and 0.06 kg P ha^{-1} $year^{-1}$, respectively. The decay constants k_j found in the same order were 0.55 and 0.68. Thus, the real returns of P (NRRP) achieved in both ecosystems were only of 0.12 and 0.04 kg P ha^{-1} $year^{-1}$.

The decomposition processes of leaf litter and nutrient release have been studied in numerous works from the methodological approach of nylon net bag technique (Bocock and Gilbert 1957) or litter bags. This method allows us to monitor changes in leaf litter mass deposited in the bag as a function of time as well as changes in nutrient content. The evolution of residual dry matter in the litter bags is traditionally adjusted to negative exponential models. One of the best known models was being proposed by Olson (1963), from the model a decomposition constant k can be obtained. This determines the speed at which the leaf litter material decomposes. In this technique, it is possible to characterize the nutrient release patterns over time: net release (mineralization), immobilization processes (biological or chemical), or enrichment of the material by external agents (fungal mycelium invasion, rainfall deposition, etc.).

Thus, under a soil P shortage, it is expected a biological immobilization of P during decomposition process. León et al. (2013) found that in degraded soils by Alluvial Gold Mining, with extremely low soil P availability, the P content of *Acacia mangium* litter increased over time up to 325% of the original amount. This P increase occurred on an excessively poor P material (0.008%), which already had restriction for the P release.

20.6 Arbuscular Mycorrhizal Fungi

The use of AMF has been considered as a viable alternative to either improve the effectiveness of P fertilizers or reduce its fertilizer dose (Khan et al. 2007). This approach is based on a more sustainable agriculture and forestry that involve environmentally friendly practices to maintain an ecological balance in soils (Vessey 2003; Borges et al. 2011). Several authors have used this biotechnological approach to enhance the effectiveness of P fertilizers (Oliveira et al. 2009; Osorio and Habte 2013).

Plant roots can form a symbiotic association with these soil fungi of the phylum *Glomeromycota* (Oehl et al. 2011). The fungal hyphae invade the cortical cells inter- and intracellularly where these form clusters of finely divided hyphae known as arbuscules (Habte 2006); the arbuscules are believed to be sites of exchange of materials between the host and the plant.

AMF absorb, via extraradical hyphae, nutrients such as N, P, K, Ca, S, Fe, Mn, Cu, and Zn from the soil solution to inside the plants' roots (Vosatka and Albrechtova 2009). The most consistent and important nutritional effect is to improve the uptake of immobile nutrients such as P. AMF are very effective in enhancing plant P uptake, particularly with plant species that lack physiological or morphological mechanisms for efficient P uptake, such as fine-branched root systems and abundant root hairs, among others (Manjunath et al. 1989; Habte and Osorio 2001).

Mycorrhizal hyphae have a higher affinity for absorbing P than roots. Schachtman et al. (1998) reported that the hyphae of *Gigaspora margarita* had an affinity constant for P (Km) of 2.5 μM (*P*: 0.077 mg L^{-1}), whereas most plants usually exhibited a Km of 6–44 μM (*P*: 0.19–1.36 mg L^{-1}), particularly those highly dependent on the mycorrhizal association (Nye and Tinker 1977; Barber 1995).

In addition, it has been claimed that rock phosphate (RP) should be applied only in acidic soils (soil pH < 5.5); because at low soil pH, this will dissolve faster (Havlin et al. 2004). However, in our experiments, the RP has been effective in increasing mycorrhizal plant P uptake and growth even in soils with pH ≥ 6.0. Several conditions can explain the better results when the mycorrhizal association is present: (1) it is clear that the elongated hyphae can capture soluble P at longer distance than the root alone; (2) the hyphae is more efficient than roots in taking up P from the soil solution; and (3) the decline of soluble P around RP particles promotes their dissolution (Manjunath et al. 1989). Presumably, the mycorrhizal hyphae exhibit a more active proton exudation than roots alone, which will favor a faster RP dissolution (Vassilev et al. 2001).

20.7 Plant Response to AMF Inoculation

In a series of greenhouse experiments, the soil was amended with grading amounts of a soluble P fertilizer (e.g., $Ca(H_2PO_4)_2$, KH_2PO_4) in order to obtain increasing soil solution P concentrations (based on soil P isotherms). In addition, the soil was either inoculated with an AMF (*Glomus fasciculatum*) or uninoculated (control). At a soil solution P of 0.02 mg L^{-1}, the shoot dry weight of mycorrhizal leucaena (*Leucaena leucocephala*) was about sevenfold higher than in non-mycorrhizal leucaena. In the case of coffee (*Coffea arabica*), banana passionfruit (*Passiflora tarminiana*), and avocado (*Persea americana*), the respective increases were 3.0, 3.7, and 1.5 times (Fig. 20.2) (Osorio and Habte 2001; Jaramillo and Osorio 2009; Montoya and Osorio 2009; Londoño 2010). In these cases the reduction in the P fertilizer dose has ranged from 50 to 79%.

Optimal response to mycorrhizal inoculation can be achieved at a soil solution P concentration of 0.02 mg L^{-1}. The amount of soluble P required to achieve such concentration can be easily determined through an isotherm of soil P fixation.

20.8 Mycorrhizal Inoculation

The inoculation consists of the process of applying around the roots AMF propagules (spores, hyphae, colonized roots) suspended in a solid sterile matrix (soil: sand) with the purpose of forming the mycorrhizal association in target plants (Habte and Osorio 2001). The inoculation is facilitated at the nursery stage; in some cases this can be applied with the seed or in the transplanting site. The dose is quite variable and largely depends on the inoculum quality (ca. 50–250 g of inoculum per site). Before inoculation AMF into a soil or substratum is necessary to know the degree of mycorrhizal dependency of the target plant. The AMF inoculation is recommended when (Osorio 2014):

- Crops require high dose of P fertilizers
- Soil exhibits a low P availability
- Native AMF population is scarce, null, or ineffective
- Soil has been subjected to intense regime of fungicides
- Soil has been eroded, degraded, or burned
- Plant species exhibits mycorrhizal dependency

Fig. 20.2 Effect of mycorrhizal inoculation on plant growth of coffee, banana passionfruit, leucaena, and avocado seedlings. In all cases the soil solution P concentration phosphate was 0.02 mg L^{-1} (Sources: Osorio and Habte (2001), Jaramillo and Osorio (2009), Montoya and Osorio (2009), Londoño (2010). Photos of N.W. Osorio, used with permission)

20.9 Phosphate-Solubilizing Microorganisms

Recent studies have demonstrated that some soil microorganisms have the ability to solubilize inorganic P compounds (Gyaneshwar et al. 2002; Bashan et al. 2012; Osorio 2011; Osorno and Osorio 2014). These P-solubilizing microorganisms release organic acids that either dissolve native or added P materials with low solubility (e.g., rock phosphate) (Reddy et al. 2002; Osorno and Osorio 2014) or desorb P that has been sorbed by the surfaces of soil minerals (He and Zhu 1998; Osorio and Habte 2012). This biotechnological approach may increase P fertilization efficiency and allow plant roots to capture P previously adsorbed in the soil (Osorio and Habte 2014).

Among the most effective bacterial PSM are species of the genera: *Pseudomonas* (Kim et al. 1998a; b; Bar-Yosef et al. 1999), *Enterobacter* (Kim et al. 1998a; b), *Bacillus* (Kim et al. 1998a; b), *Burkholderia* (Song et al. 2008; Tao et al. 2008), *Serratia* (Chen et al. 2006; Hameeda et al. 2006), *Citrobacter* (Patel et al. 2008), *Xanthomonas* (Sharan et al. 2008), *Rhizobium* (Alikhani et al. 2006), *Azospirillum* (Rodriguez et al. 2004), and *Klebsiella* (Chung et al. 2005). Effective fungal PSM belong to *Penicillium* (Reyes et al. 2002; Wakelin et al. 2004; Morales et al. 2007), *Aspergillus* (Vassilev et al. 1997; Vassileva et al. 1998; Whitelaw 2000), and *Mortierella* (Osorio and Habte 2001; Zhang et al. 2011; Osorio and Habte 2013). Also, some yeasts and actinomycetes species have been reported as effective PSM (Caroline 1994; Beauchamp and Hume 1997; Atlas and Bartha 1998; Hamdali et al. 2008).

The mechanisms for microbial phosphate solubilization consist of:

- Release of organic acids (oxalic acid, citric acid)
- Excretion of protons due to NH_4^+ assimilation (Habte and Osorio 2012)
- Formation of complexes between organic anions with cations (Al^{3+}, Fe^{2+}, Ca^{2+})
- Phosphate desorption of soils clays and oxides (He and Zhu 1998; Whitelaw 2000; Welch et al. 2002; Hameeda et al. 2006; Osorio 2011; Osorio and Habte 2013)

Although bacteria have received great attention, several authors have indicated that fungi may be consistently more effective than bacteria in solubilizing P (Arora and Gaur 1979; Kucey 1983; Alam et al. 2002; Osorio and Habte 2009). It seems that after several subcultures, bacteria PSM lose their ability to solubilize P compounds, while fungal subcultures retain this capacity (Whitelaw 2000; Rashid et al. 2004). Moreover, the elongated growth of hyphae allows fungi to have a rapid and abundant contact onto the surface of RP particles and even inside RP particles (Fig. 20.3).

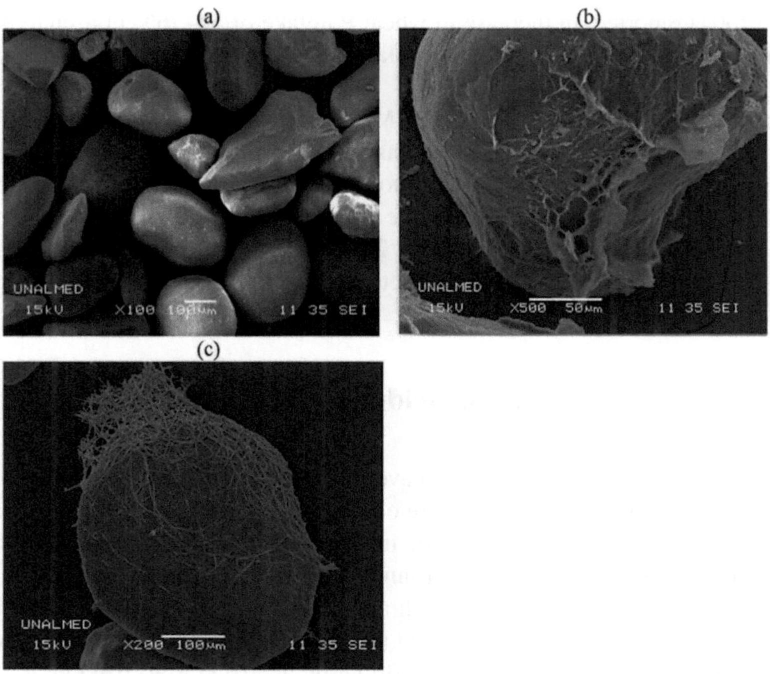

Fig. 20.3 SEM photographs showing RP particles either untreated (**a**) or treated with PSM (**b**). Notice the high degree of corrosion due to the organic acid attack on the RP surface (**b**) and the colonization of a RP particle by hyphae of a PSM (**c**) (Photos of A. Zapata and N.W. Osorio, used with permission)

The following reaction illustrates the dissolution of apatite by citric acid (produced by soil microorganisms) and the consequently release of P ions:

$$Ca_5\left(PO_4\right)_3 OH + 7\mathrm{H}^+ + 5\,\mathrm{citrate} \leftrightarrow 3\mathrm{H}_2PO_4^{\;2-} + 5\,\mathrm{citrate} - Ca^{2+} + \mathrm{H_2O} \quad K = 10^{37.9}$$

Several authors have reported beneficial effects with the PSM inoculation on plant P uptake and growth of diverse plant species grown in soils of tropical and subtropical zones (Table 20.2). The effects are higher on plant P uptake than in plant growth; there are several reasons that explain this: (1) most of these studies have been conducted with seedlings or plantlets that accumulate P in the first stages of growth; (2) plant growth depends on other factors (water and other nutrient availability, light, etc.). In general, in temperate soils, the increases with PSM on plant P uptake are higher than in tropical soils, likely due to the higher P fixation in tropical soils. However, this contrast of soil types and their influence can be also observed in the tropical zone. For instance, Osorio and Habte (2001) reported that the plant P uptake of seedlings of non-mycorrhizal leucaena increased by 14% with a PSM inoculation (*Mortierella* sp.) in a Hawaiian Oxisol (medium P fixation); in a similar experiment established in a Mollisol (low P fixation), Osorio (2008) reported an increase of 59% with the same PSM. The results reported by Duponnois et al. (2006) are higher (56–74) perhaps due to the lower P fixation expected in sandy soils. In the temperate soils, the contrast is also clear; Wakelin et al. (2004) reported an increase in wheat P uptake of 34–76% in a sandy soil of Australia (low P fixation), while Whitelaw et al. (1997) registered an increase of only 8% in an Ultisol.

On the other hand, the presence of AMF seems to have an important role in the magnitude of the plant response. For instance, the increase in plant P uptake by PSM inoculation raised from 14% in non-mycorrhizal leucaena to 40–73% with mycorrhizal leucaena (Osorio and Habte 2001; Osorio 2008; Londoño 2010) with the same PSM (*Mortierella* sp.) (Table 20.2). This synergism between AMF and PSM will be discussed below in more details.

20.10 Rock Phosphate Bioacidulation

Phosphorus-containing fertilizers have an important role in agriculture. Conventionally soluble P fertilizers are obtained from RP. The PSM can be used in a biotechnological process aiming to improve RP agronomic effectiveness and reduce both production cost and environmental pollution in making soluble P fertilizers (Stewart and Howell 2003; Smith and Moore 2005). The bioacidification of RP is a clean, environment-friendly, and innovative alternative that might make this material attractive for agricultural use (Borges et al. 2011). Bar-Yosef et al. (1999) proposed the use of a bacterial PSM to dissolve RP by acidification (gluconic acid) and thus produces a more soluble P fertilizer (superphosphate type).

Table 20.2 Increase in plant P uptake and shoot dry weight (SDW) of diverse plant species by PSM inoculation in soil of tropical and subtropical zones

PSM	Soil/site	Plant	Increase (%)		References
			P uptake	SDW	
Tropical zone					
Arthrobotrys oligospora	Sandy soil, Senegal	*Acacia holosericea*	56–74	–	Duponnois et al. (2006)
Mortierella sp.	Oxisol, Hawaii, RP added	Non-mycorrhizal *L. leucocephala*	14	19	Osorio and Habte (2001)
Mortierella sp.	Mollisol, Colombia	Non-mycorrhizal *L. leucocephala*	59	31	Osorio (2008)
Mortierella sp.	Oxisol, Hawaii, RP added	Mycorrhizal *L. leucocephala*	73	28	Osorio and Habte (2001)
Mortierella sp.	Oxisol, Colombia, RP added	Mycorrhizal *L. leucocephala*	33	24	Londoño (2010)
Mortierella sp.	Oxisol, Colombia, RP added	Mycorrhizal *L. leucocephala*	40	15	Osorio and Habte (2013)
Mortierella sp.	Oxisol, Colombia	*Vigna unguiculata*	54	22	Ramirez et al. (2013)
Unknown	Acidic soil, Taiwan (subtropics)	*L. leucocephala*	20–24	–	Young et al. (1990)

It seems that under in vitro conditions PSM can dissolve as much as 40% of the RP in only 5–7 days (Osorio 2008). Among several factors that control the efficiency of RP bioacidification are RP type and particle size, RP amount in suspension, microbial composition of culture media, type of microorganisms (PSM), stirring conditions, temperature, pH, energy sources for PSM, and incubation time (Xiao et al. 2008; Osorno 2013). For instance, it has been found that the production of these acids depends mainly on C and N sources. We have found that under in vitro conditions *Aspergillus niger* and *Mortierella* sp. are more efficient in dissolving RP if N is supplied as NH_4^+ (Fig. 20.4). The excess of NH_4^+ causes an excess of positive charge in the cytoplasm, which is an offset by increasing the H^+ pump into the external solution. Another mechanism to increase negative charge in the cytoplasm consists in diverting some organic anions (e.g., citrate) from the Krebs cycle to it, which although decreases microbial growth can improve RP bioacidification (Habte and Osorio 2012; Osorno 2013). Conversely, excessive NO_3^- uptake by the fungi is compensated by the release of HCO_3^- or OH^- to the external medium, which prevents RP dissolution. Currently, we are investigating other factors that seem to be important (glucose and ammonium concentration, type and particle size, type of microorganisms, incubation time, and cell immobilization with alginate, among others).

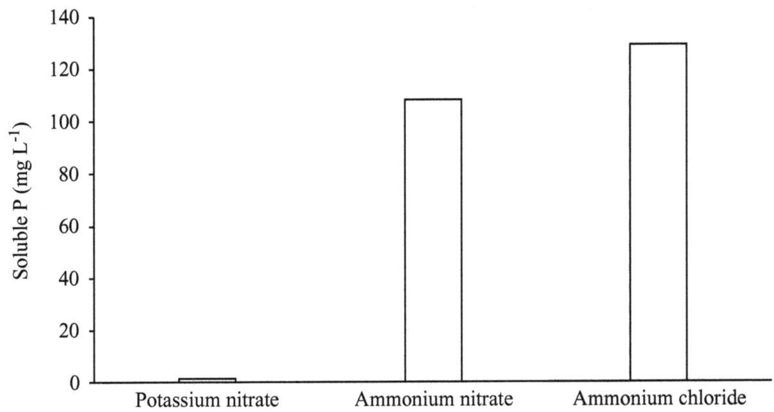

Fig. 20.4 Effect of N source in the ability of a fungus *P* solubilizer for increasing the concentration of soluble P by dissolving RP (Source: Osorio (2008))

20.11 Synergistic Effects of AMF and PSM

It has been shown that the dual inoculation with AMF and PSM can increase the RP effectiveness beyond the effect of the AMF inoculation alone. Osorio and Habte (2013) evaluated the effects of single and dual inoculation with the AMF *G. fasciculatum* and the PSM *Mortierella* sp. in plant P uptake and growth of leucaena grown in a Colombia Oxisol at Carimagua. The addition of RP increased slightly the plant growth of leucaena seedlings; however, the effect of RP addition was significantly higher when *G. fasciculatum* (AMF+) was inoculated and even higher when both microorganisms (AMF + PSM+) were concomitantly co-inoculated (Fig. 20.5a). The results were higher with a rate of 1.2 g of RP per kg of soil. At this level the AMF inoculation (AMF+) increased the shoot dry weight by 2.8 times and the dual inoculation (AMF + PSM+) by 3.2 times over the uninoculated control. The P use efficiency of non-mycorrhizal leucaena was only 0.33, but it was increased by 11 times with the mycorrhizal inoculation and by 13 times with the dual inoculation. Comparable results were obtained by Londoño (2010) in a similar experiment with leucaena grown in a Colombian Oxisol at Santander de Quilichao (Fig. 20.5b). In this case, at the RP addition rate of 1.2 g kg^{-1}, the increase in plant P uptake was 6.6 with AMF alone (AMF+) and 8.2 with both fungi (AMF + PSM+).

These synergistic effects are associated to the complementary roles of each type of microorganism. This is the PSM dissolving RP releasing thus $H_2PO_4^-$ ions into the soil solution, which are absorbed by the mycorrhizal hyphae that then transfer them into the plant roots, avoiding the P re-fixation by soil minerals.

On the other hand, some soil microorganisms are capable of releasing P from organic P compounds. They can do that via the release of phosphatase enzymes that attack ester bonds (C—O) with which P is link to organic compounds (P-lipids, P-sugars, ATP, ADN) (Alvarez et al. 2012) (Fig. 20.6). In this way P from these compounds is available for plant P uptake. Many microorganisms are capable of

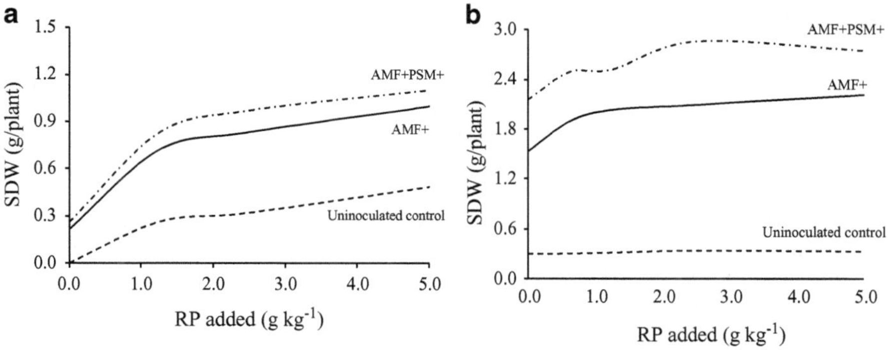

Fig. 20.5 Shoot dry weight (SDW) of *L. leucocephala* as a function of the RP level added and the inoculation with *G. fasciculatum* (AMF+) and the dual inoculation with *G. fasciculatum* and *Mortierella* sp. (AMF + PSM+) in two Colombian Oxisols (Sources: elaborated after (**a**) Osorio and Habte (2013) and (**b**) Londoño (2010))

Fig. 20.6 Phosphatase enzyme breaks the ester bond and thus releases phosphate into the soil solution

$$\text{R-O-}\underset{\text{OH}}{\overset{\text{O}}{\overset{\|}{\underset{|}{\text{P}}}}}\text{-OH} + H_2O \xrightarrow{\textit{Phosphatase}} \text{R-OH} + \text{HO-}\underset{\text{OH}}{\overset{\text{O}}{\overset{\|}{\underset{|}{\text{P}}}}}\text{-OH}$$

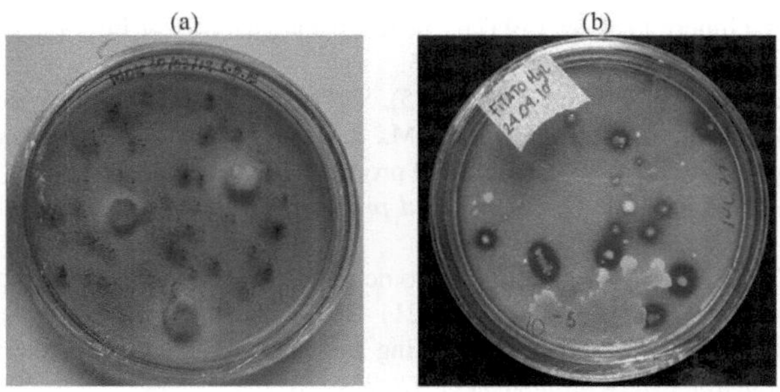

Fig. 20.7 Petri dishes contained culture media for isolation of RP-solubilizing microorganisms (**a**) and microorganisms with phytate activity (**b**). Notice the halos around the most active microbial colonies in both media (Photos of N.W. Osorio and C. Alvarez, used with permission)

producing phosphatases. A simple method using phytic acid or Ca or Na phytate as the only source of P can allow us to detect those soil microorganisms with this ability (Fig. 20.7). They can be eventually used to accelerate the release of P from organic amendments (composts, manures, leaf litter) in a way they can be used for improving plant performance. Positive results were reported by Ramirez and Kloepper (2010) using *Bacillus amyloliquefaciens* on *Brassica rapa*. More research in tropical soils needs to be done in order to know the viability of this approach under field conditions.

References

Alam, S., Khalil, S., Ayub, N., & Rashid, M. (2002). In vitro solubilization of inorganic phosphate by phosphate solubilizing microorganisms (PSM) from maize rhizosphere. *International Journal of Agriculture and Biology, 4*, 454–458.

Alikhani, H. A., Saleh-Rastin, N., & Antoun, H. (2006). Phosphate solubilization activity of rhizobia native to Iranian soils. *Plant and Soil, 287*, 35–41.

Alvarez, C. L., Marin, M., Diez, M. C., & Osorio, N. W. (2012). Molecular identification of microorganisms associated to the rhizosphere of vanilla and their potential use as biofertilizers. *Acta Horticulturae, 964*, 107–114.

Arora, D., & Gaur, A. C. (1979). Microbial solubilization of different inorganic phosphates. *Indian Journal of Experimental Biology, 17*, 1258–1261.

Atlas, R., & Bartha, R. (1998). *Microbial ecology*. New York: Addison Wesley Longman.

Barber, S. A. (1995). *Soil nutrient bioavailability. A mechanistic approach*. New York: Wiley.

Bar-Yosef, B., Rogers, R. D., Wolfram, J. H., & Richman, E. (1999). *Pseudomonas cepacia*-mediated rock phosphate solubilization in kaolinite and montmorillonite suspensions. *Soil Science Society of America Journal, 63*, 1703–1708.

Bashan, Y., Kamnev, A. A., & De-Bashan, L. E. (2012). Tricalcium phosphate is inappropriate as a universal selection factor for isolating and testing phosphate-solubilizing bacteria that enhance plant growth: A proposal for an alternative procedure. *Biology and Fertility of Soils, 49*, 465–479.

Batti, T., & Yamar, W. (2010). Bacterial solubilization of phosphorus from phosphate rock containing sulfur-mud. *Hidrometallurgy, 103*, 54–59.

Beauchamp, E. G., & Hume, D. J. (1997). Agricultural soil manipulation: The use of bacteria, manuring, and plowing. In J. D. Van Elsas, J. T. Trevors, & E. Wellintong (Eds.), *Modern soil microbiology* (pp. 643–664). New York: Marcel Dekker.

Binkley, D. (1986). *Forest Nutrition Management*. New York: Wiley.

Bocock, K. L., & Gilbert, O. J. W. (1957). The disappearance of leaf litter under different woodland conditions. *Plant and Soil, 9*, 179–185.

Bohn, H., Mcneal, B. L., & O'connor, G. (1985). *Soil chemistry*. New York: Wiley.

Borges, S. V., Bevilacqua, A. L., Magrini, G. M., & Borges, M. M. (2011). Generation of bioenergy and biofertilizer on a sustainable rural property. *Biomass and Bioenergy, 35*, 2608–2618.

Brady, N., & Weil, R. (1999). *The nature and properties of soils*. Prentice Hall: Upper Saddle River.

Campo-Alves, J. (2003). Disponibilidad y flujo de nutrimentos en una toposecuencia con bosque tropical seco en México. *Agrociencia, 37*, 211–219. Spanish.

Caroline, C. (1994). Rock phosphate solubilizing and cellulolytic actinomycete isolates of earthworm casts. *Environmental Management, 18*, 257–261.

Chen, Y. P., Rekha, P. D., Arun, A. B., Shen, F. T., Lai, W. A., & Young, C. C. (2006). Phosphate solubilizing bacteria from subtropical soil and their tricalcium phosphate solubilizing abilities. *Applied Soil Ecology, 34*, 33–41.

Chung, H., Parka, M., Madhaiyana, M., Seshadria, S., Songb, J., Chob, H., & Sa, T. (2005). Isolation and characterization of phosphate solubilizing bacteria from the rhizosphere of crop plants of Korea. *Soil Biology and Biochemistry, 37*, 1970–1974.

Collavino, M., Sansberro, P., Mroginski, L., & Aguilar, M. (2010). Comparison of *in vitro* solubilization activity of diverse phosphate-solubilizing bacteria native to acid soil and their ability to promote *Phaseolus vulgaris* growth. *Biology and Fertility of Soils, 46*, 727–738.

Cuevas, E., & Lugo, A. (1998). Dynamics of organic matter and nutrient return from litterfall in stands of ten tropical tree plantation species. *Forest Ecology and Management, 112*, 263–279.

Del Valle, J. I. (2003). Cantidad, calidad y nutrientes reciclados por la hojarasca en bosques pantanosos del Pacífico Sur Colombiano. *Interciencia, 28*, 443–449. Spanish.

Do Carmo Harta, M., & Torrent, J. (2007). Phosphorus desorption kinetics in relation to phosphorus forms and sorption properties of Portuguese acid soils. *Soil Science, 172*, 631–638.

Duponnois, R., Kisa, M., & Plenchette, C. (2006). Phosphate-solubilizing potential of the nematophagous fungus *Arthrobotrys oligospora*. *Journal of Plant Nutrition and Soil Science, 169*, 280–282.
Edwards, P. J., & Grubb, P. J. (1982). Studies of mineral cycling in a montane rain forest in New Guinea. IV. Soil characteristics and the division of mineral elements between the vegetation and soil. *Journal of Ecology, 70*, 649–666.
Fairhurst, T., Lefroy, R., Metert, E., & Batjes, N. (1999). The importance, distribution and causes of phosphorus deficiency as constraint to crop production in the tropics. *Agroforest Forum, 9*, 2–8.
Fassbender, H. W., & Grimm, V. (1981). Ciclos biogeoquímicos en un ecosistema forestal de los Andes Occidentales de Venezuela. II. Producción y descomposición de los residuos vegetales. *Turrialba, 31*, 39–47. Spanish.
Fisher, J. B., Malhi, Y., Torres, I. C., Metcalfe, D. B., Van de Weg, M. J., Meir, P., Silva-Espejo, J. E., & Huasco, W. H. (2013). Nutrient limitation in rainforests and cloud forests along a 3,000-m elevation gradient in the Peruvian Andes. *Oecologia, 172*, 889–902.
Flórez-Flórez, C. P., León-Peláez, J. D., Osorio-Vega, N. W., & Restrepo-Llano, M. F. (2013). Dinámica de nutrientes en plantaciones forestales establecidas para restauración de tierras degradadas en Colombia. *Revista de Biología Tropical, 61*, 515–529. Spanish.
Fox, R. L., & Kamprath, E. (1970). Phosphate sorption isotherms for evaluating phosphorus requirements of soils. *Soil Science Society of America Proceedings, 34*, 902–907.
Golley, F. B., JT, M. G., Clements, R. G., Child, G. I., & Deuver, M. J. (1975). *Mineral cycling in a tropical moist forest ecosystem*. Athens: University of Georgia Press.
Gyaneshwar, P., Kumar, G. N., Parekh, L. J., & Poole, P. S. (2002). Role of soil microorganisms in improving P nutrition of plants. In J. J. Adu-Gyamfi (Ed.), *Food security in nutrient-stressed environments: Exploiting plants' genetic capabilities* (pp. 133–143). Berlin: Springer.
Habte, M. (2006). The roles of arbuscular mycorrhizas in plant and soil health. In N. Uphoff et al. (Eds.), *Biological approaches to sustainable soil systems* (pp. 129–147). Boca Raton: CRC Press.
Habte, M., & Osorio, N. W. (2001). *Arbuscular mycorrhizas: Producing and applying arbuscular mycorrhizal inoculum*. Honolulu: University of Hawaii.
Habte, M., & Osorio, N. W. (2012). Effect of nitrogen form on the effectiveness of a phosphate-solubilizing fungus to dissolve rock phosphate. *Journal of Biofertilizers & Biopesticides, 3*, 127–130.
Hamdali, H., Hafidi, M., Virolle, M., & Ouhdouch, Y. (2008). Rock phosphate-solubilizing actinomycetes: Screening for plant growth-promoting activities. *World Journal of Microbiology and Biotechnology, 24*, 2565–2575.
Hameeda, B., Reddy, Y. H. K., Rupela, O. P., Kumar, G. N., & Reddy, G. (2006). Effect of carbon substrates on rock phosphate solubilization by bacteria from composts and macrofauna. *Current Microbiology, 53*, 298–302.
Havlin, J., Beaton, J., Tisdale, S. L., & Nelson, W. (2004). *Soil fertility and fertilizers. An introduction to nutrient management*. Prentice Hall: Upper Saddle River.
He, Z., & Zhu, J. (1998). Microbial utilization and transformation of phosphate adsorbed by variable charge minerals. *Soil Biology and Biochemistry, 30*, 917–923.
Herbohn, J. L., & Congdon, R. A. (1998). Ecosystem dynamics at disturbed and undisturbed sites in north Queensland wet tropical rain forest. III. Nutrient returns to the forest floor through litterfall. *Journal of Tropical Ecology, 14*, 217–229.
Hermansah, A. Z., Tsugiyuki, M., & Toshiyuki, W. (2002). Litterfall and nutrient flux in tropical rain forest, West Sumatra, Indonesia. In *17th WCSS, 2002 August 14–21, Bangkok, Thailand.*
Hue, N. V., & Fox, R. L. (2010). Predicting plant phosphorus requirements for Hawaii soils using a combination of phosphorus sorption isotherms and chemical extraction methods. *Communications in Soil Science and Plant Analysis, 41*, 133–143.
Jaramillo, S., & Osorio, N. W. (2009). Mycorrhizal dependency of coffee seedlings at three levels of soil solution phosphorus. *Revista Suelos Ecuatoriales, 39*, 143–147.

Jenny, H., Gessel, S. P., & Bingham, F. T. (1949). Comparative study of decomposition of organic matter in temperate and tropical regions. *Soil Science, 68*, 419–432.

Kaspari, M., Garcia, M., Harms, K., Santana, M., Wright, S. J., & Yavitt, J. (2008). Multiple nutrients limit litterfall and decomposition in a tropical forest. *Ecology Letters, 11*, 35–43.

Khan, M. S., Zaidi, A., & Wani, P. A. (2007). Role of phosphate-solubilizing microorganisms in sustainable agriculture—A review. *Agronomy for Sustainable Development, 27*, 29–43.

Kim, K. Y., Jordan, D., & McDonald, G. A. (1998a). Enterobacter agglomerans, phosphate solubilizing bacteria and microbial activity in soil. Effect of carbon sources. *Soil Biology and Biochemistry, 30*, 995–1003.

Kim, K. Y., McDonald, G. A., & Jordan, D. (1998b). Effect of phosphate solubilizing bacteria and vesicular-arbuscular mycorrhizae on tomato growth and soil microbial activity. *Biology and Fertility of Soils, 26*, 79–87.

Kitayama, K., & Aiba, S. I. (2002). Ecosystem structure and productivity of tropical rainforests along altitudinal gradients with contrasting soil phosphorus pools on Mount Kinabalu, Borneo. *Journal of Ecology, 90*, 37–51.

Kucey, R. M. N. (1983). Phosphate solubilising bacteria and fungi in various cultivated and virgin Alberta soils. *Canadian Journal of Soil Science, 63*, 671–678.

León, J. D., & Osorio, N. W. (2014). Role of litter turnover in soil quality in tropical degraded lands of Colombia. *The Scientific World Journal, 2014*, 1–11.

León, J. D., González, M. I., & Gallardo, J. F. (2011). Ciclos biogeoquímicos en bosques naturales y plantaciones de coníferas en ecosistemas de alta montaña de Colombia. *Revista BiologíaTtropical, 59*, 1883–1894. Spanish.

León, J. D., Castellanos, J., Casamitjana, M., Osorio, N. W., & Loaiza, J. C. (2013). Alluvial gold-mining degraded soils reclamation using *Acacia mangium* plantations: An evaluation from biogeochemistry. In R. Hai (Ed.), *Plantations biodiversity, carbon sequestration and restoration* (pp. 155–176). New York: Nova Science Publishers.

Liao, J. H., Wang, H. H., Tsai, C. H., & Hseu, Z. Y. (2006). Litter production, decomposition and nutrient return of uplifted coral reef tropical forest. *Forest Ecology and Management, 235*, 4–185.

Liu, W., Fox, E. D., & Xu, Z. (2002). Biomass and nutrient accumulation in montane evergreen broad-leaved forest (*Lithocarpus xylocarpus* type) in Ailao mountains, SW China. *Forest Ecology and Management, 158*, 223–235.

Londoño, A. (2010). *Efecto de la inoculación con un hongo micorrizal y un hongo solubilizador de fósforo en la adsorción de fosfato y el crecimiento de Leucaena en un Oxisol de la estación experimental del CIAT en Santander de Quilichao*. Thesis M.Sc., Universidad Nacional de Colombia; Medellín. Spanish.

Lugo, A. E. (1992). Comparison of tropical tree plantations with secondary forests of similar age. *Ecological Monographs, 62*, 1–41.

Lundgren, B. 1978. *Soil conditions and nutrient cycling under natural and plantation forests in Tanzanian Highlands* (Reports in Forest Ecology and Forest Soils No. 31). Uppsala: Swedish University of Agricultural Sciences.

Manjunath, A., Hue, N. V., & Habte, M. (1989). Response of *Leucaena leucocephala* to vesicular-arbuscular mycorrhizal colonization and rock phosphate fertilization in an Oxisol. *Plant and Soil, 114*, 127–133.

McDonald, M. A., & Healey, J. R. (2000). Nutrient cycling in secondary forests in the Blue Mountains of Jamaica. *Forest Ecology and Management, 139*, 257–278.

Medina, E., Cuevas, E., & Weaver, P. L. (1981). Composición foliar y transpiración de especies leñosas del Pico del Este, Sierra de Luquillo, Puerto Rico. *Acta Científica Venezolana, 32*, 159–165.

Meentemeyer, V. (1978). Macroclimate the lignin control of litter decomposition rates. *Ecology, 59*, 465–472.

Montoya, B., & Osorio, N. W. (2009). Mycorrhizal dependency of avocado at different levels of soil solution phosphorus. *Revista Suelos Ecuatoriales, 39*, 100–106.

Morales, A., Alvear, M., Valenzuela, E., Rubio, R., & Borie, F. (2007). Effect of inoculation with *Penicillium albidum*, a phosphate-solubilizing fungus, on the growth of *Trifolium pratense* cropped in a volcanic soil. *Journal of Basic Microbiology, 47*, 275–280.
Muoghalu, J. I. (2003). Contributions of throughfall, stemflow and litterfall to nutrient cycling in a secondary lowland rain forest in Ile-Ife, Nigeria. *Journal of Tropical Forest Science, 15*, 399–410.
Murphy, J., & Riley, J. P. (1962). A modified single solution method for the determination of phosphate in natural waters. *Analytica Chimica Acta, 27*, 31–35.
Nanzyo, M., Dahlgren, R., & Shoji, S. (1993). Chemical characteristics of volcanic ash soils. In S. Shoji, M. Nanzyo, & R. Dahlgren (Eds.), *Volcanic ash soils*. Amsterdam: Elsevier.
Nye, P. H. (1961). Organic matter and nutrient cycles under moist tropical forest. *Plant and Soil, 13*, 333–346.
Nye, P. H., & Tinker, P. B. (1977). *Solute movement in the soil-root system*. Oxford: Blackwell Scientific Publications.
Nzila, J. D., Bouillet, J. P., Laclaud, J. P., & Ranger, J. (2002). The effect of slash management on nutrient cycling and tree growth in *Eucalyptus* plantations in the Congo. *Forest Ecology and Management, 171*, 209–221.
Oberson, A., Bunemann, E. K., Friesen, D. K., Rao, I. M., Smithson, P. C., Turner, B. L., & Frossard, E. (2006). Improving phosphorus fertility in tropical soils through biological interventions. In N. Uphoff (Ed.), *Biological approaches to sustainable soil systems* (pp. 531–546). Boca Raton: CRC Press.
Oehl, F., Sieverding, E., Palenzuela, J., Ineichen, K., & Alves, G. (2011). Advances in Glomeromycota taxonomy and classification. *IMA Fungus, 2*, 191–199.
Oliveira, C. A., Alves, V. M. C., Marriel, I. E., Gomes, E. A., Scotti, M. R., Carneiro, N. P., Guimaraes, C. T., Schaffert, R. E., & Sa, N. M. H. (2009). Phosphate solubilizing microorganisms isolated from rhizosphere of maize cultivated in an oxisol of the Brazilian Cerrado Biome. *Soil Biology and Biochemistry, 41*, 1782–1787.
Olson, J. S. (1963). Energy storage and the balance of producers and decomposers in ecological systems. *Ecology, 44*, 322–331.
Osorio, N. W. 2008. *Effectiveness of microbial solubilization of phosphate in enhancing plant phosphate uptake in tropical soils and assessment of the mechanisms of solubilization.* Dissertation, University of Hawaii, Honolulu.
Osorio, N. W. (2011). Effectiveness of phosphate solubilizing microorganism in increasing plant phosphate uptake and growth in tropical soils. In D. K. Maheshwari (Ed.), *Bacteria in agrobiology. Plant nutrient management* (pp. 65–80). Berlin: Springer.
Osorio, N. W. (2014). *Manejo de nutrientes en suelos del trópico*. Medellin: Universidad Nacional de Colombia. Spanish.
Osorio, N. W., & Habte, M. (2001). Synergistic influence of an arbuscular mycorrhizal fungus and a P solubilizing fungus on growth and P uptake of *Leucaena leucocephala* in an Oxisol. *Arid Land Research and Management, 15*, 263–274.
Osorio, N. W., & Habte, M. (2009). Strategies for utilizing arbuscular mycorrhizal fungi and phosphate-solubilizing microorganisms for enhanced phosphate uptake and growth of plants in the soils of the tropics. In M. S. Khan, A. Zaidi, & J. Musarrat (Eds.), *Microbial strategies for crop improvement* (pp. 325–351). Berlin: Springer.
Osorio, N. W., & Habte, M. (2012). Phosphate desorption from the surface of soil mineral particles by a phosphate-solubilizing fungus. *Biology and Fertility of Soils, 49*, 481–486.
Osorio, N. W., & Habte, M. (2013). Synergistic effect of a phosphate solubilizing fungus and an arbuscular mycorrhizal fungus on Leucaena seedlings in an oxisol fertilized with rock phosphate. *Botany, 91*, 274–281.
Osorio, N. W., & Habte, M. (2014). Soil phosphate desorption induced by a phosphate-solubilizing fungus. *Communications in Soil Science and Plant Analysis, 45*, 451–460.
Osorio, N. W., Habte, M., & León, J. D. (2015). Effectiveness of a rock phosphate solubilizing fungus to increase soil solution phosphate impaired by the soil phosphate sorption capacity. *Revista Facultad Nacional de Agronomía, 68*, 7627–7636.

Osorno, L. 2013. *Biacidulación de roca fosfórica bajo condiciones in vitro*. Thesis M.Sc., Universidad Nacional de Colombia, Medellin.
Osorno, L., & Osorio, N. W. (2014). Effect of carbon and nitrogen source and concentration on rock phosphate dissolution induced by fungi. *Journal of Applied Biotechnology, 2*, 32–36.
Pande, P. K. (2004). Nutrient cycling in disturbed tropical dry deciduous teak forest of Satpura plateau, Madhya Pradesh, India. *Journal of Tropical Forest Science, 16*, 94–105.
Parrotta, J. (1999). Productivity, nutrient cycling, and succession in single and mixed species plantations of *Casuarina equisetifolia*, *Eucalyptus robusta*, and *Leucaena leucocephala* in Puerto Rico. *Forest Ecology and Management, 124*, 45–77.
Parzych, A., & Trojanowski, J. (2006). Precipitation and dust fall as natural sources of nitrogen and phosphorus for forest soils in the Slowin Ski National Park. *Baltic Coastal Zone. Journal of Ecology and Protection of the Coastline, 10*, 47–59.
Patel, D. K., Archana, G., & Kumar, G. N. (2008). Variation in the nature of organic acid secretion and mineral phosphate solubilization by *Citrobacter* sp. DHRSS in the presence of different sugars. *Current Microbiology, 56*, 168–174.
Proctor, J., Anderson, J. M., Fogden, S. C. L., & Vallack, H. W. (1983). Ecological studies in four contrasting lowland rainforest in Gunung Mulu National Park, Sarawak. II. Litterfall, litter standing crop and preliminary observations on herbivory. *Journal of Ecology, 71*, 237–260.
Ramirez, C., & Kloepper, J. (2010). Plant growth promotion by *Bacillus amyloliquefaciens* FZB45 depends on inoculum rate and P-related soil properties. *Biology and Fertility of Soils, 56*, 835–844.
Ramirez, G., Osorno, L., Osorio, N. W., & Morales, J. G. (2013). Microbiological alternatives for the improvement of cowpea growth. *Revista Facultad Nacional de Agronomía, 66*, 7035–7044.
Rashid, M., Khalil, S., Ayub, N., Alam, S., & Latif, F. (2004). Organic acids production and phosphate solubilization by phosphate solubilizing microorganisms (PSM) under *in vitro* conditions. *Pakistan Journal of Biological Sciences, 7*, 187–196.
Reddy, M. S., Kumar, S., Babita, K., & Reddy, M. S. (2002). Biosolubilization of poorly soluble rock phosphates by *Aspergillus tubingensis* and *Aspergillus niger*. *Bioresource Technology, 84*, 187–189.
Restrepo, M. F., Flórez, C. P., León, J. D., & Osorio, N. W. (2013). Passive and active restoration strategies to activate soil biogeochemical nutrient cycles in a degraded tropical dry land. *ISRN Soil Science, 2013*, 461984. 6 pages.
Reyes, I., Bernier, L., & Antoun, H. (2002). Rock phosphate solubilization and colonization of maize rhizosphere by wild and genetically modified strains of *Penicillium rugulosum*. *Microbial Ecology, 44*, 39–48.
Rodriguez, H., Gonzalez, T., Goire, I., & Bashan, Y. (2004). Gluconic acid production and phosphate solubilization by the plant growth-promoting bacterium *Azospirillum* spp. *Die Naturwissenschaften, 91*, 552–555.
Sanchez, P., & Logan, T. (1992). Myths and science about the chemistry and fertility of soils in the tropics. In R. Lal & P. Sanchez (Eds.), *Myths and science of soils of the tropics* (pp. 35–46). Madison, WI: Soil Science Society of America.
Scervino, J., Prieto, M., Della, I., Recchi, M., Sarmiento, N., & Godeas, A. (2010). Soil fungal isolates produce different organic acid patterns involved in phosphate salts solubilization. *Biology and Fertility of Soils, 46*, 755–763.
Schachtman, D. P., Reid, R., & Ayling, S. M. (1998). Phosphorus uptake by plants: From soil to cell. *Plant Physiology, 116*, 447–453.
Scott, D. A., Proctor, J., & Thompson, J. (1992). Ecological studies on a lowland evergreen rain forest on Maraca Island, Roraima, Brazil. II. Litter and nutrient cycling. *Journal of Ecology, 80*, 705–717.
Sharan, A., Shikha, & Darmwal, N. S. (2008). Efficient phosphorus solubilization by mutant strain of *Xanthomonas campestris* using different carbon, nitrogen and phosphorus sources. *World Journal of Microbiology and Biotechnology, 24*, 3087–3090.
Singh, K. P., Singh, P. K., & Tripathi, S. K. (1999). Litterfall, litter decomposition, and nutrient release patterns in four native tree species raised on coal mine spoil at Singrauli, India. *Biology and Fertility of Soils, 29*, 371–378.

Singh, R., Kumar, R., & Agrawal, M. (2004). Litter decomposition and nutrient release in relation to atmospheric deposition of S and N in a dry tropical region. *Pedobiologia, 48*, 305–311.

Smith, F. W. (2002). The phosphate uptake mechanism. *Plant and Soil, 245*, 105–114.

Smith, D. R., & Moore, J. (2005). Soil extractable phosphorus changes with time after application of fertilizer: II. Manure from swine fed modified diets. *Soil Science, 170*, 640–651.

Song, O. R., Lee, S. J., Lee, Y. S., Lee, S. C., Kim, K. K., & Choi, Y. L. (2008). Solubilization of insoluble inorganic phosphate by *Burkholderia cepacia* DA23 isolated from cultivated soils. *Brazilian Journal of Microbiology, 39*, 151–156.

Songwe, N. C., Fasehun, F. E., & Okali, D. U. U. (1988). Litterfall and productivity in a tropical rain forest, Southern Bakundu Forest Reserve, Cameroon. *Journal of Tropical Ecology, 4*, 25–37.

Stewart, B. A., & Howell, T. A. (2003). *Encyclopedia of water science*. New York: Marcel Dekker.

Tanner, E. V. J., Kapos, V., & Franco, W. (1992). Nitrogen and phosphorus fertilization effects on Venezuelan montane forest trunk growth and litterfall. *Ecology, 73*, 78–86.

Tao, G. C., Tian, S. J., Cai, M. Y., & Xie, G. H. (2008). Phosphate-solubilizing and -mineralizing abilities of bacteria isolated from soils. *Pedosphere, 18*, 515–523.

Tobón, C., Sevink, J., & Verstraten, J. M. (2004). Litterflow chemistry and nutrient uptake from the forest floor in northwest Amazonian forest ecosystems. *Biogeochemistry, 69*, 315–339.

Trolove, S. N., Hedley, M. J., Kirk, G. J. D., Bolan, N. S., & Loganathan, P. (2003). Progress in selected areas of rhizosphere research on P acquisition. *Australian Journal of Soil Research, 41*, 471–499.

Turner, B. L., Frossard, E., & Oberson, A. (2006). Enhancing phosphorus availability in low-fertility soils. In N. Uphoff et al. (Eds.), *Biological approaches to sustainable soil systems* (pp. 191–206). Boca Raton: CRC Press.

Vassilev, N., Vassileva, M., & Azcon, R. (1997). Solubilization of rock phosphate by immobilized *Aspergillus niger*. *Bioresource Technology, 59*, 1–4.

Vassilev, N., Vassileva, M., Fenice, M., & Federici, F. (2001). Immobilized cell technology applied in solubilization of insoluble inorganic (Rocks) phosphates and P plant acquisitions. *Bioresource Technology, 79*, 263–271.

Vassileva, M., Azcon, R., Barea, J., & Vassilev, N. (1998). Application of an encapsulated filamentous fungus in solubilization of inorganic phosphate. *Journal of Biotechnology, 63*, 67–72.

Veneklaas, E. J. (1991). Litterfall and nutrient fluxes in two montane tropical rain forests, Colombia. *Journal of Tropical Ecology, 7*, 319–336.

Vera, M., Cavelier, J., & Santamaría, J. (1999). Reabsorción de nitrógeno y fósforo foliar en árboles de bosques montanos en los Andes centrales de Colombia. *Revista de Biología Tropical, 47*, 33–43.

Vessey, J. K. (2003). Plant growth promoting rhizobacteria as biofertilizers. *Plant and Soil, 255*, 571–586.

Villela, D., & Proctor, J. (1999). Litterfall mass, chemistry, and nutrient retranslocation in a monodominant forest on Maracá Island, Roraima, Brazil. *Biotropica, 31*, 198–211.

Vitousek, P. M., Gerrish, G., Turner, D. R., Walker, L. R., & Mueller-Dombois, D. (1995). Litterfall and nutrient cycling in four Hawaiian montane rainforests. *Journal of Tropical Ecology, 11*, 189–203.

Vosatka, M., & Albrechtova, J. (2009). Benefits of arbuscular mycorrhizal fungi to sustainable crop production. In M. S. Khan (Ed.), *Microbial strategies for crop improvement* (pp. 205–225). Berlin: Springer.

Wakelin, A., Warren, R., Harvey, P., & Ryder, P. (2004). Phosphate solubilization by *Penicillium* spp. closely associated with wheat roots. *Biology and Fertility of Soils, 40*, 36–43.

Weerakkody, J., & Parkinson, D. (2006). Input, accumulation, and turnover of organic matter, nitrogen and phosphorus in surface organic layers of an upper montane rainforest in Sri Lanka. *Pedobiologia, 50*, 377–383.

Welch, S. A., Taunton, A. E., & Banfield, J. F. (2002). Effect of microorganisms and microbial metabolites on apatite dissolution. *Geomicrobiology Journal, 19*, 343–367.

Whitelaw, M. A. (2000). Growth promotion of plants inoculated with phosphate solubilizing fungi. *Advances in Agronomy, 69*, 99–151.

Whitelaw, M. A., Harden, T. J., & Bender, G. L. (1997). Plant growth promotion of wheat inoculated with *Penicillium radicum* sp. nov. *Australian Journal of Soil Research, 35*, 291–300.

Xiao, C. Q., Chi, R. A., Huang, X. H., Zhang, W. X., Qiu, G. Z., & Wang, D. Z. (2008). Optimization for rock phosphate solubilization by phosphate-solubilizing fungi isolated from phosphate mines. *Ecological Engineering, 33*, 187–193.

Young, C. C., Chen, C. L., & Chao, C. C. (1990). Effect of *Rhizobium*, vesicular-arbuscular mycorrhiza, and phosphate solubilizing bacteria on yield and mineral phosphorus uptake of crops in subtropical-tropical. In *14th International Congress of Soil Science. Transactions* (Vol. III). Kyoto, Japan: International Society of Soil Science.

Zhang, H., Wu, X., Li, G., & Qin, P. (2011). Interactions between arbuscular mycorrhizal fungi and phosphate-solubilizing fungus (*Mortierella* sp.) and their effects on *Kostelelzkya virginica* growth and enzyme activities of rhizosphere and bulk soils at different salinities. *Biology and Fertility of Soils, 47*, 543–554.

Chapter 21
Improving Plant Phosphorus (P) Acquisition by Phosphate-Solubilizing Bacteria

Muhammad Saleem Arif, Sher Muhammad Shahzad, Tahira Yasmeen, Muhammad Riaz, Muhammad Ashraf, Muhammad Arslan Ashraf, Muhammad Salman Mubarik, and Rizwana Kausar

Abstract Phosphorus (P) is an essential plant nutrient required for sustainable production of food and bioenergy crops. A sufficient supply of P to the crop plants is necessary in order to meet global and regional food security challenges. However, limited mobility of P in the soil and its high fixation capabilities within the soil matrix necessitate the use of P fertilizers, which are again prone to fixation, thereby reducing the availability of this crucial element for plant nutrition. Rhizosphere is an intricate zone under the influence of plant roots and harbours variety of microbial species which confer growth and nutrition benefits to the crop plants. Phosphate solubilizing bacteria (PSB) play a crucial role in solubilizing various forms of phosphorus in soil and making them available for plant uptake. The bacterial phosphate solubilization process is mainly triggered by the secretions of organic acids, siderophores, exopolysaccharides, and enzyme (phytase-phosphatase) activities. The bacterial metabolites either solubilize the inorganic forms of phosphorus or mobilize the organic sources of phosphorus through enhanced enzyme activities. In this chapter, we attempt to provide

M.S. Arif (✉) • T. Yasmeen • M. Riaz
Department of Environmental Sciences and Engineering, Government College University Faisalabad, Faisalabad 38000, Pakistan
e-mail: msarif@outlook.com

S.M. Shahzad • M. Ashraf
Department of Soil & Environmental Sciences, University College of Agriculture, University of Sargodha, Sargodha, Pakistan

M.A. Ashraf
Department of Botany, Government College University Faisalabad, Faisalabad 38000, Pakistan

M.S. Mubarik
Centre of Agricultural Biochemistry and Biotechnology (CABB), University of Agriculture, Faisalabad 38040, Pakistan

R. Kausar
Soil and Water Testing Laboratory for Research, Sargodha, Punjab, Pakistan

M. Naeem et al. (eds.), *Essential Plant Nutrients*,
DOI 10.1007/978-3-319-58841-4_21

an overview about the potential contribution of PSB in improving plant P nutrition. Moreover, we also discussed the action mechanism involving PSB and key features that make it a useful value-added product for sustainable agriculture.

Keywords Phosphorus aquaisition • Bacteria • Fertilizers • Alkaline soils

21.1 Introduction

Phosphorus (P) is a fundamental and non-substitutable nutrient element in food production system. Plant phosphorus uptake can be a difficult proposition and thus often considered an important yield limiting factor in most agriculture systems of the world (Ringeval et al. 2017). Even though phosphorus is abundant in soils (organic and inorganic forms), its availability to plant is often restricted due to the formation of insoluble phosphate complexes in soil (Adesemoye and Kloepper 2009). The application of P fertilizer is the most sought out option to counteract phosphorus limitation in crop plants. However, about 80% of applied P fertilizer can be lost due to the fixation and adsorption processes (Lopez-Bucio et al. 2000), either in the form of Fe/Al phosphate in acidic soils or in the form of Ca phosphate in neutral to alkaline soils (Bertrand et al. 2003). Therefore, most of the applied P fertilizer rapidly becomes unavailable to plants that led to reduced production potential of crop plants.

The phosphorus fertilizers are primarily produced from mined rock phosphate; a nonrenewable and geographically restricted resource. The majority of rock phosphate reserves ~85% are found in Morocco, which is the leading world phosphorus producer. Though estimated amounts are not likely running out in the short term, rock phosphate will become scarce, at least in terms of pricing due to increased demand (van de Wiel et al. 2016). In addition, anthropogenic influences such as excessive mining, growing demand, increasing price, geopolitical constraints, excessive wastage, and high discharge to water bodies tend to hinder the sustainable management of the global P resource (Chowdhury et al. 2017).

These arising concerns have led to look for other viable options of phosphorus nutrition in crop plants. One of the alternatives and emerging solution to solve this problem is to exploit the microhabitat of plant under the influence of root (rhizosphere). Rhizosphere is characterized as nutrient-rich niche of immense microbial activity. Certain culturable root zone bacteria have intrinsic ability to aggressively colonize the host plant and improve plant growth and development (plant growth-promoting rhizobacteria (PGPR)). There are number of reports that entail promising effect of these PGPR on crop productivity under various soil conditions (Arif et al. 2016a, b; da Silva et al. 2017). Among these PGPR inoculants, several bacteria are able to mobilize and/or solubilize insoluble P into soluble form by releasing acidic metabolites, chelation, and ion exchange reaction which makes P available for crop plants (Chung et al. 2005; Jorquera et al. 2008). These naturally occurring phosphate-solubilizing bacteria (PSB) are unique environmentally friendly alternative that could offer a sustainable P nutrition for various crop plants. In addition, the application of these P bioinoculant could also negate the depressing effect of P fertilizer on ecosystem health.

21.2 Rock Phosphate-P Fertilizer-Food Security: A Crucial Nexus

The main source of phosphorus fertilizer is the geological deposits of phosphate rock. Morocco holds the phosphate monopoly with approximately 85% of global rock phosphate reserves. Other important rock phosphate-producing countries are China, the USA, Russia, Brazil, and Canada. Global fertilizer sector has seen a dramatic rise (about 430%) of P fertilizer production during the past 50 years (Fig. 21.1a). Extensive population growth and diversified food demand are chief contributors behind this mammoth increase of P fertilizer production. Currently >80% of extracted rock phosphate is being utilized to manufacture P fertilizer for agriculture, which raised questions and concerns about the depletion timeline of these reserves (Ibrahim et al. 2010).

The world food production system need to produce 70 more food (FAO 2009), and securing food sufficiency by 2050 is one of the top most priority to meet the global future food demand of approximately 9.1 billion people. Increased food production

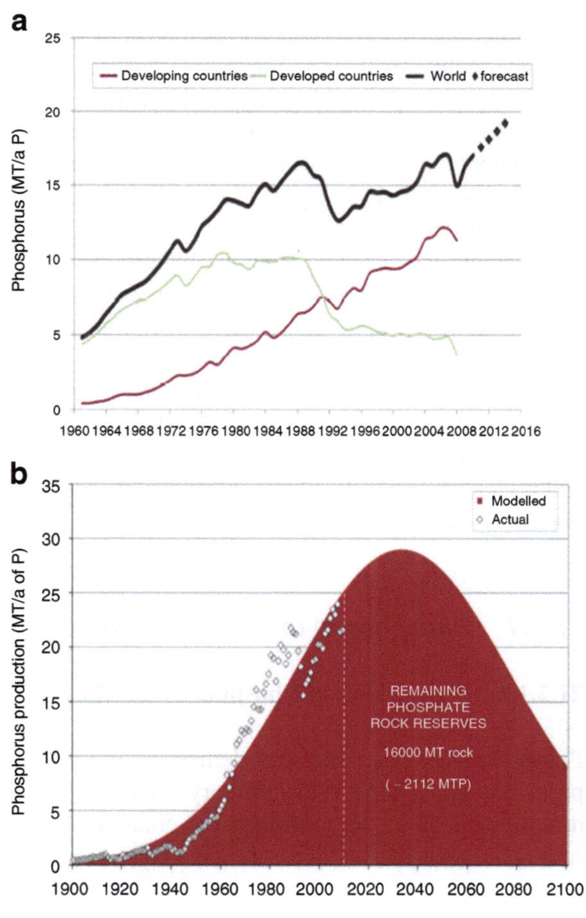

Fig. 21.1 (**a**) Global phosphorus fertilizer consumption between 1961 and 2006 (in million tons P). The figure indicates that while demand in the developed world reached a plateau and then declined around 1990, fertilizer demand has been steadily increasing in the developing world (IFA 2009). (**b**) Peak phosphorus curve indicating a peak in production by 2033, derived from the US Geological Survey and industry data (Cordell et al. 2009)

will come from higher crop yields and an expansion in harvested land, both of which will necessitate greater fertilizer use. At a time when intensive cultivation practices are depleting soil fertility, crop yield improvements continue to decelerate. Considering population growth and rising food demand, it is anticipated that the use of phosphate fertilizer will soar up manifold to achieve higher yield and quality crop product. Based on these trends, the International Fertilizer Industry Association (IFA) is projecting medium-term growth in fertilizer demand of about 1.7% per year, with total fertilizer demand exceeding 200 million metric tons of N, P_2O_5, and K_2O by 2020.

Phosphorus has received only limited attention compared to other important agricultural inputs such as nitrogen and water. Because of the vital role of phosphorus in food production, any consideration of food security needs to include an informed discussion concerning more sustainable use of P due to its limited resource base. Keeping in view the increasing global demand for phosphorus fertilizers, the ongoing debate over the long-term availability of phosphate rock, lack of adequate phosphorus accessibility by many of the world's poor farmers from developing world, shortage of P recycling system, and injudicious phosphorus fertilization warrant careful planning and consideration for P in sustainable agriculture. Moreover, detailed exploratory research is also required to provide reliable, global-scale quantification of the amount of phosphorus available for food production. A global phosphorus assessment, including further insights from scientists and other experts, policy-makers, and other stakeholders, could contribute to improving fertilizer accessibility, waste management in urban settings, and recycling of phosphorus from food waste products.

The long-term availability of phosphorus for global food production is of fundamental importance to the world population. Given the diversity of issues surrounding phosphorus, only an integrated set of policy options and technical measures can ensure its efficient and sustainable use. Environmental solutions that improve nutrient management and recycling minimize phosphorus losses due to soil erosion, and foster sustainable production and consumption also promote wise use of a finite resource. This could be the basis for fostering environmental innovation and other actions at local, national, regional, and international levels to improve phosphorus management. The future of this resource will also depend on governance with regard to its extraction and distribution around the world. There is a need for accurate information about the extent of global reserves, new technologies, infrastructure, institutions, attitudes, and policies to meet the challenge of sustainably feeding a rapidly growing global population while maintaining a healthy and productive environment.

21.2.1 Phosphorus in Soil-Plant System

21.2.1.1 Significance of Phosphorus for Plant

Phosphorus (P) is an essential element to all life forms of the earth ecosystem. In particular, phosphorus is key primary macronutrient necessary for plant growth and development along with nitrogen and potassium. P entry into plant is facilitated by root hairs, root tips, and the outermost layer of root cells. Plants typically take up P in inorganic form either as primary orthophosphates ($H_2PO_4^-$) or secondary orthophosphates (HPO_4^{2-}) ion from soil solution. The dynamic balance of P availability and its absorption in soil plant system is tightly coupled with its pH. Once P surpassed the plant roots, absorbed inorganic P is either stored in the root or transported into the aerial plant parts through various mechanisms (Schachtman et al. 1998). Phosphorus is the structural component of several fundamental macromolecules and mainly involved in genetic, regulatory, signal transduction, and metabolic processes. In certain conditions, P can get incorporated into multitude of organic compounds ranging from smaller macromolecule (nucleotide, phospholipid, sugar phosphates) to larger macromolecule (DNA, RNA, phosphoprotein, ADP/ATP) of phosphorus (Turner et al. 2002; Condron et al. 2005). High energy phosphate forms that constitute adenosine diphosphate (ADP) and triphosphate (ATP) molecules drive several biochemical process within the plant. Energy transfer through phosphate carrier ADP and ATP to other molecular component of the cell (phosphorylation) controls many key biochemical process in plants (Baginsky 2016). Phosphorus also exists as phytin, a major P reserve of seeds and fruits, required for seed formation and early developmental stages of embryonic plant. Nearly half of total P in legumes seed and two-third in cereal grain are synthesized in the form of phytin. Hence, P deficiency appeared to reduce seed size, seed number, and viability. Moreover, inorganic and organic phosphates in plants also serve as buffers in the maintenance of cellular pH.

21.2.1.2 Soil Phosphorus Cycle

Unlike carbon, oxygen, and nitrogen, global P cycle is a sedimentary cycle that originates from phosphate-bearing mineral deposit and crustal rock sediments. The global occurrence of P cycle in soil followed a dynamic flow of different P forms involving soil, plants, and microorganisms. P enters into the biosphere almost entirely from soil through numerous ways. In natural system, various soil processes control different pools and fluxes of P and subsequently drive soil P cycle in the ecosystem.

21.2.1.3 Forms of Soil Phosphorus

Soil phosphorus forms can be broadly categorized into "sorbed P," "mineral P," and "organic P," and all these P forms diffused into solution P collectively as orthophosphate for plant uptake. Sorbed P matrix is comprised of P adsorbed onto the surface of iron and aluminum oxyhydroxides and $CaCO_3$ by electrostatic and covalent bonding (Moody et al. 2013). Phosphorus also exists in mineral form as a part of the structure of a wide variety of soil minerals, such as rock phosphate, present as fluorapatite [$Ca_5(PO_4)_3F$] or hydroxyapatite [$Ca_5(PO_4)_3OH$]. Soil inorganic P forms are not found in any typical ratios and pattern; they can be formed by sorbed/precipitated P on amorphous Fe and Al oxides and hydrous oxides. In addition, soil pedological processes and/or the reaction products of added P fertilizers also favored the formation of mineral P matrix.

Phosphorus fertilizers are the major inorganic P pool in agricultural soils and contribute approximately 70–80% of inorganic P share in these soils. In agriculture system, typical P fertilizer granule soon after its application predominantly dissolved into soluble inorganic P ($H_2PO_4^-$, HPO_4^{2-}) forms by available soil moisture. Most of soluble phosphates may not be readily available for plant as they are negatively charged and rapidly immobilized by sorption onto the positively charged soil mineral surfaces (Fe^{3+}, Al^{3+}, Ca^{2+}), or fixation into soil organic matter complexes (Frossard et al. 2000; Shen et al. 2013). Organic P largely exists as an inositol polyphosphate fraction and accounts for ~50% of the total organic P in soil (Koopmans et al. 2003) Additionally, phospholipids (~1%) and nucleic acid (5–10%) and their degradation products make up the remaining organic P fraction. Phospholipids and nucleic acids that enter the soil are degraded rapidly by soil microorganisms. Inositol polyphosphates are usually associated with high molecular weight molecules extracted from the soil, suggesting that they are an important component of humus (Dao 2004).

A wide range of soil microorganisms are capable of mineralizing organic phosphates into inorganic orthophosphate via phosphate-specific enzyme activity (Dobbelaere et al. 2003). The enzymes involved in the hydrolysis of phosphate from organic P resources are collectively called as phosphatases. Microorganisms have tendency to produce both acid and alkaline phosphatases, but plant can solely secrete acid phosphatase (Tarafdar 1989). Mineralized pool of inorganic P from organic fraction enters the soil solution phase and might be taken by microbes and plants, adsorbed onto the solid matrix or rarely lost by leaching and run off. Phosphorus released from organic P fraction is highly dependent on soil moisture and temperature conditions (Adhya et al. 2015).

21.2.1.4 Phosphorus Equilibrium in Soil

Soil solution P is the ultimate source of P supply to the plant, primarily through the process of root diffusion, influenced by many factors, i.e., P concentration gradient between the root surface and the bulk soil solution, rate of P re-supply to solution P after its withdrawal, soil water content, soil P buffer capacity (change in the quantity of soil P for a change in solution P concentration), and the connectivity of water films in soil pores (tortuosity factor) (Nye 1980). Both biotic and abiotic factors control the ultimate fate of P in soil solution. Weathering of sedimentary rocks (rock phosphate) containing P minerals primarily apatite [$(Ca_5(PO_4)_3(F, Cl, OH))$] is the principle source of P to the soil. In general, apatite deposits are distributed across the globe. Individual mineral P (apatite) deposits are mostly of sedimentary origin, but some igneous reserves also exist in lesser amount (Cisse and Mrabet 2004). The dissolution of these P-bearing minerals is synergistically driven by both biotic and abiotic processes which ultimately lead to the release of mineral phosphate. The main mechanism underlying P mineral dissolution involves the release of acidic metabolites usually from microbial activity (e.g., Frossard et al. 1995; Welch et al. 2002). Solubilized phosphate is bioavailable P pool that is taken up by plants and assimilated into different plant parts and potentially can be recycled back to soil by plant residue (Damon et al. 2014).

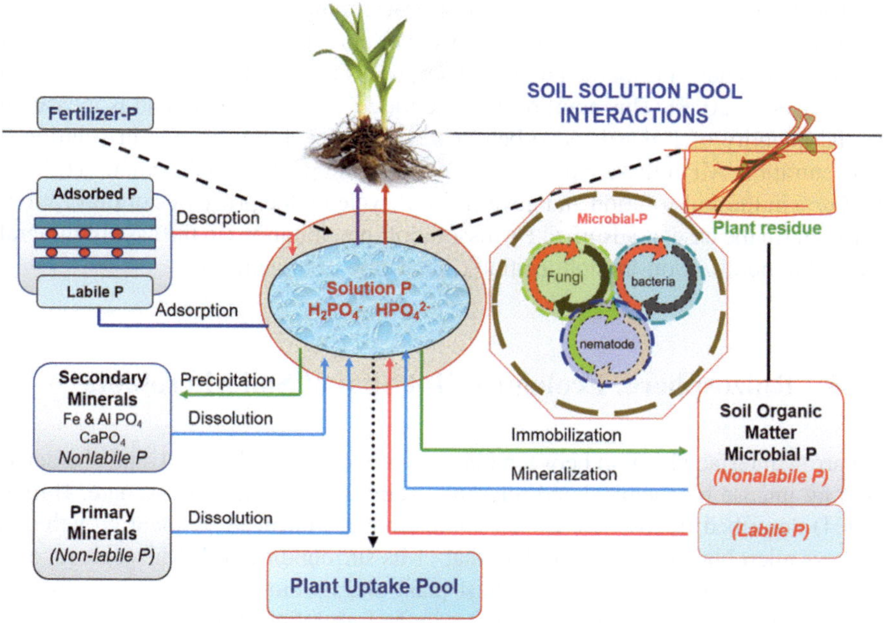

Fig. 21.2 An overview phosphorus cycle

Soil microorganisms act as sink and source of phosphorus (P) and mediate key processes in the soil P cycle, e.g., P mineralization and immobilization (Oberson and Joner 2005). P released into the soil solution from the mineralization of organic compounds might be taken up by soil microbial population, taken up by growing plants, and transferred to soil inorganic pool (Fig. 21.1). Over time, some P fraction that undergo microbial immobilization can affect P availability by removing inorganic P from the soil solution (Olander and Vitousek 2004). Most of inorganic P pool entering soil solution, either by mineralization or P fertilizer addition, is rapidly converted into insoluble P due to sorption and precipitation reactions. The sorption of inorganic P from solution is closely related to the presence of iron and aluminum oxides and hydrous oxides (Tian et al. 2016) and $CaCO_3$ (Pizzeghello et al. 2011). The P concentration of the soil solution at equilibrium state will provide maximum P for plant uptake, highest at the slightly acidic to neutral pH range and are reduced considerably in strongly acidic or alkaline soil conditions (Fig. 21.2).

In calcareous soil, the amount of $CaCO_3$ affects soil solution P equilibrium as reduced P solubility is often associated with the presence of excessive lime ($CaCO_3$) (Mahdi et al. 2011). The reaction of phosphorus with $CaCO_3$ was initially favored P sorption on these surfaces followed by rapid precipitation of soil solution P as Ca–P minerals (Sharpley et al. 1989). The formation and precipitation of these low solubility calcium phosphate compounds depressed P mobility and availability in these soils. In acidic soil, P sorption to Al and Fe oxides may be of equal or greater consideration than P sorption to $CaCO_3$ and other similar compounds. In organic P may adsorb on hydrous oxides and oxides of aluminum and iron of clay mineral surfaces (Syers et al. 2008a, b), and P is precipitated as insoluble Fe and Al phosphate complexes.

The balance between P adsorption and desorption maintains the equilibrium between solid phase and P in solution phase. The amount of P adsorbing or desorbing from surfaces depends on the number of sorption sites and the energy of adsorption (Moody et al. 2013). This equilibrium is termed P buffer capacity and is measured as the quantity of P that is adsorbed or desorbed for a unit change in solution P concentration. All of these P pools are in equilibrium with orthophosphate ($H_2PO_4^-$, HPO_4^{2-}) in the soil solution and govern soil P cycle by the processes of desorption-sorption (in the case of adsorbed P), dissolution-precipitation (in the case of mineral P), and mineralization-immobilization (in the case of organic P).

21.3 Rhizosphere: Ecological Hot Spot of Soil Microbiota

A narrow interfacial region between plant roots and surrounding soil, and characterized by intense biochemical and microbial activities, is called rhizosphere. Hiltner (1904) described rhizosphere as soil compartment influenced by plant growth that harbors microbial activity more than that of its surrounding soil. Rhizosphere, as a unique site of interaction between host plant and its biotic component, is mainly driven by the release of organic resources in the form of plant metabolites (root exudates) and plant debris (dead cells, mucilage). These organic resources served as food reserves for the microbes and support the growth and development of microbial population. These rhizodeposits represent a substantial part of photosynthetically fixed carbon (20–40%) and total plant nitrogen (10–16%) allocated to the underground root system, largely depending on plant species and plant age (Jones et al. 2009). The net sequestration of organic carbon and nitrogen by roots is thought to stimulate soil microbial multiplication in the vicinity of root tissues, because (a) most known soil bacteria are organotrophs, i.e., they derive the energy for growth from organic substrates, and (b) the accessibility and availability of organic compounds are limited in most soils (Alden et al. 2001; Demoling et al. 2007).

Rhizosphere microbiota is attracted by these organic resources and exerts numerous effects on plant growth and soil fertility (Antoun and Prevost 2005). As distinct microbial activity is usually associated with intense biochemical changes close to soil-root interface, therefore, it is also characterized as "microbial hot spot" (Reinhold-Hurek et al. 2015). These interactions are part of complex exchanges between roots and microorganisms and establish either beneficial, detrimental, or neutral relationships regulated by complex molecular signaling (Dardanelli et al. 2010). The beneficial interaction may favor plant growth, protect them from pathogens, and consequently have positive influence on crop productivity.

The rhizosphere, which is the narrow soil zone of soil that is influenced by plant roots and its released metabolites, can contain up to 10^{11} microbial cells per gram root (Egamberdiyeva et al. 2008) and more than 30,000 prokaryotic species (Mendes et al. 2011). In general, the microbial population densities in the rhizosphere are 10–100 higher than in the surrounding bulk soil (Spaepen et al. 2009). As microbial activity dwell on to the close proximity (~50 μm) of root surface, plant rhizodeposition of (approx. 50–100 mg per g root) is sufficient to support the growth of 2×10^{10} bacterial cells (Foster 1983).

An increasing body of evidence signifies the importance of this root microbiome, which consists of the entire complex of rhizosphere-associated microbes, their genetic elements and complex interactions in determining plant growth and health. The rhizosphere microbial population are usually characterized by rapid growth rate and utilize available substrates (chitin, mucilage, dead cells, and root exudates) that helps in maintaining the dynamic equilibrium of rhizosphere and controls its associated competition (Dessaux et al. 2016). Rhizosphere microbial populations are the most vibrant, highly competitive, and aggressive colonizer of the plant roots (Bouizgarne 2013; Mommer et al. 2016).

21.3.1 Plant Growth-Promoting Rhizobacteria (PGPR)

Rhizosphere microbial communities are increasingly understood to interact extensively with plants, and this association is very crucial to the overall plant health and development. A group of bacteria (PGPR) are known to establish an active synergy with plants through aggressive root colonization that subsequently confer plant growth-promoting benefits to the plants (Hartman et al. 2008; Shahzad et al. 2013). These bacteria can be found within and on roots and in soil associated with roots. The role of PGPR in promoting plant growth depends strongly on their survival and growth under variable field conditions (Rivera et al. 2008). Concerning soil heterogeneity, PGPR potency to compete, proliferate, and improve plant growth is predominantly fueled by root exudates and other organic substrates (Khalid et al. 2006; Yuan et al. 2015). Additionally, PGPR traits such as motility, chemotaxis, attachment, growth, and stress resistance also contribute to the overall competence of bacteria to survive in the rhizosphere and successfully colonize plant tissues.

A more feasible classification of PGPR is their separation as extracellular (e-PGPR) to denote those existing in the rhizosphere, on the rhizoplane, and intracellular (i-PGPR) and to denote bacteria that exist in the spaces between the cells of the root cortex or in specialized nodular structures (Gray and Smith 2005). On similar lines, Ambrosini et al. (2016) also classified soil beneficial bacteria and their association with plant roots, i.e., symbiotic (inside leguminous nodules), endophytic (intercellular spaces), associative (root surface adhered), and rhizospheric (root-soil interface associated). Unlike PGPR, certain free living soil bacteria are opportunistic in their association with plants, as they have loosely bound acclimation to the rhizosphere. These bacterial communities dwell around rhizosphere only in the presence of substantial organic substrate availability and in turn benefit plant in numerous ways. In another study, Bulgarelli and his colleagues (Bulgarelli et al. 2013) explored aboveground plant exterior surfaces as a habitat for microbes. According to their view, aerial plant parts (leave and stem surfaces) are thought to represent one of the largest but less explored microbial habitats called as phyllosphere. Compared with fungi and archaea, bacteria are the most prevalent phyllosphere-colonizing microbes, with bacterial titers averaging approximately 10^6–10^7 microbial cells per square centimeter of leaf area (Lindow and Brandl 2003). A benefiting plant-microbe interaction always involved several molecular signaling events that establish growth-promoting association with plant (Weiland-Bräuer et al. 2015). Such relationships vary according to plant genotypes and bacterial strains and with respect to the degree of proximity between the roots and surrounding soil as well as with the abilities of bacteria to improve plant growth.

There are several mechanisms by which PGPR can promote plant growth and development. Lugtenberg and Kamilova (2009) outlined tripartite contribution of PGPR toward plant growth, i.e., biofertilizer, phytostimulators, and stress controllers (Pereira and Castro 2014; Kurepin et al. 2014; Shahzad et al. 2014). PGPR use various mechanisms which may take place simultaneously or sequentially at different plant growth stages. The action mechanisms of plant growth promotion by PGPR can be grouped into two major categories, i.e., direct and indirect mode of plant growth promotion (Lugtenberg and Kamilova 2009; Bhattacharyya and Jha 2012; Ashraf et al. 2013). Direct plant growth-promoting activities mainly involve an improvement of nutrient availability to the plant by the fixation of atmospheric nitrogen, production of iron-chelating sideroaphores, organic matter mineralization (thereby meeting the nitrogen, sulfur, phosphorus nutrition of plants), and solubilization of insoluble phosphates (Martinez-Viveros et al. 2010; Chauhan et al. 2015; Etesami and Alikhani 2016). Another important direct mechanism involves the production of plant growth hormones and growth-regulating enzyme 1-aminocyclopro pane-1-carboxylate (ACC) deaminase (Shahzad et al. 2013). PGPR can also promote plant growth indirectly by inhibiting plant pathogen growth. This can also be achieved by the synthesis of enzymes that hydrolyze fungal cell walls, production of HCN, and lytic enzymes and induced systemic resistance by producing various antibiotic metabolites (Yuttavanichakul et al. 2012; Pieterse et al. 2014; Bensidhoum et al. 2016).

21.3.1.1 Phosphate-Solubilizing Bacteria (PSB)

Phosphorus is the most important nutrient after N that affects growth and metabolism processes of plant (Widawati and Suliasih 2006). The mobility of phosphate ions ($H_2PO_4^-$ and HPO_4^{2-}) in soil is very low because of their high retention in soil. But as a result of adsorption, precipitation and conversion to organic forms, only 10–30% of the applied phosphate mineral fertilizer can be recovered by the crops grown after the fertilization (Holford 1997; Syers et al. 2008a, b). The remaining 70–90% is accumulated in soil or in the form of immobile that is bound by Al or Fe in acid soils, or Ca and Mg in alkaline calcareous soils (Prochnow et al. 2006; Yang et al. 2010). While plants cannot absorb P in bound form, the P must be converted into available form. Phosphate-solubilizing bacteria (PSB) can play an important role in dissolving both of fertilizer P and bound P in the soil that is environmentally friendly and sustainable (Khan et al. 2007). The exploration of phosphate-solubilizing bacteria has been conducted by many researchers from soils

(Chen et al. 2006; Widawati and Rahmansyah 2009; Gupta et al. 2013) and rhizosphere (Poonguzhali et al. 2008; Khan et al. 2013).

Some bacterial species have mineralization and solubilization potential for organic and inorganic phosphorus, respectively (Hilda and Fraga 2000; Khiari and Parent 2005). Phosphorus-solubilizing activity is determined by the ability of microbes to release metabolites, such as organic acids, which through their hydroxyl and carboxyl groups chelate the cation bound to phosphate, the latter being converted to soluble forms (Khan et al. 2013; Sharma et al. 2013a, b). Phosphate solubilization takes place through various mechanisms including organic acid production and proton extrusion (Nahas 1996; Khan et al. 2009; Marra et al. 2011). A wide range of microbial P solubilization mechanisms exist in nature, and much of the global cycling of insoluble organic and inorganic soil phosphates is attributed to bacteria and fungi (Banik and Dey, 1982). Phosphorus solubilization is carried out by a large number of rhizobacteria acting on sparingly soluble soil phosphates, mainly by chelation-mediated mechanisms (Whitelaw, 2000; Reyes et al. 2001). Inorganic P is solubilized by the action of organic and inorganic acids secreted by PSB in which hydroxyl and carboxyl groups of acids chelate cations (Al, Fe, Ca) and decrease the pH in basic soils (Kpomblekou and Tabatabai 1994; Stevenson 2005). The PSB dissolve the soil P through the production of low molecular weight organic acids mainly gluconic and keto gluconic acids (Goldstein 1995; Deubel et al. 2000), in addition to lowering the pH of rhizosphere. The pH of rhizosphere is lowered through biotical production of proton/bicarbonate release (anion/cation balance) and gaseous (O_2/CO_2) exchanges. The phosphorus solubilization ability of PSB has direct correlation with pH of the medium. The release of root exudates such as organic ligands can also alter the concentration of P in the soil solution (Hinsinger 2001). Organic acids produced by PSB solubilize insoluble phosphates by lowering the pH, chelation of cations, and competing with phosphate for adsorption sites in the soil (Nahas, 1996). Inorganic acids, e.g., hydrochloric acid can also solubilize phosphate, but they are less effective compared to organic acids at the same pH (Kim et al. 1997). In certain cases, phosphate solubilization is induced by phosphate starvation (Gyaneshwar et al. 1999; Khan et al. 2007).

The mineralization of soil organic P (Po) plays an imperious role in P cycling of a farming system. Organic P may constitute 4–90% of the total soil P. Almost half of the microorganisms in soil and plant roots possess P mineralization potential under the action of phosphatases. Alkaline and acid phosphatases use organic phosphate as a substrate to convert it into inorganic form (Beech et al. 2001). Principal mechanism for mineralization of soil organic P is the production of acid phosphatases (Hilda and Fraga 2000). The release of organic anions and production of siderophores and acid phosphatase by plant roots/microbes (Yadaf and Tarafdar, 2001) or alkaline phosphatase (Tarafdar and Claasen 1988) enzymes hydrolyze the soil organic P or split P from organic residues. The largest portion of extracellular soil phosphatases is derived from the microbial population (Dodor and Tabatabai 2003).

21.4 Methodological Advancement for the Isolation of PSB

A diversity of growth mediums are being used in laboratories for cultivation, isolation, characterization, and subsequently selection of P-solubilizing bacteria. Most of the PSB growth media have differential chemical composition and also characterized with varied growth efficiency both on liquid and solid cultures (Table 21.1). Biological phosphate solubilization activity of root zone microbe was first described by Gerretsen (1948), who explained microbial mediated solubilization of insoluble inorganic P and its resultant effects on plant growth promotion. In the beginning, PSB isolation was primarily administered by plate screening assay and based on visual detection of clear halo zone formation as PSB colony niche in the presence of insoluble mineral P source (tricalcium phosphate/hydroxyapatite) (Pikovskaya 1948; Katznelson et al. 1962; Gupta et al. 1994). The selected method was adopted and generally considered a reliable approach for preliminary isolation and characterization of PSB (Katznelson et al. 1962; Goldstein and Liu 1987; Illmer and Schinner 1995).

However, Gupta et al. (1994) developed some modifications for PSB isolation using a Pikovskaya's medium with bromophenol blue supplementation. These modifications were primarily designed to improve the visibility of halo zone, which were formed by acidic metabolites of PSB and associated pH changes. In some cases, it appears that there was quite a distinct functional mismatch between plate halo detection and P solubilization activity in liquid culture of these so-called PSB. Several workers reported no visible clear halo zone formation on inorganic P supplement could still go on to solubilize various types of insoluble inorganic phosphate in liquid medium (Louw and Webley 1959; Das 1963). This may be because of varying diffusion rates of different organic acids secreted by an organism (Johnson 1959). Nautiyal (1999) emphasized the importance of defined media for screening-efficient PSB and associated P solubilization activity. He formulated National Botanical Research Institute's Phosphate (NBRIP) growth medium containing bromophenol blue as a pH indicator. Once the efficient PSB are selected, they are tested for their ability to solubilize insoluble P under liquid culture medium. Finally, the efficient P-solubilizing bacteria are selected and used for the advancement of inoculants whose efficacy is tested under natural environment against various crops. Similarly, Bashan et al. (2013) argued the vitality of Pikovskaya medium containing tricalcium phosphate (TCP) as the sole P source is used for screening of P-solubilizing bacteria. However, the lack of reliable evidence of TCP-P solubilizer, absence of metal-P source, and increase in the use of rock phosphate as P fertilizer emphasized the need to design a novel medium that could support the growth of P-solubilizing bacteria. They also raised halos zone-based selection concerns for screening P solubilizer as PSB colony growth is often without halos even after the media is replaced several times. They also suggested the adoption of modified liquid broth for the isolation and screening of efficient PSB solubilizer in different soil and culture conditions.

Table 21.1 Plant association and growth-promoting characteristics of PGPR in agriculture

Vegetation type	Host plant	Dominant exudation	Identified microbiota	References
Legumes	Soybean [Glycine max (L.) Merr.]	Phosphate solubilization, N-fixation, siderophores production, protease production	*Bacillus amyloliquefaciens* LL2012, *Bradyrhizobium japonicum*	Masciarelli et al. (2014)
	Chickpea (*Cicer arietinum* L.)	Siderophores production, chitinase activity, ACC-deaminase activity, exopolysaccharide production, phosphate solubilization, HCN production	*Serratia marcescens* (SF3) and *Serratia* spp. (ST9) + *M. ciceri*	Shahzad et al. (2014)
	Mung bean (*Vigna radiata* L.)	ACC-deaminase activity, Auxin production, phosphate solubilization antibiotic resistance	*Pseudomonas fragi* P5, *Pseudomonas jesseni* P10 and *Rhizobium leguminosarum* Z22	Iqbal et al. (2012)
Cereals	Wheat (*Triticum aestivum* L.),	IAA, HCN, siderophores	*Serratia marcescens*	Selvakumar et al. (2008)
	maize (*Zea mays* L)	Acid phosphatase, alkaline phosphatase, IAA production	*Azospirillum brasilense CNPSo 2083*, *Rhizobium tropici CIAT 899*	Marks et al. (2015)
	Rice (*Oryza sativa* L.)	IAA production, gibberellic acid production	*Enterobacter* spp. and *Azospirillium* spp.	Hasan et al. (2014)
	Oat (*Avena sativa* L.) and barley (*Hordeum vulgare* L.)	IAA production, siderophores production, phosphate solubilization	*Sinorhizobium meliloti L3Si*, *Pseudomonas* sp. LG *Azotobacter chroococcum* AV, *Enterobacter sp. E1*,	Stajković-Srbinović et al. (2014)
	Oat (*Avena sativa* L.)	ACC deaminase, HCN, IAA production, phosphate solubilization	*Sinorhizobium meliloti*, *Azotobacter* sp., *Pseudomonas* sp.	Delić et al. (2012)
	Sugarcane (*Saccharum officinarum* L.)	Production of IAA, phosphate solubilization, Induced systemic resistance,	*Azospirillum* sp.	Moutia et al. (2010)
	Sugarcane (*Saccharum officinarum* L.)	Phosphate solubilization, HCN production, IAA production	*Bacillus megaterium*	Sundara et al. (2002)

(continued)

Table 21.1 (continued)

Vegetation type	Host plant	Dominant exudation	Identified microbiota	References
Oil seed	Turnip mustard (*Brassica rapa* L.)	IAA, ACC deaminase, Siderophores	*Pseudomonas* sp.	Poonguzhali et al. (2008)
	Mustard (*Brassica campestris* L.)	HCN production, IAA production	*Mesorhizobium loti* MP6	Chandra et al. (2007)
	Canola (*Brassica napus* L.)	Siderophores, IAA, salicylic acid, ACC deaminase	*Dyella ginsengisoli, Burkholderia kururiensis, Pandoraea* sp. ATSB30	Anandham et al. (2008)
	Sunflower (*Helianthus annuus* L.)	Siderophores production and IAA production	*Pseudomonas fluorescens* biotype F and *Pseudomonas fluorescens* CECT 378^T	Shilev et al. (2012)
Trees	*Pinus roxburghii*	Siderophores production and IAA production	*Bacillus subtilis*	Singh et al. (2008)
	Italian stone pine (*Pinus pinea* L.)	Phosphate solubilization, IAA, exopolysaccharide production, organic acid production	*Bacillus licheniformis CECT 5106 and Bacillus pumilus CECT 5105*	Probanza et al. (2001)
	Teak (*Tectona Grandis*) and Indian redwood (*Chukrasia Tabularis*)	Nitrogen fixation, phosphate solubilization, siderophores production	*Azotobacter sp. DCU26 and Bacillus megaterium A3.3*	Aditya et al. (2009)
Grasses	Canary grass (*Phalaris minor* L.)	IAA production, Nitrogen fixation, HCN production	*Azotobacter* and *Azospirillum*	Zaefarian et al. (2012)
	Bermuda grass (*Cynodon dactylon* L.)	Phosphate solubilization, Exopolysaccharide production, ACC-deaminase activity, HCN production,	*Serratia sp.*—TRY2 and *Bacillus sp.*—TRY4	Sarathambal and Ilamurugu (2013)
	Barnyard grass (*Echinochloa crus-galli* L.), Italian ryegrass (*Lolium multiflorum* L.)	Phosphate solubilization, HCN production, IAA production, antifungal, HCN production,	*Bacillus*, *Arthrobacter*, *Stenotrophomonas*, *Acinetobacter*, and *Pseudomonas*	Sturz et al. (2001)
	Nut grass (*Cyperus rotundus* L.)	Phosphate solubilization, Organic acids production, siderophores production, HCN production	*Enterobacter* sp. Arh 1, *Pseudomonas* sp. Bro 5	Diogo et al. (2010)

(continued)

Table 21.1 (continued)

Vegetation type	Host plant	Dominant exudation	Identified microbiota	References
Vegetables	Red pepper (*Capsicum annuum* L.)	Gibberellic acid, IAA production	*Bacillus cereus* MJ-1	Joo et al. (2005)
	Mint (*Mentha piperita* L.)	Phosphate solubilization, siderophores production, IAA production	*Agrobacterium rubi* A16, *Burkholderia gladii* BA7, *P. putida* BA8, *B. subtilis* OSU142, *B. megaterium* M3	Kaymak et al. (2008)
	Cabbage (*Brassica oleracea* L.)	IAA production, Phosphate solubilization, HCN production, Organic production	*Bacillus megaterium* TV-91C, *Pantoea agglomerans* RK-92 *and B. subtilis* TV-17C	Turan et al. (2014)
	Tomato (*Solanum lycopersicum* L.)	IAA production, antagonistic behavior, HCN production, siderophores production, Gibberellic acid production	*Pseudomonas putida, P. fluorescens, Serratia marcescens, Bacillus subtilis, B. amyloliquefaciens*, and *Bacillus cereus*	Almaghrabi et al. (2013)
	Cucumber	Antagonistic effect, HCN production, siderophores production, Phosphate solubilization,	*Bacillus* sp.	Stout et al. (2002)
	Bitter gourd (*Momordica charantia* L.)	Phosphate solubilization, Nitrogen fixation, siderophores production, HCN production, ACC deaminase activity	*Azospirillum, Pseudomonas fluorescens*, and *Bacillus subtilis*	Kumar et al. (2012)

Quantitative estimation of biologically solubilized P as dissolved reactive phosphorus (most readily available P) is usually measured by molybdate colorimetric test (Murphy and Riely, 1962). This method was based on the observation that ammonium heptamolybdate and antimony potassium tartrate react with dilute orthophosphate solution in an acidic medium to form an antimony-phospho-molybdate complex. The reduction of the complex by ascorbic acid gives it an intense blue color that is proportional to the orthophosphate concentration.

21.4.1 *Mechanisms of P Solubilization*

21.4.1.1 Organic Acid Production

Phosphorus-solubilizing bacteria have characteristics ability to release acidic metabolites such as organic acids. These acidic secretions have the tendency to enhance mobility and/or solubility of inorganic P compounds (Son et al. 2006). On quantitative basis, the ability of PSB to solubilize insoluble phosphate in liquid culture medium is investigated by a number of researchers (Narula et al. 2000; Whitelaw, 2000). The solubilization of soil P in liquid medium by PSB has often been resulted due to the excretion of organic acids. In general, PSB produce variety of organic acids, i.e., acetic acid, gluconic acid, oxalic acid, citric acid, and lactic acid in liquid culture filtrates, and usually analyzed by thin layer chromatography or by high-performance liquid chromatography (HPLC). In addition, certain enzymatic methods are also employed for an accurate identification of unknown organic acids (Gyaneshwar et al. 1998).

These organic acids can either directly dissolve the mineral phosphate as a result of anion exchange of PO_4^{2-} by acid anion or can chelate both iron and aluminum ions associated with phosphate (Omar, 1998). In certain cases phosphate solubilization is induced by phosphate starvation (Gyaneshwar et al. 1999). The role of organic acids produced by PSB in solubilizing insoluble phosphate mainly attributed to the lowering of pH, chelating of cations, and competing with phosphate for adsorption sites in soil (Nahas 1996). Some inorganic acids, i.e., hydrochloric acid and sulfuric acid can also solubilize phosphate, but they are less effective compared with organic acids at the same pH (Kim et al. 1997). Effective P mobilization and/or solubilization by organic acid metabolites is related to the number and structure of the carboxyl groups, general order of carboxyl group effectiveness: tricarboxylate (e.g., $citrate^{3-}$) > dicarboxylate (e.g., $malate^{2-}$) > monocarboxylates (e.g., acetate $^{1-}$) (Ryan et al. 2001, 2012). The ability of organic acid secretions is gene-regulated mechanism but can also be influenced by prevailing environmental conditions. Soil nutrient content, i.e., C, N could affect the nature of organic acid secretions and P solubilizer (Narsian and Patel, 2000). Moreover, chelating ability of various organic acids has also been shown as an efficient mechanism of P solubility in P-deficient environment (Chapin et al. 2012).

21.4.1.2 Siderophores Production

It is well known that certain microbes secrete organic ligand to solubilize Fe from poorly available sources. Microbial siderophores are low molecular weight organic ligand produced as a scavenging agent to combat iron limitation. Siderophores production is usually not a widely investigated mechanism for phosphate solubilization. Many PSB have the ability to forage Fe from mineral complex into soluble Fe^{3+} form that is taken up by active transport carrier mechanism (Collavino et al. 2010). Siderophores production by PSB has indirect potential to improve P availability as these ligands can also extract Fe from ferric citrate and ferric phosphate (Zaidi et al. 2009). Approximately, 500 different siderophores structure are known to be produced by several gram-positive and gram-negative bacteria.

Very few works have been carried out to evaluate siderophores production as a method of P solubilization. Reid et al. (1985) showed 13-fold increments in P diffusion when compared with water. In view of mineral dissolution ascendency over ligand exchange by organic acids as a P-solubilizing mechanism, the probable siderophores contribution to improve P availability should be more pronounced.

21.4.1.3 Exopolysaccharides Production

The role of low molecular weight organic acids in the solubilization of mineral P is well documented. However, the knowledge on the role of high molecular weight microbial exudates (nonenzymatic mucilage) on P solubilization is limited. Exopolysaccharide (EPS) and biosurfactants are produced by bacteria largely in response to biofilm formation and stress. Microbial exopolysaccharides are polymers of carbohydrates excreted by bacteria on the outer side of their cell walls. The structural composition of exopolysaccharides is quite heterogenous (homo- or heteropolysaccharides) and may possess various organic and inorganic substituents (Sutherland, 2001). Some earlier studies have shown that the exopolysaccharides have the ability to form complexes with metals in soil (order of affinity to form complexes $Al^{3+} > Cu^{2+} > Zn^{2+} > Fe^{3+} > Mg^{2+} > K^{+}$) (Ochoa-Loza et al. 2001) that implicates their role of P solubilization in soil. Microbial exopolysaccharides have shown to stimulate the dissolution of tricalcium phosphate (TCP) in synergy with organic anions (Yi et al. 2008). Further the rate of dissolution was showed dependent on microbial source and concentration of EPS. Recently, Taktek et al. (2017) showed that exopolysaccharide-producing rhizobacteria solubilized igneous phosphate rock through secretions of viable biofilm cells and release of organic acids. They also found concomitant effect on plant P nutrition and yield of maize.

Phosphorus also releases from complex organic P compounds in soil by enzymatic activities; (a) phosphatases (Bandick and Dick 1999), which perform dephosphorylation of phospho-ester or phosphoanhydride bonds in organic matter; (b) phytases (Maougal et al. 2014), which particularly cause P release from phytic acid; (c) phosphonatases and C–P lyases, enzymes that exhibit C–P cleavage in organophosphonates.

21.4.1.4 Phosphatases Activity

Phosphatases are broad range of hydrolytic enzymes which showed strong affinity to catalyze the hydrolysis of both organic phosphate esters and anhydrides of H_3PO_4. Organic P is an important component of global phosphorus cycle. The P enzyme activities drive the mineralization of organic P into available inorganic form of P that easily assimilable by plants (Nannipieri 2011).

Interest in soil enzymes activity has increased manifold during the last decade because of their significance in P cycling (Beech et al. 2001; Rodríguez et al. 2006). Many bacteria having phosphatases release inorganic phosphate from organophosphate complexes. However, activities of these P-cycling enzymes largely depend upon their pH and are classified as either alkaline ($pH > 7$) or acid ($pH < 6$) phosphatase.

These enzymes catalyze the hydrolysis of many different phosphate systems including those of primary, secondary, cyclic, and sugar alcohols as well as phenols and amines.

Acid phosphatases are widely distributed hydrolase and primarily a plant origin enzyme exhibiting minimal substrate specificity (Duff et al. 1994). Several acid phosphatase genes from gram-negative bacteria have been isolated and characterized (Rossolini et al. 1998). Some of them code for acid phosphatase enzymes that are capable of performing well in soil. For example, the acpA gene isolated from *Francisella tularensis* expresses an acid phosphatase with optimum action at pH 6 with a wide range of substrate specificity (Reilly et al. 1996; Beech et al. 2001).

Alkaline phosphatases are group of hydrolases originally released by microbes under alkaline and neutral soil conditions. They can potentially hydrolyze up to 90% of total organic P into available P source in soil (Jarosch et al. 2015). This enzyme catalyzes the hydrolysis of a wide variety of phosphomonoesters and diesters excluding inositol phosphate (Kageyama et al. 2011). According to previous studies, PSB inoculation acts as an orderly stimulus with higher alkaline phosphatase activity that eventually lead to improve soil P status and plant P nutrition (Shahzad et al. 2014; Kaur and Reddy 2014).

21.4.1.5 Phytases Activity

Mostly phytases (myoinositol hexakisphosphate phosphohydrolases) belong to high molecular weight acid phosphatases. In its basic form, phytate is the primary source of inositol and the major stored form of phosphate in plant seeds and pollen. Most genetic engineering studies have focused on the search for phytases that are optimal for improving animal nutrition. Another attractive application of these enzymes that is not currently exploited is solubilization of soil organic phosphorus through phytate degradation. Phytate is the major component of organics forms of P in soil (Rodríguez et al. 2006). The ability of plants to obtain P directly from phytate is very limited. However, the growth and P nutrition of *Arabidopsis* plants supplied with phytate were improved significantly when they were genetically transformed with the phytase gene (phyA) (Richardson et al. 2001). This resulted in improved P nutrition, such that the growth and P content of the plant were equivalent to control plants supplied with inorganic phosphate. In relation to plant P availability, inoculation with phytate-mineralizing rhizobacteria improved P nutrition in cereals crop without P fertilization under Chilean Andisol (Martínez et al. 2015). Extracellular phytase-producing rhizobacteria have also been identified (Kumar et al. 2013; Li et al. 2013).

21.5 Recent Advances and Future Prospects of PSB

Based on above discussions, it is explicitly concluded that limited plant available P as well as low solubility of applied P fertilizer in soil are the major constraint in most agroecosystem. P deficiency can cause some serious concerns for overall growth, development, and yield of crop plants. However, the impediment caused by P limitation on plant growth can be dispelled and/or minimized by naturally occurring microorganisms capable of solubilizing P such as PSB.

Recently, proteomic-based techniques have emerged as an effective tool to disclose genotypic adaptation mechanisms involved in various crop plants under P deficiency. Various phosphate starvation responsive proteins have been identified in plant using these approaches, and using these proteins and their corresponding genes, it is now possible to improve plant P acquisition capacity by an upregulation of plant system for an efficient P uptake in the near future. The identification of metabolic genes regulating bioacidulation (principle mode of P solubilization) mechanism and its linked traits in PSB would advance our understanding about the underlying molecular basis of solubilized P fraction in soil. So far most of the studies involving PSB are originally in vitro which lack rigor and reliability to select an efficient PSB as a deliverable product for field trials. Thus, the work will further bridge the existing knowledge gap related to unclear role of PSB inoculant under varied soil environment. Another potential option is to develop transgenic plants encoding the genes of particular traits related to bacterial P solubilization. The literature shows that these transgenic plants have the adaptive capacity to counteract limited nutritional reserves. However, such studies are very limited in number and are at very early stage which warrants some elaborative validations before performing extensive experimentation. Moreover, information about the molecular mechanisms regulating P deficiency in crop plants is also scarce. Overall, future research should be focused (1) to mediate PSB-based metabolite engineering under P-deficient environments, (2) to explore multiple mechanistic traits of P solubilization in PSB, (3) to identify target P responsive genes for promoting growth under P starvation, and (4) the transference of targeted genes for efficient P utilization in plants through biotechnology.

21.6 Conclusion

Various modern agro-biotechnological interventions are being used to boost up plant P acquisition under P-limited conditions. One of the most emerging tools to negate P solubility and availability concerns for crop plant is the use of PSB as P bioinoculant. PSB play an important role in phosphorus nutrition by enhancing its availability to plants through solubilization and mineralization of inorganic and organic phosphates in soil. Therefore, plant beneficial microbes (including phosphate-solubilizing bacteria, i.e., PSB) and their associative interaction with host plant are key determinant of overall plant growth, development, and yield of crops under P-limited conditions. However, efficiency of PSB as a value-added P biofertilization approach further necessitates a more rigorous selection criteria for quality PSB formulation, which requires considerable attention of the scientists to overcome such challenges (Tables 21.2, 21.3, 21.4, 21.5 and 21.6).

Table 21.2 Effectiveness of PGPR for plant growth regulation under varied soil type

PGPR	Source/experimental soil	Host plant	Botanical name	Plant growth regulation	References
Bacillus firmus NCIM 2636	Heavy loam	Paddy	*Oryza* Spp.	Increased root biomass and phytohormones in plant (e.g., auxin, gibberellins, cytokinins, etc.)	Datta et al. (1982)
B. megaterium + *G. fasciculatum* and *G. fasciculatum*	Clay loam soil	Banana	*Musa paradisiaca*	Increased fresh biomass and phosphorous intake in plant as compared to uninoculated control	Patil et al. (2002)
Pseudomonas striata	Sandy loam	Chickpea, soybean	*Glycine max*	Increased the number of nodules per plant, dry and fresh weight of nodules and grain yield legumes	Son et al. (2006)
P. Fluorescence + *Bacillus megaterium*	Sandy clay loam soil	Chickpea	*Cicer arietinum*	Improving seedling growth, phosphorus uptake, total chlorophyll content, fresh and dry weight of roots	Sharma et al. (2007)
Pseudomonas putida	Heavy loam	Barley	*Hordeum vulgare*	Increased total chlorophyll contents in plant leaf	Mehrvarz et al. (2008)
Enterobacter gergoviae J107, *P. fluorescens* J108, *S. proteamaculans* J119 and *C. koseri* J120	Sandy clay loam	Chickpea	*Cicer arietinum*	Out of these, J119 strain was found to be the most effective PGPR in improving root and shoot growth, nodulation and grain yield of chickpea than control in growth pouches, pot and field trials	Shahzad et al. (2010)

PGPR	Source/experimental soil	Host plant	Botanical name	Plant growth regulation	References
G. etunicatum + *Burkholderia cepacia* BAM 6	Loam soil	Wheat	*Triticum aestivum*	Increased biomass, phosphorous uptake in plants and total grain yield	Minaxi et al. (2013)
Pseudomonas thivervalensis STF3 and *Serratia marcescens* STJ5	Sandy clay loam	Maize	*Zea mays* L.	Bacterial isolates, with 75 and 100% CF, significantly improved the growth, yield and NPK uptake in maize than control. The growth and yield promoting effect of STF3 strain with 75% CF were similar to CF alone. But with 100% CF, same strain significantly increased total biomass, grain yield and chlorophyll content than control	Shahzad et al. (2013)
Pseudomonas tolaasii IEXb, *Pseudomonas koreensis* SP28	Silt loam	Maize	*Zea mays* L.	Both strains showed a positive effect on plant growth. A significant increase in plant height (45%), shoot dry weight (40%) was observed in plants treated with IEXb, while SP28 has significantly increased P content than uninoculated control	Viruel et al. (2014)

(continued)

Table 21.2 (continued)

PGPR	Source/experimental soil	Host plant	Botanical name	Plant growth regulation	References
S. marcescens (SF3) and Serratia spp. (ST9) + M. ciceri	Sandy clay loam (irrigated area), sandy loam (rain-fed area)	Chickpea	*Cicer arietinum*	Significantly increased the nodules per plant, grain yield, protein content, P uptakes under irrigated and rain-fed conditions than sole inoculation. Integrating PEC with co-inoculation gave an additive effect on the nodulation and growth of chickpea under both farming systems	Shahzad et al. (2014)
Klebsiella sp. Br1, *Klebsiella pneumoniae* Fr1, *Bacillus pumilus* S1r1 and *Acinetobacter* sp. S3r2	Clay loam	Maize	*Zea mays* L.	Significantly improved the dry biomass of top, root and ear, total N and P content of maize roots due to PGPR inoculation. In particular, the plants inoculated with *B. pumilus* S1r1 generally out performed those with the other treatments	Kuan et al. (2016)

PGPR	Source/experimental soil	Host plant	Botanical name	Plant growth regulation	References
Bacillus megaterium RC01 and *Bacillus* M-13	loam soil	Barley	*Hordeum vulgare* L.	Seed inoculation of barley significantly increased the root weight, shoot weight and P uptakes by 32.1%, 54.2% and 39.6%, respectively, over uninoculated control	Cakmakcı et al. (2007)
Bacillus M-13	Silt clay loam	Sunflower	*Helianthus annuus* L.	Inoculation with PSB significantly enhanced the head diameter, 1000 seed weight, kernel ratio and oil content and led to seed and oil yield increases of 15.0 and 24.7% over control respectively	Ekin (2010)
Burkholderia sp. PS-01, *Bacillus* sp. PS-12, *Pseudomonas* sp. PS-32, *Flavobacterium* sp. PS-41 and *Pseudomonas* sp. PS-51	Sandy clay loam	Maize	*Zea mays* L.	Inoculation significantly increased the plant height, root length, shoot dry weight, root dry weight and grain yield were observed which were up to 16, 11, 42, 29 and 33%, respectively, over control in the presence of FYM at 16 Mg ha^{-1}	Hussain et al. (2013)

(continued)

Table 21.2 (continued)

PGPR	Source/experimental soil	Host plant	Botanical name	Plant growth regulation	References
Pseudomonas aurantiaca and *Pseudomonas fluorescens*	Sandy loam	Paddy	(*Oryza sativa* L.)	Inoculation with PSB significantly improved the plant height, shoot and root dry weight, P and N uptake and net photosynthetic rate of walnut seedlings. Application of these two PSB strains also improved soil quality, as indicated by increased activities of dehydrogenase, neutral phosphatase and urease in the soil	Yu et al. (2014)

Table 21.3 Culture media and their chemical composition used for P-solubilizing bacteria

Chemical composition of media	PVK	NBRIP	NBRIY	YEM	Ashby's
Ammonium sulfate	0.50	0.10	0.50		
Bromophenol blue (BPB)		0.03			
Calcium carbonate				2.00	5.00
Dipotassium hydrogen phosphate				0.50	0.20
Ferrous sulfate	0.00				
Glucose	10.00	10.00	10.00		
Iron(III) chloride			0.002		
Magnesium chloride hexahydrate		5.00			
Magnesium sulfate heptahydrate	0.10		0.10	0.20	0.20
Manganese sulfate	0.25	0.25	0.00		
Mannitol				10.00	20.00
Potassium chloride	0.20	0.20	0.20		
Potassium sulfate					0.10
Sodium chloride	0.20		0.20		

PVK Pikovskaya's medium (Pikovskaya, 1948), *NBRIP* National Botanical Research Institute's phosphate growth medium (Nautiyal 1999), *NBRIY* National Botanical Research Institute's phosphate growth medium devoid of yeast extract medium (Nautiyal 1999), *YEM* yeast extract mannitol broth (Holt et al. 1994), Ashby's medium (SubbaRao 1977)

Table 21.4 Effects of different types of phosphate-solubilizing bacterial inoculant on plant performance

Type of P bacteria	Name	Source	Host plant	Botanical name	Response	Conditions	References
Free living	*Arthrobacter* sp. and *Bacillus subtilis*	Soil	Wheat	*Triticum aestivum* L.	Increases dry weight up to 26% and P uptake up to 19% when co-inoculated at 2 dS $m^{-1}$of salinity level and 40 and 34% when co-inoculated at 6 dS m^{-1} of salinity level	Pot	Upadhyay et al. (2011)
	Serratia liquefaciens, *Bacillus sp.Pseudomonas sp.*	Soil	Maize	*Zea mays* L.	Grain yield increased up to 14% when they were inoculated as consortia over control. In sole inoculation, *S. liquefaciens* increased the dry weight of maize respect to control more than 10%, *Bacillus* sp. more than 7% and *Pseudomonas* sp. more than 10%	Pot (Green house)	Lalande et al. (1989)

Type of P bacteria	Name	Source	Host plant	Botanical name	Response	Conditions	References
Associative	*Pseudomonas putida* R-168, *Pseudomonas fluorescens* R-93, *P. fluorescens* DSM 50090, *A. lipoferum* DSM 1691,	Root	Maize	*Zea mays* L.	Plant height, seed weight, number of seed per ear and leaf area, shoot dry weight and phosphorus content significantly increased over control	Field	Gholami et al. (2009)
	Pseudomonas fluorescens PGPR1, PGPR2, PGPR4	Root	Peanut	*Arachis hypogaea* L.	Significantly enhanced pod yield, haulm yield, nodule dry weight and P uptake over the control	Pot and field	Dey et al. (2004)
	Enterobacter sakazakii 8MR5, *Pseudomonas sp.* 4MKS8, *Klebsiella oxytoca* 10MKR7	Root	Maize	*Zea mays* L.	Inoculation increased growth parameters and NP uptake	Pot	Babalola et al. (2003)

(continued)

Table 21.4 (continued)

Type of P bacteria	Name	Source	Host plant	Botanical name	Response	Conditions	References
Rhizobia	*Mesorhizobium* sp. RC3	Nodule	Chickpea	*Cicer arietinum* L.	Increased the dry matter accumulation, number of nodules, seed yield, P content and grain protein by 71%, 86%, 36%, 23%, and 16%, respectively, over uninoculated control plants	Pot	Wani et al. (2008)
	Bradyrhizobium MRM6	Nodule	Green gram	*Vigna radiata* (L.) Wilczek	When herbicide-tolerant *Rhizobium* strain MRM6 was used with herbicide, it increased the growth parameters at all tested concentrations of herbicides (quizalafop-p-ethyl and clodinafop)	Pot	Ahemad and Khan (2012)
	Mesorhizobium strain MRC4	Nodule	Chickpea	*Cicer arietinum* L.	Significantly increased symbiotic properties (nodulation and leghemoglobein content), root N, shoot N, root P, shoot P, seed yield, and seed protein	Pot	Ahemad and Khan (2010)

Type of P bacteria	Name	Source	Host plant	Botanical name	Response	Conditions	References
Endophytes	*Enterobacter agglomerans*	Tissue	Tomato	*Solanum lycopersicum* L.	Significantly increased the P concentration in shoot and fruit parts than control plants	Pot	Kim et al. (1998)
	Pseudomonas mutant CRPF7	Tissue	Mung bean	*Vigna radiata* L.	Significantly enhanced the nodulation, P in shoot and grain seed yield, and seed protein than uninoculated plants	Field	Das et al. (2003)
Phyllosphere	*Azotobacter chroococcum* REN 2 and *Beijerinckia indica* JN1	Leaf	Wheat	*Triticum aestivum* L.	Gave significantly increase in number of tiller, fresh biomass, grain yield, protein content, and P content in shoot and grain as compared to respective control	Field	Pati and Chandra (1981)

Table 21.5 P-solubilizing bacterial metabolites and plant growth promotion

Bacterial name	Activity	Host plant growth	Botanical name	References
Pseudomonas putida	Organic acids	Enhanced chlorophyll content and P uptake in plants	*Hordeum vulgare* L.	Mehrvarz et al. (2008)
Paenibacillus lentimorbus B-30488	Organic acids	Enhanced plant growth due to biofilm formation and phosphate solubilization	*Zea mays* L.	Khan et al. (2011)
Pseudomonas striata	Phosphatase	Increased the number of nodules, fresh and dry weights of nodules, and grain yield and improve the P uptake in grain	*Cicer arietinum* L. and *Glycine max* L.	Son et al. (2006)
Pseudomonas fluorescence + *Bacillus megaterium*	Phosphatase	Enhanced seedling growth and phosphorus content in plants in comparison to uninoculated plants	*Cicer arietinum* L.	Sharma et al. (2013a, b)
Enterobacterium with *Sinorhizobium meliloti*	Phosphatase	Increased P uptake, fresh biomass and grain yield than uninoculated plants	*Cicer arietinum* L. and *Pisum sativum* L.	Hynes et al. (2008)
Bacillus sp. with *Rhizobium*	Phytase	Co-inoculation improved pod and straw yield; increased the root length, root mass, and number of nodule and mass; and enhanced the nutrient concentration in mash plant and grains	*Vigna mungo* L.	Qureshi et al. (2012)

(continued)

Table 21.5 (continued)

Bacterial name	Activity	Host plant growth	Botanical name	References
A. chroococcum	Phytase	Enhanced plant growth due to phosphate solubilization, auxin production, and catalase activity	*Zea mays* L.	Rojas-Tapias et al. (2012)
Acinetobacter sp. (PSGB04), *Pseudomonas* (PRGBB06)	Exopolysaccharide	Increased root length, shoot length, seedling vigor, dry mass/IAA, salicylic acid, N-fixation, and P uptake in seedlings	*Brassica napus*, Tomato	Indiragandhi et al. (2008)
Pseudomonas aeruginosa	Siderophores	Improving root and shoot growth of plant under water stress and increase P uptake in plant as compared to untreated plants	Black gram	Ganesan (2008)
Enterobacter sp.	Siderophores	Increasing chlorophyll content and iron and phosphorus content in leaves than uninoculated plants	*Brassica juncea*	Kumar et al. (2008)

Table 21.6 Types and functional expression of P-solubilizing genes in plant beneficial bacteria

Identified Gene	Source	Function/mineral P solubilized	References
pKG3791	*Serratia marcescens*	PQQ biosynthesis, produce gluconic acid, and solubilizes P mineral/TCP	Krishnaraj and Goldstein (2001)
gabY	*Pseudomonas cepacia*	Produces gluconic acid and also having phosphatase activity/TCP	Babu-Khan et al. (1995)
pqqE	*Erwinia herbicola*	Produces gluconic acid and solubilizes P mineral/TCP	Vikram et al. (2007)
pqqED genes	*Rahnella aquatilis*	Produces gluconic acid and acetic acid and solubilizes P mineral/HAP	Kim et al. (1998)
Unknown	*Enterobacter agglomerans*	Produces gluconic acid and solubilizes P mineral/DCP	Kim et al. (1997)
pqqABCDEF genes	*Enterobacter intermedium*	Produces citric acid, gluconic acid, and solubilizes P mineral/HAP	Kim et al. (2003)
Ppts-gcd, P gnlA-gcd	*E. coli*	Produces gluconic acid and solubilizes P mineral/TCP	Sashidhar and Podile (2009)
gabY putative PQQ transporter	*Pseudomonas cepacia*	Produces gluconic acid	Babu-Khan et al. (1995)
gltA/citrate synthase	*E. coli* K12	Produces citric acid and solubilizes P mineral/DCP	Buchet et al. (1999)
Unknown	*Synechocystis* PCC 6803	Produces gluconic and acetic acids and solubilizes P mineral/RP	Gyaneshwar et al. (2002)
gad/gluconate dehydrogenase	*P. putida* KT2440	Produces gluconic and 2-Ketobutyric acids and solubilizes P mineral/RP	Kumar et al. (2013)
nap A	*Burkholderia cepacia IS-16*	Increased extracellular alkaline phosphatases and solubilizes P mineral/CP	Fraga et al. (2001)

References

Adesemoye, A., & Kloepper, J. (2009). Plant–microbes interactions in enhanced fertilizer use efficiency. *Applied Microbiology and Biotechnology, 85*, 1–12.

Adhya, T. K., Kumar, N., Reddy, G., Podile, A. R., Bee, H., & Samantaray, B. (2015). Microbial mobilization of soil phosphorus and sustainable P management in agricultural soils. *Current Science, 108*, 1280–1287.

Aditya, B., Abhrajyoti Ghosh, A., & Chattopadhyay, D. (2009). Co-inoculation effects of nitrogen fixing and phosphate solublising microorganisms on teak (*Tectona grandis*) and indian redwood (*Chukrasiatu bularis*). *Journal of Biological Sciences, 1*, 23–29.

Ahemad, M., & Khan, M. S. (2010). Ameliorative effects of Mesorhizobium sp. MRC4 on chickpea yield and yield components under different doses of herbicide stress. *Pesticide Biochemistry and Physiology, 98*, 183–190.

Ahemad, M., & Khan, M. S. (2012). Productivity of greengram in tebuconazole-stressed soil, by using a tolerant and plant growth–promoting *Bradyrhizobium* sp. MRM6 strain. *Acta Physiologiae Plantarum, 34*, 245–254.

Alden, L., Demoling, F., & Baath, E. (2001). Rapid method of determining factors limiting bacterial growth insoil. *Applied and Environmental Microbiology, 67*, 1830–1838.

Almaghrabi, O. A., Massoud, S. I., & Abdelmoneim, T. S. (2013). Influence of inoculation with plant growth promoting rhizobacteria (PGPR) on tomato plant growth and nematode reproduction under greenhouse conditions. *Saudi Journal of Biological Sciences, 20*, 57–61.

Ambrosini, A., Souza, R., & Passaglia, L. M. P. (2016). Ecological role of bacterial inoculants and their potential impact on soil microbial diversity. *Plant and Soil, 400*, 193–207.

Anandham, R., Gandhi, P. I., Madhaiyan, M., & Sa, T. (2008). Potential plant growth promoting traits and bioacidulation of rock phosphate by thiosulfate oxidizing bacteria isolated from crop plants. *Journal of Basic Microbiology, 48*, 439–447.

Antoun, H., & Prevost, D. (2005). Ecology of plant growth promoting rhizobacteria. In Z. A. Siddiqui (Ed.), *PGPR: Biocontrol and biofertilization* (pp. 1–38). Dordrecht: Springer.

Arif, M. S., Riaz, M., Shahzad, S. M., Yasmeen, T., Akhtar, M. J., Riaz, M. A., Jassey, V. E. J., Bragazza, L., & Buttler, A. (2016a). Associative interplay of plant growth promoting rhizobacteria (*Pseudomonas aeruginosa* QS40) with nitrogen fertilizers improves sunflower (*Helianthus annuus* L.) productivity and fertility of Aridisol. *Applied Soil Ecology, 108*, 238–247.

Arif, M. S., Riaz, M., Shahzad, S. M., Yasmeen, T., Ali, S., & Akhtar, M. J. (2016b). Phosphorus mobilizing plant growth promoting rhizobacteria (*Bacillus cereus* GS6) improved symbiotic efficiency of soybean (*Glycine max* L.) in compost amended aridisol. Pedosphere (submitted).

Ashraf, M. A., Asif, M., Zaheer, A., Malik, A., Ali, Q., & Rasool, M. (2013). Plant growth promoting rhizobacteria & sustainable agriculture: A review. *African Journal of Microbiology Research, 7*, 704–709.

Babalola, O. O., Osir, E. O., Sanni, A., Odhaimbo, G. D., & Bulimo, W. D. (2003). Amplification of 1-aminocyclopropane-1-carboxylic (ACC) deaminase from plant growth promoting rhizobacteria in Striga-infested soils. *African Journal of Biotechnology, 2*, 157–160.

Babu-Khan, S., Yeo, T. C., Martin, W. L., Duron, M. R., Rogers, R. D., & Goldstein, A. H. (1995). Cloning of a mineral phosphate-solubilizing gene from *Pseudomonas cepacia*. *Applied and Environmental Microbiology, 61*, 972–978.

Baginsky, S. (2016). Protein phosphorylation in chloroplasts–A survey of phosphorylation targets. *Journal of Experimental Botany, 67*(13), 3873–3882. doi:10.1093/jxb/erw098.

Bandick, A. K., & Dick, R. P. (1999). Field management effects on soil enzyme activities. *Soil Biology and Biochemistry, 31*, 1471–1479.

Banik, S., & Dey, B. K. (1982). Available phosphate content of an alluvial soil as influenced by inoculation of some isolated phosphate solubilizing microorganisms. *Plant and Soil, 69*, 353–364.

Bashan, Y., Kamnev, A. A., & de-Bashan, L.E. (2013). A proposal for isolating and testing phosphate-solubilizing bacteria that enhance plant growth. *Biology and Fertility of Soils, 49*, 1–2.

Bashan, Y., & Holguin, G. (1998). Proposal for the division of plant growth promoting rhizobacteria into two classifications: Biocontrol–PGPB plant growth–promoting bacteria, and PGPB. *Soil Biology and Biochemistry, 30*, 1225–1228.

Beech, I. B., Paiva, M., Caus, M., & Coutinho, C. (2001). Enzymatic activity and within biofilms of sulphate reducing bacteria. In P. G. Gilbert, D. Allison, M. Brading, J. Verran, & J. Walker (Eds.), *Biofilm community interactions: Chance or necessity?* (pp. 231–239). Cardiff, UK: BioLine.

Bensidhoum, L., Nabti, E., Tabli, N., Kupferschmied, P., Weiss, A., Rothballer, M., Schmid, M., Keel, C., & Hartmann, A. (2016). Heavy metal tolerant *Pseudomonas protegens* isolates from agricultural well water in northeastern Algeria with plant growth promoting, insecticidal and antifungal activities. *European Journal of Soil Biology, 75*, 38–46.

Bertrand, I., Holloway, R. E., Armstrong, R. D., & McLaughlin, M. J. (2003). Chemical characteristics of phosphorus in alkaline soils from southern Australia. *Australian Journal of Soil Research, 41*, 61–76.

Bhattacharyya, P. N., & Jha, D. K. (2012). Plant growth–promoting rhizobacteria (PGPR): Emergence in agriculture. *World Journal of Microbiology and Biotechnology, 28*, 1327–1350.

Bouizgarne, B. (2013). Bacteria for plant growth promotion and disease management. In D. K. Maheshwari (Ed.), *Bacteria in agrobiology: Disease management* (pp. 15–47). Berlin Heidelberg: Springer-Verlag.

Buchet, A., Nasser, W., Eichler, K., & Mandrand-Berthelot, M. A. (1999). Positive co-regulation of the *Escherichia coli* carnitine pathway *cai* and *fix* operons by CRP and the CaiF activator. *Molecular Microbiology, 34*, 562–575.

Bulgarelli, D., Schlaeppi, K., Spaepen, S., Loren, V., van Themaat, E., & Schulze-Lefert, P. (2013). Structure and functions of the bacterial microbiota of plants. *Annual Review of Plant Biology, 64*, 807–838.

Cakmakcı, R., Erat, M., Erdo, U. G., & Donmez, M. F. (2007). The influence of PGPR on growth parameters, antioxidant and pentose phosphate oxidative cycle enzymes in wheat and spinach plants. *Journal of Plant Nutrition and Soil Science, 170*, 288–295.

Chandra, S., Choure, K., Dubey, R. C., & Maheshwari, D. K. (2007). Rhizosphere competent *Mesorhizobium loti* MP6 induces root hair curling, inhibits *Sclerotiniasclerotiorum* and enhances growth of Indian mustard (*Brassica campestris*). *Brazilian Journal of Microbiology, 38*, 128–130.

Chapin, F. S., Maton, P. A., & Vitousel, P. M. (2012). *Principles of terrestrial ecosystem ecology* (p. 436). New York: Springer-Verlag.

Chauhan, H., Bagyaraj, D. J., Selvakumar, G., & Sundaram, S. P. (2015). Novel plant growth promoting rhizobacteria–prospects. *Applied Soil Ecology, 95*, 38–53.

Chen, Y. P., Rekha, P. D., Arun, A. B., Shen, F. T., Lai, W. A., & Young, C. C. (2006). Phosphate solubilizing bacteria from subtropical soil and their tricalcium phosphate abilities. *Applied Soil Ecology, 34*, 33–41.

Chowdhury, R. B., Moore, G. A., Weatherley, A. J., & Arora, M. (2017). Key sustainability challenges for the global phosphorus resource, their implications for global food security, and options for mitigation. *Journal of Cleaner Production, 140*, 945–963.

Chung, H., Park, M., Madhaiyan, M., Seshadri, S., Song, J., Cho, H., & Sa, T. (2005). Isolation and characterization of phosphate solubilizing bacteria from the rhizosphere of crop plants of Korea. *Soil Biology and Biochemistry, 37*, 1970–1974.

Cisse, L., & Mrabet, T. (2004). World phosphate production: Overview and prospects. *Phosphorus Research Bulletin, 15*, 21–25.

Condron, L. M., Turner, B. L., & Cade-Menun, B. J. (2005). Chemistry and dynamics of soil organic phosphorus. In J. T. Sims & A. N. Sharpley (Eds.), *Phosphorus: Agriculture and the environment* (pp. 87–121). Madison: American Society of Agronomy, Crop Science Society of America, Soil Science Society of America, Inc.

Collavino, M. M., Sansberro, P. A., Mroginski, L. A., & Aguilar, O. M. (2010). Comparison of in vitro solubilization activity of diverse phosphate-solubilizing bacteria native to acid soil and their ability to promote *Phaseolus vulgaris* growth. *Biology and Fertility of Soils, 46*, 727–738.

Cordell, D., Drangert, J. O., & White, S. (2009). The story of phosphorus: Global food security and food for thought. *Global Environmental Change, 19*, 292–305.
Damon, P. M., Bowden, B., Rose, T., & Rengel, Z. (2014). Crop residue contributions to phosphorus pools in agricultural soils: A review. *Soil Biology and Biochemistry, 74*, 127–137.
Dao, T. H. (2004). Ligands and phytase hydrolysis of organic phosphorus in soils amended with dairy manure. *Agronomy Journal, J96*, 1188–1195.
Dardanelli, M. S., Manyani, H., Gonzalez-Barroso, S., Rodriguez-Carvajal, M. A., Gil-Serrano, A. M., Espuny, M. R., López-Baena, F. J., Bellogin, R. A., Megías, M., & Ollero, F. J. (2010). Effect of the presence of the plant growth promoting rhizobacterium (PGPR) *Chryseobacteriumbalustinum* Aur9 and salt stress in the pattern of flavonoids exuded by soybean roots. *Plant and Soil, 328*, 483–493.
Das, A. C. (1963). Utilization of insoluble phosphates by soil fungi. *Journal of the Indian Society of Soil Science, 11*, 203–207.
Das, K., Katiyar, V., & Goel, R. (2003). P solubilization potential of plant growth promoting Pseudomonas mutants at low temperature. *Microbiological Research, 158*, 359–362.
Datta, M. S., Banik, M., & Gupta, R. K. (1982). Studies on the efficacy of a phytohormone producing phosphate solubilizing bacillus firmis in augmenting paddy yield in acid soils of Nagaland. *Plant and Soil, 69*, 365–373.
Delić, D., Stajković-Srbinović, O., Kuzmanović, D., Rasulić, N., Maksimović, S., Radović, J., & Simić, A. (2012). Influence of plant growth promoting rhizobacteria on alfalfa, Medicago sativa L. yield by inoculation of a preceding Italian ryegrass, *Lolium multiflorum* Lam. In S. Barth & D. Milbourne (Eds.), *Breeding strategies for sustainable forage and turf grass improvement* (pp. 333–339). Dordrecht, Netherlands: Springer.
Demoling, F., Figueroa, D., & Baath, E. (2007). Comparison of factors limiting bacterial growth in differentsoils. *Soil Biology and Biochemistry, 39*, 2485–2495.
Dessaux, Y., Grandclément, C., & Faure, D. (2016). Engineering the rhizosphere. *Trends in Plant Science, 21*, 266–278.
Deubel, A., Gransee, A., & Merbach, W. (2000). Transformation of organic rhizodeposits byrhizoplane bacteria and its influence on the availability of tertiary calcium phosphate. *Journal of Plant Nutrition and Soil Science, 163*, 387–392.
Dey, R., Pal, K. K., Bhatt, D. M., & Chauhan, S. M. (2004). Growth promotion and yield enhancement of peanut (*Arachish ypogaea* L.) by application of plant growth-promoting rhizobacteria. *Microbiological Research, 159*, 371–394.
Diogo, J., Korenblum, E., Casella, R., Vital, R. L., & Seldin, L. (2010). Polyphasic analysis of the bacterial community in the rhizosphere and roots of *Cyperus rotundus* L. grown in a petroleum contaminated soil. Microbiol. *Biotechnology, 20*, 862–870.
Dobbelaere, S., Vanderleyden, J., & Okon, Y. (2003). Plant growth promoting effects of diazotrophs in the rhizosphere. *Critical Reviews in Plant Sciences, 22*, 107–149.
Dodor, D. E., & Tabatabai, A. M. (2003). Effect of cropping systems on phosphatases in soils. *Journal of Plant Nutrition and Soil Science, 166*, 7–13.
Duff, S. M., Sarath, G., & Plaxton, W. C. (1994). The role of acid phosphatases in plant phosphorus metabolism. *Physiologia Plantarum, 90*, 791–800.
Egamberdiyeva, D., Kamilova, F., Validov, S., Gafurova, L., Kucharova, Z., & Lugtenberg, B. (2008). High incidence of plant growth stimulating bacteria associated with the rhizosphere of wheat grown on salinated soil in Uzbekistan. *Environmental Microbiology, 10*, 1–9.
Ekin, Z. (2010). Performance of phosphate solubilizing bacteria for improving growth and yield of sunflower (*Helianthus annuus* L.) in the presence of phosphorus fertilizer. *African Journal of Biotechnology, 9*, 3794–3800.
Etesami, H., & Alikhani, H. A. (2016). Rhizosphere and endorhiza of oilseed rape (*Brassica napus* L) plant harbor bacteria with multifaceted beneficial effects. *Biological Control, 94*, 11–24.
FAO. (2009). *How to feed the world in 2050*. Rome, Italy: Food and Agriculture Organization.
Fiske, C. H., & Subbarow, Y. (1925). A colorimetric determination of phosphorous. *The Journal of Biological Chemistry, 66*, 375–400.

Foster, R. C. (1983). The fine structure of epidermal cell mucilages of roots. *The New Phytologist, 91*, 727–740.
Fraga, R., Rodriguez, H., & Gonzalez, T. (2001). Transfer of the gene encoding the Nap A acid phosphatase from *Morganellamorganii* to a *Burkholderiacepacia* strain. *Acta Biotechnologica, 21*, 359–369.
Frossard, E., Brossard, M., Hedley, M. J., & Meterell, A. (1995). Reactions controlling the cycling of P in soils. In H. Tiessen (Ed.), *Phosphorus in the global environment* (pp. 107–137). New York: Wiley.
Frossard, E., Condron, L. M., Oberson, A., Sinaj, S., & Fardeau, J. C. (2000). Processes governing phosphorus availability in temperate soils. *Journal of Environmental Quality, 29*, 15–23.
Ganesan, V. (2008). Rhizoremediation of cadmium soil using a cadmium-resistant plant growth-promoting rhizopseudomonad. *Current Microbiology, 56*, 403–407.
Gerretsen, F. C. (1948). The influence of microorganisms on the phosphate intake by the plant. *Plant and Soil, 1*, 51–81.
Gholami, A., Shahsavani, S., & Nezarat, S. (2009). The effect of plant growth promoting rhizobacteria (PGPR) on germination, seedling growth and yield of maize. *International Journal of Biological, Biomolecular, Agricultural, Food and Biotechnological Engineering, 1*, 9–14.
Goldstein, A. H. (1995). Recent progress in understanding the molecular genetics and biochemistry of calcium phosphate solubilization by Gram–negative bacteria. *Biological Agriculture and Horticulture, 12*, 185–193.
Goldstein, A. H., & Liu, S. T. (1987). Molecular cloning and regulation of a mineral phosphate solubilizing gene from *Erwiniaherbicola. Biotechnology, 5*, 72–74.
Gray, E. J., & Smith, D. L. (2005). Intracellular and extracellular PGPR: Commonalities and distinctions in the plant–bacterium signaling processes. *Soil Biology and Biochemistry, 3*, 7395–7412.
Gupta, R., Singal, R., Sankar, A., Kuhad, R. C., & Saxena, R. K. (1994). A modified plate assay for screening phosphate solubilizing microorganisms. *The Journal of General and Applied Microbiology, 40*, 255–260.
Gupta, G., Panwar, J., & Jha, P. (2013). Natural occurrence of Pseudomonas aeruginosa, a dominant cultivable Diazotrophic endophytic bacterium colonizing *Pennisetum glaucum* (L) R. Br. *Applied Soil Ecology, 64*, 252–261.
Gyaneshwar, P., Naresh, K. P. G., & Parekh, J. L. (1998). Effect of buffering on the phosphate-solubilizing ability of microorganisms. *World Journal of Microbiology and Biotechnology, 14*, 669–673.
Gyaneshwar, P., Parekh, L. J., Archana, G., Podle, P. S., Collins, M. D., Hutson, R. A., & Naresh, K. G. (1999). Involvement of a phosphate starvation inducible glucose dehydrogenase in soil phosphate solubilization by *Enterobacter asburiae. FEMS Microbiology Letters, 171*, 223–229.
Gyaneshwar, P., Kumar, G. N., Parekh, L. J., & Poole, P. S. (2002). Role of soil microorganisms in improving P nutrition of plants. *Plant and Soil, 245*, 83–93.
Hartmann, A., Schmid, M., van Tuinen, D., & Berg, G. (2009). Plant–driven selection of microbes. *Plant and Soil, 321*, 235–257.
Hasan, M., Bano, A., Hassan, S. G., Iqbal, J., Awan, U., Rong-ji, D., & Khan, K. A. (2014). Enhancement of rice growth and production of growth–promoting phytohormones by inoculation with rhizobium and other rhizobacteria. *World Applied Sciences Journal, 31*, 1734–1743.
Hilda, R., & Fraga, R. (2000). Phosphate solubilizing bacteria and their role in plant growth promotion. *Biotechnology Advances, 17*, 319–359.
Hiltner, L. (1904). Uber neuere Erfahrungen und Probleme auf dem Gebiet der Bodenbakteriologie und unter besonderer Berucksichtigung der Grundungung und Brache. *Arbeiten der Deutschen Landwirtschaftlichen Gesellschaft, 98*, 59–78.
Hinsinger, P. (2001). Bioavailability of soil inorganic P in the rhizosphere as affected by root induced chemical changes: A review. *Plant and Soil, 237*, 173–195.
Holford, I. C. R. (1997). Soil phosphorus: Its measurements and its uptake by plants. *Australian Journal of Soil Research, 35*, 227–239.

Holt, J. G., Krieg, N. R., Sneath, P. H. A., Staley, J. T., & Willams, S. T. (1994). *Bergeys manual of determinative bacteriology* (9th ed.). Baltimore: Williams and Wilkins.

Hussain, M. I., Asghar, H. N., Akhtar, M. J., & Arshad, M. (2013). Impact of phosphate solubilizing bacteria on growth and yield of maize. *Plant, Soil and Environment, 32*, 71–78.

Hynes, L. G. C., Hirkala, D. L., & Nelson, L. M. (2008). Isolation, selection, and characterization of beneficial rhizobacteria from pea, lentil and chickpea grown in Western Canada. *Canadian Journal of Microbiology, 54*, 248–258.

Ibrahim, S. S., El-Midany, A. A., & Boulos, T. R. (2010). Economic preferences of mechanical activation over mineral beneficiation for phosphate rock direct applications. *Physicochemical Problems of Mineral Processing, 44*, 63–78.

Illmer, P., & Schinner, F. (1995). Solubilization of inorganic calcium phosphate solubilization mechanisms. *Soil Biology and Biochemistry, 27*, 257–563.

Indiragandhi, P., Anandham, R., Kim, K., Yim, W. J., Madhaiyan, M., & Sa, T. M. (2008). Induction of defense responses in tomato against *Pseudomonas syringae* pv. tomato by regulating the stress ethylene level with *Methylobacteriumoryzae* CBMB20 containing 1-aminocyclo propane-1-carboxylate deaminase. *World Journal of Microbiology and Biotechnology, 24*, 1037–1045.

International Fertilizer Industry Association (IFA). (2009). IFADATA; IFA: Paris, France. Available online: http://www.fertilizer.org/ifa/ifadata/search. Accessed 22.01.17.

Iqbal, M. A., Khalid, M., Shahzad, S. M., Ahmad, M., Soleman, N., & Akhtar, N. (2012). Integrated use of *Rhizobium leguminosarum*, plant growth promoting rhizobacteria and enriched compost for improving growth, nodulation and yield of lentil (*Lens culinaris* Medik.) *Chilean Journal of Agricultural Research, 72*, 104–110.

Iqbal, S., Khan, M. Y., Asghar, H. N., & Akhtar, M. J. (2016). Combined use of phosphate solubilizing bacteria and poultry manure to enhance the growth and yield of mung bean in calcareous soil. *Soil & Environment, 35*, 146–154.

Jarosch, K. A., Doolette, A. L., Smernik, R. J., Tamburini, F., Frossard, E., & Bünemann, E. K. (2015). Characterisation of soil organic phosphorus in NaOH-EDTA extracts: a comparison of 31P NMR spectroscopy and enzyme addition assays. *Soil Biology and Biochemistry, 91*, 298–309.

Johnson, H. W. (1959). The solubilization of "insoluble" phosphate IV the reaction between organic acids and tricalcium phosphate. *New Zealand Journal of Science, 2*, 215–218.

Jones, D. L., Nguyen, C., & Finlay, R. D. (2009). Carbon flow in the rhizosphere: Carbon trading at the soil-rootinterface. *Plant and Soil, 321*, 5–33.

Joo, G. J., Kim, Y. M., Kim, J. T., & Lee, L. I. J. (2005). Gibberellins-producing rhizobacteria increase endogenous gibberellins content and promote growth of red peppers. *Journal of Microbiology, 43*, 510–515.

Jorquera, M. A., Hernandez, M. T., Rengel, Z., Marschner, P., & de la Luz, M. M. (2008). Isolation of culturable phosphobacteria with both phytate–mineralization and phosphate–solubilization activity from the rhizosphere of plants grown in a volcanic soil. *Biology and Fertility of Soils, 44*, 1025–1034.

Kageyama, H., Tripathi, K., Rai, A. K., Chaum, S., Waditee-Sirisattha, R., & Takabe, T. (2011). An alkaline phosphatase/phosphodiesterase, PhoD, induced by salt stress and secreted out of the cells of Aphanothece halophytica, a halotolerant cyanobacterium. *Applied and Environmental Microbiology, 77*, 5178–5183.

Katznelson, H., Peterson, E. A., & Rovatt, J. W. (1962). Phosphate dissolving microoganisms on seed and in the root zone of plants. *Canadian Journal of Botany, 40*, 1181–1186.

Kaur, G., & Reddy, M. S. (2014). Influence of P-solubilizing bacteria on crop yield and soil fertility at multi locational sites. *European Journal of Soil Biology, 61*, 35–40.

Kaymak, H. C., Yarali, F., Guvenc, I., & Donmez, M. F. (2008). The effect of inoculation with Plant Growth Promoting Rhizobacteria (PGPR) on root formation of mint (*Mentha piperita* L.) cuttings. *African Journal of Biotechnology, 7*, 4479–4483.

Khalid, A., Arshad, M., & Zahir, Z. A. (2006). Phytohormones: Microbial production and applications. In N. Uphoff, A. S. Ball, E. Fernandes, H. Herren, O. Husson, M. Laing, C. Palm,

J. Pretty, P. Sanchez, N. Sanginga, & J. Thies (Eds.), *Biological approaches to sustainable soil systems* (pp. 207–220). Boca Raton: Taylor and Francis.

Khan, M. S., Zaidi, A., & Wani, P. A. (2007). Role of phosphate solubilizing microorganisms in sustainable agriculture. A review. *Agronomy for Sustainable Development, 27*, 29–43.

Khan, M. S., Zaidi, A., Wani, P. A., Ahemad, M., & Oves, M. (2009). Functional diversity among plant growth–promoting rhizobacteria. In M. S. Khan, A. Zaidi, & J. Musarrat (Eds.), *Microbial strategies for crop improvement* (pp. 105–132). Berlin: Springer.

Khan, N., Khan, N. W., & Khan, S. A. (2011). Combined effect of nitrogen fertilizers and herbicides upon maize production in Peshawar. *Journal of Animal and Plant Sciences, 21*, 1001–1006.

Khan, M. S., Ahmad, E., Zaidi, A., & Oves, M. (2013). Functional aspect of phosphate–solubilizing bacteria. Importance in crop production. In D. K. Maheshwari, M. Saraf, & A. Aeron (Eds.), *Bacteria in agrobiology: Crop productivity* (pp. 237–265). Berlin: Springer.

Khiari, L., & Parent, L. E. (2005). Phosphorus transformations in acid light–textured soils treated with dry swine manure. *Canadian Journal of Soil Science, 85*, 75–87.

Kim, K. Y., Jordan, D., & McDonald, G. A. (1997). Solubilization of hydroxyapatite by *Enterobacter agglomerans*and cloned *Escherichia coli* in culture medium. *Biology and Fertility of Soils, 24*, 347–352.

Kim, K. Y., Jordan, D., & Krishnan, H. B. (1998). Expression of genes from Rahnellaaquatilis that are necessary for mineral phosphate solubilization in *Escherichia coli. FEMS Microbiology Letters, 159*, 121–127.

Kim, C. H., Han, S. H., Kim, K. Y., Cho, B. H., Kim, Y. H., Koo, B. S., & Kim, C. Y. (2003). Cloning and expression of pyrroloquinoline quinone (PQQ) genes from a phosphate-solubilizing bacterium *Enterobacter intermedium. Current Microbiology, 47*, 457–461.

Koopmans, G. F., Chardon, W. J., Dolfing, J., Oenema, O., Van der Meer, P., & Van Riemsdijk, W. H. (2003). Wet chemical phosphorus-31 nuclear magnetic resonance analysis of phosphorus speciation in a sand soil receiving long-term fertiliser or animal manure applications. *Journal of Environmental Quality, 32*, 287–295.

Kpomblekou, K., & Tabatabai, M. A. (1994). Effect of organic acids on release of phosphorus from phosphate rocks. *Soil Science, 158*, 442–453.

Krishnaraj, P. U., & Goldstein, A. H. (2001). Cloning of a *Serratia marcescens* DNA fragment that induces quinoprotein glucose dehydrogenase–mediated gluconic acid production in Escherichia coli in the presence of stationary phase *Serratia marcescens. FEMS Microbiology Letters, 205*, 215–220.

Kuan, K. B., Othman, R., Rahim, K. A., & Shamsuddin, Z. H. (2016). Plant growth–promoting rhizobacteria inoculation to enhance vegetative growth, nitrogen fixation and nitrogen remobilisation of maize under greenhouse conditions. *PloS One, 11*, e0152478. doi:10.1371/journal. pone. 0152478.

Kumar, K. V., Singh, N., Behl, H. M., & Srivastava, S. (2008). Influence of plant growth promoting bacteria and its mutant on heavy metal toxicity in *Brassica juncea* grown in fly ash amended soil. *Chemosphere, 72*, 678–683.

Kumar, A., Kumar, A., Devi, S., Patil, S., Payal, C., & Negi, S. S. (2012). Isolation, screening and characterization of bacteria from rhizospheric soils for different plant growth promotion (PGP) activities: An in vitro study. *Recent Research in Science and Technology, 4*, 01–05.

Kumar, V., Singh, P., Jorquera, M. A., Sangwan, P., Kumar, P., Verma, A. K., & Agrawal, S. (2013). Isolation of phytaseproducing bacteria from Himalayan soils and their effect on growth and phosphorus uptake of Indian mustard (*Brassica juncea*). *World Journal of Microbiology and Biotechnology, 29*, 1361–1365.

Kurepin, L. V., Zaman, M., & Pharis, R. P. (2014). Phytohormonal basis for the plant growth promoting action of naturally occurring biostimulators. *Journal of the Science of Food and Agriculture, 94*, 1715–1722.

Lalande, R., Bissonnette, N., Coutlée, D., & Antoun, H. (1989). Identification of rhizobacteria from maize and determination of their plant-growth promoting potential. *Plant and Soil, 115*, 7–11.

Li, G. E., Wu, X. Q., Ye, J. R., Hou, L., Zhou, A. D., & Zhao, L. (2013). Isolation and identification of phytate-degrading rhizobacteria with activity of improving growth of poplar and Masson pine. *World Journal of Microbiology and Biotechnology, 29*, 2181–2193.

Lindow, S. E., & Brandl, M. T. (2003). Microbiology of the phyllosphere. *Applied and Environmental Microbiology, 69*, 1875–1883.

Lopez-Bucio, J., de la Vega, O. M., Guevara-García, A., & Herrera-Estrella, L. (2000). Enhanced phosphorus uptake in transgenic tobacco plants that overproduce citrate. *Nature Biotechnology, 18*, 450–453.

Louw, H. A., & Webley, D. M. (1959). A study of soil bacteria dissolving certain phosphate fertilizers and related compounds. *The Journal of Applied Bacteriology, 22*, 227–233.

Lugtenberg, B., & Kamilova, F. (2009). Plant–growth–promoting rhizobacteria. *Annual Review of Microbiology, 63*, 541–556.

Mahdi, S. S., Hassan, G. I., Hussain, A., & Rasool, F. (2011). Phosphorus availability issue—Its fixation and role of phosphate solubilizing bacteria in phosphate solubilization. *Research Journal of Agricultural Sciences, 2*, 174–179.

Maougal, R. T., Brauman, A., Plassard, C., Abadie, J., Djekoun, A., & Drevon, J. J. (2014). Bacterial capacitiesto mineralize phytate increase in the rhizosphere of nodulated common bean (*Phaseolus vulgaris*) under P deficiency. *European Journal of Soil Biology, 62*, 8–14.

Marks, B. B., Megías, M., Ollero, F. J., Nogueira, M. A., Araujo, R. S., & Hungria, M. (2015). Maize growth promotion by inoculation with *Azospirillum brasilense* and metabolites of *Rhizobium tropici* enriched on lipochito oligosaccharides (LCOs). *AMB Express, 5*, 71–79. doi:10.1186/s13568-015-0154-z.

Marra, L. M., de Oliveira, S. M., Soares, C. R. F. S., & de Souza Moreira, F. M. (2011). Solubilisation of inorganic phosphates by inoculants strains from tropical legumes. *Scientia Agricola, 68*, 603–609.

Martinez-Viveros, O., Jorquera, M., Crowley, D. E., Gajardo, G., & Mora, M. L. (2010). Mechanisms and practical considerations involved in plant growth promotion by rhizobacteria. *Journal of Soil Science and Plant Nutrition, 10*, 293–319.

Martínez, O. A., Crowley, D. E., Mora, M. L., & Jorquera, M. A. (2015). Short-term study shows that phytatemineralizing rhizobacteria inoculation affects the biomass, phosphorus (P) uptake and rhizosphere properties of cereal plants. *Journal of Soil Science and Plant Nutrition, 15*, 153–166.

Masciarelli, O., Llanes, A., & Luna, V. (2014). A new PGPR co-inoculated with *Bradyrhizobium japonicum* enhances soybean nodulation. *Microbiological Research, 169*, 609–615.

Mehrvarz, S., Chaichi, M. R., & Alikhani, H. A. (2008). Effects of phosphate solubilizing microorganisms and phosphorus chemical fertilizer on yield and yield components of Barely (*Hordeum vulgare* L.) *American-Eurasian Journal of Agricultural & Environmental Sciences, 3*, 822–828.

Mendes, R., Kruijt, M., de Bruijn, I., Dekkers, E., van der Voort, M., Schneider, J. H. M., Piceno, Y. M., DeSantis, T. Z., Andersen, G. L., Bakker, P. A. H. M., & Raaijmakers, J. M. (2011). Deciphering the rhizosphere microbiome for disease–suppressive bacteria. *Science, 332*, 1097–1100.

Minaxi, Saxena, J., Chandra, S., & Nain, L. (2013). Synergistic effect of phosphate solubilizing rhizobacteria and *arbuscular mycorrhiza* on growth and yield of wheat plants. *Journal of Soil Science and Plant Nutrition, 13*, 511–525. doi:10.4067/S0718–95162013005000040.

Mommer, L., Kirkegaard, J., & van Ruijven, J. (2016). Root-root interactions: Towards a rhizosphere framework. *Trends in Plant Science, 21*, 209–217.

Moody, P. W., Speirs, S. D., Scott, B. J., & Mason, S. D. (2013). Soil phosphorus tests I: What soil phosphorus pools and processes do they measure? *Crop & Pasture Science, 64*, 461–468.

Moutia, J. F. Y., Saumtally, S., Spaepen, S., & Vanderleyden, J. (2010). Plant growth promotion by *Azospirillum* sp. in sugarcane is influenced by genotype and drought stress. *Plant and Soil, 337*, 233–242.

Murphy, J., & Riely, J. P. (1962). A modified single solution method for the determination of phosphate in natural waters. *Analytica Chimica Acta, 27*, 31–36.

Nahas, E. (1996). Factors determining rock phosphate solubilization by microorganism isolated from soil. *World Journal of Microbiology and Biotechnology, 12*, 18–23.
Nannipieri, P. (2011). Potential impact of climate change on microbial function in soil. In T. J. Sauer, J. M. Norman, & M. V. K. Sivakumar (Eds.), *Sustaining soil productivity in response to global climate change* (1st ed., pp. 199–209). New York: Wiley.
Narsian, V., & Patel, H. H. (2000). *Aspergillus aculeatus* as a rock phosphate solubilizer. *Soil Biology and Biochemistry, 32*, 559–565.
Narula, N., Kumar, V., Behl, R. K., Duebel, A. A., Gransee, A., & Merbach, W. (2000). Effect of P solubilizing *Azotobacter chroococcum* on N, P, K uptake in P responsive wheat genotypes grown under greenhouse conditions. *Journal of Plant Nutrition and Soil Science, 163*, 393–398.
Nautiyal, C. S. (1999). An efficient microbiological growth medium for screening phosphate solubilizing microorganism. *FEMS Microbiology Letters, 170*, 265–270.
Nye, P. H. (1980). Diffusion of ions and uncharged solutes in soils and soil clays. *Advances in Agronomy, 31*, 225–272.
Oberson, A., & Joner, E. J. (2005). Microbial turnover of phosphorus in soil. In B. L. Turner, E. Frossard, & D. S. Baldwin (Eds.), *Organic phosphorus in the environment* (pp. 133–164). Wallingford: CABI.
Ochoa-Loza, F. J., Artiola, J. F., & Maier, R. M. (2001). Stability constants for the complexation of various metals with a rhamnolipid biosurfactant. *Journal of Environmental Quality, 30*, 479–485.
Olander, L. P., & Vitousek, P. M. (2004). Biological and geochemical sinks for phosphorus in soil from a wet tropical forest. *Ecosystems, 7*, 404–419.
Omar, S. A. (1998). The role of rock phosphate solubilizing fungi and vesicular arbuscular mycorrhiza (VAM) in growth of wheat plants fertilized with rock phosphate. *World Journal of Microbiology and Biotechnology, 2*, 211–218.
Pati, B. R., & Chandra, A. K. (1981). Effect of spraying nitrogen-fixing phyllospheric bacterial isolates on wheat plants. *Plant and Soil, 61*, 419–427.
Patil, M. G., Sayyed, R. Z., Chaudhari, A. B., & Chincholkar, S. B. (2002). Phosphate solubilizing microbes: A potential bioinoculant for efficient use of phosphate fertilizers. In S. M. Reddy, S. R. Reddy, & S. Grisham (Eds.), *Bioinoculants for sustainable agriculture and forestry* (pp. 127–138). Jodhpur: Scientific Publisher.
Pereira, S. I. A., & Castro, P. M. L. (2014). Phosphate solubilizing rhizobacteria enhance *Zea mays* growth in agricultural P–deficient soils. *Ecological Engineering, 73*, 526–535.
Pieterse, C. M. J., Zamioudis, C., Berendsen, R. L., Weller, D. M., Van Wees, S. C. M., & Bakker, P. A. H. M. (2014). Induced systemic resistance by beneficial microbes. *Annual Review of Phytopathology, 52*, 347–375.
Pikovskaya, R. I. (1948). Mobilization of phosphorus in soil in connection with the vital activity of some microbial species. *Mikrobiologiya, 17*, 362–370.
Pizzeghello, D., Berti, A., Nardi, S., & Morari, F. (2011). Phosphorus forms and P–sorption properties in three alkaline soils after long–term mineral and manure applications in north–eastern Italy. *Agriculture, Ecosystems & Environment, 141*, 58–66.
Poonguzhali, S., Madhaiyan, M., & Sa, T. (2008). Isolation and identification of phosphate solubilizing bacteria from Chinese cabbage and their effect on growth and phosphorus utilization of plants. *Journal of Microbiology and Biotechnology, 18*, 773–777.
Probanza, A., Mateos, J. L., Lucas-Garcia, J. A., Ramos, B., de Felipe, M. R., & Gutierrez Manero, F. J. (2001). Effects of inoculation with PGPR Bacillus and *Pisolithustinctorius* on *Pinus pinea* L. growth, bacterial rhizosphere colonization, and mycorrhizal infection. *Microbial Ecology, 41*, 140–148.
Prochnow, L. I., Fernando, J., Quispe, S., Artur, E., Francisco, B., & Braga, G. (2006). Effectiveness of phosphate fertilizers of different water solubilities in relation to soil phosphorus adsorption. *Agronomy Journal, 95*, 293–302.
Qureshi, M. A., Iqbal, A., Akhtar, N., Shakir, M. A., & Khan, A. (2012). Co-inoculation of phosphate solubilizing bacteria and rhizobia in the presence of L-tryptophan for the promotion of mash bean (*Vigna mungo* L.) *Soil and Environment, 31*, 47–54.

Reid, R. K., Reid, C. P. P., & Szaniszlo, P. J. (1985). Effects of synthetic and microbially produced chelates on the diffusion of iron and phosphorus to a simulated root in soil. *Biology and Fertility of Soils, 1*, 45–52.

Reilly, T. J., Baron, G. S., Nano, F., & Kuhlenschmidt, M. S. (1996). Characterization and sequencing of a respiratory burst inhibiting acid phosphatase from *Francisella tularensis*. *The Journal of Biological Chemistry, 271*, 10973–10983.

Reinhold-Hurek, B., Bünger, W., Burbano, C. S., Sabale, M., & Hurek, T. (2015). Roots shaping their microbiome: Global hotspots for microbial activity. *Annual Review of Phytopathology, 53*, 403–424.

Reyes, I., Baziramakenga, R., Bernier, L., & Antoun, H. (2001). Solubilization of phosphate rocks and minerals by a wild–type strain and two UV induced mutants of *Penicillium rugulosum*. *Soil Biology and Biochemistry, 33*, 1741–1747.

Richardson, A. E., Hadobas, P. A., & Hayes, J. E. (2001). Extracellular secretion of *Aspergillus phytase* from Arabidopsis roots enables plants to obtain phosphorous from phytate. *The Plant Journal, 25*, 641–649.

Ringeval, B., Augusto, L., Monod, H., van Apeldoorn, D., Bouwman, L., Yang, X., Achat, D. L., Chini, L. P., Van Oost, K., Guenet, B., Wang, R., Decharme, B., Nesme, T., & Pellerin, S. (2017). Phosphorus in agricultural soils: Drivers of its distribution at the global scale. *Glob Change Biology*. doi:10.1111/gcb.13618.

Rivera, C. M., Trujillo, N. A., Córdova, B. G., Kohler, J., Caravaca, F., & Roldan, A. (2008). Poultry manure and banana waste are effective biofertilizer carriers for promoting plant growth and soil sustainability in banana crops. *Soil Biology and Biochemistry, 40*, 3092–3095.

Rodríguez, H., Fraga, R., Gonzalez, T., & Bashan, T. (2006). Genetics of phosphate solubilization and itspotential applications for improving plant growth promoting bacteria. *Plant and Soil, 287*, 15–21.

Rojas-Tapias, D., Moreno-Galván, A., Pardo-Díaz, S., Obando, M., Rivera, D., & Bonilla, R. (2012). Effect of inoculation with plant growth-promoting bacteria (PGPB) on amelioration of saline stress in maize (*Zea mays*). *Applied Soil Ecology, 61*, 264–272.

Rossolini, G. M., Shippa, S., Riccio, M. L., Berlutti, F., Macaskie, L. E., & Thaller, M. C. (1998). Bacterial nonspecific acid phosphatases: Physiology, evolution and use as tools in microbial biotechnology. *Cellular and Molecular Life Sciences, 54*, 833–850.

Ryan, P. R., Delhaize, E., & Jones, D. L. (2001). Function and mechanism of organic anion exudation from plant roots. *Annual Review of Plant Physiology and Plant Molecular Biology, 52*, 527–560.

Ryan, M. H., Tibbett, M., Edmonds-Tibbett, T., Suriyagoda, L. D. B., Lambers, H., Cawthray, G. R., & Pang, J. (2012). Carbon trading for phosphorus gain: The balance between rhizosphere carboxylates and arbuscular mycorrhizal symbiosis in plant phosphorus acquisition. *Plant, Cell & Environment, 35*, 2170–2180.

Sarathambal, C., & Ilamurugu, K. (2013). Saline tolerant plant growth promoting diazotrophs from rhizosphere of bermuda grass and their effect on rice. *Indian Journal of Weed Science, 45*, 80–85.

Sashidhar, B., & Podile, A. R. (2009). Transgenic expression of glucose dehydrogenase in *Azotobacter vinelandii* enhances mineral phosphate solubilization and growth of sorghum seedlings. *Microbial Biotechnology, 2*, 521–529.

Schachtman, D. P., Robert, J., & Reid, A. S. M. (1998). Phosphorus uptake by plants: From soil to cell. *Plant Physiology, 116*, 447–453.

Selvakumar, G., Mohan, M., Kundu, S., Gupta, A. D., Joshi, P., Nazim, S., & Gupta, H. S. (2008). Cold tolerance and plant growth promotion potential of *Serratia marcescens* strain SRM (MTCC 8708) isolated from flowers of summer squash (*Cucurbita pepo*). *Letters in Applied Microbiology, 46*, 171–175.

Shahzad, S. M., Khalid, A., Arshad, M., Tahir, J., & Mahmood, T. (2010). Improving nodulation, growth and yield of Cicer arietinum L. through bacterial ACC–deaminase induced changes in root architecture. *European Journal of Soil Biology, 46*, 342–347.

Shahzad, S. M., Arif, M. S., Riaz, M., Ashraf, M., & Iqbal, Z. (2013). PGPR with varied ACC–deaminase activity induced different growth and yield response in maize (*Zea mays* L) under fertilized conditions. *European Journal of Soil Biology, 57*, 27–34.

Shahzad, S. M., Khalid, A., Arif, M. S., Riaz, M., Ashraf, M., & Iqbal, Z. (2014). Co-inoculation integrated with P-enriched compost improved nodulation and growth of chickpea (*Cicer arietinum* L.) under irrigated and rainfed farming systems. *Biology and Fertility of Soils, 50*, 1–12.

Sharma, K. K., Mathur, P. B., & Jatanand, B. (2007). Chickpea (*Cicer arietinum* L). In K. Wang (Ed.), *Agrobacterium protocol, volume 1, Methods in molecular biology* (Vol. 343, 2nd ed.). Tootowa NJ: Humana Inc.

Sharma, S., Upadhyaya, H. D., Roorkiwal, M., Varshney, R. K., & Gowda, C. L. L. (2013a). Chickpea. In M. Singh, H. D. Upadhyaya, & I. S. Bisht (Eds.), *Genetic and genomic resources of grain legume improvement* (pp. 81–104). London: Elsevier Inc..

Sharma, S. B., Sayyed, R. Z., Trivedi, M. H., & Gobi, T. A. (2013b). Phosphate solubilizing microbes: Sustainable approach for managing phosphorus deficiency in agricultural soils. *Springerplus, 2*, 587–600.

Sharpley, A. N., Singh, U., Uehara, G., & Kimble, J. (1989). Modeling soil and plant phosphorus dynamics in calcareous and highly weathered soils. *Soil Science Society of America Journal, 53*, 153–158.

Shen, J., Li, C., Mi, G., Li, L., Yuan, L., Jiang, R., & Zhang, F. (2013). Maximizing root/rhizosphere efficiency to improve crop productivity and nutrient use efficiency in intensive agriculture of China. *Journal of Experimental Botany, 64*, 1181–1192.

Shilev, S., Sancho, E. D., & Benlloch-González, M. (2012). Rhizospheric bacteria alleviate salt-produced stress in sunflower. *Journal of Environmental Management, 95*, 37–41.

da Silva, T. F., Vollú, R. E., do Carmo Dias, B., de Lacerda, J. R. M., Marques, J. M., Nishikawa, M. M., de Vasconcelos Goulart, F. R., Alviano, C. S., & Seldin, L. (2017). Cultivable bacterial communities associated with roots of rose-scented geranium (*Pelargonium graveolens*) with the potential to contribute to plant growth. *Applied Soil Ecology, 111*, 123–128.

Singh, N., Pandey, P., Dubey, R. C., & Maheshwari, D. K. (2008). Biological control of root rot fungus *Macrophomina phaseolina* and growth enhancement of *Pinus roxburghii* (Sarg.) by rhizosphere competent *Bacillus subtilis* BN1. *World Journal of Microbiology and Biotechnology, 24*, 1669–1679.

Son, H. J., Park, G. T., Cha, M. S., & Heo, M. S. (2006). Solubilization of insoluble inorganic phosphates by a novel salt- and pH tolerant *Pantoea agglomerans* R-42 isolated from soybean rhizosphere. *Bioresource Technology, 97*, 204–210.

Spaepen, S., Vanderleyden, J., & Okon, Y. (2009). Plant growth-promoting actions of rhizobacteria. In L. C. van Loon, J. C. Ed Kader, & M. Delseny (Eds.), *Advances in botanical research* (Vol. 51, pp. 283–320). Amsterdam, The Netherlands: Elsevier.

Stajković-Srbinović, O., Delić, D., Kuzmanović, D., Protić, N., Rasulić, N., & Knežević-Vukčević, J. (2014). Growth and nutrient uptake in oat and barley plants as affected by rhizobacteria. *Romanian Biotechnology Letters, 19*, 9429–9436.

Stevenson, F. J. (2005). *Cycles of soil: Carbon, nitrogen, phosphorus, sulfur, micronutrients* (p. 448). New York: Wiley.

Stout, M. J., Zehnder, G. W., & Baur, M. E. (2002). Potential for the use of elicitors of plant defense in arthropod management programs. *Archives of Insect Physiology and Biochemistry, 51*, 222–235.

Sturz, A. V., Matheson, B. G., Arsenault, W., & Christie, L. B. R. (2001). Weeds as a source of plant growth promoting rhizobacteria in agricultural soils. *Canadian Journal of Microbiology, 47*, 1013–1024.

SubbaRao (Ed.). (1977). *soil microorganisms and plant growth*. India: Oxford and IBH Publishing Co.

Sundara, B., Natarajan, V., & Hari, K. (2002). Influence of phosphorus solubilizing bacteria on the changes in soil available phosphorus and sugarcane and sugar yields. *Field Crops Research, 77*, 43–49.

Sutherland, I. (2001). Biofilm exopolysaccharides: a strong and sticky framework. *Microbiology, 147*, 3–9.

Syers, J. K., Johnston, A. E., & Curtin, D. (2008a). *Efficiency of soil and fertilizer phosphorus use FAO*. Rome, Italy: Fertilizer and Plant Nutrition Bulletin 18.

Syers, J. K., Johnston, A. E., & Curtin, D. (2008b). *Efficiency of soil and fertilizer phosphorus: Reconciling changing concepts of soil phosphorus behaviour with agronomic information* (pp. 27–44). Rome, Italy: FAO Fertilizer and Plant Nutrition Bulletin 18.

Taktek, S., St-Arnaud, M., Piché, Y., Fortin, J. A., & Antoun, H. (2017). Igneous phosphate rock solubilization by biofilm forming mycorrhizo bacteria and hyphobacteria associated with *Rhizoglomus irregular* DAOM 197198. *Mycorrhiza, 27*, 13–22.

Tarafdar, J. C. (1989). Use of electrofocussing technique for characterizing the phosphatases in the soil and root exudates. *Journal of the Indian Society of Soil Science, 37*, 393–395.

Tarafdar, J. C., & Claasen, N. (1988). Organic phosphorus compounds as a phosphorus source for higher plants through the activity of phosphatases produced by plant roots and microorganisms. *Biology and Fertility of Soils, 5*, 308–312.

Tian, G., Cox, A. E., Kumar, K., Granato, T. C., O'Connor, G. A., & Elliott, H. A. (2016). Assessment of plant availability and environmental risk of biosolids–phosphorus in a US midwest corn–belt soil. *Journal of Environmental Management, 172*, 171–176.

Turan, M., Ekinci, M., Yildirim, E., & Dursun, A. (2014). Plant growth-promoting rhizobacteria improved growth, nutrient, and hormone content of cabbage (*Brassica oleracea*) seedlings. *Turkish Journal of Agriculture and Forestry, 38*, 327–333.

Turner, B. L., Papházy, M. J., & Haygarth, P. M. (2002). Inositol phosphates in the environment. *Philosophical Transactions of the Royal Society of London. Series B, Biological Sciences, 357*, 449–469.

Upadhyay, S. K., Singh, J. S., Saxena, A. K., & Singh, D. P. (2011). Impact of PGPR inoculation on growth and antioxidant status of wheat under saline conditions. *Plant Biology, 14*, 605–611.

Van de Wiel, C. C. M., van der Linden, C. G., & Scholten, O. E. (2016). Improving phosphorus use efficiency in agriculture: Opportunities for breeding. *Euphytica, 207*, 1–22.

Vikram, A., Ajjanna, R., Alagawadi, A. P. U., Krishnaraj, A. K. S., & Kumar, M. (2007). Transconjugation studies in *Azospirillum* sp. negative to mineral phosphate solubilization. *World Journal of Microbiology and Biotechnology, 23*, 1333–1337.

Viruel, E., Erazzú, L. E., Calsina, L. M., Ferrero, M. A., Lucca, M. E., & Siñeriz, F. (2014). Inoculation of maize with phosphate solubilizing bacteria: Effect on plant growth and yield. *Journal of Soil Science and Plant Nutrition, 14*, 819–831.

Wani, P. A., Khan, M. S., & Zaidi, A. (2008). Chromium-reducing and plant growth-promoting Mesorhizobium improves chickpea growth in chromium-amended soil. *Biotechnology Letters, 30*, 159–163.

Weiland-Bräuer, N., Pinnow, N., & Schmitz, R. A. (2015). Novel reporter for identification of interference with acyl homoserine lactone and autoinducer-2 quorum sensing. *Applied and Environmental Microbiology, 81*, 1477–1489.

Welch, S. A., Taunton, A. E., & Banfield, J. F. (2002). Effect of microorganisms and microbial metabolites on apatite dissolution. *Geomicrobiology Journal, 19*, 343–367.

Whitelaw, M. A. (2000). Growth promotion of plants inoculated with phosphate solubilizing fungi. *Advances in Agronomy, 69*, 99–151.

Widawati, S., & Rahmansyah, M. (2009). The influence of bacteria inoculation to *jarakpagar* (*Jatropha curcas* L) growth. *Jurnal Biologi Indonesia, 6*, 107–117.

Widawati, S., & Suliasih. (2006). Augmentation of potential phosphate solubilizing bacteria (PSB) stimulate growth of green mustard (*Brassica caventis* Ocd) in marginal soil. *Biodiversitas, 7*, 10–14.

Yadaf, R. S., & Tarafdar, J. C. (2001). Influence of organic and inorganic phosphorus supply on the maximum secretion of acid phosphatase by plants. *Biology and Fertility of Soils, 34*, 140–143.

Yang, M., Ding, G., Shi, L., Xu, F., & Meng, J. (2010). Detection of QTL for phosphorus efficiency at vegetative stage in *Brassica napus*. *Plant and Soil, 339*, 97–111.

Yi, Y., Huang, W., & Ying, G. (2008). Exopolysaccharide: A novel important factor in the microbial dissolution of tricalcium phosphate. *World Journal of Microbiology and Biotechnology, 24*, 1059–1065.

Yu, X., Liu, X., & Zhu, T. (2014). Walnut growth and soil quality after inoculating soil containing rock phosphate with phosphate–solubilizing bacteria. *Science Asia, 40*, 21–27.

Yuan, J., Zhang, N., & Huang, Q. (2015). Organic acids from root exudates of banana help root colonization of PGPR strain *Bacillus amylo liquefaciens* NJN–6. *Scientific Reports, 5*, 134–438.

Yuttavanichakul, W., Lawongsa, P., Wongkaew, S., Teaumroong, N., Boonkerd, N., Nomura, N., & Tittabutr, P. (2012). Improvement of peanut rhizobial inoculant by incorporation of plant growth promoting rhizobacteria (PGPR) as biocontrol against the seed borne fungus, *Aspergillus niger*. *Biological Control, 63*, 87–97.

Zaidi, A., Khan, M. S., Ahemad, M., & Oves, M. (2009). Plant growth promotion by phosphate solubilizing bacteria. *Acta Microbiologica et Immunologica Hungarica, 56*, 263–284.

Zaefarian, F., Vahidzadeh, S., Rahdari, P., Rezvani, M., & Zadeh, H. G. (2012). Effectiveness of plant growth promoting rhizobacteria in facilitating lead and nutrient uptake by little seed canary grass. *Brazilian Journal of Botany, 35*, 241–248.

Index

M. Naeem et al. (eds.), *Essential Plant Nutrients*,
DOI 10.1007/978-3-319-58841-4

D

N

MIX
Papier aus verantwortungsvollen Quellen
Paper from responsible sources
FSC® C105338

If you have any concerns about our products,
you can contact us on
ProductSafety@springernature.com

In case Publisher is established outside the EU,
the EU authorized representative is:
Springer Nature Customer Service Center GmbH
Europaplatz 3, 69115 Heidelberg, Germany

Printed by Libri Plureos GmbH
in Hamburg, Germany